AF335077

Biochemistry of Atherosclerosis

ADVANCES IN BIOCHEMISTRY IN HEALTH AND DISEASE

Series Editor: Naranjan S. Dhalla, PhD, MD (Hon), DSc (Hon) Winnipeg, Manitoba, Canada

Editorial Board: A. Angel, Toronto, Canada; I.M.C. Dixon, Winnipeg, Canada; L.A. Kirshenbaum, Winnipeg, Canada; Dennis B. McNamara, New Orleans, Louisiana; M.A.Q. Siddiqui, Brooklyn, New York; A.K. Srivastava, Montreal, Canada

Volume 1: S. K. Cheema (ed.), *Biochemistry of Atherosclerosis*

Biochemistry of Atherosclerosis

Edited by

Sukhinder Kaur Cheema, PhD
*Memorial University of Newfoundland, St. John's
NL, Canada*

 Springer

Sukhinder Kaur Cheema, PhD
Associate Professor
Department of Biochemistry
Memorial University of Newfoundland
St. John's, NL
Canada A1B 3X9
skaur@mun.ca

Library of Congress Control Number: 2005938503

ISBN-10: 0-387-31252-8
ISBN-13: 978-0387-31252-1

Printed on acid-free paper.

Printed in the United States of America.   (SPI/MVY)

9 8 7 6 5 4 3 2 1

springer.com

Preface

Atherosclerosis and its complications are the major cause of morbidity and mortality in both developed and developing nations. There is an urgent need to understand the pathogenesis and progression of atherosclerosis, and to develop a strategy to prevent an epidemic episode. Crucial advances in our understanding of the pathogenesis of atherosclerosis and its complications have been achieved in recent years. This text, "Biochemistry of Atherosclerosis", is a compilation of contributions from world-renowned scientists, who are at the forefront of atherosclerosis research. This compendium will be highly valuable for individuals in the healthcare field and in basic research as it covers a variety of research topics on the onset, progression, and management of atherosclerosis. Each contribution deals in detail with the biochemical processes involved and provides in-depth information in specific areas.

Atherosclerosis comes from the Greek words; "athero" meaning gruel or paste and "sclerosis" meaning hardness. Atherosclerosis begins with damage to the artery caused by elevated levels of cholesterol and triglycerides in the blood, as well as high blood pressure. Several other factors are also associated with the onset and progression of atherosclerosis, i.e. hyperglycemia, hyper-homocysteinemia, disruption of the immune system, glycation end products, and infectious agents. The first two sections of this book are dedicated to the association of hyperlipidaemia, diabetes, and hypertension with atherosclerosis. Section I addresses recent advances in the regulation of lipid and cholesterol metabolism, and how various biochemical pathways are involved in the development and progression of atherosclerosis. Many of these chapters cover recent research that employed transgenic and genetically altered mice. Section II concentrates on diabetes and hypertension. Type 2 diabetes, associated with insulin resistance and obesity, is on a rapid rise in the North American population, not only in the adults but also in children. The risk of heart disease in people with diabetes is two to four times higher than in the nondiabetic population. This section highlights recent advances in these areas.

Homocysteine is an amino acid that is found in the blood; elevated circulating levels are related to a higher risk of heart disease. Section III deals with the regulation of homocysteine metabolism and the factors that influence this regulation. Section IV is directed towards other factors that are associated with the onset and progression of atherosclerosis, i.e., immune function, infection, and endothelial dysfunction. Immunological responses and infectious agents also play an important role in transplant arteriopathy, which is a major cause of death in long-term survivors after heart transplantation. This additional information provides an excellent bridge between an understanding of the regulation of metabolic pathways and the clinical implications. The complications of atherosclerosis become acute when the plaque ruptures and blocks blood flow (thrombosis). Stability of the plaque is maintained to a large extent by the composition of the plaque. New insights are provided in this section on the significance of inflammation to the vulnerability of plaque to rupture. Lastly, with the rapid increase in the risk of atherosclerosis and associated risk factors, emphasis needs to be placed on the prevention of atherosclerosis. Thus, the last section addresses dietary and lifestyle interventions as prevention and management strategies for atherosclerosis.

I sincerely appreciate the support from authors in helping to bring together this book. This compendium provides a breadth of knowledge as well as new insights into all aspects of atherosclerosis. These contributions from around the world indicate that heart disease is a worldwide problem. It is clear from these contributions that we need to further our understanding of the onset, progression, and management of atherosclerosis.

Sukhinder Kaur Cheema

Contents

Section I. Hyperlipidaemia and Atherosclerosis   **1**

Chapter 1: Reverse Cholesterol Transport   3
Jim W. Burgess, Philip A. Sinclair,
Christophe M. Chretien, Jonathon Boucher
and Daniel L. Sparks

Chapter 2: The Role of LCAT in Atherosclerosis   23
Dominic S. Ng

Chapter 3: Howdy Partner: Apolipoprotein A-I-ABCA1   39
Interactions Mediating HDL Particle
Formation
Sereyrath Ngeth and Gordon A. Francis

Chapter 4: Role of the Scavenger Receptor Class B   53
Type I in Lipoprotein Metabolism and
Atherosclerosis: Insights from Genetically Altered
Mice
Bernardo Trigatti

Chapter 5: The Role of Scavenger Receptors in   70
Signaling, Inflammation and Atherosclerosis
Daisy Sahoo and Victor A. Drover

Chapter 6: ABC Transporters and Apolipoprotein E: 92
Critical Players in Macrophage
Cholesterol Efflux and Atherosclerosis
Kathryn E. Naus and Cheryl L. Wellington

Chapter 7: Provision of Lipids for Very Low-Density   121
Lipoprotein Assembly
Dean Gilham and Richard Lehner

Chapter 8: Oxidatively Modified Low-Density   150
Lipoproteins and Thrombosis
Garry X. Shen

Chapter 9:   The Clinical Significance of Small, Dense Low-Density Lipoproteins   168
Manfredi Rizzo and Kaspar Berneis

Chapter 10:   Bile Acids: At the Crossroads of Sterol, Fat and Carbohydrate Metabolism   186
Luis B. Agellon

Section II. Diabetes and Hypertension Induced Atherosclerosis   **203**

Chapter 11:   Lipoprotein Metabolism in Insulin-Resistant States   205
Rita Kohen Avramoglu, Heather Basciano and Khosrow Adeli

Chapter 12:   The Roles of Protein Glycation, Glycoxidation, and Advanced Glycation End-Product Formation in Diabetes-Induced Atherosclerosis   247
Imran Rashid, Bronwyn E. Brown, David M. van Reyk and Michael J. Davies

Chapter 13:   Molecular and Cellular Mechanisms by Which Diabetes Mellitus Promotes the Development of Atherosclerosis   284
Geoff H. Werstuck

Chapter 14:   Hypertension and Atherosclerosis: Advanced Glycation End Products— A Common Link   305
Sudesh Vasdev and Vicki Gill

Section III. Hyperhomocysteinemia and Atherosclerosis   **327**

Chapter 15:   Homocysteine Metabolism   329
Enoka P. Wijekoon, Margaret E. Brosnan and John T. Brosnan

Chapter 16:   Role of Hyperhomocysteinemia in Atherosclerosis   358
Stephen M. Colgan, Donald W. Jacobsen and Richard C. Austin

Chapter 17:   Molecular and Biochemical Mechanisms of Hyperhomocysteinemia-Induced Cardiovascular Disorders   380
Karmin O, Yaw L. Siow, Patrick C. Choy and Grant M. Hatch

Section IV. Other Factors in Atherosclerosis and the Associated Complications   **395**

    Chapter 18: The Role of the Immune System in Atherosclerosis: Lessons Learned from Using Mouse Models of the Disease   397
Stewart C. Whitman and Tanya A. Ramsamy

    Chapter 19: The Role of Infectious Agents in Atherogenesis   413
Paul K. M. Cheung and Grant N. Pierce

    Chapter 20: Transplant Arteriopathy: Role of Nitric Oxide Synthase   435
Nandini Nair, Hannah Valantine and John P. Cooke

    Chapter 21: Insights into the Molecular Mechanisms of Plaque Rupture and Thrombosis   455
P.K. Shah and Behrooz Sharifi

Section V. Management of Atherosclerosis   **473**

    Chapter 22: Modification of Biochemical and Cellular Processes in the Development of Atherosclerosis by Red Wine   475
Harjot K. Saini, Parambir Dhami, Yan-Jun Xu, Sukhinder Kaur Cheema, Amarjit S. Arneja and Naranjan S. Dhalla

    Chapter 23: Endothelial Dysfunction and Atherosclerosis: Role of Dietary Fats   495
Kanta Chechi, Pratibha Dubey and Sukhinder Kaur Cheema

    Chapter 24: Oxidized LDL and Antioxidants in Atherosclerosis   519
Lesley MacDonald-Wicks and Manohar Garg

    Chapter 25: Dietary Fatty Acids and Stroke   542
Gene R. Herzberg

Index   **553**

Contributors

Khosrow Adeli, Department of Pediatric Laboratory Medicine, Hospital for Sick Children, University of Toronto, Toronto, Ontario, Canada
Email: khosrow.adeli@sickkids.ca

Luis B. Agellon, Department of Biochemistry, University of Alberta, Edmonton, Alberta, Canada
Email: luis.agellon@ualberta.ca

Amarjit S. Arneja, Department of Internal Medicine, Faculty of Medicine, University of Manitoba, Winnipeg, Canada

Richard C. Austin, Department of Pathology and Molecular Medicine, McMaster University and the Henderson Research Centre, Hamilton, Ontario, Canada
Email: raustin@thrombosis.hhscr.org

Rita Kohen Avramoglu, Department of Pediatric Laboratory Medicine, Hospital for Sick Children, University of Toronto, Toronto, Ontario, Canada

Heather Basciano, Department of Pediatric Laboratory Medicine, Hospital for Sick Children, University of Toronto, Toronto, Ontario, Canada

Kaspar Berneis, Division of Endocrinology and Diabetology, University Hospital Zurich, Switzerland
Email: kkberneis@hotmail.com

Jonathon Boucher, The Lipoprotein and Atherosclerosis Research Group, University of Ottawa, 40 Ruskin Street, Ottawa, Ontario, Canada

John T. Brosnan, Department of Biochemistry, Memorial University of Newfoundland, St. John's, Newfoundland, Canada
Email: jbrosnan@mun.ca

Margaret E. Brosnan, Department of Biochemistry, Memorial University of Newfoundland, St. John's, Newfoundland, Canada

Bronwyn E. Brown, Free Radical Group, The Heart Research Institute, 114 Pyrmont Bridge Road, Camperdown, Sydney, NSW 2050, Australia

Jim W. Burgess, Liponex, Inc., 1740 Woodroffe Avenue, Ottawa, Ontario, Canada, K2G 3R8

Kanta Chechi, Department of Biochemistry, Memorial University of Newfoundland, St. John's, Newfoundland, Canada

Sukhinder Kaur Cheema, Department of Biochemistry, Memorial University of Newfoundland, St. John's, Newfoundland, Canada
Email: skaur@mun.ca

Paul K.M. Cheung, Division of Stroke and Vascular Disease, St. Boniface General Hospital Research Centre, Department of Physiology, University of Manitoba, Winnipeg, Manitoba, Canada

Patrick C. Choy, Centre for Research and Treatment of Atherosclerosis, St. Boniface Hospital Research Centre, University of Manitoba, Winnipeg, Manitoba, Canada

Christophe M. Chretien, Liponex, Inc., 1740 Woodroffe Avenue, Ottawa, Ontario, Canada, K2G 3R8

Stephen M. Colgan, Department of Pathology and Molecular Medicine, McMaster University and the Henderson Research Centre, Hamilton, Ontario, Canada

John P. Cooke, Division of Cardiovascular Medicine, Stanford University School of Medicine, 300 Pasteur Drive, Stanford, California, USA
Email: john.cooke@stanford.edu

Michael J. Davies, Free Radical Group, The Heart Research Institute, 114 Pyrmont Bridge Road, Camperdown, Sydney, NSW 2050, Australia
Email: m.davies@hri.org.au

Naranjan S. Dhalla, Institute of Cardiovascular Sciences, St. Boniface General Hospital Research Centre and Department of Physiology, Faculty of Medicine, University of Manitoba, Winnipeg, Canada

Parambir Dhami, Institute of Cardiovascular Sciences, St. Boniface General Hospital Research Centre and Department of Physiology, Faculty of Medicine, University of Manitoba, Winnipeg, Canada

Victor A. Drover, Department of Pharmacological Sciences, State University of New York, Stony Brook, New York, USA

Pratibha Dubey, Department of Biochemistry, Memorial University of Newfoundland, St. John's, Newfoundland, Canada

Gordon A. Francis, CIHR Group on Molecular and Cell Biology of Lipids, Departments of Medicine and Biochemistry, University of Alberta, Edmonton, Alberta, Canada
Email: gordon.francis@ualberta.ca

Manohar Garg, School of Biomedical Sciences, Faculty of Health, University of Newcastle, Callaghan, NSW 2308, Australia
Email: Manohar.Garg@newcastle.edu.au

Dean Gilham, Departments of Pediatrics and Cell Biology, University of Alberta, Edmonton, Alberta, Canada

Vicki Gill, Discipline of Medicine, Faculty of Medicine, Health Sciences Centre, Memorial University of Newfoundland, St. John's, Newfoundland, Canada

Grant M Hatch, Department of Pharmacology and Therapeutics, Faculty of Medicine, Centre for Research and Treatment of Atherosclerosis, St. Boniface Hospital Research Centre, University of Manitoba, Winnipeg, Manitoba, Canada
Email: hatchgm@Ms.UManitoba.CA

Gene R. Herzberg, Department of Biochemistry, Memorial University of Newfoundland, St. John's, Newfoundland, Canada
Email: geneherz@mun.ca

Donald W. Jacobsen, Department of Cell Biology, Lerner Research Institute Department of Cardiovascular Medicine, Division of Medicine, Cleveland Clinic Foundation and Department of Molecular Medicine, Cleveland Clinic Lerner College of Medicine, Case Western Reserve University, Cleveland, Ohio USA

Richard Lehner, Departments of Pediatrics and Cell Biology, University of Alberta, Edmonton, Alberta, Canada
Email: richard.lehner@ualberta.ca

Lesley MacDonald-Wicks, School of Health Sciences, Faculty of Health, University of Newcastle, Callaghan, NSW 2308, Australia

Nandini Nair, Division of Cardiovascular Medicine, Stanford University School of Medicine, 300 Pasteur Drive, Stanford, California, USA

Kathryn E. Naus, Department of Pathology and Laboratory Medicine, University of British Columbia, Vancouver, British Columbia, Canada

Dominic S. Ng, Department of Medicine, St. Michael's Hospital, West Annex 2-015, 38 Shuter Street, Toronto, Ontario, Canada
Email: ngd@smh.toronto.on.ca

Sereyrath Ngeth, CIHR Group on Molecular and Cell Biology of Lipids, Departments of Medicine and Biochemistry, University of Alberta, Edmonton, Alberta, Canada

Grant N. Pierce, Division of Stroke and Vascular Disease, St. Boniface General Hospital Research Centre, Department of Physiology, University of Manitoba, Winnipeg, Manitoba, Canada
Email: gpierce@sbrc.ca

Karmin O, Department of Animal Science, Faculty of Agricultural and Food Sciences, Department of Physiology, Faculty of Medicine, Centre for Research and Treatment of Atherosclerosis, National Centre for Agri-Food Research in Medicine, St. Boniface Hospital Research Centre, University of Manitoba, Winnipeg, Manitoba, Canada

Tanya A. Ramsamy, Departments of Pathology and Laboratory Medicine & Cellular and Molecular Medicine, University of Ottawa Heart Institute, Ottawa, Ontario, Canada

Imran Rashid, Free Radical Group, The Heart Research Institute, 114 Pyrmont Bridge Road, Camperdown, Sydney, NSW 2050, Australia

Manfredi Rizzo, Department of Clinical Medicine and Emerging Diseases, University of Palermo, Italy
Email: rizzomv@yahoo.it

Daisy Sahoo, Department of Pharmacological Sciences, State University of New York, Stony Brook, New York, USA
Email: daisysahoo@hotmail.com

Harjot K. Saini, Institute of Cardiovascular Sciences, St. Boniface General Hospital Research Centre and Department of Physiology, Faculty of Medicine, University of Manitoba, Winnipeg, Canada

P.K. Shah, Division of Cardiology and Atherosclerosis Research Center, Burns and Allen Research Institute and Department of Medicine, Cedars Sinai Medical Center and UCLA School of Medicine, Los Angeles, California, USA

Behrooz Sharifi, Division of Cardiology and Atherosclerosis Research Center, Burns and Allen Research Institute and Department of Medicine, Cedars Sinai Medical Center and UCLA School of Medicine, Los Angeles, California, USA
Email: sharifi@cshs.org

Garry X. Shen, Departments of Internal Medicine and of Physiology, University of Manitoba, Winnipeg, Manitoba, Canada
Email: gshen@ms.umanitoba.ca

Philip A. Sinclair, Liponex, Inc., 1740 Woodroffe Avenue, Ottawa, Ontario, Canada, K2G 3R8

Yaw L. Siow, Department of Physiology, Faculty of Medicine, Centre for Research and Treatment of Atherosclerosis, National Centre for Agri-Food Research in Medicine, St. Boniface Hospital Research Centre, University of Manitoba, Winnipeg, Manitoba, Canada

Daniel L. Sparks, Liponex, Inc., 1740 Woodroffe Avenue, Ottawa, Ontario, Canada, K2G 3R8 and The Lipoprotein and Atherosclerosis Research Group, University of Ottawa, 40 Ruskin Street, Ottawa, Ontario, Canada
Email: DSparks@ottawaheart.ca

Bernardo Trigatti, Department of Biochemistry and Biomedical Sciences, McMaster University, Hamilton, Ontario, Canada
Email: trigatt@mcmaster.ca

Hannah Valantine, Division of Cardiovascular Medicine, Stanford University School of Medicine, 300 Pasteur Drive, Stanford, California, USA

David M. van Reyk, Department of Health Sciences, University of Technology Sydney, PO Box 123, Broadway, NSW, 2007, Australia.

Sudesh Vasdev, Discipline of Medicine, Faculty of Medicine, Health Sciences Centre, Memorial University of Newfoundland, St. John's, Newfoundland, Canada
Email: svasdev@mun.ca

Cheryl L.Wellington, Department of Pathology and Laboratory Medicine, University of British Columbia, Vancouver, British Columbia, Canada
Email: cheryl@cmmt.ubc.ca

Geoff H. Werstuck, Departments of Biochemistry and Biomedical Sciences, Medicine, and the Henderson Research Centre, McMaster University, Hamilton, Ontario, Canada
Email: gwerstuck@thrombosis.hhscr.org

Stewart C. Whitman, Departments of Pathology and Laboratory Medicine & Cellular and Molecular Medicine, University of Ottawa Heart Institute, Ottawa, Ontario, Canada
Email: swhitman@ottawaheart.ca

Enoka P. Wijekoon, Department of Biochemistry, Memorial University of Newfoundland, St. John's, Newfoundland, Canada

Yan-Jun Xu, Institute of Cardiovascular Sciences, St. Boniface General Hospital Research Centre and Department of Physiology, Faculty of Medicine, University of Manitoba, Winnipeg, Canada

Section I

Hyperlipidaemia and Atherosclerosis

1

Reverse Cholesterol Transport

JIM W. BURGESS, PHILIP A. SINCLAIR, CHRISTOPHE M. CHRETIEN, JONATHON BOUCHER, AND DANIEL L. SPARKS

Abstract

Atherosclerosis, the accumulation of cholesterol in the arteries resulting in heart attacks and strokes, is the leading cause of death in the USA and most other industrialized countries in the world. Plasma levels of high-density lipoprotein (HDL) cholesterol are invariably found to be inversely associated with the risk of atherosclerosis. This protective effect has classically been ascribed to HDL-mediated reverse cholesterol transport (RCT). In this process, nascent HDL in the circulation removes unesterified (free) cholesterol from peripheral cells, such as macrophages, through the transfer of cholesterol across the cell membrane by the ATP-binding cassette (ABCA1) transporter protein. The mature, spherical HDL particle, which contains these cholesterol esters in its core, then transfers the cholesterol to the liver through receptor-mediated processes. HDL cholesterol that is taken up by the liver is then excreted in the form of bile acids and cholesterol, completing the process of RCT. Because HDL-mediated RCT reduces serum cholesterol levels and is associated with a reduction in the risk of cardiovascular disease, increasing HDL is now being realized as a promising therapeutic end point.

Keywords: ABCA1; apolipoproteins; bile acids; cholesterol ester transfer protein; high-density lipoprotein; lipase; scavenger receptor BI

Abbreviations: ABC, ATP-binding cassette proteins; ACAT, acyl coenzyme A-cholesterol acyltransferase; apoA-I, apolipoprotein A-I; apoB, apolipoprotein B; AP1, transcription factor AP1; CE, cholesterol ester; CETP, cholesterol ester transfer protein; CYP7A1, cytochrome P450, family 7, subfamily A, polypeptide 1; EL, endothelial lipase; FXR, farnesoid X receptor; HDL, high-density lipoprotein; HNF-4, hepatocyte nuclear factor-4; HL, hepatic lipase; HMGCoA, hydroxymethylglutaryl coenzyme A; JAK2, Janus kinase 2; LCAT, lecithin cholesterol acyltransferase; LXR, liver X receptor; LDL, low-density lipoprotein; LPL, lipoprotein lipase; LRP, LDL-receptor related protein; nCEH, neutral cholesterol ester hydrolase; PPAR, peroxisome proliferator activated receptor; RCT, reverse cholesterol transport; RXR, retinoid X receptor; SR-BI, scavenger receptor class B type I; Sp1, transcription factor Sp1; SREBP, sterol regulatory element binding protein; TG, triglyceride; StAR, steroidogenic acute regulatory protein; VLDL, very low-density lipoprotein

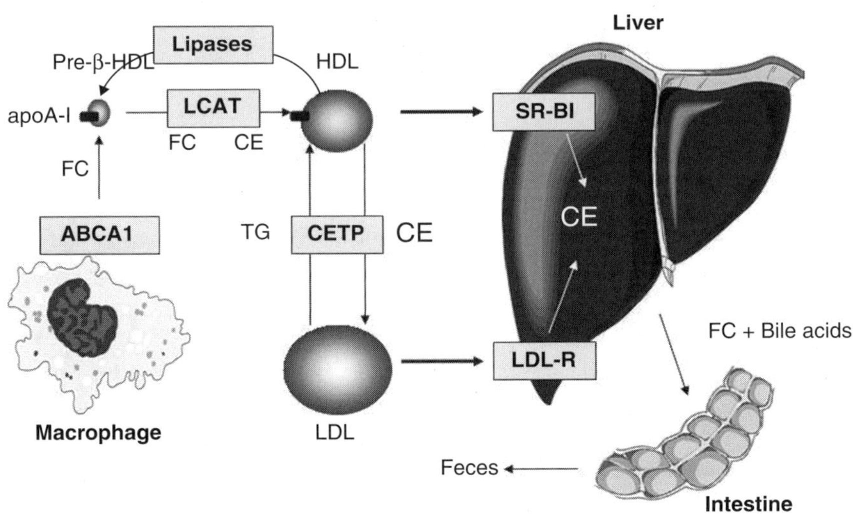

FIGURE 1.1. Schematic of the reverse cholesterol transport pathway. Cholesterol efflux to nascent HDL is regulated by ABCA1. Within the HDL pool, LCAT esterifies FC to CE and then CETP transfers CE from HDL to the LDL pool. FC and CE are taken up by the liver and recycled into the lipoprotein pool or secreted into the bile. ABCA1, ATP-binding cassette protein A1; apoA-I, apolipoprotein A-I; CE, cholesterol ester; CETP, cholesterol ester transfer protein; FC, free cholesterol; HDL, high-density lipoprotein; LCAT, lecithin cholesterol acyltransferase; LDL-R, low-density lipoprotein receptor; SR-BI, scavenger receptor class BI; TG, triglyceride.

Cholesterol that is synthesized in extrahepatic tissues or acquired from lipoproteins returns to the liver for excretion in a process called reverse cholesterol transport (RCT). RCT is a multistep process involving (i) cholesterol efflux from peripheral cells, (ii) maturation of the pre-β-high-density lipoprotein (pre-β-HDL) particle to spherical HDL, (iii) transfer of HDL-associated cholesterol to the liver, (iv) hepatic recycling or excretion of excess cholesterol, and (v) regeneration of the cholesterol acceptor pre-β-HDL. These individual steps are shown in Fig. 1.1 and will be reviewed separately in this chapter.

Cholesterol Efflux

The first step in RCT involves the transfer of cholesterol from cell membranes to acceptor lipoprotein particles in the extravascular fluids. Two mechanisms control this transfer of cholesterol. In the aqueous diffusion model, cholesterol molecules spontaneously disassociate from cell membranes, traverse the intervening aqueous space by diffusion, and then incorporate into acceptor particles. This process is driven by the gradient of free cholesterol (FC) between cells and the aqueous medium and does not require energy or interaction of the cholesterol acceptor with a cell surface receptor [1]. The second mechanism requires interaction between lipid-poor HDL and cell surface ABCA1. ABCA1, an ATP-binding cassette transporter mutated in Tangier disease [2], promotes cellular phospholipid and cholesterol efflux by loading free apolipoprotein A-I (apoA-I) with these lipids. This process involves binding of apoA-I to the cell surface and the energy-dependent transfer of phospholipid and cholesterol by ABCA1 to apoA-I [3]. Cross-linking studies have suggested a very close association of apoA-I and ABCA1 at the cell surface [4, 5], however, there is also evidence that the mechanisms by which apoA-I associates with the plasma membrane and ABCA1 vary depending on cell type. Macrophages demonstrate a requirement for the C-terminal domain of apoA-I for optimal binding and efflux [6]. This may result from a binding site for the C-terminal domain of apoA-I on the macrophage extracellular matrix [7]. In contrast, apoA-I binding and efflux to wild-type apoA-I and several deletion mutants including those lacking residues 187–243 of the C-terminal tail [6] is equally efficient in human fibroblasts. Coincidentally, fibroblasts lack the extracellular matrix binding site [7]. By analogy with multidrug resistance proteins, also members of the ATP-binding cassette transporter family, it has been suggested that ABCA1 forms a channel within the plasma membrane through which phospholipids and cholesterol are transferred to lipid-poor particles [8].

In addition to stimulating lipid efflux from cells, the interaction of apoA-I and cell surface ABCA1 also turns on signal transduction pathways and their downstream effectors. A schematic model illustrating these processes is shown in Fig. 1.2. In fibroblasts, HDL or apoA-I has been shown to activate members of the Rho family of G-proteins [9] and downstream processes including adenylate cyclase and phospholipases including phosphatidylinositol-specific phospholipase C and phosphatidylcholine-specific phospholipase D [10, 11]. Downstream protein kinases including protein kinase A (PKA) [12], protein kinase C (PKC) [13, 14], and Janus kinase 2 (JAK2) [15] are also activated. ApoA-I binding to human fibroblasts and consequent PKC activation is reported to mobilize intracellular cholesterol for efflux from an acyl coenzyme A-cholesterol acyltransferase (ACAT) accessible compartment [16]. Direct modulation of ABCA1 activity has been reported as a result of its phosphorylation. PKC directly phosphorylates ABCA1, which is reported to protect the protein against proteolytic degradation [17]. PKA activation by analogs of cAMP increase ABCA1 phosphorylation and cellular cholesterol efflux [18, 19]. Conversely, apoA-I-mediated release of cellular lipids is sensitive to inhibitors of PKA and impaired when PKA phosphorylation sites in ABCA1 are mutated [19]. Naturally occurring mutations of ABCA1 from Tangiers patients also demonstrated severely reduced apoA-I-mediated cAMP production, ABCA1 phosphorylation, apoA-I binding, and lipid efflux [18]. The protein-tyrosine kinase, JAK2, is required for optimum ABCA1-dependent lipid loading of apolipoproteins and appears to modulate

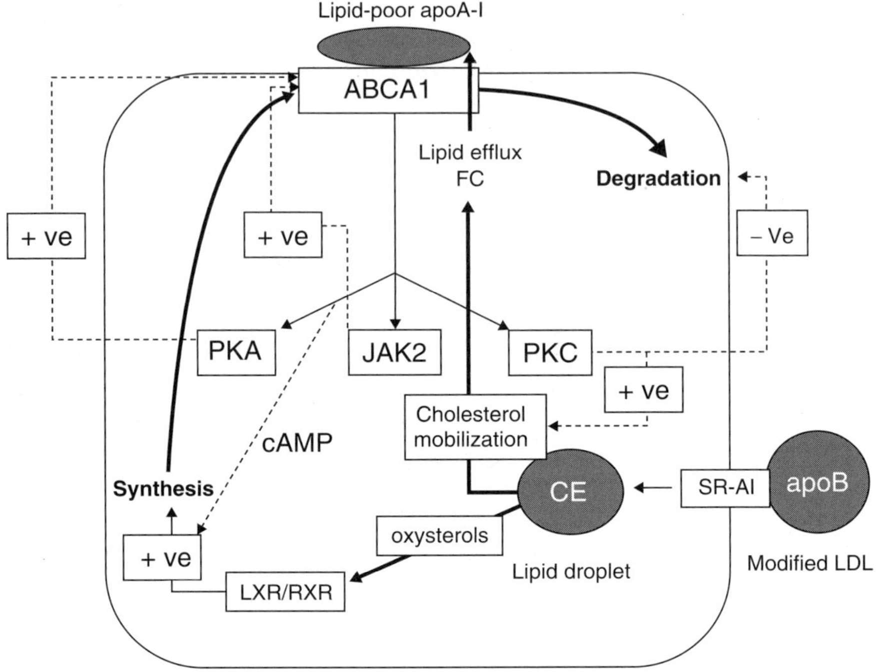

FIGURE 1.2. Regulation of the cholesterol efflux pathway. Binding of lipid-poor apoA-I to cell surface ABCA1 stimulates the efflux of FC and phospholipids and the activation of several protein kinases including PKA, PKC, and JAK2. PKA-mediated phosphorylation of ABCA1 optimizes the interaction between apoA-I and ABCA1 and increases the cholesterol translocase activity of ABCA1. Activation of JAK2 also enhances ABCA1 translocase activity. PKC-mediated phosphorylation of ABCA1 reduces degradation of ABCA1 and increases FC levels by mobilizing intracellular stores of cholesterol. The levels of ABCA1 are tightly controlled in cells. Synthesis is increased by metabolic products of lipid degradation that are ligands for the nuclear LXR/RXR receptor. Activated LXR/RXR increases transcription of the ABCA1 gene. ApoB, apolipoprotein B; JAK2, Janus kinase 2; LXR, liver X receptor; PKC, protein kinase C; PKA, protein kinase A; SR-AI, scavenger receptor class AI.

the apolipoprotein interactions with the ABCA1 transporter [15]. However, the lack of detectable phosphotyrosines in ABCA1 implies that the target of JAK2 is probably not ABCA1 [15].

The level of ABCA1 in cells is highly regulated. ApoA-I binding increases ABCA1 protein levels in mouse primary hepatocytes and peritoneal macrophages without affecting ABCA1 mRNA levels. This indicates that the rates of catabolism exert some control over ABCA1 levels in cells [20]. At the gene level the ABCA1 promoter contains binding motifs for several transcription factors, including the sterol regulatory element binding protein

(SREBP), the peroxisome proliferator activated receptor-α (PPAR-α), the liver X receptor (LXR), retinoid X receptor (RXR), and transcription factors Sp1 and AP1 binding sites [21]. These factors allow regulation through many diverse metabolites, including cAMP [5], sterols [22], *cis*-retinoic acid [23], PPAR agonists [24], and interferon-γ [25]. PPAR-α is a nuclear receptor activated by fatty acid derivatives and hypolipidemic drugs of the fibrate class. PPAR-α is expressed in monocytes, macrophages, and foam cells, suggesting a role for this receptor in macrophage lipid homeostasis, foam cell formation, and the consequent development of atherosclerosis [26].

Maturation of Nascent Pre-β-HDL

The transport of cholesterol from peripheral tissues to the liver by HDL (Fig. 1.1) is considered to be stimulated by the plasma enzymes lecithin cholesterol acyltransferase (LCAT) and cholesterol ester transfer protein (CETP). LCAT is synthesized in the liver and associates predominantly with HDL in the bloodstream. LCAT is activated by apoA-I [27] and converts HDL-associated cholesterol to cholesterol ester (CE). Nascent pre-β-HDL particles eventually become mature spherical particles after accumulation of sufficient CE. Spherical HDL is then acted on by CETP, which facilitates the transport of CE to the storage lipoproteins, low-density lipoprotein (LDL) and very low-density lipoprotein (VLDL), in exchange for TG [28]. LCAT and CETP therefore regulate the plasma storage of cholesterol and may contribute to RCT by channeling cholesterol from extrahepatic tissues to apolipoproteins B (apoB) that are subsequently catabolized in the liver [29]. Conversely, there is also data to suggest that inhibition of cholesterol storage pathways may stimulate RCT. Injection of phosphatidylinositol into the bloodstream of rabbits has been shown to inhibit LCAT activity and to directly stimulate RCT [30].

LCAT activity does not appear to be controlled at the gene level but is more affected by the composition and electrostatic properties of the plasma lipoproteins. In humans and some animal models on atherogenic diets, treatment with cholesterol-reducing drugs does not have a significant impact on LCAT expression [31]. In contrast, the human CETP promoter is under the control of several regulatory elements including the orphan nuclear receptor ARP1 [32], the CCAAT/enhancer-binding protein [33], *trans*-retinoic acid [34], the SREBP [35] and the liver receptor homolog-1 [36]. This allows several metabolic factors to modulate CETP levels including dietary cholesterol [37], fatty acids [38], and corticosteroids [39]. Numerous studies have reported that CETP modifies HDL cholesterol (HDL-C) and apoA-I levels, which are decreased by overexpression of CETP [40]. Human genetic deficiency of CETP results in significantly increased plasma HDL levels [41]. This has led to a considerable interest in CETP inhibitors as drugs to elevate HDL-C levels.

Lipase activity in plasma also appears to exert effects on RCT. Increased hydrolysis of TG-rich lipoproteins by lipoprotein lipase (LPL) results in elevated levels of HDL in the bloodstream and may positively impact RCT [42, 43]. In agreement with these findings, low levels of LPL activity, through knockout or mutation in animal models or human LPL deficiency, are correlated with reduced levels of HDL [44, 45]. LPL may indirectly impact RCT through the generation of lipoprotein remnants that transport cholesterol to the liver. The generation of remnants produces substrates for CETP-mediated transfer of CE from HDL [42] thereby providing additional carriers for the transport of cholesterol to the liver. Thus, LPL may stimulate RCT by increasing HDL levels and enhancing CE transport to the liver.

Hepatic lipase (HL) may also play a role in the movement of cholesterol to the liver. HL is expressed primarily in the liver and is found bound to the sinusoidal endothelium [46]. HL hydrolyzes HDL phospholipids and triglycerides (TGs) as well as chylomicron and VLDL lipids [47]. A portion of nascent HDL is derived from lipase-mediated hydrolysis of TG-rich lipoproteins and lipolysis of chylomicron and VLDL lipids is correlated with an increase in circulating HDL [42]. It is assumed that high-LPL/low-HL activity is related to high-HDL levels and that enhanced TG-rich hydrolysis is associated with a stimulation of cholesterol transport [42]. Studies have shown that hypertriglyceridemia is associated with low-HDL levels and high-HL activity [48]. This appears to suggest that a low LPL/high HL may be causative to high-TG and low-HDL levels. This phenotype is also commonly associated with increased risk of atherosclerosis. HL may also catalyze the actions of CETP to promote RCT. TG-enrichment of HDL by CETP followed by HL-mediated hydrolysis has been shown to enhance uptake of CE by scavenger receptor class B type I (SR-BI) [49].

Transfer of HDL-Associated Cholesterol to the Liver

Cholesterol is stored and transported in the bloodstream in HDL and the apoB-containing lipoproteins almost exclusively as CE. More than half of the apoB-associated cholesterol is internalized by hepatic LDL-receptors, which are located in clathrin-coated pits on the hepatocyte cell surface. The receptor–ligand complex is delivered to lysosomes by the endosomal pathway [50]. The LDL-receptor is efficiently returned to the cell surface by recycling endosomes and returned to coated pits for another round of binding and internalization. ApoB-associated cholesterol can also be internalized by the multifunctional LDL-receptor related protein (LRP) [51]. Holoparticle uptake by this path is complex and involves interactions between LRP and cell surface heparin sulfate proteoglycans [52]. These long unbranched highly polyanionic molecules are required for LRP-mediated lipoprotein uptake and act by forming a binding matrix with lipoproteins through associations with apoE, HL, or LPL [53]. Regardless of the receptor and pathways involved, internalized apoB-containing lipoproteins are delivered to lysosomes and completely degraded to amino acids, FC, and free fatty acids (Fig. 1.3).

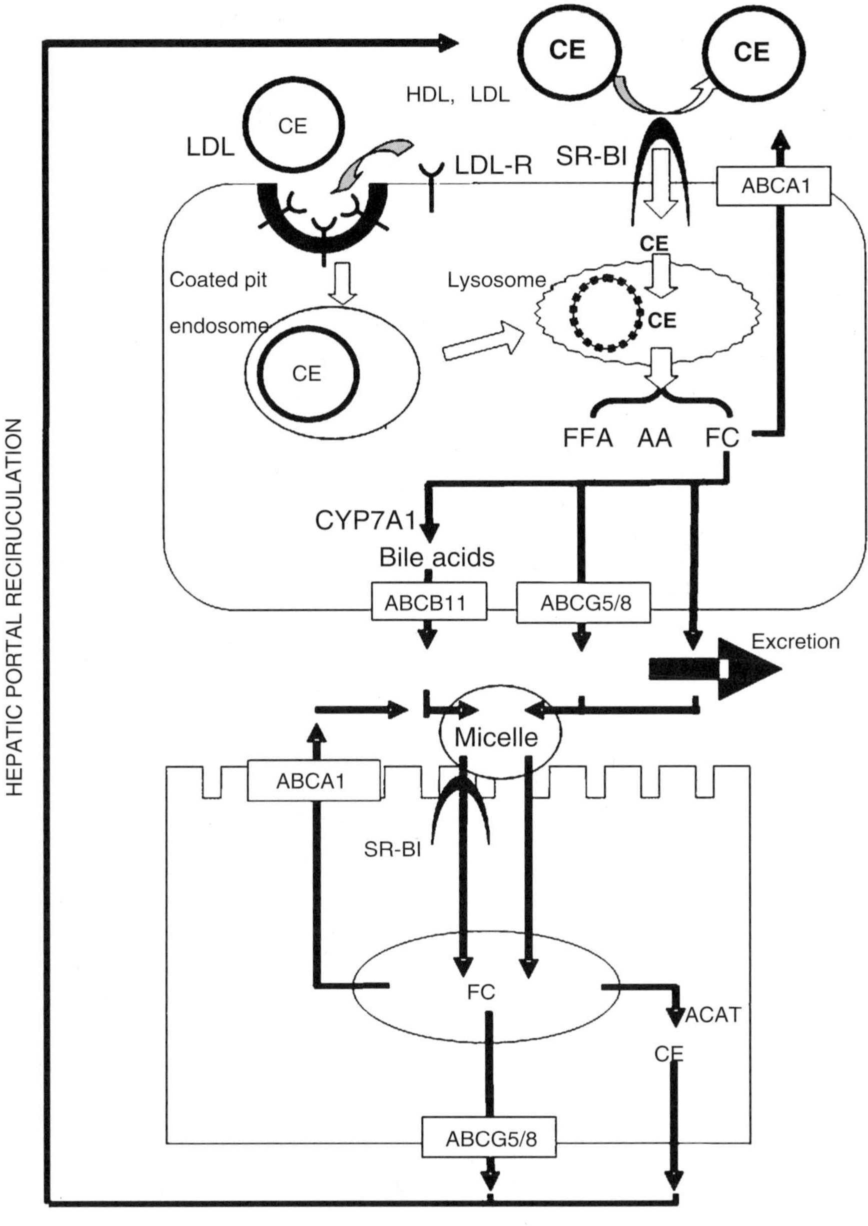

FIGURE 1.3. Cholesterol is taken up and metabolized within the liver hepatocyte through two distinct pathways. LDL is taken up by the LDL receptor (LDL-R) and undergoes endocytosis and degradation within lysosomes. HDL and LDL cholesterol can also be taken up by SR-BI, through a selective uptake of cholesterol and CE with the recycling of a cholesterol-poor lipoprotein. CE is stored within the cell or hydrolyzed to cholesterol and secreted intact into the bile duct or as a bile acid. In the gut lumen, cholesterol is absorbed by the enterocyte, which can excrete cholesterol in the lumen by the action of ABCA1. Dietary cholesterol is converted to CE and directed to the hepatic portal circulation.

HDL cholesterol can be taken into cells without parallel apolipoprotein uptake and degradation. This process, termed selective uptake, is believed to be an important step in RCT. SR-BI has been shown to mediate this process in rodents. SR-BI drives selective uptake by a poorly understood two-step process: binding of HDL via apoA-I followed by selective transfer of cholesterol to cells. Wang et al. [54] suggested that the action of HL on apoA-I-containing lipoproteins may facilitate the SRBI-mediated uptake of HDL lipid. This was later confirmed by another group that demonstrated the bridging properties of HL and proteoglycans with SR-BI [55]. Bridging of cells and lipoproteins has also been recently suggested to induce the SR-BI-independent selective uptake of CE [56]. CD36, an SR-BI-related protein and best known for its ability to mediate uptake of oxidized LDL also mediates selective uptake, although to a lesser extent than SR-BI [57].

Three models have been proposed to describe the process of selective uptake. In the first model, SR-BI mediates the fusion between HDL and the plasma membrane, allowing the lipid transfer to the plasma membrane down a concentration gradient [58]. Rodrigueza et al. [59] have suggested that SR-BI forms a nonaqueous channel in the plasma membrane, through which cholesterol can move down a concentration gradient into the cell. Finally, a retroendocytic mechanism has been proposed in which SR-BI-associated HDL is internalized in caveolae and cholesterol is removed intracellularly followed by resecretion of cholesterol-poor HDL [60].

In rats, 60–70% of HDL-C is removed by SR-BI-mediated selective uptake [61]. The human homolog of SR-BI, termed CLA-1, exhibits similar tissue distribution [62], binding properties for a wide spectrum of plasma lipoproteins, and identical cholesterol transfer capacities as those of murine SR-BI [63]. Data from a human hepatic cell model demonstrate that CLA-1 is responsible for the selective uptake of cholesterol from HDL and LDL particles [64]. Furthermore, recent epidemiological studies have identified single-nucleotide polymorphisms in the CLA-1 gene that are associated with abnormal plasma lipid levels and lipoprotein composition [65]. These studies indicate that SR-BI/CLA-1 play a pivotal role in RCT and the metabolism of the cholesterol component of both HDL and apoB-lipoproteins in humans.

Cholesterol homeostasis in the liver is tightly controlled at the receptor and synthesis level. High-intracellular concentrations of cholesterol downregulates the LDL receptor [66] and hydroxymethylglutaryl coenzyme A (HMGCoA) reductase (the rate-limiting step in cholesterol synthesis) expression [67]. This process is regulated by SREBP [68]. The regulation of LRP expression is poorly understood although it has been reported that certain growth factors, for example insulin and nerve growth factors as well as cytokines like interleukin-β and tumor necrosis factor-α, upregulate the synthesis of LRP in HepG2 cells [69]. The expression of SR-BI gene is modulated by the activation of multiple signaling pathways, mediated by the involvement of SREBP, LXR, and PPAR-α [70].

Enterohepatic Metabolism of Cholesterol

Both the liver and the intestine are essential organs in the regulation of cholesterol metabolism. Cholesterol that is transported to the liver via lipoproteins becomes a substrate for bile acid synthesis and can be secreted into the bile in the form of bile acids and cholesterol. This cholesterol–bile acid pool is released into the large intestine where it is recycled and reabsorbed via the enterohepatic circulatory pathway. However, approximately 50% of cholesterol that is secreted into the gastrointestinal tract is not absorbed and is excreted [71]. Thus, the enterohepatic pathway (Fig. 1.3) provides the major route for cholesterol removal from the body.

Cholesterol that is taken up by the liver is made available for storage, membrane function, hormone and vitamin synthesis, and bile acid synthesis. FC that is taken into the hepatocyte can be esterified by ACAT [72], which is an intracellular enzyme that catalyzes the formation of cholesteryl esters from cholesterol and fatty acyl coenzyme A. Cholesterol hydrolases such as neutral cholesterol ester hydrolase (nCEH), also known as hormone-sensitive lipase, catalyze the hydrolytic cleavage of CEs to form FC [73]. The process by which cholesterol is mobilized from the lipid droplet is uncertain. It is likely that transfer proteins such as caveolin [74] or the steroidogenic acute regulatory protein (StAR) [75] stimulate CE hydrolysis and mobilization to the endoplasmic reticulum for bile acid synthesis or to the mitochondria for steroidogenesis.

Bile acids are end products of cholesterol metabolism and function to emulsify exogenous and endogenous lipids. HDL has been implicated as the preferential source of cholesterol for bile acid synthesis such that most of the nonesterified cholesterol from HDL is converted to bile acids [76]. The conversion of cholesterol into bile acids is initiated by the hepatic P450 enzyme, cholesterol 7α-hydroxylase (CYP7A1) [77]. This is followed by multiple hydroxylations of the cholesterol steroid ring structure and cleavage of the side chain. Thus, the bile acids retain both lipophilic and hydrophilic properties and this amphipathic property allows for the formation of mixed micelles with phospholipids and cholesterol in the bile.

In the hepatocyte, cholesterol transport into the bile involves multiple factors, all of which influence cholesterol homeostasis. The ability to transport bile acids and cholesterol across the canalicular plasma membrane of the hepatocyte is a controlling factor in biliary lipid secretion/excretion. A specific membrane transporter belonging to the ABC-binding cassette family of proteins, termed the bile salt export pump (ABCB11), is responsible for bile acid transport across the canalicular membrane [78]. Changes in the expression of hepatic ABCG5 and ABCG8 transporters has been shown to be correlated with changes in biliary cholesterol secretion [79]. Thus, ABCG5 and ABCG8 may help to facilitate the excretion of FC into the bile. Once secreted into the bile, the bile acids form very strong molecular attractions to phospholipids and aid in solubilizing FC for transport into the intestine.

The majority of bile acids in the intestinal tract are reabsorbed. This is confirmed by the finding that relatively low levels (approximately 5%) of bile acids are present in the feces [80]. Thus, up to 95% of bile acid is reabsorbed by the small intestine into the portal vein and returned to the liver, where it either functions to regulate bile acid synthesis or is recirculated into the bile. This enterohepatic circulation recycles approximately ten times per day and has a major influence on the level of bile acid synthesis and hence, cholesterol homeostasis. Biliary cholesterol is also regulated by the enterohepatic circulation. Approximately 50% of biliary cholesterol is reabsorbed; however, this value can range between 20 and 80% depending on differences between individuals.

The regulation of the enterohepatic metabolism of cholesterol is of key interest, as the cholesterol and bile acids that are not reabsorbed constitute the major route for cholesterol excretion from the body. The enterohepatic metabolism of cholesterol is a process that is largely regulated at the molecular level. Many nuclear hormone receptors have been identified that play significant roles in the transcriptional control of the genes involved in bile acid synthesis and cholesterol absorption. Two nuclear hormone receptors that are involved in cholesterol homeostasis are the LXR and the farnesoid X receptor (FXR) [81]. Upon ligand binding, these receptors form a heterodimer with RXR and interact with a response element within the proximal promoter regions of genes involved in cholesterol metabolism. In the liver, LXR regulates the 7α-hydroxylation of cholesterol by inducing the expression of CYP7A1, and thus enhances the catabolism/excretion of cholesterol [82]. However, as a feedback control, FXR can bind to bile acids, followed by interaction with a negative response element on the CYP7A1 gene, causing an inhibition of bile acid synthesis [83]. In addition to FXR, the hepatocyte nuclear factor-4α (HNF-4α) is also implicated in the downregulation of CYP7A1 [84]. Thus, bile acids returning to the liver through the enterohepatic circulation downregulate the transcription of CYP7A1. These receptor systems therefore work in tandem to control cholesterol homeostasis through their ability to enhance or inhibit bile acid synthesis.

The LXR receptor is also involved in the regulation of cholesterol transport proteins such as the ABC transporters at extrahepatic locations such as the intestine. Upon binding cholesterol and/or its metabolites, LXR upregulates the transcription of ABCA1, ABCG5, and ABCG8, leading to increased efflux of intracellular cholesterol [85]. ABCA1 activity is also suggested to be enhanced by LXR interacting with PPAR [86]. In the liver, PPAR-α has been shown to decrease CYP7A1 expression [87] while increasing hepatic SR-BI levels [88] leading to increased hepatic cholesterol uptake.

Nuclear receptors are highly involved in the regulation of RCT, bile acid synthesis, and cholesterol excretion. Thus, alterations in receptor-dependent mechanisms could lead to differences in the amount of mRNA that is transcribed. Based on this concept, these nuclear receptors are likely targets for therapeutic interventions for the treatment of atherosclerosis.

Regeneration of the Cholesterol Acceptor Pre-β-HDL

The factors regulating the regeneration of cholesterol accepting pre-β-HDL are shown in Fig. 1.1. HL plays a major role in HDL remodeling by hydrolyzing phospholipid in the larger HDL_2 subfractions to produce the smaller, denser HDL_3 and by generating pre-β-HDL from the hydrolysis of spherical HDL [89]. Overexpression of HL in LCAT-transgenic mice results in significant reductions in total cholesterol, HDL-C, and phospholipids [90]. In addition, overexpression of a catalytically inactive HL also results in a decrease in HDL-C [91]. Thus, the HDL-lowering potential of HL may involve a component that is not related to its enzymatic activity. In contrast, low-HL activity from gene mutations or in HL-deficient patients causes elevated HDL levels [92].

The recently discovered endothelial lipase (EL) hydrolyzes almost exclusively HDL phospholipid *in vivo* [93, 94] and appears to play a very important and antagonistic role in the regulation of HDL levels. In EL overexpression animal models, HDL-C levels were drastically reduced [93]. In contrast, EL knockout animals showed elevated HDL-C [95]. Likewise, genetic variation in the human EL gene has been correlated with HDL levels [96]. Characterization of human EL is less comprehensive than in animal models, however, studies do show that SNPs in the EL gene (LIPG gene) are linked to changes in HDL-C levels, suggesting that it is a potential antiatherogenic target [97]. By lowering HDL, EL may block the initiation of the RCT pathway. Transgenic mice with the double apoE and EL knockout had much lower HDL levels compared to wild type and apoE knockout and a 70% reduction in atherosclerosis [96]. EL activity also appears to affect efflux of cholesterol via SR-BI and ABCA1 in human apoA-I transgenic mice. Overexpression of EL caused an approximately 91% reduction in HDL-C and total cholesterol. It also caused a 90% decrease in efflux potential of the serum via SR-BI and a 63% increase in the efflux potential of the serum via ABCA1 [98]. Yancey et al. [98] suggest that the type of lipase acting on HDL *in vivo* will determine whether HDL is effluxed via the SR-BI or ABCA1 pathway. Further, EL-modified HDL displays a reduced ability to efflux cholesterol via the SR-BI pathway, while not affecting ABCA1-mediated cholesterol efflux. Studies have shown that EL-hydrolysis of HDL increases the negative charge of HDL, rendering HDL less able to bind to SR-BI on the cell surface [99]. Thus, EL lowers HDL levels, reduces the efflux potential of HDL particles, and decreases SR-BI-mediated selective uptake of CE.

Drugs that Affect HDL Levels and RCT

The large number of drugs in development to increase HDL cholesterol levels reflect a new focus for the pharmaceutical industry and the importance of HDL as a therapeutic target [100]. Of the currently available drugs to raise HDL levels [101], niacin has shown considerable clinical benefits by elevating HDL cholesterol by 15–30% [102]. Among the various mechanisms characterized are inhibition of lipolysis in adipose tissue, inhibition of TG synthesis, and inhibition of apoA-I catabolism [102]. Unfortunately the adverse effects associated with this drug make compliance poor. Raising HDL can also be achieved using fibrates or certain statins such as simvastatin, however these effects are fairly modest (usually <10%). Fibrates, such as gemfibrozil and fenofibrate, are best known for their TG-lowering effects. Their mechanism of action is also complex. They belong to the class of PPAR agonists [101, 102]. PPAR agonists are a family of nuclear hormone receptors that are widespread throughout the body. Fibrates stimulate PPAR-α in the liver and other tissues, leading to the expression of multiple genes involved in lipoprotein metabolism. Fibrates stimulate apoA-I and ABCA1 synthesis and decrease the synthesis of TGs. Fibrates are also known to enhance lipase synthesis and activity, suggesting that conventional drugs therapies to treat dyslipidemia act, at least in part, by stimulating LPL, HL, and EL. Fibrates also increase SR-BI levels in human monocytes and macrophages, although an opposite effect is observed in murine liver at a posttranscriptional level [103].

Some drugs under development have been designed to directly promote RCT. Among the more novel approaches is administration of apoA-I or related peptides (artificial HDL), either as long-term oral therapy or by infusion for subacute treatment. In a recent study involving a naturally occurring apoA-I variant (apoA-I Milano) administered intravenously [104], coronary atheroma volume decreased by 1% after 5 weeks. This approach represents a paradigm shift in thinking about lipid therapy, which is generally considered to reduce risk gradually over months to years. Intravenous administration of these compounds is thought to rapidly mobilize cholesterol from arteries for disposal by the liver, thereby stabilizing vulnerable plaque.

Inhibition of CETP is another approach to raise HDL levels, although it is unclear whether it will positively affect RCT. CETP inhibitors such as torcetrapib and JTT-705 are currently in clinical trials and have shown promising elevations in apoA-I and HDL [105, 106]. A vaccine against CETP antigen is also being evaluated in clinical trials [100]. Phosphatidylinositol is being tested as an HDL-elevating therapeutic in clinical trials and early efficacy results suggest that the compound can significantly elevate both apoA-I and HDL cholesterol levels [107]. Phosphatidylinositol also stimulates multiple steps in the RCT pathway. This naturally occurring lipid was shown to increase cholesterol efflux from cholesterol-loaded macrophages through the inositol-signaling cascade and to increase cholesterol transport to the liver and feces [30]. Drugs that inhibit HDL apoA-I catabolism or the remodeling of HDL particles [108] (e.g., EL inhibitors) represent another promising approach.

The enterohepatic circulation plays an important role in the excretion of cholesterol derived from RCT and a number of strategies have been designed to pharmacologically inhibit cholesterol absorption (or reabsorption) in the gut. The micellar solubilization of cholesterol can be disrupted with the use of agents such as orlistat (Xenical), a lipase inhibitor, which forms an oil phase in the intestinal lumen which entraps cholesterol, allowing for its fecal excretion [109]. In addition, bile acid resins have also been used to interfere with cholesterol absorption. Bile acid resins, such as colesevelam, bind to both cholesterol and bile acids in the intestinal tract, forming an insoluble matrix, which allows for their subsequent elimination in the feces [110]. A disadvantage of bile acid resins is their potential ability to additionally block fat-soluble vitamin absorption. More recently, a major focus has been directed towards ezetimibe, a cholesterol absorption inhibitor that interrupts the absorption of dietary cholesterol and biliary cholesterol without affecting the absorption of fat-soluble vitamins. Ezetimibe is rapidly absorbed and is recycled enterohepatically multiple times [111].

References

1. Yokoyama S: Release of cellular cholesterol: molecular mechanism for cholesterol homeostasis in cells and in the body. Biochim Biophys Acta 1529: 231–244, 2000.
2. Oram JF: Tangier disease and ABCA1. Biochim Biophys Acta 1529: 321–330, 2000.
3. Oram JF, Vaughan AM: ABCA1-mediated transport of cellular cholesterol and phospholipids to HDL apolipoproteins. Curr Opin Lipidol 11: 253–260, 2000.
4. Wang N, Silver DL, Costet P, Tall AR: Specific binding of apoA-I, enhanced cholesterol efflux, and altered plasma membrane morphology in cells expressing ABC1. J Biol Chem 275: 33053–33058, 2000.
5. Oram JF, Lawn RM, Garvin MR, Wade DP: ABCA1 is the cAMP-inducible apolipoprotein receptor that mediates cholesterol secretion from macrophages. J Biol Chem 275: 34508–34511, 2000.
6. Burgess JW, Frank PG, Franklin V, Liang P, McManus DC, Desforges M, Rassart E, Marcel YL: Deletion of the C-terminal domain of apolipoprotein A-I impairs cell surface binding and lipid efflux in macrophage. Biochemistry 38: 14524–14533, 1999.
7. Burgess JW, Kiss RS, Zheng H, Zachariah S, Marcel YL: Trypsin-sensitive and lipid-containing sites of the macrophage extracellular matrix bind apolipoprotein A-I and participate in ABCA1-dependent cholesterol efflux. J Biol Chem 277: 31318–31326, 2002.
8. Oram JF, Lawn RM: ABCA1. The gatekeeper for eliminating excess tissue cholesterol. J Lipid Res 42: 1173–1179, 2001.
9. Nofer JR, Feuerborn R, Levkau B, Sokoll A, Seedorf U, Assmann G: Involvement of Cdc42 signaling in apoA-I-induced cholesterol efflux. J Biol Chem 278: 53055–53062, 2003.
10. Nofer JR, Fobker M, Hobbel G, Voss R, Wolinska I, Tepel M, Zidek W, Junker R, Seedorf U, von Eckardstein A, Assmann G, Walter M: Activation of phosphatidylinositol-specific phospholipase C by HDL-associated lysosphingolipid. Involvement in mitogenesis but not in cholesterol efflux. Biochemistry 39: 15199–15207, 2000.
11. Walter M, Reinecke H, Gerdes U, Nofer JR, Hobbel G, Seedorf U, Assmann G: Defective regulation of phosphatidylcholine-specific phospholipases C and D in

a kindred with Tangier disease. Evidence for the involvement of phosphatidyl-choline breakdown in HDL-mediated cholesterol efflux mechanisms. J Clin Invest 98: 2315–2323, 1996.

12. Haidar B, Denis M, Marcil M, Krimbou L, Genest J Jr: Apolipoprotein A-I activates cellular cAMP signaling through the ABCA1 transporter. J Biol Chem 279: 9963–9969, 2004.

13. Theret N, Delbart C, Aguie G, Fruchart JC, Vassaux G, Ailhaud G: Cholesterol efflux from adipose cells is coupled to diacylglycerol production and protein kinase C activation. Biochem Biophys Res Commun 173: 1361–1368, 1990.

14. Mendez AJ, Oram JF, Bierman EL: Protein kinase C as a mediator of high density lipoprotein receptor-dependent efflux of intracellular cholesterol. J Biol Chem 266: 10104–10111, 1991.

15. Tang C, Vaughan AM, Oram JF: Janus kinase 2 modulates the apolipoprotein interactions with ABCA1 required for removing cellular cholesterol. J Biol Chem 279: 7622–7628, 2004.

16. Yamauchi Y, Chang CC, Hayashi M, Abe-Dohmae S, Reid PC, Chang TY, Yokoyama S: Intracellular cholesterol mobilization involved in the ABCA1/apolipoprotein-mediated assembly of high density lipoprotein in fibroblasts. J Lipid Res 45: 1943–1951, 2004.

17. Yamauchi Y, Hayashi M, Abe-Dohmae S, Yokoyama S: Apolipoprotein A-I activates protein kinase C alpha signaling to phosphorylate and stabilize ATP binding cassette transporter A1 for the high density lipoprotein assembly. J Biol Chem 278: 47890–47897, 2003.

18. Haidar B, Denis M, Krimbou L, Marcil M, Genest J Jr: cAMP induces ABCA1 phosphorylation activity and promotes cholesterol efflux from fibroblasts. J Lipid Res 43: 2087–2094, 2002.

19. See RH, Caday-Malcolm RA, Singaraja RR, Zhou S, Silverston A, Huber MT, Moran J, James ER, Janoo R, Savill JM, Rigot V, Zhang LH, Wang M, Chimini G, Wellington CL, Tafuri SR, Hayden MR: Protein kinase A site-specific phos-phorylation regulates ATP-binding cassette A1 (ABCA1)-mediated phospho-lipid efflux. J Biol Chem 277: 41835–41842, 2002.

20. Wang N, Chen W, Linsel-Nitschke P, Martinez LO, Agerholm-Larsen B, Silver DL, Tall AR: A PEST sequence in ABCA1 regulates degradation by calpain protease and stabilization of ABCA1 by apoA-I. J Clin Invest 111: 99–107, 2003.

21. Yang XP, Freeman LA, Knapper CL, Amar MJ, Remaley A, Brewer HB Jr, Santamarina-Fojo S: The E-box motif in the proximal ABCA1 promoter mediates transcriptional repression of the ABCA1 gene. J Lipid Res 43: 297–306, 2002.

22. Repa JJ, Turley SD, Lobaccaro JA, Medina J, Li L, Lustig K, Shan B, Heyman RA, Dietschy JM, Mangelsdorf DJ: Regulation of absorption and ABC1-mediated efflux of cholesterol by RXR heterodimers. Science 289: 1524–1529, 2000.

23. Costet P, Lalanne F, Gerbod-Giannone MC, Molina JR, Fu X, Lund EG, Gudas LJ, Tall AR: Retinoic acid receptor-mediated induction of ABCA1 in macrophages. Mol Cell Biol 23: 7756–7766, 2003.

24. Chinetti G, Lestavel S, Bocher V, Remaley AT, Neve B, Torra IP, Teissier E, Minnich A, Jaye M, Duverger N, Brewer HB, Fruchart JC, Clavey V, Staels B: PPAR-alpha and PPAR-gamma activators induce cholesterol removal from human macrophage foam cells through stimulation of the ABCA1 pathway. Nat Med 7: 53–58, 2001.

25. Panousis CG, Zuckerman SH: Interferon-gamma induces downregulation of Tangier disease gene (ATP-binding-cassette transporter 1) in macrophage-derived foam cells. Arterioscler Thromb Vasc Biol 20: 1565–1571, 2000.

26. Cabrero A, Cubero M, Llaverias G, Jove M, Planavila A, Alegret M, Sanchez R, Laguna JC, Carrera MV: Differential effects of peroxisome proliferator-activated receptor activators on the mRNA levels of genes involved in lipid metabolism in primary human monocyte-derived macrophages. Metabolism 52: 652–657, 2003.

27. Jonas A: Lecithin cholesterol acyltransferase. Biochim Biophys Acta 1529: 245–256, 2000.

28. Tall AR: Plasma high density lipoproteins. Metabolism and relationship to atherogenesis. J Clin Invest 86: 379–384, 1990.

29. Barter PJ, Hopkins GJ, Calvert GD: Transfers and exchanges of esterified cholesterol between plasma lipoproteins. Biochem J 208: 1–7, 1982.

30. Burgess JW, Boucher J, Neville TA, Rouillard P, Stamler C, Zachariah S, Sparks DL: Phosphatidylinositol promotes cholesterol transport and excretion. J Lipid Res 44: 1355–1363, 2003.

31. Jonas A: Regulation of lecithin cholesterol acyltransferase activity. Prog Lipid Res 37: 209–234, 1998.

32. Gaudet F, Ginsburg GS: Transcriptional regulation of the cholesteryl ester transfer protein gene by the orphan nuclear hormone receptor apolipoprotein AI regulatory protein-1. J Biol Chem 270: 29916–29922, 1995.

33. Agellon LB, Zhang P, Jiang XC, Mendelsohn L, Tall AR: The CCAAT/enhancer-binding protein trans-activates the human cholesteryl ester transfer protein gene promoter. J Biol Chem 267: 22336–22339, 1992.

34. Jeoung NH, Jang WG, Nam JI, Pak YK, Park YB: Identification of retinoic acid receptor element in human cholesteryl ester transfer protein gene. Biochem Biophys Res Commun 258: 411–415, 1999.

35. Chouinard RA Jr, Luo Y, Osborne TF, Walsh A, Tall AR: Sterol regulatory element binding protein-1 activates the cholesteryl ester transfer protein gene *in vivo* but is not required for sterol up-regulation of gene expression. J Biol Chem 273: 22409–22414, 1998.

36. Luo Y, Liang CP, Tall AR: The orphan nuclear receptor LRH-1 potentiates the sterol-mediated induction of the human CETP gene by liver X receptor. J Biol Chem 276: 24767–24773, 2001.

37. Jiang XC, Agellon LB, Walsh A, Breslow JL, Tall A: Dietary cholesterol increases transcription of the human cholesteryl ester transfer protein gene in transgenic mice. Dependence on natural flanking sequences. J Clin Invest 90: 1290–1295, 1992.

38. Hirano R, Igarashi O, Kondo K, Itakura H, Matsumoto A: Regulation by long-chain fatty acids of the expression of cholesteryl ester transfer protein in HepG2 cells. Lipids 36: 401–406, 2001.

39. Masucci-Magoulas L, Moulin P, Jiang XC, Richardson H, Walsh A, Breslow JL, Tall A: Decreased cholesteryl ester transfer protein (CETP) mRNA and protein and increased high density lipoprotein following lipopolysaccharide administration in human CETP transgenic mice. J Clin Invest 95: 1587–1594, 1995.

40. Marotti KR, Castle CK, Boyle TP, Lin AH, Murray RW, Melchior GW: Severe atherosclerosis in transgenic mice expressing simian cholesteryl ester transfer protein. Nature 364: 73–75, 1993.

41. Brown ML, Inazu A, Hesler CB, Agellon LB, Mann C, Whitlock ME, Marcel YL, Milne RW, Koizumi J, Mabuchi H, Takeda R, Tall AR: Molecular basis of lipid transfer protein deficiency in a family with increased high-density lipoproteins. Nature 342: 448–451, 1989.
42. Miller NE, Nanjee MN: Evidence that reverse cholesterol transport is stimulated by lipolysis of triglyceride-rich lipoproteins. FEBS Lett 285: 132–134, 1991.
43. Navab M, Hama SY, Hough GP, Hedrick CC, Sorenson R, La Du BN, Kobashigawa JA, Fonarow GC, Berliner JA, Laks H, Fogelman AM: High density associated enzymes: their role in vascular biology. Curr Opin Lipidol 9: 449–456, 1998.
44. Coleman T, Seip RL, Gimble JM, Lee D, Maeda N, Semenkovich CF: COOH-terminal disruption of lipoprotein lipase in mice is lethal in homozygotes, but heterozygotes have elevated triglycerides and impaired enzyme activity. J Biol Chem 270: 12518–12525, 1995.
45. Parrott CL, Alsayed N, Rebourcet R, Santamarina-Fojo S: ApoC-IIParis2: a premature termination mutation in the signal peptide of apoC-II resulting in the familial chylomicronemia syndrome. J Lipid Res 33: 361–367, 1992.
46. Santamarina-Fojo S, Haudenschild C, Amar M: The role of hepatic lipase in lipoprotein metabolism and atherosclerosis. Curr Opin Lipidol 9: 211–219, 1998.
47. Thuren T: Hepatic lipase and HDL metabolism. Curr Opin Lipidol 11: 277–283, 2000.
48. Soderlund S, Soro-Paavonen A, Ehnholm C, Jauhiainen M, Taskinen MR: Hypertriglyceridemia is associated with prebeta-HDL levels in subjects with familial low HDL. J Lipid Res 2005.
49. Collet X, Tall AR, Serajuddin H, Guendouzi K, Royer L, Oliveira H, Barbaras R, Jiang XC, Francone OL: Remodeling of HDL by CETP *in vivo* and by CETP and hepatic lipase *in vitro* results in enhanced uptake of HDL CE by cells expressing scavenger receptor B-I. J Lipid Res 40: 1185–1193, 1999.
50. Meddings JB, Dietschy JM: Regulation of plasma levels of low-density lipoprotein cholesterol: interpretation of data on low-density lipoprotein turnover in man. Circulation 74: 805–814, 1986.
51. Ji ZS, Dichek HL, Miranda RD, Mahley RW: Heparan sulfate proteoglycans participate in hepatic lipase and apolipoprotein E-mediated binding and uptake of plasma lipoproteins, including high density lipoproteins. J Biol Chem 272: 31285–31292, 1997.
52. Williams KJ, Fless GM, Petrie KA, Snyder ML, Brocia RW, Swenson TL: Mechanisms by which lipoprotein lipase alters cellular metabolism of lipoprotein(a), low density lipoprotein, and nascent lipoproteins. Roles for low density lipoprotein receptors and heparan sulfate proteoglycans. J Biol Chem 267: 13284–13292, 1992.
53. Rhainds D, Brissette L: Low density lipoprotein uptake: holoparticle and cholesteryl ester selective uptake. Int J Biochem Cell Biol 31: 915–931, 1999.
54. Wang N, Weng W, Breslow JL, Tall AR: Scavenger receptor BI (SR-BI) is up-regulated in adrenal gland in apolipoprotein A-I and hepatic lipase knock-out mice as a response to depletion of cholesterol stores. *In vivo* evidence that SR-BI is a functional high density lipoprotein receptor under feedback control. J Biol Chem 271: 21001–21004, 1996.
55. Lambert G, Chase MB, Dugi K, Brewer J, Santamarina-Fojo S, Bensadoun A, Lambert G: Hepatic lipase promotes the selective uptake of high density

lipoprotein-cholesteryl esters via the scavenger receptor B1. J Lipid Res 40: 1294–1303, 1999.

56. Brundert M, Greten H, Rinninger F, Heeren J: Hepatic lipase mediates an increase in selective uptake of HDL-associated cholesteryl esters by cells in culture independent from SR-BI. J Lipid Res 44: 1020–1032, 2003.

57. Connelly MA, Klein SM, Azhar S, Abumrad NA, Williams DL: Comparison of class B scavenger receptors, CD36 and scavenger receptor BI (SR-BI), shows that both receptors mediate high density lipoprotein-cholesteryl ester selective uptake but SR-BI exhibits a unique enhancement of cholesteryl ester uptake. J Biol Chem 274: 41–47, 1999.

58. Gu X, Trigatti B, Xu S, Acton S, Babitt J, Krieger M: The efficient cellular uptake of high density lipoprotein lipids via scavenger receptor class B type I requires not only receptor-mediated surface binding but also receptor-specific lipid transfer mediated by its extracellular domain. J Biol Chem 273: 26338–26348, 1998.

59. Rodrigueza WV, Thuahnai ST, Temel RE, Lund-Katz S, Phillips MC, Williams DL: Mechanism of scavenger receptor class B type I-mediated selective uptake of cholesteryl esters from high density lipoprotein to adrenal cells. J Biol Chem 274: 20344–20350, 1999.

60. Silver DL, Jiang XC, Arai T, Bruce C, Tall AR: Receptors and lipid transfer proteins in HDL metabolism. Ann NY Acad Sci 902: 103–111, 2000.

61. Pittman RC, Steinberg D: Sites and mechanisms of uptake and degradation of high density and low density lipoproteins. J Lipid Res 25: 1577–1585, 1984.

62. Calvo D, Gomez-Coronado D, Lasuncion MA, Vega MA: CLA-1 is an 85-kD plasma membrane glycoprotein that acts as a high-affinity receptor for both native (HDL, LDL, and VLDL) and modified (OxLDL and AcLDL) lipoproteins. Arterioscler Thromb Vasc Biol 17: 2341–2349, 1997.

63. Calvo D, Gomez-Coronado D, Suarez Y, Lasuncion MA, Vega MA: Human CD36 is a high affinity receptor for the native lipoproteins HDL, LDL, and VLDL. J Lipid Res 39: 777–788, 1998.

64. Rhainds D, Brodeur M, Lapointe J, Charpentier D, Falstrault L, Brissette L: The role of human and mouse hepatic scavenger receptor class B type I (SR-BI) in the selective uptake of low-density lipoprotein-cholesteryl esters. Biochemistry 42: 7527–7538, 2003.

65. Acton S, Osgood D, Donoghue M, Corella D, Pocovi M, Cenarro A, Mozas P, Keilty J, Squazzo S, Woolf EA, Ordovas JM: Association of polymorphisms at the SR-BI gene locus with plasma lipid levels and body mass index in a white population. Arterioscler Thromb Vasc Biol 19: 1734–1743, 1999.

66. Brown MS, Goldstein JL: A receptor-mediated pathway for cholesterol homeostasis. Science 232: 34–47, 1986.

67. Saucier SE, Kandutsch AA, Gayen AK, Swahn DK, Spencer TA: Oxysterol regulators of 3-hydroxy-3-methylglutaryl-CoA reductase in liver. Effect of dietary cholesterol. J Biol Chem 264: 6863–6869, 1989.

68. Brown MS, Goldstein JL: A proteolytic pathway that controls the cholesterol content of membranes, cells, and blood. Proc Natl Acad Sci USA 96: 11041–11048, 1999.

69. Kumar A, Middleton A, Chambers TC, Mehta KD: Differential roles of extracellular signal-regulated kinase-1/2 and p38(MAPK) in interleukin-1beta- and tumor necrosis factor-alpha-induced low density lipoprotein receptor expression in HepG2 cells. J Biol Chem 273: 15742–15748, 1998.

70. Rigotti A, Miettinen HE, Krieger M: The role of the high-density lipoprotein receptor SR-BI in the lipid metabolism of endocrine and other tissues. Endocr Rev 24: 357–387, 2003.

71. Bosner MS, Lange LG, Stenson WF, Ostlund RE Jr: Percent cholesterol absorption in normal women and men quantified with dual stable isotopic tracers and negative ion mass spectrometry. J Lipid Res 40: 302–308, 1999.

72. Sturley SL: Molecular aspects of intracellular sterol esterification: the acyl coenzyme A: cholesterol acyltransferase reaction. Curr Opin Lipidol 8: 167–173, 1997.

73. Ghosh S, Mallonee DH, Hylemon PB, Grogan WM: Molecular cloning and expression of rat hepatic neutral cholesteryl ester hydrolase. Biochim Biophys Acta 1259: 305–312, 1995.

74. Miller WL, Strauss JF III: Molecular pathology and mechanism of action of the steroidogenic acute regulatory protein, StAR. J Steroid Biochem Mol Biol 69: 131–141, 1999.

75. van Meer G: Caveolin, cholesterol, and lipid droplets? J Cell Biol 152: F29–F34, 2001.

76. Botham KM, Bravo E: The role of lipoprotein cholesterol in biliary steroid secretion. Studies with *in vivo* experimental models. Prog Lipid Res 34: 71–97, 1995.

77. Russell DW: The enzymes, regulation, and genetics of bile acid synthesis. Annu Rev Biochem 72: 137–174, 2003.

78. Gerloff T, Stieger B, Hagenbuch B, Madon J, Landmann L, Roth J, Hofmann AF, Meier PJ: The sister of P-glycoprotein represents the canalicular bile salt export pump of mammalian liver. J Biol Chem 273: 10046–10050, 1998.

79. Yu L, Li-Hawkins J, Hammer RE, Berge KE, Horton JD, Cohen JC, Hobbs HH: Overexpression of ABCG5 and ABCG8 promotes biliary cholesterol secretion and reduces fractional absorption of dietary cholesterol. J Clin Invest 110: 671–680, 2002.

80. Small DM, Dowling RH, Redinger RN: The enterohepatic circulation of bile salts. Arch Intern Med 130: 552–573, 1972.

81. Chen HC: Molecular mechanisms of sterol absorption. J Nutr 131: 2603–2605, 2001.

82. Repa JJ, Mangelsdorf DJ: Nuclear receptor regulation of cholesterol and bile acid metabolism. Curr Opin Biotechnol 10: 557–563, 1999.

83. Lu TT, Makishima M, Repa JJ, Schoonjans K, Kerr TA, Auwerx J, Mangelsdorf DJ: Molecular basis for feedback regulation of bile acid synthesis by nuclear receptors. Mol Cell 6: 507–515, 2000.

84. Hayhurst GP, Lee YH, Lambert G, Ward JM, Gonzalez FJ: Hepatocyte nuclear factor 4alpha (nuclear receptor 2A1) is essential for maintenance of hepatic gene expression and lipid homeostasis. Mol Cell Biol 21: 1393–1403, 2001.

85. Repa JJ, Turley SD, Lobaccaro JA, Medina J, Li L, Lustig K, Shan B, Heyman RA, Dietschy JM, Mangelsdorf DJ: Regulation of absorption and ABC1-mediated efflux of cholesterol by RXR heterodimers. Science 289: 1524–1529, 2000.

86. Bocher V, Pineda-Torra I, Fruchart JC, Staels B: PPARs: transcription factors controlling lipid and lipoprotein metabolism. Ann NY Acad Sci 967: 7–18, 2002.

87. Hunt MC, Yang YZ, Eggertsen G, Carneheim CM, Gafvels M, Einarsson C, Alexson SE: The peroxisome proliferator-activated receptor alpha (PPARalpha) regulates bile acid biosynthesis. J Biol Chem 275: 28947–28953, 2000.

88. Malerod L, Sporstol M, Juvet LK, Mousavi A, Gjoen T, Berg T: Hepatic scavenger receptor class B, type I is stimulated by peroxisome proliferator-activated

receptor gamma and hepatocyte nuclear factor 4alpha. Biochem Biophys Res Commun 305: 557–565, 2003.

89. Connelly PW: The role of hepatic lipase in lipoprotein metabolism. Clin Chim Acta 286: 243–255, 1999.

90. Dugi KA, Vaisman BL, Sakai N, Knapper CL, Meyn SM, Brewer HB Jr, Santamarina-Fojo S: Adenovirus-mediated expression of hepatic lipase in LCAT transgenic mice. J Lipid Res 38: 1822–1832, 1997.

91. Dugi KA, Amar MJ, Haudenschild CC, Shamburek RD, Bensadoun A, Hoyt RF Jr, Fruchart-Najib J, Madj Z, Brewer HB Jr, Santamarina-Fojo S: *In vivo* evidence for both lipolytic and nonlipolytic function of hepatic lipase in the metabolism of HDL. Arterioscler Thromb Vasc Biol 20: 793–800, 2000.

92. Hegele RA, Little JA, Vezina C, Maguire GF, Tu L, Wolever TS, Jenkins DJ, Connelly PW: Hepatic lipase deficiency. Clinical, biochemical, and molecular genetic characteristics. Arterioscler Thromb 13: 720–728, 1993.

93. Jaye M, Lynch KJ, Krawiec J, Marchadier D, Maugeais C, Doan K, South V, Amin D, Perrone M, Rader DJ: A novel endothelial-derived lipase that modulates HDL metabolism. Nat Genet 21: 424–428, 1999.

94. Hirata K, Dichek HL, Cioffi JA, Choi SY, Leeper NJ, Quintana L, Kronmal GS, Cooper AD, Quertermous T: Cloning of a unique lipase from endothelial cells extends the lipase gene family. J Biol Chem 274: 14170–14175, 1999.

95. Cilingiroglu M, Ballantyne C: Endothelial lipase and cholesterol metabolism. Curr Atheroscler Rep 6: 126–130, 2004.

96. Ishida T, Choi SY, Kundu RK, Spin J, Yamashita T, Hirata K, Kojima Y, Yokoyama M, Cooper AD, Quertermous T: Endothelial lipase modulates susceptibility to atherosclerosis in apolipoprotein-E-deficient mice. J Biol Chem 279: 45085–45092, 2004.

97. Mank-Seymour AR, Durham KL, Thompson JF, Seymour AB, Milos PM: Association between single-nucleotide polymorphisms in the endothelial lipase (LIPG) gene and high-density lipoprotein cholesterol levels. Biochim Biophys Acta 1636: 40–46, 2004.

98. Yancey PG, Kawashiri MA, Moore R, Glick JM, Williams DL, Connelly MA, Rader DJ, Rothblat GH: *In vivo* modulation of HDL phospholipid has opposing effects on SR-BI- and ABCA1-mediated cholesterol efflux. J Lipid Res 45: 337–346, 2004.

99. Gauster M, Oskolkova OV, Innerlohinger J, Glatter O, Knipping G, Frank S: Endothelial lipase-modified high-density lipoprotein exhibits diminished ability to mediate SR-BI (scavenger receptor B type I)-dependent free-cholesterol efflux. Biochem J 382: 75–82, 2004.

100. Bays H, Stein EA: Pharmacotherapy for dyslipidaemia—current therapies and future agents. Expert Opin Pharmacother 4: 1901–1938, 2003.

101. Sacks FM: The role of high-density lipoprotein (HDL) cholesterol in the prevention and treatment of coronary heart disease: expert group recommendations. Am J Cardiol 90: 139–143, 2002.

102. Kashyap ML, Tavintharan S, Kamanna VS: Optimal therapy of low levels of high density lipoprotein-cholesterol. Am J Cardiovasc Drugs 3: 53–65, 2003.

103. Mardones P, Altay M, Miquel JF, Rigotti A, Pilon A, Bouly M, Duran D, Luc G, Clavey V, Staels B, Nishimoto T, Arai H, Kozarsky KF: Fibrates down-regulate hepatic scavenger receptor class B type I protein expression in mice. J Biol Chem 278: 7884–7890, 2003.

104. Nissen SE, Tsunoda T, Tuzcu EM, Schoenhagen P, Cooper CJ, Yasin M, Eaton GM, Lauer MA, Sheldon WS, Grines CL, Halpern S, Crowe T, Blankenship JC, Kerensky R: Effect of recombinant apoA-I Milano on coronary atherosclerosis in patients with acute coronary syndromes: a randomized controlled trial. JAMA 290: 2292–2300, 2003.
105. Clark RW, Sutfin TA, Ruggeri RB, Willauer AT, Sugarman ED, Magnus-Aryitey G, Cosgrove PG, Sand TM, Wester RT, Williams JA, Perlman ME, Bamberger MJ: Raising high-density lipoprotein in humans through inhibition of cholesteryl ester transfer protein: an initial multidose study of torcetrapib. Arterioscler Thromb Vasc Biol 24: 490–497, 2004.
106. de Grooth GJ, Zerba KE, Huang SP, Tsuchihashi Z, Kirchgessner T, Belder R, Vishnupad P, Hu B, Klerkx AH, Zwinderman AH, Jukema JW, Sacks FM, Kastelein JJ, Kuivenhoven JA: The cholesteryl ester transfer protein (CETP) TaqIB polymorphism in the cholesterol and recurrent events study: no interaction with the response to pravastatin therapy and no effects on cardiovascular outcome: a prospective analysis of the CETP TaqIB polymorphism on cardiovascular outcome and interaction with cholesterol-lowering therapy. J Am Coll Cardiol 43: 854–857, 2004.
107. Burgess JW, Neville TA, Rouillard P, Harder Z, Beanlands DS, Sparks DL: Phosphatidylinositol increases HDL-C levels in humans. J Lipid Res 46: 350–355, 2005.
108. Morgan JM, Carey CM, Lincoff A, Capuzzi DM: The effects of niacin on lipoprotein subclass distribution. Prev Cardiol 7: 182–187, 2004.
109. Mittendorfer B, Ostlund RE Jr, Patterson BW, Klein S: Orlistat inhibits dietary cholesterol absorption. Obes Res 9: 599–604, 2001.
110. Davidson MH, Dillon MA, Gordon B, Jones P, Samuels J, Weiss S, Isaacsohn J, Toth P, Burke SK: Colesevelam hydrochloride (cholestagel): a new, potent bile acid sequestrant associated with a low incidence of gastrointestinal side effects. Arch Intern Med 159: 1893–1900, 1999.
111. Altmann SW, Davis HR Jr, Yao X, Laverty M, Compton DS, Zhu LJ, Crona JH, Caplen MA, Hoos LM, Tetzloff G, Priestley T, Burnett DA, Strader CD, Graziano MP: The identification of intestinal scavenger receptor class B, type I (SR-BI) by expression cloning and its role in cholesterol absorption. Biochim Biophys Acta 1580: 77–93, 2002.

2

The Role of LCAT in Atherosclerosis

DOMINIC S. NG

Abstract

Lecithin–cholesterol acyltransferase (LCAT) is one of the major modulators of plasma high-density lipoprotein cholesterol (HDL-C) and plays a central role in the reverse cholesterol transport (RCT) process. Clinically, partial LCAT deficiency is common in a number of chronic disorders at risk for coronary heart diseases (CHD) but the role of LCAT in atherosclerosis remains controversial. Patients with monogenic causes of complete LCAT deficiency appear not to be prone to premature CHD. On the other hand, recent studies on patients with monogenic-based partial LCAT deficiency suggest they may be at increased atherogenic risk. Animal models with transgenic overexpression of the LCAT genes showed variable degrees of antiatherogenic properties except in one transgenic mouse model. LCAT knockout mouse models from different laboratories showed conflicting findings in their predisposition to aortic atherosclerosis. On the other hand, a number of studies using the LCAT knockout mice revealed significant impact of LCAT deficiency on not only lipoprotein metabolism but also in systemic oxidative stress, intrahepatic lipid, and glucose metabolism, each of which may individually modulate atherogenesis. Much remain to be learned with respect to the impact of LCAT deficiency on various proatherogenic pathways to better delineate its pathophysiologic link to atherosclerosis. This is of particular significance in patients suffering from common atherosclerosis-prone disorders that are known to be associated with partial LCAT deficiency, including diabetes and renal insufficiencies.

Keywords: atherosclerosis; high-density lipoprotein (HDL); glomerulopathy; lecithin–cholesterol acyltransferase (LCAT); lipoprotein-X (LpX); triglycerides, glucose

Abbreviations: LCAT, lecithin–cholesterol acyltransferase; HDL, high-density lipoprotein; LDL, low-density lipoprotein; CETP, cholesteryl ester transfer protein; PLTP, phospholipids transfer protein; ABCA1, ATP-binding cassette A1; PC, phosphatidylcholine; SR-BI, scavenger receptor class B type I

Introduction

High-Density Lipoprotein and Atherosclerosis— Epidemiologic Evidence

Numerous large-scale epidemiological studies persistently demonstrated an inverse relationship between plasma high-density lipoprotein cholesterol (HDL-C) level and the risk of coronary heart disease (CHD). An aggregate analysis of four of the largest US studies, which include the Framingham Heart Study [1], the Lipid Research Clinic Prevalence Mortality Follow-up Study, Lipid Research Clinic Primary Prevention Trial, and Multiple Risk Factor Intervention Trial [2], estimated that each 1 mg/dL (0.02 mmol/L) elevation of HDL-C is associated with a 2–3% reduction in CHD risk, a magnitude comparable to that for low-density lipoprotein (LDL) lowering. While this quantitative relationship may be valid in an epidemiologic context, and may also reflect some biological function of the circulating lipoprotein fraction, the use of a single HDL-C level as a risk predictor require caution. The heterogeneity in the HDL–CHD relationship is best exemplified by studying kindreds with monogenic causes of low HDL-C, namely with mutations in the apolipoprotein apoA-I, lecithin–cholesterol acyltransferase (LCAT), and ATP-binding cassette A1 (ABCA1) genes. Individuals homozygous for mutations in these genes uniformly develop severe HDL deficiency but the affected are not equally at risk of premature CHD. This is particularly the case with patients with complete LCAT deficiency.

The Role of LCAT in HDL Metabolism and Reverse Cholesterol Transport

LCAT was first described as a plasma enzyme, which mediates the transfer of fatty acids at the *sn-2* position from phosphatidylcholine (PC) to free cholesterol (FC), forming the neutral lipid cholesterol ester (CE) and lysophosphatidylcholine (LPC) [3]. It is a 416 amino acid glycoprotein synthesized and secreted primarily by the liver. mRNA messages of the LCAT gene have also been detected in testes and the brain [4] but their physiological significance are not well understood.

Upon secretion by the liver, LCAT circulates in plasma bound primarily to HDL but it is also found in apoB-containing lipoproteins, especially the LDL. Kinetic studies in humans have provided an estimate that approximately 70% of plasma CE are formed in the HDL and 30% in the apoB-containing particles [5]. Esterification of FC on apoB-containing particles and HDL by LCAT are coined by the terms β- and α-activity, respectively, and each plays different role in lipoprotein metabolism.

By way of its primary enzymatic action, LCAT plays a major role in HDL metabolism, especially in the reverse cholesterol transport (RCT) pathway, a multistep process by which cholesterol in the peripheral tissues is transferred back to the liver for eventual elimination from the body (Fig. 2.1). To date, the RCT pathway continues to be considered as one of the major mechanisms by which HDL confers its cardioprotective effects [6]. The first step of RCT entails the efflux of cellular cholesterol, along with phospholipids, onto the cholesterol acceptors, with the lipid-free apoA-I and the disc-shaped pre-β-HDL being most avid [7]. Esterification of tissue-derived FC by LCAT transfers the neutral CE into the lipoprotein core, hence sustaining a chemical gradient to accept more FC from tissues. The accumulation of core CE converts the disc-shaped HDL to spherical particles. The mature HDL can further acquire other apoproteins from triglyceride-rich lipoprotein particles. The mature HDL particles can deliver its CE content directly into the liver through the selective uptake process mediated by scavenger receptor class B type I (SR-BI). On the other hand, the tissue-derived CE in HDL can also be transferred to the apoB-containing lipoproteins, in exchange for TG, through the action of

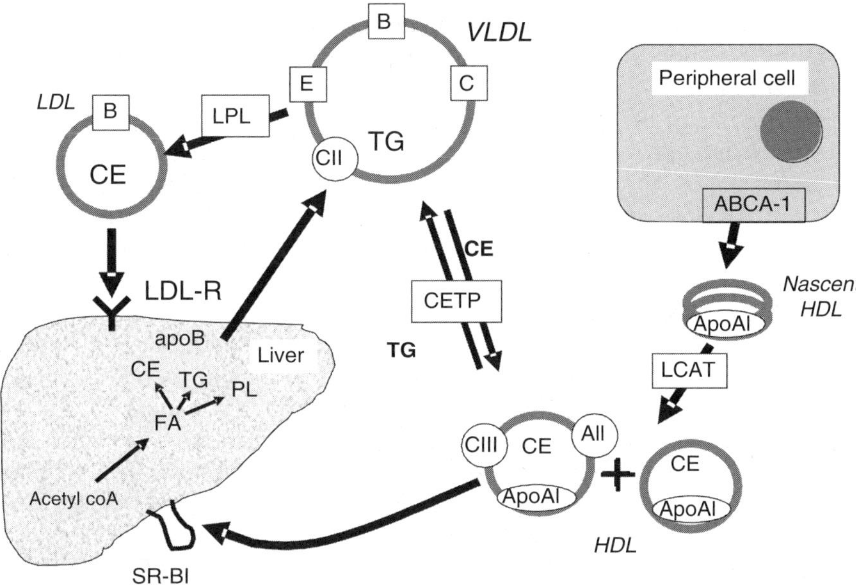

FIGURE 2.1. Schematics of the reverse cholesterol transport (RCT) pathway and the role of LCAT in HDL metabolism. Key steps in the RCT pathway include (1) cholesterol efflux: the transport of free cholesterol from peripheral cells (e.g., arterial wall macrophages) onto the circulating nascent HDL or lipid-free apoA-I (not shown)—mediated by the ATP-binding cassette AI (ABCA1) transporter; (2) cholesterol esterification: cell-derived cholesterol are esterified by lecithin–cholesterol acyltransferase (LCAT), generating neutral esterified cholesterol (CE) which enters the core; (3) lipid transfer: exchange of CE from HDL with triglyceride-rich lipoproteins for TG—mediated by cholesteryl ester transfer protein (CETP). The CE of the mature HDL can also be taken up directly into the liver—mediated by scavenger receptor class B type I (SR-BI). LPL, lipoprotein lipase; LDL-R, low-density lipoprotein receptor.

cholesteryl ester transfer protein (CETP). When the apoB-containing lipoproteins are taken up by the liver, the tissue-derived cholesterol can be channeled for elimination from the body through the biliary cholesterol pathway [8].

More recently, Nakamura et al. [9] examined the differential efficiency of LCAT activity in the generation of CE in various HDL fractions. These authors reported LCAT is associated with various size fractions of HDL and the smallest HDL identified in this experimental regime is consistent with a pre-β-HDL/LCAT complex. Among the various size fractions, the small pre-β-HDL-associated LCAT was responsible for the majority of the LCAT-mediated esterification of CE in HDL. Furthermore, these authors also examined the role of HDL size on the activity of CETP and reported a preferential transfer of CE from the small HDL fractions, including those derived from the pre-β-HDL. Continued activity of LCAT leads to the "maturation" and formation of the α-migrating HDL. Meanwhile, the larger HDL are subject to the modulatory action of phospholipids transfer protein (PLTP), replenishing the pool of pre-β-HDL. In this paradigm, one could envision a pool of HDL that would recycle between the smaller "metabolically active" form and the more mature form, and the cycle will continue to fuel the net movement of cell-derived cholesterol towards the liver for removal.

LCAT is one of the several major modulators of the plasma HDL-C levels, the other being apoA-I, ABCA1, and CETP. In humans subjects with genetic deficiency of LCAT, those who are homozygous a functional LCAT gene mutation develop severe HDL deficiency and the ones heterozygous for a defective LCAT gene develop intermediate levels of HDL deficiency [10, 11]. The marked reduction in HDL-C in the LCAT deficient subjects is attributed primarily to an accelerated catabolism of the HDL particles, with those containing apoA-I and apoA-II being more rapidly cleared than those containing apoA-I alone [12]. The residual circulating lipoproteins in the HDL density range include (a) the discoidal particles containing apoA-I, apoA-II, and occasionally apoE which often form rouleaux and (b) the small apoA-I containing particles that are rich in FC and PL. Interestingly, plasma from LCAT deficient subjects have been demonstrated to be equally effective in mediated cholesterol efflux in fibroblasts [13] but this finding was not shared by other [14, 15]. It is tempting to postulate that the residual HDL in LCAT deficient subjects are functionally sufficient for the majority of the esterification of the cell-derived cholesterol [9] although confirmatory evidence is lacking.

LCAT and Atherosclerosis—Clinical Studies

Monogenic Disorders—Humans

The human LCAT gene is located in the q12–22 region of chromosome 16. Up until recently, 40 mutations have been reported (HGMD; http://uwcmmll1s.uwcm.ac.uk/mg/search/119359.html). Calabresi et al. [11] reported an additional

15 novel mutations more recently. In addition to a marked reduction in HDL-C, other lipoprotein phenotypes in homozygotes and compound heterozygotes further distinguish the LCAT mutations into two distinct syndromes, the complete LCAT deficiency (CLD) and the fish-eye disease (FED). The CLD is characterized by a complete or near-complete deficiency of LCAT activity in the plasma with the absence of cholesterol esterification in all lipoprotein classes. Furthermore, marked increase in total plasma FC/CE ratio and an accumulation of the phospholipid precursor are also hallmarks of the CLD syndrome. Clinically, despite the disruption of the RCT pathway and the severe low level of HDL-C, CLD subjects are paradoxically not particularly predisposed to premature CHD [10, 16–18]. Instead, there is a high prevalence of glomerulopathy in these subjects [19]. Other phenotypes include modest hypertriglyceridemia (HTG), presence of LpX vesicles and mild anemia [20]. In FED, LCAT activity is absent selectively in the HDL fractions. In these subjects, HDL-C is markedly reduced but cholesterol esterification in the apoB-containing lipoprotein particles is relatively preserved. FED patients are not associated with renal complications. However, a number of FED mutations had been found to be associated with premature coronary artery disease (CAD) [21, 22] but the underlying pathogenesis remains obscure.

Although subjects with complete LCAT deficiency appear to be associated with a paradoxical absence of premature CAD in spite of severe HDL deficiency, little is known about the predisposition of partial LCAT deficient subjects based on earlier studies. More recently, Ayyobi et al. [23] reported a 25-year longitudinal follow-up of a large Canadian LCAT deficient kindred concerning their vascular risk. Carotid ultrasound for intima-medial thickness (IMT) and brachial artery flow-mediated dilatation (FMD) were used as surrogate markers for cardiovascular endpoints. These authors reported that, based on two homozygotes, nine heterozygotes, and four unaffected family members, the heterozygotes are associated with pronounced IMT abnormalities including detection of atherosclerotic plaques whereas for the homozygotes, the IMT are only minimally increased. A similar vascular evaluation by IMT in 68 carriers of various known FED mutations showed significant increase in IMT progression in both heterozygotes and homozygotes [24].

Putting it all together, data from families with genetic causes of LCAT deficiency suggest that FED mutations may confer higher risk of CAD than those with CLD. On the other hand, subjects homozygous for CLD mutations are generally not at risk of accelerated CAD but the heterozygotes with partial LCAT deficiency, appear to be more prone. In light of the clinical observations that partial LCAT deficiency being detected in a variety of common diseases known to be at high risk of premature CAD, e.g., diabetes [25, 26], uremia [27], cigarette smoking [28] etc., the proatherogenic potential of partial LCAT deficiency require confirmation as this would form the basis for LCAT being a potential therapeutic target.

The LCAT–Atherosclerosis Controversy

Transgenic Mouse Models

To investigate the role of LCAT in atherosclerosis, a number of animal models have been generated. LCAT transgenic mice have been created by a number of laboratories. As expected, a modest overexpression of the human LCAT gene in mice resulted in a moderate elevation of total plasma HDL-C levels and an increase in the HDL particle size. In case of the human LCAT gene being coexpressed with the human apoA-I transgene, a mouse strain showed a humanlike polymodal size distribution of HDL, modest particle size increase was observed in each individual HDL subfractions attributable to increased accumulation of CE [29]. Meanwhile, there was a modest reduction in fasting plasma TG. These data demonstrated that overexpression of LCAT *in vivo* resu *in vivo* a "favorable" lipid profile. However, transgenic mice with a modest overproduction of LCAT failed to impact on aortic atherosclerotic lesions [30, 31]. On the other hand, high-level overexpression (>100-fold) of human LCAT resulted in not only a marked increase in plasma HDL-C, but there was also appearance of a large-sized, HDL1-like subfraction [32]. Surprisingly, this strain of high-expressor mice showed a paradoxical increase in diet-induced aortic atherosclerosis [33]. On the other hand, the same level of overexpression of human LCAT in rabbits was found associated with a significant reduction in aortic atherosclerosis. More detailed analyses of the lipoprotein changes in the two animal models suggested that the absence of CETP activity in mice may be responsible for the presence of the large HDL fraction and the increased aortic lesions. To test this hypothesis, Foger et al. [34] examined the effect of a concurrent overexpression of CETP in the LCAT transgenic mice in atherosclerosis and found that CETP expression partially attenuated the increase in atherosclerosis seen in the LCAT transgenic mice.

The enzymatic action of LCAT is a critical determinant of the lipid composition of circulating lipoproteins. A tenfold overexpression of the human LCAT gene in mice fed an atherogenic diet (15% of calories from palm oil, 1.0% cholesterol, and 0.5% cholic acid) resulted in a twofold increase in the ratio of saturated+monounsaturated to polyunsaturated CE species in the apoB-containing lipoproteins. However, there was no difference in aortic atherosclerosis determined on the basis of aortic CE content despite a doubling of the saturation index of the apoB-containing particles [31].

In a different experimental system, a short-term overexpression of the LCAT genes appears beneficial when administered to the LDL receptor/leptin double deficient mice. This double mutant serves as a model for combined severe hyperlipidemia, insulin resistance, and leptin deficient-obesity. Furthermore, the mice were also found to have reduced serum platelet-activating factor acetylhydrolase (PAF-AH) activities and paraoxonase 1 (PON1) activities, two oxidative stress markers. Meanwhile, the plasma titre of oxidized LDL autoantibody, homing of the macrophages and aortic atherosclerosis were increased. Adenovirus-mediated gene transfer of LCAT resulted in a reduction of the titre for the autoantibody to the oxidized LDL and the aortic plaque volume in only 6 weeks. Unlike previous transgenic models, the short-term overexpression of the LCAT gene in this experiment, despite a 64% raise in the LCAT activity, did not result in any significant change in HDL-C, the plasma FC/CE ratio, or the ability of the plasma to increase efflux of cholesterol from cells [35]. Therefore, the observed reduction in atherosclerosis as a result of LCAT-gene transfer was attributed to LCAT playing a significant role in modulating the oxidative stress.

Collectively, hepatic overproduction of LCAT in mouse models have resulted in mixed results in the context of its role in modulating atherogenicity. In the specific situations where beneficial effects were observed, correction of the elevated oxidative stress appeared to be the major mechanism. In the rabbit models, the observed cardioprotective effects of LCAT overproduction may be attributable to the decrease in the proatherogenic apoB-containing lipoprotein particles and the increase in HDL-C. However, changes in oxidative stress markers have not been reported in the rabbit models.

LCAT Knockout Mouse Models

In patients with complete LCAT deficiency, there is no significant increase in the risk of atherosclerosis in spite of the disruption of the RCT pathway and a marked reduction in plasma levels of HDL-C and apoA-I. Plasma very-low-density lipoprotein (VLDL) and LDL fractions are both altered morphologically and compositionally. The apoB-containing lipoproteins are relatively TG-enriched and CE-poor and there is a high prevalence of fasting HTG in LCAT deficient subjects. However, how these dyslipidemic changes might modulate atherogenicity of the lipoproteins remain to be fully elucidated.

A mouse model for LCAT deficiency has been generated by homologous recombination of the murine LCAT gene [36, 37]. The lipoprotein abnormalities in these mice showed remarkable resemblance to those seen in LCAT deficient humans, especially in the changes in HDL. In the homozygous knockout mice (LCAT–/–), plasma CE was markedly reduced to 9.4% of control and there was a 11.7-fold increase in the FC/CE ratio. Both HDL-C and apoA-I levels were reduced in the LCAT deficient mice in a gene dose-dependent manner but the reduction in CE content being dramatically more so than the FC content. The HDL fraction ($d = 1.063–1.21$ g/mL) was characterized by the presence of particles with a wide distribution in sizes but dominated by a small, 7.6-nm peak. Electron microscopy revealed rouleaux formation from the disc-shaped HDL particles, remarkably similar to those seen in humans [38]. In the $d < 1.063$ g/mL fractions, vesicle-like particles have also been detected [36, 37]. The presence of compositionally and morphologically altered particles in the LDL and VLDL range is consistent with a direct action of LCAT on these particles.

LCAT knockout mice have been studied to explore the role of LCAT in atherosclerosis as well as to elucidate the possible mechanism for the paradoxical absence of accelerated atherosclerosis in LCAT deficient humans. A recent study by Lambert et al. [39] reported significant reductions in aortic atherosclerosis attributable to LCAT deficiency in a number of different dyslipidemic backgrounds, including wild type, LDL receptor knockout (LDLR–/–), and CETP transgenic mice, fed a cholate-containing, high cholesterol/high fat diet, and in apoE$^{-/-}$ background fed a chow diet. In this study, the authors observed a uniform reduction in the level of apoB-containing lipoproteins in the LCAT-deficient mice across all dyslipidemic backgrounds examined and proposed that the apoB-containing particle levels play a key role in modulating atherosclerosis risk in LCAT deficiency.

A study by Furbee et al. [40] using apoE$^{-/-}$xLCAT–/– and LDLR–/–xLCAT–/– mice fed a cholate-free, atherogenic diet yielded opposite results regarding the impact of LCAT deficiency on atherosclerosis. In this study, LDLR–/–xLCAT–/– double knockout mice and their LDLR–/–xLCAT+/+ single knockout control have similar apoB-containing lipoprotein levels but the LDL–CE in the former are more enriched in the saturated, and putatively more proatherogenic, fatty acids. On the other hand, the apoE$^{-/-}$xLCAT–/– double knockout mice have not only more saturated LDL–CE, but also higher levels of apoB-containing lipoproteins. The reason for the disparate findings between these two studies are not apparent. However, the difference in the composition of the diets used in the studies, and the methods of aortic lesion quantification may play a key role.

More recently, Ng et al. [41] reported the effect of LCAT deficiency on atherosclerosis using the hyperlipidemic apoE$^{-/-}$ as background. The apoE$^{-/-}$xLCAT–/– and apoE$^{-/-}$xLCAT+/+ mice were fed a regular chow diet and the aortic atherosclerosis lesions were compared at 8–9 months of age. The lipoprotein profile by Fast Protein Liquid Chromatography (FPLC) revealed, in the apoE$^{-/-}$xLCAT–/– double knockout mice, a marked reduction in HDL-C and a modest reduction in apoB-containing particles in all VLDL and LDL/IDL (intermediate-density lipoprotein) fractions. These findings are in agreement with those reported by Lambert et al. [39]. In their investigation of the mechanism of the impact of LCAT on atherosclerosis, the authors first observed a significant increase in two independently determined oxidative stress markers, namely the plasma F2-isoprostane level and the aortic superoxide production rate, in both the LCAT–/– mice and the apoE$^{-/-}$ mice. The marked increase in oxidative stress status in the LCAT–/– mice is consistent with a previous study showing that, in this model, there is a significant reduction in plasma PON1 activity and PAF-AH activity, in part due to the marked reduction in plasma HDL level [42]. Surprisingly, the oxidative stress markers in the apoE$^{-/-}$xLCAT–/– were completely normalized to the same level as the wild-type mice despite persistent severe hyperlipidemia. This normalization of the oxidative markers was found associated with a 50% reduction in aortic atherosclerosis lesion size. Intriguingly, the complete normalization of the two oxidative markers was also found to be associated with a restoration of the total plasma level of PON1 arylesterase activity to that of the wild-type mice, despite a persistent severe HDL deficiency. Detailed FPLC fractionation revealed that the PON1 arylesterase activity is no longer restricted to the HDL fractions but is broadly redistributed to all the non-HDL fractions, the latter accounting for nearly 50% of the total plasma PON1 activity.

The mechanism by which PON1 redistribute in the plasma of the LCAT deficient mice remains unclear. A recent study by Sorenson et al. [43] demonstrated that PON1 can bind to phospholipid vesicles via its hydrophobic N-terminal signal sequence peptide and remains active enzymatically without apoA-I. The authors further demonstrated that HDL-associated PON1 can be transferred to these PL vesicles. The physical characteristics of such synthetic vesicles are not dissimilar to that of LpX, a FC- and PL-rich vesicles frequently found in LCAT deficient humans as has been shown to be present in modest quantity in the apoE$^{-/-}$xLCAT$-$/$-$ mice. However, the possibility of an association of PON1 with other apoB-containing lipoproteins cannot be ruled out. Similarly, the role of the retained PON1 in the observed dramatic reversal of oxidative markers is also unknown. In light of the previous reports on the impact of PON1 on oxidative stress, especially the *in vivo* murine models [44–46], it is conceivable that the paradoxical normalization of the oxidative markers in the apoE$^{-/-}$xLCAT$-$/$-$ double knockout mice may, at least in part, be attributed to the redistribution and retention of serum PON1. Similar PON1 arylesterase activity redistribution profile by FPLC is also detected in the LDLR$-$/$-$xLCAT$-$/$-$ double knockout mice (Ng, unpublished data). More importantly, PON1 activity has been detected in the non-HDL fractions from a subject homozygous for an LCAT mutation but is absent from his unaffected siblings (Ng, Connelly, and Frohlich, unpublished data).

Collectively, data to date based on murine models of the impact of LCAT deficiency in atherosclerosis are also mixed but are suggestive of strong gene–gene and possibly gene–diet interactions. They also suggest that oxidative modifications of lipids may play an important role in mediating the effect of LCAT on atherogenesis. In addition to the observed significant alterations in PON1 abundance and distribution as a result of changes in the dyslipidemic background and LCAT gene dose, it is conceivable that other enzymes pertinent to oxidation and perhaps inflammation may also be altered.

The Role of LCAT in the Metabolism of apoB-Containing Lipoproteins

Several lines of experimental evidence suggest that LCAT is active not only on HDL particles, but a small fraction of the total plasma LCAT activity is detected in the LDL particles. In humans, several mutations in the LCAT gene led to a selective loss of transesterification activity in the HDL particles, resulting in clinically a marked reduction in HDL-C levels but relative normal

LDL levels, compositions, and the FC/CE ratio. These subjects with selective LCAT deficiency do not develop renal disease but tend to have more severe clouding of the cornea, hence the term FED [10]. To further address the role of LCAT in the CE content of apoB-containing particles, Furbee et al. [47] examined the lipoprotein compositions of two strains of double knockout mice, namely the apoE$^{-/-}$xLCAT–/– and LDLR–/–xLCAT–/– mice. The removal of functional LCAT in these two strains of hyperlipidemic mice revealed a differential impact on the fatty acid composition of CE in the apoB-containing particles. In the case of LCAT deficiency in the LDLR–/– background, there was a significant loss of linoleate and a complete loss of polyunsaturated fatty acids (PUFA) in the LDL–CE. On the other hand, in the apoE$^{-/-}$ background, LCAT deficiency resulted in only a modest alteration in the LDL–CE fatty acid content. These authors concluded that LCAT contributes to the CE fatty acid pool of apoB-containing lipoproteins and is the only source of plasma long-chain polyunsaturated CE in these mice. In a follow-up study, Zhao et al. [48] found that, in apoE$^{-/-}$xapoA-I–/– mice, the dramatic reduction in plasma cholesterol esterificaton rate (CER) is disproportionate to the more modest reduction in the LCAT mass. However, incubation of the VLDL from these double knockout mice with apoE resulted in a threefold increase in CER whereas a similar incubation with apoA-I only resulted in a modest 80% increase. The authors concluded that apoE is a more significant physiologic activator of LCAT than apoA-I-containing lipoproteins.

The Role of LCAT in Triglyceride and Glucose Metabolism

HTG is a risk marker for CHD based on epidemiologic studies but the pathogenesis of how TG-rich lipoproteins contribute to the atherosclerosis is frequently confounded by other associated risk factors. HTG is a common feature in a number of highly proatherogenic conditions, namely type 2 diabetes mellitus, metabolic syndrome, and familial combined hyperlipidemia [49], in which low HDL-C, insulin resistance, central obesity are frequently present. The dyslipidemia in these conditions share the cardinal feature of a hepatic overproduction of VLDL/apoB as contributor to the HTG, which in turn promotes the generation of small dense LDL and lowering of HDL-C through CETP. HTG has also been documented in LCAT deficient subjects, both in homozygotes and heterozygotes but due to the small number of subjects in this rare syndrome [11, 50, 51], the pathophysiologic relevance and pathogenesis of HTG remain poorly understood. In animal models of either LCAT overexpression or deficiency, an inverse association between the LCAT gene dose and fasting TG levels has begun to emerge. Transgenic mice [29] and rabbits [52] overexpressing the human LCAT gene have been reported to have a modest reduction in fasting TG levels. Gene-targeted mice deficient in

LCAT activities are associated with elevated TG in a gene dose-dependent manner [36, 37]. Collectively, these data are in support of a mechanistic link between LCAT deficiency and HTG. In humans, a partial reduction in postheparin lipase activity has been observed in association with a fasting HTG [53], other possible contributions have not been explored.

The possible mechanism of HTG in complete LCAT deficient mice was recently reported by Ng et al. [54] using the LDLR–/–xLCAT–/– double knockout mice, which showed modest fasting HTG when compared with the LDLR–/–xLCAT+/+ single knockout control. FPLC lipid profile showed accumulation of VLDL particles in the double knockout mice and the apoB-containing lipoproteins (VLDL/IDL/LDL fractions) are all relatively TG-enriched. Ultracentrifuge-isolated VLDL fractions are, at least semiquantitatively, better substrate for the exogenous (bovine) lipoprotein lipase. On the other hand, postheparin lipase activity assay revealed a significant isolated reduction in the LPL activity, consistent with those seen in human subjects. More intriguingly, the double knockout mice were found to have a marked increase in hepatic TG production rate, a metabolic alteration not seen in other models of primary severe low HDL syndromes. This was further found to be associated with an upregulation of the hepatic SREBP1 gene, which was further linked to a coordinated upregulation of a number of target genes encoding for enzymes in lipogenesis on the one hand, and a downregulation of PEPCK1, a gene for the rate-limiting enzyme for gluconeogenesis on the other. The change in the fasting PEPCK mRNA level was further found to be associated with a reduction in fasting glucose and fasting insulin levels in the double knockout mice, consistent with the notion of an improvement in hepatic insulin sensitivity and reduced gluconeogenesis. Although both *srebp1* and *pepck1* gene are known as target genes for the transcription factor liver X receptor α (LXRα) [55, 56], survey of the expression levels of a number of other specific LXRα target genes suggests that the SREBP1 upregulation in this LCAT deficiency mouse model is unlikely to be driven by LXR. On the other hand, the LDLR–/–xlcat–/– mouse LDL were recently shown to be markedly depleted of linoleic acid moiety and are virtually completely devoid of other PUFA [47]. In light of the suppressive effect of PUFA on *srebp1* mRNA level shown by many laboratories [57, 58], the authors postulated that the lack of PUFA in LDL in the double knockout mice might have resulted in a reduction in the abundance of PUFA in the liver, leading to less suppression of the SREBP1 gene expression. In short, the findings presented here suggest that LCAT deficiency may constitute a unique metabolic milieu in which an increase in hepatic TG production, possibly through an induction of hepatic SREBP1 expression and the downstream lipogenesis, acts in concert with the reduction in postheparin lipase activity to cause HTG. This pathway represents first to link a primary HDL deficiency state to HTG through hepatic TG overproduction and, putatively through increased lipogenesis, may extend our mechanistic view of the yin/yang relationship between HTG and low HDL-C seen in many proatherogenic states.

Conclusion

LCAT plays a central role in RCT and is a major modulator of plasma HDL-C levels as well as compositions and levels of apoB-containing lipoproteins. However, its role in atherosclerosis continues to be controversial. Studies in LCAT deficient humans with monogenic LCAT gene defects continue to be consistent with the notion of a lack of accelerated atherosclerosis in those with complete LCAT deficiency. On the other hand, based on recent studies with larger cohorts of subjects who are either heterozygotes for CLD mutations or both heterozygotes and homozygotes for FED mutations, suggest that partial LCAT deficiency may be proatherogenic. Data from animal models are also conflicting. LCAT knockout mice have been shown to be proatherogenic and antiatherogenic. Although these studies were carried out by different laboratories using slightly different experimental protocols, the contradictions are not easily reconciled. On the other hand, the significant impact of LCAT deficiency on oxidative status seen in the LCAT knockout mice underscores the complex functional impact of LCAT on not only lipoprotein metabolism but also in oxidative stress, the latter likely plays important roles in atherogenesis. Likewise, the recent discoveries of how LCAT deficiency in the murine model significantly impact on intrahepatic TG, fatty acid, and glucose metabolism further exemplifies the multiplicity in the phenotypic changes in response to a reduction in the enzymatic activity. In light of the common occurrence of partial LCAT deficiency in a number of common proatherogenic disorders like diabetes, renal insufficiency etc., a better understanding of the pathophysiologic impact of this metabolic disorder at the molecular level would enable improved and targeted therapeutic strategies.

Acknowledgments: This work was in part funded by the Heart and Stroke Foundation of Ontario grant-in-aids, CIHR New Investigator Award, and CIHR Operating grants.

References

1. Gordon T, Castelli WP, Hjortland MC, Kannel WB, Dawber TR: High density lipoprotein as a protective factor against coronary heart disease. The Framingham Study. Am J Med 62: 707–714, 1977.
2. Gordon DJ, Probstfield JL, Garrison RJ, et al: High-density lipoprotein cholesterol and cardiovascular disease. Four prospective American studies. Circulation 79: 8–15, 1989.
3. Glomset JA: The mechanism of the plasma cholesterol esterification reaction: plasma fatty acid transferase. Biochim Biophys Acta 65: 128–135, 1962.
4. Warden CH, Langner CA, Gordon JI, Taylor BA, McLean JW, Lusis AJ: Tissue-specific expression, developmental regulation, and chromosomal mapping of the lecithin:cholesterol acyltransferase gene. Evidence for expression in brain and testes as well as liver. J Biol Chem 264: 21573–21581, 1989.

5. Lacko AG, Pritchard PH: International Symposium on Reverse Cholesterol Transport. Report on a meeting. J Lipid Res 31: 2295–2299, 1990.

6. Rader DJ: Regulation of reverse cholesterol transport and clinical implications. Am J Cardiol 92(4A): 42J–49J, 2003.

7. Knight BL: ATP-binding cassette transporter A1: regulation of cholesterol efflux. Biochem Soc Trans 32(Pt 1): 124–127, 2004.

8. Zhang Y, Zanotti I, Reilly MP, Glick JM, Rothblat GH, Rader DJ: Overexpression of apolipoprotein A-I promotes reverse transport of cholesterol from macrophages to feces *in vivo*. Circulation 108: 661–663, 2003.

9. Nakamura Y, Kotite L, Gan Y, Spencer TA, Fielding CJ, Fielding PE: Molecular mechanism of reverse cholesterol transport: reaction of pre-beta-migrating high-density lipoprotein with plasma lecithin/cholesterol acyltransferase. Biochemistry 43(46): 14811–14820, 2004.

10. Kuivenhoven JA, Pritchard H, Hill J, Frohlich J, Assmann G, Kastelein J: The molecular pathology of lecithin:cholesterol acyltransferase (LCAT) deficiency syndromes. J Lipid Res 38: 191–205, 1997.

11. Calabresi L, Pisciotta L, Costantin A, Frigerio I, Eberini I, Alessandrini P, Arca M, Bittolo Bon G, Boscutti G, Busnach G, Frasca G, Gesualdo L, Gigante M, Lupattelli G, Montali A, Pizzolitto S, Rabbone I, Rolleri M, Ruotolo G, Sampietro T, Sessa A, Vaudo G, Cantafora A, Veglia F, Calandra S, Bertolini S, Franceschini G: The molecular basis of lecithin:cholesterol acyltransferase deficiency syndromes. A comprehensive study of molecular and biochemical findings in 13 unrelated Italian families. Arterioscler Thromb Vasc Biol 25: 1972–1978, 2005.

12. Rader DJ, Ikewaki K, Duverger N, et al: Markedly accelerated catabolism of apolipoprotein A-II (ApoA-II) and high density lipoproteins containing ApoA-II in classic lecithin:cholesterol acyltransferase deficiency and fish-eye disease. J Clin Invest 93: 321–330, 1994.

13. Berard AM, Clerc M, Brewer B Jr, Santamarina-Fojo S: A normal rate of cellular cholesterol removal can be mediated by plasma from a patient with familial lecithin–cholesterol acyltransferase (LCAT) deficiency. Clin Chim Acta 314(1–2): 131–139, 2001.

14. Ohta T, Nakamura R, Ikeda Y, Frohlich J, Takata K, Saito Y, Horiuchi S, Matsuda I: Evidence for impaired cellular cholesterol removal mediated by apo A-I containing lipoproteins in patients with familial lecithin: cholesterol acyltransferase deficiency. Biochim Biophys Acta 1213(3): 295–301, 1994.

15. von Eckardstein A, Huang Y, Wu S, Funke H, Noseda G, Assmann G: Reverse cholesterol transport in plasma of patients with different forms of familial HDL deficiency. Arterioscler Thromb Vasc Biol 15(5): 691–703, 1995.

16. Assmann G, von Eckardstein A, Funke H: Lecithin:cholesterol acyltransferase deficiency and fish-eye disease. Curr Opin Lipidol 2: 110–117, 1991.

17. Glomset JA, Assmann G, Gjone E, Norum KR: Lecithin:cholesterol acyltransferase deficiency and fish eye disease. In: Scriver CR, Beaudet AL, Sly WS, Stanbury JB, Wyngaarden, Fredrickson DS (eds), The Metabolic and Molecular Bases of Inherited Disease. McGraw-Hill, New York, 1995, pp. 1933–1951.

18. Funke H, von Eckardstein A, Pritchard PH, Hornby AE, Wiebusch H, Motti C, Hayden MR, Dachet C, Jacotot B, Gerdes U, Faergeman O, Albers JJ, Colleoni N, Catapano A, Frohlich J, Assmann G: Genetic and phenotypic heterogeneity in familial lecithin:cholesterol acyltransferase (LCAT) deficiency. Six newly

identified defective alleles further contribute to the structural heterogeneity in this disease. J Clin Invest 91(2): 677–683, 1993.

19. Borysiewicz LK, Soutar AK, Evans DJ, Thompson GR, Rees AJ: Renal failure in familial lecithin: cholesterol acyltransferase deficiency. Q J Med 51: 411–426, 1982.

20. Santamarina-Fojo S, Lambert G, Hoeg JM, Brewer HB Jr: Lecithin–cholesterol acyltransferase: role in lipoprotein metabolism, reverse cholesterol transport and atherosclerosis. Curr Opin Lipidol 11: 267–275, 2000.

21. Kuivenhoven JA, van Voorst tot Voorst EJ, Wiebusch H, Marcovina SM, Funke H, Assmann G, Pritchard PH, Kastelein JJ: A unique genetic and biochemical presentation of fish-eye disease. J Clin Invest 96(6): 2783–2791, 1995.

22. Kuivenhoven JA, Stalenhoef AF, Hill JS, Demacker PN, Errami A, Kastelein JJ, Pritchard PH: Two novel molecular defects in the LCAT gene are associated with fish eye disease. Arterioscler Thromb Vasc Biol 16(2): 294–303, 1996.

23. Ayyobi AF, McGladdery SH, Chan S, John Mancini GB, Hill JS, Frohlich JJ: Lecithin:cholesterol acyltransferase (LCAT) deficiency and risk of vascular disease: 25 year follow-up. Atherosclerosis 177(2): 361–366, 2004.

24. Hovingh GK, de Groot E, van der Steeg W, Boekholdt SM, Hutten BA, Kuivenhoven JA, Kastelein JJ: Inherited disorders of HDL metabolism and atherosclerosis. Curr Opin Lipidol 16(2): 139–145, 2005.

25. Kiziltunc A, Akcay F, Polat F, Kuskay S, Sahin YN: Reduced lecithin:cholesterol acyltransferase (LCAT) and Na+, K+, ATPase activity in diabetic patients. Clin Biochem 30(2): 177–182, 1997.

26. Weight MJ, Coetzee HS, Smuts CM, Marais MP, Maritz JS, Hough FS, Benade AJ, Taljaard JJ: Lecithin:cholesterol acyltransferase activity and high-density lipoprotein subfraction composition in type 1 diabetic patients with improving metabolic control. Acta Diabetol 30(3): 159–165, 1993.

27. Gillett MP, Teixeira V, Dimenstein R: Decreased plasma lecithin:cholesterol acyltransfer and associated changes in plasma and red cell lipids in uraemia. Nephrol Dial Transplant 8(5): 407–411, 1993.

28. Freeman DJ, Caslake MJ, Griffin BA, Hinnie J, Tan CE, Watson TD, Packard CJ, Shepherd J: The effect of smoking on post-heparin lipoprotein and hepatic lipase, cholesteryl ester transfer protein and lecithin:cholesterol acyl transferase activities in human plasma. Eur J Clin Invest 28(7): 584–591, 1998.

29. Francone OL, Gong EL, Ng DS, Fielding CJ, Rubin EM: Expression of human lecithin–cholesterol acyltransferase in transgenic mice. Effect of human apolipoprotein AI and human apolipoprotein all on plasma lipoprotein cholesterol metabolism. J Clin Invest 96:1440–1448, 1995.

30. Mehlum A, Gjernes E, Solberg LA, Hagve TA, Prydz H: Overexpression of human lecithin:cholesterol acyltransferase in mice offers no protection against diet-induced atherosclerosis. APMIS 108: 336–342, 2000.

31. Furbee JW Jr, Parks JS: Transgenic overexpression of human lecithin:cholesterol acyltransferase (LCAT) in mice does not increase aortic cholesterol deposition. Atherosclerosis.165: 89–100, 2002.

32. Vaisman BL, Klein HG, Rouis M, Berard AM, Kindt MR, Talley GD, Meyn SM, Hoyt RF Jr, Marcovina SM, Albers JJ, et al: Overexpression of human lecithin cholesterol acyltransferase leads to hyperalphalipoproteinemia in transgenic mice. J Biol Chem 270: 12269–12275, 1995.

33. Berard AM, Foger B, Remaley A, Shamburek R, Vaisman BL, Talley G, Paigen B, Hoyt RF Jr, Marcovina S, Brewer HB Jr, Santamarina-Fojo S: High plasma

HDL concentrations associated with enhanced atherosclerosis in transgenic mice overexpressing lecithin–cholesteryl acyltransferase. Nat Med 3: 744–749, 1997.

34. Foger B, Chase M, Amar MJ, Vaisman BL, Shamburek RD, Paigen B, Fruchart-Najib J, Paiz JA, Koch CA, Hoyt RF, Brewer HB Jr, Santamarina-Fojo S: Cholesteryl ester transfer protein corrects dysfunctional high density lipoproteins and reduces aortic atherosclerosis in lecithin cholesterol acyltransferase transgenic mice. J Biol Chem 274: 36912–36920, 1999.

35. Mertens A, Verhamme P, Bielicki JK, Phillips MC, Quarck R, Verreth W, Stengel D, Ninio E, Navab M, Mackness B, Mackness M, Holvoet P: Increased low-density lipoprotein oxidation and impaired high-density lipoprotein antioxidant defense are associated with increased macrophage homing and atherosclerosis in dyslipidemic obese mice: LCAT gene transfer decreases atherosclerosis. Circulation 107: 1640–1646, 2003.

36. Ng DS, Francone OL, Forte TM, Zhang J, Haghpassand M, Rubin EM: Disruption of the murine lecithin:cholesterol acyltransferase gene causes impairment of adrenal lipid delivery and up-regulation of scavenger receptor class B type I. J Biol Chem 272: 15777–15781, 1997.

37. Sakai N, Vaisman BL, Koch CA, Hoyt RF Jr, Meyn SM, Talley GD, Paiz JA, Brewer HB Jr, Santamarina-Fojo S: Targeted disruption of the mouse lecithin:cholesterol acyltransferase (LCAT) gene. Generation of a new animal model for human LCAT deficiency. J Biol Chem 272: 7506–7510, 1997.

38. Forte T, Nichols A, Glomset J, Norum K: The ultrastructure of plasma lipoproteins in lecithin:cholesterol acyltransferase deficiency. Scand J Clin Lab Invest Suppl 137: 121–132, 1974.

39. Lambert G, Sakai N, Vaisman BL, et al: Analysis of glomerulosclerosis and atherosclerosis in lecithin cholesterol acyltransferase-deficient mice. J Biol Chem 276: 15090–15098, 2001.

40. Furbee JW Jr, Sawyer JK, Parks JS: Lecithin:cholesterol acyltransferase deficiency increases atherosclerosis in the low density lipoprotein receptor and apolipoprotein E knockout mice. J Biol Chem 277: 3511–3519, 2002.

41. Ng DS, Maguire GF, Wylie J, Ravandi A, Xuan W, Ahmed Z, Eskandarian M, Kuksis A, Connelly PW: Oxidative stress is markedly elevated in lecithin:cholesterol acyltransferase-deficient mice and is paradoxically reversed in the apolipoprotein E knockout background in association with a reduction in atherosclerosis. J Biol Chem 277: 11715–11720, 2002.

42. Forte TM, Oda MN, Knoff L, Frei B, Suh J, Harmony JA, Stuart WD, Rubin EM, Ng DS: Targeted disruption of the murine lecithin:cholesterol acyltransferase gene is associated with reductions in plasma paraoxonase and platelet-activating factor acetylhydrolase activities but not in apolipoprotein J concentration. J Lipid Res 40: 1276–1283, 1999.

43. Sorenson RC, Bisgaier CL, Aviram M, Hsu C, Billecke S, La Du BN: Human serum paraoxonase/arylesterase's retained hydrophobic N-terminal leader sequence associates with HDLs by binding phospholipids: apolipoprotein A-I stabilizes activity. Arterioscler Thromb Vasc Biol 19: 2214–2225, 1999.

44. Shih DM, Gu L, Xia YR, Navab M, Li WF, Hama S, Castellani LW, Furlong CE, Costa LG, Fogelman AM, Lusis AJ: Mice lacking serum paraoxonase are susceptible to organophosphate toxicity and atherosclerosis. Nature 394: 284–287, 1998.

45. Tward A, Xia YR, Wang XP, Shi YS, Park C, Castellani LW, Lusis AJ, Shih DM: Decreased atherosclerotic lesion formation in human serum paraoxonase transgenic mice. Circulation 106: 484–490, 2002.

46. Rozenberg O, Rosenblat M, Coleman R, Shih DM, Aviram M: Paraoxonase (PON1) deficiency is associated with increased macrophage oxidative stress: studies in PON1-knockout mice. Free Radic Biol Med 34: 774–784, 2003.
47. Furbee JW Jr, Francone O, Parks JS: *In vivo* contribution of LCAT to apolipoprotein B lipoprotein cholesteryl esters in LDL receptor and apolipoprotein E knockout mice. J Lipid Res 43: 428–437, 2002.
48. Zhao Y, Thorngate FE, Weisgraber KH, Williams DL, Parks JS: Apolipoprotein E is the major physiological activator of lecithin–cholesterol acyltransferase on apolipoproteins B lipoprotein particles. Biochemistry 44: 1013–1025, 2004.
49. Brunzell JD, Ayyobi AF: Dyslipidemia in the metabolic syndrome and type 2 diabetes mellitus. Am J Med 115 (Suppl 8A): 24S–28S, 2003.
50. Kodama T, Akanuma Y, Okazaki M, Aburatani H, Itakura H, Takahashi K, Sakuma M, Takaku F, Hara I: Abnormalities in plasma lipoprotein in familial partial lecithin:cholesterol acyltransferase deficiency. Biochim Biophys Acta 752: 407–415, 1983.
51. Frohlich J, McLeod R, Pritchard PH, Fesmire J, McConathy W: Plasma lipoprotein abnormalities in heterozygotes for familial lecithin:cholesterol acyltransferase deficiency. Metabolism 37: 3–8, 1988.
52. Hoeg JM, Vaisman BL, Demosky SJ Jr, Meyn SM, Talley GD, Hoyt RF Jr, Feldman S, Berard AM, Sakai N, Wood D, Brousseau ME, Marcovina S, Brewer HB Jr, Santamarina-Fojo S: Lecithin:cholesterol acyltransferase overexpression generates hyperalpha-lipoproteinemia and a nonatherogenic lipoprotein pattern in transgenic rabbits. J Biol Chem 271: 4396–4402, 1996.
53. Blomhoff JP, Holme R, Sauar J, Gjone E: Familial lecithin:cholesterol acyltransferase deficiency. Further studies on plasma lipoproteins and plasma postheparin lipase activity of a patient with normal renal function. Scand J Clin Lab Invest Suppl 150: 177–182, 1978.
54. Ng DS, Xie C, Maguire GF, Zhu X, Ugwu F, Lam E, Connelly PW: Hypertriglyceridemia in lecithin–cholesterol acyltransferase-deficient mice is associated with hepatic overproduction of triglycerides, increased lipogenesis, and improved glucose tolerance. J Biol Chem 279: 7636–7642, 2004.
55. Cao G, Liang Y, Broderick CL, Oldham BA, Beyer TP, Schmidt RJ, Zhang Y, Stayrook KR, Suen C, Otto KA, Miller AR, Dai J, Foxworthy P, Gao H, Ryan TP, Jiang XC, Burris TP, Eacho PI, Etgen GJ: Antidiabetic action of a liver X receptor agonist mediated by inhibition of hepatic gluconeogenesis. J Biol Chem 278: 1131–1136, 2003.
56. Pawar A, Botolin D, Mangelsdorf DJ, Jump DB: The role of liver X receptor-alpha in the fatty acid regulation of hepatic gene expression. J Biol Chem 278: 40736–40743, 2003.
57. Sekiya M, Yahagi N, Matsuzaka T, Najima Y, Nakakuki M, Nagai R, Ishibashi S, Osuga J, Yamada N, Shimano H: Polyunsaturated fatty acids ameliorate hepatic steatosis in obese mice by SREBP-1 suppression. Hepatology 38: 1529–1539, 2003.
58. Field FJ, Born E, Murthy S, Mathur SN: Polyunsaturated fatty acids decrease the expression of sterol regulatory element-binding protein-1 in $CaCO_2$ cells: effect on fatty acid synthesis and triacylglycerol transport. Biochem J 368: 855–864, 2002.

3

Howdy Partner: Apolipoprotein A-I–ABCA1 Interactions Mediating HDL Particle Formation

SEREYRATH NGETH* AND GORDON A. FRANCIS

Abstract

High-density lipoproteins (HDL) are the fraction of cholesterol-containing particles in plasma that mediate the removal of excess cholesterol from tissues, thereby protecting against atherosclerosis. The interaction between the main protein of HDL, apolipoprotein A-I (apoA-I), and the membrane transporter ATP-binding cassette transporter A1 (ABCA1) is the rate-limiting step in HDL particle formation. Evidence exists for both indirect and direct interactions between apoA-I and ABCA1 at the cell surface that mediate phospholipid and cholesterol removal to generate nascent HDL particles. In addition, apoA-I alters the phosphorylation status and stability of ABCA1, further enhancing lipid removal and HDL particle formation by cells. This chapter reviews the current literature concerning the critical partnership between apoA-I and ABCA1 in mediating this key step in HDL metabolism.

Keywords: ABCA1; apolipoprotein A-I; atherosclerosis; cholesterol efflux; HDL; reverse cholesterol transport

Abbreviations: apo, apolipoprotein; ABCA1, ATP-binding cassette transporter A1; HDL, high-density lipoproteins; LXR, liver X receptor; PM, plasma membrane; PS, phosphatidylserine

Introduction

Atherosclerosis in the form of coronary artery and cerebrovascular disease is the leading cause of death in the world in adults [1, 2]. A major risk factor in the development of atherosclerosis is a low level of the protective class of cholesterol-containing particles in plasma, high-density lipoproteins (HDL) [3]. Protection by HDL is believed to result largely from the ability of these particles to remove cholesterol from cells and to deliver it back to the liver to be excreted into bile, referred to as reverse cholesterol transport (RCT) [4].

---

*Recipient of the CIHR/HSFC/Merck Frosst/Schering Trainee Award in Stroke, Cardiovascular, Obesity, Lipid, and Atherosclerosis Research (SCOLAR).

The active interaction of HDL proteins, particularly apolipoprotein A-I (apoA-I), with cells to remove cellular lipids is the critical step in the initial production of HDL, without which no HDL is available for further passive removal of cell cholesterol or subsequent steps of the RCT pathway.

Studies of a rare genetic cause of extremely low HDL levels, Tangier disease, indicated phospholipid [5] and cholesterol [5, 6] efflux to apoA-I and other HDL apoproteins [7] was impaired from Tangier disease cells. The defect in Tangier disease was subsequently identified as mutations in the ATP-binding cassette transporter AI (ABCA1) [8–12]. ABCA1 is a member of the large superfamily of ABC membrane transporters [13], and is present on the surface of most cells to stimulate the delivery of cell phospholipids and cholesterol to apoA-I and other HDL apolipoproteins including apoA-II and apoE [14]. The critical importance of functional ABCA1 in determining plasma HDL cholesterol (HDL-C) levels is suggested by the near-absence of HDL particles in individuals with mutations in both ABCA1 alleles (Tangier disease), and approximately 1/2 normal plasma HDL-C levels in individuals with one functional ABCA1 allele [15, 16]. It therefore appears that subsequent steps in the RCT pathway cannot compensate for an initial decrease in ABCA1 activity to normalize HDL-C levels. Tangier disease homozygotes and heterozygotes are consequently at increased risk of developing atherosclerosis [17, 18].

ABCA1 is a full ABC transporter containing two transmembrane domains and two ATP-binding domains [13]. ABCA1 expression is regulated at multiple levels (reviewed in Ref. [14]). Cellular cholesterol accumulation leads to increased ABCA1 expression [11, 19] through oxysterol-dependent activation of liver X receptor (LXR), a nuclear receptor that binds to the promoter region of the ABCA1 gene to activate its expression [20–22]. ABCA1 expression is therefore an additional way by which cells attempt to maintain cholesterol balance. Increased ABCA1 expression correlates with increased cholesterol efflux to HDL apoproteins [23, 24], as well as cell surface binding of apoA-I [25, 26].

ApoA-I is an amphipathic protein containing ten α-helical domains [27]. Lipid-free/lipid-poor apoA-I as well as other amphipathic HDL apolipoproteins including apoA-II, apoA-IV, apoC, and apoE have been shown to simulate ABCA1-mediated lipid efflux to generate nascent HDL particles [24]. However, the nature of this critical interaction between HDL apoproteins and ABCA1, and whether this definitely requires a direct apoA-I–ABCA1 interaction, remains unclear. Here we summarize recent literature regarding the nature of the apoA-I–ABCA1 interaction, and the regulation of ABCA1 activity by apoA-I-dependent and -independent phosphorylation mechanisms. Although apoA-I is the most abundant HDL apoprotein, and the protein used in most *in vitro* studies of apoprotein–ABCA1 interactions, the results of these studies in many cases have also been found to apply to the interactions of ABCA1 with other lipid-poor HDL apoproteins.

ApoA-I–ABCA1 Interactions

Currently, there are two schools of thought concerning the mechanism by which apoA-I interacts with ABCA1 to mediate lipid efflux. The first model suggests apoA-I interacts with plasma membrane (PM) lipid domains created by ABCA1, but not with ABCA1 directly [28]. The second proposes that apoA-I directly and physically binds ABCA1 in a receptor–ligand type interaction to promote phospholipid and cholesterol efflux.

Indirect Association/Lipid Domain Model

In the first model (Fig. 3.1A), functional ABCA1 is thought to generate an unstable lipid domain by translocating phospholipids, primarily phosphatidylserine (PS), to the exofacial leaflet of the PM [29]. It has been observed that ABCA1-expressing cells have membrane structure perturbations, which may be enriched in phospholipids and cholesterol in the outer leaflet of the PM [30, 31]. Such an unstable microenvironment on the cell surface is proposed to allow the amphipathic apoA-I to interact more readily with the PM and thereby solubilize membrane lipids, leading to cellular lipid removal upon dissociation of the lipidated protein from the membrane [32, 33]. Indeed, there is a strong correlation between the lipid binding affinity of apoA-I peptide variants and their ability to remove cellular phospholipids and cholesterol [34]. Another study using a fluorescence photobleaching technique demonstrated that membrane-associated apoA-I has diffusional

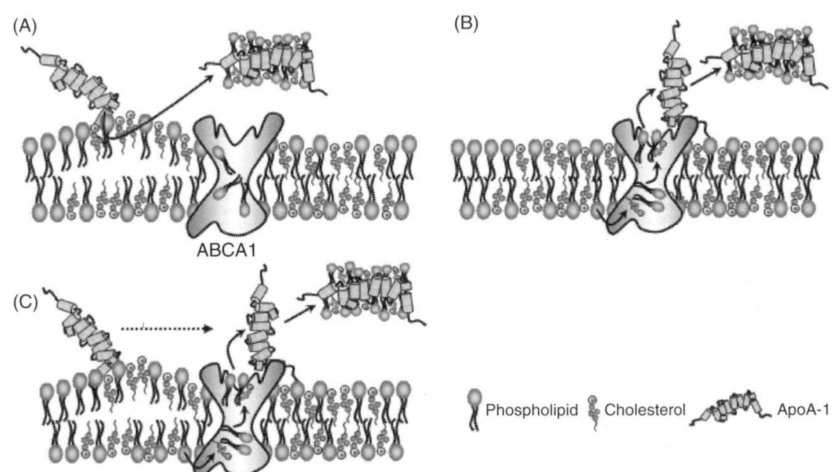

FIGURE 3.1. ApoA-1–ABCA1 interaction models. (A) Indirect association/lipid domain model, (B) direct association/receptor–ligand model, and (C) hybrid model. See text for details.

properties suggestive of an interaction with membrane lipids, rather than a membrane protein receptor [35].

Studies with ABCA1 mutants have also shone light on the nature of the molecular interaction between apoA-I and ABCA1. Chambenoit et al. [35] have shown that increased ABCA1 expression does not linearly correlate to increased apoA-I binding and that apoA-I cell surface binding is dependent on a fully functional ABCA1 protein because cells expressing an ATPase-defective ABCA1 fail to bind apoA-I. The conclusion was that cells lacking functional ABCA1 would not be able to flip PS to the outer leaflet of the PM, thereby inhibiting the generation of a suitable lipid environment required for apoA-I docking. In support of this is the observation that the expression of ABCA1 correlates with the amount of exofacial PS exposed on the cell surface [35] and the requirement for PS flipping to release cellular phospholipids to apoA-I [36]. Another study, however, showed that apoptosis-induced exofacial PS flipping was not sufficient to promote lipid efflux to apoA-I, and that annexin V, a PS-binding protein, did not compete for apoA-I cell surface binding [37]. A more recent study showed treatment of ABCA1-expressing human fibroblasts with phosphatidylcholine-specific phospholipase C or sphingomyelinase had no effect on the ability of apoA-I to associate with ABCA1, suggesting alterations in PM phospholipid content by ABCA1 do not predict apoA-I binding to cells [38].

Direct Association/Receptor–Ligand Model

Studies using chemical cross-linkers have suggested that apoA-I binds directly to cell surface ABCA1 to form a high affinity complex that is required for ABCA1-mediated lipid efflux [25, 31, 39] (Fig. 3.1B). ApoA-I has been shown to cross-link to ABCA1 in an aqueous environment [40] and at a distance as little as 3 Å away, indicating a very close association between the two proteins [41]. One ABCA1 transporter seems to bind to one molecule of apoA-I; however, ABCA1 has been proposed to oligomerize into a homotetrameric complex binding four molecules of apoA-I [42]. In support of the direct association model, inhibition of ABCA1 by glibenclamide has been shown to inhibit phospholipid and cholesterol efflux as well as cross-linking of apoA-I to ABCA1, suggesting a close relationship between apoA-I cross-linking and lipid efflux [26].

Studies with ABCA1 mutants have also provided mounting evidence for the direct binding of apoA-I to ABCA1. Fitzgerald et al. [40] have shown that four naturally occurring ABCA1 mutants harboring missense mutations in the two extracellular loops of ABCA1 fail to promote cholesterol efflux and had diminished ability to cross-link to apoA-I, even though the ABCA1 mutant proteins were properly targeted to the PM [40]. This would imply that a direct interaction between apoA-I and ABCA1 is required for efflux. Interestingly, one other ABCA1 mutant (W590S) that failed to promote lipid efflux retained full cross-linking ability with apoA-I, suggesting that apoA-I binding is necessary but not sufficient to stimulate cholesterol efflux [40]. Further studies with the W590S mutant revealed that apoA-I binds and dissociates from the defective ABCA1 transporter at a similar rate to wild-type ABCA1, but the dissociated apoA-I was without lipids, implying that dissociation is not dependent on lipid transfer [39]. An explanation for the inability of the W590S mutant to efflux lipids may be its' inability to drive the flipping of PS to the exofacial side of the PM, proposed to be a requirement for lipid efflux in the indirect association model [36]. Hence, the apoA-I–ABCA1 interaction is complex and may involve the association of apoA-I with lipid domains and ABCA1.

Panagotopulos and colleagues observed that helix 10 of apoA-I is crucial for apolipoprotein-mediated cholesterol efflux, and that two apoA-I mutants containing intact helix 10 exhibited markedly reduced lipid-binding affinity, but promoted cholesterol efflux much like wild-type apoA-I [43]. To explain this observation, the authors propose a hybrid model integrating the protein–protein and lipid domain interaction hypotheses. They speculate that helix 10 functions to tether apoA-I to the ABCA1-generated lipid domain such that apoA-I can diffuse laterally until it comes into contact with ABCA1 to form a productive complex, leading to the lipidation of apoA-I [43]. This hybrid model (Fig. 3.1C) would explain why apoA-I does not readily cross-link to ABCA1 at lower temperatures, as the lack of membrane fluidity would prevent the initial insertion and lateral diffusion of apoA-I [40].

The actual mechanism of ABCA1 in flipping lipids to the cell surface remains unknown. It may act primarily as a phospholipid flippase between the inner and outer PM leaflets, with cholesterol moving secondarily to maintain a constant phospholipid:cholesterol ratio in the outer leaflet, or ABCA1 might facilitate the simultaneous movement of both phospholipids and cholesterol. Further studies are required to resolve this question.

ApoA-I–ABCA1 Interaction Domains

ABCA1 is primarily found on the cell surface, consistent with its function as a cell surface lipid transporter, but it has also been found to localize to the Golgi and endolysosomal compartments [31, 44, 45]. Due to its presence in intracellular compartments as well as at the cell surface, it has been proposed that ABCA1-mediated lipid efflux may include ABCA1 internalization, possibly along with apoA-I, to intracellular sites and back to the PM for release of nascent HDL. Consistent with this idea was the report that ABCA1 preferentially mediates the removal of cholesterol from late endosomes–lysosomes [46] that would otherwise be esterified by acyl CoA:cholesterol acyltransferase [47]. As well, Golgi to PM vesicular transport is increased during apoA-I-mediated cholesterol efflux [48] and cellular lipid export from the Golgi to the PM is defective in ABCA1-deficient cells [45]. Alternatively, intracellular ABCA1 recycling may be mainly a mechanism to regulate

ABCA1 degradation [14, 44]. ApoA-I has been shown to be internalized and resecreted by phagocytic cells [49], consistent with a possible role in late endocytic vesicular trafficking and mobilization of intracellular lipid pools [50].

Regardless of the mechanism of ABCA1–apoA-I-dependent lipid efflux, the initial interaction of apoA-I with ABCA1 likely occurs at the PM. Since ABCA1 is localized mainly to the PM, it is thought that it functions primarily at the cell surface to efflux lipids, consistent with a direct role in the binding of apoA-I to ABCA1 at the cell surface. For this interaction to occur, ABCA1 must be properly trafficked to the PM in order to bind and facilitate lipid efflux to apoA-I. ABCA1 mutants (R587W, Q597R, L1379F, V1704D) that are retained in the endoplasmic reticulum and fail to localize to the PM do not efflux lipids or bind apoA-I [36, 51, 52]. Studies with ABCA1 mutants have revealed that amino acid residues in the two large extracellular loops of ABCA1 may form critical apoA-I-binding domains. Indeed, many of the Tangier disease-causing ABCA1 missense mutations have been mapped to these two domains [8–10, 18, 53]. Fitzgerald et al. [40] observed that four of five mutants carrying a mutation in either of the extracellular loops were properly translocated to the PM, but lacked the ability to cross-link with apoA-I, implying that these extracellular loops play an important role in mediating apoA-I binding. Another ABCA1-dependent requirement for apoA-I binding is functional ATPase domains. Engineered ATPase-deficient mutants that are correctly folded and targeted to the PM do not elicit apoA-I cell surface binding as assessed by fluorescence photobleaching [35]. This lack of apoA-I binding may be due to the inability of the ABCA1 mutant to create a PM lipid domain required for apoA-I docking [35]. Alternatively, ATP binding/hydrolysis may be required to induce a conformational change of ABCA1 that enables the direct binding of apoA-I to ABCA1, as suggested by lack of cross-linking of apoA-I to an ATPase-deficient ABCA1 mutant [26]. In agreement with this idea is the observation that the W590S ABCA1 mutant that fails to promote lipid efflux but is properly targeted to the PM and had a functional ATPase domain exhibited enhanced cross-linking activity to apoA-I [40, 51].

ABCA1 has been shown to promote lipid efflux to other exchangeable apolipoproteins associated with HDL particles, namely apoA-II, apoA-IV, apoC-I, apoC-II, apoC-III, and apoE [24]. This is seemingly implausible if apolipoproteins interact with ABCA1 in a classical receptor–ligand type complex, considering that the apolipoproteins do not share a similar primary amino acid sequence. Several of these proteins, however, have been shown to compete effectively with apoA-I for cross-linking and are themselves cross-linked to ABCA1 [39, 54]. The common feature between these apolipoproteins is that they contain multiple amphipathic helical domains, suggesting that the secondary structure rather than the primary amino acid sequence is the structural element that allows an apolipoprotein to interact with ABCA1. In support of this theory, Fitzgerald and colleagues showed that synthetic amphipathic apolipoproteins were able to block the cross-linking of apoA-I to ABCA1, whereas nonamphipathic lipid binding proteins did not [39]. Similar to this theory, the scavenger receptor class B type I has been shown to bind various apolipoproteins and the major motif found to mediate the interaction was amphipathic helices [55]. Hence, it is proposed that apolipoproteins may interact with ABCA1 by way of a hydrophobic groove within the ABCA1 transporter, an interaction that is seen between protein kinase A (PKA) and its amphipathic helical anchoring proteins [39]. ApoA-I likely interacts with a hydrophobic domain of ABCA1; however, some of the contact points between apoA-I and ABCA1 may occur in an aqueous environment because cross-linking was achievable with a water-soluble agent [40].

Interestingly, ABCA1 contains a cytoplasmic domain that is critical for promoting cholesterol efflux and apoA-I binding. Alterations in the highly conserved C-terminal VFVNFA motif resulted in defective lipid efflux as well as impaired apoA-I cross-linking [56]. Since the VFVNFA domain faces the cytosol, the authors speculate that this domain may be involved in protein–protein interactions that are responsible for recruiting additional factors to form an efflux complex [56]. ABCA1-expressing cells treated with a peptide mimetic of the VFVNFA motif inhibited ABCA1-mediated cholesterol efflux [56]. Clearly, less is known about the apoA-I–ABCA1 interaction domain than the molecular association of apoA-I to ABCA1. Additional studies designed to map and identify the critical amino acid residues on apolipoproteins and ABCA1 that interact will be necessary to fully understand the role of apolipoprotein binding to the ABCA1 transporter in lipid efflux. Perhaps obtaining the crystal structure of ABCA1 could best answer the question of what are the interaction domains between apoA-I and ABCA1.

Phosphorylation Events Modifying ABCA1 Activity and the Role of HDL Apolipoproteins

Constitutive and ApoA-I-Dependent Phosphorylation of ABCA1

In addition to the indirect or direct interactions between ABCA1 and apoA-I mediating cellular lipid efflux at the PM, numerous phosphorylation and dephosphorylation events affect ABCA1 activity, several of which are dependent on apoA-I binding to cells. See et al. [57] reported that PKA-dependent phosphorylation of serines 1042 and 2054 in the nucleotide binding domains of ABCA1 occurs constitutively (i.e., independent of apoA-I binding), and that decreased phosphorylation of ABCA1 through mutation of serine 2054 reduces apoA-I-mediated phospholipid efflux but not apoA-I binding or ABCA1 protein stability. Lawn et al. [11] showed that cyclic AMP (cAMP) could increase ABCA1 expression and cell surface levels of ABCA1 protein. Haidar et al. [58] subsequently showed that treatment of human fibroblasts with 8-bromo cAMP increased ABCA1 phosphorylation, and concluded that abnormal folding of ABCA1 caused by mutations might prevent phosphorylation of critical serine–threonine residues [58]. Further studies from this group suggested apoA-I also activates ABCA1 phosphorylation through a cAMP/PKA-dependent pathway under conditions where ABCA1 is present in high levels (following treatment with an LXR agonist), an effect that was reduced in Tangier disease fibroblasts [59]. These results suggested binding of apoA-I to ABCA1 may also activate cAMP signaling through a G protein-coupled ABCA1 transporter [59].

Earlier results showed that apoA-I-dependent lipid release is dependent on activation of protein kinase C (PKC) [60–62]. Yamauchi et al. [63] have recently provided evidence that PKC inhibitors suppress the stabilization of ABCA1 and cellular lipid release mediated by apoA-I. They showed that digestion of sphingomyelin increased ABCA1 protein levels, and that inhibition of phosphatidylcholine phospholipase C (PC-PLC) inhibits this effect. They also proposed PKC-dependent phosphorylation of ABCA1 by apoA-I, as suggested by inhibition of this effect by PKC-specific inhibitors [63]. They conclude that apoA-I activates PKC following depletion of the PM of sphingomyelin, thereby activating PC-PLC and generating diacylglycerol that activates PKC, leading to the phosphorylation and stabilization of ABCA1.

Yet another phosphorylation mechanism activated by the apoA-I interaction with cells is an increase in the tyrosine kinase Janus kinase 2 (JAK2) activity [64]. JAK2 activity was required for apoA-I binding and apoA-I-dependent efflux of cellular lipids, but did not directly affect ABCA1 phosphorylation or cholesterol translocase activity of ABCA1 [64]. These results suggest JAK2 is activated by apoA-I, and that JAK2 promotes the productive interaction between apoA-I and ABCA1 to mobilize cellular lipids, possibly by enhancing the activity of another partner protein that facilitates the binding of apoA-I to ABCA1.

Lastly, phosphorylation sites in conserved regions of a cytoplasmic domain of ABCA1 downstream of the first nucleotide domain have been identified that are phosphorylated by protein kinase 2, leading to decreased apoA-I and apoA-II binding, ABCA1 flippase activity, and phospholipid plus cholesterol efflux [65]. These results demonstrate an inhibitory mechanism of ABCA1 phosphorylation independent of, but subsequently inhibiting, apoA-I binding.

Protection by apoA-I against Calpain-Dependent Proteolysis of ABCA1

Another remarkable facet of the apoA-I–ABCA1 interaction is the demonstration that HDL apoproteins serve to protect ABCA1 against proteolysis, thereby increasing ABCA1 activity and further facilitating apoprotein-mediated lipid efflux [66–68]. Arakawa and Yokoyama [66] demonstrated that incubation of apoA-I with cultured THP-1 cells increased ABCA1 protein levels without increasing ABCA1 message, and that free apoA-II did the same thing, however intact HDL particles did not [66]. Thiol protease inhibitors could also prevent ABCA1 degradation and enhanced cholesterol efflux, suggesting a possible mechanism by which the apoproteins could be stabilizing ABCA1. Wang and colleagues subsequently reported that calpain-dependent proteolysis is involved in the degradation of ABCA1 through a proline, glutamate, serine, and threonine (PEST)-rich cytoplasmic sequence of the transporter, and that deletion of this PEST sequence or treatment of cells with apoA-I could inhibit this proteolysis and increase cell surface ABCA1 [67]. They also demonstrated that infusion of apoA-I increased ABCA1 protein levels in the liver and peritoneal macrophages of C57BL/6J mice [67]. This group subsequently reported that phosphorylation of threonine residues within this PEST sequence is required for calpain-dependent proteolysis of ABCA1, and that phosphorylation of this sequence is reduced by apoA-I. In this way, apoA-I modulates yet another mechanism of ABCA1 phosphorylation, by somehow decreasing phosphorylation of the PEST sequence, to increase ABCA1 activity. Further studies suggest the PEST sequence of ABCA1 is required for the internalization of the transporter, to mediate efflux of the portion of total cholesterol efflux derived from intracellular pools [69].

Conclusions

The interactions of apoA-I with ABCA1 that mediate nascent HDL particle formation are the rate-limiting step in the RCT pathway. Evidence is presented for both an indirect and direct interaction of apoA-I with the cell surface, mediated by ABCA1, which facilitates the delivery of cellular phospholipids and cholesterol to apoA-I to generate HDL particles. The complexity of this partnership is demonstrated by the numerous phosphorylation events mediated in part by apoA-I interactions with cells that enhance or reduce ABCA1 activity. Further research into this critical interaction will enhance our understanding of HDL particle formation and our ability to utilize this pathway for the treatment and prevention of atherosclerosis.

Acknowledgments: This work is supported by CIHR operating grant MOP-12660. Gordon. A. Francis is a Senior Scholar of the Alberta Heritage Foundation for Medical Research.

References

1. Murray CJ and Lopez AD: Mortality by cause for eight regions of the world: Global Burden of Disease Study. Lancet 349: 1269–1276, 1997.
2. Okrainec K, Banerjee DK, and Eisenberg MJ: Coronary artery disease in the developing world. Am Heart J 148: 7–15, 2004.
3. Castelli WP: Cholesterol and lipids in the risk of coronary artery disease—the Framingham Heart Study. Can J Cardiol 4 (Suppl A): 5A–10A, 1988.

4. Glomset JA: The plasma lecithin: cholesterol acyltransferase reaction. J Lipid Res 9: 155–167, 1968.

5. Francis GA, Knopp RH, and Oram JF: Defective removal of cellular cholesterol and phospholipids by apolipoprotein A-I in Tangier disease. J Clin Invest 96: 78–87, 1995.

6. Walter M, Gerdes U, Seedorf U, and Assmann G: The high density lipoprotein- and apolipoprotein A-I-induced mobilization of cellular cholesterol is impaired in fibroblasts from Tangier disease subjects. Biochem Biophys Res Commun 205: 850–856, 1994.

7. Remaley AT, Schumacher UK, Stonik JA, Farsi BD, Nazih H, and Brewer HB Jr: Decreased reverse cholesterol transport from Tangier disease fibroblasts. Acceptor specificity and effect of brefeldin on lipid efflux. Arterioscler Thromb Vasc Biol 17: 1813–1821, 1997.

8. Bodzioch M, Orso E, Klucken J, Langmann T, Bottcher A, Diederich W, Drobnik W, Barlage S, Buchler C, Porsch-Ozcurumez M, Kaminski WE, Hahmann HW, Oette K, Rothe G, Aslanidis C, Lackner KJ, and Schmitz G: The gene encoding ATP-binding cassette transporter 1 is mutated in Tangier disease. Nat Genet 22: 347–351, 1999.

9. Brooks-Wilson A, Marcil M, Clee SM, Zhang LH, Roomp K, van Dam M, Yu L, Brewer C, Collins JA, Molhuizen HO, Loubser O, Ouelette BF, Fichter K, Ashbourne-Excoffon KJ, Sensen CW, Scherer S, Mott S, Denis M, Martindale D, Frohlich J, Morgan K, Koop B, Pimstone S, Kastelein JJ, and Hayden MR: Mutations in ABC1 in Tangier disease and familial high-density lipoprotein deficiency. Nat Genet 22: 336–345, 1999.

10. Rust S, Rosier M, Funke H, Real J, Amoura Z, Piette JC, Deleuze JF, Brewer HB, Duverger N, Denefle P, and Assmann G: Tangier disease is caused by mutations in the gene encoding ATP-binding cassette transporter 1. Nat Genet 22: 352–355, 1999.

11. Lawn RM, Wade DP, Garvin MR, Wang X, Schwartz K, Porter JG, Seilhamer JJ, Vaughan AM, and Oram JF: The Tangier disease gene product ABC1 controls the cellular apolipoprotein-mediated lipid removal pathway. J Clin Invest 104: R25–R31, 1999.

12. Brousseau ME, Schaefer EJ, Dupuis J, Eustace B, Van Eerdewegh P, Goldkamp AL, Thurston LM, FitzGerald MG, Yasek-McKenna D, O'Neill G, Eberhart GP, Weiffenbach B, Ordovas JM, Freeman MW, Brown RH Jr, and Gu JZ: Novel mutations in the gene encoding ATP-binding cassette 1 in four tangier disease kindreds. J Lipid Res 41: 433–441, 2000.

13. Dean M, Hamon Y, and Chimini G: The human ATP-binding cassette (ABC) transporter superfamily. J Lipid Res 42: 1007–1017, 2001.

14. Oram JF: ATP-binding cassette transporter A1 and cholesterol trafficking. Curr Opin Lipidol 13: 373–381, 2002.

15. Assmann G, Von Eckardstein A, and Brewer HB: Familial Analphalipoproteinemia: Tangier Disease. In: Scriver CR, Beaudet AL, Sly WS, and Valle D (eds), The Metabolic and Molecular Bases of Inherited Disease, 8th edn. McGraw-Hill, New York, 2001, pp. 2937–2980.

16. Asztalos BF, Brousseau ME, McNamara JR, Horvath KV, Roheim PS, and Schaefer EJ: Subpopulations of high density lipoproteins in homozygous and heterozygous Tangier disease. Atherosclerosis 156: 217–225, 2001.

17. Serfaty-Lacrosniere C, Civeira F, Lanzberg A, Isaia P, Berg J, Janus ED, Smith MP, Pritchard PH, Frohlich J, Lees RS, Barnard GF, Ordovas JM, and Schaefer EJ: Homozygous Tangier disease and cardiovascular disease. Atherosclerosis 107: 85–98, 1994.

18. Clee SM, Kastelein JJ, van Dam M, Marcil M, Roomp K, Zwarts KY, Collins JA, Roelants R, Tamasawa N, Stulc T, Suda T, Ceska R, Boucher B, Rondeau C, DeSouich C, Brooks-Wilson A, Molhuizen HO, Frohlich J, Genest J Jr, and Hayden MR: Age and residual cholesterol efflux affect HDL cholesterol levels and coronary artery disease in ABCA1 heterozygotes. J Clin Invest 106: 1263–1270, 2000.

19. Langmann T, Klucken J, Reil M, Liebisch G, Luciani MF, Chimini G, Kaminski WE, and Schmitz G: Molecular cloning of the human ATP-binding cassette transporter 1 (hABC1): evidence for sterol-dependent regulation in macrophages. Biochem Biophys Res Commun 257: 29–33, 1999.

20. Costet P, Luo Y, Wang N, and Tall AR: Sterol-dependent transactivation of the ABC1 promoter by the liver X receptor/retinoid X receptor. J Biol Chem 275: 28240–28245, 2000.

21. Schwartz K, Lawn RM, and Wade DP: ABC1 gene expression and ApoA-I-mediated cholesterol efflux are regulated by LXR. Biochem Biophys Res Commun 274: 794–802, 2000.

22. Francis GA, Annicotte JS, and Auwerx J: Liver X receptors: Xcreting Xol to combat atherosclerosis. Trends Mol Med 8: 455–458, 2002.

23. Bortnick AE, Rothblat GH, Stoudt G, Hoppe KL, Royer LJ, McNeish J, and Francone OL: The correlation of ATP-binding cassette 1 mRNA levels with cholesterol efflux from various cell lines. J Biol Chem 275: 28634–28640, 2000.

24. Remaley AT, Stonik JA, Demosky SJ, Neufeld EB, Bocharov AV, Vishnyakova TG, Eggerman TL, Patterson AP, Duverger NJ, Santamarina-Fojo S, and Brewer HB Jr: Apolipoprotein specificity for lipid efflux by the human ABCAI transporter. Biochem Biophys Res Commun 280: 818–823, 2001.

25. Oram JF, Lawn RM, Garvin MR, and Wade DP: ABCA1 is the cAMP-inducible apolipoprotein receptor that mediates cholesterol secretion from macrophages. J Biol Chem 275: 34508–34511, 2000.

26. Wang N, Silver DL, Thiele C, and Tall AR: ATP-binding cassette transporter A1 (ABCA1) functions as a cholesterol efflux regulatory protein. J Biol Chem 276: 23742–23747, 2001.

27. Frank PG and Marcel YL: Apolipoprotein A-I: structure–function relationships. J Lipid Res 41: 853–872, 2000.

28. Oram JF: HDL apolipoproteins and ABCA1: partners in the removal of excess cellular cholesterol. Arterioscler Thromb Vasc Biol 23: 720–727, 2003.

29. Hamon Y, Broccardo C, Chambenoit O, Luciani MF, Toti F, Chaslin S, Freyssinet JM, Devaux PF, McNeish J, Marguet D, and Chimini G: ABC1 promotes engulfment of apoptotic cells and transbilayer redistribution of phosphatidylserine. Nat Cell Biol 2: 399–406, 2000.

30. Lin G and Oram JF: Apolipoprotein binding to protruding membrane domains during removal of excess cellular cholesterol. Atherosclerosis 149: 359–370, 2000.

31. Wang N, Silver DL, Costet P, and Tall AR: Specific binding of ApoA-I, enhanced cholesterol efflux, and altered plasma membrane morphology in cells expressing ABC1. J Biol Chem 275: 33053–33058, 2000.

32. Vaughan AM and Oram JF: ABCA1 redistributes membrane cholesterol independent of apolipoprotein interactions. J Lipid Res 44: 1373–1380, 2003.
33. Gillotte KL, Davidson WS, Lund-Katz S, Rothblat GH, and Phillips MC: Removal of cellular cholesterol by pre-beta-HDL involves plasma membrane microsolubilization. J Lipid Res 39: 1918–1928, 1998.
34. Gillotte KL, Zaiou M, Lund-Katz S, Anantharamaiah GM, Holvoet P, Dhoest A, Palgunachari MN, Segrest JP, Weisgraber KH, Rothblat GH, and Phillips MC: Apolipoprotein-mediated plasma membrane microsolubilization. J Biol Chem 274: 2021–2028, 1999.
35. Chambenoit O, Hamon Y, Marguet D, Rigneault H, Rosseneu M, and Chimini G: Specific docking of apolipoprotein A-I at the cell surface requires a functional ABCA1 transporter. J Biol Chem 276: 9955–9960, 2001.
36. Rigot V, Hamon Y, Chambenoit O, Alibert M, Duverger N, and Chimini G: Distinct sites on ABCA1 control distinct steps required for cellular release of phospholipids. J Lipid Res 43: 2077–2086, 2002.
37. Smith JD, Waelde C, Horwitz A, and Zheng P: Evaluation of the role of phosphatidylserine translocase activity in ABCA1-mediated lipid efflux. J Biol Chem 277: 17797–17803, 2002.
38. Denis M, Haidar B, Marcil M, Bouvier M, Krimbou L, and Genest J Jr: Molecular and cellular physiology of apolipoprotein A-I lipidation by the ATP-binding cassette transporter A1 (ABCA1). J Biol Chem 279: 7384–7394, 2004.
39. Fitzgerald ML, Morris AL, Chroni A, Mendez AJ, Zannis VI, and Freeman MW: ABCA1 and amphipathic apolipoproteins form high-affinity molecular complexes required for cholesterol efflux. J Lipid Res 45: 287–294, 2004.
40. Fitzgerald ML, Morris AL, Rhee JS, Andersson LP, Mendez AJ, and Freeman MW: Naturally occurring mutations in the largest extracellular loops of ABCA1 can disrupt its direct interaction with apolipoprotein A-I. J Biol Chem 277: 33178–33187, 2002.
41. Chroni A, Liu T, Fitzgerald ML, Freeman MW, and Zannis VI: Cross-linking and lipid efflux properties of apoA-I mutants suggest direct association between apoA-I helices and ABCA1. Biochemistry 43: 2126–2139, 2004.
42. Denis M, Haidar B, Marcil M, Bouvier M, Krimbou L, and Genest J: Characterization of oligomeric human ATP binding cassette transporter A1. Potential implications for determining the structure of nascent high density lipoprotein particles. J Biol Chem 279: 41529–41536, 2004.
43. Panagotopulos SE, Witting SR, Horace EM, Hui DY, Maiorano JN, and Davidson WS: The role of apolipoprotein A-I helix 10 in apolipoprotein-mediated cholesterol efflux via the ATP-binding cassette transporter ABCA1. J Biol Chem 277: 39477–39484, 2002.
44. Neufeld EB, Remaley AT, Demosky SJ, Stonik JA, Cooney AM, Comly M, Dwyer NK, Zhang M, Blanchette-Mackie J, Santamarina-Fojo S, and Brewer HB Jr: Cellular localization and trafficking of the human ABCA1 transporter. J Biol Chem 276: 27584–27590, 2001.
45. Orso E, Broccardo C, Kaminski WE, Bottcher A, Liebisch G, Drobnik W, Gotz A, Chambenoit O, Diederich W, Langmann T, Spruss T, Luciani MF, Rothe G, Lackner KJ, Chimini G, and Schmitz G: Transport of lipids from Golgi to plasma membrane is defective in Tangier disease patients and Abc1-deficient mice. Nat Genet 24: 192–196, 2000.

46. Chen W, Sun Y, Welch C, Gorelik A, Leventhal AR, Tabas I, and Tall AR: Preferential ATP-binding cassette transporter A1-mediated cholesterol efflux from late endosomes/lysosomes. J Biol Chem 276: 43564–43569, 2001.
47. Yamauchi Y, Chang CC, Hayashi M, Abe-Dohmae S, Reid PC, Chang TY, and Yokoyama S: Intracellular cholesterol mobilization involved in the ABCA1/apolipoprotein-mediated assembly of high density lipoprotein in fibroblasts. J Lipid Res 45: 1943–1951, 2004.
48. Zha X, Gauthier A, Genest J, and McPherson R: Secretory vesicular transport from the Golgi is altered during ATP-binding cassette protein A1 (ABCA1)-mediated cholesterol efflux. J Biol Chem 278: 10002–10005, 2003.
49. Takahashi Y and Smith JD: Cholesterol efflux to apolipoprotein AI involves endocytosis and resecretion in a calcium-dependent pathway. Proc Natl Acad Sci U S A 96: 11358–11363, 1999.
50. Neufeld EB, Stonik JA, Demosky SJ Jr, Knapper CL, Combs CA, Cooney A, Comly M, Dwyer N, Blanchette-Mackie J, Remaley AT, Santamarina-Fojo S, and Brewer HB Jr: The ABCA1 transporter modulates late endocytic trafficking: insights from the correction of the genetic defect in Tangier disease. J Biol Chem 279: 15571–15578, 2004.
51. Tanaka AR, Abe-Dohmae S, Ohnishi T, Aoki R, Morinaga G, Okuhira K, Ikeda Y, Kano F, Matsuo M, Kioka N, Amachi T, Murata M, Yokoyama S, and Ueda K: Effects of mutations of ABCA1 in the first extracellular domain on subcellular trafficking and ATP binding/hydrolysis. J Biol Chem 278: 8815–8819, 2003.
52. Albrecht C, Baynes K, Sardini A, Schepelmann S, Eden ER, Davies SW, Higgins CF, Feher MD, Owen JS, and Soutar AK: Two novel missense mutations in ABCA1 result in altered trafficking and cause severe autosomal recessive HDL deficiency. Biochim Biophys Acta 1689: 47–57, 2004.
53. Schmitz G, Kaminski WE, and Orso E: ABC transporters in cellular lipid trafficking. Curr Opin Lipidol 11: 493–501, 2000.
54. Krimbou L, Denis M, Haidar B, Carrier M, Marcil M, and Genest J Jr: Molecular interactions between apoE and ABCA1: impact on apoE lipidation. J Lipid Res 45: 839–848, 2004.
55. Williams DL, de La Llera-Moya M, Thuahnai ST, Lund-Katz S, Connelly MA, Azhar S, Anantharamaiah GM, and Phillips MC: Binding and cross-linking studies show that scavenger receptor BI interacts with multiple sites in apolipoprotein A-I and identify the class A amphipathic alpha-helix as a recognition motif. J Biol Chem 275: 18897–18904, 2000.
56. Fitzgerald ML, Okuhira K, Short GF III, Manning JJ, Bell SA, and Freeman MW: ATP-binding cassette transporter A1 contains a novel C-terminal VFVNFA motif that is required for its cholesterol efflux and ApoA-I binding activities. J Biol Chem 279: 48477–48485, 2004.
57. See RH, Caday-Malcolm RA, Singaraja RR, Zhou S, Silverston A, Huber MT, Moran J, James ER, Janoo R, Savill JM, Rigot V, Zhang LH, Wang M, Chimini G, Wellington CL, Tafuri SR, and Hayden MR: Protein kinase A site-specific phosphorylation regulates ATP-binding cassette A1 (ABCA1)-mediated phospholipid efflux. J Biol Chem 277: 41835–41842, 2002.
58. Haidar B, Denis M, Krimbou L, Marcil M, and Genest J Jr: cAMP induces ABCA1 phosphorylation activity and promotes cholesterol efflux from fibroblasts. J Lipid Res 43: 2087–2094, 2002.

59. Haidar B, Denis M, Marcil M, Krimbou L, and Genest J Jr: Apolipoprotein A-I activates cellular cAMP signaling through the ABCA1 transporter. J Biol Chem 279: 9963–9969, 2004.
60. Mendez AJ, Oram JF, and Bierman EL: Protein kinase C as a mediator of high density lipoprotein receptor-dependent efflux of intracellular cholesterol. J Biol Chem 266: 10104–10111, 1991.
61. Li Q and Yokoyama S: Independent regulation of cholesterol incorporation into free apolipoprotein-mediated cellular lipid efflux in rat vascular smooth muscles cells. J Biol Chem 270: 26216–26223, 1995.
62. Li Q, Tsujita M, and Yokoyama S: Selective down-regulation by protein kinase C inhibitors of apolipoprotein-mediated cellular cholesterol efflux in macrophages. Biochemistry 36: 12045–12052, 1997.
63. Yamauchi Y, Hayashi M, Abe-Dohmae S, and Yokoyama S: Apolipoprotein A-I activates protein kinase C alpha signaling to phosphorylate and stabilize ATP binding cassette transporter A1 for the high density lipoprotein assembly. J Biol Chem 278: 47890–47897, 2003.
64. Tang C, Vaughan AM, and Oram JF: Janus kinase 2 modulates the apolipoprotein interactions with ABCA1 required for removing cellular cholesterol. J Biol Chem 279: 7622–7628, 2004.
65. Roosbeek S, Peelman F, Verhee A, Labeur C, Caster H, Lensink MF, Cirulli C, Grooten J, Cochet C, Vandekerckhove J, Amoresano A, Chimini G, Tavernier J, and Rosseneu M: Phosphorylation by protein kinase CK2 modulates the activity of the ATP binding cassette A1 transporter. J Biol Chem 279: 37779–37788, 2004.
66. Arakawa R and Yokoyama S: Helical apolipoproteins stabilize ATP-binding cassette transporter A1 by protecting it from thiol protease-mediated degradation. J Biol Chem 277: 22426–22429, 2002.
67. Wang N, Chen W, Linsel-Nitschke P, Martinez LO, Agerholm-Larsen B, Silver DL, and Tall AR: A PEST sequence in ABCA1 regulates degradation by calpain protease and stabilization of ABCA1 by apoA-I. J Clin Invest 111: 99–107, 2003.
68. Martinez LO, Agerholm-Larsen B, Wang N, Chen W, and Tall AR: Phosphorylation of a pest sequence in ABCA1 promotes calpain degradation and is reversed by ApoA-I. J Biol Chem 278: 37368–37374, 2003.
69. Chen W, Wang N, and Tall AR: A PEST deletion mutant of ABCA1 shows impaired internalization and defective cholesterol efflux from late endosomes. J Biol Chem 280: 29277–29281, 2005.

4

Role of the Scavenger Receptor Class B Type I in Lipoprotein Metabolism and Atherosclerosis: Insights from Genetically Altered Mice

BERNARDO TRIGATTI

Abstract

The scavenger receptor class B type I (SR-BI) is a cell surface multiligand receptor that can bind high-density lipoproteins and mediate exchange of lipids with cells. It is expressed abundantly in liver and in cells of the vascular wall. Analyses of genetically manipulated mice with altered levels of SR-BI expression in different tissues have yielded important insight into its roles in HDL metabolism and atherosclerosis.

Keywords: atherosclerosis; eNOS; HDL; knockout; liver; macrophage; PDZK1; SR-BI; transgenic

Abbreviations: ABC, ATP-binding cassette transporter; AGE-BSA, advance glycation end-product modified bovine serum albumin; CETP, cholesteryl ester transfer protein; eNOS, endothelial nitric oxide synthase; HDL, high-density lipoprotein; KO, knockout; LCAT, lecithin cholesterol acyltransferase; LDL, low-density lipoprotein; MDCK, Madin–Darby canine kidney; PAF-AH, platelet activating factor acylhydrolase; PDZ, Postsynaptic density protein-95/Drosophila discs-large/ZO1; PKA, protein kinase A; PL, phospholipid; PON, paraoxonase; SPAP, small PDZK1-associated protein; SAA, serum amyloid alpha; SR-BI, scavenger receptor class B type I; VLDL, very low-density lipoprotein

Introduction

Clinical, epidemiological, and basic research points to a role for high-density lipoproteins (HDL) in protection against atherosclerosis and heart disease [1–3]. Because of this relationship, treatments aimed at increasing HDL function have garnered great interest [3]. Efforts over the past several years have been aimed at attempting to understand the molecular and physiological mechanisms by which HDL protects against atherosclerosis. A variety of pathways appear to be involved. For example, HDL is a carrier of antioxidant vitamins (such as α-tocopherol—vitamin E), which help protect cells from harmful effects of reactive oxygen species [4, 5]. HDL is also associated with enzymes such as paraoxonase (PON) and platelet activating factor acylhydrolase (PAF-AH) that may play important roles in the removal of oxidized lipids from other lipoproteins such as low-density lipoprotein (LDL) [6–8]. HDLs also trigger signaling events within cells leading to diverse consequences, including the activation of endothelial nitric oxide synthase (eNOS) and protection of cells in the vascular wall against apoptosis, processes which have been associated with protection against atherosclerosis [9–18].

HDL also plays a key role in reverse cholesterol transport. This is the removal of cholesterol from cells in peripheral tissues (e.g., macrophages and foam cells in the artery wall), and its delivery to the liver. HDL-associated cholesterol is converted to cholesteryl ester by lecithin cholesterol acyltransferase (LCAT) [19, 20]. HDL-associated cholesteryl ester can be transferred to very low-density lipoprotein (VLDL) in exchange for triglycerides by cholesteryl ester transfer protein (CETP) [21]. Alternatively, cholesteryl ester that remains associated with HDL can be taken up by liver hepatocytes via a pathway called selective lipid uptake. This involves the net internalization of only the lipid portion of the HDL particle, without the net internalization and degradation of the HDL particle itself (reviewed in [22]). Internalized cholesterol can be secreted into bile either as cholesterol or after conversion to bile acids or be repackaged into nascent VLDL and secreted from the hepatocyte into the circulation (reviewed in [22]). HDL has also been implicated in the reverse transport of oxidized lipids, derived from cells or from other lipoproteins, to the liver for clearance [23]. Therefore, the "reverse cholesterol transport" pathway may play important roles in the clearance of diverse lipids in addition to cholesterol, highlighting the importance of this pathway to protection against atherosclerosis.

The Scavenger Receptor Class B Type I

Almost a decade ago, Krieger and coworkers [24] identified the scavenger receptor class B type I (SR-BI) as an HDL receptor that could mediate the selective uptake of HDL lipids. SR-BI is a member of the CD36 family of proteins (see Chapter 5) and contains two transmembrane domains, short N-terminal and C-terminal cytoplasmic regions, and a large extracellular region that is heavily glycosylated (reviewed in [22], see Fig. 4.1). SR-BI is also palmitoylated on two Cys residues near the junction between the C-terminal cytoplasmic and transmembrane regions [25, 26]. It is expressed most abundantly in cells with a high capacity for selective HDL lipid uptake—steroidogenic cells and hepatocytes (reviewed in [22]). This suggested that SR-BI may play an important role in the later stages of reverse

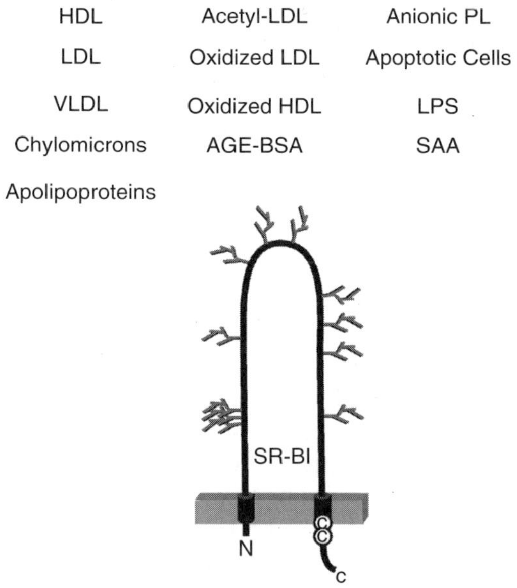

FIGURE 4.1. SR-BI and known ligands. SR-BI is represented with two transmembrane regions residing in the membrane (gray bar), cytoplasmic C-terminus and N-terminus, and a large extracellular region. Approximate locations of N-linked glycosylation sites are shown in gray, as are the locations of two cysteine residues that are sites of palmitoylation. Known SR-BI ligands are listed at the top and include lipoproteins, chemically modified proteins, and lipoproteins including advance glycation end-product modified BSA (AGE-BSA), liposomes containing anionic phospholipids (PL), apoptotic cells, bacterial lipopolysaccharide, and the acute phase protein serum amyloid alpha (SAA) (either free or associated with lipoproteins) (reviewed in [27] and [86]).

cholesterol transport and transport of HDL cholesterol to steroidogenic tissues. SR-BI is also expressed in a variety of other cell types (reviewed in [22, 27, 28]) including macrophages and endothelial cells [29–32] and could therefore influence atherosclerosis locally at the vessel wall (explained later).

When overexpressed in transfected cells, SR-BI can mediate both increased selective uptake of HDL lipids and increased efflux of free cholesterol tracer to HDL acceptors [24, 33, 34]. This has led to the notion that SR-BI can mediate the bidirectional flux of lipids between cells and bound lipoproteins and the suggestion that the direction of net flux is dependent on the concentration gradient of the lipid [35]. For example, in Chinese hamster ovary-derived cells, which have relatively low levels of intracellular cholesterol, the net flux appears to be in the direction of the cell [24]. The observation that SR-BI can mediate the efflux of unesterified cholesterol tracer from cells has led to the suggestion that it may play an important role in HDL-dependent cholesterol efflux from macrophages (see [33, 34]). Consequently, SR-BI has been implicated in both early steps (i.e., cholesterol efflux from macrophages) and late steps (hepatic selective HDL cholesterol uptake) in reverse cholesterol transport.

SR-BI has also been implicated in mediating HDL-dependent signaling events in endothelial cells (see Chapter 5). For example, the HDL-dependent stimulation of eNOS appears to be dependent on SR-BI [9, 11, 12, 14]. This may occur via the transfer of bioactive lipids (estradiol, ceramide, and/or sphingosine-1-phosphate) with signaling activity and/or through the generation of intracellular signaling events including the activation PKB/Akt pathways [9, 11, 12, 14, 17]. The ability of SR-BI to mediate HDL-dependent eNOS activation may also be related to SR-BI-mediated cholesterol efflux to HDL [36]. The role of SR-BI in cholesterol efflux from cells appears to be controversial, however. Overexpression of SR-BI in transfected cells has been shown quite clearly to increase the efflux of cholesterol tracer to HDL [33, 34, 37]. In contrast, several studies have indicated that SR-BI is not required for cholesterol efflux to HDL in different cells including endothelial cells [38] and macrophages [39–41].

Certain tissues, including the liver, contain an alternatively spliced transcript giving rise to an isoform called SR-BII, which differs from SR-BI in its C-terminal cytoplasmic region [42, 43]. SR-BI appears to be considerably more active than SR-BII in mediating lipid transfer between cells and lipoproteins [44]. This may be the result of a different subcellular distribution of the two isoforms, with SR-BI demonstrating a greater distribution on the plasma membrane than does SR-BII [44].

Recently, Smart and coworkers [18] have reported that SR-BI expression induces apoptosis in cells in the absence of ligand stimulation. Both HDL interaction with SR-BI and eNOS activation were able to inhibit SR-BI-triggered apoptosis. This suggests the possibility that SR-BI could contribute to apoptosis of endothelial cells or other cells in an HDL- and eNOS-regulated manner. This process could contribute to endothelial dysfunction and initiation or development of atherosclerosis, and HDL and eNOS might act, at least in part, by halting this process [18].

In addition to mediating the binding of HDL and HDL-dependent lipid exchange with cells, SR-BI also binds a variety of other ligands (Fig. 4.1). These include other lipoproteins (LDL, VLDL, and chylomicrons), modified proteins/lipoproteins (e.g., advance glycation end-product modified bovine serum albumin (AGE-BSA), oxidized or acetylated LDL), the acute phase protein serum amyloid alpha (SAA—either free or bound to HDL), phospholipid (PL) vesicles containing anionic PL, apoptotic cells, and bacterial cell surface components (e.g., lipopolysaccharide) (see [27] for recent review). Therefore in addition to playing a role in HDL metabolism, SR-BI may play roles in a variety of other physiological processes, many of which may affect the development or progression of atherosclerosis.

SR-BI and Atherosclerosis: Studies of Gene-Targeted Mice

Studies of genetically manipulated mice have provided substantial insight into SR-BI's physiological role in HDL metabolism and atherosclerosis. SR-BI knockout (KO) mice, in which the gene encoding SR-BI (*Scarb1*) has been inactivated by targeted mutagenesis, exhibit no detectable SR-BI expression in tissues examined [45]. These mice have increased cholesterol levels associated with enlarged HDL particles with altered compositions [45–48]. The enlarged HDL particles contain increased apoE and nonesterified/esterified cholesterol ratios accompanied by decreased levels of LCAT activity [45, 47–49]. These alterations in plasma HDL composition and structure are accompanied by reduced lipid storage in steroidogenic tissues and reduced biliary cholesterol secretion [45, 46, 50]. Mice that are heterozygous for the targeted mutation express half of normal SR-BI levels in tissues that have been analyzed, including liver and adrenal glands [45]. These mice have slightly increased HDL cholesterol, and ~50% reductions in adrenal cholesterol levels [45]. SR-BI KO mice do not, however, exhibit altered hepatic cholesterol levels or cholesterol biosynthesis [49, 50]. SR-BI KO mice have been reported to exhibit increased levels of hepatic expression of ABCG1, an ATP-binding cassette transporter (ABC) implicated in cholesterol efflux to HDL, and mice carrying one or two copies of the inactivating mutation exhibit reduced hepatic ABCG5 and ABCG8, other ABC family members which have been implicated in regulating hepatic cholesterol secretion into bile [49, 51, 52].

SR-BI KO mice have reduced hepatic and adrenal uptake of HDL cholesterol and exhibit reduced selective clearance of HDL and LDL cholesterol from plasma [53–55]. These studies suggest that SR-BI is required for selective uptake of cholesterol from HDL and LDL by liver and steroidogenic cells *in vivo* [45, 46, 50, 53–56]. SR-BI KO mice have also been reported to exhibit a greater increase in plasma triglycerides after an intragastric fat load whereas adenovirus-infected mice with hepatic overexpression of SR-BI show a reduced response, relative to controls [57, 58]. This combined with *in vitro* studies, demonstrating reduced binding of chylomicron-like particles to hepatocytes from SR-BI KO mice relative to controls, suggest that SR-BI may also be involved in chylomicron metabolism *in vivo* [57, 58].

SR-BI KO mice develop increased atherosclerosis either when challenged with high-fat/atherogenic diets or when SR-BI deficiency is combined with deficiency in genes encoding either LDL receptor or apoE [39, 46, 49]. Mice deficient in SR-BI alone develop increased diet-induced atherosclerosis compared to either heterozygous or homozygous SR-BI-expressing mice [49]. Examination of gene expression in aortas from these mice has revealed increased expression of genes involved in monocyte adhesion to endothelial cells, including P- and E-selectins and vascular cell adhesion molecule (VCAM)-1; similar increases were seen in CD68, apoE, ABCA1, and ABCG1, most likely reflecting increased macrophage content [49]. These results suggest increased inflammation in SR-BI KO mice at least when fed a high fat, atherogenic diet.

Elimination of SR-BI expression in the context of inactivating mutations in the genes encoding either LDL receptor or apoE has more profound effects on atherosclerosis [39, 46, 59]. Elimination of SR-BI in LDL receptor-deficient mice resulted in substantially increased Western diet-induced atherosclerosis relative to LDL receptor KO controls [39]. Elimination of SR-BI in normal chow-fed apoE KO mice resulted in increased aortic sinus atherosclerosis at an age (5–7 weeks) when apoE KO mice show no signs of atherosclerotic plaque [46]. Both Western diet fed SR-BI/LDL receptor and normal chow-fed SR-BI/apoE double KO mice exhibited decreased levels of plasma apoB containing lipoproteins relative to corresponding SR-BI-expressing controls. Cholesterol associated with HDL-like particles (increased in size relative to normal HDL) was increased relative to that in SR-BI-expressing mice. Thus the increased atherosclerosis in these mice appeared to be accompanied by impaired hepatic HDL cholesterol clearance [46].

In addition to increased aortic atherosclerosis, SR-BI/apoE double KO mice also exhibited occlusive coronary artery disease, myocardial infarction, and death between 5 and 8 weeks of age [46, 60]. These mice exhibit impaired heart functional and conductance properties [47, 60] and represent a unique mouse model of coronary heart disease. The antioxidant drug, probucol, has been shown to be effective in overcoming the effects of a lack of SR-BI expression on both plasma lipoprotein structure and composition and delays the onset of coronary heart disease in SR-BI/apoE double KO mice [47]. Exactly how probucol works is still not clear. Whether probucol enhances selective HDL cholesteryl ester clearance in an SR-BI-independent manner in SR-BI-deficient mice remains to be determined [61, 62]. A low level of apoE expression is sufficient to prevent the development of coronary heart disease, at least in chow-fed mice, since elimination of SR-BI in mice homozygous for a hypomorphic apoE allele (express very low levels of apoE) did not induce atherosclerosis or coronary heart disease [63]. When these mice were switched to a high-fat diet, however, atherosclerosis and occlusive coronary artery disease were induced [63]. These lead to myocardial infarction, reduced heart function and conductance abnormalities, and death with similar kinetics to those seen in SR-BI/apoE double KO mice [63]. These studies suggest that SR-BI and apoE may be involved in parallel lipoprotein metabolic pathways that normally prevent occlusive coronary artery atherosclerosis.

Role of Hepatic SR-BI in Protection against Atherosclerosis

As discussed above, *in vitro* studies in cells and analyses of SR-BI KO mice suggest that SR-BI plays an important role in hepatic selective HDL cholesterol uptake and reverse cholesterol transport. Confirmation of this has come from studies in which hepatic SR-BI levels have been either overexpressed or suppressed in a tissue-specific manner. Hepatic overexpression of SR-BI in either transgenic mice or using adenovirus-mediated gene transfer reduced plasma HDL cholesterol levels, increased biliary cholesterol levels, and reduced atherosclerosis in susceptible strains ([64–67] and references cited therein). Decreased atherosclerosis was associated with decreased HDL cholesterol levels in LDL receptor-deficient mice overexpressing adenovirus-delivered SR-BI in a liver-specific manner, suggesting that decreased atherosclerosis resulted from increased clearance of HDL cholesterol and increased HDL-dependent reverse cholesterol transport [64, 66]. Alternatively, in another study, liver-specific transgenic SR-BI overexpression in LDL receptor +/− mice reduced atherosclerosis in a manner that was correlated with decreased plasma LDL cholesterol, suggesting that overexpression of SR-BI increased LDL cholesterol clearance [65]. *In vitro* studies suggest that SR-BI mediates selective uptake of LDL cholesterol by hepatocytes and *in vivo* studies in SR-BI KO mice indicate that SR-BI participates in LDL cholesterol clearance from plasma [54, 68, 69], supporting this possibility. In another study of different lines of SR-BI transgenic mice (driven by an apoA1 promoter), only moderate (twofold) increased levels of hepatic SR-BI overexpression reduced diet-induced atherosclerosis in apoB transgenic mice [67, 70]. In contrast, high levels (tenfold) of SR-BI overexpression did not reduce diet-induced atherosclerosis in apoB transgenic mice relative to controls even though the level of reduction of HDL cholesterol was dependent on the level of SR-BI overexpression [67]. In this case, the increased atherosclerosis in high-SR-BI-expressing relative to low-SR-BI-expressing transgenic mice may have been related to reduced levels of plasma HDL and altered HDL particle composition [67].

Mice with liver-restricted reduction in SR-BI gene expression have been generated by the insertion of a neomycin gene cassette into the promoter region of the mouse *Scarb1* gene [71]. Mice with both alleles containing the insertion (called SR-BI attenuated or SR-BI[att] mice) express ~50% of normal levels of SR-BI in liver and normal levels of SR-BI in other tissues examined (e.g., adrenal gland) [71]. SR-BI[att] mice exhibit slightly increased levels of plasma HDL cholesterol, reduced clearance of HDL cholesterol from blood, and reduced hepatic selective uptake of HDL cholesterol [71]. These studies, together with those of SR-BI KO mice discussed above, suggest that hepatic SR-BI is solely responsible for hepatic selective HDL cholesteryl ester uptake in mice, and plays a major role in hepatic reverse cholesterol transport. LDL receptor KO mice that carry two copies of the SR-BI[att] insertion (SR-BI[att]/LDL receptor KO mice) have also been generated and exhibit increased levels of LDL cholesterol, supporting the idea that normal levels of hepatic SR-BI expression are required for normal clearance of LDL cholesterol [59]. SR-BI[att]/LDL receptor KO mice also develop increased diet-induced atherosclerosis relative to LDL receptor KO mice with normal *Scarb1* alleles, consistent with hepatic SR-BI protecting against atherosclerosis [59].

Regulation of Hepatic SR-BI Protein Expression

A distinct mouse model of tissue-restricted SR-BI suppression has been generated. These mice contain a targeted mutation in the gene encoding an adaptor protein called PDZK1, which contains multiple postsynaptic density protein-95/Drosophila discs-large/ZO1 (PDZ) protein–protein interaction domains (reviewed in [72, 73]). PDZK1 (also called CLAMP) was identified as a protein that interacted with the C-terminal three amino acids (AKL) of SR-BI via one of its four PDZ protein–protein interaction domains. SR-BI homologs from all mammalian species identified to date contain a potential PDZ target sequence (see Fig. 4.2 and [93]) though PDZK1 interaction has been verified only for mouse SR-BI [74–77]. An intact PDZK1 interaction domain has been demonstrated to be essential for the proper trafficking of SR-BI to the cell surface in hepatocytes but not in other cells examined (such as Chinese hamster ovary cells) [72, 74, 75]. Similarly, expression of PDZK1 is required for trafficking of SR-BI to the cell surface in hepatocytes [77, 78]. In the absence of PDZK1, SR-BI in hepatocytes is degraded resulting in a substantial decrease in SR-BI expression at the protein level [77, 78]. SR-BI is also moderately reduced in the small intestine but not in steroidogenic tissues including the adrenal gland, testes, and ovary [77]. PDZK1 KO mice exhibit altered plasma HDL cholesterol levels and enlarged HDL particles of altered composition, similar to those found in SR-BI KO mice ([77], reviewed in [73]). Treatment of mice with fibrates has also been shown to result in decreased expression of both SR-BI and PDZK1 protein, although the mechanisms involved remain uncertain [78, 79]. Mice fed diets supplemented with fibrates exhibit alterations in plasma HDL cholesterol levels and HDL structure similar to those seen in SR-BI KO mice [79].

A PDZK1-interacting protein, called small PDZK1-associated protein (SPAP) [76] has been identified by virtue of coimmunoprecipitation with PDZK1 from mouse liver and kidney extracts. Transgenic mice overexpressing SPAP in a liver-specific manner exhibit drastically reduced levels of hepatic PDZK1 and SR-BI protein [76]. These mice also exhibit increased levels of plasma cholesterol associated with large HDL particles reminiscent

```
Mouse      - - -KGTVLQE[AKL]
Human      - - -KGSVLQEAKL
Rat        - - -KGTVLQEAKL
CHO        - - -KGTVLQEAKL
Rabbit     - - -KGTVLQEARL
Pig        - - -KGTVLQEARL
Bovine     - - -KGTVLQEARL
```

FIGURE 4.2. Comparison of SR-BI C-terminal sequences from different species. Shown are the sequences of the last ten amino acids of SR-BI from different species. The PDZ recognition sequence identified in mouse SR-BI is indicated with a box. CHO is Chinese hamster ovary. Amino acid sequences are based on cDNA sequences [24, 87–92].

of those seen in SR-BI KO mice [76]. Overexpression of SPAP in cultured hepatocytes increases PDZK1 protein-degradation via an as yet uncharacterized proteasome-independent pathway [76]. It is not yet clear if SPAP at physiological levels of expression are involved in the regulation of PDZK1 and SR-BI.

Hepatocytes are polarized cells, with distinct basolateral and apical membranes. PDZK1 appears to be required for regulation of SR-BI cell surface expression only in polarized hepatocytes [75]. When hepatocytes are cultured under conditions in which they are not polarized, SR-BI expression is no longer dependent on the presence of PDZK1 [75]. SR-BI is expressed basolaterally and continuously internalized and recycled in polarized hepatocytes and other polarized cells, such as Madin–Darby canine kidney (MDCK) cells [75, 80]. It is not known whether SR-BI stability and trafficking to the cell surface are dependent on PDZK1 in other polarized cells. SR-BI's polarized distribution in MDCK cells is disrupted by overexpression, depletion of cellular cholesterol, or activation of protein kinase A (PKA) [80]. It is not known if the distribution of SR-BI in polarized hepatocytes is also subjected to similar regulation, or if PDZK1 is involved in these processes.

Role of SR-BI in Cells in the Vascular Wall

SR-BI is expressed in both macrophages and vascular endothelial cells [29–32], and therefore can have local effects on the development of atherosclerosis. As discussed above, SR-BI in endothelial cells can mediate the HDL-stimulated activation of eNOS. This appears to be important physiologically since HDL is able to stimulate the relaxation of arteries from SR-BI-expressing mice, whereas HDL-stimulated relaxation was impaired if arteries were from SR-BI KO mice [14]. Normal levels of eNOS activity have been proposed to protect against the development of atherosclerosis [81–85]. Alterations in eNOS expression clearly affect atherogenesis in mouse models, however, contrary to expectation, inactivation of eNOS reduced atherosclerosis and overexpression of eNOS increased atherosclerosis in apoE KO mice [81, 84, 85]. This may have been due to an imbalance between eNOS activity and the amount of tetrahydrobiopterin, an eNOS cofactor required for its proper function (NO production) and without which, eNOS is dysfunctional leading to the production of superoxide radical [81]. In fact increased endothelial tetrahydrobiopterin production in transgenic mice overexpressing GTP-cyclohydrolase I, an enzyme required for its synthesis, results in reduced atherosclerosis in apoE KO mice [82]. This suggests that increased functional eNOS may protect against atherosclerosis. Therefore, a lack of SR-BI-dependent HDL stimulation of eNOS in SR-BI KO mice could contribute to the development of atherosclerosis through reduced functional eNOS. Confirmation of this awaits the generation of mice with endothelial cell-specific alterations in SR-BI expression.

In atherosclerotic plaques, SR-BI appears most abundant in macrophages [31, 32]. As indicated above, SR-BI overexpression in transfected cells has been shown to confer increased cholesterol efflux to HDL and other cholesterol acceptors [33, 34, 37]. In contrast, cholesterol efflux to HDL and other acceptors does not appear to be substantially affected in macrophages from SR-BI KO mice compared to those from SR-BI-expressing controls [39–41]. This suggests that SR-BI may not be required for cholesterol efflux from macrophages in atherosclerotic plaque. To test whether SR-BI in macrophages and other leukocytes affect the development of atherosclerosis, bone marrow transplantation has been used to generate chimeric mice with impaired SR-BI expression in bone marrow-derived cells (including monocyte-derived macrophages and other leukocytes) but normal SR-BI expression in non-hematopoietic cells [39–41]. These mice developed increased atherosclerosis relative to controls receiving bone marrow from SR-BI-expressing mice [39–41]. These studies demonstrated that leukocyte SR-BI is normally required for protection against atherosclerosis at least in advanced stages. Curiously, the sizes of small lesions at early stages of atherosclerosis appeared to be decreased in mice with leukocyte-specific SR-BI gene inactivation, suggesting that leukocyte SR-BI might promote early steps in plaque formation but prevent later steps [41].

Conclusion

SR-BI plays an important role in HDL metabolism, HDL cholesterol transport, and noncholesterol transport properties of HDL. Its activity in both liver and cells in the vascular wall affects the development and progression of atherosclerosis in mouse models. Several studies have identified sequence variations and single nucleotide polymorphisms in the human gene encoding SR-BI. Associations of these sequence variants with a variety of parameters including HDL and LDL cholesterol, postprandial triglyceride, body mass index, and incidence of coronary artery disease (reviewed in [86]) suggests that variations in SR-BI function/levels may contribute to variations in lipoprotein metabolism and heart disease in the general population. Research in mouse models of atherosclerosis will continue to provide insights into the role of SR-BI in this disease process and will complement our understanding of its involvement in human disease.

Acknowledgements: Research in the Trigatti lab is supported by grants from the Canadian Institutes of Health Research, The Heart and Stroke Foundation of Ontario, and the Natural Sciences and Engineering Research Council of Canada. B.T. is a Heart and Stroke Foundation of Canada New Investigator.

References

1. Gordon DJ, Rifkind BM: High-density lipoprotein—the clinical implications of recent studies. N Engl J Med 321: 1311–1316, 1989.
2. Kwiterovich PO Jr: The antiatherogenic role of high-density lipoprotein cholesterol. Am J Cardiol 82: 13Q–21Q, 1998.
3. Nicholls SJ, Rye KA, Barter PJ: High-density lipoproteins as therapeutic targets. Curr Opin Lipidol 16: 345–349, 2005.
4. Herrera E, Barbas C: Vitamin E: action, metabolism and perspectives. J Physiol Biochem 57: 43–56, 2001.
5. Mardones P, Strobel P, Miranda S, Leighton F, Quinones V, Amigo L, Rozowski J, Krieger M, Rigotti A: Alpha-tocopherol metabolism is abnormal in scavenger receptor class B type I (SR-BI)-deficient mice. J Nutr 132: 443–449, 2002.
6. Tselepis AD, Chapman MJ: Inflammation, bioactive lipids and atherosclerosis: potential roles of a lipoprotein-associated phospholipase A2, platelet activating factor-acetylhydrolase. Atherosclerosis (Suppl 3): 57–68, 2002.
7. Mackness MI, Durrington PN, Mackness B: The role of paraoxonase 1 activity in cardiovascular disease: potential for therapeutic intervention. Am J Cardiovasc Drugs 4: 211–217, 2004.
8. Connelly PW, Draganov D, Maguire GF: Paraoxonase-1 does not reduce or modify oxidation of phospholipids by peroxynitrite. Free Radic Biol Med 38: 164–174, 2005.
9. Mineo C, Shaul PW: HDL stimulation of endothelial nitric oxide synthase: a novel mechanism of HDL action. Trends Cardiovasc Med 13: 226–231, 2003.
10. Mineo C, Yuhanna IS, Quon MJ, Shaul PW: High density lipoprotein-induced endothelial nitric-oxide synthase activation is mediated by Akt and MAP kinases. J Biol Chem 278: 9142–9149, 2003.
11. Gong M, Wilson M, Kelly T, Su W, Dressman J, Kincer J, Matveev SV, Guo L, Guerin T, Li XA, Zhu W, Uittenbogaard A, Smart EJ: HDL-associated estradiol stimulates endothelial NO synthase and vasodilation in an SR-BI-dependent manner. J Clin Invest 111: 1579–1587, 2003.
12. Li XA, Titlow WB, Jackson BA, Giltiay N, Nikolova-Karakashian M, Uittenbogaard A, Smart EJ: High density lipoprotein binding to scavenger receptor, class B, type I activates endothelial nitric-oxide synthase in a ceramide-dependent manner. J Biol Chem 277: 11058–11063, 2002.
13. Uittenbogaard A, Shaul PW, Yuhanna IS, Blair A, Smart EJ: High density lipoprotein prevents oxidized low density lipoprotein-induced inhibition of endothelial nitric-oxide synthase localization and activation in caveolae. J Biol Chem 275: 11278–11283, 2000.
14. Yuhanna IS, Zhu Y, Cox BE, Hahner LD, Osborne-Lawrence S, Lu P, Marcel YL, Anderson RG, Mendelsohn ME, Hobbs HH, Shaul PW: High-density lipoprotein binding to scavenger receptor-BI activates endothelial nitric oxide synthase. Nat Med 7: 853–857, 2001.
15. Nofer JR, Junker R, Pulawski E, Fobker M, Levkau B, von Eckardstein A, Seedorf U, Assmann G, Walter M: High density lipoproteins induce cell cycle entry in vascular smooth muscle cells via mitogen activated protein kinase-dependent pathway. Thromb Haemost 85: 730–735, 2001.

16. Nofer JR, Levkau B, Wolinska I, Junker R, Fobker M, von Eckardstein A, Seedorf U, Assmann G: Suppression of endothelial cell apoptosis by high density lipoproteins (HDL) and HDL-associated lysosphingolipids. J Biol Chem 276: 34480–34485, 2001.

17. Nofer JR, van der Giet M, Tolle M, Wolinska I, von Wnuck Lipinski K, Baba HA, Tietge UJ, Godecke A, Ishii I, Kleuser B, Schafers M, Fobker M, Zidek W, Assmann G, Chun J, Levkau B: HDL induces NO-dependent vasorelaxation via the lysophospholipid receptor S1P3. J Clin Invest 113: 569–581, 2004.

18. Li XA, Guo L, Dressman JL, Asmis R, Smart EJ: A novel ligand-independent apoptotic pathway induced by scavenger receptor class B, type I and suppressed by endothelial nitric-oxide synthase and high density lipoprotein. J Biol Chem 280: 19087–19096, 2005.

19. Breslow J: Familial disorders of high density lipoprotein metabolism. In: Scriver CR, Beaudet AL, Sly WS, Valle D (eds), The Metabolic and Molecular Bases of Inherited Diseases. McGraw-Hill, New York, 1995, pp 2031–2052.

20. Ng DS: Insight into the role of LCAT from mouse models. Rev Endocr Metab Disord 5: 311–318, 2004.

21. Bruce C, Chouinard RA Jr, Tall AR: Plasma lipid transfer proteins, high-density lipoproteins, and reverse cholesterol transport. Annu Rev Nutr 18: 297–330, 1998.

22. Rigotti A, Trigatti B, Babitt J, Penman M, Xu S, Krieger M: Scavenger receptor BI—a cell surface receptor for high density lipoprotein. Curr Opin Lipidol 8: 181–188, 1997.

23. Fluiter K, Sattler W, De Beer MC, Connell PM, van der Westhuyzen DR, van Berkel TJ: Scavenger receptor BI mediates the selective uptake of oxidized cholesterol esters by rat liver. J Biol Chem 274: 8893–8899, 1999.

24. Acton S, Rigotti A, Landschulz KT, Xu S, Hobbs HH, Krieger M: Identification of scavenger receptor SR-BI as a high density lipoprotein receptor. Science 271: 518–520, 1996.

25. Babitt J, Trigatti B, Rigotti A, Smart EJ, Anderson RG, Xu S, Krieger M: Murine SR-BI, a high density lipoprotein receptor that mediates selective lipid uptake is N-glycosylated and fatty acylated and colocalizes with plasma membrane caveolae. J Biol Chem 272: 13242–13249, 1997.

26. Gu X, Trigatti B, Xu S, Acton S, Babitt J, Krieger M: The efficient cellular uptake of high density lipoprotein lipids via scavenger receptor class B type I requires not only receptor-mediated surface binding but also receptor-specific lipid transfer mediated by its extracellular domain. J Biol Chem 273: 26338–26348, 1998.

27. Trigatti B, Covey S, Rizvi A: Scavenger receptor class B type I in high-density lipoprotein metabolism, atherosclerosis and heart disease: lessons from gene-targeted mice. Biochem Soc Trans 32: 116–120, 2004.

28. Krieger M: Charting the fate of the "good cholesterol": identification and characterization of the high-density lipoprotein receptor SR-BI. Annu Rev Biochem 68: 523–558, 1999.

29. Hatzopoulos AK, Rigotti A, Rosenberg RD, Krieger M: Temporal and spatial pattern of expression of the HDL receptor SR-BI during murine embryogenesis. J Lipid Res 39: 495–508, 1998.

30. Fluiter K, van der Westhuijzen DR, van Berkel TJ: *In vivo* regulation of scavenger receptor BI and the selective uptake of high density lipoprotein cholesteryl esters in rat liver parenchymal and Kupffer cells. J Biol Chem 273: 8434–8438, 1998.

31. Hirano K, Yamashita S, Nakagawa Y, Ohya T, Matsuura F, Tsukamoto K, Okamoto Y, Matsuyama A, Matsumoto K, Miyagawa J, Matsuzawa Y: Expression of human scavenger receptor class B type I in cultured human monocyte-derived macrophages and atherosclerotic lesions. Circ Res 85: 108–116, 1999.
32. Chinetti G, Gbaguidi FG, Griglio S, Mallat Z, Antonucci M, Poulain P, Chapman J, Fruchart JC, Tedgui A, Najib-Fruchart J, Staels B: CLA-1/SR-BI is expressed in atherosclerotic lesion macrophages and regulated by activators of peroxisome proliferator-activated receptors. Circulation 101: 2411–2417, 2000.
33. Jian B, de la Llera-Moya M, Ji Y, Wang N, Phillips MC, Swaney JB, Tall AR, Rothblat GH: Scavenger receptor class B type I as a mediator of cellular cholesterol efflux to lipoproteins and phospholipid acceptors. J Biol Chem 273: 5599–5606, 1998.
34. Ji Y, Jian B, Wang N, Sun Y, Moya ML, Phillips MC, Rothblat GH, Swaney JB, Tall AR: Scavenger receptor BI promotes high density lipoprotein-mediated cellular cholesterol efflux. J Biol Chem 272: 20982–20985, 1997.
35. Yancey PG, Bortnick AE, Kellner-Weibel G, de la Llera-Moya M, Phillips MC, Rothblat GH: Importance of different pathways of cellular cholesterol efflux. Arterioscler Thromb Vasc Biol 23: 712–719, 2003.
36. Assanasen C, Mineo C, Seetharam D, Yuhanna IS, Marcel YL, Connelly MA, Williams DL, de la Llera-Moya M, Shaul PW, Silver DL: Cholesterol binding, efflux, and a PDZ-interacting domain of scavenger receptor-BI mediate HDL-initiated signaling. J Clin Invest 115: 969–977, 2005.
37. Gu X, Kozarsky K, Krieger M: Scavenger receptor class B, type I-mediated [3H]cholesterol efflux to high and low density lipoproteins is dependent on lipoprotein binding to the receptor. J Biol Chem 275: 29993–30001, 2000.
38. O'Connell BJ, Denis M, Genest J: Cellular physiology of cholesterol efflux in vascular endothelial cells. Circulation 110: 2881–2888, 2004.
39. Covey SD, Krieger M, Wang W, Penman M, Trigatti BL: Scavenger receptor class B type I-mediated protection against atherosclerosis in LDL receptor-negative mice involves its expression in bone marrow-derived cells. Arterioscler Thromb Vasc Biol 23: 1589–1594, 2003.
40. Zhang W, Yancey PG, Su YR, Babaev VR, Zhang Y, Fazio S, Linton MF: Inactivation of macrophage scavenger receptor class B type I promotes atherosclerotic lesion development in apolipoprotein E-deficient mice. Circulation 108: 2258–2263, 2003.
41. Van Eck M, Bos IS, Hildebrand RB, Van Rij BT, Van Berkel TJ: Dual role for scavenger receptor class B, type I on bone marrow-derived cells in atherosclerotic lesion development. Am J Pathol 165: 785–794, 2004.
42. Webb NR, de Villiers WJ, Connell PM, de Beer FC, van der Westhuyzen DR: Alternative forms of the scavenger receptor BI (SR-BI). J Lipid Res 38: 1490–1495, 1997.
43. Webb NR, Connell PM, Graf GA, Smart EJ, de Villiers WJ, de Beer FC, van der Westhuyzen DR: SR-BII, an isoform of the scavenger receptor BI containing an alternate cytoplasmic tail, mediates lipid transfer between high density lipoprotein and cells. J Biol Chem 273: 15241–15248, 1998.
44. Eckhardt ER, Cai L, Sun B, Webb NR, van der Westhuyzen DR: High density lipoprotein uptake by scavenger receptor SR-BII. J Biol Chem 279: 14372–14381, 2004.

45. Rigotti A, Trigatti BL, Penman M, Rayburn H, Herz J, Krieger M: A targeted mutation in the murine gene encoding the high density lipoprotein (HDL) receptor scavenger receptor class B type I reveals its key role in HDL metabolism. Proc Natl Acad Sci USA 94: 12610–12615, 1997.

46. Trigatti B, Rayburn H, Vinals M, Braun A, Miettinen H, Penman M, Hertz M, Schrenzel M, Amigo L, Rigotti A, Krieger M: Influence of the high density lipoprotein receptor SR-BI on reproductive and cardiovascular pathophysiology. Proc Natl Acad Sci USA 96: 9322–9327, 1999.

47. Braun A, Zhang S, Miettinen HE, Ebrahim S, Holm TM, Vasile E, Post MJ, Yoerger DM, Picard MH, Krieger JL, Andrews NC, Simons M, Krieger M: Probucol prevents early coronary heart disease and death in the high-density lipoprotein receptor SR-BI/apolipoprotein E double knockout mouse. Proc Natl Acad Sci USA 100: 7283–7288, 2003.

48. Ma K, Forte T, Otvos JD, Chan L: Differential additive effects of endothelial lipase and scavenger receptor-class B type I on high-density lipoprotein metabolism in knockout mouse models. Arterioscler Thromb Vasc Biol 25: 149–154, 2005.

49. Van Eck M, Twisk J, Hoekstra M, Van Rij BT, Van der Lans CA, Bos IS, Kruijt JK, Kuipers F, Van Berkel TJ: Differential effects of scavenger receptor BI deficiency on lipid metabolism in cells of the arterial wall and in the liver. J Biol Chem 278: 23699–23705, 2003.

50. Mardones P, Quinones V, Amigo L, Moreno M, Miquel JF, Schwarz M, Miettinen HE, Trigatti B, Krieger M, VanPatten S, Cohen DE, Rigotti A: Hepatic cholesterol and bile acid metabolism and intestinal cholesterol absorption in scavenger receptor class B type I-deficient mice. J Lipid Res 42: 170–180, 2001.

51. Yu L, Gupta S, Xu F, Liverman AD, Moschetta A, Mangelsdorf DJ, Repa JJ, Hobbs HH, Cohen JC: Expression of ABCG5 and ABCG8 is required for regulation of biliary cholesterol secretion. J Biol Chem 280: 8742–8747, 2005.

52. Klett EL, Lu K, Kosters A, Vink E, Lee MH, Altenburg M, Shefer S, Batta AK, Yu H, Chen J, Klein R, Looije N, Oude-Elferink R, Groen AK, Maeda N, Salen G, Patel SB: A mouse model of sitosterolemia: absence of Abcg8/sterolin-2 results in failure to secrete biliary cholesterol. BMC Med 2: 5, 2004.

53. Out R, Hoekstra M, Spijkers JA, Kruijt JK, van Eck M, Bos IS, Twisk J, Van Berkel TJ: Scavenger receptor class B type I is solely responsible for the selective uptake of cholesteryl esters from HDL by the liver and the adrenals in mice. J Lipid Res 45: 2088–2095, 2004.

54. Brodeur MR, Luangrath V, Bourret G, Falstrault L, Brissette L: Physiological importance of SR-BI in the *in vivo* metabolism of human HDL and LDL in male and female mice. J Lipid Res 46: 687–696, 2005.

55. Brundert M, Ewert A, Heeren J, Behrendt B, Ramakrishnan R, Greten H, Merkel M, Rinninger F: Scavenger receptor class B type I mediates the selective uptake of high-density lipoprotein-associated cholesteryl ester by the liver in mice. Arterioscler Thromb Vasc Biol 25: 143–148, 2005.

56. Temel RE, Trigatti B, DeMattos RB, Azhar S, Krieger M, Williams DL: Scavenger receptor class B, type I (SR-BI) is the major route for the delivery of high density lipoprotein cholesterol to the steroidogenic pathway in cultured mouse adrenocortical cells. Proc Natl Acad Sci USA 94: 13600–13605, 1997.

57. Out R, Kruijt JK, Rensen PC, Hildebrand RB, de Vos P, Van Eck M, Van Berkel TJ: Scavenger receptor BI plays a role in facilitating chylomicron metabolism. J Biol Chem 279: 18401–18406, 2004.

58. Out R, Hoekstra M, de Jager SC, de Vos P, van der Westhuyzen DR, Webb NR, Van Eck M, Biessen EA, Van Berkel TJ: Adenovirus-mediated hepatic overexpression of scavenger receptor class B type I accelerates chylomicron metabolism in C57BL/6J mice. J Lipid Res 46: 1172–1181, 2005.

59. Huszar D, Varban ML, Rinninger F, Feeley R, Arai T, Fairchild-Huntress V, Donovan MJ, Tall AR: Increased LDL cholesterol and atherosclerosis in LDL receptor-deficient mice with attenuated expression of scavenger receptor BI. Arterioscler Thromb Vasc Biol 20: 1068–1073, 2000.

60. Braun A, Trigatti BL, Post MJ, Sato K, Simons M, Edelberg JM, Rosenberg RD, Schrenzel M, Krieger M: Loss of SR-BI expression leads to the early onset of occlusive atherosclerotic coronary artery disease, spontaneous myocardial infarctions, severe cardiac dysfunction, and premature death in apolipoprotein E-deficient mice. Circ Res 90: 270–276, 2002.

61. Rinninger F, Wang N, Ramakrishnan R, Jiang XC, Tall AR: Probucol enhances selective uptake of HDL-associated cholesteryl esters *in vitro* by a scavenger receptor B-I-dependent mechanism. Arterioscler Thromb Vasc Biol 19: 1325–1332, 1999.

62. Miettinen HE, Rayburn H, Krieger M: Abnormal lipoprotein metabolism and reversible female infertility in HDL receptor (SR-BI)-deficient mice. J Clin Invest 108: 1717–1722, 2001.

63. Zhang S, Picard MH, Vasile E, Zhu Y, Raffai RL, Weisgraber KH, Krieger M: Diet-induced occlusive coronary atherosclerosis, myocardial infarction, cardiac dysfunction, and premature death in scavenger receptor class B type I-deficient, hypomorphic apolipoprotein ER61 mice. Circulation 111: 3457–3464, 2005.

64. Kozarsky KF, Donahee MH, Rigotti A, Iqbal SN, Edelman ER, Krieger M: Overexpression of the HDL receptor SR-BI alters plasma HDL and bile cholesterol levels. Nature 387: 414–417, 1997.

65. Arai T, Wang N, Bezouevski M, Welch C, Tall AR: Decreased atherosclerosis in heterozygous low density lipoprotein receptor-deficient mice expressing the scavenger receptor BI transgene. J Biol Chem 274: 2366–2371, 1999.

66. Kozarsky KF, Donahee MH, Glick JM, Krieger M, Rader DJ: Gene transfer and hepatic overexpression of the HDL receptor SR-BI reduces atherosclerosis in the cholesterol-fed LDL receptor-deficient mouse. Arterioscler Thromb Vasc Biol 20: 721–727, 2000.

67. Ueda Y, Gong E, Royer L, Cooper PN, Francone OL, Rubin EM: Relationship between expression levels and atherogenesis in scavenger receptor class B, type I transgenics. J Biol Chem 275: 20368–20373, 2000.

68. Stangl H, Hyatt M, Hobbs HH: Transport of lipids from high and low density lipoproteins via scavenger receptor-BI. J Biol Chem 274: 32692–32698, 1999.

69. Swarnakar S, Temel RE, Connelly MA, Azhar S, Williams DL: Scavenger receptor class B, type I, mediates selective uptake of low density lipoprotein cholesteryl ester. J Biol Chem 274: 29733–29739, 1999.

70. Ueda Y, Royer L, Gong E, Zhang J, Cooper PN, Francone O, Rubin EM: Lower plasma levels and accelerated clearance of high density lipoprotein (HDL) and non-HDL cholesterol in scavenger receptor class B type I transgenic mice. J Biol Chem 274: 7165–7171, 1999.

71. Varban ML, Rinninger F, Wang N, Fairchild-Huntress V, Dunmore JH, Fang Q, Gosselin ML, Dixon KL, Deeds JD, Acton SL, Tall AR, Huszar D: Targeted mutation reveals a central role for SR-BI in hepatic selective uptake of high density lipoprotein cholesterol. Proc Natl Acad Sci USA 95: 4619–4624, 1998.

72. Silver DL: SR-BI and protein–protein interactions in hepatic high density lipoprotein metabolism. Rev Endocr Metab Disord 5: 327–333, 2004.
73. Yesilaltay A, Kocher O, Rigotti A, Krieger M: Regulation of SR-BI-mediated high-density lipoprotein metabolism by the tissue-specific adaptor protein PDZK1. Curr Opin Lipidol 16: 147–152, 2005.
74. Ikemoto M, Arai H, Feng D, Tanaka K, Aoki J, Dohmae N, Takio K, Adachi H, Tsujimoto M, Inoue K: Identification of a PDZ-domain-containing protein that interacts with the scavenger receptor class B type I. Proc Natl Acad Sci USA 97: 6538–6543, 2000.
75. Silver DL: A carboxyl-terminal PDZ-interacting domain of scavenger receptor B, type I is essential for cell surface expression in liver. J Biol Chem 277: 34042–34047, 2002.
76. Silver DL, Wang N, Vogel S: Identification of small PDZK1-associated protein, DD96/MAP17, as a regulator of PDZK1 and plasma high density lipoprotein levels. J Biol Chem 278: 28528–28532, 2003.
77. Kocher O, Yesilaltay A, Cirovic C, Pal R, Rigotti A, Krieger M: Targeted disruption of the PDZK1 gene in mice causes tissue-specific depletion of the high density lipoprotein receptor scavenger receptor class B type I and altered lipoprotein metabolism. J Biol Chem 278: 52820–52825, 2003.
78. Lan D, Silver DL: Fenofibrate induces a novel degradation pathway for scavenger receptor B-I independent of PDZK1. J Biol Chem 280: 23390–23396, 2005.
79. Mardones P, Pilon A, Bouly M, Duran D, Nishimoto T, Arai H, Kozarsky KF, Altayo M, Miquel JF, Luc G, Clavey V, Staels B, Rigotti A: Fibrates down-regulate hepatic scavenger receptor class B type I protein expression in mice. J Biol Chem 278: 7884–7890, 2003.
80. Burgos PV, Klattenhoff C, de la Fuente E, Rigotti A, Gonzalez A: Cholesterol depletion induces PKA-mediated basolateral-to-apical transcytosis of the scavenger receptor class B type I in MDCK cells. Proc Natl Acad Sci USA 101: 3845–3850, 2004.
81. Kawashima S, Yokoyama M: Dysfunction of endothelial nitric oxide synthase and atherosclerosis. Arterioscler Thromb Vasc Biol 24: 998–1005, 2004.
82. Alp NJ, McAteer MA, Khoo J, Choudhury RP, Channon KM: Increased endothelial tetrahydrobiopterin synthesis by targeted transgenic GTP-cyclohydrolase I overexpression reduces endothelial dysfunction and atherosclerosis in ApoE-knockout mice. Arterioscler Thromb Vasc Biol 24: 445–450, 2004.
83. Shaul PW: Endothelial nitric oxide synthase, caveolae and the development of atherosclerosis. J Physiol 547: 21–33, 2003.
84. Ozaki M, Kawashima S, Yamashita T, Hirase T, Namiki M, Inoue N, Hirata K, Yasui H, Sakurai H, Yoshida Y, Masada M, Yokoyama M: Overexpression of endothelial nitric oxide synthase accelerates atherosclerotic lesion formation in apoE-deficient mice. J Clin Invest 110: 331–340, 2002.
85. Shi W, Wang X, Shih DM, Laubach VE, Navab M, Lusis AJ: Paradoxical reduction of fatty streak formation in mice lacking endothelial nitric oxide synthase. Circulation 105: 2078–2082, 2002.
86. Trigatti BL: Hepatic high-density lipoprotein receptors: roles in lipoprotein metabolism and potential for therapeutic modulation. Curr Atheroscler Rep 7: 344–350, 2005.
87. Acton SL, Scherer PE, Lodish HF, Krieger M: Expression cloning of SR-BI, a CD36-related class B scavenger receptor. J Biol Chem 269: 21003–21009, 1994.

88. Calvo D, Vega MA: Identification, primary structure, and distribution of CLA-1, a novel member of the CD36/LIMPII gene family. J Biol Chem 268: 18929–18935, 1993.
89. Rajapaksha WR, McBride M, Robertson L, O'Shaughnessy PJ: Sequence of the bovine HDL-receptor (SR-BI) cDNA and changes in receptor mRNA expression during granulosa cell luteinization *in vivo* and *in vitro*. Mol Cell Endocrinol 134: 59–67, 1997.
90. Shiratsuchi A, Kawasaki Y, Ikemoto M, Arai H, Nakanishi Y: Role of class B scavenger receptor type I in phagocytosis of apoptotic rat spermatogenic cells by sertoli cells. J Biol Chem 274: 5888–5894, 1999.
91. Ritsch A, Tancevski I, Schgoer W, Pfeifhofer C, Gander R, Eller P, Foeger B, Stanzl U, Patsch JR: Molecular characterization of rabbit scavenger receptor class B types I and II: portal to central vein gradient of expression in the liver. J Lipid Res 45: 214–222, 2004.
92. Kim JG, Vallet JL, Nonneman D, Christenson RK: Molecular cloning and endometrial expression of porcine high density lipoprotein receptor SR-BI during the estrous cycle and early pregnancy. Mol Cell Endocrinol 222: 105–112, 2004.
93. Hung AY, Sheng M: PDZ domains: structural modules for protein complex assembly. J Biol Chem 277: 5699–5702, 2002.

5

The Role of Scavenger Receptors in Signaling, Inflammation, and Atherosclerosis

DAISY SAHOO AND VICTOR A. DROVER

Abstract

The scavenger receptor (SR) family of cell surface, integral membrane proteins has been extensively characterized as mediators of modified lipoprotein binding and internalization. The role of these receptors in atherogenesis is highlighted by their ability to regulate the uptake of cholesterol both from modified low-density lipoproteins in peripheral tissues and cells such as macrophages, as well as from high density lipoprotein in tissues such as the liver and adrenal. However, it is becoming increasingly apparent that like other cell surface receptors, SRs appear to be associated with a plethora of ligand-induced intracellular signaling events that may modulate their impact on atherosclerosis, independent of their effects on cholesterol homeostasis. This chapter will describe the current understanding of these processes for the following SRs: macrophage scavenger receptor-1 (SR-A), CD36, scavenger receptor class B type I (SR-BI), and lectin-like oxLDL receptor-1 (LOX-1).

Keywords: atherosclerosis; CD36; LOX-1; scavenger receptor; signaling; SR-A; SR-BI

Abbreviations: acLDL, acetylated LDL; BAEC, bovine arterial endothelial cells; eNOS, endothelial NO synthase; FH, familial hypercholesterolemia; HCAEC, human coronary artery endothelial cells; HDL, high-density lipoprotein; IL, interleukin; LDL, low-density lipoprotein; LDL-R, LDL receptor; LOX-1, lectin-like oxLDL receptor-1; LPS, lipopolysaccharide; MAPK, mitogen-activated protein kinase; mLDL, modified LDL; NF-κB, nuclear factor-κB; NO, nitric oxide; oxLDL, oxidized LDL; PKC, protein kinase C; PTK, protein tyrosine kinase; ROS, reactive oxygen species; SAA, serum amyloid A; SMC, smooth muscle cell; SR, scavenger receptor; SR-A, macrophage scavenger receptor-1; SR-BI, scavenger receptor class B type I; TLR, toll-like receptor; TNF, tumor necrosis factor; uPA, urokinase-type plasminogen activator; WT, wild type

Introduction

Atherosclerosis is a chronic inflammatory state characterized by the formation of arterial plaques—lesions resulting from the accumulation of cholesterol-rich macrophages in the intimal space of the vessel wall. Left untreated, atherosclerosis can lead to hypertension, myocardial infarction, stroke, and death. Scavenger receptors (SRs) are membrane-bound proteins and play a critical role in the pathology of atherosclerosis via their ability to bind lipoproteins and internalize their lipid cargo. In addition to cholesterol transport, SRs can also mediate intracellular signaling and inflammatory responses. Given the increasingly important role of these processes in the development and progression of atherosclerosis, characterizing the specific signaling functions of SRs is crucial to a thorough understanding of disease progression. We begin with a brief description of the pathology of atherosclerosis. We will then describe the potential roles of SRs in regulating intracellular signaling pathways involved in atherogenesis.

Atherosclerosis: A Chronic Inflammatory State

A hallmark feature of atherosclerosis is the presence of white, lipid-laden "fatty streaks" in the arterial wall. Fatty streaks or plaques are thickenings of the intima, the innermost layer of the artery wall, and consist primarily of neutral lipids (such as cholesteryl esters), apoptotic cellular debris, connective tissue, and inflammatory/immune cells including macrophages and T cells. These lesions are commonly found in young persons and, with age, may disappear or develop into advanced plaques. A developing lesion can grow large enough to protrude into the arterial lumen, reduce blood flow, and thus cause hypertension. If the plaque ruptures, the resultant thrombosis can completely block the artery leading to myocardial infarction, stroke, and/or death.

Once thought to be a simple disorder of lipid metabolism, we now recognize that atherosclerosis is a complex process whereby lipoproteins interact with inflammatory cells in the intima to produce foam cells rich in cholesteryl esters. The following sections will describe these interactions in four stages: lesion initiation, monocyte infiltration and differentiation, lesion progression, and plaque rupture.

Lesion Initiation

Lesion formation in the intimal space is thought to occur at sites of endothelial perturbation such as vessel injury, turbulent blood flow, and lipoprotein infiltration [1]. While controversy exists as to the whether arterial inflammation precedes the lipid accumulation or vice versa [2]—a matter which is further complicated by the fact that hypercholesterolemia is itself associated with inflammation [3]—it is clear that a close relationship exists between these two processes during lesion formation.

Since lipoprotein penetration is directly related to the low-density lipoprotein (LDL)-associated cholesterol concentration in the plasma, LDL levels are a major indicator for the development of atherosclerosis [4]. When LDL penetration occurs at sites of hemodynamic strain (such as vessel bifurcation), the inflammatory process begins and LDL can be oxidatively or enzymatically modified (reviewed in [5]). LDL modification results in the release of oxidized phospholipids and lysophospholipids in the intimal space. These bioactive lipids activate the endothelial cells lining the arterial wall, resulting in the expression of adhesion molecules, chemotactic proteins, and cytokines (reviewed in [6]).

Monocyte Infiltration and Differentiation

Monocytes adhere to the endothelium via receptors for adhesion molecules and chemotactic proteins and readily enter the intima [7]. Cytokines present in the intimal space promote (i) monocyte differentiation into macrophages, (ii) SR expression, and (iii) the internalization of modified LDL (mLDL) [8, 9]. Like the uptake of native LDL through the classical LDL receptor (LDL-R) pathway [10], the cholesterol derived via the SR pathway exerts feedback repression on additional cholesterol synthesis and uptake via the LDL-R. However, the expression of SRs is not downregulated by excess cellular cholesterol. Rather, SR expression is increased by oxidized LDL (oxLDL) [11] and macrophages in the intimal space accumulate massive amounts of cholesterol/cholesteryl esters as a result. These cholesterol-loaded macrophages are called foam cells and are the main constituents of fatty streaks.

Lesion Progression

The inflammatory environment of the fatty streak/early plaque activates T cells to produce a host of pro- and anti-inflammatory cytokines, which potentiate atherosclerosis [12, 13]. The transition from a fatty streak to a complex plaque depends upon a vicious inflammatory cycle of T cell and macrophage activation. This in turn causes smooth muscle cell (SMC) migration, additional foam cell formation, and the production of extracellular matrix proteins which form a fibrous cap over the plaque protruding into the lumen of the artery. Cellular lipid accumulation in the increasingly inflammatory environment of the intimal space triggers apoptosis and necrosis of macrophage- and SMC-derived foam cells, producing a core of acellular debris. As a result, advanced lesions contain a number of cell types not abundant in early fatty steaks including SMCs, mast cells, and T cells. In addition, a late-stage plaque has a core of insoluble lipid and apoptotic cellular debris covered by a fibrous cap.

Plaque Rupture

Although arterial narrowing may cause some symptoms of atherosclerosis (such as hypertension), myocardial infarction and stroke likely result from plaque rupture (reviewed in [14]). Once the thrombotic material below the fibrous cap (i.e., plaque lipids and tissue factor) come into contact with the blood, the coagulation cascade begins. Subsequent platelet adhesion and thrombus formation eventually occludes blood flow. Advanced plaques may also develop through a number of smaller plaque ruptures and thrombotic events, which successively narrow the artery prior to a major ischemic rupture.

The stability of an atherosclerotic lesion is very much dependent on the integrity of the fibrous cap [14]. Thus, lesions with thin fibrous caps or large necrotic cores are particularly susceptible to rupture. The "shoulder region" where the edges of the fibrous cap meet the arterial wall is a frequent site of plaque rupture and a relative accumulation of foam cells in this area is commonly observed. The secretion of matrix metalloproteinases by foam cells close to the shoulder region may result in local degradation of the fibrous cap, thus leading to plaque rupture.

Scavenger Receptor Classification, Functions, and Regulation of Inflammation/Signaling

The identification of SRs has its roots in the characterization of the LDL-R pathway by Goldstein and Brown (reviewed in [10]). LDL, the most abundant source of plasma cholesterol in humans, is internalized and degraded via the cell surface LDL-R. Cholesterol acquired by this pathway exerts feedback repression on de novo cholesterol synthesis as well as additional LDL-R-mediated cholesterol uptake. Mutations in the LDL-R result in familial hypercholesterolemia (FH), a disease that presents with abnormally high levels of cholesterol in the plasma [15]. Additional symptoms include delayed clearance of LDL from the plasma and heart attacks in early childhood. Importantly, cholesterol still accumulates in macrophages, kidney, skin, and other tissues in FH patients, suggesting an alternate "scavenging" cholesterol transport pathway.

One hypothesis put forth from the work described above was that the delayed clearance of LDL in FH patients might result in a chemical modification of the lipoprotein, thus making it available to the scavenger pathway *in vivo*. Based on this premise, acetylated LDL (acLDL) was used to identify a high-affinity binding site in liver and macrophages that did not recognize native LDL [16]. The uptake of acLDL via this alternate pathway caused massive cellular cholesterol deposition due to a lack of feedback regulation. AcLDL binding was then used to isolate the first SR [17] and the partial protein sequence was used to clone the cDNA [17, 18].

After more than a decade of research, mLDL binding remains the unifying trait of the SR family, which is now composed of a diverse group of integral membrane proteins divided into at least eight classes (reviewed in [19]). In addition to binding mLDL and/or other lipoproteins, a wide range of additional ligands exists for this protein family such as fatty acids, oxidized phospholipids, apoptotic cells, and bacterial lipopolysaccharides (LPS; see Table 5.1). Together, the SRs exert significant effects on cholesterol and fat metabolism as well as innate immune function. While these proteins have been studied primarily in regard to their ability to mediate ligand clearance, it is now becoming clear that intracellular signaling pathways initiated by these ligands extend the functionality of SRs into many metabolic processes. However, due to their recent classification or identification, little or no information is available for some receptors in regard to signaling, inflammation, or atherosclerosis. Thus, this chapter will focus on some of the most well-described SRs—macrophage scavenger receptor-1 (SR-A), CD36, scavenger receptor class B type I (SR-BI), and lectin-like oxLDL receptor-1 (LOX-1)—which may be involved in the development and progression of atherosclerosis. The impact of intracellular signaling on the expression of SRs will not be discussed so that we can examine the converse: how SR expression regulates inflammation and related signaling pathways. In addition, much of the early characterization of the SR family utilized "specific" ligands to make conclusions about a particular SR. We now recognize that many SR ligands (e.g., oxLDL) can engage multiple receptors. Therefore, we will restrict our discussion to experimental techniques, such as targeted gene disruption or mutagenesis, which clearly identify the appropriate receptor.

TABLE 5.1. Properties of the scavenger receptor family.

				Ligands			
Class	Members	Topology	Primary membrane	Modified LDL[a]	Apoptotic cells	Bacterial/ LPS	Oxidized lipids
A	SR-AI/II, MARCO, SRCL	Type II	PM	+	+	+	NR
B	CD36, SR-B, LIMP-II	Type III	PM	+	+	+	+
C	dSR-CI	Type I	PM	+	NR	+	NR
D	CD68, LAMP-1, -2, -3	Type I	Lysosome	+	NR	NR	NR
E	LOX-1	Type II	PM	+	+	+	NR
F	SREC-I, SREC-II	Type I	PM	+	+	NR	NR
G	SR-PSOX/ CXCL16	Type I	PM	+	NR	+	NR
H	FEEL-1, FEEL-2	Type I	PM/ endosome	+	NR	+	NR

+, ligand binding reported.
NR, ligand binding not reported.
[a]Includes oxidized or acetylated LDL.

Macrophage Scavenger Receptor-1 (SR-A)

SR-A was the first scavenger receptor to be identified and cloned [17, 18]. It is a homotrimeric, type II integral membrane protein (single transmembrane domain with a cytoplasmic N-terminus) expressed in macrophages, endothelial cells, and smooth muscle tissue. It can bind mLDL/oxLDL and may account for as much as 30% of oxLDL uptake by macrophages [20]. SR-A also binds bacterial components and is an important factor in controlling endotoxemia [21].

The contribution of SR-A to the development of atherosclerosis is currently a controversial subject despite the *in vitro* evidence that mLDL can engage and be internalized by SR-A [20, 22]. Babaev et al. [23] demonstrated that SR-A$^{-/-}$ C57BL/6 mice on a butterfat diet had smaller lesions in the proximal aorta without a significant change in plasma cholesterol levels [23]. In addition, lethally irradiated C57BL/6 or LDL-R$^{-/-}$ mice reconstituted with SR-A$^{-/-}$ fetal liver cells were protected from lesion formation compared to mice reconstituted with SR-A$^{+/+}$ control cells. These data corroborate an earlier finding in apoE$^{-/-}$ mice [17] and suggest that SR-A expression is atherogenic due to its ability to mediate the uptake of mLDL. In contrast, Herijgers et al. [24] found that SR-A had no effect on lesion area in LDL-R$^{-/-}$ mice fed a "Western-type" diet and transplanted with bone marrow from mice overexpressing SR-A [24]. The observation that macrophage-specific overexpression of SR-A decreases atherosclerosis in LDL-R$^{-/-}$ mice fed a diet enriched in fat and cholesterol also contradicts the notion that SR-A is atherogenic [25], as does a recent report using apoE$^{-/-}$ SR-A$^{-/-}$ double-null mice [22]. Thus, it seems likely that SR-A can mediate both pro- and antiatherogenic processes *in vivo* and the overall susceptibility to lesion formation depends on diet and genetic background of the animal model.

Although the mechanisms which underlie the ability of SR-A to promote or inhibit lesion formation require further study, a small body of work suggests that SR-A may achieve some of its functions via intracellular signaling processes. Sequence analysis of SR-A revealed the presence of three conserved phosphorylation sites in the cytoplasmic domains [26], giving the first indication that SR-A was part of a signaling pathway. In addition, phosphorylation at serine 21 or 49 of mouse SR-A appears to modulate the efficiency of mLDL internalization and metabolism in COS-7 cells [27]. Mice, which lack c-Jun N-terminal kinase 2, display reduced foam cell formation, reduced uptake of mLDL by macrophages, and reduced SR-A phosphorylation [28]. However, SR-A is not only a target of intracellular signaling. In rat peritoneal macrophages, acLDL (a signature SR-A ligand) stimulated tumor necrosis factor (TNF)-α secretion and increased intracellular calcium levels [29]. The authors showed that TNF-α production could be blocked by inhibitors to heterotrimeric G-proteins, protein kinase C (PKC), and protein tyrosine kinases (PTK). Inhibiting heterotrimeric G-proteins with pertussis toxin also appears to reduce acLDL uptake, thus indicating a positive feedback pathway between G-proteins and SR-A [30]. Such a pathway may partially explain the continued uptake of mLDL via the SR pathway despite high levels of intracellular cholesterol.

Perhaps the best evidence for a direct link between SR-A and intracellular signaling pathways comes from the plasminogen activator field. Urokinase-type plasminogen activator (uPA) is a protease expressed by macrophages in the arterial wall; in human arterial lesions, uPA levels are elevated and correlate directly with the severity of atherosclerotic lesion formation [31]. In THP-1 macrophages, acLDL increases the steady-state levels of uPA in a PTK- and PKC-dependent fashion [32]. AcLDL also stimulates a dramatic increase in cellular protein tyrosine phosphorylation in these cells as well as in Bowes human melanoma cells transfected with SR-A. In contrast, the parental Bowes cells, which do not express SR-A, showed limited protein tyrosine phosphorylation upon acLDL incubation [32]. The requirement of SR-A expression for protein tyrosine phosphorylation and elevated uPA levels strongly implicate SR-A in intracellular signaling pathways. Interestingly, the exact role of uPA in lesion formation is not clear due to discrepancies between various studies similar to the discrepancies reported for SR-A (discussed earlier). One might speculate that the atherogenic potential of SR-A is related to its ability to induce or repress uPA levels. However, prospective studies are required to confirm this notion.

SR-A also appears to mediate the activation of the Src-family of PTKs. Miki et al. [33] have reported that acLDL rapidly induces the phosphorylation and activation of Lyn in THP-1 macrophages. Under identical conditions, Lyn was coimmunoprecipitated with an α-SR-A antibody. The same group more recently reported that Lyn$^{-/-}$ mice have reduced atherosclerosis despite elevated plasma cholesterol [34], but did not test SR-A function directly. It is imperative to determine whether the atheroprotection observed in this strain is due to SR-A dysfunction, and whether Lyn-mediated signaling pathways are altered in SR-A$^{-/-}$ mice.

Finally, SR-A plays an important but apparently indirect role in mediating inflammation induced by bacterial glycolipids such as endotoxin or LPS. SR-A$^{-/-}$ mice are more susceptible to LPS-induced death compared to wild-type (WT) mice [21]. This phenotype appears to be due to an elevated inflammatory response as serum TNF-α and interleukin (IL)-6 levels are higher in the absence of SR-A and survivability of SR-A$^{-/-}$ mice injected with LPS is restored when pretreated with a TNF-α blocking antibody. The mechanism for the sensitivity to endotoxic shock is unclear due to conflicting reports on the half-life of serum LPS in WT and SR-A$^{-/-}$ mice [35, 36]. Thus, it is not known what role SR-A has, if any, on modulating the downstream signaling events following LPS activation.

CD36

CD36, also known as FAT, GPIV, and SCARB3, is a type III (multiple transmembrane domains) membrane glycoprotein expressed in numerous tissues (adipose, heart, muscle, intestine) and cell types (monocytes/macrophages, microvascular endothelium, dendritic cells, retinal epithelia, SMCs). It is predicted to have a "hairpin" topology (a large extracellular loop and two transmembrane domains with both the C- and N-termini projecting into the cytoplasm) and has been implicated in numerous cellular functions including atherosclerosis (reviewed in [37]). This broad range of functions is reflected by the numerous ligands for this receptor, which include thrombospondin, anionic phospholipids, β-amyloid, long-chain fatty acids, native lipoproteins, and oxLDL.

In an atherogenic background, CD36$^{-/-}$ mice appear to be protected from plaque formation and atherosclerosis [38] despite only subtle changes in plasma cholesterol [39]. Bone marrow transplantation studies have also shown that the proatherogenic effects of CD36 are dependent upon bone marrow-derived cells (i.e., monocytes/macrophages) [40]. *In vitro* experiments demonstrating that CD36 is the major oxLDL receptor in macrophages are consistent with an atherogenic role of CD36 [20, 41, 42]. However, a recent report has called into question the relevance of CD36 in atherogenesis [22] and will likely spark renewed interest in this area to resolve the controversy.

Although the mechanism of atheroprotection in CD36$^{-/-}$ mice likely involves reduced oxLDL/cholesterol uptake by macrophages, a growing body of evidence suggests that CD36 may also play a role in regulating inflammatory responses via the transcription factor nuclear factor-κB (NF-κB; see Fig. 5.1 for a description of the NF-κB pathway). Indirect evidence for this link comes from the observation that oxLDL can modulate NF-κB activity (reviewed in [43]). The first direct evidence was reported by Lipsky et al. [44] using Chinese hamster ovary cells transfected with an expression vector encoding CD36. Preincubating the cells with oxLDL increased the amount of activated NF-κB following exposure to TNF-α. However, this was not observed when cells were transfected with a vector encoding a mutant CD36 lacking the last (3′) six codons. Janabi et al. [45] further showed that oxLDL induction of NF-κB activation was defective in monocyte-derived macrophages from CD36-deficient human patients. Thus, it is reasonable to conclude that an intracellular protein interacting with the C-terminal cytoplasmic tail of CD36 mediates the effects of oxLDL on NF-κB. In line with this, Bull et al. [46] reported a physical interaction between CD36 and Src-related PTKs Fyn, Lyn, and Yes in human dermal microvascular endothelial cells. Moore et al. [47] independently verified these coimmunoprecipitation studies between CD36 and Lyn in elicited murine peritoneal macrophages treated with a nonlipid CD36 ligand and further showed that this interaction was required for subsequent, ligand-induced mitogen-activated protein kinase (MAPK) activation [47]. In COS-7 cells, an interaction between CD36

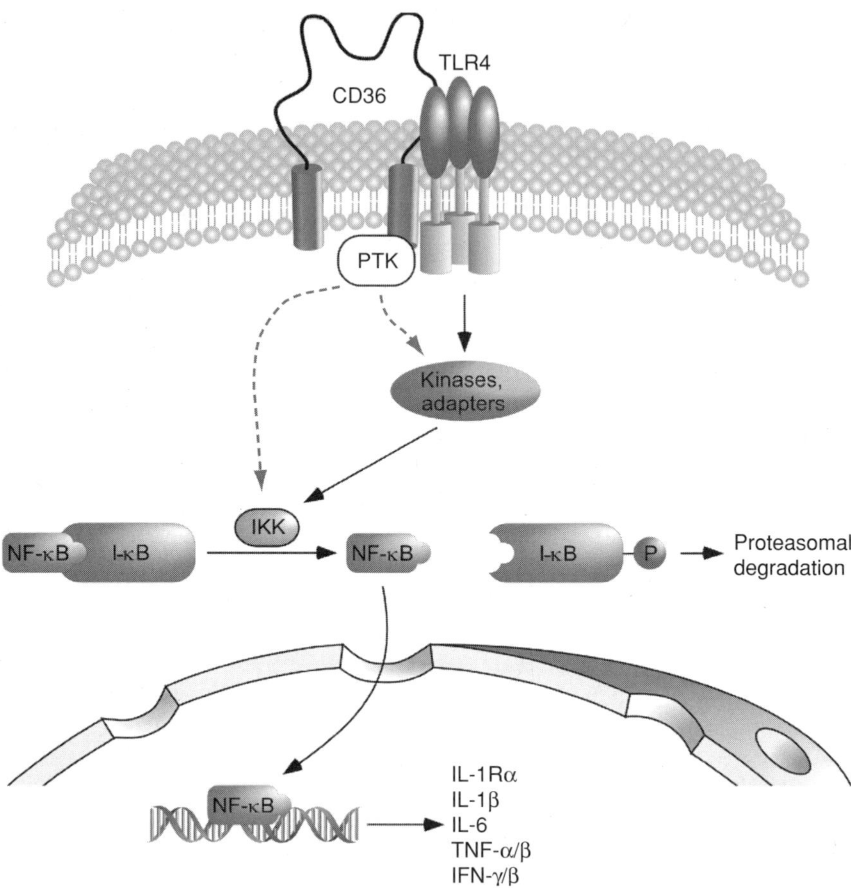

FIGURE 5.1. Hypothetical model of CD36–TLR4 interactions and NF-κB activation. NF-κB is an evolutionarily conserved transcription factor required for many aspects of the immune and inflammatory response. In atherosclerosis, NF-κB coordinates the expression of a wide array of inflammatory genes (reviewed in [100]) and is thus central to the pathology of this disease. The activity of NF-κB is tightly regulated by I-κB inhibitory proteins, which bind to NF-κB, sequestering it in the cytoplasm. When the NF-κB pathway is activated, I-κBs are phosphorylated by kinases (IKKs) leading to the dissociation of the NF-κB:I-κB complex and NF-κB translocation to the nucleus. In the nucleus, NF-κB binds to the promoter of target genes and stimulates gene expression. An important regulator of the NF-κB pathway is the toll-like receptor (TLR) family of pattern recognition receptors. TLRs are crucial to the innate host response to infectious agents such as bacteria due to their ability to bind endotoxin/LPS. The growing body of data linking such pathogens with atherosclerosis has cast a new light on the role of TLRs in lesion development and progression. In particular, the TLR4 pathway has garnered much attention (reviewed in [101]). Ligand engagement of TLR4 triggers the recruitment of intracellular adapter proteins such as MyD88, which results in the activation of numerous kinases, phosphorylation of I-κB and thus elevated NF-κB activity. We hypothesize that CD36–TLR4 aggregation results in the recruitment of CD36-associated protein tyrosine kinases (PTKs), which further enhance IKK and NF-κB activity. Potential sites of PTK activity are indicated by dashed arrows as are some of the genes which may be affected by CD36 expression levels [45].

and c-Src, has also been suggested [48]. Together, the data strongly argue for a functional effect of CD36 on NF-κB activation, perhaps via intracellular kinases associated with the C-terminal cytoplasmic tail of CD36.

Physical interactions between CD36 and a major regulatory protein in the NF-κB pathway have also been detected. As noted in Fig. 5.1, the toll-like receptor (TLR)-4 is an important mediator of NF-κB and has been well characterized as a major receptor for LPS, a glycolipid found on the outer membrane of gram-negative bacteria [49–51]. Similar to other immunoreceptors, the first step in TLR4 signaling is ligand-dependent receptor aggregation, which initiates signaling cascades and thus leads to the appropriate inflammatory response. CD36 appears to participate in this receptor aggregation. In human monocytes, stimulation with LPS induced the aggregation of TLR4 and CD36 with CD14 [52] (another cell surface LPS receptor which facilitates interactions between LPS and TLR4). Receptor aggregation/colocalization between CD36 and TLR4 was also detected in retinal pigment epithelial cells [53, 54], the cells that lie between the retina and the choroids (the vascular bed which oxygenates the retina), and are adjacent to the retinal photoreceptors. These examples of CD36–TLR4 interactions provide one possible mechanism whereby CD36 may regulate the NF-κB pathway.

The data above can be summarized as follows: (i) CD36 physically interacts with Src-related PTKs, (ii) CD36 expression modulates the magnitude of oxLDL-mediated NF-κB activation, and (iii) ligand engagement of TLR4 (which mediates NF-κB activation by LPS) induces aggregation with CD36. In addition, we have observed that CD36$^{-/-}$ mice display a muted inflammatory response following LPS injection (V.A. Drover and N.A. Abumrad, unpublished results). Thus, we propose a model (Fig. 5.1) whereby CD36 and its associated PTKs aggregate with TLR4 upon ligand binding. The PTKs may modulate TLR4 signaling to NF-κB via phosphorylation of downstream effector molecules or even the regulatory kinases directly as has been demonstrated previously [55, 56]. The report that CD36 is a signaling bridge for microbial diacylglyceride-stimulated TLR2 [57] supports this model. Further studies in lesions and macrophages from WT and CD36$^{-/-}$ mice are required to test the validity of this model and the potential role of CD36 in mediating atherosclerosis and inflammation.

Scavenger Receptor BI

Like CD36, SR-BI is a class B receptor and a type III transmembrane protein. SR-BI appears to have a hairpin topology, is heavily glycosylated, and binds numerous ligands (reviewed in [58]). SR-BI has a broad expression pattern including intestine, macrophages, endothelial cells, and SMCs but is most highly expressed in the liver and steroidogenic tissues where it plays a major role in cholesterol catabolism and hormone synthesis, respectively.

SR-BI is best known as the most physiologically relevant receptor for high-density lipoprotein (HDL) particles [59]. SR-BI-mediated cholesterol uptake from HDL occurs by a process called selective uptake; the cholesteryl ester in the core of the HDL particle is selectively extracted and delivered to the cell for subsequent hydrolysis and/or storage. Mice lacking SR-BI have elevated plasma cholesterol, increased HDL particle size, and reduced cholesterol disposal in the liver [60–62]. *In vitro*, SR-BI also facilitates cholesterol efflux to lipid acceptors such as small unilamellar vesicles and small HDL particles [17, 63]. Together, the biochemical data would predict that SR-BI plays an atheroprotective role by promoting cholesterol efflux into the plasma as HDL and then facilitating hepatic HDL cholesterol disposal. Indeed, numerous studies in atherogenic strains of mice show that SR-BI deficiency promotes atherosclerosis [64–66] while hepatic overexpression is protective [67] (reviewed in [68]; see also Chapter 4).

While structure–function relationships in SR-BI have been extensively characterized, the role of SR-BI in cell signaling and inflammation remains poorly described. However, a few reports use various techniques to directly implicate SR-BI in cell signaling and inflammatory pathways, which may be relevant to atherosclerosis. For instance, Vishnyakova et al. [69] demonstrated that SR-BI is also a receptor for LPS in murine RAW and HeLa cells. Like SR-A, SR-BI may be involved in the clearance of LPS and attenuating the inflammatory response, thus protecting against atherosclerosis.

Serum amyloid A (SAA), a major acute-phase reactant typically transported on HDL, appears to be a ligand for SR-BI. SAA engagement of SR-BI in transfected HeLa cells results in enhanced secretion of IL-8 [70], a proinflammatory and atherogenic chemokine (reviewed in [71]). This SR-BI-dependent, SAA-induced increase in IL-8 secretion requires the activation of ERK1/2 and p38 MAPK [70] and suggests a role for SR-BI in these signaling pathways. The observation that ERK1/2 activation by HDL is reduced by SR-BI neutralizing antibodies [72] further support this notion. One scenario whereby this pathway may be physiologically relevant is the ability of dietary cholesterol to raise serum SAA levels [73], which in turn can increase inflammation/IL-8 secretion and thus contribute to lesion formation. However, it is difficult to reconcile the concept that SR-BI mediates an inflammatory role in SAA/MAPK signaling with the observations from genetically modified mice illustrating an atheroprotective role for SR-BI *in vivo* (discussed earlier). One possibility is that SR-BI is atherogenic in the early stages of lesion development where inflammation may play a more prominent role in disease pathology. This is in agreement with a recent bone marrow transplantation study showing that SR-BI increased lesion development in LDL-R-deficient mice fed a Western diet for only 4 weeks [74].

The best evidence for SR-BI-dependent cell signaling in atherogenesis comes from the nitric oxide (NO) field. NO is critical for arterial tonicity and is an important mediator of atherosclerosis since chronic inhibition of NO production enhances lesion development in hypercholesterolemic rabbits [75]. NO is produced by endothelial NO synthase (eNOS) and several groups have reported that HDL stimulates eNOS activity in both epithelial and CHO cells in an SR-BI-dependent fashion [76, 77]. As a physiologic measure of NO production, Yuhanna et al. [77] measured arterial relaxation in aortic rings ex vivo and found that HDL-mediated relaxation was highly attenuated in aortic rings from SR-BI$^{-/-}$ mice. Although the mechanism of action is unclear, intracellular ceramide levels covary with eNOS activation [76], suggesting that SR-BI may be mediating eNOS activity by regulating levels of this sphingolipid.

Cross talk between SR-BI-mediated eNOS activation and apoptotic cell death adds another layer of complexity to the role of SR-BI in cellular signaling mechanisms. Li et al. [78] recently reported that in the absence of ligand, SR-BI expression activates caspase-8-mediated apoptosis in CHO cells, mouse embryonic fibroblasts, and endothelial cells. Cotransfection with eNOS or incubation with HDL reversed this SR-BI-dependent cell death. The authors suggest that in healthy cells, HDL-SR-BI ligation inhibits inappropriate apoptosis via NO production. Cells in stressful, inflammatory environments have reduced eNOS activity and apoptosis occurs in an attempt to reduce further inflammation and damage to neighboring cells. Elevated mRNA levels of inflammatory markers in the artery in SR-BI$^{-/-}$ mice [79] are consistent with uncontrolled inflammation and support the authors' hypothesis.

The Lectin-Like oxLDL Receptor-1 (LOX-1)

Also known as OLR1, this SR displays a type II membrane topology and is expressed on endothelial cells, macrophages, SMCs, and platelets. In addition to mLDL, bacteria, and apoptotic cells, LOX-1 can also bind advanced glycation end products and fibronectin, thus suggesting a role in cell adhesion (discussed later). Ligand engagement to LOX-1 appears to be linked to a host of intracellular signaling processes (discussed later) that may affect atherosclerotic progression but prospective studies in genetically modified mice have not been reported. However, the observations that (i) humans and rabbits exhibit LOX-1 accumulation in atherosclerotic lesions [17, 80] and (ii) LOX-1 mutations in humans are associated with increased cardiovascular disease [55] suggest that LOX-1 is an important mediator of atherosclerosis. As mice lacking LOX-1 are not currently available, most of the definitive studies on the role of LOX-1 in inflammation/signaling utilize a mixed approach of heterologous gene expression, antisense knockdown technology, and specific α-LOX-1 antibodies which block oxLDL binding to the receptor.

One mechanism whereby LOX-1 may mediate atherosclerotic lesion development is through its ability to stimulate leukocyte adhesion following exposure to inflammatory stimuli. In human coronary artery endothelial cells (HCAEC), oxLDL-induced LOX-1 expression caused elevated monocyte chemoattractant protein-1 expression and monocyte adhesion. These effects were attenuated by transfection of the cells with antisense LOX-1, which abrogated LOX-1 expression [81]. Similarly, CHO-K1 cells transfected with a plasmid encoding LOX-1 exhibited increased leukocyte and THP-1 cell adhesion which could be blocked by oxLDL or an α-LOX-1 antibody [82]. In rats, leukocytes accumulated at sites of inflammation/damage in a LOX-1-dependent fashion with concomitant increases in adhesion molecule expression [83–85]. Although the mechanism(s) responsible for these LOX-1-mediated effects are not known, both MAPK (p38 and p42/p44) and PKC-β pathways appear to be involved in modulating changes in the expression of adhesion molecules [81, 84, 86].

The ability of oxLDL to stimulate other signaling pathways may also be involved in the activation of endothelial cells. Most notable is the ability of oxLDL to elevate levels of reactive oxygen species (ROS). Cominacini et al. [87] documented an oxLDL and LOX-1-dependent increase in intracellular ROS in bovine arterial endothelial cells (BAEC). Levels of NF-κB in the nucleus were also elevated by oxLDL exposure as determined by a qualitative mobility shift assay. Although a causal relationship between ROS production and nuclear NF-κB translocation was not shown in this study, the concept that ROS are upstream signals for NF-κB activation is well documented in the literature [88–92]. The additional observation that ROS levels following oxLDL stimulation rise well before NF-κB activation is detected [87] forms a strong circumstantial case for a LOX-1/ROS-mediated link between oxLDL and the NF-κB pathway in endothelial cells and other cell types such as articular chondrocytes [93].

The LOX-1-dependent rise of ROS in oxLDL-treated BAECs also results in a concomitant reduction of intracellular NO [94]. The authors speculate that oxLDL ligation of LOX-1 in atherosclerotic lesions may explain the impaired endothelium-dependent vasodilation commonly observed in this type of vascular disease. Since ROS are thought to increase oxLDL production *in vivo*, we see the potential for a vicious loop of LDL oxidation, ROS formation, NO inactivation, and NF-κB-dependent inflammation—all dependent on endothelial LOX-1. Perhaps not surprisingly, the situation may be further exacerbated by the feed-forward upregulation of LOX-1 expression by oxLDL [95, 96].

In addition to ROS signaling, LOX-1 appears to play some role in a myriad of signaling pathways that may be relevant to atherosclerosis. In addition to the role of p38 MAPK in adhesion molecule expression (discussed earlier), oxLDL induces apoptosis in neonatal rat cardiac myocytes in a LOX-1- and p38 MAPK-dependent fashion [97]. In cultured rat chondrocytes and HCAECs, oxLDL induces cell death via PI3 kinase/Akt [98] and NF-κB pathways [99], respectively; in each instance, cell death could be blocked by α-LOX-1 antibodies or abrogating LOX-1 expression. Given the complexity of the signaling pathways stimulated by oxLDL via LOX-1, significant cross talk appears likely.

Since LOX-1 is thought to be a major SR on endothelial cells, much of the work described herein has used primary or cultured endothelial cells. However, it is important to note that many of the pathways implicated in LOX-1 signaling are present in macrophages. LOX-1 is expressed on macrophages and oxLDL ligation may affect inflammation/apoptosis in this cell type in addition to endothelial cell activation. It is also likely that some of these signaling pathways terminate at the genes encoding adhesion molecules and other cell type-specific genes in the inflammatory/atherosclerotic expression profile (Fig. 5.2).

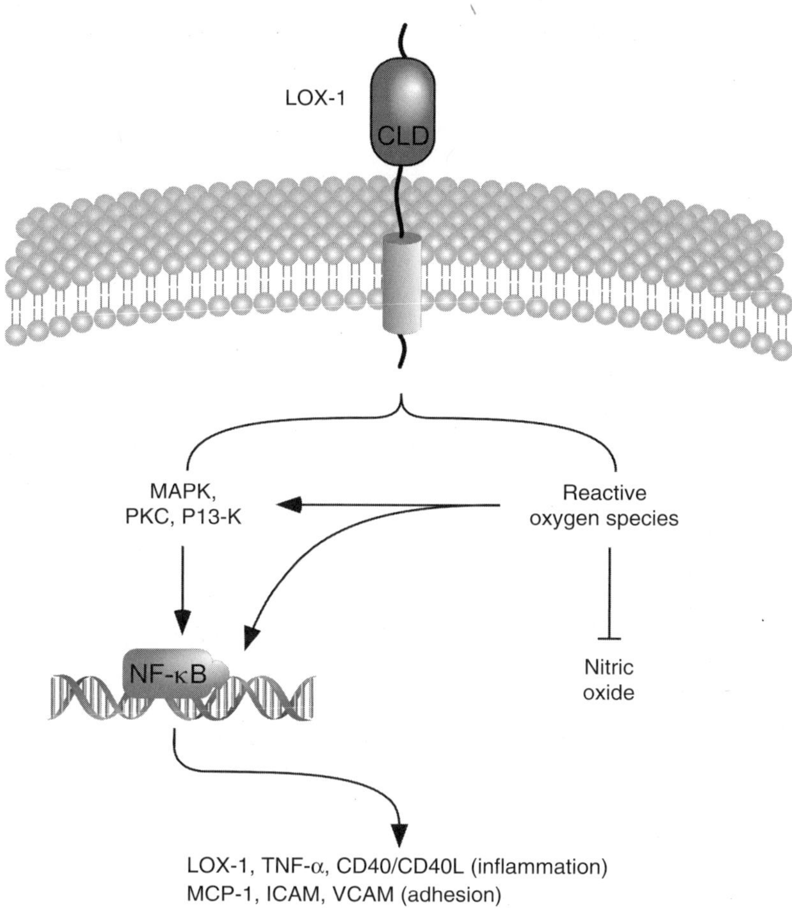

FIGURE 5.2. Intracellular signaling pathways following LOX-1 engagement by oxLDL in endothelial cells. LOX-1 is a type II integral membrane protein and contains a C-type lectin-like domain (CLD). Oxidized LDL engagement of LOX-1 causes the increased expression of numerous genes whose protein products are involved in inflammation, monocyte adhesion, and plaque rupture. These changes in gene expression occur as a result of the activation of numerous intracellular signaling pathways including mitogen-activated protein kinase (MAPK), protein kinase C (PKC), and PI3 kinase (PI3-K), as well as through the production of reactive oxygen species (ROS). It is likely that many of these signaling pathways act directly or indirectly via the NF-κB pathway. In addition, increased ROS production inactivates intracellular nitric oxide (NO), which may result in impaired vasodilation commonly observed in patients with atherosclerosis.

A major hurdle in the elucidation of these pathways is the lack of mice with a disrupted LOX-1 gene. Tissue/cell type-specific gene ablation may be necessary to disentangle the plethora of intracellular signals initiated by LOX-1 ligand engagement.

Concluding Remarks

As the field of SR biology expands, our understanding of the functions of these proteins moves beyond the clearance and utilization of a particular ligand. As should now be evident, SRs are as promiscuous in their use of intracellular signaling pathways as they are in their cell surface ligands. Our understanding of these pathways and their role in the pathology of atherosclerosis is still a developing area of research, especially as many SRs have only recently been classified in this protein family. In addition, determining the net effect of engaging multiple SRs with a common ligand such as mLDL/oxLDL will be a challenging task. This will be further complicated by the fact that multiple SRs can affect the same signaling pathways. For instance, LOX-1 shares a functional link to NF-κB and NO with CD36 and SR-BI, respectively. Similarly, CD36 and SR-A both interact with Src-family PTKs.

While examining the effects of multilocus gene disruption may provide an overall measure of the relative contribution of the SRs to atherogenesis, determining the relevant mechanisms will require careful experimental design and attention to the effective cell type. In addition to the relative distribution of the SRs in critical cell types such as endothelial cells and monocytes/macrophages, the ability of a ligand or nutritional regimen to regulate SR levels and concomitant signaling pathways must be considered. The continued use of genetically modified mice and tissue-specific gene ablation are thus likely to play important roles in answering the remaining questions in this field.

References

1. Skalen K, Gustafsson M, Rydberg EK, Hulten LM, Wiklund O, Innerarity TL, Boren J: Subendothelial retention of atherogenic lipoproteins in early atherosclerosis. Nature 417: 750–754, 2002.
2. Stocker R, Keaney JF Jr: Role of oxidative modifications in atherosclerosis. Physiol Rev 84: 1381–1478, 2004.
3. Stokes KY, Cooper D, Tailor A, Granger DN: Hypercholesterolemia promotes inflammation and microvascular dysfunction: role of nitric oxide and superoxide. Free Radic Biol Med 33: 1026–1036, 2002.
4. Kannel WB: Status of risk factors and their consideration in antihypertensive therapy. Am J Cardiol 59: 80A–90A, 1987.
5. Abumrad N, Coburn C, Ibrahimi A: Membrane proteins implicated in longchain fatty acid uptake by mammalian cells: CD36, FATP and FABPm. Biochim Biophys Acta 1441: 4–13, 1999.
6. Leitinger N: Oxidized phospholipids as modulators of inflammation in atherosclerosis. Curr Opin Lipidol 14: 421–430, 2003.
7. Eriksson EE, Xie X, Werr J, Thoren P, Lindbom L: Importance of primary capture and L-selectin-dependent secondary capture in leukocyte accumulation in inflammation and atherosclerosis *in vivo*. J Exp Med 194: 205–218, 2001.
8. Janeway CA Jr, Medzhitov R: Innate immune recognition. Annu Rev Immunol 20: 197–216, 2002.

9. Peiser L, Mukhopadhyay S, Gordon S: Scavenger receptors in innate immunity. Curr Opin Immunol 14: 123–128, 2002.

10. Goldstein JL, Brown MS: The low-density lipoprotein pathway and its relation to atherosclerosis. Annu Rev Biochem 46: 897–930, 1977.

11. Nagy L, Tontonoz P, Alvarez JG, Chen H, Evans RM: Oxidized LDL regulates macrophage gene expression through ligand activation of PPARgamma. Cell 93: 229–240, 1998.

12. Frostegard J, Ulfgren AK, Nyberg P, Hedin U, Swedenborg J, Andersson U, Hansson GK: Cytokine expression in advanced human atherosclerotic plaques: dominance of pro-inflammatory (Th1) and macrophage-stimulating cytokines. Atherosclerosis 145: 33–43, 1999.

13. Uyemura K, Demer LL, Castle SC, Jullien D, Berliner JA, Gately MK, Warrier RR, Pham N, Fogelman AM, Modlin RL: Cross-regulatory roles of inter-leukin(IL)-12 and IL-10 in atherosclerosis. J Clin Invest 97: 2130–2138, 1996.

14. Falk E, Shah PK, Fuster V: Coronary plaque disruption. Circulation 92: 657–671, 1995.

15. Brown AM, Baker PW, Gibbons GF: Changes in fatty acid metabolism in rat hepatocytes in response to dietary n-3 fatty acids are associated with changes in the intracellular metabolism and secretion of apolipoprotein B-48. J Lipid Res 38: 469–481, 1997.

16. Goldstein JL, Ho YK, Basu SK, Brown MS: Binding site on macrophages that mediates uptake and degradation of acetylated low density lipoprotein, pro-ducing massive cholesterol deposition. Proc Natl Acad Sci USA 76: 333–337, 1979.

17. Suzuki H, Kurihara Y, Takeya M, Kamada N, Kataoka M, Jishage K, Ueda O, Sakaguchi H, Higashi T, Suzuki T, Takashima Y, Kawabe Y, Cynshi O, Wada Y, Honda M, Kurihara H, Aburatani H, Doi T, Matsumoto A, Azuma S, Noda T, Toyoda Y, Itakura H, Yazaki Y, Horiuchi S, Takahashi K, Kruijt JK, van Berkel TJ, Steinbrecher UP, Ishibashi S, Maeda N, Gordon S, Kodama T et al: A role for macrophage scavenger receptors in atherosclerosis and susceptibility to infec-tion. Nature 386: 292–296, 1997.

18. Rohrer L, Freeman M, Kodama T, Penman M, Krieger M: Coiled-coil fibrous domains mediate ligand binding by macrophage scavenger receptor type II. Nature 343: 570–572, 1990.

19. Murphy JE, Tedbury PR, Homer-Vanniasinkam S, Walker JH, Ponnambalam S: Biochemistry and cell biology of mammalian scavenger receptors. Atherosclerosis 182: 1–15, 2005.

20. Kunjathoor VV, Febbraio M, Podrez EA, Moore KJ, Andersson L, Koehn S, Rhee JS, Silverstein R, Hoff HF, Freeman MW: Scavenger receptors class A-I/II and CD36 are the principal receptors responsible for the uptake of modified low density lipoprotein leading to lipid loading in macrophages. J Biol Chem 277: 49982–49988, 2002.

21. Haworth R, Platt N, Keshav S, Hughes D, Darley E, Suzuki H, Kurihara Y, Kodama T, Gordon S: The macrophage scavenger receptor type A is expressed by activated macrophages and protects the host against lethal endotoxic shock. J Exp Med 186: 1431–1439, 1997.

22. Moore KJ, Kunjathoor VV, Koehn SL, Manning JJ, Tseng AA, Silver JM, McKee M, Freeman MW: Loss of receptor-mediated lipid uptake via scavenger receptor A or CD36 pathways does not ameliorate atherosclerosis in hyperlipi-demic mice. J Clin Invest 115: 2192–2201, 2005.

23. Babaev VR, Gleaves LA, Carter KJ, Suzuki H, Kodama T, Fazio S, Linton MF: Reduced atherosclerotic lesions in mice deficient for total or macrophage-specific expression of scavenger receptor-A. Arterioscler Thromb Vasc Biol 20: 2593–2599, 2000.

24. Herijgers N, de Winther MP, Van Eck M, Havekes LM, Hofker MH, Hoogerbrugge PM, Van Berkel TJ: Effect of human scavenger receptor class A overexpression in bone marrow-derived cells on lipoprotein metabolism and atherosclerosis in low density lipoprotein receptor knockout mice. J Lipid Res 41: 1402–1409, 2000.

25. Whitman SC, Rateri DL, Szilvassy SJ, Cornicelli JA, Daugherty A: Macrophage-specific expression of class A scavenger receptors in LDL receptor(–/–) mice decreases atherosclerosis and changes spleen morphology. J Lipid Res 43: 1201–1208, 2002.

26. Ashkenas J, Penman M, Vasile E, Acton S, Freeman M, Krieger M: Structures and high and low affinity ligand binding properties of murine type I and type II macrophage scavenger receptors. J Lipid Res 34: 983–1000, 1993.

27. Fong LG, Le D: The processing of ligands by the class A scavenger receptor is dependent on signal information located in the cytoplasmic domain. J Biol Chem 274: 36808–36816, 1999.

28. Ricci R, Sumara G, Sumara I, Rozenberg I, Kurrer M, Akhmedov A, Hersberger M, Eriksson U, Eberli FR, Becher B, Boren J, Chen M, Cybulsky MI, Moore KJ, Freeman MW, Wagner EF, Matter CM, Luscher TF: Requirement of JNK2 for scavenger receptor A-mediated foam cell formation in atherogenesis. Science 306: 1558–1561, 2004.

29. Pollaud-Cherion C, Vandaele J, Quartulli F, Seguelas MH, Decerprit J, Pipy B: Involvement of calcium and arachidonate metabolism in acetylated-low-density-lipoprotein-stimulated tumor-necrosis-factor-alpha production by rat peritoneal macrophages. Eur J Biochem 253: 345–353, 1998.

30. Whitman SC, Daugherty A, Post SR: Regulation of acetylated low density lipoprotein uptake in macrophages by pertussis toxin-sensitive G proteins. J Lipid Res 41: 807–813, 2000.

31. Kienast J, Padro T, Steins M, Li CX, Schmid KW, Hammel D, Scheld HH, van de Loo JC: Relation of urokinase-type plasminogen activator expression to presence and severity of atherosclerotic lesions in human coronary arteries. Thromb Haemost 79: 579–586, 1998.

32. Hsu HY, Hajjar DP, Khan KM, Falcone DJ: Ligand binding to macrophage scavenger receptor-A induces urokinase-type plasminogen activator expression by a protein kinase-dependent signaling pathway. J Biol Chem 273: 1240–1246, 1998.

33. Miki S, Tsukada S, Nakamura Y, Aimoto S, Hojo H, Sato B, Yamamoto M, Miki Y: Functional and possible physical association of scavenger receptor with cytoplasmic tyrosine kinase Lyn in monocytic THP-1-derived macrophages. FEBS Lett 399: 241–244, 1996.

34. Miki S, Horikawa K, Nishizumi H, Suemura M, Sato B, Yamamoto M, Takatsu K, Yamamoto T, Miki Y: Reduction of atherosclerosis despite hypercholesterolemia in Lyn-deficient mice fed a high-fat diet. Genes Cells 6: 37–42, 2001.

35. Amersfoort ESV, Van Berkel TJC, Kuiper J: Clearance of lipopolysaccharide in scavenger receptor knockout mice. J Leukocyte Biol 210 (Suppl): 48, 1996.

36. Yanai H, Chiba H, Morimoto M, Abe K, Fujiwara H, Fuda H, Hui SP, Takahashi Y, Akita H, Jamieson GA, Kobayashi K, Matsuno K: Human CD36

deficiency is associated with elevation in low-density lipoprotein-cholesterol. Am J of Med Gen 93: 299–304, 2000.

37. Febbraio M, Hajjar DP, Silverstein RL: CD36: a class B scavenger receptor involved in angiogenesis, atherosclerosis, inflammation, and lipid metabolism. J Clin Invest 108: 785–791, 2001.

38. Febbraio M, Podrez EA, Smith JD, Hajjar DP, Hazen SL, Hoff HF, Sharma K, Silverstein RL: Targeted disruption of the class B scavenger receptor CD36 protects against atherosclerotic lesion development in mice. J Clin Invest 105: 1049–1056, 2000.

39. Febbraio M, Abumrad NA, Hajjar DP, Sharma K, Cheng W, Pearce SF, Silverstein RL: A null mutation in murine CD36 reveals an important role in fatty acid and lipoprotein metabolism. J Biol Chem 274: 19055–19062, 1999.

40. Watt MJ, Southgate RJ, Holmes AG, Febbraio MA: Suppression of plasma free fatty acids upregulates peroxisome proliferator-activated receptor (PPAR) alpha and delta and PPAR coactivator 1 alpha in human skeletal muscle, but not lipid regulatory genes. J Mol Endocrinol 33: 533–544, 2004.

41. Malaud E, Hourton D, Giroux LM, Ninio E, Buckland R, McGregor JL: The terminal six amino-acids of the carboxy cytoplasmic tail of CD36 contain a functional domain implicated in the binding and capture of oxidized low-density lipoprotein. Biochem J 364: 507–515, 2002.

42. Puente Navazo MD, Daviet L, Ninio E, McGregor JL: Identification on human CD36 of a domain (155–183) implicated in binding oxidized low-density lipoproteins (ox-LDL). Arterioscler Thromb Vasc Biol 16: 1033–1039, 1996.

43. Robbesyn F, Salvayre R, Negre-Salvayre A: Dual role of oxidized LDL on the NF-kappaB signaling pathway. Free Radic Res 38: 541–551, 2004.

44. Lipsky RH, Eckert DM, Tang Y, Ockenhouse CF: The carboxyl-terminal cytoplasmic domain of CD36 is required for oxidized low-density lipoprotein modulation of NF-kappaB activity by tumor necrosis factor-alpha. Recept Signal Transduct 7: 1–11, 1997.

45. Janabi M, Yamashita S, Hirano K, Sakai N, Hiraoka H, Matsumoto K, Zhang Z, Nozaki S, Matsuzawa Y: Oxidized LDL-induced NF-kappa B activation and subsequent expression of proinflammatory genes are defective in monocyte-derived macrophages from CD36-deficient patients. Arterioscler Thromb Vasc Biol 20: 1953–1960, 2000.

46. Bull HA, Brickell PM, Dowd PM: Src-related protein tyrosine kinases are physically associated with the surface antigen CD36 in human dermal microvascular endothelial cells. FEBS Lett 351: 41–44, 1994.

47. Moore KJ, El Khoury J, Medeiros LA, Terada K, Geula C, Luster AD, Freeman MW: A CD36-initiated signaling cascade mediates inflammatory effects of beta-amyloid. J Biol Chem 277: 47373–47379, 2002.

48. Lee H, Woodman SE, Engelman JA, Volonte D, Galbiati F, Kaufman HL, Lublin DM, Lisanti MP: Palmitoylation of caveolin-1 at a single site (Cys-156) controls its coupling to the c-Src tyrosine kinase: targeting of dually acylated molecules (GPI-linked, transmembrane, or cytoplasmic) to caveolae effectively uncouples c-Src and caveolin-1 (TYR-14). J Biol Chem 276: 35150–35158, 2001.

49. Hoshino K, Takeuchi O, Kawai T, Sanjo H, Ogawa T, Takeda Y, Takeda K, Akira S: Cutting edge: Toll-like receptor 4 (TLR4)-deficient mice are hyporesponsive to lipopolysaccharide: evidence for TLR4 as the Lps gene product. J Immunol 162: 3749–3752, 1999.

50. Medzhitov R, Preston-Hurlburt P, Janeway CA Jr: A human homologue of the Drosophila Toll protein signals activation of adaptive immunity. Nature 388: 394–397, 1997.

51. Qureshi ST, Lariviere L, Leveque G, Clermont S, Moore KJ, Gros P, Malo D: Endotoxin-tolerant mice have mutations in Toll-like receptor 4 (TLR4). J Exp Med 189: 615–625, 1999.

52. Pfeiffer A, Bottcher A, Orso E, Kapinsky M, Nagy P, Bodnar A, Spreitzer I, Liebisch G, Drobnik W, Gempel K, Horn M, Holmer S, Hartung T, Multhoff G, Schutz G, Schindler H, Ulmer AJ, Heine H, Stelter F, Schutt C, Rothe G, Szollosi J, Damjanovich S, Schmitz G: Lipopolysaccharide and ceramide docking to CD14 provokes ligand-specific receptor clustering in rafts. Eur J Immunol 31: 3153–3164, 2001.

53. Finnemann SC, Silverstein RL: Differential roles of CD36 and alphavbeta5 integrin in photoreceptor phagocytosis by the retinal pigment epithelium. J Exp Med 194: 1289–1298, 2001.

54. Kindzelskii AL, Elner VM, Elner SG, Yang D, Hughes BA, Petty HR: Toll-like receptor 4 (TLR4) of retinal pigment epithelial cells participates in transmembrane signaling in response to photoreceptor outer segments. J Gen Physiol 124: 139–149, 2004.

55. Huang WC, Chen JJ, Chen CC: c-Src-dependent tyrosine phosphorylation of IKKbeta is involved in TNF-alpha-induced intercellular adhesion molecule-1 expression. J Biol Chem 278(11): 9944–9952, 2003.

56. Liu R, Aupperle K, Terkeltaub R: Src family protein tyrosine kinase signaling mediates monosodium urate crystal-induced IL-8 expression by monocytic THP-1 cells. J Leukoc Biol 70: 961–968, 2001.

57. Hoebe K, Georgel P, Rutschmann S, Du X, Mudd S, Crozat K, Sovath S, Shamel L, Hartung T, Zahringer U, Beutler B: CD36 is a sensor of diacylglycerides. Nature 433: 523–527, 2005.

58. Williams DL, Connelly MA, Temel RE, Swarnakar S, Phillips MC, de la Llera-Moya M, Rothblat GH: Scavenger receptor BI and cholesterol trafficking. Curr Opin Lipidol 10: 329–339, 1999.

59. Acton S, Rigotti A, Landschulz KT, Xu S, Hobbs HH, Krieger M: Identification of scavenger receptor SR-BI as a high density lipoprotein receptor. Science 271: 518–520, 1996.

60. Rigotti A, Trigatti BL, Penman M, Rayburn H, Herz J, Krieger M: A targeted mutation in the murine gene encoding the high density lipoprotein (HDL) receptor scavenger receptor class B type I reveals its key role in HDL metabolism. Proc Natl Acad Sci USA 94: 12610–12615, 1997.

61. Trigatti B, Rayburn H, Vinals M, Braun A, Miettinen H, Penman M, Hertz M, Schrenzel M, Amigo L, Rigotti A, Krieger M: Influence of the high density lipoprotein receptor SR-BI on reproductive and cardiovascular pathophysiology. Proc Natl Acad Sci USA 96: 9322–9327, 1999.

62. Varban ML, Rinninger F, Wang N, Fairchild-Huntress V, Dunmore JH, Fang Q, Gosselin ML, Dixon KL, Deeds JD, Acton SL, Tall AR, Huszar D: Targeted mutation reveals a central role for SR-BI in hepatic selective uptake of high density lipoprotein cholesterol. Proc Natl Acad Sci USA 95: 4619–4624, 1998.

63. Jian B, de la Llera-Moya M, Ji Y, Wang N, Phillips MC, Swaney JB, Tall AR, Rothblat GH: Scavenger receptor class B type I as a mediator of cellular cholesterol

efflux to lipoproteins and phospholipid acceptors. J Biol Chem 273: 5599–5606, 1998.

64. Chen M, Yang Y, Braunstein E, Georgeson KE, Harmon CM: Gut expression and regulation of FAT/CD36: possible role in fatty acid transport in rat enterocytes. Am J Physiol Endocrinol Metab 281: E916–E923, 2001.

65. Covey SD, Krieger M, Wang W, Penman M, Trigatti BL: Scavenger receptor class B type I-mediated protection against atherosclerosis in LDL receptor-negative mice involves its expression in bone marrow-derived cells. Arterioscler Thromb Vasc Biol 23: 1589–1594, 2003.

66. Zhang W, Yancey PG, Su YR, Babaev VR, Zhang Y, Fazio S, Linton MF: Inactivation of macrophage scavenger receptor class B type I promotes atherosclerotic lesion development in apolipoprotein E-deficient mice. Circulation 108: 2258–2263, 2003.

67. Kozarsky KF, Donahee MH, Glick JM, Krieger M, Rader DJ: Gene transfer and hepatic overexpression of the HDL receptor SR-BI reduces atherosclerosis in the cholesterol-fed LDL receptor-deficient mouse. Arterioscler Thromb Vasc Biol 20: 721–727, 2000.

68. Trigatti BL, Krieger M, Rigotti A: Influence of the HDL receptor SR-BI on lipoprotein metabolism and atherosclerosis. Arterioscler Thromb Vasc Biol 23: 1732–1738, 2003.

69. Vishnyakova TG, Bocharov AV, Baranova IN, Chen Z, Remaley AT, Csako G, Eggerman TL, Patterson AP: Binding and internalization of lipopolysaccharide by Cla-1, a human orthologue of rodent scavenger receptor B1. J Biol Chem 278: 22771–22780, 2003.

70. Baranova IN, Vishnyakova TG, Bocharov AV, Kurlander R, Chen Z, Kimelman ML, Remaley AT, Csako G, Thomas F, Eggerman TL, Patterson AP: Serum amyloid A binding to CLA-1 (CD36 and LIMPII analogous-1) mediates serum amyloid A protein-induced activation of ERK1/2 and p38 mitogen-activated protein kinases. J Biol Chem 280: 8031–8040, 2005.

71. Boisvert WA, Curtiss LK, Terkeltaub RA: Interleukin-8 and its receptor CXCR2 in atherosclerosis. Immunol Res 21: 129–137, 2000.

72. Grewal T, de Diego I, Kirchhoff MF, Tebar F, Heeren J, Rinninger F, Enrich C: High density lipoprotein-induced signaling of the MAPK pathway involves scavenger receptor type BI-mediated activation of Ras. J Biol Chem 278: 16478–16481, 2003.

73. Lewis KE, Kirk EA, McDonald TO, Wang S, Wight TN, O'Brien KD, Chait A: Increase in serum amyloid A evoked by dietary cholesterol is associated with increased atherosclerosis in mice. Circulation 110: 540–545, 2004.

74. Van Eck M, Bos IS, Hildebrand RB, Van Rij BT, Van Berkel TJ: Dual role for scavenger receptor class B, type I on bone marrow-derived cells in atherosclerotic lesion development. Am J Pathol 165: 785–794, 2004.

75. Cayatte AJ, Palacino JJ, Horten K, Cohen RA: Chronic inhibition of nitric oxide production accelerates neointima formation and impairs endothelial function in hypercholesterolemic rabbits. Arterioscler Thromb 14: 753–759, 1994.

76. Li XA, Titlow WB, Jackson BA, Giltiay N, Nikolova-Karakashian M, Uittenbogaard A, Smart EJ: High density lipoprotein binding to scavenger receptor, class B, type I activates endothelial nitric-oxide synthase in a ceramide-dependent manner. J Biol Chem 277: 11058–11063, 2002.

77. Yuhanna IS, Zhu Y, Cox BE, Hahner LD, Osborne-Lawrence S, Lu P, Marcel YL, Anderson RG, Mendelsohn ME, Hobbs HH, Shaul PW: High-density lipoprotein binding to scavenger receptor-BI activates endothelial nitric oxide synthase. Nat Med 7: 853–857, 2001.

78. Li XA, Guo L, Dressman JL, Asmis R, Smart EJ: A novel ligand-independent apoptotic pathway induced by scavenger receptor class B, type I and suppressed by endothelial nitric-oxide synthase and high density lipoprotein. J Biol Chem 280: 19087–19096, 2005.

79. Van Eck M, Twisk J, Hoekstra M, Van Rij BT, Van der Lans CA, Bos IS, Kruijt JK, Kuipers F, Van Berkel TJ: Differential effects of scavenger receptor BI deficiency on lipid metabolism in cells of the arterial wall and in the liver. J Biol Chem 278: 23699–23705, 2003.

80. Chen M, Kakutani M, Minami M, Kataoka H, Kume N, Narumiya S, Kita T, Masaki T, Sawamura T: Increased expression of lectin-like oxidized low density lipoprotein receptor-1 in initial atherosclerotic lesions of Watanabe heritable hyperlipidemic rabbits. Arterioscler Thromb Vasc Biol 20: 1107–1115, 2000.

81. Li D, Mehta JL: Antisense to LOX-1 inhibits oxidized LDL-mediated upregulation of monocyte chemoattractant protein-1 and monocyte adhesion to human coronary artery endothelial cells. Circulation 101: 2889–2895, 2000.

82. Hayashida K, Kume N, Minami M, Kita T: Lectin-like oxidized LDL receptor-1 (LOX-1) supports adhesion of mononuclear leukocytes and a monocyte-like cell line THP-1 cells under static and flow conditions. FEBS Lett 511: 133–138, 2002.

83. Honjo M, Nakamura K, Yamashiro K, Kiryu J, Tanihara H, McEvoy LM, Honda Y, Butcher EC, Masaki T, Sawamura T: Lectin-like oxidized LDL receptor-1 is a cell-adhesion molecule involved in endotoxin-induced inflammation. Proc Natl Acad Sci USA 100: 1274–1279, 2003.

84. Li D, Williams V, Liu L, Chen H, Sawamura T, Antakli T, Mehta JL: LOX-1 inhibition in myocardial ischemia-reperfusion injury: modulation of MMP-1 and inflammation. Am J Physiol Heart Circ Physiol 283: H1795–H1801, 2002.

85. Nakagawa T, Akagi M, Hoshikawa H, Chen M, Yasuda T, Mukai S, Ohsawa K, Masaki T, Nakamura T, Sawamura T: Lectin-like oxidized low-density lipoprotein receptor 1 mediates leukocyte infiltration and articular cartilage destruction in rat zymosan-induced arthritis. Arthritis Rheum 46: 2486–2494, 2002.

86. Li D, Liu L, Chen H, Sawamura T, Ranganathan S, Mehta JL: LOX-1 mediates oxidized low-density lipoprotein-induced expression of matrix metalloproteinases in human coronary artery endothelial cells. Circulation 107: 612–617, 2003.

87. Cominacini L, Pasini AF, Garbin U, Davoli A, Tosetti ML, Campagnola M, Rigoni A, Pastorino AM, Lo Cascio V, Sawamura T: Oxidized low density lipoprotein (ox-LDL) binding to ox-LDL receptor-1 in endothelial cells induces the activation of NF-kappaB through an increased production of intracellular reactive oxygen species. J Biol Chem 275: 12633–12638, 2000.

88. Chan EL, Murphy JT: Reactive oxygen species mediate endotoxin-induced human dermal endothelial NF-kappaB activation. J Surg Res 111: 120–126, 2003.

89. Robbesyn F, Garcia V, Auge N, Vieira O, Frisach MF, Salvayre R, Negre-Salvayre A: HDL counterbalance the proinflammatory effect of oxidized LDL by inhibiting intracellular reactive oxygen species rise, proteasome activation, and subsequent NF-kappaB activation in smooth muscle cells. FASEB J 17: 743–745, 2003.

90. Schreck R, Meier B, Mannel DN, Droge W, Baeuerle PA: Dithiocarbamates as potent inhibitors of nuclear factor kappa B activation in intact cells. J Exp Med 175: 1181–1194, 1992.

91. Schreck R, Rieber P, Baeuerle PA: Reactive oxygen intermediates as apparently widely used messengers in the activation of the NF-kappa B transcription factor and HIV-1. EMBO J 10: 2247–2258, 1991.

92. Weber C, Erl W, Pietsch A, Strobel M, Ziegler-Heitbrock HW, Weber PC: Antioxidants inhibit monocyte adhesion by suppressing nuclear factor-kappa B mobilization and induction of vascular cell adhesion molecule-1 in endothelial cells stimulated to generate radicals. Arterioscler Thromb 14: 1665–1673, 1994.

93. Yoshizumi T, Nozaki S, Fukuchi K, Yamasaki K, Fukuchi T, Maruyama T, Tomiyama Y, Yamashita S, Nishimura T, Matsuzawa Y: Pharmacokinetics and metabolism of 123I-BMIPP fatty acid analog in healthy and CD36-deficient subjects. J Nucl Med 41: 1134–1138, 2000.

94. Cominacini L, Rigoni A, Pasini AF, Garbin U, Davoli A, Campagnola M, Pastorino AM, Lo Cascio V, Sawamura T: The binding of oxidized low density lipoprotein (ox-LDL) to ox-LDL receptor-1 reduces the intracellular concentration of nitric oxide in endothelial cells through an increased production of superoxide. J Biol Chem 276: 13750–13755, 2001.

95. Aoyama T, Fujiwara H, Masaki T, Sawamura T: Induction of lectin-like oxidized LDL receptor by oxidized LDL and lysophosphatidylcholine in cultured endothelial cells. J Mol Cell Cardiol 31: 2101–2114, 1999.

96. Mehta JL, Li DY: Identification and autoregulation of receptor for ox-LDL in cultured human coronary artery endothelial cells. Biochem Biophys Res Commun 248: 511–514, 1998.

97. Iwai-Kanai E, Hasegawa K, Sawamura T, Fujita M, Yanazume T, Toyokuni S, Adachi S, Kihara Y, Sasayama S: Activation of lectin-like oxidized low-density lipoprotein receptor-1 induces apoptosis in cultured neonatal rat cardiac myocytes. Circulation 104: 2948–2954, 2001.

98. Nakagawa T, Yasuda T, Hoshikawa H, Shimizu M, Kakinuma T, Chen M, Masaki T, Nakamura T, Sawamura T: LOX-1 expressed in cultured rat chondrocytes mediates oxidized LDL-induced cell death-possible role of dephosphorylation of Akt. Biochem Biophys Res Commun 299: 91–97, 2002.

99. Li D, Mehta JL: Upregulation of endothelial receptor for oxidized LDL (LOX-1) by oxidized LDL and implications in apoptosis of human coronary artery endothelial cells: evidence from use of antisense LOX-1 mRNA and chemical inhibitors. Arterioscler Thromb Vasc Biol 20: 1116–1122, 2000.

100. Kutuk O, Basaga H: Inflammation meets oxidation: NF-kappaB as a mediator of initial lesion development in atherosclerosis. Trends Mol Med 9: 549–557, 2003.

101. Michelsen KS, Doherty TM, Shah PK, Arditi M: TLR signaling: an emerging bridge from innate immunity to atherogenesis. J Immunol 173: 5901–5907, 2004.

6

ABC Transporters and Apolipoprotein E: Critical Players in Macrophage Cholesterol Efflux and Atherosclerosis

KATHRYN E. NAUS AND CHERYL L. WELLINGTON

Abstract

Macrophages are key mediators of atherosclerosis. Circulating monocytes adhere to activated endothelial cells at lesion-prone sites within large arteries, and migrate into the subendothelial space where they differentiate into macrophages. Within the arterial intima, macrophages mobilize the uptake of oxidized lipoproteins through scavenger receptors, and also ingest cell debris found within the forming fatty streak. Lipids derived from internalized lipoproteins or the membranes of phagocytosed cells are transferred from the endosome to the lysosome where acid lipase hydrolyses the internalized cholesterol esters (CE). The free cholesterol (FC) thus generated is released into the cytoplasm and is rapidly delivered to the plasma membrane. From there, cholesterol is either effluxed from the cell, or it is re-esterified by acyl-coenzyme A:cholesterol acyltransferase (ACAT) and stored as CE lipid droplets that give macrophages their characteristic foamy appearance in the fatty streak. The appearance of foam cells is a visual cue that the ability of the macrophage to remove excess lipid and maintain cholesterol homeostasis has been exceeded. Furthermore, foam cells play a major role in the evolution of fatty streaks into more complex atherosclerotic lesions. Therefore, improving the ability of macrophages to promote cholesterol efflux may lead to substantial therapeutic promise for the prevention and treatment of atherosclerosis.

This review will focus on the pathways by which macrophages can efflux lipids, with a particular emphasis on the role of ABC transporters and apolipoprotein E (apoE) in cholesterol efflux. We will begin with an overview of lipid metabolism and atherosclerosis, followed by a brief summary of the major pathways that macrophages use to maintain cholesterol homeostasis. We will then focus our attention on recent studies of two of the ABC transporters that play critical roles in macrophage cholesterol efflux, ABCA1 and ABCG1. Finally, we will attempt to integrate these observations and end with some new questions to ponder.

Keywords: ABCA1; ABCG1; apoAI; apoE; atherosclerosis; cholesterol; efflux; HDL; macrophage

Abbreviations: CE, cholesterol ester; FC, free cholesterol; ACAT, acyl-coenzyme A:cholesterol acyltransferase; LDL, low-density lipoprotein; HDL, high-density lipoprotein; RCT, reverse cholesterol transport; LCAT, lecithin:cholesterol acyltransferase; CETP, cholesterol ester transfer protein; EL, endothelial lipase; SRBI, scavenger receptor BI; PPAR, peroxisome proliferator activated receptor; RXR, retinoid X receptor; LXR, liver X receptor; GGPP, geranylgeranyl pyrophosphate; PKC, protein kinase C; PKA, protein kinase A; SCD, stearoyl CoA-desaturase; BMT, bone marrow transplantation; LTR, long terminal repeat; WT, wild-type

Lipid Metabolism and Atherosclerosis

Nearly half of the deaths of persons in the Western world are caused by atherosclerotic processes, the most common of which manifest as coronary artery disease and stroke [1–4]. Aberrant regulation of lipid and lipoprotein homeostasis can lead to atherosclerosis [5–19]. Well-established risk factors include elevated plasma low-density lipoprotein (LDL) and depressed high-density lipoprotein (HDL) levels [4, 15, 20–27]. Conversely, high levels of HDL protect from atherosclerosis even in the presence of high LDL levels [2, 26, 27]. Statins, drugs that lower LDL levels by inhibiting the rate-limiting step in cholesterol biosynthesis, are widely used to control LDL levels and to reduce the risk of cardiovascular disease [19]. However, the clinical event rate in patients on statins is approximately 70% of the rate in subjects treated with a placebo, suggesting that additional therapeutic strategies are still needed to reduce the burden of cardiovascular disease [19]. Therapeutic approaches that increase the levels of HDL, augment its antiatherogenic properties, and promote reverse cholesterol transport (RCT) are considered particularly desirable [19, 21].

RCT is the process by which excess cholesterol is extracted from peripheral cells and transported to the liver in the form of HDL particles for elimination as bile acids [21, 28, 29]. HDL itself exhibits several antiatherogenic properties including stimulating nitric oxide production, inhibiting endothelial cell adhesion molecule expression, inhibiting the oxidation of LDL, preventing endothelial cell apoptosis, and inhibiting platelet aggregation [19, 21, 27]. HDL is the smallest and densest circulating lipoprotein whose major function is to transfer excess lipids from extrahepatic tissues to the liver. HDL is a heterogenous lipoprotein consisting mainly of phospholipid, cholesterol, and protein. Biogenesis of HDL occurs predominately in the liver and intestine, and requires the action of the ATP-binding cassette transporter ABCA1 to transfer cholesterol and phospholipids onto lipid-free apoAI to form discoidal pre-β-HDL, which then matures into α-migrating spherical HDL particles upon acquisition of additional lipids and esterification of cholesterol into cholesterol esters (CE) by lecithin:cholesterol acyltransferase (LCAT) [4, 26].

Currently used agents that can raise HDL levels include nicotinic acid, fibrates, statins, thiazolidinediones, estrogen, and ethanol, but these typically result in relatively modest increases of HDL [19]. Strategies that target specific genes involved in HDL metabolism, including apoAI, cholesterol ester transfer protein (CETP), and endothelial lipase (EL), show promise as potential therapies. Weekly intravenous injections of recombinant apoAI-Milano, a single base-pair variant of apoAI, resulted in significant regression of coronary atherosclerosis over a 6-week treatment period [30]. Additionally, inhibition of CETP-mediated movement of cholesterol from HDL to LDL in rabbits increases plasma HDL and reduces atherosclerosis, and CETP inhibition also increases HDL levels in humans [31–33]. Recently, overexpression of EL in mice was shown to reduce HDL levels by approximately 19%, whereas inactivation of EL increases HDL levels by approximately 50% [34–36].

Although absolute levels of circulating HDL are important determinants of atherosclerotic risk, the ability of the macrophage to efflux accumulated cholesterol within the arterial wall plays a crucial role in the development and progression of atherosclerotic lesions. Strategies to augment the process of macrophage cholesterol efflux, and thus retard foam cell formation and the initiation of inflammatory pathways that exacerbate lesion progression, may also yield novel therapeutic agents that slow the development of advanced atherosclerotic lesions and reduce the incidence of plaque rupture and infarction.

Pathways for Cholesterol Efflux from Macrophages

Although mammalian cells can synthesize cholesterol, they do not have the capacity to degrade the sterol ring. In order to remove excess cholesterol, macrophages and other cell types must export cholesterol to extracellular acceptors that transport excess cholesterol to the liver for eventual excretion in bile. It is generally accepted that there are several pathways by which macrophages are thought to efflux excess cholesterol (reviewed in Refs. [37, 38]). For the purposes of this review, we will classify these into two major subdivisions: passive/facilitated aqueous diffusion and transporter-mediated active efflux.

Passive/Facilitated Aqueous Diffusion

Passive diffusion involves the desorption of membrane cholesterol down a concentration gradient into the aqueous environment followed by absorption onto a phospholipid-containing acceptor [39, 40]. Cholesterol efflux by passive diffusion does not consume metabolic energy and is independent of the cellular growth state and cholesterol content. Rather, the rate of cholesterol transfer is influenced both by the structure and lipid composition of the plasma membrane on the donor cell and properties of the acceptor molecule including concentration, size, and lipid composition. Because the aqueous layer between donor and acceptor represents a significant diffusion barrier, cholesterol efflux by passive diffusion is inefficient and slow.

However, passive diffusion can be facilitated by expression of the scavenger receptor BI (SRBI), which is the first membrane protein shown to function in the bidirectional flux of free cholesterol (FC). SRBI is expressed in many tissues including macrophages, and the highest levels of SRBI are found in the liver and steroidogenic tissues such as adrenals, testis, and ovary. A major function of SRBI is to promote the selective uptake of esterified cholesterol from HDL, a unidirectional process that occurs without endocytosis of the entire HDL particle [41–43]. Expression of SRBI also stimulates the bidirectional flux of FC, with the net direction of transfer dictated by the concentration gradient between donor and acceptor [44, 45]. In a number of cell types, the rate of FC efflux to acceptors correlates well with expression levels of SRBI [46]. Although many acceptors bind to SRBI, including HDL, LDL, oxidized and acetylated LDL, small unilamellar vesicles, and lipid-free apoAI [47, 48], only those acceptors that contain phospholipids will accept cholesterol via SRBI. Furthermore, both phospholipid content and species influence SRBI-mediated efflux. For example, it is well established that enrichment of HDL or serum with phosphotidylcholine increases SRBI-mediated efflux, whereas depletion of phosphotidylcholine decreases SRBI-mediated efflux [45, 49]. Enrichment of HDL with sphingomyelin also results in a net increase in SRBI–cholesterol efflux, but this occurs by inhibiting SRBI-mediated FC influx rather than by enhancing FC efflux [45].

SRBI is thought to facilitate passive diffusion by reorganizing membrane microdomains into cholesterol-rich lipid rafts that supply the majority of cholesterol during passive diffusion [48, 50–52]. This mechanism may be particularly important at high HDL concentrations when SRBI receptors are saturated, and increased efflux is independent of HDL binding. However, under conditions of low HDL concentrations, binding of HDL to SRBI influences cholesterol efflux. In this case, it appears that SRBI and the acceptor molecule must be precisely aligned into a productive complex in order to promote efficient FC efflux [53]. For example, SRBI-mediated FC efflux to HDL was found to be severely decreased in the presence of two apoAI mutations even though binding of the mutant HDL particle to SRBI was not affected. Intriguingly, efficient FC efflux to HDL containing the mutated apoAI was rescued in the presence of mutations within SRBI, presumably by restoring the alignment between donor and acceptor [53].

Transporter-Mediated Active Efflux

Several members of the ATP-binding cassette transporter superfamily are key regulators of cholesterol efflux, RCT, and HDL metabolism. There are 48 known human ABC transporters that are grouped into seven classes [54]. Members of the ABCA and ABCG classes are believed to form a functional network of lipid transporters that are critical gatekeepers of cell and sterol homeostasis [54–56]. Two of these transporters, ABCA1 and ABCG1, are highly induced in cholesterol-loaded macrophages and mediate lipid efflux to apoAI and HDL, respectively. These observations suggest a prominent role for ABCA1 and ABCG1 in the coordinated regulation of macrophage lipid homeostasis.

ABCA1

ABCA1 is the prototype of the full-sized ABCA class of transporters and contains two membrane spanning regions interspersed with two ATP-binding domains [55, 56]. Deficiency of ABCA1 results in Tangier disease, which is characterized by extremely low plasma HDL levels, tissue deposition of CE, and an increased risk of atherosclerosis [57–60]. ABCA1 catalyses the ATP-dependent transport of cholesterol and phospholipids from the plasma membrane to lipid-free apoAI to form immature, pre-β-HDL, which is considered to be a key step in RCT [57, 59, 61, 62]. Mature, α-HDL is formed by further accumulation of lipids on pre-β-HDL followed by esterification of cholesterol by LCAT [29, 63–65]. Cholesterol from α-HDL is taken up by the liver through a number of pathways including SRBI-mediated selective uptake, holoprotein uptake by apoE receptors, or indirect pathways involving CETP, hepatic lipase, and EL. Within hepatocytes, cholesterol is converted to bile acids and excreted from the body in bile [41, 42, 66]. Because ABCA1 has a singular role in the biogenesis of HDL and the initiation of RCT, it is therefore an attractive candidate molecule on which to base atheroprotective therapeutic strategies. Additional support for ABCA1 as a potential therapeutic target has come from studies demonstrating that overexpression of ABCA1 in mice increases plasma HDL levels [67, 68] and dramatically reduces the progression of atherosclerotic lesions on an apoE-deficient background [69].

Transcription of ABCA1 is subject to several layers of control (recently reviewed in Refs. [70–74]). In many cell types including macrophages, the most effective direct transcriptional modulators of ABCA1 include members of the retinoid X receptor (RXR) and liver X receptor (LXR) nuclear orphan receptors, which act as obligate heterodimers upon binding their respective agonists [75–78]. Ligands for RXRs include 9-*cis*-retinoic acid, and ligands for LXRs are oxysterols such as 22-hydroxycholesterol, 20-hydroxycholesterol, 24-hydroxycholesterol, 24,25-epoxycholesterol, and 27-hydroxycholesterol. Although LXR and RXR agonists can each induce ABCA1 expression independently, the most potent increase in ABCA1 mRNA levels occurs when LXR–RXR agonists act synergistically. It is generally believed that the induction of ABCA1 expression in lipid-loaded macrophages occurs primarily through the LXR–RXR pathway, which is activated upon conversion of accumulated sterols to endogenous oxysterols such as 27-hydroxycholesterol. In J774 and primary murine macrophages, ABCA1 mRNA is also induced by cAMP treatment [79]. Peroxisome proliferator-activated receptors (PPARs) are nuclear receptors that are activated by fatty acid derivatives and heterodimerize with RXRs to regulate lipid and glucose metabolism [80]. PPARs (PPARα, PPARBβ/δ, and PPARγ) have been shown to upregulate ABCA1 as well as ABCG1 expression in macrophages [81, 82]. Although this has been thought to employ an indirect mechanism via induction of LXRα expression by PPAR–RXR heterodimers [81], recent studies have shown that ABCA1 expression and apoAI-dependent cholesterol efflux can be induced by the selective PPARδ agonist GW50516 in an LXRα-independent manner in THP-1 cells, intestinal cells, and fibroblasts [83]. Furthermore, conditional inactivation of PPARγ impaired expression of ABCA1, ABCG1, and apoE in macrophages and led to reduced cholesterol efflux [84]. Recently, PPARα and PPARγ agonists were found to strongly inhibit atherosclerosis in the LDLR–/– model, which was mediated through ABCG1 rather than ABCA1 expression [85]. These observations further support a role for PPARs in macrophage lipid homeostasis and regulation of ABC transporter expression.

ABCA1 mRNA synthesis can be repressed by geranylgeranyl pyrophosphate (GGPP), an intermediate in the mevalonate pathway that reduces the transactivation potential of LXRs [86]. In addition, lipopolysaccharide has recently been shown to downregulate ABCA1 expression in an NFκB-dependent manner [87, 88], and the proinflammatory cytokines IL-1, INFγ, and TNFα each suppress ABCA1 mRNA levels in macrophages [89–91]. These observations suggest that the ability of inflammatory mediators to promote atherosclerosis lesion formation may in part be mediated by suppressing ABCA1 expression and thereby reducing cholesterol efflux. Thyroid hormone (T3) can also inhibit ABCA1 transcription, which occurs when thyroid hormone–RXR complexes compete for LXR–RXR heterodimers at the ABCA1 promoter [92].

ABCA1 protein levels are also subject to regulation through posttranscriptional mechanisms. ABCA1 contains a PEST sequence from amino acids 1283–1306 that modulates protein degradation by calpain [93]. Residues T1286 and T1305 within the PEST sequence are normally constitutively phosphorylated, which results in a rapid turnover (half-life of 1–2 h) of ABCA1 in the absence of apoAI [93, 94]. However, these residues become dephosphorylated in the presence of apoAI, which inhibits calpain-mediated degradation of ABCA1 [93, 94]. This may occur through activation of the protein kinase Cα (PKCα) signaling pathway by lipid-poor apoAI [95]. In contrast, protein kinase A (PKA) appears not to be involved in the stabilization of ABCA1 protein [94], but may promote cholesterol efflux by phosphorylating residues S1044 and S2054 on ABCA1 [96]. In cAMP-stimulated J774 macrophage cells, ABCA1 protein levels are also destabilized by unsaturated but not by saturated fatty acids [97]. In contrast, ABCA1 is destabilized by both saturated and unsaturated fatty acids in LXR–RXR-treated cells [98]. The differential effect of fatty acids on ABCA1 stability is thought to be due to the induction of stearoyl CoA-desaturase (SCD) specifically in LXR–RXR-stimulated cells, resulting in the formation of unsaturated fatty acids from their saturated precursors thereby triggering ABCA1 degradation [98]. ABCA1 is also degraded by the ubiquitin-mediated proteasomal pathway in macrophages loaded with FC [99]. This pathway may be particularly important during lesion progression when there is a progressive increase in the FC:CE ratio in lipid-laden macrophages. Additionally, because this pathway requires a functional Neimann–Pick type C gene [99], it may also contribute to impaired ABCA1-dependent lipid efflux and reduced HDL levels in NPC disease [100]. Recently, Albrecht et al. [101] showed that ABCA1 protein levels are dramatically reduced in advanced atherosclerotic lesions despite high levels of ABCA1 mRNA, suggesting that the microenvironment surrounding the atherosclerotic plaque triggers signaling pathways that regulate ABCA1 abundance. This study demonstrates that posttranscriptional regulation of ABCA1 occurs *in vivo*, and that a deeper understanding of these pathways may lead to an improved ability to retard foam cell formation and reduce the development of atherosclerotic lesions.

Lack of ABCA1 in humans and mice results in a nearly complete absence of plasma HDL, demonstrating that apoAI requires functional ABCA1 activity to acquire lipids and contribute to HDL biogenesis [57–59, 102–104]. In addition to apoAI, however, ABCA1 can also deliver lipids to other exchangeable lipoproteins including apoAIV, apoE, and apoC [79, 105], suggesting that ABCA1 can efflux lipids to proteins that contain amphipathic α-helices. The ability of synthetic amphipathic helices to accept lipids from ABCA1 further supports this hypothesis [106–108]. Exactly how ABCA1 transfers lipids to apoAI remains a subject of considerable investigation. One model proposes that contact between ABCA1 and helices 9/10 of apoAI induces the net transfer of phospholipids and cholesterol [109–112]. Recently, the interaction with apoAI has been suggested to occur with dimeric and tetrameric, but not with monomeric ABCA1 [113], and the formation of higher order ABCA1 structures could explain previous observations of dominant negative effects of truncation mutations on ABCA1 protein levels, which implies a physical interaction among ABCA1 molecules [114]. A second model proposes that ABCA1 translocates lipids to the exofacial leaflet of the plasma membrane independent of apoAI, generating a membrane microenvironment required for subsequent apoAI docking and lipid efflux [115–117]. Whether transfer of phospholipids and cholesterol onto apoAI occur simultaneously [118–121] or sequentially [122–125] remains to be fully elucidated.

ABCA1 is broadly expressed in many tissues, and liver and macrophages are key cell types involved in sterol homeostasis [126, 127]. Liver ABCA1 is the major regulator of plasma HDL levels, as selective inactivation of hepatic ABCA1 reduces plasma HDL by approximately 80% [128], whereas specific overexpression of ABCA1 in the liver increases plasma HDL levels [129, 130]. Within the liver, ABCA1 is most abundant in parenchymal and Kupffer cells compared to endothelial cells [131], and hepatocyte ABCA1 has been shown to play a critical role in the lipidation of newly synthesized apoAI [132]. In contrast to the liver, macrophage ABCA1 plays a minimal role in determination of plasma HDL-C levels but critically regulates atherosclerotic lesion development, as described below.

ABCG1

ABCG1 is the founding member of the ABCG subclass of ABC transporters [54]. ABCG1 is a half-size transporter with one ATPase-binding cassette [54–56, 133–136], and therefore requires either homodimerization or heterodimerization to provide functional transporter activity. Recent reports suggest that ABCG1 can function as a homodimer, and there are conflicting reports whether ABCG1 may also form functional heterodimers with ABCG4. In one study, HEK 293 cells transiently transfected with either ABCG1 or ABCG4 demonstrated that separately, each promoted cholesterol efflux to HDL but not to lipid-free apoAI [137]. However, coexpression of ABCG1 and ABCG4 did not potentiate lipid efflux compared to cells individually expressing either transporter, suggesting that ABCG1–ABCG4 heterodimers, if formed, may not display an enhanced ability to efflux lipid. In contrast, another report recently suggested that ATPase activity increased in insect cells expressing either ABCG1 or ABCG4, and that ATPase activity was augmented by coexpression of both genes [138]. This study concluded that ABCG1 and ABCG4 may function as homodimers or heterodimers, at least with respect to ATPase activity [138].

The human ABCG1 gene contains 23 known exons and spans approximately 98 kb [134–136, 139]. The murine ABCG1 gene contains regions of similarity to the human gene from exons 5–23. Thus far, human ABCG1 transcripts have been studied primarily in macrophages, where at least 11 different mRNAs have been described that result from complex patterns of alternative transcription initiation, alternative splicing, and translation initiation [134–136, 139]. At least seven different ABCG1 protein isoforms with amino termini comprised of various combinations of exons 1–10 spliced to exons 11–23 have been predicted in macrophages alone [134–136, 139]. The presence of multiple alternative amino termini for ABCG1 suggests that exons 1–10 may encode regulatory domains and that exons 11–23, which encode the nucleotide binding and transmembrane domains, may constitute the catalytic portion of ABCG1. Multiple ABCG1 transcripts are responsive to LXR–RXR stimulation [135, 136]. Two LXR responsive elements conserved between human and murine ABCG1 are located within intron 7 and have been shown to regulate transcription of ABCG1 [135]. In addition, an RXR response element located approximately 1.6 kb upstream of exon 1 has been postulated, but not proven, to regulate ABCG1 transcription [136]. Recently, the murine ABCG1 gene has also been shown to encode multiple transcripts [140].

ABCG1 mRNA is most abundant in brain and macrophage-rich tissues such as spleen and lung, as well as in placenta, intermediate in heart and muscle, and low in liver [140, 141]. Mice deficient in ABCG1 have recently been shown to accumulate lipids in tissues when challenged with a high fat diet [142], suggesting that ABCG1 does participate in the regulation of body sterol homeostasis.

ABCG1 facilitates cholesterol efflux to HDL2 and HDL3 but not to lipid-free apoA1 in transiently transfected HEK 293 cells [137, 140]. Furthermore, antisense inhibition of ABCG1 expression reduces cholesterol efflux in a dose-dependent manner [137, 143]. These observations link ABCG1 to pathways relevant for atherosclerosis and RCT. Macrophage ABCG1 mRNA is highly induced by lipid loading as well as by LXR–RXR agonists [137, 144], and efflux of cholesterol from lipid-laden macrophages suppresses ABCG1 expression [143]. ABCG1 is also upregulated in macrophages isolated from individuals with Tangier disease [145], suggesting the possibility of a compensatory mechanism to enhance macrophage ABCG1 function in the absence of ABCA1. Recently, ABCG1 in human monocyte-derived macrophages was shown to be extraordinarily sensitive to cholesterol loading or exposure to LXR–RXR agonists, and was more closely associated with cholesterol efflux than ABCA1 levels [146]. These observations suggest that ABCG1 may also mediate lipid homeostasis and lipid efflux in macrophages, although this remains to be determined experimentally.

Macrophage-Specific Influences on Atherosclerosis

Because macrophages are derived from hematopoietic stem cells, bone marrow transplantation (BMT) is a popular approach for the study of macrophage-specific effects on cholesterol efflux and atherosclerosis (for review see Ref. [147]). After irradiation, donation of marrow to a recipient mouse leads to the repopulation of the recipient animal with macrophages of donor origin. A second method to obtain macrophage-specific expression *in vivo* is with the visna virus long terminal repeat (LTR). A construct that contains the visna virus LTR will allow directed expression of a gene of interest to macrophages [148].

The two most commonly used animal models of atherosclerosis are apoE$^{-/-}$ and LDLR$-/-$ mice. Both of these models develop atherosclerotic lesions and have elevated plasma cholesterol; apoE$^{-/-}$ mice develop atherosclerosis spontaneously, whereas LDLR$-/-$ mice require feeding with an atherosclerotic or Western-type diet (for reviews see Refs. [149, 150]). Feeding a Western-type diet to apoE$^{-/-}$ mice is often used to induce a more severe atherosclerotic pathology at a younger age. In addition, gender differences are apparent in both models, with LDLR$-/-$ males having significantly greater lesion areas than females. Similar trends are observed in the apoE$^{-/-}$ model [151].

Macrophage apoE and Atherosclerosis

ApoE plays a crucial role in atherosclerosis [152, 153]. ApoE is synthesized predominantly by the liver, but is also made in several peripheral tissues including macrophages [152, 153]. Although a major function of apoE is to facilitate the hepatic uptake of triglyceride-rich lipoprotein particles, macrophage-derived apoE has potent antiatherogenic activities that appear to act independently of its effects on plasma lipoproteins. *In vitro* lipid loading of macrophages stimulates synthesis and secretion of phospholipid-bound apoE, which in turn facilitates cholesterol efflux [146, 154–159]. The local source of apoE not only supplies phospholipids as fatty acid donors for esterification of cholesterol, but transfer of apoE to HDL expands its surface and increases its capacity to accept cholesterol. Both human monocyte-derived macrophages [160, 161] and mouse peritoneal macrophages [159] are known to have the capacity to use endogenously secreted apoE as a cholesterol acceptor in the absence of other exogenous factors. Intriguingly, macrophage apoE-mediated efflux is dependent on culture volume; large culture volumes dilute the secreted apoE and result in low level efflux mediated only by apoE in the immediate juxtacellular space [159]. In contrast, at low culture volumes where apoE levels are saturating, cholesterol efflux is robust and is mediated by extracellular as well as juxtacellular apoE [159]. These observations suggest that endogenous apoE has both autocrine and paracrine effects that may become particularly important in the context of the vessel wall [159].

In addition to promoting local cholesterol efflux, macrophage-derived apoE has several other antiatherogenic properties including modulation of the inflammatory response within the vessel wall (reviewed in Refs. [162–164]). ApoE inhibits the expression of vascular cell adhesion molecule-1 on endothelial cells through its interaction with cell surface apoE receptor 2, which results in stimulation of nitric oxide synthase in endothelial cells [165, 166]. ApoE also inhibits lipid oxidation, T cell activation and proliferation, the migration and proliferation of smooth muscle cells, and platelet aggregation [167–174]. Thus, reductions in apoE levels or functional properties can affect lesion development through a number of mechanisms.

Many BMT and viral experiments done to determine the role of macrophage apoE in plasma lipid metabolism and atherosclerosis involved reconstitution of apoE$^{-/-}$ or wild-type (WT) mice with bone marrow from donors with varying levels of apoE (Table 6.1) (reviewed in Ref. [175]). Many groups have shown that reconstitution of WT bone marrow in hypercholesterolemic apoE$^{-/-}$ recipients resulted in a decrease in plasma total cholesterol, even when challenged with an atherogenic diet [176–182], demonstrating that apoE secreted by the WT macrophages can influence lipid metabolism. Furthermore, this decrease in cholesterol was dependent on the gene dosage, as apoE+/− donor marrow gave an intermediate level of plasma total cholesterol [178, 182]. Consistent with an antiatherogenic role of macrophage apoE, mean lesion area was reduced in mice that received WT compared to apoE$^{-/-}$ bone marrow [176, 177, 180–183]. Additionally, Zhu et al. [179] found that macrophage cholesterol efflux was increased in animals that received WT marrow. These antiatherogenic changes were obtained despite the fact that plasma apoE levels were in all cases <10% of the WT murine levels, indicating the

TABLE 6.1. Effect of macrophage apoE on atherosclerosis.

Group	Donor	Recipient	Plasma apoE (% of WT)*	Diet	mΦ Cholesterol efflux	Total cholesterol	Mean lesion area
Bellosta et al. [176]	$hTgE+/0$ $hTgE+/+$	$apoE^{-/-}$	$hTgE+/0$ (3–5%) $hTgE+/+$ (8%)	C	NM	$hTgE+/0$ ($\downarrow$ 31%) $hTgE+/+$ ($\downarrow$ 60%)	$hTgE+/0$ ($\downarrow$ 3×)
Boisvert et al. [177]	$apoE+/+$ $apoE^{-/-}$	$apoE^{-/-}$	NM	C	NM	$apoE+/+$ ($\downarrow$50%) $apoE^{-/-}$ (no change)	
				A/W		$apoE^{-/-}$ $\uparrow$ 4× above $apoE+/+$	$apoE+/+$ ($\downarrow$ not quantified)
Linton et al. [178]	$apoE+/+$ $apoE+/-$ $apoE^{-/-}$	$apoE^{-/-}$	NM	C	NM	$apoE+/+$ ($\downarrow$ 74%) $apoE+/-$ ($\downarrow$ 51%) $apoE^{-/-}$ (no change)	
				A/W		$apoE^{-/-}$ $\uparrow$ 3× above $apoE+/+$	$apoE+/+$ ($\downarrow$ 52×)
Fazio et al. [183]	$apoE+/+$ $apoE^{-/-}$	WT (♀)	No change	C A/W	NM	No change No change	$apoE+/+$ ($\downarrow$ 10×)
Van Eck et al. [182]	$apoE+/+$ $apoE+/-$ $apoE^{-/-}$	$apoE^{-/-}$	$apoE+/+$ (3.5%) $apoE+/-$ (1.9%)	C	NM	$apoE+/+$ ($\downarrow$ 91%) apoE+/- ($\downarrow$ 82%) $apoE^{-/-}$ ($\downarrow$ 13%)	
				A/W		$apoE^{-/-}$ $\uparrow$ 1.4× above $apoE+/+$	$apoE+/+$ ($\downarrow$ not quantified)
Zhu et al. [179]	$apoE+/+$ $hTgE+/0$ $apoE^{-/-}$	$apoE^{-/-}$ (?)	$hTgE+/0$ (7%)	C	$apoE+/+$ ($\uparrow$ 48–74%) $hTgE+/0$ ($\uparrow$ 37–49%)	$apoE+/+$ ($\downarrow$ 82%) $hTgE+/0$ (no change)	NM
Boisvert et al. [184]	$apoE+/+$ $apoE^{-/-}$	WT (♂)	No change	C A/W	NM	No change No change	$apoE+/+$ ($\uparrow$ 2×)
Van Eck et al. [185]	$apoE+/+$ $apoE^{-/-}$	WT (♀)	No change	C A/W	$apoE+/+$ ($\uparrow$ 42%)	No change $apoE^{-/-}$ FC $\uparrow$ 2× above $apoE+/+$	$apoE+/+$ ($\downarrow$ 4×)

| Van Eck et al. [181] | *apoE+/+*
h-apoE2
h-apoE3-
Leiden | *apoE⁻ᐟ⁻* | *h-apoE2* (14%)
h-apoE3-Leiden
(19%) of *apoE+/+* | C | No change | *apoE+/+* (↓ 87%)
h-apoE2, h-apoE3-Leiden
(transient ↓) | *apoE+/+* (↓ 23×)
h-apoE2, h-apoE3-
Leiden (no change) |

All comparisons made relative to baseline recipient measurements unless stated otherwise. NM, not measured; C, chow diet; A/W, atherosclerotic/Western-type diet; FC, free cholesterol; h-apoE, human apoE.

*% of WT murine apoE levels.

atherogenic importance of locally produced macrophage apoE even in the absence of plasma apoE [176, 179, 182].

Conversely, transplantation of apoE$^{-/-}$ bone marrow into atherosusceptible C57B/6 WT recipients creates a model with deletion of apoE solely in macrophages. These mice display no appreciable change in plasma cholesterol levels [183–185], indicating that the selective deficiency of macrophage apoE is not sufficient to affect plasma lipid metabolism when apoE is synthesized normally elsewhere. However, following several weeks on an atherogenic diet, lesion area was increased in WT mice that received apoE$^{-/-}$ bone marrow. Since these differences in lesion areas are seen without changes in plasma cholesterol levels, the protective effects of apoE clearly extend beyond its ability to lower cholesterol [179].

Taken together, these studies strongly support an antiatherogenic role for macrophage apoE. This conclusion is not without a caveat, however, as locally secreted apoE would be predicted to encourage cholesterol influx into macrophages via lipoprotein uptake by apoE receptors [186], and Boisvert et al. [184] found that WT mice reconstituted with apoE-producing macrophages had increased lesions compared to those with apoE$^{-/-}$ macrophages. This unexpected result may be due to differences in gender, diet, or feeding schedules compared to other similarly designed BMT experiments. However, the majority of BMT experiments strongly support an atheroprotective role for macrophage apoE, wherein even low levels of this apolipoprotein are capable of altering the atherosclerotic fate of the animals.

Macrophage ABCA1 and Atherosclerosis

Using BMT experiments, much insight has been gained about the importance of ABCA1 in the macrophage (Table 6.2). Like humans, mice with a complete disruption of ABCA1 have decreased levels of plasma total cholesterol and HDL [102, 103]. However, selective deficiency of macrophage ABCA1 does not decrease plasma lipid or apoAI levels, and conversely, reconstitution of ABCA1–/– animals with WT bone marrow does not restore plasma HDL levels, although very small increases in plasma HDL and apoAI levels have been observed [187]. These studies show that macrophage ABCA1 has only a minimal impact on the circulating pool of mature plasma HDL.

Breeding ABCA1–/– mice to either the apoE$^{-/-}$ or LDLR–/– background was found to reduce plasma lipids in both models on both a chow and high fat high cholesterol diet, as expected due to the role of ABCA1 in HDL regulation [188]. Unexpectedly, however, aortic lesions in ABCA1–/–/apoE$^{-/-}$ mice did not advance beyond fatty streaks on either a chow or an atherogenic diet despite the presence of severe xanthomas and foam cells in the skin [188]. Although more complex lesions were observed in the LDLR–/– compared to the apoE$^{-/-}$ model, lesion size did not differ between LDLR–/– mice that contained or lacked total ABCA1 [188]. These findings suggested that the less atherogenic plasma lipid profile observed in the complete absence of ABCA1

TABLE 6.2. Effect of macrophage ABCA1 on atherosclerosis.

Group	Donor	Recipient (sex)	Plasma apoE (% of WT)*	Plasma apoAI (% of WT)*	Diet	mΦ Cholesterol efflux	Total cholesterol	Mean lesion area
Haghpassand et al. [187]	*ABCA1+/+* *ABCA1–/–*	*WT* (♀)	NM	No change	C	*ABCA1–/–* (↓ 7%)	*ABCA1+/+* (↑19%) *ABCA1–/–* (↑ 24%)	NM
					A/W		HDL-C: *ABCA1+/+* (↑2×) *ABCA1–/–* (no change)	
	ABCA1+/+ *ABCA1–/–*	*ABCA1–/–*	NM	*ABCA1–/–* (↓ 50%)	C	*ABCA1–/–* (↓11%)	*ABCA1+/+* (↑ 20%) *ABCA1–/–* (↑ 45%)	NM
					A/W		No change in HDL-C	
Aiello et al. [188]	*apoEABCA 1–/– apoE⁻ᐟ⁻*	*apoE⁻ᐟ⁻*	NM	NM	C	NM	No change in HDL-C	*apoEABC* A1–/– ↑ 3.3× above *apoE⁻ᐟ⁻*
Van Eck et al. [189]	*ABCA1 +/+* *ABCA1–/–*	*LDLR–/–*	*ABCA1–/–* (↓ 56%)	*ABCA1–/–* (↓ 24%)	C	*ABCA1–/–* (↓ 4%)	*ABCA1–/–* (↓ 14%; TG ↑ 74%)	
					A/W		*ABCA1–/–* ↓ 1.4× below *ABCA1+/+*	*ABCA1–/–* (↑ 1.6×)

All comparisons made relative to baseline recipient measurements or animals reconstituted with WT marrow (unless stated otherwise). NM, not measured; C, chow diet; A/W, atherosclerotic/Western-type diet; TG, triglycerides.

may have counterbalanced the proposed proatherogenic effect of macrophage ABCA1 deficiency, resulting in no net change in lesion size. To specifically test the influence of macrophage ABCA1 on atherosclerosis, Van Eck et al. reconstituted apoE$^{-/-}$ mice with apoE$^{-/-}$/ABCA1–/– bone marrow, and observed no change in plasma lipids but a significant increase in lesion area [188]. Similarly, reconstitution of LDLR–/– mice with ABCA1–/– bone marrow had no effect on plasma HDL levels, but resulted in an increased lesion area as well as the presence of more advanced lesions [189]. These studies clearly demonstrated that selective deficiency of macrophage ABCA1 leads to greater development of atherosclerotic lesions, but this effect is independent of circulating HDL.

Thus far, there are conflicting reports on whether overexpression of ABCA1 throughout the body protects from atherosclerosis. Overexpression of ABCA1 under the control of the apoE promoter in atherosusceptible C57Bl/6 mice resulted in decreased atherosclerosis when challenged with a proatherogenic diet [190]. However, these mice unexpectedly developed increased lesions when crossed onto the apoE$^{-/-}$ model [190]. In contrast, Singaraja et al. [69] demonstrated that ABCA1 BAC transgenic mice, which express excess ABCA1 under endogenous regulatory signals, exhibit dramatically reduced lesion size when crossed onto the apoE$^{-/-}$ background. These observations demonstrate that appropriate physiological regulation of ABCA1 is crucial for its antiatherogenic effects *in vivo*. Whether selective overexpression of appropriately regulated ABCA1 in macrophages is sufficient to protect from atherosclerotic lesion development remains to be determined.

Conclusions

Macrophages play crucial roles in the development of atherosclerosis. One vital aspect of macrophage physiology in atherosclerosis is the efflux of cholesterol accumulated during the uptake of oxidized lipoproteins. Failure of macrophages to rid themselves of accumulated cholesterol is a critical step in lesion progression. Within the context of the vessel wall, local synthesis and lipidation of apoE may also enhance the efficiency of cholesterol efflux from macrophages, and thus retard foam cell formation. In this review, we have highlighted aspects of ABCA1, ABCG1, and apoE function within macrophages and their roles in cholesterol efflux and atherosclerosis. However, many questions remain. For example, ABCA1 is required for efficient secretion of apoE from macrophages [191], and Van Eck et al. [189] have shown that plasma apoE and apoAI levels are reduced by the lack of macrophage ABCA1, providing further evidence that macrophage ABCA1 can affect the metabolism of apoE *in vivo*. Because macrophage-secreted apoE may have an important role in facilitating cholesterol efflux within the vessel wall, it is possible that ABCA1 may also affect the ability of endogenous apoE to promote RCT under conditions of minimal exposure to plasma

HDL or apoAI [157, 158, 160, 161, 192]. Although the experiments of Singaraja et al. [69] clearly demonstrated that apoE is not required for ABCA1 to promote cholesterol efflux and reduce lesion progression *in vivo*, these findings do not rule out the possibility that a greater reduction in lesion size or progression may be observed if ABCA1 is overexpressed in the presence of macrophage apoE, particularly if overexpression of ABCA1 is itself restricted to macrophages. Additionally, although Huang et al. [193, 194] have shown that cholesterol efflux in apoE-expressing J774 cells does not depend on ABCA1 but rather is modulated by SRBI expression, ABCA1 and ABCG1 do affect apoE secretion and spontaneous apoE-dependent lipid efflux in human monocyte-derived macrophages [146, 191]. These studies highlight the importance of further investigation into the interplay of the efflux pathways mediated by ABCA1, ABCG1, apoE, and SRBI in relevant model systems. Studies using recently described ABCG1-deficient mice should soon reveal whether or not ABCG1 participates in atherosclerotic lesion development. Because ABCG1 has been reported to be upregulated in ABCA1-deficient macrophages [145], it will also be important to determine the extent to which compensatory gene expression pathways may contribute to macrophage cholesterol efflux and atherosclerosis *in vivo*. Through these and other investigations, a deeper understanding of how the functions of all of these molecules are integrated within the macrophage may lead to novel methods to preserve optimal macrophage function within the arterial wall and reduce the burden of atherothrombotic disease.

Acknowledgments: We are grateful to the colleagues who have provided critical feedback on this review, including Veronica Hirsch-Reinshagen, Luis Maia, Marcia MacDonald, and Liam Brunham. CLW is supported by a New Investigator Award from the Canadian Institutes of Health Research (CIHR) and operating grants from CIHR and the Heart and Stroke Foundation of Canada.

References

1. Murray CJL, Lopez AD: Alternative projections of mortality and disability by cause 1990–2020: Global Burden of Disease Study. Lancet 349: 1498–1504, 1997.
2. Murray CJL, Lopez AD: Global mortality, disability, and the contribution of risk factors: Global Burden of Disease Study. Lancet 349: 1436–1442, 1997.
3. Gordon T, Kannel WB, Castelli WP, Dawber TR: Lipoproteins, cardiovascular disease and death. Arch Intern Med 141: 1128–1131, 1981.
4. Gordon DJ, Rifkind BM: High density lipoprotein—the clinical implications of recent studies. N Engl J Med 321: 1311–1316, 1989.
5. Hodis HN, Mack WJ: Triglyceride-rich lipoproteins and progression of atherosclerosis. Eur Heart J 19: A40–A44, 1998.

6. Zilversmit DB: Atherogenesis: A postprandial phenomenon. Circulation 60: 473–485, 1979.

7. Zilversmit DB: Atherogenic nature of triglycerides, postprandial lipemia, and triglyceride-rich remnant lipoproteins. Clin Chem 41: 153–158, 1995.

8. Napoli C, D'Armiento FP, Mancini FP, Postiglione A, Witztum JL, Palumbo G, Palinski W: Fatty streak formation occurs in human fetal aortas and is greatly enhanced by maternal hypercholesterolemia. J Clin Invest 100: 2680–2690, 1997.

9. Criqui MH, Golomb BA: Epidemiological aspects of lipid abnormalities. Am J Med 105: 48S–57S, 1998.

10. Criqui MH: Triglycerides and cardiovascular disease. Eur Heart J 19: A36–A39, 1998.

11. Stary HC, Chandler AB, Glagov S, Guyton JR, Insull W Jr, Rosenfeld ME, Schaffer SA, Schwartz CJ, Wagner WD, Wissler RW: A definition of initial, fatty streak and intermediate lesions of atherosclerosis. Arterioscler Thromb 14: 840–856, 1994.

12. Stary HC: Natural history and histological classification of atherosclerotic lesions. Arterioscler Thromb Vasc Biol 20: 1177–1178, 2000.

13. Solberg LA, Strong JP: Risk factors and atherosclerotic lesions. Arteriosclerosis 3: 187–198, 1983.

14. Kannel WB, Wilson PWF: Efficacy of lipid profiles in prediction of coronary disease. Am Heart J 124: 768–774, 1992.

15. Kannel WB: High-density lipoproteins: epidemiologic profile and risks of coronary artery disease. Am J Cardiol 52: 9B–12B, 1983.

16. Kwiterovich PO: The antiatherogenic role of high-density lipoprotein cholesterol. Am J Cardiol 82: 13Q–21Q, 1998.

17. Hoeg JM: Lipoproteins and atherogenesis. Endocrinol Metabol Clin 27: 569–584, 1998.

18. Lusis AJ: Atherosclerosis. Nature 407: 233–241, 2000.

19. Meyers CD, Kashyap ML: Pharmacologic elevation of high-density lipoproteins: recent insights on mechanism of action and atherosclerosis protection. Curr Opin Cardiol 19: 366–373, 2004.

20. Gordon DJ, Probstfield JL, Garrison RJ, Neaton JD, Castelli WP, Knoke JD, Jacobs DR, Bangdiwala S, Tyroler HA: High-density lipoprotein cholesterol and cardiovascular disease. Four prospective American studies. Circulation 79: 8–15, 1989.

21. Toth PP, Davidson MH: Therapeutic interventions targeted at the augmentation of reverse cholesterol transport. Curr Opin Cardiol 19: 374–379, 2004.

22. Ross R, Glomset JA: The pathogenesis of atherosclerosis. N Engl J Med 295: 369–377, 1976.

23. Gaziano JM, Hennekens CH, O'Donnell CJ, Breslow JL, Buring JE: Fasting triglycerides, high density lipoprotein, and risk of myocardial infarction. Circulation 96: 2520–2525, 1997.

24. Stampfer MJ, Sacks FM, Salvini S, Willett WC, Hennekens CH: A prospective study of cholesterol, apolipoproteins, and their risk of myocardial infarction. N Engl J Med 325: 373–381, 1991.

25. Stein O, Stein Y: Atheroprotective mechanisms of HDL. Atheroscler 144: 285–301, 1999.

26. Eisenberg S: High density lipoprotein metabolism. J Lipid Res 25: 1017–1058, 1984.
27. Asztalos BF: High-density lipoprotein metabolism and progression of atherosclerosis: new insights from the HDL Atherosclerosis Treatment Study. Curr Opin Cardiol 19: 385–391, 2004.
28. Fielding CJ, Fielding PE: Molecular physiology of reverse cholesterol transport. J Lipid Res 36: 211–228, 1995.
29. von Eckardstein A, Assmann G: High density lipoproteins and reverse cholesterol transport: Lessons from mutations. Atherosclerosis 137: S7–S11, 1998.
30. Nissen SE, Tsunoda T, Tuzcu EM, Schoenhagen P, Cooper CJ, Yasin M, Eaton GM, Lauer MA, Sheldon WS, Grines CL, Halpern S, Crowe T, Blankenship JC, Kerensky R: Effect of recombinant apoA-I Milano on coronary atherosclerosis in patients with acute coronary syndromes. JAMA 290: 2292–2300, 2003.
31. Kobayashi J, Okamoto H, Otabe M, Bujo H, Saito Y: Effect of HDL, from Japanese white rabbit administered a new cholesterol ester transfer protein inhibitor JTT-705, on cholesterol ester accumulation induced by acetylated low density lipoprotein in J774 macrophage. Atherosclerosis 162: 131–135, 2002.
32. Okamoto H, Yonemori F, Wakitani K, Minowa T, Maeda K, Shinkai H: A cholesterol ester transfer protein inhibitor attenuates atherosclerosis in rabbits. Nature 406: 203–207, 2000.
33. de Grooth GJ, Kuivenhoven JA, Stalenhoef AFH, de Graaf J, Zwinderman AH, Posma JL, van Tol A, Kastelein JJP: Efficacy and safety of a novel cholesterol ester transfer protein inhibitor, JTT-705, in humans. Circulation 105: 2159–2165, 2002.
34. Ma K, Cilingioroglu M, Otvos JD, Ballantyne CM, Marian AJ, Chan L: Endothelial lipase is a major genetic determinant for high-density lipoprotein concentration, structure, and metabolism. Proc Natl Acad Sci USA 100: 2748–2753, 2002.
35. Jin W, Millar JS, Broedl U, Glick JM, Rader DJ: Inhibition of endothelial lipase causes increased HDL cholesterol levels *in vivo*. J Clin Invest 111: 357–362, 2003.
36. Ishida T, Choi S, Kundu RK, Irata K, Rubin EM, Cooper AD, Quertermous T: Endothelial lipase is a major determinant of HDL level. J Clin Invest 111: 347–355, 2003.
37. Yancey PG, Bortnick AE, Kellner-Weibel G, de la Llera-Moya M, Phillips MC, Rothblat GH: Importance of different pathways of cellular cholesterol efflux. Arterioscler Thromb Vasc Biol 23: 712–719, 2003.
38. Singaraja RR, Brunham LR, Visscher H, Kastelein JJ, Hayden MR: Efflux and atherosclerosis: The clinical and biochemical impact of variations in the ABCA1 gene. Arterioscler Thromb Vasc Biol 23: 1322–1332, 2003.
39. Phillips MC, Johnson WJ, Rothblat GH: Mechanisms and consequences of cellular cholesterol exchange and transfer. Biochim Biophys Acta 906: 223–276, 1987.
40. Rothblat GH, de la Llera-Moya M, Atger V, Kellner-Weibel G, Williams DL, Phillips MC: Cell cholesterol efflux: integration of old and new observations provides new insights. J Lipid Res 40: 781–796, 1999.
41. Acton S, Rigotti A, Landschulz KT, Xu S, Hobbs HH, Krieger M: Identification of scavenger receptor SR-BI as a high density lipoprotein receptor. Science 518–520, 1996.
42. Krieger M: The "best" of cholesterols, the "worst" of cholesterols: A tale of two receptors. Proc Natl Acad Sci USA 95: 4077–4080, 1998.

43. Landschulz KT, Pathak RK, Rigotti A, Krieger M, Hobbs HH: Regulation of scavenger receptor, class B, type I, a high density lipoprotein receptor, in liver and steroidogenic tissues of the rat. J Clin Invest 98: 984–995, 1996.

44. de la Llera-Moya M, Connelly MA, Drazul D, Klein SM, Favari E, Yancey PG, Williams DA, Rothblat GH: Scavenger receptor class B type I affects cholesterol homeostasis by magnifying cholesterol flux between cells and HDL. J Lipid Res 42: 1969–1978, 2001.

45. Yancey PG, de la Llera-Moya M, Swarnaker S, Monzo P, Klein SM, Connelly MA, Johnson WJ, Williams DL, Rothblat GH: HDL phospholipid composition is a major determinant of the bi-directional flux and net movement of cellular free cholesterol mediated by scavenger receptor-BI (SR-BI). J Biol Chem 275: 36596–36604, 2000.

46. Ji Y, Jian B, Wang N, Sun Y, de la Llera-Moya M, Phillips MC, Rothblat GH, Swaney JB, Tall AR: Scavenger receptor BI promotes high density lipoprotein-mediated cellular cholesterol efflux. J Biol Chem 272: 20982–20985, 1997.

47. Rigotti A, Acton S, Krieger M: The class B scavenger receptors SR-BI and CD36 are receptors for anionic phospholipids. J Biol Chem 270: 16221–16224, 1995.

48. de la Llera-Moya M, Rothblat GH, Connelly MA, Kellner-Weibel G, Sakr SW, Phillips MC, Williams DL: Scavenger receptor BI (SRBI) mediates free cholesterol flux independently of HDL tethering to the cell surface. J Lipid Res 40: 575–580, 1999.

49. Jian B, de la Llera-Moya M, Royer L, Rothblat G, Francone O, Swaney JB: Modification of the cholesterol efflux properties of human serum by enrichment with phospholipid. J Lipid Res 38: 734–744, 1997.

50. Kellner-Weibel G, de la Llera-Moya M, Connelly MA, Stoudt G, Christian AE, Haynes MP, Williams DL, Rothblat GH: Expression of scavenger receptor BI in COS-7 cells alters cholesterol content and distribution. Biochemistry 39: 221–229, 2000.

51. Babitt J, Trigatti B, Rigotti A, Smart EJ, Anderson RGW, Xu S, Krieger M: SR-BI, a high density lipoprotein receptor that mediates selective lipid uptake, is N-glycosylated and fatty acylated and colocalizes with plasma membrane caveolae. J Biol Chem 272: 13242–13249, 1997.

52. Graf GA, Connell PM, van der Westhuyzen DR, Smart EJ: The class B, type I scavenger receptor promotes the selective uptake of high density lipoprotein cholesterol ethers into caveolae. J Biol Chem 274: 12043–12048, 1999.

53. Liu T, Krieger M, Kan H-Y, Zannis VI: The effects of mutations in helices 4 and 6 of apoA-I on scavenger receptor class B type I (SR-BI)-mediated cholesterol efflux suggest that formation of a productive complex between reconstituted high density lipoproteins and SR-BI is required for efficient lipid transport. J Biol Chem 277: 21576–21584, 2002.

54. Dean M, Hamon Y, Chimini G: The human ATP-binding cassette (ABC) transporter superfamily. J Lipid Res 42: 1007–1017, 2001.

55. Dean M, Allikmets R: Complete characterization of the human ABC gene family. J Bioenerg Biomemb 33: 475–479, 2001.

56. Schmitz G, Kaminski WE, Orsó E: ABC transporters in cellular lipid trafficking. Curr Opin Lipidol 11: 493–501, 2000.

57. Brooks-Wilson A, Marcil M, Clee SM, Zhang L, Roomp K, van Dam M, Yu L, Brewer C, Collins JA, Molhuizen HOF, Loubser O, Ouellette BFF, Fichter K, Ashbourne Excoffon KJD, Sensen CW, Scherer S, Mott S, Denis M, Martindale D,

Frohlich J, Morgan K, Koop B, Pimstone SN, Kastelein JJP, Genest J Jr, Hayden MR: Mutations in ABC1 in Tangier disease and familial high-density lipoprotein deficiency. Nat Genet 22: 336–345, 1999.

58. Rust S, Rosier M, Funke H, Amoura Z, Piette J-C, Deleuze J-F, Brewer HB Jr, Duverger N, Denèfle P, Assmann G: Tangier disease is caused by mutations in the gene encoding ATP-binding cassette transporter 1. Nat Genet 22: 352–355, 1999.

59. Bodzioch M, Orsó E, Klucken J, Langmann T, Böttcher A, Diederich W, Drobnik W, Barlage S, Büchler C, Porsch-Özcürümez M, Kaminski WE, Hahmann HW, Oette K, Rothe G, Aslanidis C, Lackner KJ, Schmitz G: The gene encoding ATP-binding cassette transporter 1 is mutated in Tangier disease. Nat Genet 22: 347–351, 1999.

60. Hayden MR, Clee SM, Brooks-Wilson A, Genest J Jr, Attie A, Kastelein JJP: Cholesterol efflux regulatory protein, Tangier disease and familial high-density lipoprotein deficiency. Curr Opin Lipidol 11: 117–122, 2000.

61. Tall AR, Wang N: Tangier disease as a test of the reverse cholesterol transport hypothesis. J Clin Invest 106: 1263–1270, 2000.

62. Rust S, Walter M, Funke H, von Eckardstein A, Cullen P, Kroes HY, Hordijk R, Geisel J, Kastelein J, Molhuizen HOF, Schreiner M, Mischke A, Hahmann HW, Assmann G: Assignment of Tangier disease to chromosome 9q31 by a graphical linkage exclusion strategy. Nat Genet 20: 96–98, 1998.

63. Kuivenhoven JA, Pritchard H, Hill J, Frohlich J, Assmann G, Kastelein J: The molecular pathology of lecithin:cholesterol acyltransferase (LCAT) deficiency syndromes. J Lipid Res 38: 191–205, 1997.

64. Glomset JA, Norum KR: The metabolic role of lecithin:cholesterol acyltransferase: Perspectives from pathology. Adv Lipid Res 11: 1–65, 1973.

65. Glomset JA: The plasma lecithin:cholesterol acyltransferase reaction. J Lipid Res 9: 155–167, 1968.

66. Liu B, Krieger M: Highly purified scavenger receptor class B, type I reconstituted into phosphatidylcholine/cholesterol liposomes mediates high affinity HDL binding and selective lipid uptake. J Biol Chem 277: 34125–34135, 2002.

67. Singaraja RR, Bocher V, James ER, Clee SM, Zhang L-H, Leavitt BR, Tan B, Brooks-Wilson A, Kwok A, Bissada N, Yang Y-Z, Liu G, Tafuri SR, Fievet C, Wellington CL, Staels B, Hayden MR: Human ABCA1 BAC transgenic mice show increased HDL-C and ApoA1-dependent efflux stimulated by an internal promoter containing LXREs in intron 1. J Biol Chem 276: 33969–33979, 2001.

68. Vaisman BL, Lambert G, Amar M, Joyce C, Ito T, Shamburek RD, Cain WJ, Fruchart-Najib J, Neufeld ED, Remaley AT, Brewer HB Jr, Santamarinao-Fojo S: ABCA1 overexpression leads to hyperalphalipoproteinemia and increased biliary cholesterol excretion in transgenic mice. J Clin Invest 108: 303–309, 2001.

69. Singaraja R, Fievet C, Castro G, Jamers ER, Hennuyer N, Clee SM, Bissada N, Choy JC, Fruchart J-C, McManus BM, Hayden MR: Increased ABCA1 activity protects against atherosclerosis. J Clin Invest 110: 35–42, 2002.

70. Schmitz G, Langmann T: Transcriptional regulatory networks in lipid metabolism control ABCA1 expression. Biochim Biophys Acta, 1735: 1–19.

71. Fitzgerald ML, Moore KJ, Freeman MW: Nuclear hormone receptors and cholesterol trafficking: the orphans fine a new home. J Mol Med 80: 271–281, 2002.

72. Edwards PA, Kast HR, Anisfeld AM: BAREing it all: the adoption of LXR and FXR and their roles in lipid metabolism. J Lipid Res 43: 2–12, 2002.

73. Santamarina-Fojo S, Remaley AT, Neufeld EB, Brewer HB Jr: Regulation and intracellular trafficking of the ABCA1 transporter. J Lipid Res 42: 1339–1345, 2001.

74. Schultz JR, Tu H, Luk A, Repa JJ, Medina JC, Li L, Schwendner S, Wang S, Thoolen M, Mangelsdorf DJ, Lustig KD, Shan B: Role of LXRs in control of lipogenesis. Genes Devel 14: 2831–2838, 2000.

75. Costet P, Luo Y, Wang N, Tall AR: Sterol-dependent transactivation of the ABC1 promoter by the liver X receptor/retinoid X receptor. J Biol Chem 275: 28240–28245, 2000.

76. Repa JJ, Turley SD, Lobaccaro JA, Medina J, Li L, Lustig K, Shan B, Heyman RA, Dietschy JM, Mangelsdorf DJ: Regulation of absorption and ABC1-mediated efflux of cholesterol by RXR heterodimers. Science 289: 1524–1529, 2000.

77. Venkateswaran A, Laffitte BA, Joseph SB, Mak PA, Wilpitz DC, Edwards PA, Tontonoz P: Control of cellular cholesterol efflux by the nuclear oxysterol receptor LXRα. Proc Natl Acad Sci USA 97: 12097–12102, 2000.

78. Schwartz K, Lawn RM, Wade DP: ABC1 gene expression and ApoA-1-mediated cholesterol efflux are regulated by LXR. Biochem Biophys Res Commun 274: 794–802, 2000.

79. Bortnick AE, Rothblat GH, Stoudt G, Hoppe KL, Royer LJ, McNeish J, Francone OL: The correlation of ATP-binding cassette 1 mRNA levels with cholesterol efflux from various cell lines. J Biol Chem 275: 28634–28640, 2000.

80. Barbier O, Torra IP, Duguay Y, Blanquart C, Fruchart J-C, Glineur C, Staels B: Pleiotropic actions of peroxisome proliferator-activated receptors in lipid metabolism and atherosclerosis. Arterioscler Thromb Vasc Biol 22: 717–726, 2002.

81. Chawla A, Boisvert WA, Lee CH, Laffitte BA, Barak Y, Joseph SB, Liao D, Nagy P, Edwards PA, Curtiss LK, Evans RM, Tontonoz P: A PPARγ-LXR-ABCA1 pathway in macrophages is involved in cholesterol efflux and atherogenesis. Mol Cell 7: 161–171, 2001.

82. Chinetti G, Fruchart J-C, Staels B: Peroxisome proliferator-activated receptors: new targets for the pharmacological modulation of macrophage gene expression and function. Curr Opin Lipidol 14: 459–468, 2003.

83. Oliver WR, Shenk JL, Snaith MR, Russell CS, Plunket KD, Bodkin NL, Lewis MC, Winegar DA, Sznaidman ML, Lambert MH, Xu HE, Sternbach DD, Kliewer SA, Hansen BC, Willson TM: A selective peroxisome proliferator-activated receptor δ agonist promotes reverse cholesterol transport. Proc Natl Acad Sci USA 98: 5306–5311, 2001.

84. Akiyama TE, Sakai S, Lambert G, Nichol CJ, Matsusue K, Pimprale S, Lee Y-H, Ricote M, Glass CK, Brewer HB, Gonzalez FJ: Conditional disruption of the peroxisome proliferator-activated receptor γ gene in mice results in lowered expression of ABCA1, ABCG1, and apoE in macrophages and reduced cholesterol efflux. Mol Cell Biochem 22: 2607–2619, 2002.

85. Li AC, Binder CJ, Gutierrez A, Brown KK, Plotkin CR, Pattison JW, Valledor AF, Davis RA, Willson TM, Witztum JL, Palinski W, Glass CK: Differential inhibition of macrophage foam-cell formation and atherosclerosis in mice by PPARα, β/δ, and γ. J Clin Invest 114: 1564–1576, 2004.

86. Gan X, Kaplan R, Menke JG, MacNaul K, Chen Y, Sparrow CP, Zhou G, Wright SD, Cai TQ: Dual mechanisms of ABCA1 regulation by geranylgeranyl pyrophosphate. J Biol Chem 276: 48702–48708, 2001.

87. Baranova I, Vishnyakova T, Bocharov A, Chen Z, Remaley AT, Stonik J, Eggerman TL, Patterson AP: Lipopolysaccharide down regulates both scav-

enger receptor B1 and ATP binding cassette transporter A1 in RAW cells. Infect Immun 70: 2995–3003, 2002.

88. Khovidhunkit W, Moser AH, Shigenaga JK, Grunfeld C, Feingold KR: Endotoxin down-regulates ABCG5 and ABCG8 in mouse liver and ABCA1 and ABCG1 in J774 murine macrophages: differential role of LXR. J Lipid Res 44: 1728–1736, 2003.

89. Panousis CG, Zuckerman SH: Interferon-gamma induces down-regulation of Tangier disease gene (ATP-binding cassette transporter 1) in macrophage-derived foam cells. Arterioscler Thromb Vasc Biol 20: 1565–1571, 2000.

90. Panousis CG, Evans GA, Zuckerman SH: TGF-beta increases cholesterol efflux and ABC-1 expression in macrophage-derived foam cells: opposing the effects of IFN-gamma. J Lipid Res 42: 856–863, 2001.

91. Wang XQ, Panousis CG, Alfaro ML, Evans GF, Zuckerman SH: Interferon-gamma-mediated downregulation of cholesterol efflux and ABC1 expression is by the Stat1 pathway. Arterioscler Thromb Vasc Biol 290: 891–897, 2002.

92. Huuskonen J, Vishnu M, Pullinger CR, Fielding PE, Fielding CJ: Regulation of ATP-binding cassette transporter A1 transcription by thyroid hormone receptor. Biochemistry 43: 1626–1632, 2004.

93. Wang N, Chen W, Linsel-Nitschke P, Martinez LO, Agerholm-Larsen B, Silver DL, Tall AR: A PEST sequence in ABCA1 regulates degradation by calpain protease and stabilization of ABCA1 by apoA-I. J Clin Invest 111: 99–107, 2003.

94. Martinez LO, Agerholm-Larsen B, Wang N, Chen W, Tall AR: Phosphorylation of a Pest sequence in ABCA1 promotes calpain degradation and is reversed by apoA-I. J Biol Chem 278: 37368–37374, 2003.

95. Yamauchi Y, Hayashi M, Abe-Dohmae S, Yokoyama S: Apolipoprotein A-I activates protein kinase Cα signaling to phosphorylate and stabilize ATP binding cassette transporter A1 for the high density lipoprotein assembly. J Biol Chem 48: 47890–47897, 2003.

96. See RH, Caday-Malcolm RA, Singaraja RR, Zhou S, Silverston A, Huber MT, Moran J, James ER, Janoo R, Savill JM, Rigot V, Zhang L-H, Wang M, Chimini G, Wellington CL, Tafuri SR, Hayden MR: Protein kinase A site-specific phosphorylation regulates ATP-binding cassette A1 (ABCA1)-mediated phospholipid efflux. J Biol Chem 277: 41835–41842, 2002.

97. Wang Y, Oram JF: Unsaturated fatty acids inhibit cholesterol efflux from macrophages by increasing degradation of ATP-binding cassette transporter A1. J Biol Chem 277: 5692–5697, 2002.

98. Wang Y, Kurdi-Haidar B, Oram JF: LXR-mediated activation of macrophage stearoyl-CoA desaturase generates unsaturated fatty acids that destabilize ABCA1. J Lipid Res 45: 972–980, 2004.

99. Feng B, Tabas I: ABCA1-mediated cholesterol efflux is defective in free cholesterol-loaded macrophages. J Biol Chem 277: 43271–43280, 2002.

100. Choi HY, Karten B, Chan T, Vance JE, Greer WL, Heidenreich RA, Garver WS, Francis GA: Impaired ABCA1-dependent lipid efflux and hypoalphalipoproteinemia in human Niemann–Pick type C disease. J Biol Chem 278: 32569–32577, 2003.

101. Albrecht C, Soumian S, Amey JS, Sardini A, Higgins CF, Davies AH, Gibbs RG: ABCA1 expression in carotid atherosclerotic plaques. Stroke 35: 2801–2806, 2004.

102. McNeish J, Aiello RJ, Guyot D, Turi T, Gabel C, Aldinger C, Hoppe KL, Roach ML, Royer LJ, de Wet J, Broccardo C, Chimini G, Francone OL: High density lipoprotein deficiency and foam cell accumulation in mice with targeted disruption of ATP-binding cassette transporter-1. Proc Natl Acad Sci USA 97: 4245–4250, 2000.

103. Christiansen-Weber TA, Voland JR, Wu Y, Ngo K, Roland BL, Nguyen S, Peterson PA, Fung-Leung WP: Functional loss of ABCA1 in mice causes severe placental malformation, aberrant lipid distribution, and kidney glomerulonephritis as well as high-density lipoprotein cholesterol deficiency. Am J Pathol 157: 1017–1029, 2000.

104. Orsó E, Broccardo C, Kaminski WE, Böttcher A, Liebisch G, Drobnik W, Götz A, Chambenoit O, Diederich W, Langmann T, Spruss T, Luciani M-F, Rothe G, Lackner KJ, Chimini G, Schmitz G: Transport of lipids from Golgi to plasma membrane is defective in Tangier disease patients and ABC1-deficient mice. Nat Genet 24: 192–196, 2000.

105. Remaley AT, Stonick JA, Demosky SJ, Neufeld EB, Bocharov AV, Vishnyakova TG, Eggerman TL, Patterson AP, Duverger NJ, Santamarinao-Fojo S, Brewer HB: Apolipoprotein specificity for lipid efflux by the human ABCA1 transporter. Biochem Biophys Res Commun 280: 818–823, 2001.

106. Mendez AJ, Anantharamaiah GM, Segrest JP, Oram JF: Synthetic amphipathic helical peptides that mimic apolipoprotein A-I in clerning cellular cholesterol. J Clin Invest 94: 1698–1705, 1994.

107. Yancey PG, Bielicki JK, Johnson WJ, Lund-Katz S, Palgunachari MN, Anantharamaiah GM, Segrest JP, Phillips MC, Rothblat GH: Efflux of cellular cholesterol and phospholipid to lipid-free apolipoproteins and class A amphipathic peptides. Biochemistry 34: 7955–7965, 1995.

108. Remaley AT, Thomas F, Stonick JA, Demosky SJ, Bark SE, Neufeld EB, Bocharov AV, Vishnyakova TG, Patterson AP, Eggerman TL, Santamarina-Fojo S, Brewer HB: Synthetic amphipathic helical peptides promote lipid efflux from cells by an ABCA1-dependent and an ABCA1-independent pathway. J Lipid Res 44: 828–836, 2003.

109. Wang N, Silver DL, Costet P, Tall AR: Specific binding of ApoA-I, enhanced cholesterol efflux, and altered plasma membrane morphology in cells expressing ABC1 [In Process Citation]. J Biol Chem 275: 33053–33058, 2000.

110. Fitzgerald ML, Morris AL, Rhee JS, Andersson LP, Mendez AJ, Freeman MW: Naturally occurring mutations in the largest extracellular loops of ABCA1 can disrupt its direct interaction with apolipoprotein A-I. J Biol Chem 277: 33178–33187, 2002.

111. Natarajan P, Forte TM, Chu B, Phillips MC, Oram JF, Bielicki JK: Identification of an apolipoprotein A-I structural element that mediates cellular cholesterol efflux and stabilizes ATP binding cassette transporter A1. J Biol Chem 279: 24044–24052, 2004.

112. Oram JF, Lawn RM, Garvin MR, Wade DP: ABCA1 is the cAMP-inducible apolipoprotein receptor that mediates cholesterol secretion from macrophages. J Biol Chem 275: 34508–34511, 2000.

113. Denis M, Haidar B, Marcil M, Bouvier M, Krimbou L, Genest J: Characterization of oligomeric human ATP binding cassette transporter A1: Potential implications for determining the structure of nascent high density lipoprotein particles. J Biol Chem 279: 41529–41536, 2004.

114. Wellington CL, Yang Y-Z, Zhou S, Clee SM, Tan B, Hirano K, Zwarts K, Kwok A, Gelfer A, Marcil M, Newman S, Roomp K, Singaraja R, Collins J, Zhang LH, Groen AK, Hovingh K, Brownlie A, Tafuri S, Genest J Jr, Kastelein JJ, Hayden MR: Truncation mutations in ABCA1 suppress normal upregulation of full-

length ABCA1 by 9-cis-retinoic acid and 22-R-hydroxycholesterol. J Lipid Res 43: 1939–1949, 2002.

115. Chambenoit O, Hamon Y, Marguet D, Rigneault H, Rosseneu N, Chimini G: Specific docking of apolipoprotein A-1 at the cell surface requires a functional ABCA1 transporter. J Biol Chem, 276: 9955–9960, 2001.

116. Vaughan AM, Oram JF: ABCA1 redistributes membrane cholesterol independently of apolipoprotein interactions. J Lipid Res 44: 1373–1380, 2003.

117. Panagotopulos SE, Witting SR, Horace EM, Hui DY, Maiorano JN, Davidson WS: The role of apolipoprotein A-I helix 10 in apolipoprotein-mediated cholesterol efflux via the ATP-binding cassette transporter ABCA1. J Biol Chem 277: 39477–39484, 2002.

118. Gillotte KL, Zaiou M, Lund-Katz S, Anantharamaiah GM, Holvoet P, Dhoest A, Palgunachari MN, Segrest JP, Weisgraber KH, Rothblat GH, Phillips MC: Apolipoprotein-mediated plasma membrane microsolubilization: role of lipid affinity and membrane penetration in the efflux of cellular cholesterol and phospholipid. J Biol Chem 274: 2021–2028, 1999.

119. Gillotte KL, Davidson WS, Lund-Katz S, Rothblat GH, Phillips MC: Removal of cellular cholesterol by pre-β-HDL involves plasma membrane microsolubilization. J Lipid Res 39: 1918–1928, 1998.

120. Neufeld EB, Remaley AT, Demosky SJ, Stonik JA, Cooney AM, Comly M, Dwyer NK, Zhang M, Blanchette-Mackie JE, Santamarina-Fojo S, Brewer HB Jr: Cellular localization and trafficking of the human ABCA1 transporter. J Biol Chem 276: 27584–27590, 2001.

121. Rothblat GH, de la Llera-Moya M, Favari E, Yancey PG, Kellner-Weibel G: Cellular cholesterol flux studies: methodological considerations. Atherosclerosis 163: 1–8, 2002.

122. Wang N, Silver DL, Theile C, Tall AR: ATP-binding cassette transporter A1 (ABCA1) functions as a cholesterol efflux regulatory protein. J Biol Chem 276: 23742–23747, 2001.

123. Yamauchi Y, Abe-Dohmae S, Yokoyama S: Differential regulation of apolipoprotein A-I/ATP binding cassette transporter A1-mediated cholesterol and phospholipid release. Biochim Biophys Acta 1585: 1–10, 2002.

124. Fielding PE, Nagao K, Hakamata H, Chimini G, Fielding CJ: A two-step mechanism for free cholesterol and phospholipid efflux from human vascular cells to apolipoprotein A-I. Biochemistry 39: 14113–14120, 2000.

125. Chen W, Sun Y, Welch C, Gorelik A, Leventhal AR, Tabas I, Tall AR: Preferential ATP-binding cassette transporter A1-mediated cholesterol efflux from late endosomes/lysosomes. J Biol Chem 276: 43564–43569, 2001.

126. Wellington CL, Walker EK, Suarez A, Kwok A, Bissada N, Singaraja R, Yang Y-Z, Zhang LH, James E, Wilson JE, Francone O, McManus BM, Hayden MR: ABCA1 mRNA and protein distribution patterns predict multiple different roles and levels of regulation. Lab Invest 82: 273–283, 2002.

127. Lawn RM, Wade DP, Couse TL, Wilcox JN: Localization of human ATP-binding cassette transporter 1 (ABC1) in normal and atherosclerotic tissues. Arterioscler Thromb Vasc Biol 21: 378–385, 2001.

128. Timmins JM, Lee JY, Boudyguina E, Kluckman KD, Brunham LR, Mulya A, Gebre AK, Coutinho JM, Colvin PL, Smith TL, Hayden MR, Maeda N, Parks JS: Targeted inactivation of hepatic Abca1 causes profound hypoalphalipopro-

teinemia and kidney hypercatabolism of apoA-I. J Clin Invest 115: 1333–1342, 2005.

129. Wellington CL, Brunham LR, Zhou S, Singaraja R, Visscher H, Gelfer A, Ross C, James E, Liu G, Huber MT, Yang YZ, Parks RJ, Groen A, Fruchart-Najib J, Hayden MR: Alterations in plasma lipids in mice via adenoviral mediated hepatic overexpression of human ABCA1. J Lipid Res 44: 1470–1480, 2003.

130. Basso F, Freeman L, Knapper CL, Remaley A, Stonik J, Neufeld EB, Tansey T, Amar MJA, Fruchart-Najib J, Duverger N, Santamarina-Fojo S, Brewer HB Jr: Role of the hepatic ABCA1 transporter in modulating intrahepatic cholesterol and plasma HDL-cholesterol concentrations. J Lipid Res 44: 296–302, 2003.

131. Hoekstra M, Kruijt JK, Van Eck M, van Berkel TJC: Specific gene expression of ATP-binding cassette transporters and nuclear hormone receptors in rat liver parenchymal, endothelial, and Kupffer cells. J Biol Chem 278: 25448–25453, 2003.

132. Kiss RS, McManus DC, Franklin V, Tan WL, McKenzie A, Chimini G, Marcel YL: The lipidation by hepatocytes of human apolipoprotein A-I occurs by both ABCA1-dependent and -independent pathways. J Biol Chem 278: 10119–10127, 2003.

133. Borst P, Zelcer N, van Helvoort A: ABC transporters in lipid transport. Biochim Biophys Acta 1486: 128–144, 2000.

134. Chen H, Rossier MD, Lalioti A, Lynn A, Charkravarti A, Perrin G, Antonarakis SE: Cloning of the cDNA for a human homologue of the *Drosophila white* gene and mapping to chromosome 21q22.3. Am J Hum Genet 59: 66–75, 1996.

135. Kennedy MA, Venkateswaran A, Tarr PT, Xenarios I, Kudoh J, Shimizu N, Edwards PA: Characterization of the human ABCG1 gene. J Biol Chem 276: 39438–39447, 2001.

136. Lorkowski S, Rust S, Engel T, Jung E, Tegelkamp K, Galinski EA, Assmann G, Cullen P: Genomic sequence and structure of the human ABCG1 (ABC8) gene. Biochem Biophys Res Commun 280: 121–131, 2001.

137. Wang N, Lan D, Chen W, Matsuura F, Tall AR: ATP-binding cassette transporters G1 and G4 mediate cellular cholesterol efflux to high-density lipoproteins. Proc Natl Acad Sci USA 101: 9774–9779, 2004.

138. Cserepes J, Szentpetery Z, Seres L, Ozvegy-Laczka C, Langmann T, Schmitz G, Glavinas H, Klein I, Homolya L, Varadi A, Sarkadi B, Elkind NB: Functional expression and characterization of the human ABCG1 and ABCG1 proteins: indications for heterodimerization. Biochem Biophys Res Commun 320: 860–867, 2004.

139. Langmann T, Porsch-Özcürümez M, Unkelbach U, Klucken J, Schmitz G: Genomic organization and characterization of the promoter of the human ATP-binding cassette transporter-G1 (ABCG1) gene. Biochim Biophys Acta 1494: 175–180, 2000.

140. Nakamura K, Kennedy MA, Baldán A, Bohanic DD, Lyons K, Edwards PA: Expression and regulation of multiple murine ATP-binding cassette transporter G1 mRNAs/isoforms that stimulate cellular cholesterol efflux to high density lipoprotein. J Biol Chem August 19 Epub: 2004.

141. Croop JM, Tiller GE, Fletcher JA, Lux E, Raab D, Goldenson D, Son S, Arciniegas S, Wu RL: Isolation and characterization of a mammalian homolog of the *Drosophila white* gene. Gene 185: 77–85, 1997.

142. Kennedy MA, Barrera GC, Nakamura K, Baldán A, Tarr P, Fishbein MC, Frank J, Francone OL, Edwards PA: ABCG1 has a critical role in mediating cholesterol efflux to HDl and preventing cellular lipid accumulation. Cell Metabol 1: 121–131, 2005.

143. Klucken J, Büchler C, Orsó E, Kaminski WE, Porsch-Özcürümez M, Liebisch G, Kapinsky M, Diederich W, Drobnik W, Dean M, Allikmets R, Schmitz G: ABCG1 (ABC8), the human homologue of the *Drosophila* white gene, is a regulator of macrophage cholesterol and phospholipid transport. Proc Natl Acad Sci USA 97: 817–822, 2000.
144. Venkateswaran AJ, Repa JJ, Lobaraccaro JM, Bronson A, Magnelsorf DJ, Edwards PA: Human white/murine ABC8 mRNA levels are highly induced in lipid-loaded macrophages. A transcriptional role for specific oxysterols. J Biol Chem 275: 14700–14707, 2000.
145. Lorkowski S, Kratz M, Wenner C, Schmidt R, Weitkamp B, Fobker M, Reinhardt J, Rauterberg J, Galinski EA, Cullen P: Expression of the ATP-binding cassette transporter gene ABCG1 (ABC8) in Tangier disease. Biochem Biophys Res Commun 283: 821–830, 2001.
146. Cignarella A, Engel T, von Eckardstein A, Kratz M, Lorkowski S, Lueken A, Assmann G, Cullen P: Pharmacological regulation of cholesterol efflux in human monocyte-derived macrophages in the absence of exogenous cholesterol acceptors. Atheroscler 179: 229–236, 2005.
147. Fazio S, Linton MF: Murine bone marrow transplantation as a novel approach to studying the role of macrophages in lipoprotein metabolism and atherogenesis. Trends Cardiovasc Med 6: 58–65, 1996.
148. Small JA, Bieberich C, Ghotbi Z, Hess J, Scangos GA, Clements JE: The visna virus long terminal repeat directs expression of a reporter gene in activated macrophages, lymphocytes, and the central nervous systems of transgenic mice. J Virol 63: 1891–1896, 1989.
149. Daugherty A: Mouse models of atherosclerosis. Am J Med Sci 323: 3–10, 2002.
150. Smith JD, Breslow JL: The emergence of mouse models of atherosclerosis and their relevance to clinical research. J Intern Med 242: 99–109, 1997.
151. Tangirala RK, Rubin EM, Palinski W: Quantitation of atherosclerosis in murine models: correlation between lesions in the aortic origin and in the entire aorta, and differences in the extent of lesions between sexes in LDL receptor-deficient and apolipoprotein E-deficient mice. J Lipid Res 36: 2320–2328, 1995.
152. Mahley RW, Rall SC Jr: Apolipoprotein E: far more than a lipid transport protein. Ann Rev Genom Hum Genet 1: 507–537, 2000.
153. Greenow K, Pearce NJ, Ramji DP: The key role of apolipoprotein E in atherosclerosis. J Mol Med 83: 329–342, 2005.
154. Mazzone T: Apolipoprotein E secretion by macrophages: its potential functions. Curr Opin Lipidol 7: 303–310, 1996.
155. von Eckardstein A: Cholesterol efflux from macrophages and other cells. Curr Opin Lipidol 7: 308–319, 1996.
156. Langer C, Huang Y, Cullen P, Mahley RW, Assmann G, von Eckardstein A: Endogenous apolipoprotein E modulates cholesterol efflux and cholesterol ester hydrolysis mediated by high density lipoprotein$_3$ and lipid-free apolipoproteins in mouse peritoneal macrophages. J Mol Med 77: 614–622, 2000.
157. Basu SK, Ho YK, Brown MS, Bilheimer DW, Anderson RGW, Goldstein JL: Biochemical and genetic studies of the apolipoprotein E secreted by mouse macrophages and human monocytes. J Biol Chem 257: 9788–9795, 1982.
158. Basu SK, Goldstein JL, Brown ML: Independent pathways for secretion of cholesterol and apolipoprotein E by macrophages. Science 219: 871–873, 1983.

159. Dove DE, Linton MF, Fazio S: ApoE-mediated cholesterol efflux from macrophages: separation of autocrine and paracrine effects. Am J Physiol Cell Physiol 288: C586–C592, 2005.

160. Zhang W-Y, Gaynor PM, Kruth HS: Apolipoprotein E produced by human monocyte-derived macrophages mediates cholesterol efflux that occurs in the absence of added cholesterol acceptors. J Biol Chem 271: 28641–28646, 1996.

161. Kruth HS, Skarlatos SI, Gaynor PM, Gamble W: Production of cholesterol-enriched nascent high density lipoproteins by human monocyte-derived macrophages is a mechanism that contributes to macrophage cholesterol efflux. J Biol Chem 269: 24511–24518, 1994.

162. Curtiss LK, Boisvert WA: Apolipoprotein E and atherosclerosis. Curr Opin Lipidol 11: 243–251, 2000.

163. Linton MF, Fazio S: Macrophages, inflammation, and atherosclerosis. Int J Obesity 27: 535–540, 2003.

164. Li AC, Glass CK: The macrophage foam cell as a target for therapeutic intervention. Nat Med 8: 1235–1242, 2002.

165. Stannard AK, Riddell DR, Sacre SM, Tagalakis AD, Langer C, von Eckardstein A, Cullen P, Athanasopouos T, Dickson G, Owen JS: Cell-derived apolipoprotein E (apoE) particles inhibit vascular cell adhesion molecule-1 (VCAM) expression in human endothelial cells. J Biol Chem 246: 46011–46016, 2001.

166. Sacre SM, Stannard AK, Owen JS: Apolipoprotein E isoforms differentially induce nitric oxide production in endothelial cells. FEBS Lett 540: 181–187, 2003.

167. Kothapalli D, Fuki I, Ali K, Stewart SA, Zhao L, Yahil R, Kwiatkowski D, Hawthorne EA, FitzGerald GA, Phillips MC, Lund-Katz S, Puré E, Rader DJ, Assoian RK: Antimitogenic effects of HDL and apoE mediated by Cox-2 dependent IP activation. J Clin Invest 113: 609–618, 2004.

168. Miyata M, Smith JD: Apolipoprotein E alle-specific antioxidant activity and effects on cytotoxicity by oxidative insults and beta-amyloid peptides. Nat Genet 14: 55–61, 1996.

169. Kelly ME, Clay MA, Mistry MJ, Hsieh-Li HM, Harmony JA: Apolipoprotein E inhibition of mitogen-activated T-lymphocytes: production of interleukin 2 with reduced biological activity. Cell Immunol 159: 124–139, 1994.

170. Mistry MJ, Clay MA, Kelly ME, Steiner MA, Harmony JA: Apolipoprotein E restricts interleukin-dependent T-lymphocyte proliferation at the G1A/G1B boundary. Cell Immunol 160: 14–23, 1995.

171. Ishigami M, Swertferger DK, Granholm NA, Hui DY: Apolipoprotein E inhibits platelet-derived growth factor-induced vascular smooth muscle cell migration and proliferation by suppressing signal transduction and preventing cell entry to G1 phase. J Biol Chem 273: 20156–20161, 1998.

172. Swertfeger DK, Hui DY: Apolipoprotein E receptor binding versus heparan sulfate proteoglycan binding in its regulation of smooth muscle cell migration and proliferation. J Biol Chem 276: 25043–25048, 2001.

173. Riddell DR, Vinogradov DV, Stannard AK, Chadwick N, Owen JS: Identification and characterisation of LRP8 (apoER2) in human blood platelets. J Lipid Res 40: 1925–1930, 1999.

174. Riddell DR, Graham A, Owen JS: Apolipoprotein E inhibits platelet aggregation through the L-arginine:nitric oxide pathway. Implications for vascular disease. J Biol Chem 272: 89–95, 1997.

175. Linton MF, Fazio S: Macrophages, lipoprotein metabolism, and atherosclerosis: insights from murine bone marrow transplantation studies. Curr Opin Lipidol 10: 97–105, 1999.
176. Bellosta S, Mahley RW, Sanan DA, Murata J, Newland DL, Taylor JM, Pitas RE: Macrophage-specific expression of human apolipoprotein E reduces atherosclerosis in hypercholesterolemic apolipoprotein E-null mice. J Clin Invest 96: 2170–2179, 1995.
177. Boisvert WA, Spangenberg J, Curtiss LK: Treatment of severe hypercholesterolemia in apolipoprotein E-deficient mice by bone marrow transplantation in mice. J Clin Invest 96: 1118–1124, 1995.
178. Linton MF, Atkinson JB, Fazio S: Prevention of atherosclerosis in apolipoprotein E-deficient mice by bone marrow transplantation. Science 267: 1034–1037, 1995.
179. Zhu Y, Bellosta S, Langer C, Bernini F, Pitas RE, Mahley RW, Assmann G, von Eckardstein A: Low-dose expression of a human apolipoprotein E transgene in macrophages restores cholesterol efflux capacity of apolipoprotein E-deficient mouse plasma. Proc Natl Acad Sci USA 95: 7585–7590, 1998.
180. Van Eck M, Herijgers N, Van Dijk KW, Havekes LM, Hofker MH, Groot PHE, Van Berkel TJ: Effect of macrophage-derived mouse apoE, human apoE3-Leiden, and human apoE2 (Arg158-Cys) on cholesterol levels and atherosclerosis in apoE-deficient mice. Arteriosclerosis 20: 119–127, 2000.
181. Van Eck M, Herijgers N, Van Dijk KW, Havekes LM, Hofker MH, Groot PH, Van Berkel TJ: Effect of macrophage-derived mouse ApoE, human ApoE3-Leiden, and human ApoE2 (Arg158→Cys) on cholesterol levels and atherosclerosis in ApoE-deficient mice. Arterioscler Thromb Vasc Biol 20: 119–127, 2000.
182. Van Eck M, Herijgers N, Yates J, Pearce NJ, Hoogerbrugge PM, Groot PH, Van Berkel TJ: Bone marrow transplantation in apolipoprotein E-deficient mice. Effect of ApoE gene dosage on serum lipid concentrations, (beta)VLDL catabolism, and atherosclerosis. Arterioscler Thromb Vasc Biol 17: 3117–3126, 1997.
183. Fazio S, Babaev VR, Murray AB, Hasty AH, Carter KJ, Gleaves LA, Atkinson JB, Linton MF: Increased atherosclerosis in mice reconstituted with apolipoprotein E null macrophages. Proc Natl Acad Sci USA 94: 4647–4652, 1997.
184. Boisvert WA, Curtiss LK: Elimination of macrophage-specific apolipoprotein E reduces diet-induced atherosclerosis in C57Bl/6J male mice. J Lipid Res 40: 806–813, 1999.
185. Van Eck M, Herijgers N, Vidgeon-Hart M, Pearce NJ, Hoogerbrugge PM, Groot PHE, Van Berkel TJ: Accelerated atherosclerosis in C57Bl/6 mice transplanted with ApoE-deficient bone marrow. Atherosclerosis 150: 71–80, 2000.
186. Ishibashi S, Nobuhiro Y, Shimano H, Mori N, Mokuno H, Gotohda T, Kawakami M, Murata J: Apolipoprotein E and lipoprotein lipase secreted from human monocyte-derived macrophages modulate very low density lipoprotein uptake. J Biol Chem 265: 3040–3047, 1990.
187. Haghpassand M, Bourassa P-AK, Francone OL, Aiello RJ: Monocyte/macrophage expression of ABCA1 has minimal contribution to plasma HDL levels. J Clin Invest 108: 1315–1320, 2001.
188. Aiello RJ, Brees D, Bourassa P-AK, Royer L, Lindsey S, Coskran T, Haghpassand M, Francone O: Increased atherosclerosis in hyperlipidemic mice with inactivation of ABCA1 in macrophages. Arterioscler Thromb Vasc Biol 22: 630–637, 2002.
189. Van Eck M, Bos ST, Kaminski WE, Orsó E, Rothe G, Twisk J, Van Amersfoort ES, Christiansen-Weber TA, Fung-Leung WP, van Berkel TJC, Schmitz G:

Leukocyte ABCA1 controls susceptibility to atherosclerosis and macrophage recruitment into tissues. Proc Natl Acad Sci USA 99: 6298–6303, 2002.

190. Joyce CW, Amar MJA, Lambert G, Vaisman BL, Paigen B, Najib-Fruchart J, Hoyt RF Jr Neufeld ED, Remaley AT, Fredrickson DS, Brewer HB Jr: The ATP binding cassette transporter A1 (ABCA1) modulates the development of aortic atherosclerosis in C57BL/6 and apoE-knockout mice. Proc Natl Acad Sci USA 99: 407–412, 2002.

191. von Eckardstein A, Langer C, Engel T, Schaukal I, Cignarella A, Reinhardt J, Lorkowski S, Li Z, Zhou X, Cullen P, Assmann G: ATP binding cassette transporter ABCA1 modulates the secretion of apolipoprotein E from human monocyte-derived macrophages. FASEB J 15: 1555–1561, 2001.

192. Koo C, Innerarity TL, Mahley RW: Obligatory role of cholesterol and apolipoprotein E in the formation of large cholesterol-enriched and receptor-active high density lipoproteins. J Biol Chem 260: 11934–11943, 1985.

193. Huang ZH, Lin CY, Oram JF, Mazzone T: Sterol efflux mediated by endogenous macrophage apoE expression is independent of ABCA1. Arterioscler Thromb Vasc Biol 21: 2019–2025, 2001.

194. Huang ZH, Mazzone T: ApoE-dependent sterol efflux from macrophages is modulated by scavenger receptor class B type I expression. J Lipid Res 43: 375–382, 2002.

7

Provision of Lipids for Very Low-Density Lipoprotein Assembly

DEAN GILHAM AND RICHARD LEHNER

Abstract

Atherosclerosis is the principal cause of heart attack and stroke in the Western world. The relationship between high levels of low-density lipoprotein (LDL) and atherosclerosis has been known for several decades. LDL is derived from very low-density lipoprotein (VLDL) in the blood via a complex series of reactions involving hydrolases and transfer of lipids and apoproteins among the circulating lipoproteins. Hepatic VLDL assembly and secretion is dependent on lipid availability. Thus stimulation of hepatic triacylglycerol (TG), phospholipid, and cholesterol synthesis increases the secretion of VLDL while inhibition of lipid synthesis has the opposite effect. Several lines of evidence suggest that the assembly of VLDL takes place in two or more steps. The first step involves the generation of dense nascent lipoprotein particles in the endoplasmic reticulum by cotranslational lipidation of apolipoprotein B and these particles mature into a VLDL by posttranslational addition of core lipids such as TG and cholesteryl ester (CE). De novo synthesized TG accounts for only a minor fraction of the TG secreted with VLDL. The majority (60–70%) of VLDL-TG is derived from lipolysis of stored TG to partial acylglycerols and fatty acids followed by reesterification by the ER-localized acyltransferases. Recently, the genes of a number of enzymes involved in the provision of lipids for VLDL assembly have been cloned and in some cases mice in which these genes were ablated have been obtained. This chapter discusses recent advances in VLDL assembly with emphasis on the lessons learned from genetically modified mouse models.

Keywords: acyltransferase; lipolysis; liver; triacylglycerol; triacylglycerol hydrolysis; VLDL

Abbreviations: ACAT, acyl coenzyme A:cholesterol acyltransferase; AGPAT, 1-acylglycerol-3-phosphate acyltransferase; apo, apolipoprotein; CCT, cytidine triphosphate:phosphocholine cytidylyltransferase; CHAPS, [(3-cholamidopropyl) dimethylammonio]-1-propanesulfonate; CE, cholesteryl ester; CVD, cardiovascular disease; DG, diacylglycerol; DGAT, diacylglycerol acyltransferase; DGTA, diacylglycerol transacylase; E600, diethyl-*p*-nitrophenyl phosphate; ER; endoplasmic reticulum; GPAT, glycerol-3-phosphate acyltransferase; HDL, high-density lipoprotein; HSL, hormone-sensitive lipase; LDL, low-density lipoprotein; MG, monoacylglycerol; MGAT, monoacylglycerol acyltransferase; MTP, microsomal triglyceride transfer protein; PC, phosphatidylcholine; PEMT, phosphatidylethanolamine *N*-methyltransferase; PAPH, phosphatidic acid phosphohydrolase; PPAR, peroxisomal proliferator-activated receptor; SCD, stearoyl-CoA desaturase; SREBP, sterol regulatory elements binding protein; TG, triacylglycerol; TGH, triacylglycerol hydrolase; VLDL, very low-density lipoprotein.

Triacylglycerol Metabolism is Linked to Several Human Diseases

The ability to store neutral lipids plays a critical role in the ability of an organism to withstand fuel deprivation. Triacylglycerol (TG) is the most concentrated form of energy available to biological tissues. It can provide over double the amount of energy by metabolic oxidation than an equal mass of either carbohydrates or proteins [1]. TG is composed of a glycerol backbone to which three fatty acids are attached via ester bonds. Most cells and tissues have the ability to synthesize and store TG to some degree. TG metabolism is of tremendous clinical relevance. Dysregulation of TG lipid synthesis, transport, or storage contributes to the development of disease. Consequently there is tremendous clinical and pharmacological relevance for enzymes that metabolize TG. Obesity alone is associated with type 2 diabetes, hyperlipidemia, cardiovascular disease, stroke, gallbladder disease, and some cancers [2–7]. Excessive TG storage within muscle and pancreatic β-cells correlates with insulin resistance, leading ultimately to diabetes [8, 9]. Excessive hepatic TG secretion contributes to high levels of circulating TG in the blood, which is a risk factor for cardiovascular disease (CVD). Plasma TG levels >150 mg/dL or 1.70 mM in a fasted state are considered elevated [10].

Components and Classification of Lipoproteins

Neutral lipids such as TG and CE are hydrophobic in nature, and hence have very low solubility in aqueous environments including blood. Therefore, in order to be transported throughout the body, where they are required for various cellular functions, hydrophobic lipids are packaged into complexes with specific proteins known as apolipoproteins (apos) as well as partly polar phospholipids. All mature TG-rich lipoproteins consist of a hydrophobic core of nonpolar lipids including cholesteryl esters (CE) and TG surrounded by a phospholipid monolayer. Free or unesterified cholesterol can also exist within the monolayer of phospholipid. Apos are also amphipathic and confer solubility to the lipoprotein particle as well as influence trafficking via interactions with various cell surface receptors, hydrolytic enzymes, and lipid-transferring proteins [11]. Most apos are exchangeable; they can detach from the surface of a lipoprotein then reattach to another. Hence the complement of apos changes during the lifetime of these particles. The exception to apos that exchange is more hydrophobic and very large apoB, which is integral and remains bound to the particle it is initially assembled with [12, 13].

Lipoproteins are classified based on their density, which is determined by their lipid and protein composition and is inversely proportional to their diameter. Chylomicrons are the largest and least dense class of TG-rich lipoproteins, with a very TG-rich core (75–1200 nm in diameter) [14]. They are synthesized by the small intestine utilizing ingested dietary lipids, and are carried by the lymphatic system to the bloodstream. Very low-density lipoprotein (VLDL) is the second largest lipoprotein (30–100 nm), and is synthesized and secreted by the liver via an intricate secretory pathway [15, 16]. VLDL is also rich in TG and is catabolized in the circulatory system via a complex series of reactions involving lipases and the transfer of lipids and apos among lipoproteins [17] to produce intermediate (IDL) and low-density lipoproteins (LDL; LDL is approximately 18–25 nm in diameter).

The final class is the smallest (25 nm in diameter or smaller), and these particles are known as high-density lipoproteins (HDL). HDL does not carry apoB, but rather contains at least one molecule of apoA-I among other apos [11]. HDL has a relatively CE-rich core. It is important for the removal of excess cholesterol from peripheral (nonhepatic) cells and its transport back to the liver for recycling or excretion as bile. The collection of cholesterol from the periphery and delivery back to the liver is known as the reverse cholesterol transport pathway. The cardioprotective nature of HDL has been shown in numerous epidemiological studies worldwide over the last 50 years. A striking inverse correlation between HDL cholesterol and incidence of CVD has been observed [18–21], strongly suggesting that a major risk factor for CVD development is low-plasma HDL [22, 23]. Although the detailed mechanisms of how cells efflux cholesterol to HDL and how HDL protects against atherosclerosis are not fully elucidated, potential strategies to treat and prevent CVD exist in raising plasma HDL levels [24].

VLDL Assembly and Secretion

The major apo component of chylomicrons, VLDL, and the sole apo of LDL is a single apoB molecule per particle [12, 17]. Several excellent reviews regarding the mechanism of VLDL assembly and secretion have been published [16, 25, 26]. Gene targeting experiments in mice revealed that apoB is the sole apo required for the secretion of a VLDL particle [27]. ApoB is encoded by a single gene, though it occurs in two forms due to mRNA editing by a deaminase that converts codon 2153 from CAA (glutamine) to a stop codon [28, 29]. The larger form, termed apoB100, is the sole form of apoB secreted by the human liver and includes 100% of the mature protein. Human apoB100 has 4536 amino acids and a calculated molecular weight of over 512 kDa [30, 31]. The shorter form, apoB48, represents the N-terminal 48% of apoB100 with 2152 amino acids and a calculated molecular weight of over 264 kDa [13]. ApoB48 is secreted by human intestine on chylomicrons for transport of dietary lipids. While apoB100 is capable of an interaction with the LDL receptor to mediate cellular uptake via endocytosis, apoB48 lacks the LDL-receptor interacting domain. It is initially counterintuitive that chylomicrons can be up to ten times larger than VLDL and carry more lipids, while at the same time they contain the shorter apoB protein. This means that the C-terminal 52% of apoB100 is not necessary for acquisition of large amounts of lipid during lipoprotein assembly. Both forms of apoB are secreted by the liver of rats and mice [27], making these animals a more challenging model in which to investigate lipoprotein metabolism as it applies to humans.

Triacylglycerol Biosynthesis

The fatty acids present within the TG molecule are provided by dietary lipids or are generated de novo. Variations in the composition and quantity of the diet affect the rate of de novo fatty acid and TG synthesis to satisfy the body's demands for lipid (or energy) storage [32]. The source of the fatty acids for TG synthesis varies depending upon the tissue [33]. For example, the enterocyte takes up dietary fatty acids and 2-monacylglycerols (MG) and resynthesizes TG from these substrates primarily through the 2-monoacylglycerol (MG) pathway. Hepatocytes take up fatty acids derived from remnant lipoproteins and free fatty acids released from adipocytes. It is generally accepted that the synthesis of TG is controlled primarily by the amount of fatty acid available. Thus, de novo TG synthesis is enhanced only when fatty acid synthesis or uptake exceeds the cellular need for energy and phospholipid synthesis. In hepatocytes, as well as adipocytes, de novo TG synthesis is primarily achieved through the glycerol-3-phosphate pathway (Fig. 7.1).

The Glycerol-3-Phosphate Pathway of TG Biosynthesis

The glycerol-3-phosphate pathway is the main route of TG biosynthesis in nonenteric tissues. Briefly, it begins with the acylation of glycerol-3-phosphate by glycerol-3-phosphate acyltransferase (GPAT) at the *sn*-1 position to form 1-acylglycerol-3-phosphate. Subsequently, the acylation of 1-acylglycerol-3-phosphate at the *sn*-2 position by 1-acylglycerol-3-phosphate acyltransferase (AGPAT) (also referred to as lysophosphatidic acid acyltransferase) synthesizes phosphatidic acid (PA). The hydrolysis of PA by phosphatidate phosphohydrolase (PAPH) generates *sn*-1,2-diacylglycerol (DG). DG is then acylated to form TG via the action of diacylglycerol acyltransferase (DGAT). The utilization of molecular cloning and gene targeting techniques has elucidated many new insights regarding these enzymes and their regulation. The

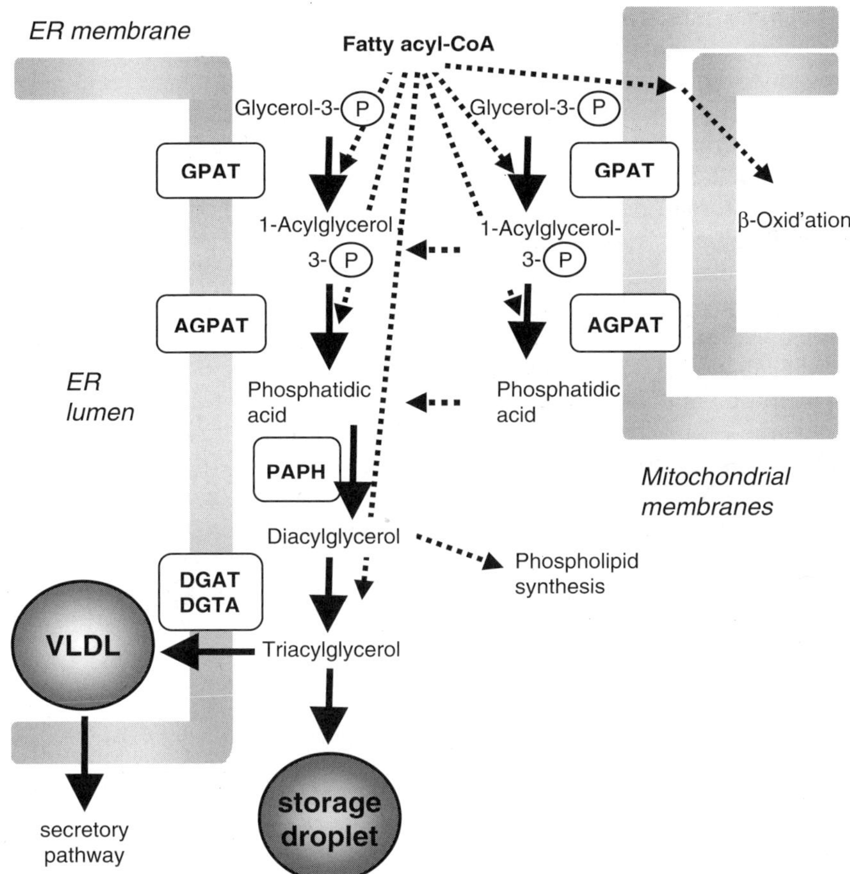

FIGURE 7.1. The glycerol phosphate pathway of triacylglycerol synthesis. Glycerol-3-phosphate is acylated to form 1-acyl-glycerol-3-phosphate by glycerol-3-phosphate acyltransferase (GPAT). A subsequent acylation is accomplished by 1-acyl-glycerol-3-phosphate acyltransferase (AGPAT), resulting in phosphatidic acid. Hydrolysis of the *sn*-3-phosphate through the action of phosphatidic acid phosphohydrolase (PAPH) produces diacylglycerol. Diacylglycerol may be converted to triacylglycerol by diacylglycerol acyltransferase (DGAT) or diacylglycerol transacylase (DGTA). Alternately, it can be diverted towards phospholipid biosynthesis. In liver, triacylglycerol is stored in droplets or secreted on VLDL.

regulation of TG synthesis and its breakdown determines the quantity of TG stored within a cell.

Glycerol-3-Phosphate Acyltransferase

GPAT represents the committed step for the synthesis of all glycerolipids. Two isoforms for this enzyme exist. A mitochondrial GPAT has been cloned and demonstrated to be tightly regulated [34]. A microsomal GPAT activity also exists, though a mammalian cDNA has not been cloned and its activity appears to be largely unregulated [34]. The murine mitochondrial GPAT cDNA is primarily expressed in liver, adipose, and muscle tissues [35].

The nutritional and hormonal regulation of the murine mitochondrial GPAT is well characterized and occurs primarily at the transcriptional level. Posttranslational reduction of mitochondrial GPAT activity has been reported to occur upon stimulation of adenosine monophosphate-activated kinase, a sensor of cellular fuel deprivation [36].

GPAT activity has strong links to obesity. For example, Zucker rats are a commonly used model for obesity as well as insulin resistance. These animals exhibit enhanced GPAT activity in both hepatic and adipose tissues [37]. The regulation of mitochondrial GPAT is consistent with a function in reducing TG synthesis and increasing oxidation when nutrients are limiting, and conversely to direct fatty acids toward TG synthesis and away from oxidation when energy is in excess [38]. The localization of this GPAT activity on the cytosolic surface of mitochondria also suggests a role in preventing fatty acids (more accurately fatty acyl-CoAs) from entering this organelle and undergoing β-oxidation under conditions of metabolic excess by converting it to 1-acylglycerol-3-phosphate and thus directing the fatty acid towards TG synthesis (Fig. 7.1).

Acylglycerol-3-Phosphate Acyltransferase

This enzyme is also referred to as lysophosphatidic acid acyltransferase [34]. Several AGPAT cDNA sequences have been identified. Human and mouse cDNAs of 2 and 1.8 kb, respectively that encode proteins with AGPAT activity were identified and designated AGPAT-1 [39, 40]. 3T3-L1 cells are a fibroblastic preadipocyte cell line that can be differentiated into cells with an adipocyte phenotype. AGPAT-1 overexpression in 3T3-L1 adipocytes inhibited fatty acid release while promoting its uptake and TG storage [41]. Expression of AGPAT-1 mRNA has been observed in all tissues examined, with highest expression in skeletal muscle [39, 40]. AGPAT-2 mRNA is expressed at highest levels in the liver, heart, and adipose tissues, and much less in other tissues [42, 43]. The protein encoded by the AGPAT-2 cDNA shares 48% homology with AGPAT-1 and is expressed at twofold higher levels than AGPAT-1 in the adipose tissue, similar levels in liver and AGPAT-1 is expressed at 1.8-fold greater levels than AGPAT-2 in skeletal muscle [44]. AGPAT-1 and -2 were localized to the endoplasmic reticulum (ER) membrane [45, 46]. Several individuals with congenital generalized lipodystrophy were discovered to have a variety of different mutations in the AGPAT-2 gene that led to the production of a dysfunctional protein [44]. These observations suggest that AGPAT-2 is necessary for TG synthesis in adipocytes. Three additional AGPAT cDNAs have been designated AGPAT-3, -4, and -5, respectively. The corresponding mRNAs are barely detectable in liver, adipose, and skeletal muscle and do not yet have a clearly identified function [44].

Phosphatidic Acid Phosphohydrolase

Two forms of phosphatidic acid phosphohydrolase (PAPH) activity exist, PAPH-1 and PAPH-2 (also called lipid phosphate phosphohydrolase). Several cDNAs for the PAPH-2 family or lipid phosphate phosphohydrolases have been cloned. These enzymes appear to play a role in signal transduction pathways by generating the lipid second messenger, DG [47]. PAPH-1 activity exists in both cytosol and microsomal fractions. It has been proposed that long-chain fatty acids stimulate the translocation of PAPH-1 from the cytosol to the ER membrane where PA is hydrolyzed to DG [48, 49]. A cDNA encoding an enzyme with PAPH-1 activity has not been cloned, but this activity is considered as the rate-limiting step for de novo TG synthesis. There is a strong correlation between the activity of membrane-associated PAPH-1 and the rate of TG synthesis [50]. PAPH-1 activity is subjected to dietary- and hormone-induced changes [50]. PAPH-1 activity was increased 2–3-fold by glucagon, cyclic adenosine monophosphate, and dexamethasone in primary rat hepatocytes [51–53]. It was hypothesized that enhanced PAPH-1 activity enables the liver to increase its ability to synthesize TG during the period of maximum feeding.

Diacylglycerol Acyltransferase

DGAT catalyzes the final step in TG synthesis. Two DGAT cDNAs have been cloned. DGAT-1 is ubiquitously expressed in human and murine tissues, with the highest levels of expression in the small intestine, colon, testis, thymus, heart, and skeletal muscle [54]. The mammalian DGAT-2 cDNA is unrelated to the mammalian DGAT-1 cDNA and was cloned based on its homology with a fungal DGAT cDNA [55]. Both proteins have been localized to microsomal membranes. DGAT-2 is highly expressed in liver and adipose, which correlates with a high capacity to synthesize TG in these tissues. Analysis of the predicted amino acid sequence from the DGAT-2 cDNA suggests the presence of a single transmembrane segment and a domain with homology to the glycerol-binding domain of GPAT and AGPAT. The glycerol-binding domain of DGAT-2 is predicted to be within the ER lumen where the synthesis of TG would be ideally linked to the site of VLDL assembly. There is an experimental evidence for the existence of overt (cytosolic) and latent (lumenal) DGAT activities in microsomes [56]. Hence, different acyltransferases (or transacylases), acting in different subcellular compartments could be involved in the synthesis of TG, and the extent of their respective activities could channel TG preferentially for storage within a cytosolic storage pool or towards the secretory pathway via assembly into lipoproteins. Whether the DGAT-1 or -2 enzyme is responsible for directing TG towards VLDL versus storage has been investigated.

Recently, adenovirus constructs were used to overexpress DGAT-1 or DGAT-2 in mice [57]. Intriguingly the data showed that DGAT-1 overexpression led to increased VLDL, while DGAT-2 overexpression led to TG accumulation in the liver. This supports the hypothesis that DGAT-1 activity directs TG synthesis to the lumen of the ER where it is then available for VLDL assembly, while DGAT-2 activity directs TG to cytoplasmic storage pools. The precise contribution of DGAT-1 to TG synthesis was also examined in mice that had the *Dgat*-1 gene disrupted. The targeted deletion of the *Dgat*-1 gene failed to eliminate the synthesis of TG in many tissues and the level of plasma TG was normal [58, 59]. *Dgat1–/–* mice had dry fur that did not repel water, and female mice had a defect in their ability to lactate [58]. In addition, DGAT-1 null mice had a 50% reduction in fat pad content and were resistant to weight gain when fed a high-fat diet that was associated with a 15% increase in daily total energy expenditure. DGAT-1 null mice had increased sensitivity to insulin and leptin. When *Dgat1–/–* mice were crossed into the background of the obese leptin-deficient *ob/ob* mouse, DGAT-1 deficiency did not affect energy and glucose metabolism [59]. In addition, DGAT-2 expression was increased in *ob/ob* mice, suggesting that the leptin pathway directly downregulates DGAT-2 expression. In the absence of leptin, DGAT-2 sufficiently compensated for the loss of DGAT-1-mediated TG synthesis [59, 60]. Further, others have observed a threefold increase in overt and latent DGAT activities in liver microsomes from obese *ob/ob* mice relative to lean controls [61]. DGAT-1 was not necessary for intestinal TG absorption and chylomicron synthesis, although a high-fat diet caused accumulation of cytosolic TG within the enterocytes of *Dgat1–/–* mice [62]. Together, these results suggest that DGAT-2 (or possibly diacylglycerol transacylase (DGTA)—discussed later) activities sufficiently compensated for the targeted deletion of the DGAT-1 gene in many tissues.

Disruption of the *Dgat*-2 gene resulted in lethality shortly after birth [63], revealing a developmental requirement of this gene product that is not compensated by DGAT-1. Surprising insights into a role of DGAT-2 in provision of lipids involved in regulating the permeability of the skin were also revealed. The skin defects in these mice were much more profound than those in DGAT-1 null mice. The phenotypes demonstrate that DGAT-1 and -2 have different roles in TG metabolism. Tissue-specific disruption of *Dgat*-2 may contribute to a better understanding of how this enzyme participates in various cellular processes.

Diacylglycerol Transacylase

As indicated in Fig. 7.1, fatty acyl coenzyme A (acyl-CoA) plays a central role in TG biosynthesis. Acyl-CoA-independent mechanisms for TG formation are also present in animal tissues. The isolation of a DGTA from enteric microsomes demonstrated that the transacylation between two 1,2-DG molecules to form TG and MG occurs [64]. The relative contribution of this pathway to TG synthesis is unknown because enzymes of the MG and PA pathways exist in the same subcellular compartment and must be dissected apart to discern these details. DGTA activity is low when compared to total DGAT activity [64, 65], however, the presence of DGTA may be important for the synthesis of TG within the ER lumen, since transport of acyl-CoA across the ER membrane would not be required for this route of TG synthesis [66].

Monoacylglycerol Pathway of TG Biosynthesis

Within the intestinal lumen, lipases hydrolyze TG to free fatty acids and 2-MG. MG is taken up by enterocytes and undergoes resynthesis back to TG. This process is initially catalyzed by monoacylglycerol acyltransferase (MGAT) to synthesize DG, which then serves as a substrate for DGAT or DGTA. Most of the TG generated by the intestine is secreted as chylomicron particles.

Monoacylglycerol Acyltransferase

MGAT activity in the small intestine is primarily responsible for the acylation of dietary MGs. In liver, MGAT activity is higher in diabetic animals, but low in obese Zucker rats [37, 67]. Significant levels of MGAT activity are detected in white adipose tissue and kidney [37, 68]. The different chromatographic and inhibition profiles of the rat intestinal and hepatic MGAT activities suggested that MGAT activity represents tissue-specific isoenzymes [69, 70]. A murine cDNA was identified as encoding a protein that possessed MGAT activity when expressed in insect cells [71]. This gene, designated as MGAT-1, is expressed in stomach, kidney, liver, and adipose tissues of the mouse. Interestingly, MGAT-1 was not expressed in the small intestine, again suggesting the existence of an additional MGAT gene. Following this, MGAT-2 and -3 have been identified, and their respective cDNAs have been cloned [72, 73]. MGAT-2 shows highest expression in the small intestine, with lower amounts of the mRNA detectable in kidney, adipose, and stomach. This enzyme was described as also for possessing a weak DGAT activity in transfected cells. MGAT-3 expression appears to be exclusive to the small intestine and is most abundant in the ileum. Both MGAT-2 and -3 are thus implicated in the process of dietary fat absorption.

Subcellular Distribution of Fatty Acyl-CoA for Lipid Synthesis

The subcellular localization of activated fatty acid that is fatty acyl-CoA, at the ER membrane for utilization by various acyltransferases including DGAT and MGAT is presently unclear. One possibility is that microsomal acyltransferases (which are typically transmembrane proteins) utilize acyl-CoA strictly from the cytosolic pool. A second possibility is the transfer of fatty acids across the ER membrane as acylcarnitine derivatives. Microsomal carnitine acyltransferases have been demonstrated to exist [74, 75]. Since tolbutamide inhibits microsomal carnitine acyltransferase and suppresses

VLDL secretion by hepatocytes, it seems reasonable to hypothesize that the microsomal carnitine acyltransferase system is an important component for the delivery of fatty acids to the ER lumen [76, 77]. However, more recent studies suggest that the observed microsomal carnitine acyltransferase activities may have been a result of contamination of microsomes with mitochondria [78, 79].

Phosphatidylcholine Biosynthesis

The primary phospholipid found on lipoproteins is phosphatidylcholine (PC) [14]. Two biochemical pathways for PC synthesis exist. One pathway is essentially liver specific and involves the conversion of phosphatidylethanolamine to PC via three consecutive methylation reactions that are all catalyzed by the enzyme phosphatidylethanolamine *N*-methyltransferase (PEMT) [80]. *S*-Adenosylmethionine is utilized by this enzyme as the donor of the methyl group to produce *S*-adenosylhomocysteine, which is subsequently converted to homocysteine. Plasma homocysteine levels are an independent risk factor for the development of CVD [81]. Disruption of the *Pemt* gene resulted in a 50% reduction in plasma homocysteine in mice [82], demonstrating that the PEMT reaction has a very significant impact on plasma levels of this molecule, and is important with respect to CVD development beyond VLDL production. The PEMT null mice also rapidly develop liver failure on a choline-deficient diet compared to wild-type mice [83]. The mechanism for this is related to a sensitivity towards the loss of PC in bile [84].

The predominant pathway for PC synthesis is found in all nucleated cells [85]. The process begins with the uptake of dietary choline, as choline is not manufactured in animal cells other than via methylation of phosphatidylethanolamine to PC followed by hydrolysis of the choline group [86]. Following uptake, choline is rapidly phosphorylated to phosphocholine by choline kinase [87]. Phosphocholine is combined with cytidine triphosphate (rather than adenosine triphosphate encountered in many biological reactions) in a rate-limiting reaction catalyzed by cytidine triphosphate:phosphocholine cytidylyltransferase (CCT) to form cytidine diphosphocholine (CDP-choline) [86]. There are two mammalian genes that encode isoforms of CCT, and the two genes exhibit tissue-specific expression. One gene encodes CCT-α, which is expressed in most tissues [88], with high expression in testis, lung, liver, and ovary. Subcellularly, CCT-α is found most abundantly in the nucleus [89], with lower amounts associated with the ER and the cytosol [88]. The second gene encodes CCT-β, which is most predominant in the brain [90], and is localized to the ER. Alternate splicing of the CCT-β gene results in at least two mRNAs termed CCT-β2 and CCT-β3. Current biochemical data have not distinguished a unique, nonredundant role for the CCT-β isoforms in cell or tissue function, although disruption of CCT-β2 produced gonadal dysfunction in both male and female mice [91]. No overt defects were found in the brain. An additional splice variation of the CCT-β gene gives rise to a CCT-β1 transcript found in the expressed sequence tag database [90], however, the protein has not been detected in either mouse or human tissues. Therefore, the existence of this protein remains speculative.

In a final step towards this phospholipid synthesis, CDP-choline is combined with 1,2-DG to form PC in a reaction catalyzed by CDP-choline:1,2-diacylglycerol cholinephosphotransferase. This enzyme and PEMT are localized to the ER where the majority of phospholipid synthesis occurs. It is currently unclear which pathway is of primary importance with respect to VLDL assembly, as evidence for both has been presented [92–94].

Summary of Cholesterol Biosynthesis

Cholesterol and CE are relatively minor, but important components of VLDL [14]. Beyond lipoproteins, cholesterol is used for membrane biogenesis, cell growth, steroid hormones, and bile acids (reviewed in [95]). The cholesterol biosynthetic pathway is complex and involves numerous enzymes, which are localized to the cytosol, ER, and peroxisomes [96]. The ultimate synthesis of this molecule occurs in the ER. The entire cholesterol biosynthetic route will not be reviewed here. The rate-limiting step in cholesterol biosynthesis occurs early in the pathway, and is catalyzed by 3-hydroxy-3-methylglutarly-CoA reductase. This enzyme's activity is regulated by cellular cholesterol levels. Chemical inhibition of this enzyme by the statin family of drugs has been used therapeutically to reduce plasma cholesterol levels [97]. Owing to their ability to lower plasma LDL cholesterol, the use of statins has shown a marked reduction in coronary events, especially in people with hypercholesterolemia.

Because cholesterol in membranes affects their fluidity, cholesterol is esterified for storage in lipid droplets. Esterification is mediated by acyl-CoA:cholesterol acyltransferase (ACAT) [98]. ACAT is thought to play an important role in membrane biology by maintaining the free sterol content of membranes within ranges optimal for proper cell function. Two isoforms of ACAT are encoded by separate genes [99]. ACAT-1 is found ubiquitously throughout the body, while ACAT-2 is expressed in liver and intestine. Both isoforms are localized to the ER membrane.

Provision of Lipid for VLDL Assembly

The liver has the ability to store neutral lipids or secrete them into the circulation as VLDL particles, thereby regulating storage and secretion with the energy needs of the body [16, 25]. Exogenous fatty acids taken up by hepatocytes are not directly utilized for secretion as VLDL-associated TG, but enter an intracellular storage pool of TG [100, 101]. In the postabsorptive state, stored TG is released by the liver as VLDL [100]. On average, a normal adult human liver stores about 5 μmol TG per gram of liver weight, though the liver has the capacity to store much larger quantities of TG [102].

A plethora of studies have been performed in hepatocytes to investigate VLDL assembly and secretion. These include the hepatoma cell lines McArdle RH7777 and HepG2 as well as primary hepatocytes isolated from rats, hamsters, and mice. ApoB, the main protein component of VLDL, is translated on ribosomes associated with the rough ER. The transcription and translation of apoB is continuous so that newly synthesized apoB is always available for assembly with lipids for secretion [103]. Lipid availability determines the percentage of apoB that is assembled into secretion-competent lipoprotein particles versus that which is misfolded and targeted for degradation [15, 103–105]. The relative importance of each core lipid in the assembly of VLDL is a fundamental issue that has been explored using a variety of methods. There is general agreement that stimulation of hepatic TG synthesis will also augment the secretion of TG and apoB [15, 106, 107]. In contrast, there is debate over whether or not CE biosynthesis alters VLDL and apoB secretion. Mice virtually devoid of hepatic ACAT activity, the enzyme catalyzing CE formation, still synthesize and secrete apoB-containing lipoproteins, although the particles were of smaller size than those produced from wild-type mice [108]. Counter to these studies, however, the overexpression of ACAT in McArdle RH7777 cells resulted in increased synthesis, cellular accumulation, and secretion of CE [109]. Decreased intracellular degradation and increased secretion of apoB were also observed. Overexpression of DGAT-1 in this cell line had similar effects in that increased TG levels and apoB secretion were encountered, reflecting the critical importance of neutral lipid availability in VLDL production.

It is generally accepted that the assembly of VLDL particles takes place in two or more steps [15, 110]. Studies using McArdle RH7777 cells and primary rat hepatocytes demonstrated that the newly translated apoB is not integrated into microsomal membranes, but is initially associated with the lumenal side of the ER membrane in a peripheral fashion [111, 112]. In the first step of VLDL assembly, apoB is partially lipidated with small quantities of TG, phospholipid, and cholesterol forming a dense primordial particle. Since apoB is a large hydrophobic protein, chaperone-assisted folding of apoB appears to be necessary during its translation and translocation into the ER lumen [113, 114]. The subsequent addition of the bulk of the neutral lipid to the small dense apoB-containing entity represents the second step and results in a very low-density apoB-containing particle that may be secreted from the hepatocyte and into the circulatory system.

There is currently debate in the literature about the location and mode of assembly of VLDL particles. Several studies demonstrated that both steps in VLDL assembly are complete before the particle enters the Golgi apparatus in primary hepatocytes and in McArdle RH7777 cells [115–117]. Alternately, completion of VLDL assembly may occur in post-ER or in the Golgi compartment [118, 119].

Regulation of VLDL Secretion

In hepatoma cells, active synthesis of lipids drives VLDL secretion. Oleic acid promotes apoB secretion by hepatoma cells through increased TG synthesis and lipid availability [120, 121]. Triacsin D blocked oleic acid-stimulated apoB secretion through the inhibition of TG synthesis without affecting CE synthesis [122]. It has been suggested that CE synthesis has a regulatory role in VLDL synthesis [123, 124]. However targeted deletion of the ACAT genes in the mouse failed to demonstrate that CE synthesis was essential for lipoprotein production [108, 125]. PC synthesis, a quantitatively small component of VLDL, is a potential regulator of VLDL assembly and secretion [92, 94, 126].

Stearoyl-CoA desaturase (SCD) catalyzes the synthesis of monounsaturated fatty acids from saturated fatty acids. SCD-1-deficient mice have impaired TG and CE synthesis [127]. Furthermore, these mice secrete lower levels of TG on VLDL [127, 128]. SCD-1 expression is tightly regulated by insulin and carbohydrates [129]. High-carbohydrate diets enhance the synthesis and secretion of TG by the liver and the hepatic secretion of TG on VLDL by SCD-1-deficient mice do not respond to a high-carbohydrate diet, suggesting that SCD-1 can regulate VLDL secretion [129, 130].

Origin of TG for VLDL Assembly and the Role of Microsomal Triglyceride Transfer Protein

A lingering debate addresses the location of the storage pool of TG that is utilized for lipidation of primordial apoB-containing particles (the second step of apoB lipidation). Fatty acids can be synthesized or taken up by hepatocytes, but enter an intracellular storage pool as TG and are not secreted directly [77]. The bulk of hepatic TG is stored in cytoplasmic lipid droplets. It remains unclear how TG from the cytosol-facing ER-associated TG droplet is recruited to the lumen of the ER for VLDL assembly. A non-apoB-associated neutral lipid droplet has been observed by electron microscopy within the lumen of the smooth ER, and hypothesized to be the source of the bulk of the lipid for the primordial VLDL particle [131]. Non-apoB-associated neutral lipid droplets in the ER lumen of hepatocytes has subsequently been demonstrated in several reports [132, 133]. Since TG cannot diffuse across the ER bilayer en bloc, the lumenal lipid droplet is presumably derived from both de novo synthesized TG and from cytoplasmic TG stores in a process that involves microsomal triglyceride transfer protein (MTP). MTP is composed of a 97-kDa subunit that is complexed with the ER-folding chaperone protein disulfide isomerase [134, 135]. MTP has the ability to transfer lipid between liposomes. This activity appears to be essential for transferring the bulk of triglycerides into the lumen of the ER for VLDL assembly [136]. MTP-dependent formation of apoB-free TG-rich droplets within the ER lumen may fuse with the primordial apoB-containing particle [137]. While the formation of the lipid droplet within the ER lumen is thought to be dependent upon MTP, the bulk lipidation of the apoB particle may be MTP independent [111, 132, 136, 138]. Inhibition of MTP caused apoB to be cotranslationally degraded at an early stage in lipoprotein assembly [139].

MTP has been identified as the defective gene in abetalipoproteinemia [140, 141]. Abetalipoproteinemia is a rare disease associated with the absence of apoB-containing lipoproteins in plasma and malabsorption of fat-soluble vitamins causing severe spinocerebellar and retinal degeneration. In mice, homozygous disruption of the *Mtp* gene results in embryonic lethality [142]. Liver-specific gene disruption of *Mtp* resulted in striking reductions in plasma apoB, VLDL triglycerides, and large reductions in both VLDL/LDL and HDL-associated cholesterol [136]. MTP inhibitors have been used as therapeutic agents in clinical trials for hypercholesterolemia in humans (reviewed in [143]). Although very impressive results were achieved with respect to plasma lipids, safety concerns arose regarding absorption of fat-soluble vitamins as well as steatosis of the liver.

Stored TG Undergoes Lipolysis and Reesterification Prior to VLDL Assembly

Several studies from various laboratories have demonstrated that only a minor fraction of TG in VLDL originates from de novo synthesis, while the majority (60–70%) is derived from preformed TG storage pools (reviewed by [102]). Further, several groups, using different experimental approaches, have quantitatively determined that stored TG within hepatocytes undergoes a cycle of lipolysis followed by reesterification prior to secretion on VLDL particles. A proportion of the TG is secreted as VLDL, though the majority of the reesterified TG gets returned to storage depots.

Dual radioisotope labeling studies in primary rat hepacytes by Wiggins and Gibbons [77] yielded two significant observations: (i) a 70% of the VLDL-TG is derived from the storage pool and (ii) the quantity of hydrolyzed TG that returned to the intracellular storage pool amounted to 1 pool per day, which was estimated to be 2–3 times greater than required to maintain TG secretion. The majority of TG that undergoes lipolysis and reesterification was returned to storage pools.

Lankester et al. [144] also used a dual radiolabeling technique to differentiate between the incorporation of acyl chains into TG that were derived from exogenous and endogenous fatty acids. The stored TG was prelabeled with [^{3}H] oleate and then cultured a further 3 h with [^{14}C] oleate. The authors observed exogenous [^{14}C] fatty acids contributed only approximately 17% of total acyl chains secreted as TG, indicating that the majority of the acyl chains were derived from the prelabeled TG stores. Furthermore, the authors demonstrated that the TG was not hydrolyzed completely to fatty acid and glycerol. Since the

[^{3}H] fatty acids derived from radiolabeled TG was not available to the same extent as exogenous [^{14}C] fatty acids for oxidation, [^{3}H]-labeled TG undergoes incomplete lipolysis to DG and is available for resynthesis as secreted TG.

Yang et al. [145] used chiral and reverse-phase high-pressure liquid chromatography with mass spectrometry to reveal similarities in positional distribution and molecular association in the 1,2-DG acyl chains of secreted TG on VLDL and stored hepatic TG. However, the 2,3-DG acyl chains of secreted TG on VLDL were different from the stored hepatic TG. These authors calculated that 60% of secreted TG was derived via lipolysis to DG followed by reesterification and 40% of secreted TG could have been derived from de novo TG synthesis. These authors added further evidence through the analyses of secreted TG following labeling of stored TG by [^{3}H] fatty acids or glycerol. The authors deduced that 30–40% of the glycerol and fatty acids in TG on VLDL are not direct products of TG stored within the liver [146]. Taken together, the data is consistent with a proportion of the TG stored within the liver undergoing lipolysis to DG/MG, which is reesterified to TG that is available for secretion on VLDL or returned to storage pools in a futile cycle.

Enzymes Handling TG for VLDL should Localize to the ER

Since apoB-containing lipoproteins are at least partly assembled within the ER lumen, it is likely that the synthesis of TG for VLDL assembly must be directed towards that compartment. The molecular mechanism and intracellular location of the enzymes responsible for the lipolysis and resynthesis of TG for VLDL secretion remain obscure, but requires ER-localized DGAT and possibly MGAT. The TG lipolysis may take place at sites where the ER membrane is in contact with cytosolic lipid droplets [147, 148]. At these contact points, lipolytic products would have increased solubility within the ER membrane and would encounter the TG-synthetic enzymes located within the ER membrane as opposed to being directed towards the oxidative pathway in the mitochondria [102]. This ER localization would promote resynthesis of TG at or in proximity to the site of VLDL assembly. TG that is not assembled onto apoB may be returned to storage pools. There has been substantial progress towards identifying and characterizing the lipase responsible for mobilizing stored TG for VLDL assembly. It has been established that this lipolysis is not catalyzed by lysosomal (acidic) lipase [77]. Additionally, the well-described cytosolic hormone-sensitive lipase (HSL) [149] is not found in appreciable quantities in the liver. Importantly, expression of HSL in HepG2 cells directed fatty acids into the oxidative pathway as opposed to the secretory pathway [150], supporting the concept that the subcellular localization of any TG lipase is crucial in directing the fate of the products of lipolysis.

Triacylglycerol Hydrolase

It was an important advance when triacylglycerol hydrolase (TGH) was puri-fied from porcine liver microsomes and characterized [151]. TGH accounts for approximately 70% of hepatic microsomal lipolytic activity. The purified pro-tein was isolated by solubilization by the zwitterionic detergent, 3-[(3-cholami-dopropyl)dimethylammonio]-1-propanesulfonate (CHAPS) and subsequent column chromatography to yield a single protein band with an apparent molec-ular mass of ~60 kDa in SDS-PAGE. TGH hydrolyzes long-, medium- and short-chain TGs. It does not hydrolyze phospholipids or acyl-CoA thioesters. Divalent cations are not required for optimal lipolytic activity. The enzyme has a neutral pH optimum and is inactivated by the lipase inhibitors tetrahydrolip-statin, diethyl-*p*-nitrophenyl phosphate (E600) and diisopropyl fluorophos-phate, indicating that it is a serine esterase [151, 152]. The N-terminal sequences of purified porcine TGH were found to be identical to that of porcine proline-β-naphthylamidase [153]. TGH expression was immunodetected in livers from humans, rats, mice, hamsters, and cows [152]. In rodents, TGH is highly expressed in liver and adipose tissue with lesser levels found in heart, kidney, and small intestine, while in humans TGH is mainly expressed in the liver, adi-pose, and small intestine [154]. TGH was subsequently purified from human and murine liver microsomes, and cDNAs for rat, mouse, and human TGH have been cloned [155–157]. These cDNAs are predicted to encode proteins of 565 amino acids for rat and mouse and 568 amino acids for human. In addi-tion to mobility on SDS-PAGE, the calculated molecular masses of these pro-teins are also ~60 kDa from the amino acid sequences derived from the cDNAs. The murine and human TGH proteins share 92% identity and rat and human proteins share 93% identity at the amino acid level [154]. Sequence analysis showed that TGH has minimal amino acid identity to previously identified lipases, and is more closely related to the family of mammalian car-boxylesterases (EC 3.1.1.1) [154]. Both the TGH mRNA and protein are expressed in mouse and rat liver toward the end of the suckling period [152, 158], which coincides with ontogeny of VLDL secretion. This altered TGH expression appears to be related to dietary changes at the time of weaning and independent of hepatic differentiation because TGH expression was unchanged in regenerating livers that undergo dedifferentiation and acquire fetal and neonatal features following partial hepatectomy [152]. The transcrip-tion factor, Sp1 has been implicated in the dramatic increase in hepatic TGH mRNA and protein observed during the suckling/weaning transition in mice [158]. The TGH genes are located within a cluster of carboxylesterase genes on mouse and human chromosomes 8 and 16, respectively. Both the murine and human TGH genes span approximately 30 kb and contain 14 exons [154].

Localization of TGH

If TGH activity were a component of the pathway for assembly and secretion of apoB-containing lipoprotein particles, then it would be anticipated that

TGH would be expressed in tissues that have the ability to synthesize and secrete apoB-containing lipoproteins. The high level of TGH expression in the liver is consistent with its involvement in the mobilization of intracellular TG stores for VLDL secretion [154, 159]. Immunocytochemical studies localize TGH expression exclusively to adult hepatocytes surrounding the capillary vessels leading to the central vein [152]. This region of the liver is most likely to be active in lipoprotein production and secretion. TGH cosediments with ER- and mitochondria-associated membranes [152], which have been demonstrated to contain enzymatic activities required for synthesis and assembly of lipoproteins [160]. Rat TGH could only be detected in liver parenchymal cells, but not Kupffer or endothelial cells [161].

Mechanism of How TGH Participates in VLDL Assembly

In rodents, TGH is highly expressed in liver and adipose tissue with lesser levels found in heart, kidney, and small intestine, while in humans TGH is mainly expressed in the liver, adipose, and small intestine [154]. TGH hydrolyzes stored TG and, in the liver, the lipolytic products are made available for VLDL synthesis [155, 162]. A similar function has been hypothesized for TGH in the small intestine regarding chylomicron assembly, although this remains to be demonstrated. A role for TGH in basal TG lipolysis in adipocytes has been described [163].

In a model of TG mobilization for VLDL assembly, TGH acts to hydrolyze TG in lipid-storage droplets that are associated with the ER. The lumenal lipid droplet is derived from cytoplasmic stores in a process that involves MTP. Lipolysis of TG to DG/MG would make the lipolytic products more soluble in the membrane, thereby facilitating the efficient transfer of the acylglycerols for resynthesis to TG by lumenally oriented acyltransferases and availability to a developing apoB-containing entity. TG that is not assembled onto apoB may be returned to either cytosolic or lumenal storage pools in a futile cycle. Importantly, TGH does not hydrolyze apoB-associated lipids within the ER, which suggest a vectored movement of lipids from ER-localized TG to nascent apoB lipoproteins [164].

Regulation of Hepatic TG Lipolysis and Reesterification

Since it has been demonstrated that intracellular TG lipolysis and reesterification was necessary for the efficient recruitment of stored TG for VLDL assembly, then altering the rate of hepatic TG lipolysis and reesterification could control the rate of VLDL secretion. Glucose increased hepatic TG secretion and increased the dilution of the glycerol label in prelabeled TG, consistent with increased lipolysis and reesterification [130, 165, 166]. In addition, it was found that glucose phosphorylation was a necessary event for increased lipolysis and reesterification since mannoheptulose, an inhibitor of glucose phosphorylation, abolished the stimulatory effect of glucose [166].

TGH expression was not affected by supplementation of the diet with 20% fatty acid, regardless of the degree of saturation [167]. Therefore, it appears that a high-fat diet does not alter the level of hepatic TGH-mediated lipolysis. Including cholesterol with the high-fat diet, however, increased murine hepatic TGH expression approximately twofold. This induction of TGH expression is consistent with the observed induction of the orthologous human TGH expression by incubation of human macrophages with exogenous LDL [168, 169]. Although sequences that bear some similarity to sterol response elements exist in the murine TGH proximal promoter, the functionality of these sequences has not been determined [158]. Further, the enforced expression of a nuclear form of sterol regulatory elements binding protein (SREBP)-1a in transgenic mice did not alter TGH mRNA expression in the liver (R. Lehner and D.E. Vance, unpublished observation). It is currently unclear whether the cholesterol-mediated regulation of TGH expression is due to SREBPs, oxysterol nuclear receptors, or an indirect mechanism. It is important to note that hypercholesterolemic $ApoE^{-/-}$ mice also have a threefold higher level of hepatic TGH expression than wild-type mice [170], supporting a role for cholesterol in influencing TGH expression in liver.

Because the action of TGH generates fatty acids from TG, and the murine TGH promoter has a peroxisomal proliferator-activated receptor (PPAR)-like response element, an investigation into whether PPARs are involved in TGH expression was performed using PPAR agonists [167]. The data indicate that PPAR-α and PPAR-γ agonists can alter TGH expression, but the changes reflect secondary responses, and PPARs are not directly involved in TGH expression [167]. This suggests that TGH expression is not directly regulated by PPARs or fatty acids both in the liver and adipose tissue.

Conclusions and Future Perspectives

The cloning and characterization of enzymes involved in TG synthesis and intracellular lipolysis constitute a major advance in this field. Further research using animal models in which the various genes coding for hepatic acyltransferases and TGH are ablated should address detailed correlations of the enzymes in the mobilization of hepatic lipid for VLDL assembly. Concerted cell biology and proteomics efforts should delineate interactions of the various proteins involved in the vectorial movement of lipids towards lipidation of apoB in the ER.

References

1. Brindley DN: Metabolism of triacylglycerols. In: Vance DE, Vance JE (eds), Biochemistry of Lipids, Lipoproteins and Membranes. Elsevier, Amsterdam, 1991, pp 171–204.
2. American Heart Association (AHA): Heart Disease and Stroke Statistics—2003 Update. American Heart Association, Dallas, TX, 2004.

3. National Institutes of Health (NIH): Clinical Guidelines on the Identification, Evaluation, and Treatment of Overweight and Obesity in Adults. National Institutes of Health, Bethesda, MD. Available: www.nhlbi.nih.gov/guidelines/obesity/ob_home.htm. 1998.

4. World Health Organization (WHO): Obesity: Preventing and Managing the Global Epidemic. Report of a WHO Consultation on Obesity, Geneva, June 3–5, 1997. 1998.

5. Husten L: Global epidemic of cardiovascular disease predicted. Lancet 352: 1530, 1998.

6. Ross R: The pathogenesis of atherosclerosis: a perspective for the 1990s. Nature 362: 801–809, 1993.

7. Zimmet P, Alberti KG, Shaw J: Global and societal implications of the diabetes epidemic. Nature 414: 782–787, 2001.

8. Shulman GI: Cellular mechanisms of insulin resistance. J Clin Invest 106: 171–176, 2000.

9. Kahn BB, Flier JS: Obesity and insulin resistance. J Clin Invest 106: 473–481, 2000.

10. The National Cholesterol Education Program (NCEP): Executive summary of the third report of The National Cholesterol Education Program (NCEP) expert panel on detection, evaluation, and treatment of high blood cholesterol in adults (Adult Treatment Panel III). JAMA 285: 2486–2497, 2001.

11. Mahley RW, Innerarity TL, Rall SC Jr, Weisgraber KH: Plasma lipoproteins: apolipoprotein structure and function. J Lipid Res 25: 1277–1294, 1984.

12. Elovson J, Chatterton JE, Bell GT, Schumaker VN, Reuben MA, Puppione DL, Reeve JR Jr, Young NL: Plasma very low density lipoproteins contain a single molecule of apolipoprotein B. J Lipid Res 29: 1461–1473, 1988.

13. Chan L: Apolipoprotein B, the major protein component of triglyceride-rich and low density lipoproteins. J Biol Chem 267: 25621–25624, 1992.

14. Davis RA, Vance JE: Structure, assembly and secretion of lipoproteins. In: Vance DE, Vance JE (eds), Biochemistry of Lipids, Lipoproteins and Membranes. Elsevier, Amsterdam, 1996, pp 473–495.

15. Fisher EA, Ginsberg HN: Complexity in the secretory pathway: the assembly and secretion of apolipoprotein B-containing lipoproteins. J Biol Chem 277: 17377–17380, 2002.

16. Shelness GS, Sellers JA: Very-low-density lipoprotein assembly and secretion. Curr Opin Lipidol 12: 151–157, 2001.

17. Fielding PE, Fielding CJ: Dynamics of lipoprotein transport in the human circulatory system. In: Vance DE, Vance JE (eds), Biochemistry of Lipids, Lipoproteins, and Membranes. Elsevier, Amsterdam, 1996, pp 494–516.

18. Gordon T, Castelli WP, Hjortland MC, Kannel WB, Dawber TR: High density lipoprotein as a protective factor against coronary heart disease. The Framingham Study. Am J Med 62: 707–714, 1977.

19. Castelli WP, Garrison RJ, Wilson PW, Abbott RD, Kalousdian S, Kannel WB: Incidence of coronary heart disease and lipoprotein cholesterol levels. The Framingham Study. JAMA 256: 2835–2838, 1986.

20. Abbott RD, Wilson PW, Kannel WB, Castelli WP: High density lipoprotein cholesterol, total cholesterol screening, and myocardial infarction. The Framingham Study. Arteriosclerosis 8: 207–211, 1988.

21. Husten L: More data reported for HDL's role in heart disease. Lancet 352: 1603, 1998.

22. Gordon DJ, Probstfield JL, Garrison RJ, Neaton JD, Castelli WP, Knoke JD, Jacobs DR Jr, Bangdiwala S, Tyroler HA: High-density lipoprotein cholesterol and cardiovascular disease. Four prospective American studies. Circulation 79: 8–15, 1989.

23. Wilson PW, Abbott RD, Castelli WP: High density lipoprotein cholesterol and mortality. The Framingham Heart Study. Arteriosclerosis 8: 737–741, 1988.

24. Linsel-Nitschke P, Tall AR: HDL as a target in the treatment of atherosclerotic cardiovascular disease. Nat Rev Drug Discov 4: 193–205, 2005.

25. Olofsson SO, Asp L, Boren J: The assembly and secretion of apolipoprotein B-containing lipoproteins. Curr Opin Lipidol 10: 341–346, 1999.

26. Gibbons GF, Brown AM, Wiggins D, Pease R: The roles of insulin and fatty acids in the regulation of hepatic very-low-density lipoprotein assembly. J R Soc Med 95(Suppl 42): 23–32, 2002.

27. Veniant MM, Kim E, McCormick S, Boren J, Nielsen LB, Raabe M, Young SG: Insights into apolipoprotein B biology from transgenic and gene-targeted mice. J Nutr 129: 451S–455S, 1999.

28. Powell LM, Wallis SC, Pease RJ, Edwards YH, Knott TJ, Scott J: A novel form of tissue-specific RNA processing produces apolipoprotein-B48 in intestine. Cell 50: 831–840, 1987.

29. Chen SH, Habib G, Yang CY, Gu ZW, Lee BR, Weng SA, Silberman SR, Cai SJ, Deslypere JP, Rosseneu M, Gotto AM, Li WH, Chan L: Apolipoprotein B-48 is the product of a messenger RNA with an organ-specific in-frame stop codon. Science 238: 363–366, 1987.

30. Chen SH, Yang CY, Chen PF, Setzer D, Tanimura M, Li WH, Gotto AM Jr, Chan L: The complete cDNA and amino acid sequence of human apolipoprotein B-100. J Biol Chem 261: 12918–12921, 1986.

31. Law SW, Grant SM, Higuchi K, Hospattankar A, Lackner K, Lee N, Brewer HB Jr: Human liver apolipoprotein B-100 cDNA: complete nucleic acid and derived amino acid sequence. Proc Natl Acad Sci USA 83: 8142–8146, 1986.

32. Coleman RA, Lee DP: Enzymes of triacylglycerol synthesis and their regulation. Prog Lipid Res 43: 134–176, 2004.

33. Lehner R, Kuksis A: Biosynthesis of triacylglycerols. Prog Lipid Res 35: 169–201, 1996.

34. Coleman RA, Lewin TM, Muoio DM: Physiological and nutritional regulation of enzymes of triacylglycerol synthesis. Annu Rev Nutr 20: 77–103, 2000.

35. Paulauskis JD, Sul HS: Cloning and expression of mouse fatty acid synthase and other specific mRNAs. Developmental and hormonal regulation in 3T3-L1 cells. J Biol Chem 263: 7049–7054, 1988.

36. Muoio DM, Seefeld K, Witters LA, Coleman RA: AMP-activated kinase reciprocally regulates triacylglycerol synthesis and fatty acid oxidation in liver and muscle: evidence that sn-glycerol-3-phosphate acyltransferase is a novel target. Biochem J 338(Pt 3): 783–791, 1999.

37. Jamdar SC, Cao WF: Triacylglycerol biosynthetic enzymes in lean and obese Zucker rats. Biochim Biophys Acta 1255: 237–243, 1995.

38. Ericsson J, Jackson SM, Kim JB, Spiegelman BM, Edwards PA: Identification of glycerol-3-phosphate acyltransferase as an adipocyte determination and differentiation factor 1 and sterol regulatory element-binding protein-responsive gene. J Biol Chem 272: 7298–7305, 1997.

39. Stamps AC, Elmore MA, Hill ME, Kelly K, Makda AA, Finnen MJ: A human cDNA sequence with homology to non-mammalian lysophosphatidic acid acyltransferases. Biochem J 326(Pt 2): 455–461, 1997.

40. Kume K, Shimizu T: cDNA cloning and expression of murine 1-acyl-*sn*-glycerol-3-phosphate acyltransferase. Biochem Biophys Res Commun 237: 663–666, 1997.

41. Ruan H, Pownall HJ: Overexpression of 1-acyl-glycerol-3-phosphate acyltransferase-α enhances lipid storage in cellular models of adipose tissue and skeletal muscle. Diabetes 50: 233–240, 2001.

42. Eberhardt C, Gray PW, Tjoelker LW: Human lysophosphatidic acid acyltransferase. cDNA cloning, expression, and localization to chromosome 9q34.3. J Biol Chem 272: 20299–20305, 1997.

43. Eberhardt C, Gray PW, Tjoelker LW: cDNA cloning, expression and chromosomal localization of two human lysophosphatidic acid acyltransferases. Adv Exp Med Biol 469: 351–356, 1999.

44. Agarwal AK, Arioglu E, De Almeida S, Akkoc N, Taylor SI, Bowcock AM, Barnes RI, Garg A: AGPAT2 is mutated in congenital generalized lipodystrophy linked to chromosome 9q34. Nat Genet 31: 21–23, 2002.

45. Aguado B, Campbell RD: Characterization of a human lysophosphatidic acid acyltransferase that is encoded by a gene located in the class III region of the human major histocompatibility complex. J Biol Chem 273: 4096–4105, 1998.

46. Leung DW: The structure and functions of human lysophosphatidic acid acyltransferases. Front Biosci 6: D944–D953, 2001.

47. Brindley DN, Waggoner DW: Mammalian lipid phosphate phosphohydrolases. J Biol Chem 273: 24281–24284, 1998.

48. Gomez-Munoz A, Hatch GM, Martin A, Jamal Z, Vance DE, Brindley DN: Effects of okadaic acid on the activities of two distinct phosphatidate phosphohydrolases in rat hepatocytes. FEBS Lett 301: 103–106, 1992.

49. Martin A, Hales P, Brindley DN: A rapid assay for measuring the activity and the Mg^{2+} and Ca^{2+} requirements of phosphatidate phosphohydrolase in cytosolic and microsomal fractions of rat liver. Biochem J 245: 347–355, 1987.

50. Tijburg LB, Geelen MJ, van Golde LM: Regulation of the biosynthesis of triacylglycerol, phosphatidylcholine and phosphatidylethanolamine in the liver. Biochim Biophys Acta 1004: 1–19, 1989.

51. Pittner RA, Fears R, Brindley DN: Effects of cyclic AMP, glucocorticoids and insulin on the activities of phosphatidate phosphohydrolase, tyrosine aminotransferase and glycerol kinase in isolated rat hepatocytes in relation to the control of triacylglycerol synthesis and gluconeogenesis. Biochem J 225: 455–462, 1985.

52. Pittner RA, Fears R, Brindley DN: Interactions of insulin, glucagon and dexamethasone in controlling the activity of glycerol phosphate acyltransferase and the activity and subcellular distribution of phosphatidate phosphohydrolase in cultured rat hepatocytes. Biochem J 230: 525–534, 1985.

53. Pittner RA, Bracken P, Fears R, Brindley DN: Spermine antagonises the effects of dexamethasone, glucagon and cyclic AMP in increasing the activity of phosphatidate phosphohydrolase in isolated rat hepatocytes. FEBS Lett 207: 42–46, 1986.

54. Cases S, Smith SJ, Zheng YW, Myers HM, Lear SR, Sande E, Novak S, Collins C, Welch CB, Lusis AJ, Erickson SK, Farese RV Jr: Identification of a gene encoding

an acyl CoA:diacylglycerol acyltransferase, a key enzyme in triacylglycerol synthesis. Proc Natl Acad Sci USA 95: 13018–13023, 1998.

55. Cases S, Stone SJ, Zhou P, Yen E, Tow B, Lardizabal KD, Voelker T, Farese RV Jr: Cloning of DGAT2, a second mammalian diacylglycerol acyltransferase, and related family members. J Biol Chem 276: 38870–38876, 2001.

56. Owen MR, Corstorphine CC, Zammit VA: Overt and latent activities of diacylglycerol acytransferase in rat liver microsomes: possible roles in very-low-density lipoprotein triacylglycerol secretion. Biochem J 323(Pt 1): 17–21, 1997.

57. Yamazaki T, Sasaki E, Kakinuma C, Yano T, Miura S, Ezaki O: Increased VLDL secretion and gonadal fat mass in mice over-expressing liver acyl CoA:diacylglycerol acyltransferase 1. J Biol Chem 280: 21506–21514, 2005.

58. Smith SJ, Cases S, Jensen DR, Chen HC, Sande E, Tow B, Sanan DA, Raber J, Eckel RH, Farese RV Jr: Obesity resistance and multiple mechanisms of triglyceride synthesis in mice lacking Dgat. Nat Genet 25: 87–90, 2000.

59. Chen HC, Smith SJ, Ladha Z, Jensen DR, Ferreira LD, Pulawa LK, McGuire JG, Pitas RE, Eckel RH, Farese RV Jr: Increased insulin and leptin sensitivity in mice lacking acyl CoA:diacylglycerol acyltransferase 1. J Clin Invest 109: 1049–1055, 2002.

60. Chen HC, Smith SJ, Tow B, Elias PM, Farese RV Jr: Leptin modulates the effects of acyl CoA:diacylglycerol acyltransferase deficiency on murine fur and sebaceous glands. J Clin Invest 109: 175–181, 2002.

61. Waterman IJ, Zammit VA: Activities of overt and latent diacylglycerol acyltransferases (DGATs I and II) in liver microsomes of *ob/ob* mice. Int J Obes Relat Metab Disord 26: 742–743, 2002.

62. Buhman KK, Smith SJ, Stone SJ, Repa JJ, Wong JS, Knapp FF Jr, Burri BJ, Hamilton RL, Abumrad NA, Farese RV Jr: DGAT1 is not essential for intestinal triacylglycerol absorption or chylomicron synthesis. J Biol Chem 277: 25474–25479, 2002.

63. Stone SJ, Myers HM, Watkins SM, Brown BE, Feingold KR, Elias PM, Farese RV Jr: Lipopenia and skin barrier abnormalities in DGAT2-deficient mice. J Biol Chem 279: 11767–11776, 2004.

64. Lehner R, Kuksis A: Triacylglycerol synthesis by an *sn*-1,2(2,3)-diacylglycerol transacylase from rat intestinal microsomes. J Biol Chem 268: 8781–8786, 1993.

65. Lehner R, Kuksis A: Utilization of 2-monoacylglycerols for phosphatidylcholine biosynthesis in the intestine. Biochim Biophys Acta 1125: 171–179, 1992.

66. Abo-Hashema KA, Cake MH, Power GW, Clarke D: Evidence for triacylglycerol synthesis in the lumen of microsomes via a lipolysis-esterification pathway involving carnitine acyltransferases. J Biol Chem 274: 35577–35582, 1999.

67. Mostafa N, Bhat BG, Coleman RA: Increased hepatic monoacylglycerol acyltransferase activity in streptozotocin-induced diabetes: characterization and comparison with activities from adult and neonatal rat liver. Biochim Biophys Acta 1169: 189–195, 1993.

68. Coleman RA, Haynes EB: Hepatic monoacylglycerol acyltransferase. Characterization of an activity associated with the suckling period in rats. J Biol Chem 259: 8934–8938, 1984.

69. Manganaro F, Kuksis A: Rapid isolation of a triacylglycerol synthetase complex from rat intestinal mucosa. Can J Biochem Cell Biol 63: 107–114, 1985.

70. Manganaro F, Kuksis A: Purification and preliminary characterization of 2-monoacylglycerol acyltransferase from rat intestinal villus cells. Can J Biochem Cell Biol 63: 341–347, 1985.

71. Yen CL, Stone SJ, Cases S, Zhou P, Farese RV Jr: Identification of a gene encoding MGAT1, a monoacylglycerol acyltransferase. Proc Natl Acad Sci USA 99: 8512–8517, 2002.

72. Cao J, Lockwood J, Burn P, Shi Y: Cloning and functional characterization of a mouse intestinal acyl-CoA:monoacylglycerol acyltransferase, MGAT2. J Biol Chem 278: 13860–13866, 2003.

73. Cheng D, Nelson TC, Chen J, Walker SG, Wardwell-Swanson J, Meegalla R, Taub R, Billheimer JT, Ramaker M, Feder JN: Identification of acyl coenzyme A:monoacylglycerol acyltransferase 3, an intestinal specific enzyme implicated in dietary fat absorption. J Biol Chem 278: 13611–13614, 2003.

74. Murthy MS, Pande SV: Malonyl-CoA-sensitive and -insensitive carnitine palmitoyltransferase activities of microsomes are due to different proteins. J Biol Chem 269: 18283–18286, 1994.

75. Broadway NM, Saggerson ED: Microsomal carnitine acyltransferases. Biochem Soc Trans 23: 490–494, 1995.

76. Broadway NM, Saggerson ED: Inhibition of liver microsomal carnitine acyltransferases by sulphonylurea drugs. FEBS Lett 371: 137–139, 1995.

77. Wiggins D, Gibbons GF: The lipolysis/esterification cycle of hepatic triacylglycerol. Its role in the secretion of very-low-density lipoprotein and its response to hormones and sulphonylureas. Biochem J 284(Pt 2): 457–462, 1992.

78. Broadway NM, Pease RJ, Birdsey G, Turner NA, Shayeghi M, Saggerson ED: Microsomal malonyl-CoA-sensitive carnitine acyltransferase. Biochem Soc Trans 29: 267–271, 2001.

79. Broadway NM, Pease RJ, Birdsey G, Shayeghi M, Turner NA, David Saggerson E: The liver isoform of carnitine palmitoyltransferase 1 is not targeted to the endoplasmic reticulum. Biochem J 370: 223–231, 2003.

80. Vance DE, Walkey CJ, Cui Z: Phosphatidylethanolamine *N*-methyltransferase from liver. Biochim Biophys Acta 1348: 142–150, 1997.

81. Refsum H, Ueland PM, Nygard O, Vollset SE: Homocysteine and cardiovascular disease. Annu Rev Med 49: 31–62, 1998.

82. Noga AA, Stead LM, Zhao Y, Brosnan ME, Brosnan JT, Vance DE: Plasma homocysteine is regulated by phospholipid methylation. J Biol Chem 278: 5952–5955, 2003.

83. Walkey CJ, Yu L, Agellon LB, Vance DE: Biochemical and evolutionary significance of phospholipid methylation. J Biol Chem 273: 27043–27046, 1998.

84. Vance JE, Vance DE: Metabolic insights into phospholipid function using gene-targeted mice. J Biol Chem 280: 10877–10880, 2005.

85. Kent C: CTP:phosphocholine cytidylyltransferase. Biochim Biophys Acta 1348: 79–90, 1997.

86. Vance DE: Gylcerolipid biosynthesis in eukaryotes. In: Vance DE, Vance JE (eds), Biochemistry of Lipids, Lipoproteins and Membranes. Elsevier, Amsterdam, 1996, pp 153–182.

87. Uchida T, Yamashita S: Molecular cloning, characterization, and expression in *Escherichia coli* of a cDNA encoding mammalian choline kinase. J Biol Chem 267: 10156–10162, 1992.

88. Lykidis A, Baburina I, Jackowski S: Distribution of CTP:phosphocholine cytidylyltransferase (CCT) isoforms. Identification of a new CCTβ splice variant. J Biol Chem 274: 26992–27001, 1999.

89. Wang Y, Sweitzer TD, Weinhold PA, Kent C: Nuclear localization of soluble CTP:phosphocholine cytidylyltransferase. J Biol Chem 268: 5899–5904, 1993.

90. Karim M, Jackson P, Jackowski S: Gene structure, expression and identification of a new CTP:phosphocholine cytidylyltransferase β isoform. Biochim Biophys Acta 1633: 1–12, 2003.

91. Jackowski S, Rehg JE, Zhang YM, Wang J, Miller K, Jackson P, Karim MA: Disruption of CCTβ2 expression leads to gonadal dysfunction. Mol Cell Biol 24: 4720–4733, 2004.

92. Jacobs RL, Devlin C, Tabas I, Vance DE: Targeted deletion of hepatic CTP:phosphocholine cytidylyltransferase α in mice decreases plasma high density and very low density lipoproteins. J Biol Chem 279: 47402–47410, 2004.

93. Noga AA, Vance DE: A gender-specific role for phosphatidylethanolamine N-methyltransferase-derived phosphatidylcholine in the regulation of plasma high density and very low density lipoproteins in mice. J Biol Chem 278: 21851–21859, 2003.

94. Noga AA, Zhao Y, Vance DE: An unexpected requirement for phosphatidylethanolamine N-methyltransferase in the secretion of very low density lipoproteins. J Biol Chem 277: 42358–42365, 2002.

95. Liscum L, Munn NJ: Intracellular cholesterol transport. Biochim Biophys Acta 1438: 19–37, 1999.

96. Olivier LM, Krisans SK: Peroxisomal protein targeting and identification of peroxisomal targeting signals in cholesterol biosynthetic enzymes. Biochim Biophys Acta 1529: 89–102, 2000.

97. Ganesh SK, Nass CM, Blumenthal RS: Anti-atherosclerotic effects of statins: lessons from prevention trials. J Cardiovasc Risk 10: 155–159, 2003.

98. Farese RV Jr: Acyl CoA:cholesterol acyltransferase genes and knockout mice. Curr Opin Lipidol 9: 119–123, 1998.

99. Joyce C, Skinner K, Anderson RA, Rudel LL: Acyl-coenzyme A:cholesteryl acyltransferase 2. Curr Opin Lipidol 10: 89–95, 1999.

100. Diraison F, Beylot M: Role of human liver lipogenesis and reesterification in triglycerides secretion and in FFA reesterification. Am J Physiol 274: E321–E327, 1998.

101. Gibbons GF, Bartlett SM, Sparks CE, Sparks JD: Extracellular fatty acids are not utilized directly for the synthesis of very-low-density lipoprotein in primary cultures of rat hepatocytes. Biochem J 287(Pt 3): 749–753, 1992.

102. Gibbons GF, Islam K, Pease RJ: Mobilisation of triacylglycerol stores. Biochim Biophys Acta 1483: 37–57, 2000.

103. Pullinger CR, North JD, Teng BB, Rifici VA, Ronhild de Brito AE, Scott J: The apolipoprotein B gene is constitutively expressed in HepG2 cells: regulation of secretion by oleic acid, albumin, and insulin, and measurement of the mRNA half-life. J Lipid Res 30: 1065–1077, 1989.

104. Borchardt RA, Davis RA: Intrahepatic assembly of very low density lipoproteins. Rate of transport out of the endoplasmic reticulum determines rate of secretion. J Biol Chem 262: 16394–16402, 1987.

105. White AL, Graham DL, LeGros J, Pease RJ, Scott J: Oleate-mediated stimulation of apolipoprotein B secretion from rat hepatoma cells. A function of the ability of apolipoprotein B to direct lipoprotein assembly and escape presecretory degradation. J Biol Chem 267: 15657–15664, 1992.

106. Boren J, Rustaeus S, Wettesten M, Andersson M, Wiklund A, Olofsson SO: Influence of triacylglycerol biosynthesis rate on the assembly of apoB-100-containing lipoproteins in HepG2 cells. Arterioscler Thromb 13: 1743–1754, 1993.

107. Yao Z, McLeod RS: Synthesis and secretion of hepatic apolipoprotein B-containing lipoproteins. Biochim Biophys Acta 1212: 152–166, 1994.
108. Buhman KK, Accad M, Novak S, Choi RS, Wong JS, Hamilton RL, Turley S, Farese RV Jr: Resistance to diet-induced hypercholesterolemia and gallstone formation in ACAT2-deficient mice. Nat Med 6: 1341–1347, 2000.
109. Liang JJ, Oelkers P, Guo C, Chu PC, Dixon JL, Ginsberg HN, Sturley SL: Overexpression of human diacylglycerol acyltransferase 1, acyl-CoA:cholesterol acyltransferase 1, or acyl-CoA:cholesterol acyltransferase 2 stimulates secretion of apolipoprotein B-containing lipoproteins in McA-RH7777 cells. J Biol Chem 279: 44938–44944, 2004.
110. Innerarity TL, Boren J, Yamanaka S, Olofsson SO: Biosynthesis of apolipoprotein B48-containing lipoproteins. Regulation by novel post-transcriptional mechanisms. J Biol Chem 271: 2353–2356, 1996.
111. Rustaeus S, Stillemark P, Lindberg K, Gordon D, Olofsson SO: The microsomal triglyceride transfer protein catalyzes the post-translational assembly of apolipoprotein B-100 very low density lipoprotein in McA-RH7777 cells. J Biol Chem 273: 5196–5203, 1998.
112. Hebbachi AM, Gibbons GF: Microsomal membrane-associated apoB is the direct precursor of secreted VLDL in primary cultures of rat hepatocytes. J Lipid Res 42: 1609–1617, 2001.
113. Linnik KM, Herscovitz H: Multiple molecular chaperones interact with apolipoprotein B during its maturation. The network of endoplasmic reticulum-resident chaperones (ERp72, GRP94, calreticulin, and BiP) interacts with apolipoprotein B regardless of its lipidation state. J Biol Chem 273: 21368–21373, 1998.
114. Chen Y, Le Caherec F, Chuck SL: Calnexin and other factors that alter translocation affect the rapid binding of ubiquitin to apoB in the Sec61 complex. J Biol Chem 273: 11887–11894, 1998.
115. Glaumann H, Bergstrand A, Ericsson JL: Studies on the synthesis and intracellular transport of lipoprotein particles in rat liver. J Cell Biol 64: 356–377, 1975.
116. Cartwright IJ, Higgins JA: Intracellular events in the assembly of very-low-density-lipoprotein lipids with apolipoprotein B in isolated rabbit hepatocytes. Biochem J 310(Pt 3): 897–907, 1995.
117. Yamaguchi J, Gamble MV, Conlon D, Liang JS, Ginsberg HN: The conversion of apoB100 low density lipoprotein/high density lipoprotein particles to apoB100 very low density lipoproteins in response to oleic acid occurs in the endoplasmic reticulum and not in the Golgi in McA RH7777 cells. J Biol Chem 278: 42643–42651, 2003.
118. Gusarova V, Brodsky JL, Fisher EA: Apolipoprotein B100 exit from the endoplasmic reticulum (ER) is COPII-dependent, and its lipidation to very low density lipoprotein occurs post-ER. J Biol Chem 278: 48051–48058, 2003.
119. Tran K, Thorne-Tjomsland G, DeLong CJ, Cui Z, Shan J, Burton L, Jamieson JC, Yao Z: Intracellular assembly of very low density lipoproteins containing apolipoprotein B100 in rat hepatoma McA-RH7777 cells. J Biol Chem 277: 31187–31200, 2002.
120. Sakata N, Wu X, Dixon JL, Ginsberg HN: Proteolysis and lipid-facilitated translocation are distinct but competitive processes that regulate secretion of apolipoprotein B in HepG2 cells. J Biol Chem 268: 22967–22970, 1993.
121. Boren J, Rustaeus S, Olofsson SO: Studies on the assembly of apolipoprotein B-100- and B-48-containing very low density lipoproteins in McA-RH7777 cells. J Biol Chem 269: 25879–25888, 1994.

122. Wu X, Sakata N, Lui E, Ginsberg HN: Evidence for a lack of regulation of the assembly and secretion of apolipoprotein B-containing lipoprotein from HepG2 cells by cholesteryl ester. J Biol Chem 269: 12375–12382, 1994.

123. Zhang Z, Cianflone K, Sniderman AD: Role of cholesterol ester mass in regulation of secretion of apoB100 lipoprotein particles by hamster hepatocytes and effects of statins on that relationship. Arterioscler Thromb Vasc Biol 19: 743–752, 1999.

124. Spady DK, Willard MN, Meidell RS: Role of acyl-coenzyme A:cholesterol acyltransferase-1 in the control of hepatic very low density lipoprotein secretion and low density lipoprotein receptor expression in the mouse and hamster. J Biol Chem 275: 27005–27012, 2000.

125. Meiner VL, Cases S, Myers HM, Sande ER, Bellosta S, Schambelan M, Pitas RE, McGuire J, Herz J, Farese RV Jr: Disruption of the acyl-CoA:cholesterol acyltransferase gene in mice: evidence suggesting multiple cholesterol esterification enzymes in mammals. Proc Natl Acad Sci USA 93: 14041–14046, 1996.

126. Yao ZM, Vance DE: The active synthesis of phosphatidylcholine is required for very low density lipoprotein secretion from rat hepatocytes. J Biol Chem 263: 2998–3004, 1988.

127. Miyazaki M, Kim YC, Gray-Keller MP, Attie AD, Ntambi JM: The biosynthesis of hepatic cholesterol esters and triglycerides is impaired in mice with a disruption of the gene for stearoyl-CoA desaturase 1. J Biol Chem 275: 30132–30138, 2000.

128. Miyazaki M, Kim YC, Ntambi JM: A lipogenic diet in mice with a disruption of the stearoyl-CoA desaturase 1 gene reveals a stringent requirement of endogenous monounsaturated fatty acids for triglyceride synthesis. J Lipid Res 42: 1018–1024, 2001.

129. Shimomura I, Shimano H, Korn BS, Bashmakov Y, Horton JD: Nuclear sterol regulatory element-binding proteins activate genes responsible for the entire program of unsaturated fatty acid biosynthesis in transgenic mouse liver. J Biol Chem 273: 35299–35306, 1998.

130. Boogaerts JR, Malone-McNeal M, Archambault-Schexnayder J, Davis RA: Dietary carbohydrate induces lipogenesis and very-low-density lipoprotein synthesis. Am J Physiol 246: E77–E83, 1984.

131. Alexander CA, Hamilton RL, Havel RJ: Subcellular localization of B apoprotein of plasma lipoproteins in rat liver. J Cell Biol 69: 241–263, 1976.

132. Wang Y, Tran K, Yao Z: The activity of microsomal triglyceride transfer protein is essential for accumulation of triglyceride within microsomes in McA-RH7777 cells. A unified model for the assembly of very low density lipoproteins. J Biol Chem 274: 27793–27800, 1999.

133. Kulinski A, Rustaeus S, Vance JE: Microsomal triacylglycerol transfer protein is required for lumenal accretion of triacylglycerol not associated with apoB, as well as for apoB lipidation. J Biol Chem 277: 31516–31525, 2002.

134. Gordon DA, Jamil H: Progress towards understanding the role of microsomal triglyceride transfer protein in apolipoprotein-B lipoprotein assembly. Biochim Biophys Acta 1486: 72–83, 2000.

135. Berriot-Varoqueaux N, Aggerbeck LP, Samson-Bouma M, Wetterau JR: The role of the microsomal triglygeride transfer protein in abetalipoproteinemia. Annu Rev Nutr 20: 663–697, 2000.

136. Raabe M, Veniant MM, Sullivan MA, Zlot CH, Bjorkegren J, Nielsen LB, Wong JS, Hamilton RL, Young SG: Analysis of the role of microsomal triglyceride transfer protein in the liver of tissue-specific knockout mice. J Clin Invest 103: 1287–1298, 1999.

137. Hamilton RL, Wong JS, Cham CM, Nielsen LB, Young SG: Chylomicron-sized lipid particles are formed in the setting of apolipoprotein B deficiency. J Lipid Res 39: 1543–1557, 1998.

138. Gordon DA, Jamil H, Gregg RE, Olofsson SO, Boren J: Inhibition of the microsomal triglyceride transfer protein blocks the first step of apolipoprotein B lipoprotein assembly but not the addition of bulk core lipids in the second step. J Biol Chem 271: 33047–33053, 1996.

139. Zhou M, Fisher EA, Ginsberg HN: Regulated co-translational ubiquitination of apolipoprotein B100. A new paradigm for proteasomal degradation of a secretory protein. J Biol Chem 273: 24649–24653, 1998.

140. Gregg RE, Wetterau JR: The molecular basis of abetalipoproteinemia. Curr Opin Lipidol 5: 81–86, 1994.

141. Wetterau JR, Aggerbeck LP, Bouma ME, Eisenberg C, Munck A, Hermier M, Schmitz J, Gay G, Rader DJ, Gregg RE: Absence of microsomal triglyceride transfer protein in individuals with abetalipoproteinemia. Science 258: 999–1001, 1992.

142. Raabe M, Flynn LM, Zlot CH, Wong JS, Veniant MM, Hamilton RL, Young SG: Knockout of the abetalipoproteinemia gene in mice: reduced lipoprotein secretion in heterozygotes and embryonic lethality in homozygotes. Proc Natl Acad Sci USA 95: 8686–8691, 1998.

143. Sudhop T, von Bergmann K: Cholesterol absorption inhibitors for the treatment of hypercholesterolaemia. Drugs 62: 2333–2347, 2002.

144. Lankester DL, Brown AM, Zammit VA: Use of cytosolic triacylglycerol hydrolysis products and of exogenous fatty acid for the synthesis of triacylglycerol secreted by cultured rat hepatocytes. J Lipid Res 39: 1889–1895, 1998.

145. Yang LY, Kuksis A, Myher JJ, Steiner G: Origin of triacylglycerol moiety of plasma very low density lipoproteins in the rat: structural studies. J Lipid Res 36: 125–136, 1995.

146. Yang LY, Kuksis A, Myher JJ, Steiner G: Contribution of de novo fatty acid synthesis to very low density lipoprotein triacylglycerols: evidence from mass isotopomer distribution analysis of fatty acids synthesized from [^{2}H$_6$]ethanol. J Lipid Res 37: 262–274, 1996.

147. Blanchette-Mackie EJ, Dwyer NK, Barber T, Coxey RA, Takeda T, Rondinone CM, Theodorakis JL, Greenberg AS, Londos C: Perilipin is located on the surface layer of intracellular lipid droplets in adipocytes. J Lipid Res 36: 1211–1226, 1995.

148. Horton JD, Shimano H, Hamilton RL, Brown MS, Goldstein JL: Disruption of LDL receptor gene in transgenic SREBP-1a mice unmasks hyperlipidemia resulting from production of lipid-rich VLDL. J Clin Invest 103: 1067–1076, 1999.

149. Strålfors P, Olsson H, Belfrage P: Hormone-sensitive lipase. In: Boyer PD, Krebs EG (eds), The Enzymes. Academic Press, Orlando, USA, 1987, pp 147–178.

150. Pease RJ, Wiggins D, Saggerson ED, Tree J, Gibbons GF: Metabolic characteristics of a human hepatoma cell line stably transfected with hormone-sensitive lipase. Biochem J 341(Pt 2): 453–460, 1999.

151. Lehner R, Verger R: Purification and characterization of a porcine liver microsomal triacylglycerol hydrolase. Biochemistry 36: 1861–1868, 1997.
152. Lehner R, Cui Z, Vance DE: Subcellullar localization, developmental expression and characterization of a liver triacylglycerol hydrolase. Biochem J 338(Pt 3): 761–768, 1999.
153. Matsushima M, Inoue H, Ichinose M, Tsukada S, Miki K, Kurokawa K, Takahashi T, Takahashi K: The nucleotide and deduced amino acid sequences of porcine liver proline-β-naphthylamidase. Evidence for the identity with carboxylesterase. FEBS Lett 293: 37–41, 1991.
154. Dolinsky VW, Gilham D, Alam M, Vance DE, Lehner R: Triacylglycerol hydrolase: role in intracellular lipid metabolism. Cell Mol Life Sci 61: 1633–1651, 2004.
155. Lehner R, Vance DE: Cloning and expression of a cDNA encoding a hepatic microsomal lipase that mobilizes stored triacylglycerol. Biochem J 343(Pt 1): 1–10, 1999.
156. Dolinsky VW, Sipione S, Lehner R, Vance DE: The cloning and expression of a murine triacylglycerol hydrolase cDNA and the structure of its corresponding gene. Biochim Biophys Acta 1532: 162–172, 2001.
157. Alam M, Ho S, Vance DE, Lehner R: Heterologous expression, purification, and characterization of human triacylglycerol hydrolase. Protein Expr Purif 24: 33–42, 2002.
158. Douglas DN, Dolinsky VW, Lehner R, Vance DE: A role for Sp1 in the transcriptional regulation of hepatic triacylglycerol hydrolase in the mouse. J Biol Chem 276: 25621–25630, 2001.
159. Gilham D, Lehner R: The physiological role of triacylglycerol hydrolase in lipid metabolism. Rev Endocr Metab Disord 5: 303–309, 2004.
160. Rusinol AE, Cui Z, Chen MH, Vance JE: A unique mitochondria-associated membrane fraction from rat liver has a high capacity for lipid synthesis and contains pre-Golgi secretory proteins including nascent lipoproteins. J Biol Chem 269: 27494–27502, 1994.
161. Gaustad R, Berg T, Fonnum F: Heterogeneity of carboxylesterases in rat liver cells. Biochem Pharmacol 44: 827–829, 1992.
162. Gilham D, Ho S, Rasouli M, Martres P, Vance DE, Lehner R: Inhibitors of hepatic microsomal triacylglycerol hydrolase decrease very low density lipoprotein secretion. FASEB J 17: 1685–1687, 2003.
163. Soni KG, Lehner R, Metalnikov P, O'Donnell P, Semache M, Gao W, Ashman K, Pshezhetsky AV, Mitchell GA: Carboxylesterase 3 (EC 3.1.1.1) is a major adipocyte lipase. J Biol Chem 279: 40683–40689, 2004.
164. Gilham D, Alam M, Gao W, Vance DE, Lehner R: Triacylglycerol hydrolase is localized to the endoplasmic reticulum by an unusual retrieval sequence where it participates in VLDL assembly without utilizing VLDL lipids as substrates. Mol Biol Cell 16: 984–996, 2005.
165. Durrington PN, Newton RS, Weinstein DB, Steinberg D: Effects of insulin and glucose on very low density lipoprotein triglyceride secretion by cultured rat hepatocytes. J Clin Invest 70: 63–73, 1982.
166. Brown AM, Wiggins D, Gibbons GF: Glucose phosphorylation is essential for the turnover of neutral lipid and the second stage assembly of triacylglycerol-rich apoB-containing lipoproteins in primary hepatocyte cultures. Arterioscler Thromb Vasc Biol 19: 321–329, 1999.

167. Dolinsky VW, Gilham D, Hatch GM, Agellon LB, Lehner R, Vance DE: Regulation of triacylglycerol hydrolase expression by dietary fatty acids and peroxisomal proliferator-activated receptors. Biochim Biophys Acta 1635: 20–28, 2003.
168. Becker A, Bottcher A, Lackner KJ, Fehringer P, Notka F, Aslanidis C, Schmitz G: Purification, cloning, and expression of a human enzyme with acyl coenzyme A: cholesterol acyltransferase activity, which is identical to liver carboxylesterase. Arterioscler Thromb 14: 1346–1355, 1994.
169. Langmann T, Becker A, Aslanidis C, Notka F, Ullrich H, Schwer H, Schmitz G: Structural organization and characterization of the promoter region of a human carboxylesterase gene. Biochim Biophys Acta 1350: 65–74, 1997.
170. Huang GS, Yang SM, Hong MY, Yang PC, Liu YC: Differential gene expression of livers from apoE deficient mice. Life Sci 68: 19–28, 2000.

8

Oxidatively Modified Low-Density Lipoproteins and Thrombosis

GARRY X. SHEN

Abstract

The atherothrombogenecity of low-density lipoprotein (LDL) is considerably increased following oxidative modification by a group of chemical, physical, or biological agents. Multiple lines of evidence suggest the presence of oxidized LDL (oxLDL) in human tissues and circulation. Elevated levels of oxLDL or its antibodies were detected in patients with atherosclerotic coronary artery disease or venous thrombosis. OxLDL or minimally modified LDL, including glycated LDL, upregulates the expression of factors promoting coagulation, antifibrinolysis, or platelet aggregation in vascular cells. The prothrombotic effects of oxidatively modified LDL may partially result from their stimulatory effects on inflammatory regulators or stress response mediators. Large clinical trials demonstrated that 3-hydroxy-3-methylglutaryl coenzyme A (HMG-CoA) reductase inhibitors, statins, effectively reduced ischemic events or deaths in cardiac or diabetic patients with or without hypercholesterolemia. The cardiac beneficial effect of statins may be partially due to the pleiotropic effects of statins on thrombosis-related factors or oxidative stress. Angiotensin-converting enzymes (ACE) inhibitors and angiotensin II (AII) receptor antagonists inhibit the oxidation of LDL, which may contribute to their effects on reducing platelet activation, thrombosis, and endothelial dysfunction. Accumulating data suggest the crucial role of oxidatively modified LDL in the development of thrombosis in addition to atherosclerosis. Pharmacological treatment with statins or AII antagonists potentially prevents thrombosis through their negative impact on the oxidation of LDL.

Keywords: angiotensin II antagonists; endothelial dysfunction; oxidized LDL; inflammation; statins; thrombosis

Abbreviations: LDL: low-density lipoprotein, IHD: ischemic heart disease, oxLDL: oxidized LDL, SR: scavenger receptor, SMC: smooth muscle cells, EC: endothelial cells, ACS: acute coronary syndrome, MI: myocardial infarction, HDL: high-density lipoprotein, APS: antiphospholipid syndrome, SLE: systematic lupus erythematosus, TF: tissue factor, TFPI: tissue factor pathway inhibitor, APC: activated protein C, EPCR: endothelial cell protein C receptor, tPA: tissue plasminogen activator, uPA: urokinase plasminogen activator, PAI-1: plasminogen activator inhibitor-1, VLDL: very low-density lipoprotein, vWF: von Willebrand factor, NO: nitric oxide, MCP-1: monocyte chemoattractant protein-1, TNF-α: tumor necrosis factor-α, ICAM-1: intracellular adhesion molecule-1, VCAM-1: vascular cell adhesion molecule-1, HSP: heat shock protein, HMG-CoA: 3-hydroxy-3-methylglutaryl coenzyme A, Statins: HMG-CoA reductase inhibitors, F1+2: prothrombin fragment 1+2, AII: angiotension II, ACE: angiotensin-converting enzyme, AT1: AII receptor type 1

Introduction

Low-density lipoprotein (LDL) is an independent risk factor for ischemic heart disease (IHD) [1]. LDL is removed from blood circulation via the LDL receptor [2]. LDL may be oxidatively modified by a group of biological factors or cells *in vitro* or *in vivo*. Particles of oxidized LDL (oxLDL) are recognized by various scavenger receptors (SR) on the surface of vascular smooth muscle cells (SMC), endothelial cells (EC), and macrophages [3, 4]. Oxidative modification substantially increases the atherogenecity of LDL [5]. Growing amounts of evidence suggest that oxLDL also affects the processes of coagulation, fibrinolysis, platelet activation, and thrombosis [6, 7]. This chapter summarizes up-to-date information on the role of oxLDL in thrombogenesis and potential pharmacological management.

Thrombosis and Cardiovascular Events

Thrombosis at the sites of atherosclerotic lesions is a common cause of acute coronary syndrome (ACS). ACS is the leading cause of morbidity and mortality in North America [8]. Thrombus has been frequently detected in coronary arteries of patients with acute myocardial infarction (MI), unstable angina, or sudden death [9]. Vulnerable plaques have been considered as triggers for thrombogenesis and ACS [10]. Exposure of blood components to rough surfaces of plaques may trigger platelet aggregation, coagulation, and the formation of fibrin clots. Under normal condition, fibrin clots are quickly cleared through fibrinolysis. When the activity of fibrinolysis is reduced, fibrin clots may develop to thrombus [11]. Intravascular thrombosis may cause ischemia in tissues or organs where blood supply relies on the involved artery.

LDL and Cardiovascular Diseases

LDL is the major carrier of cholesterol in blood circulation. Apolipoprotein B-100 is the sole apolipoprotein in LDL. The lipid core of LDL is composed of cholesterol, cholesteryl esters, triglycerides, and phospholipids. LDL may be subclassed to small dense and large LDL based on their sizes, density, and lipid components. Small dense LDL contains relatively abundant amounts of triglycerides compared to large LDL. The levels of small dense LDL are closely associated with the incidences of IHD and diabetes [12]. LDL transports cholesterol to peripheral tissues for the synthesis of cellular membrane and steroid hormones. Peripheral cells internalize LDL through the LDL receptors and clathrin vesicle-dependent endocytosis [13]. The expression of the LDL receptor is negatively regulated by intracellular cholesterol [14]. Increased levels of LDL and accelerated development of atherosclerosis were found in humans and animals with deficiency or mutation of LDL receptor [15, 16].

Oxidative Modification of LDL

LDL modified by copper ion or acetylation may be incorporated into macrophages or SMC [17]. Biological products (hypochloride, peroxynitrite, or glucose), prolonged exposure to cells (EC, SMC, or monocytes), or physical factors (ultraviolet or radiation) may also oxidatively modify LDL [18, 19]. Increased peroxidation products are detected in glycated or cell-modified LDL, which have been known as minimally modified LDL [20, 21]. Glycation increases the susceptibility of LDL to oxidation [22]. Increased levels of glycated LDL have been detected in diabetic patients [23]. Modified LDL particles are recognized and incorporated into macrophages, SMC, or EC via SR or other receptors. SR-AI/II mediates the uptake of oxLDL, but not acetyl-LDL. SR-BI mediates the uptake of oxLDL, acetyl-LDL, and high-density lipoprotein (HDL) [24]. CD36 is a type of SR-B, but it is expressed in EC in addition to macrophages [25]. Lectin-like oxLDL receptor-1 is another type of receptor mediating the uptake of oxLDL into EC [26]. Receptors specifically mediating the uptake of glycated LDL have not been identified. The oxidative derivative of glycated LDL is potentially recognized by SR or receptor for advanced glycation end products [27].

Existence of OxLDL in Human Body

The presence of oxLDL in tissue or blood circulation of humans has been a controversial issue for decades. Antigens of oxLDL were detected in diet-induced atherosclerotic animals and human atherosclerotic lesions in late 1980s [28]. OxLDL is associated with SMC- or macrophage-derived foam cells and necrotic lipid core in atherosclerotic lesions [29]. Increased levels of autoantibody against malondialdehyde-LDL were detected in plasma of patients with IHD [30]. Increased levels of oxLDL antigen were recently detected in blood circulation of patients with cardiac events [31].

OxLDL and Thrombosis

OxLDL has been considered as a risk factor for atherothrombosis [32]. In comparison to atherosclerosis, the role of oxLDL in thrombosis has been less well understood. Since increased levels of LDL and oxidative stress often coexist with other risk factors (diabetes, hypercholesterolemia, hyperglycemia, hypertriglyceridemia, and low HDL-cholesterol), it is often difficult to separate the effect of oxLDL from the other risk factors. High incidences of thrombosis are characteristic in patients with antiphospholipid syndrome (APS) or systemic lupus erythematosus (SLE). Elevated levels of autoantibody against oxLDL were correlated with venous thrombosis in SLE patients in a cohort study [33]. Significantly increased levels of oxLDL and its antibody were detected in APS patients compared to healthy subjects, or in APS patients with thrombosis compared to those without [34]. Circulating oxLDL may form complexes with β2-glycoprotein I. High levels of oxLDL-β2-glycoprotein I complex were associated with arterial thrombosis in APS patients [35]. These findings suggest that oxLDL may cause thrombosis in patients with immunological disorders. The effects of oxLDL on thrombosis-related activities are discussed in the following sections.

Effects of oxLDL on Coagulation

The activation of tissue factor (TF) is crucial for triggering the extrinsic coagulation cascade, which leads to the formation of thrombin and fibrin clot. TF may activate factor X and IX. This process is inhibited by tissue factor pathway inhibitor (TFPI). Thrombomodulin is a binding site on EC surface for thrombin. Both oxLDL and minimally modified LDL stimulate the expression of thrombomodulin in EC [36, 37], which may partially compensate the increased activity of coagulation. OxLDL increases the activity of protein C in vascular EC [38]. Protein C is activated by the complex of thrombin and thrombomodulin to generate activated protein C (APC) [39]. The formation of APC is increased when protein C is bound to endothelial cell protein C receptor (EPCR). APC becomes anticoagulant after it dissociates from EPCR and binds to protein S. Protein S–APC complex inactivates the activated forms of factor V and VIII, which inhibits coagulation (Fig. 8.1). OxLDL enhances the expression of TF in monocytes induced by endotoxin [40]. Both LDL and oxLDL induce the expression of TF in vascular SMC [41]. OxLDL increases the binding sites for factor VIII, and the activation of factor VIII and IX on the surface of macrophages and SMC [42]. OxLDL does not affect the expression of TFPI but promotes the degradation of TFPI in EC [43]. These findings suggest that oxLDL activates both extrinsic and intrinsic coagulation pathways.

Role of oxLDL in Platelet Activation

Treatment with oxLDL increases the aggregation and adhesion of platelets [44], and reduces the membrane fluidity of platelets [45]. OxLDL enhances the synthesis of thromboxane A2, one of the strongest biological agonists for platelet aggregation, in platelets [46]. OxLDL increases the production of prostacyclin, a strong inhibitor for platelet aggregation after a short period of incubation with EC, but inhibits the production of prostacyclin following

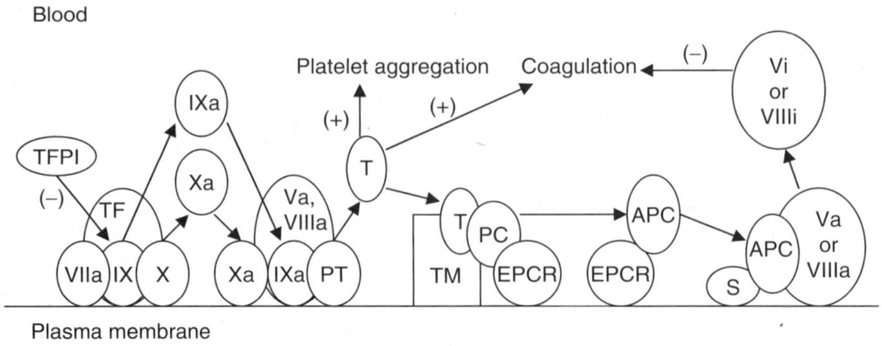

154   Garry X. Shen

FIGURE 8.1. Simplified scheme for coagulation and protein C pathway. TF, tissue factor; TFPI, tissue factor pathway inhibitor; VIIa, activated factor VII; X, factor X; Xa, activated X; IX, factor IX; IXa, activated IX; PT, prothrombin; Va, activated factor V; T, thrombin; TM, thrombomodulin; PC, protein C; EPCR, endothelial protein C receptor; V, factor V; VIII, factor VIII; Va, activated V; VIIIa, activated VIII; APC, activated protein C; S, protein S; Vi, inactivated factor V; VIIIi, inactivated factor VIII. (+), stimulation; (−), inhibition.

longer-term incubations [47, 48]. Glycated LDL promotes platelet aggregation that is associated with increased intracellular calcium concentration and nitric oxide (NO) production in platelets [49]. Hypochlorite-modified LDL induces greater platelet aggregation than copper ion-oxLDL [50]. The phosphorylation of p38 kinase mediates the activation of platelet by oxLDL or hypochloride-modified LDL [51]. CD36 mediates oxLDL-induced platelet activation [52]. Thus, oxidatively modified LDL may activate platelets through multiple mechanisms.

Impact of oxLDL on Fibrinolysis

Imbalance between coagulation and fibrinolysis leads to intravascular thrombosis. The biologically active product of fibrinolytic system is plasmin, which functions in dissolving fibrin clot and maintains the fluency of blood flow. The precursor of plasmin is plasminogen, which is synthesized in liver and is abundant in most biological fluids. The generation of plasmin is mainly regulated by tissue and urokinase plasminogen activators (tPA and uPA) (Fig. 8.2). The activity of tPA and uPA is regulated by their major physiological inhibitor, plasminogen activator inhibitor-1 (PAI-1) [53, 54]. Tremoli et al. [55] reported that LDL and oxLDL increased the release of PAI-1 from EC via a LDL receptor-independent pathway. The activity and mRNA levels of PAI-1 in EC are also elevated by oxLDL treatment [56]. The release of tPA from EC is attenuated by incubation with oxLDL [57]. LDL increases the generation of PAI-1 and decreases the release of tPA from EC. Augmentation of the formation of conjugated diene is detected in previously unmodified

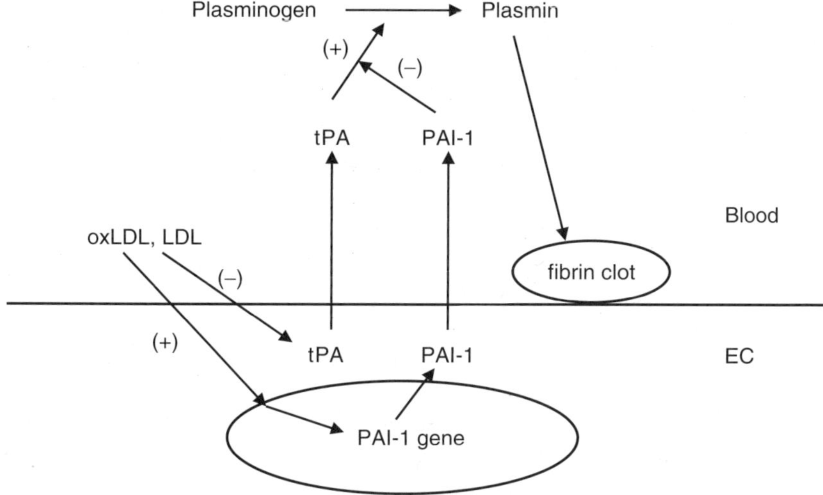

FIGURE 8.2. Simplified scheme for fibrinolytic system and effect of oxLDL and LDL. tPA, tissue plasminogen activator; PAI-1, plasminogen activator inhibitor-1; EC, endothelial cells; LDL, low-density lipoprotein; oxLDL, oxidized LDL. (+), stimulation; (−), inhibition.

LDL following prolonged exposure to EC [58]. The activation of protein kinase C-β is required for oxLDL-induced overproduction of PAI-1 in EC [59]. Glycation enhances the effects of LDL on PAI-1 production and further reduces the release of tPA from EC [58]. LDL isolated from type 2 diabetic patients with suboptimal glucose control stimulates the transcription and generation of PAI-1 from EC [60]. Beside LDL and its modified forms, very low-density lipoprotein (VLDL) and lipoprotein(a) also stimulate the generation of PAI-1 from EC [56, 61, 62]. A VLDL-responsive element has been detected in −672/−657 bp of the PAI-1 promoter [63]. OxLDL and LDL activate a responsive element independent from the VLDL-responsive element in the PAI-1 promoter [64]. The findings from above studies suggest that oxidatively modified LDL may reduce fibrinolytic activity in blood circulation by enhancing the production of PAI-1 and decreasing the generation of tPA in vascular EC.

Endothelial Dysfunction and Thrombosis

Intact endothelium provides an anticoagulant and antiplatelet deposition surface, which is critical for preventing thrombosis in vascular lumen. Endothelial dysfunction may be assessed from abnormal secretion of endothelial products, vascular tone, or the loss of EC [65]. EC synthesize a large numbers of products, including protein C, protein S, tPA, PAI-1, prostacyclin, TF, TFPI, von Willebrand factor (vWF), thrombomodulin, heparin, and factor V. vWF is a coagulation factor mainly produced by EC. OxLDL stimulates the release of vWF from EC [66]. Results from epidemiological studies suggest that elevated level of vWF is a predictor of the progression of atherosclerosis [67]. Prostacyclin is predominantly produced in EC. It is not only an inhibitor of platelet aggregation, but also a potent vasodilator [68]. Extensive studies on endothelium-derived relaxing factor or NO provides additional insight into the importance of endothelium in the regulation of vascular tone [69]. OxLDL reduces the generation of NO and impairs vascular relaxation [70, 71]. Vasoconstriction results in turbulence of blood flow, which may activate platelet and coagulation cascade. Flow-mediated dilation may be detected in brachial artery using 2D ultrasound scan with pharmacological provocation [72]. This technique provides a non-invasive tool to assess endothelial vasodilation *in vivo*, although the accuracy and reproducibility of the method requires to be improved. Severe endothelial injury may increase the amounts of EC in blood circulation. Circulating EC in blood may be assessed using antibody against CD146 [73]. Increased levels of circulating EC have been detected in clinical situations associated with endothelial injury, including ACS, peripheral vascular disease, sickle cell anemia, and inflammatory diseases [74–77]. Impairment of endothelial layer leads to a prothrombotic surface in vascular lumen. Clinical significance of increased levels of circulating EC remains unclear.

Inflammation and Thrombosis

Inflammation plays a crucial role at the initial stage of atherosclerosis. Monocytes and neutrophils migrate from blood circulation and penetrate into endothelium via adhesion molecules, including P-selectin, E-selectin, monocyte chemoattractant protein-1 (MCP-1), tumor necrosis factor-α (TNF-α), intracellular adhesion molecule-1 (ICAM-1), and vascular cell adhesion molecule-1 (VCAM-1) [78, 79]. Inflammation creates a procoagulant surface on vascular intima, which activates coagulation cascade and promotes the generation of thrombin. Thrombin elicits multiple inflammatory responses, including the upregulation of the expression of inflammatory mediators [80]. OxLDL is a strong agonist for the expression of many inflammatory mediators, including P-selectin [81], ICAM-1, VCAM-1 [82], MCP-1 [83], and TNF-α [84]. Anticoagulation complex, APC–EPCR, has anti-inflammatory effects *in vivo* [85], oxLDL increases the formation of APC [38]. OxLDL plays multifunctional roles in the modulation of inflammation, which indirectly promote thrombogenesis.

Stress and Thrombosis

Stress-mediated response regulates the folding and processing of a number of cellular proteins during various biological stresses, including heat shock, inflammation, oxidative, shear, and restrain stresses. Heat shock factor is activated during many stresses, which mediates the transcription of heat shock protein (HSP) genes [86]. OxLDL induces the expression of HSP-70 in EC [87]. Hyperthemia increases the expression of PAI-1 in EC [88]. Restrain stress induced the expression of PAI-1 in mice in a tissue- and cell-type-specific manner. Increased PAI-1 mRNA and thrombosis was more evident in aged mice than in young mice [89]. The findings suggest that stress may promote thrombosis partially resulting from its inhibition on fibrinolysis. The relationship between oxLDL, stress and the expression of thrombosis-related factors remains to be exploited.

Thrombosis, Endothelial Functions, and Diabetes

Increased thrombotic vascular diseases were detected in patients with diabetes [90]. Elevated levels of vWF, factor VII, TF, fibrinogen or PAI-1, and decreased levels of antithrombin III, protein C, or tPA were associated with diabetes [91, 92]. Increased platelet activation was found in diabetic patients [93]. Endothelial injury and oxidative stress are potential underlying mechanisms for the increased thrombosis in diabetes. Hyperglycemia, advanced glycation end products, and dyslipidemia may impair endothelial function and increase oxidative stress. LDL from type 2 diabetic patients or glycated LDL increase the production of PAI-1 in EC and decrease the release of tPA from EC compared to unmodified LDL from healthy donors [58, 60]. Endothelial dysfunction is an early finding of diabetic vascular abnormalities. Reduction in flow-mediated dilation has been detected in children with diabetes [94]. Hyperglycemia increases susceptibility of LDL to oxidation. Substantial increases in the generation of hydrogen peroxide are detected in EC treated with glycated LDL or oxLDL for 1–2 h [95]. Insulin resistance is one of the underlying mechanisms for the pathogenesis of type 2 diabetes. Insulin resistance is associated with an upregulation of CD36, a receptor for oxLDL, in macrophages of experimental obesity [96]. Increased uptake of oxLDL in vascular wall may elicit the production of oxidative stress, impairs endothelium, and promote thrombosis-related processes in diabetic patients.

Effects of Statins on Thrombosis

3-Hydroxy-3-methylglutaryl coenzyme A (HMG-CoA) reductase is a rate-limiting enzyme for cholesterol synthesis, which regulates the generation of mevalonate from acetyl CoA. Mevalonate is a precursor of a group of active intracellular mediators (farnesyl and geranyl pyrophosphates) in addition to cholesterol. HMG-CoA reductase inhibitors (statins) were designed for lowering cholesterol synthesis [97]. Results from a recent meta-analysis demonstrated that statin treatment reduced coronary event by 27%, stroke by 18%, and all-causes of mortality by 15% in hypercholesterolemic patients in comparison to placebo controls [98]. The beneficial effects of statins have been seen in IHD or diabetic patients with normal cholesterol levels [99]. Mechanism for pleiotropic effects of statins has been actively studied [100]. Growing amounts of evidence demonstrate that statins are helpful in the management of thrombotic, inflammatory, autoimmune, and neurodegenerative diseases as well as organ transplantations, in addition to atherosclerotic cardiovascular diseases. Statin treatment decreases oxLDL–LDL ratio in plasma and the thickness of carotid intima in patients with carotid artery atherosclerosis [101]. Statins reduce the levels of TF, fibrinogen, PAI-1, or platelet aggregation in patients with hypercholesterolemia or diabetes [102–105]. In a recent study conducted by our group, simvastatin reduced the levels of PAI-1 and prothrombin fragment 1 and 2 (F1+2) in plasma of type 2 diabetic patients. F1+2 is a marker of thrombin generation. Positive correlations were found between PAI-1, but not F1+2, and total or LDL-cholesterol levels in diabetic patients [106]. The findings suggest that certain antithrombotic activities of statins may be secondary to cholesterol lowering, but others may be independent from cholesterol metabolism. The mechanism for statin-induced antithrombotic effects remains unclear. Results from experimental studies suggest that Ras and other prenylated G-proteins, including Rho, Rac, and Rap, are implicated in statin-mediated cellular activities [107, 108]. Statins have outstanding safety profiles. The major side effect of statins, myotoxicity, may be largely prevented by avoidance of coadministration of certain medications [109]. Statins are potential medications for preventing thrombosis in patients with hypercholesterolemia, diabetes, or increased oxidative stress.

Renin–Angiotensin Antagonists and Thrombosis

The renin–angiotensin system plays a critical role in the development of atherosclerosis, endothelial dysfunction, thrombosis, and diabetic nephropathy. Angiotensin II (AII), the major biological activator of the system, is converted from angiotensin I by angiotensin-converting enzyme (ACE). AII is a potent vasoconstrictor and a stimulator of aldosterone, an adrenal salt–water-retention hormone. AII also stimulates the proliferation of vascular cells and predisposes to vascular inflammation, thrombosis, and oxidative stress. Most biological effects of AII are mediated through AII receptor-1 (AT1) on cell surface [110]. ACE inhibitors are commonly used in the treatment of hypertension, heart failure, diabetic nephropathy, and metabolic syndrome [111]. Captopril, the first generation of ACE inhibitors, reduces platelet deposition and thrombus formation in patients with MI [112]. Ramipril reduced PAI-1 antigen and activity but did not significantly affect tPA level in MI patients in HEART study [113]. Enalapril reduced the levels of TF and MCP-1, but not TFPI, in patients with acute MI [114]. AT1 antagonists inhibit platelet aggregation and thrombus formation [115, 116]. ACE inhibitors and AT1 antagonists reduced the oxidation of LDL in humans and in several types of animal models including apolipoprotein E knockout mice [117–122]. The inhibitory effect of AII antagonists on the oxidation of LDL may contribute, at least in part, to their antithrombotic effects.

Conclusion

Oxidative modification of LDL plays a critical role in the development of thrombosis in addition to atherosclerosis. Oxidized or minimally modified LDL cause inflammation on endothelium, which leads to a prothrombotic surface in vascular lumen. Oxidized or minimally modified LDL may activate platelet and coagulation, but reduces fibrinolytic and anticoagulation activity in vasculature (Fig. 8.3). Reduction of oxidative stress or the oxidation of LDL may be an additional therapeutic target for the prevention of thrombosis as well as atherosclerosis. The results from large trials on the effects of antioxidant vitamins for preventing cardiovascular events were not encouraging. Statins and AII antagonists may prevent thrombosis in patients with high risks through inhibiting oxidative modification of LDL.

Acknowledgments: The author thanks for the collaborations from Drs. John W. Fenton II, George King, Liam J. Murphy, Peter Cattini, and Sora Ludwig, contributions from students, postdoctoral fellows, and other staff working in his laboratories to relevant studies and grant supports from Canadian Institute of Health Research, Canadian Diabetes Association, Heart Stroke Foundation of Canada, Manitoba Medical Service Foundation, Manitoba Health Research Council, University of Manitoba, and Dr. Paul Tholackson Foundation, and research space provided by Health Sciences Centre Foundation.

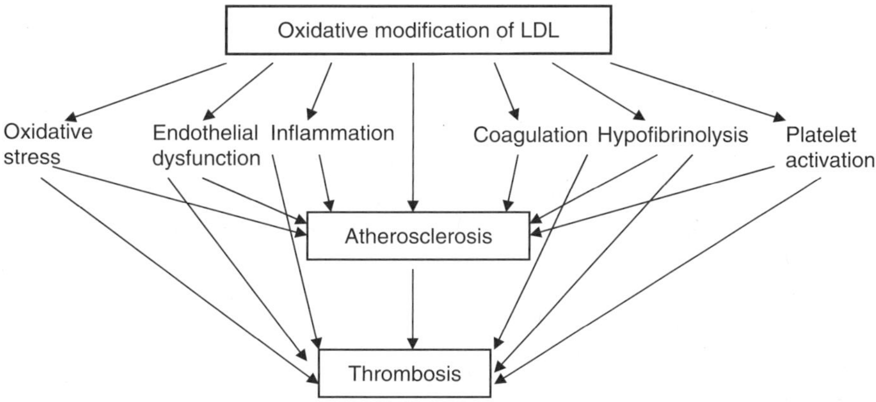

FIGURE 8.3. Scheme for relationships between oxidatively modified LDL, atherosclerosis, and thrombosis.

References

1. Castelli WP, Anderson K. A population at risk. Prevalence of high cholesterol levels in hypertensive patients in the Framingham Study. Am J Med 80:23–32, 1986.
2. Brown MS, Goldstein JL. Regulation of the activity of the low density lipoprotein receptor in human fibroblasts. Cell 6:307–316, 1975.
3. Freeman MW. Scavenger receptors in atherosclerosis. Curr Opin Hematol 4:41–47, 1997.
4. Platt N, Gordon S. Scavenger receptors: diverse activities and promiscuous binding of polyanionic ligands. Chem Biol 5:R193–R203, 1998.
5. Steinberg D. Low density lipoprotein oxidation and its pathobiological significance. J Biol Chem 272:20963–20966, 1997.
6. Holvoet P, Collen D. Oxidized lipoproteins in atherosclerosis and thrombosis. FASEB J 8:1279–1284, 1994.
7. Relou IA, Hackeng CM, Akkerman JW, Malle E. Low-density lipoprotein and its effect on human blood platelets. Cell Mol Life Sci 60:961–971, 2003.
8. Fischer A, Gutstein DE, Fuster V. Thrombosis and coagulation abnormalities in the acute coronary syndromes. Cardiol Clin 17:283–294, 1999.
9. Arbustini E, Grasso M, Diegoli M, Pucci A, Bramerio M, Ardissino D, Angoli L, de Servi S, Bramucci E, Mussini A, et al. Coronary atherosclerotic plaques with and without thrombus in ischemic heart syndromes: a morphologic, immuno-histochemical, and biochemical study. Am J Cardiol 68:36B–50B, 1991.
10. Bertolet BD, Dinerman J, Hartke R Jr, Conti CR. Unstable angina: relationship of clinical presentation, coronary artery pathology, and clinical outcome. Clin Cardiol 16:116–122, 1993.
11. Kluft C. The fibrinolytic system and thrombotic tendency. Pathophysiol Haemost Thromb 33:425–429, 2003–2004.
12. Krauss RM. Dense low density lipoproteins and coronary artery disease. Am J Cardiol 75:53B–57B, 1995.
13. Schneider WJ. Lipoprotein receptor. In: Vance DE, Vance JE (eds), Biochemistry of Lipids, Lipoproteins and Membranes, 4th edn. Elsevier, Amsterdam, 2002, pp. 553–572.
14. Brown MS, Ho YK, Goldstein JL. The low-density lipoprotein pathway in human fibroblasts: relation between cell surface receptor binding and endocytosis of low-density lipoprotein. Ann N Y Acad Sci 275:244–257, 1976.
15. Brown MS, Goldstein JL. The LDL receptor locus and the genetics of familial hypercholesterolemia. Annu Rev Genet 13:259–289, 1979.
16. Ishibashi S, Herz J, Maeda N, Goldstein JL, Brown MS. The two-receptor model of lipoprotein clearance: tests of the hypothesis in "knockout" mice lacking the low density lipoprotein receptor, apolipoprotein E, or both proteins. Proc Natl Acad Sci U S A 91:4431–4435, 1994.
17. Sparrow CP, Parthasarathy S, Steinberg D. A macrophage receptor that recognizes oxidized low density lipoprotein but not acetylated low density lipoprotein. J Biol Chem 264:2599–2604, 1989.
18. Aviram M. Modified forms of low density lipoprotein and atherosclerosis. Atherosclerosis 98:1–9, 1993.
19. Tabas I. Nonoxidative modifications of lipoproteins in atherogenesis. Annu Rev Nutr 19:123–139, 1999.

20. Berliner JA, Territo MC, Sevanian A, Ramin S, Kim JA, Bamshad B, Esterson M, Fogelman AM. Minimally modified low density lipoprotein stimulates monocyte endothelial interactions. J Clin Invest 85:1260–1266, 1990.
21. Lyons TJ, Li W, Wojciechowski B, Wells-Knecht MC, Wells-Knecht KJ, Jenkins AJ. Aminoguanidine and the effects of modified LDL on cultured retinal capillary cells. Invest Ophthalmol Vis Sci 41:1176–1180, 2000.
22. Jenkins AJ, Klein RL, Chassereau CN, Hermayer KL, Lopes-Virella MF. LDL from patients with well-controlled IDDM is not more susceptible to *in vitro* oxidation. Diabetes 45:762–767, 1996.
23. Lyons TJ, Baynes JW, Patrick JS, Colwell JA, Lopes-Virella MF. Glycosylation of low density lipoprotein in patients with type 1 (insulin-dependent) diabetes: correlations with other parameters of glycaemic control. Diabetologia 29:685–689, 1986.
24. Dhaliwal BS, Steinbrecher UP. Scavenger receptors and oxidized low density lipoproteins. Clin Chim Acta 286:191–205, 1999.
25. Endemann G, Stanton LW, Madden KS, Bryant CM, White RT, Protter AA. CD36 is a receptor for oxidized low density lipoprotein. J Biol Chem 268: 11811–11816, 1993.
26. Nagase M, Hirose S, Sawamura T, Masaki T, Fujita T. Enhanced expression of endothelial oxidized low-density lipoprotein receptor (LOX-1) in hypertensive rats. Biochem Biophys Res Commun 237:496–498, 1997.
27. Schmidt AM, Hori O, Cao R, Yan SD, Brett J, Wautier JL, Ogawa S, Kuwabara K, Matsumoto M, Stern D. RAGE: a novel cellular receptor for advanced glycation end products. Diabetes 45 (Suppl 3):S77–S80, 1996.
28. Yla-Herttuala S, Palinski W, Rosenfeld ME, Parthasarathy S, Carew TE, Butler S, Witztum JL, Steinberg D. Evidence for the presence of oxidatively modified low density lipoprotein in atherosclerotic lesions of rabbit and man. J Clin Invest 84:1086–1095, 1989.
29. Holvort P. Oxidized LDL and coronary heart disease. Acta Cardiol 59:479–484, 2004.
30. Salonen JT, Yla-Herttuala S, Yamamoto R, Butler S, Korpela H, Salonen R, Nyyssonen K, Palinski W, Witztum JL. Autoantibody against oxidised LDL and progression of carotid atherosclerosis. Lancet 339:883–887, 1992.
31. Shimada K, Mokuno H, Matsunaga E, Miyazaki T, Sumiyoshi K, Miyauchi K, Daida H. Circulating oxidized low-density lipoprotein is an independent predictor for cardiac event in patients with coronary artery disease. Atherosclerosis 174:343–347, 2004.
32. Mertens A, Holvoet P. Oxidized LDL and HDL: antagonists in atherothrombosis. FASEB J 15:2073–2084, 2001.
33. Hayem G, Nicaise-Roland P, Palazzo E, de Bandt M, Tubach F, Weber M, Meyer O. Anti-oxidized low-density-lipoprotein (OxLDL) antibodies in systemic lupus erythematosus with and without antiphospholipid syndrome. Lupus 10:346–351, 2001.
34. Zhao D, Ogawa H, Wang X, Cameron GS, Baty DE, Dlott JS, Triplett DA. Oxidized low-density lipoprotein and autoimmune antibodies in patients with antiphospholipid syndrome with a history of thrombosis. Am J Clin Pathol 116:760–767, 2001.
35. Kobayashi K, Kishi M, Atsumi T, Bertolaccini ML, Makino H, Sakairi N, Yamamoto I, Yasuda T, Khamashta MA, Hughes GR, Koike T, Voelker DR,

Matsuura E. Circulating oxidized LDL forms complexes with beta2-glycoprotein I: implication as an atherogenic autoantigen. J Lipid Res 44:716–726, 2003.

36. Oida K, Tohda G, Ishii H, Horie S, Kohno M, Okada E, Suzuki J, Nakai T, Miyamori I. Effect of oxidized low density lipoprotein on thrombomodulin expression by THP-1 cells. Thromb Haemost 78:1228–1233, 1997.

37. Kim JA, Tran ND, Berliner JA, Fisher MJ. Minimally oxidized low-density lipoprotein regulates hemostasis factors of brain capillary endothelial cells. J Neurol Sci 217:135–141, 2004.

38. Weis JR, Pitas RE, Wilson BD, Rodgers GM. Oxidized low-density lipoprotein increases cultured human endothelial cell tissue factor activity and reduces protein C activation. FASEB J 5:2459–2465, 1991.

39. Riewald M, Petrovan RJ, Donner A, Mueller BM, Ruf W. Activation of endothelial cell protease activated receptor 1 by the protein C pathway. Science 296:1880–1882, 2002.

40. Brand K, Banka CL, Mackman N, Terkeltaub RA, Fan ST, Curtiss LK. Oxidized LDL enhances lipopolysaccharide-induced tissue factor expression in human adherent monocytes. Arterioscler Thromb 14:790–797, 1994.

41. Cui MZ, Penn MS, Chisolm GM. Native and oxidized low density lipoprotein induction of tissue factor gene expression in smooth muscle cells is mediated by both Egr-1 and Sp1. J Biol Chem 274:32795–32802, 1999.

42. Ananyeva NM, Kouiavskaia DV, Shima M, Saenko EL. Intrinsic pathway of blood coagulation contributes to thrombogenicity of atherosclerotic plaque. Blood 99:4475–4485, 2002.

43. Horie S, Hiraishi S, Hirata Y, Kazama M, Matsuda J. Oxidized low-density lipoprotein impairs the anti-coagulant function of tissue-factor-pathway inhibitor through oxidative modification by its high association and accelerated degradation in cultured human endothelial cells. Biochem J 352 (part 2): 277–285, 2000.

44. Dardik R, Varon D, Tamarin I, Zivelin A, Salomon O, Shenkman B, Savion N. Homocysteine and oxidized low density lipoprotein enhanced platelet adhesion to endothelial cells under flow conditions: distinct mechanisms of thrombogenic modulation. Thromb Haemost 83:338–344, 2000.

45. Vlasova II. The effect of oxidatively modified low-density lipoproteins on platelet aggregability and membrane fluidity. Platelets 11:406–414, 2000.

46. Mahfouz MM, Kummerow FA. Oxidized low-density lipoprotein (LDL) enhances thromboxane A(2) synthesis by platelets, but lysolecithin as a product of LDL oxidation has an inhibitory effect. Prostaglandins Other Lipid Mediat 62:183–200, 2000.

47. Triau JE, Meydani SN, Schaefer EJ. Oxidized low density lipoprotein stimulates prostacyclin production by adult human vascular endothelial cells. Arteriosclerosis 8:810–818, 1988.

48. Myers DE, Huang WN, Larkins RG. Lipoprotein-induced prostacyclin production in endothelial cells and effects of lipoprotein modification. Am J Physiol 271:C1504–C1511, 1996.

49. Ferretti G, Rabini RA, Bacchetti T, Vignini A, Salvolini E, Ravaglia F, Curatola G, Mazzanti L. Glycated low density lipoproteins modify platelet properties: a compositional and functional study. J Clin Endocrinol Metab 87:2180–2184, 2002.

50. Volf I, Roth A, Cooper J, Moeslinger T, Koller E. Hypochlorite modified LDL are a stronger agonist for platelets than copper oxidized LDL. FEBS Lett 483:155–159, 2000.

51. Coleman LG Jr, Polanowska-Grabowska RK, Marcinkiewicz M, Gear AR. LDL oxidized by hypochlorous acid causes irreversible platelet aggregation when combined with low levels of ADP, thrombin, epinephrine, or macrophage-derived chemokine (CCL22). Blood 104:380–389, 2004.

52. Korporaal SJ, Gorter G, van Rijn HJ, Akkerman JW. Effect of oxidation on the platelet-activating properties of low-density lipoprotein. Arterioscler Thromb Vasc Biol 25:867–872, 2005.

53. Plow EF, Herren T, Redlitz A, Miles LA, Hoover-Plow JL. The cell biology of the plasminogen system. FASEB J 9:939–945, 1995.

54. Shen GX. Vascular cell-derived fibrinolytic regulators and atherothrombotic vascular disorders. Int J Mol Med 1:399–408, 1998.

55. Tremoli E, Camera M, Maderna P, Sironi L, Prati L, Colli S, Piovella F, Bernini F, Corsini A, Mussoni L. Increased synthesis of plasminogen activator inhibitor-1 by cultured human endothelial cells exposed to native and modified LDLs. An LDL receptor-independent phenomenon. Arterioscler Thromb 13:338–346, 1993.

56. Ren S, Man RY, Angel A, Shen GX. Oxidative modification enhances lipoprotein(a)-induced overproduction of plasminogen activator inhibitor-1 in cultured vascular endothelial cells. Atherosclerosis 128:1–10, 1997.

57. Kugiyama K, Sakamoto T, Misumi I, Sugiyama S, Ohgushi M, Ogawa H, Horiguchi M, Yasue H. Transferable lipids in oxidized low-density lipoprotein stimulate plasminogen activator inhibitor-1 and inhibit tissue-type plasminogen activator release from endothelial cells. Circ Res 73:335–343, 1993.

58. Zhang J, Ren S, Sun D, Shen GX. Influence of glycation on LDL-induced generation of fibrinolytic regulators in vascular endothelial cells. Arterioscler Thromb Vasc Biol 18:1140–1148, 1998.

59. Ren S, Shatadal S, Shen GX. Protein kinase C-beta mediates lipoprotein-induced generation of PAI-1 from vascular endothelial cells. Am J Physiol Endocrinol Metab 278:E656–E662, 2000.

60. Ren S, Lee H, Hu L, Lu L, Shen GX. Impact of diabetes-associated lipoproteins on generation of fibrinolytic regulators from vascular endothelial cells. J Clin Endocrinol Metab 87:286–291, 2002.

61. Stiko-Rahm A, Wiman B, Hamsten A, Nilsson J. Secretion of plasminogen activator inhibitor-1 from cultured human umbilical vein endothelial cells is induced by very low density lipoprotein. Arteriosclerosis 10:1067–1073, 1990.

62. Etingin OR, Hajjar DP, Hajjar KA, Harpel PC, Nachman RL. Lipoprotein (a) regulates plasminogen activator inhibitor-1 expression in endothelial cells. A potential mechanism in thrombogenesis. J Biol Chem 266:2459–2465, 1991.

63. Eriksson P, Nilsson L, Karpe F, Hamsten A. Very-low-density lipoprotein response element in the promoter region of the human plasminogen activator inhibitor-1 gene implicated in the impaired fibrinolysis of hypertriglyceridemia. Arterioscler Thromb Vasc Biol 18:20–26, 1998.

64. Shen G, Zhao R, Lu L, Ma X, Hu L. Oxidized low density lipoprotein induces interaction between a novel response element in plasminogen activator inhibitor-1 promoter and a nuclear protein from vascular endothelial cells. Atherosclerosis 4/2:185, 2003 (abstract).

65. Blann AD. Assessment of endothelial dysfunction: focus on atherothrombotic disease. Pathophysiol Haemost Thromb 33:256–261, 2003–2004.

66. Blann AD, Burrows G, McCollum CN. Oxidised and native low-density lipoproteins induce the release of von Willebrand factor from human endothelial cells *in vitro*. Br J Biomed Sci 60:155–160, 2003.

67. Blann AD, Waite MA. von Willebrand factor and soluble E-selectin in hypertension: influence of treatment and value in predicting the progression of atherosclerosis. Coron Artery Dis 7:143–147, 1996.
68. Hyman AL, Kadowitz PJ. Pulmonary vasodilator activity of prostacyclin (PGI2) in the cat. Circ Res 45:404–409, 1979.
69. Furchgott RF, Zawadzki JV. The obligatory role of endothelial cells in the relaxation of arterial smooth muscle by acetylcholine. Nature 288:373–376, 1980.
70. Jacobs M, Plane F, Bruckdorfer KR. Native and oxidized low-density lipoproteins have different inhibitory effects on endothelium-derived relaxing factor in the rabbit aorta. Br J Pharmacol 100:21–26, 1990.
71. Liu SY, Lu X, Choy S, Dembinski TC, Hatch GM, Mymin D, Shen X, Angel A, Choy PC, Man RY. Alteration of lysophosphatidylcholine content in low density lipoprotein after oxidative modification: relationship to endothelium dependent relaxation. Cardiovasc Res 28:1476–1481, 1994.
72. Corretti MC, Anderson TJ, Benjamin EJ, Celermajer D, Charbonneau F, Creager MA, Deanfield J, Drexler H, Gerhard-Herman M, Herrington D, Vallance P, Vita J, Vogel R; International Brachial Artery Reactivity Task Force. Guidelines for the ultrasound assessment of endothelial-dependent flow-mediated vasodilation of the brachial artery: a report of the International Brachial Artery Reactivity Task Force. J Am Coll Cardiol 39:257–265, 2002.
73. Dignat-George F, Sampol J. Circulating endothelial cells in vascular disorders: new insights into an old concept. Eur J Haematol 65:215–220, 2000.
74. Mutin M, Canavy I, Blann A, Bory M, Sampol J, Dignat-George F. Direct evidence of endothelial injury in acute myocardial infarction and unstable angina by demonstration of circulating endothelial cells. Blood 93:2951–2958, 1999.
75. Makin AJ, Blann AD, Chung NA, Silverman SH, Lip GY. Assessment of endothelial damage in atherosclerotic vascular disease by quantification of circulating endothelial cells. Relationship with von Willebrand factor and tissue factor. Eur Heart J 25:371–376, 2004.
76. Solovey A, Lin Y, Browne P, Choong S, Wayner E, Hebbel RP. Circulating activated endothelial cells in sickle cell anemia. N Engl J Med 337:1584–1590, 1997.
77. Woywodt A, Schroeder M, Gwinner W, Mengel M, Jaeger M, Schwarz A, Haller H, Haubitz M. Elevated numbers of circulating endothelial cells in renal transplant recipients. Transplantation 76:1–4, 2003.
78. Libby P, Sukhova G, Lee RT, Liao JK. Molecular biology of atherosclerosis. Int J Cardiol 62 (Suppl 2):S23–S29, 1997.
79. Dong ZM, Chapman SM, Brown AA, Frenette PS, Hynes RO, Wagner DD. The combined role of P- and E-selectins in atherosclerosis. J Clin Invest 102:145–152, 1998.
80. Esmon CT. Coagulation and inflammation. J Endotoxin Res 9:192–198, 2003.
81. Gebuhrer V, Murphy JF, Bordet JC, Reck MP, McGregor JL. Oxidized low-density lipoprotein induces the expression of P-selectin (GMP140/PADGEM/CD62) on human endothelial cells. Biochem J 306 (Pt 1):293–298, 1995.
82. Amberger A, Maczek C, Jurgens G, Michaelis D, Schett G, Trieb K, Eberl T, Jindal S, Xu Q, Wick G. Co-expression of ICAM-1, VCAM-1, ELAM-1 and Hsp60 in human arterial and venous endothelial cells in response to cytokines and oxidized low-density lipoproteins. Cell Stress Chaperones 2:94–103, 1997.
83. Wang GP, Deng ZD, Ni J, Qu ZL. Oxidized low density lipoprotein and very low density lipoprotein enhance expression of monocyte chemoattractant protein-1 in rabbit peritoneal exudate macrophages. Atherosclerosis 133:31–36, 1997.

84. Jovinge S, Ares MP, Kallin B, Nilsson J. Human monocytes/macrophages release TNF-alpha in response to Ox-LDL. Arterioscler Thromb Vasc Biol 16:1573–1579, 1996.

85. Joyce DE, Gelbert L, Ciaccia A, DeHoff B, Grinnell BW. Gene expression profile of antithrombotic protein C defines new mechanisms modulating inflammation and apoptosis. J Biol Chem 276:11199–11203, 2001.

86. Morano KA, Thiele DJ. Heat shock factor function and regulation in response to cellular stress, growth, and differentiation signals. Gene Expr 7:271–282, 1999.

87. Wojta J, Holzer M, Hufnagl P, Christ G, Hoover RL, Binder BR. Hyperthermia stimulates plasminogen activator inhibitor type 1 expression in human umbilical vein endothelial cells *in vitro*. Am J Pathol 139:911–919, 1991.

88. Zhu W, Roma P, Pellegatta F, Catapano AL. Oxidized-LDL induce the expression of heat shock protein 70 in human endothelial cells. Biochem Biophys Res Commun 200:389–394, 1994.

89. Yamamoto K, Takeshita K, Shimokawa T, Yi H, Isobe K, Loskutoff DJ, Saito H. Plasminogen activator inhibitor-1 is a major stress-regulated gene: implications for stress-induced thrombosis in aged individuals. Proc Natl Acad Sci U S A 99:890–895, 2002.

90. Petrauskiene V, Falk M, Waernbaum I, Norberg M, Eriksson JW. The risk of venous thromboembolism is markedly elevated in patients with diabetes. Diabetologia Mar 19 (Epub), 2005.

91. Moreno PR, Fuster V. New aspects in the pathogenesis of diabetic atherothrombosis. J Am Coll Cardiol 44:2293–2300, 2004.

92. Carr ME. Diabetes mellitus: a hypercoagulable state. J Diabetes Complications 15:44–54, 2001.

93. Petersen HD, Gormsen J. Platelet aggregation in diabetes mellitus. Acta Med Scand 203:125–130, 1978.

94. Jarvisalo MJ, Raitakari M, Toikka JO, Putto-Laurila A, Rontu R, Laine S, Lehtimaki T, Ronnemaa T, Viikari J, Raitakari OT. Endothelial dysfunction and increased arterial intima-media thickness in children with type 1 diabetes. Circulation 109:1750–1755, 2004.

95. Zhao R, Shen GX. Functional modulation of antioxidant enzymes in vascular endothelial cells by glycated LDL. Atherosclerosis 179:277–284, 2005.

96. Liang CP, Han S, Okamoto H, Carnemolla R, Tabas I, Accili D, Tall AR. Increased CD36 protein as a response to defective insulin signaling in macrophages. J Clin Invest 113:764–773, 2004.

97. Witiak DT, Kuwano E, Feller DR, Baldwin JR, Newman HA, Sankrappa SK. Synthesis and antilipidemic properties of cis-7-chloro-3a, 8b-dihydro-3a-methylfuro[3,4-b]benzofuran-3(1H)-one, a tricyclic clofibrate related lactone having a structural resemblance to mevalonolactone. J Med Chem 19:1214–1220, 1976.

98. Cheung BM, Lauder IJ, Lau CP, Kumana CR. Meta-analysis of large randomized controlled trials to evaluate the impact of statins on cardiovascular outcomes. Br J Clin Pharmacol 57:640–651, 2004.

99. Corsini, A. The safety of HMG-CoA reductase inhibitors in special populations at high cardiovascular risk. Cardiovasc Drugs Ther 17:265–285, 2003.

100. Davignon J. Beneficial cardiovascular pleiotropic effects of statins. Circulation 109:III39–43, 2004.

101. Vasankari T, Ahotupa M, Toikka J, Mikkola J, Irjala K, Pasanen P, Neuvonen K, Raitakari O, Viikari J. Oxidized LDL and thickness of carotid intima-media

are associated with coronary atherosclerosis in middle-aged men: lower levels of oxidized LDL with statin therapy. Atherosclerosis 155:403–412, 2001.

102. Ferro D, Basili S, Alessandri C, Mantovani B, Cordova C, Violi F. Simvastatin reduces monocyte-tissue-factor expression type IIa hypercholesterolaemia. Lancet 350:1222, 1997.

103. Davignon J, Laaksonen R. Low-density lipoprotein-independent effects of statins. Curr Opin Lipidol 10:543–559, 1999.

104. Balk EM, Lau J, Goudas LC, Jordan HS, Kupelnick B, Kim LU, Karas RH. Effects of statins on nonlipid serum markers associated with cardiovascular disease: a systematic review. Ann Intern Med 139:670–682, 2003.

105. van Nieuw Amerongen GP, Vermeer MA, Negre-Aminou P, Lankelma J, Emeis JJ, van Hinsbergh VW. Simvastatin improves disturbed endothelial barrier function. Circulation 102:2803–2809, 2000.

106. Ludwig S, Dharmalingam S, Erickson-Nesmith S, Ren S, Ma G, Zhu F, Zhao R, Fenton II JW, Ofosu F, te Velthuis H, Van Mierlo G, Shen GX. Impact of simvastatin on hemostatic and fibrinolytic regulators in type 2 diabetes mellitus. Diabet Res Clin Prac, 70: 110–118, 2005.

107. Takemoto M, Liao JK. Pleiotropic effects of 3-hydroxy-3-methylglutaryl coenzyme a reductase inhibitors. Arterioscler Thromb Vasc Biol 21:1712–1719, 2001.

108. Comparato C, Altana C, Bellosta S, Baetta R, Paoletti R, Corsini A. Clinically relevant pleiotropic effects of statins: drug properties or effects of profound cholesterol reduction? Nutr Metab Cardiovasc Dis 11:328–343, 2001.

109. Moghadasian MH. A safety look at currently available statins. Expert Opin Drug Saf 1:269–274, 2002.

110. O'Keefe JH, Lurk JT, Kahatapitiya RC, Haskin JA. The renin–angiotensin–aldosterone system as a target in coronary disease. Curr Atheroscler Rep 5:124–130, 2003.

111. McFarlane SI, Kumar A, Sowers JR. Mechanisms by which angiotensin-converting enzyme inhibitors prevent diabetes and cardiovascular disease. Am J Cardiol 91:30H–37H, 2003.

112. Zurbano MJ, Anguera I, Heras M, Roig E, Lozano M, Sanz G, Escolar G. Captopril administration reduces thrombus formation and surface expression of platelet glycoprotein IIb/IIa in early postmyocardial infarction stage. Arterioscler Thromb Vasc Biol 19:1791–1795, 1999.

113. Vaughan DE, Rouleau JL, Ridker PM, Arnold JM, Menapace FJ, Pfeffer MA. Effects of ramipril on plasma fibrinolytic balance in patients with acute anterior myocardial infarction. HEART Study Investigators. Circulation 96:442–447, 1997.

114. Soejima H, Ogawa H, Yasue H, Kaikita K, Takazoe K, Nishiyama K, Misumi K, Miyamoto S, Yoshimura M, Kugiyama K, Nakamura S, Tsuji I. Angiotensin-converting enzyme inhibition reduces monocyte chemoattractant protein-1 and tissue factor levels in patients with myocardial infarction. J Am Coll Cardiol 34:983–988, 1999.

115. Li P, Fukuhara M, Diz DI, Ferrario CM, Brosnihan KB. Novel angiotensin II AT(1) receptor antagonist irbesartan prevents thromboxane A(2)-induced vasoconstriction in canine coronary arteries and human platelet aggregation. J Pharmacol Exp Ther 292:238–246, 2000.

116. Lopez-Farre A, Sanchez de Miguel L, Monton M, Jimenez A, Lopez-Bloya A, Gomez J, Nunez A, Rico L, Casado S. Angiotensin II AT(1) receptor antagonists and platelet activation. Nephrol Dial Transplant 16:45–49, 2001.

117. Rachmani R, Lidar M, Brosh D, Levi Z, Ravid M. Oxidation of low-density lipoprotein in normotensive type 2 diabetic patients. Comparative effects of enalapril versus nifedipine: a randomized cross-over over study. Diabetes Res Clin Pract 48:139–145, 2000.
118. de Nigris F, D'Armiento FP, Somma P, Casini A, Andreini I, Sarlo F, Mansueto G, De Rosa G, Bonaduce D, Condorelli M, Napoli C. Chronic treatment with sulfhydryl angiotensin-converting enzyme inhibitors reduce susceptibility of plasma LDL to *in vitro* oxidation, formation of oxidation-specific epitopes in the arterial wall, and atherogenesis in apolipoprotein E knockout mice. Int J Cardiol 81:107–115, 2001.
119. Napoli C, Cicala C, D'Armiento FP, Roviezzo F, Somma P, de Nigris F, Zuliani P, Bucci M, Aleotti L, Casini A, Franconi F, Cirino G. Beneficial effects of ACE-inhibition with zofenopril on plaque formation and low-density lipoprotein oxidation in watanabe heritable hyperlipidemic rabbits. Gen Pharmacol 33:467–477, 1999.
120. Rachmani R, Levi Z, Zadok BS, Ravid M. Losartan and lercanidipine attenuate low-density lipoprotein oxidation in patients with hypertension and type 2 diabetes mellitus: a randomized, prospective crossover study. Clin Pharmacol Ther 72:302–307, 2002.
121. Hayek T, Attias J, Coleman R, Brodsky S, Smith J, Breslow JL, Keidar S. The angiotensin-converting enzyme inhibitor, fosinopril, and the angiotensin II receptor antagonist, losartan, inhibit LDL oxidation and attenuate atherosclerosis independent of lowering blood pressure in apolipoprotein E deficient mice. Cardiovasc Res 44:579–587, 1999.
122. Strawn WB, Chappell MC, Dean RH, Kivlighn S, Ferrario CM. Inhibition of early atherogenesis by losartan in monkeys with diet-induced hypercholesterolemia. Circulation 101:1586–1593, 2000.

9

The Clinical Significance of Small Dense Low-Density Lipoproteins

MANFREDI RIZZO AND KASPAR BERNEIS

Abstract

Peak size of low-density lipoproteins (LDL) in humans does not show a normal, but a bimodal distribution and can be separated into two phenotypes, that differ in size, density, physicochemical composition, metabolic behavior, and atherogenicity. These phenotypes have been assigned as pattern A when larger, more buoyant LDL and pattern B when smaller, more dense LDL predominate. Small dense LDL correlates negatively with plasma HDL levels and positively with plasma triglyceride concentrations and is associated with the metabolic syndrome and increased risk for cardiovascular disease. LDL size seems to be an important predictor of cardiovascular events and progression of coronary heart disease (CHD). Recently the predominance of small dense LDL has been accepted as an emerging cardiovascular risk factor by the National Cholesterol Education Program Adult Treatment Panel III. In addition, several studies have suggested that therapeutic modulation of specific LDL subclasses may be of great benefit in reducing the atherosclerotic risk. Therefore, LDL size measurement may be of potential value in the clinical assessment and management of patients at high risk of CHD, a category that comprise individuals with both coronary and noncoronary forms of atherosclerosis. Screening for the presence of small dense LDL in such patients may potentially identify those with even higher vascular risk and may contribute in directing specific antiatherosclerotic treatments to prevent new vascular events in the same or another district.

Keywords: atherosclerosis; coronary heart disease; prevention; small dense LDL

Abbreviations: LDL, low-density lipoproteins; HDL, high-density lipoproteins; VLDL, very low-density lipoprotein; IDL, intermediate-density lipoprotein; LpL, lipoprotein lipase; CETP, cholesteryl ester transfer protein; CAD, coronary artery disease; CHD, coronary heart disease.

Introduction

Peak size of low-density lipoproteins (LDL) in humans does not show a normal, but a bimodal distribution and can be separated into two phenotypes, that differ in size, density, physicochemical composition, metabolic behavior, and atherogenicity. These phenotypes have been assigned as pattern A when larger, more buoyant LDL and pattern B when smaller, more dense LDL predominate [1–5].

LDL size correlates positively with plasma high-density lipoproteins (HDL) levels and negatively with plasma triglyceride concentrations and the combination of small dense LDL, decreased HDL-cholesterol, and increased triglycerides has been called as the "atherogenic lipoprotein phenotype" [6] (Fig. 9.1). This partially heritable trait is a feature of the metabolic syndrome and is associated with increased cardiovascular risk.

LDL size seems also to be an important predictor of cardiovascular events and progression of coronary artery disease (CAD) and the predominance of small dense LDL have been accepted as an emerging cardiovascular risk factor by the National Cholesterol Education Program Adult Treatment Panel III [1, 2, 7].

Genetic and Environmental Influences

The predominance of small dense LDL is approximately 30% in adult men, 5–10% in young men and women less than 20 years, and approximately 15–25% in postmenopausal women [1, 2]. It has been shown that LDL size is genetically influenced with a heritability ranging from 35% to 45% based on an autosomal dominant or codominant model with varying additive and polygenic effects [8]. Thus, nongenetic and environmental factors influence

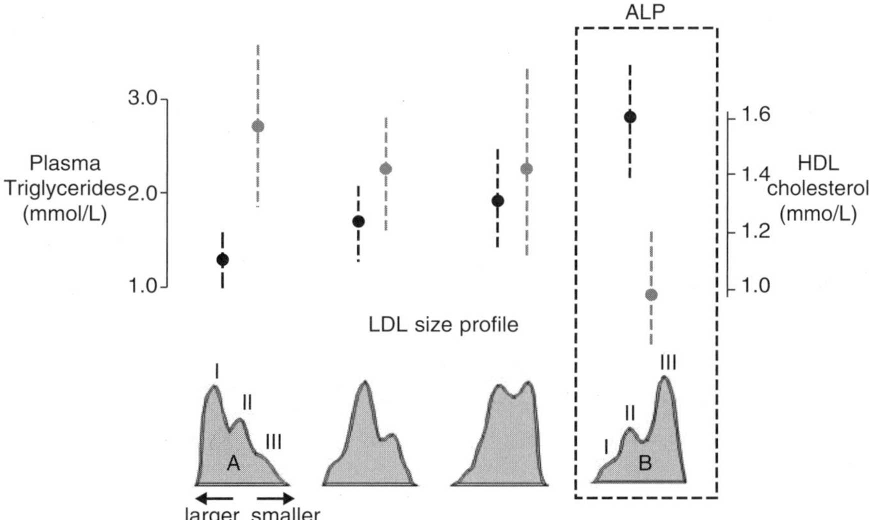

FIGURE 9.1. LDL heterogeneity and plasma triglyceride and HDL-cholesterol concentrations [2, 26]. ALP, atherogenic lipoprotein phenotype.

the expression of this phenotype and an increase of small dense LDL has been shown for abdominal adiposity and oral contraceptive use [9–11].

Dietary factors are also important. It has been shown that a very low-fat, high-carbohydrate diet can induce the pattern B phenotype in persons genetically predisposed to this phenotype [12]. In addition, the predominance of small dense LDL is commonly found in conjunction with familial disorders of lipoprotein metabolism that are associated with increased risk of premature CAD, including familial combined hyperlipidemia, hyperbetalipoproteinemia, and hypoalphalipoproteinemia [13–15].

Heterogeneity of apoB-Containing Particles

It is commonly accepted today that apolipoprotein B (apoB) particles do not comprise a population with continuously variable size; instead there are a multiple subclasses with discrete size and density, different physicochemical composition, and different metabolic behavior. Based on their characteristic appearance in analytical ultracentrifugation and gradient gel electrophoresis distinct subclasses of very low-density lipoprotein (VLDL), intermediate-density lipoprotein (IDL), and LDL particles have been defined [3, 16–32].

There are at least two subclasses of VLDL, two subclasses of IDL, and seven subspecies of LDL from large LDL I to very small LDL IVB [3]. Size differences of VLDL particles are mainly due to changes in core triglycerides while the coat consisting of apolipoproteins, free cholesterol, and phospholipids is of relatively constant thickness. LDL subclasses show differences in the surface lipid content and certain features of the apolipoprotein B100 structure probably contribute to changes in the size of these particles [3].

Metabolism

It has been suggested that there are parallel metabolic channels within the delipidation cascade from VLDL to LDL [3]. A metabolic relationship between large VLDL particles and small LDL particles has been demonstrated using stable isotopes in subjects with a predominance of small dense LDL [33]. Kinetic analysis of tracer studies in humans demonstrate that LDL particles show an initial rapid plasma decay which is due to both intra–extravascular exchange and catabolism of LDL These studies have not yet identified the specific precursors of individual LDL subclasses, however there are data from animal models suggesting that separate pathways may be responsible for the generation of distinct LDL particles [3, 34]. Inverse correlation's of changes in large LDL (LDL-I) and small LDL (LDL-III) and of changes of medium-sized LDL (LDL II) and very small LDL (LDL IV) in dietary intervention studies raise the possibility of precursor–product relationships between distinct LDL subclasses [3].

Activity of lipolytic enzymes is related to the size of the LDL particles. A significant inverse relationship between postheparin lipoprotein lipase (LpL) activity and small dense LDL has been demonstrated and increase of LpL by high-fat diet was associated with an increase of large LDL and decrease of small dense LDL. Reduced activity of LpL and increased activity of hepatic lipase (HL) has been shown in subjects with the pattern B phenotype [35].

HL has a higher affinity for LDL than LpL and is positively correlated with plasma triglycerides, apoB, mass of large VLDL and small dense LDL, but not with the mass of large LDL [36], suggesting an important role for HL in the lipolytic conversion of these particles [34]. The strong relationship of LDL size and triglycerides is based on their importance as substrates for the size reduction of LDL particles by exchange of cholesteryl esters with triglycerides, LDL and HDL can become triglyceride-enriched and can be further processed by lipases.

Profound changes in the physicochemical composition of both LDL and HDL particles with increasing triglyceridemia, while core cholesterol esters are progressively depleted and replaced by triglyceride molecules have been described [37]. In addition, the production of large triglyceride-rich VLDL 1 is dependent on triglyceride availability and VLDL 1 is associated with smaller denser LDL particles [3].

Cholesteryl ester transfer protein (CETP) probably has an important role in the remodeling of larger to smaller LDL particles by mediating triglyceride enrichment of IDL and large LDL [37]. In type 2 diabetes patients it has been demonstrated that CETP contributes significantly to the increased levels of small dense LDL by preferential cholesteryl ester transfer from HDL to small dense LDL, as well as through an indirect mechanism involving enhanced CE transfer from HDL to VLDL 1 [38].

As we already reported [2], the formation of small dense LDL particles is mostly observed in presence of a hypertriglyceridemic state. Indeed, hypertriglyceridemia from a variety of causes is associated with an increased exchange of triglycerides from triglyceride-rich lipoproteins to LDL and HDL particles in exchange of cholesteryl esters through the action of the CETP. This phenomenon results in the generation of VLDL particles enriched in cholesteryl esters and to smaller triglyceride-rich LDL and HDL particles. These smaller lipoproteins are good substrates for HL that has a higher binding affinity for small lipoproteins. Therefore, the lipolysis of the triglycerides in these LDL particles will lead to the formation of highly atherogenic small, dense, cholesteryl ester-depleted LDL particles (Fig. 9.2).

LDL Size Measurement

Particle size distribution of plasma LDL subfractions may be measured by different laboratory techniques [39]; however, the most common procedure is represented by the 2–16% gradient gel electrophoresis at 10°C using a Tris

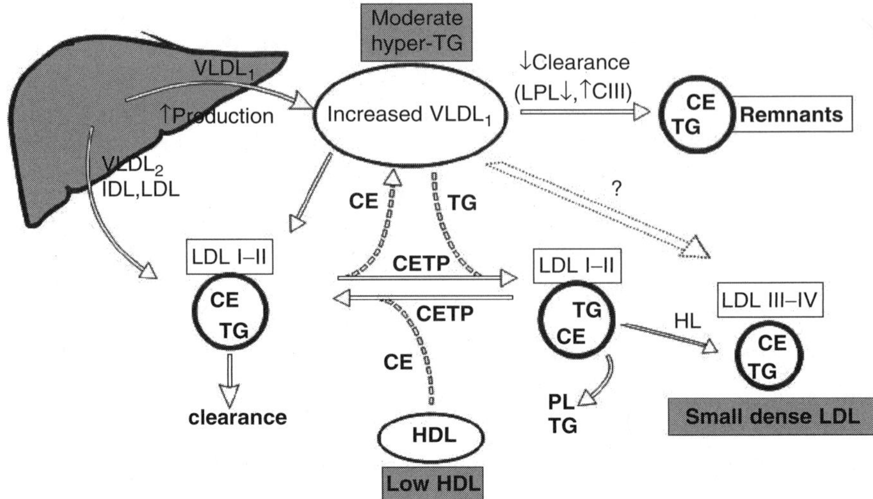

FIGURE 9.2. The metabolic origins of atherogenic lipoprotein phenotype [2, 5]. VLDL, very low-density lipoprotein; IDL, intermediate-density lipoprotein; CE, cholesterol esters; TG, triglycerides; LpL, lipoprotein lipase; CIII, apolipoprotein CIII; CETP, cholesteryl ester transfer protein; HL, hepatic lipase.

(0.09 M)–boric acid (0.08 M)–Na₂ EDTA(0.003 M) buffer (pH 8.3) [1, 2]. In detail, plasma is adjusted to 20% sucrose, and 3 to 10 µL is applied to the gel. Potentials are set at 40 (15 min), 80 (15 min), and 125 mV (24 h). Gels are fixed and stained for lipids in a solution containing oil red O in 60% ethanol at 55°C, for proteins in a solution containing 0.1% Coomassie brilliant blue R-250, 50% ethanol and 9% acetic acid and then scanned at 530 nm with a densitometer. Molecular diameters are determined on the basis of migration distance by comparison with standards of known diameter [4, 9, 34].

Assignment of LDL subclass phenotypes is based on particle diameter of the major plasma LDL peak: LDL phenotype A (larger, more buoyant LDL) is defined as an LDL subclass pattern with the major gradient gel peak at a particle diameter of 258 Å or greater, whereas the major peak of LDL phenotype B (small, dense LDL) is at a particle diameter of less than 258 Å [1, 2, 9, 40] (Figs. 9.3 and 9.4).

Atherogenicity of Small Dense LDL

Several reasons have been suggested for atherogenicity of small dense LDL: smaller denser LDL are taken up more easily by arterial tissue than larger LDL [41], suggesting greater transendothelial transport of smaller particles. In addition, smaller LDL particles may also have decreased receptor-mediated uptake and increased proteoglycan binding [42]. Sialic acid,

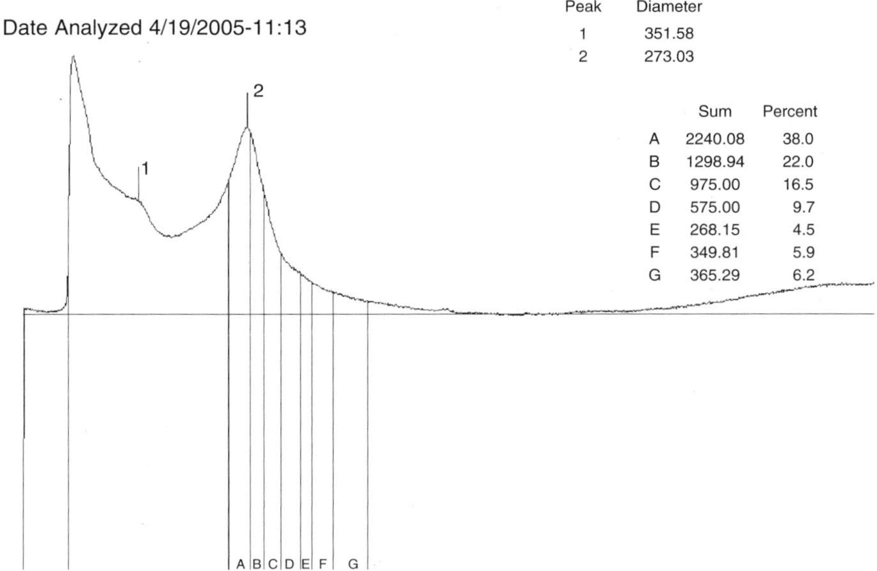

FIGURE 9.3. A representative densitometric scan of lipid stained 2–16% nondenaturing gradient gel electrophoresis of whole plasma from a subject with a predominance of large, buoyant LDL (LDL phenotype "pattern A").

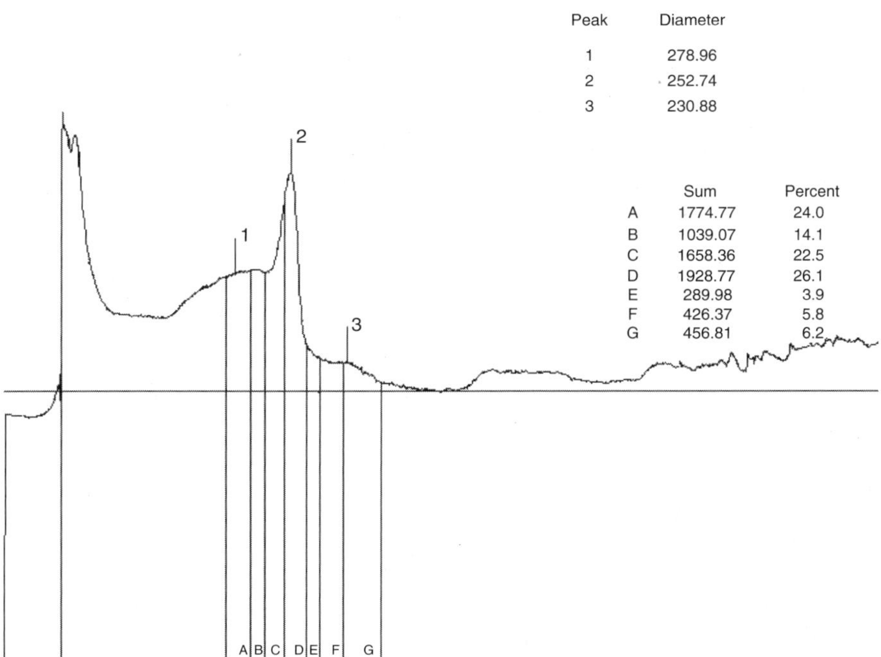

FIGURE 9.4. A representative densitometric scan of lipid stained 2–16% nondenaturing gradient gel electrophoresis of whole plasma from a subject with a predominance of small dense LDL (LDL phenotype "pattern B").

perhaps because of its exposure at the LDL surface, plays a determinant role in the *in vitro* association of LDL with the polyanionic proteoglycans [43] and it has been shown that sialic acid content of LDL particles of subjects with the pattern B phenotype is reduced.

Further, it has been shown that oxidative susceptibility increases and antioxidant concentrations decreases with decreasing LDL size [44]. Altered properties of the surface lipid layer associated with reduced content of free cholesterol [45] and increased content of polyunsaturated fatty acids [46] might contribute to enhanced oxidative susceptibility of small dense LDL.

Recently [47] we have chosen the model of apoB transgenic mice to evaluate the kinetic behavior of human LDL particles of different size *in vivo* in a genetically homogeneous recipient avoiding other metabolic differences that could influence LDL metabolism. We found that small LDL particles have intrinsic features that lead to retarded metabolism and decreased intra–extravascular equilibration compared to medium-sized LDL; these properties could contribute to greater atherogenicity of small dense LDL

LDL size and Coronary Heart Disease

The predominance of atherogenic small dense LDL has been associated with an approximately threefold increased risk for CAD [48]. This has been demonstrated in case–control studies of myocardial infarction [48–50] and of angiographically documented coronary disease [51]. However, in most, but not all [50] of these studies, the disease risk associated with small dense LDL was no longer significant after adjusting for covariates, including triglycerides [48, 51] or other cardiovascular risk factors [49].

In a nested case–control study of myocardial infarction during 7 years in patients of the physicians health study cases had significantly smaller LDL size than controls matched for age and smoking. However, LDL size was not an independent risk predictor after adjustment for triglycerides [52].

In the prospective Stanford Five City Project the association of the incidence of fatal and nonfatal CAD with LDL diameter has been investigated. The significant difference in LDL size between cases and controls was independent of levels of HDL-cholesterol, non-HDL-cholesterol, triglycerides, smoking, systolic blood pressure, and body mass index, but not independent from the ratio of total cholesterol to HDL-cholesterol. In this study LDL size was the best differentiator of CAD status in logistic regression analysis [53].

In the Quebec Cardiovascular Study the association between LDL particle size and incident ischemic heart disease has been analyzed on the basis of data from the entire population-based, prospective cohort of men initially free from coronary heart disease (CHD) with a follow up of 5 years. In this study, small dense LDL particles predicted the rate of CHD independent of LDL-cholesterol, triglycerides, HDL-cholesterol, apoB and the total cholesterol to HDL-cholesterol ratio. Further, the increase in cardiovascular risk attributed to lipid risk factors was modulated to a significant extent by variations in LDL particle size [54].

In addition, the cholesterol concentration in small dense LDL particle may give even more precise information. Again, in the Quebec heart study, St-Piere et al. [55] demonstrated that the cholesterol concentration in small dense LDL particles showed the strongest association with the risk of CHD. Multivariate logistic and survival models indicated that the relationship between LDL-cholesterol levels in particles with a diameter less than 255 Å and CHD risk was independent of all nonlipid risk factors and of LDL-cholesterol, HDL-cholesterol, triglycerides, and lipoprotein(a) level [55]. These data suggest that the cholesterol within small dense LDL is particularly harmful. Measurement of LDL particle size and possibly in addition cholesterol content within these particles may enhance our capability to predict cardiovascular events.

In a population of 98 men less than 50 years and 100 women less than 50 years who underwent elective diagnostic coronary arteriography, smaller denser LDL were associated with CAD independently of traditional risk factors (age, sex, smoking, diabetes, LDL and HDL-cholesterol concentrations), other than plasma triglycerides. These results stress the importance of triglycerides and small dense LDL in premature CAD [51]. Taken together these studies suggest that LDL size is an important predictor of CAD. However, in most studies LDL size was not completely independent of traditional lipid especially triglycerides. This is not surprising as these parameters are obviously metabolically linked.

Mykkanen et al. [56] found that LDL size was not a predictor of CHD events in elderly men and women after controlling for diabetes status. The main reason for the discrepancies might be a survival bias due to old age in these Finnish subjects. Further, Finnish subjects had relatively high LDL-cholesterol and total cholesterol. Therefore the power to detect effects of LDL size on CHD events might have been diminished.

Interestingly, a recent study analyzing data from the CARE trial in a prospective nested case–control study found that larger LDL size after adjustment for other variables was an independent predictor of recurrent coronary events in a population with CAD [57]. However, in this study cases and controls were closely matched for prevalence of LDL subclass pattern B (approximately 40%). Thus, the population was one in which the atherogenic lipoprotein phenotype did not discriminate risk for recurrent events, and in this context a strong risk associated with larger LDL was detected.

Interestingly, significantly larger LDL and HDL particles have been found in Ashkenazi Jews with exceptional longevity compared to an age-matched control group. Larger LDL particles in this study were associated with a lower prevalence of hypertension, cardiovascular disease, and metabolic syndrome [58].

Finally, we have recently found that small dense LDL may be associated with premature CAD [59] and with the extension of coronary atherosclerosis [60].

LDL Size and Acute Myocardial Infarction

Acute myocardial infarction is accompanied by profound plasma lipid and lipoprotein modifications that have a great clinical relevance since they must be taken into account in making therapeutic decisions [61].

The common lipid alterations observed during the acute phase include a rise of triglyceride levels and a fall of total, LDL- and HDL-cholesterol concentrations [62–64]. These lipoprotein modifications may be linked to other changes occurring in lipid metabolism during the acute phase of myocardial infarction: either up- or downregulation of cholesterol synthesis, an increased LDL-receptor activity, and an impairment of lipolytic enzymes [65–68]. Also glucose metabolism is severely impaired [69].

Therefore, the acute myocardial infarction and the atherogenic lipoprotein phenotype seems to share a similar array of interrelated metabolic aberrations, including increased levels of triglyceride-rich lipoproteins, reduced levels of HDL, and a relative resistance to insulin-mediated glucose uptake. However, despite a number of data regarding the modifications of total plasma lipoprotein fractions during a myocardial infarction, it is less defined whether or not the LDL peak particle size is also modified by the acute phase and therefore the best time to measure it [62–64].

We recently found [70] in a group of subjects admitted to hospital for a myocardial infarction, and followed until the discharge and 3 months after the event, that the reduction of LDL peak particle size is premature and persist during the hospitalization, with a significant increase 3 months after the myocardial infarction. In addition, the timing of these changes seems to precede those of all other lipoproteins. Therefore, we suggest that the evaluation of the LDL peak particle size should be extended to patients with acute myocardial infarction as much as possible.

LDL Size and Vascular Diseases

It has been recently stated by the National Cholesterol Education Program Adult Treatment Panel III that clinical forms of noncoronary atherosclerosis carry a risk for CHD equal to those with established CHD [7]. These conditions include peripheral arterial disease, symptomatic (transient ischemic attack or stroke of carotid origin) and asymptomatic (>50% stenosis on angiography or ultrasound) carotid artery disease, and abdominal aortic aneurysm [7].

However, despite a huge number of data on the relation between LDL size and atherosclerosis in patients with CHD, very few authors have investigated such relation in patients with noncoronary forms of atherosclerosis and we need to wait for further studies with larger number of patients. Yet, the available data suggest a strong association between small dense LDL and noncoronary forms of atherosclerosis.

It has been found that smaller denser LDL particles represent a risk factor for peripheral arterial disease, in absence or presence of diabetes [71]. In addition, some studies have showed that common features of peripheral arterial disease are represented by increased triglyceride levels and lower HDL-cholesterol concentrations [71] and it is known that patients with such lipid abnormalities consist primarily of subjects with predominantly atherogenic small dense LDL particles [1–3].

The association between carotid disease and small dense LDL has been found in healthy subjects [72–76] as well as in different categories of patients: with familial combined hyperlipidemia, familial hypercholesterolemia, vascular dementia, Alzheimer's disease, insulin resistance, or type 2 diabetes [77–81]. In addition, it has been recently suggested that carotid atherosclerosis regression or progression may be linked to LDL size [78, 82].

Regarding patients with abdominal aortic aneurysms, no published studies directly examined the association of LDL size with the presence of such disease. However, it has been recently found [83] that patients with abdominal aortic aneurysms show an elevated prevalence of the metabolic syndrome, which is known to be associated with the predominance of small dense LDL [1, 2, 6] and, therefore, it is likely that such patients may have reduced LDL size.

LDL Size and Diabetes

Patients with diabetes represent a category of high-risk individuals that need stronger measures of lipid-lowering therapies [7, 84, 85]. Hypertriglyceridemia, low HDL-cholesterol, and an increased fraction of small dense LDL particles characterize diabetic dyslipidemia [86–88], whereas LDL or total cholesterol are generally not increased in diabetes patients except for a slight increase of LDL-cholesterol in women (UKPDS) [89]. A predominance of small dense LDL particles has been identified as being associated with a higher risk of CHD in the Physician's Health Study [52], the population based on Stanford Five Cities Project [53] and the Quebec Cardiovascular Study [54, 55].

In addition, small dense LDL is associated with the cluster of risk factors that characterize the insulin resistance syndrome [90] and it has been suggested that the small dense LDL particles can be added to the group of changes described as the metabolic syndrome [87]. Interestingly, subjects with predominance of small dense LDL have a greater than twofold increased risk for developing type 2 diabetes mellitus, independent from age, sex, glucose tolerance, and body mass index. An increase of peak LDL size was associated with a 16% decrease in the risk of developing type 2 diabetes mellitus [91].

Recently, we investigated the clinical significance of LDL size and LDL subclasses in diabetes type 2 patients [81]. Diabetes patients with manifest CHD had decreased LDL particle sizes and altered LDL subclass distributions, i.e., specifically more LDL III B and less LDL I and LDL II as compared to diabetes patients without established CHD. Multivariate analysis revealed that LDL size was the strongest marker of CHD as compared to nine other established cardiovascular risk factors, including plasma lipids and lipoproteins.

Increased intima media thickness (IMT) is considered a reliable surrogate marker of early atherosclerosis and it has been demonstrated that small dense LDL is independently related to CCA IMT in 50-year-old men [73]. IMT has been shown to correlate significantly with the presence of CHD and to predict coronary events [92–95]. In addition, significant relationships of IMT with other lipid parameters such as LDL-cholesterol [96] and apoB [97] have been demonstrated. In the study described above LDL size was significantly associated with carotid IMT in diabetes type 2 patients and LDL size was the second strongest predictor of IMT, after smoking when compared to nine other cardiovascular risk factors and the strongest of all lipid parameters.

In summary LDL size is a marker of clinical apparent (CHD) and nonapparent (IMT) atherosclerosis in type 2 diabetes. However a potential superior clinical value to predict cardiovascular events in this population needs to be shown in prospective studies, before routine laboratory measurements of LDL size can be recommended.

Effects of Hypolipidemic Treatment on LDL Size

Hypolipidemic treatment is capable of altering LDL subclass distribution. Particularly medication with triglyceride-lowering effects will shift LDL peak size from smaller denser to larger more buoyant particles. As explained in more detail earlier, reduced availability of triglyceride-rich VLDL particles lead to a reduction in the production of small dense LDL. This has been shown for fibrates and niacin: these substances lower preferentially small dense LDL, so that the LDL peak size shifts to larger particles. Statins potentially lower large, medium, and small LDL particles; therefore, the beneficial net effect of statins on LDL size is often none or only moderate. However, a strong variation has been noticed among the different statin molecules [1, 98–101].

Conclusions

Genetic and environmental factors influence the expression of small dense LDL, which is not completely independent of traditional lipids and, in fact, correlates negatively with plasma HDL concentrations and positively with plasma triglyceride levels. Small dense LDL are associated with the metabolic syndrome and with increased risk for cardiovascular disease and diabetes mellitus. LDL size seems also to be an important predictor of cardiovascular events and progression of CAD and the predominance of small dense LDL has been accepted as an emerging cardiovascular risk factor by the National Cholesterol Education Program Adult Treatment Panel III [7].

In addition, patients with acute myocardial infarction show an early reduction of LDL size, which persist during hospitalization and seems to precede all other plasma lipoprotein modifications. It has been recently shown that even angina itself (on the background of coronary artery spasm) without atherosclerosis may lower LDL size [102]. Hypolipidemic treatment is able to increase LDL particles size [1–3] and this increment correlates with regression of coronary stenosis [1–3]. However, it is still on debate whether to measure the LDL size routinely and in which categories of patients [39].

Recently, the Coordinating Committee of the National Cholesterol Education Program suggested that very high-risk patients may benefit from stronger measures of lipid-lowering therapies [84]. Since the therapeutic modulation of distinct LDL subspecies is of great benefit in reducing the risk of cardiovascular events, LDL size measurement should be extended in primary prevention as much as possible to patients at high risk of cardiovascular diseases. In addition, screening for the presence of small dense LDL in patients with coronary or noncoronary forms of atherosclerosis may also identify those with even higher vascular risk and may contribute in directing specific antiatherosclerotic treatments (secondary prevention) in order to prevent new vascular events in the same or another district.

Acknowledgments: The authors wish to thank Prof. Ronald M. Krauss (Senior Scientist, Lawrence Berkeley National Laboratory, University of California, Berkeley, USA and Director of Atherosclerosis Research, Children's Hospital Oakland Research Institute, Oakland, California, USA) and Prof. Alberto Notarbartolo (Head, Department of Clinical Medicine and Emerging Diseases, University of Palermo, Italy) for their strong, valuable, and unique support on the authors' researches in this field. Kaspar Berneis was supported by a research grant from the Swiss National Foundation 3200B0-105258/1.

References

1. Berneis K, Rizzo M. LDL size: does it matter? Swiss Med Weekly 134:720–724, 2004.
2. Rizzo M, Berneis K. Should we measure routinely the LDL peak particle size? Int J Card 107: 147–151, 2006.
3. Berneis KK, Krauss RM. Metabolic origins and clinical significance of LDL heterogeneity. J Lipid Res 43:1363–1379, 2002.
4. Krauss RM, Blanche PJ. Detection and quantification of LDL subfractions. Curr Opin Lipidol 3:377–383, 1992.
5. Packard CJ, Shepherd J. Lipoprotein heterogeneity and apolipoprotein B metabolism. Arterioscler Thromb Vasc Biol 17:3542–3556, 1997.
6. Austin MA, King MC, Vranizan KM, Krauss RM. Atherogenic lipoprotein phenotype. A proposed genetic marker for coronary heart disease risk. Circulation 82:495–506, 1990.
7. National Cholesterol Education Program (NCEP). Expert Panel on Detection, Evaluation, and Treatment of High Blood Cholesterol in Adults (Adult Treatment Panel III). Third Report of the National Cholesterol Education Program (NCEP) Expert Panel on Detection, Evaluation, and Treatment of

High Blood Cholesterol in Adults (Adult Treatment Panel III) final report. Circulation 106:3143–421, 2002.

8. Austin MA. Genetic epidemiology of low-density lipoprotein subclass phenotypes. Ann Med 24:477–481, 1992.

9. Rizzo M, Barbagallo CM, Severino M, Polizzi F, Onorato F, Noto D, Cefalu AB, Pace A, Marino G, Notarbartolo A, Averna RM. Low-density-lipoprotein peak particle size in a Mediterranean population. Eur J Clin Invest 33:126–133, 2003.

10. Terry RB, Wood PD, Haskell WL, Stefanick ML, Krauss RM. Regional adiposity patterns in relation to lipids, lipoprotein cholesterol, and lipoprotein subfraction mass in men. J Clin Endocrinol Metab 68:191–199, 1989.

11. de Graaf J, Swinkels DW, Demacker PN, de Haan AF, Stalenhoef AF. Differences in the low density lipoprotein subfraction profile between oral contraceptive users and controls. J Clin Endocrinol Metab 76:197–202, 1993.

12. Dreon DM, Fernstrom HA, Williams PT, Krauss RM. LDL subclass patterns and lipoprotein response to a low-fat, high-carbohydrate diet in women. Arterioscler Thromb Vasc Biol 17:707–714, 1997.

13. Austin MA, Brunzell JD, Fitch WL, Krauss RM. Inheritance of low density lipoprotein subclass patterns in familial combined hyperlipidemia. Arteriosclerosis 10:520–530, 1990.

14. Teng B, Thompson GR, Sniderman AD, Forte TM, Krauss RM, Kwiterovich PO Jr. Composition and distribution of low density lipoprotein fractions in hyperapobetalipoproteinemia, normolipidemia, and familial hypercholesterolemia. Proc Natl Acad Sci U S A 80:6662–6666, 1983.

15. Genest J Jr, Bard JM, Fruchart JC, Ordovas JM, Schaefer EJ. Familial hypoalphalipoproteinemia in premature coronary artery disease. Arterioscler Thromb 13:1728–1737, 1993.

16. Gofman JW, Lindgren F. The role of lipids and lipoproteins in atherosclerosis. Science 111:166–171, 1950.

17. Lindgren FT, Nichols AV, Hayes TL, Freeman NK, Gofman JW. Structure and homogeneity of the low-density serum lipoproteins. Ann N Y Acad Sci 72: 826–844, 1959.

18. Gofman JW, Young W, Tandy R. Ischemic heart disease, atherosclerosis, and longevity. Circulation 34: 679–697, 1966.

19. Sata T, Havel RJ, Jones AL. Characterization of subfractions of triglyceride-rich lipoproteins separated by gel chromatography from blood plasma of normolipemic and hyperlipemic humans. J. Lipid Res 13:757–768, 1972.

20. Lee DM, Alaupovic P. Composition and concentration of apolipoproteins in very-low- and low-density lipoproteins of normal human plasma. Atherosclerosis 19:501–520, 1974.

21. Packard CJ, Shepherd J, Joerns S, Gotto AM Jr, Taunton OD. Very low density and low density lipoprotein subfractions in type III and type IV hyperlipoproteinemia. Chemical and physical properties. Biochim Biophys Acta 572:269–282, 1979.

22. Krauss RM, Burke D. Identification of multiple subclasses of plasma low density lipoproteins in normal humans. J. Lipid Res 23:97–104, 1982.

23. Musliner TA, Giotas C, RM Krauss. Presence of multiple subpopulations of lipoproteins of intermediate density in normal subjects. Arteriosclerosis 6:79–87, 1986.

24. Swinkels DW, Hak-Lemmers HL, Demacker PN. Single spin density gradient ultracentrifugation method for the detection and isolation of light and heavy low density lipoprotein subfractions. J Lipid Res 28:1233–1239, 1987.
25. Krauss RM. Relationship of intermediate and low-density lipoprotein subspecies to risk of coronary artery disease. Am Heart J 113:578–582, 1987.
26. Griffin BA, Caslake MJ, Yip B, Tait GW, Packard CJ, Shepherd J. Rapid isolation of low density lipoprotein (LDL) subfractions from plasma by density gradient ultracentrifugation. Atherosclerosis 83:59–67, 1990.
27. Alaupovic P. Apolipoprotein composition as the basis for classifying plasma lipoproteins. Characterization of ApoA- and ApoB-containing lipoprotein families. Prog Lipid Res 30:105–138, 1991.
28. Kandoussi A, Cachera C, Parsy D, Bard JM, Fruchart JC. Quantitative determination of different apolipoprotein B containing lipoproteins by an enzyme linked immunosorbent assay: apoB with apoC-III and apoB with apoE. J Immunoassay 12:305–323, 1991.
29. Alaupovic P. The lipoprotein family concept and its clinical significance. Nutr Metab Cardiovasc Dis 2:52–59, 1992.
30. Otvos JD, Jeyarajah EJ, Bennett DW, Krauss RM. Development of a proton nuclear magnetic resonance spectroscopic method for determining plasma lipoprotein concentrations and subspecies distributions from a single, rapid measurement. Clin Chem 38:1632–1638, 1992.
31. Meyer BJ, Caslake MJ, McConnell MM, Packard CJ. Two subpopulations of intermediate density lipoprotein and their relationship to plasma triglyceride and cholesterol levels. Atherosclerosis 153:355–362, 2000.
32. Segrest JP, Jones MK, De Loof H, Dashti N. Structure of apolipoprotein B-100 in low density lipoproteins. J Lipid Res 42:1346–1367, 2001.
33. Krauss RM, Hellerstein MK, Neese RA, Blanche PJ, La Belle M, Shames DM. Altered metabolism of large low density lipoproteins in subjects with predominance of small low density lipoproteins. Circulation 92:1–102, 1995.
34. Rizzo M, Taylor JM, Barbagallo CM, Berneis K, Blanche PJ, Krauss RM. Effects on lipoprotein subclasses of combined expression of human hepatic lipase and human apoB in transgenic rabbits. Arterioscler Thromb Vasc Biol 24:141–146, 2004.
35. Jansen H, Hop W, van Tol A, Bruschke AV, Birkenhager JC. Hepatic lipase and lipoprotein lipase are not major determinants of the low density lipoprotein subclass pattern in human subjects with coronary heart disease. Atherosclerosis 107:45–54, 1994.
36. Campos H, Dreon DM, Krauss RM. Associations of hepatic and lipoprotein lipase activities with changes in dietary composition and low density lipoprotein subclasses. J Lipid Res 36:462–472, 1995.
37. Deckelbaum RJ, Granot E, Oschry Y, Rose L, Eisenberg S. Plasma triglyceride determines structure-composition in low and high density lipoproteins. Arteriosclerosis 4:225–231, 1984.
38. Guerin M, Le Goff W, Lassel TS, Van Tol A, Steiner G, Chapman MJ. Atherogenic role of elevated CE transfer from HDL to VLDL(1) and dense LDL in type 2 diabetes: impact of the degree of triglyceridemia. Arterioscler Thromb Vasc Biol 21:282–288, 2001.
39. Sacks FM, Campos H. Low-density lipoprotein size and cardiovascular disease: a reappraisal. J Clin Endocr Metab 88:4525–4532, 2003.

40. Georgieva AM, van Greevenbroek MMJ, Krauss RM, Brouwers MCGJ, Vermeulen VM M-J, Robertus-Teunissen MG, van der Kallen CJH, de Bruin TWA. Subclasses of low-density lipoprotein and very low-density lipoprotein in familial combined hyperlipidemia: relationship to multiple lipoprotein phenotype. Arterioscler Thromb Vasc Biol 24:1–7, 2004.

41. Bjornheden T, Babyi A, Bondjers G, Wiklund O. Accumulation of lipoprotein fractions and subfractions in the arterial wall, determined in an *in vitro* perfusion system. Atherosclerosis 123:43–56, 1996.

42. Galeano NF, Al-Haideri M, Keyserman F, Rumsey SC, Deckelbaum RJ. Small dense low density lipoprotein has increased affinity for LDL receptor-independent cell surface binding sites: a potential mechanism for increased atherogenicity. J Lipid Res 39:1263–1273, 1998.

43. Camejo G, Lopez A, Lopez F, Quinones J. Interaction of low density lipoproteins with arterial proteoglycans. The role of charge and sialic acid content. Atherosclerosis 55:93–105, 1985.

44. Tribble DL, Rizzo M, Chait A, Lewis DM, Blanche PJ, Krauss RM. Enhanced oxidative susceptibility and reduced antioxidant content of metabolic precursors of small, dense low-density lipoproteins. Am J Med 110:103–110, 2001.

45. Tribble DL, Holl LG, Wood PD, Krauss RM. Variations in oxidative susceptibility among six low density lipoprotein subfractions of differing density and particle size. Atherosclerosis 93:189–199, 1992.

46. de Graaf J, Hak-Lemmers HL, Hectors MP, Demacker PN, Hendriks JC, Stalenhoef AF. Enhanced susceptibility to *in vitro* oxidation of the dense low density lipoprotein subfraction in healthy subjects. Arterioscler Thromb 11:298–306, 1991.

47. Berneis K, Shames DM, Blanche PJ, La Belle M, Rizzo M, Krauss RM. Plasma clearance of human low-density lipoprotein in human apolipoprotein B transgenic mice is related to particle diameter. Metabolism 53:483–487, 2004.

48. Austin MA, Breslow JL, Hennekens CH, Buring JE, Willett WC, Krauss RM. Low-density lipoprotein subclass patterns and risk of myocardial infarction. JAMA 260:1917–1921, 1988.

49. Campos H, Genest JJ Jr, Blijlevens E, McNamara JR, Jenner JL, Ordovas JM, Wilson PW, Schaefer EJ. Low density lipoprotein particle size and coronary artery disease. Arterioscler Thromb 12:187–195, 1992.

50. Griffin BA, Freeman DJ, Tait GW, Thomson J, Caslake MJ, Packard CJ, Shepherd J. Role of plasma triglyceride in the regulation of plasma low density lipoprotein (LDL) subfractions: relative contribution of small dense LDL to coronary heart disease risk. Atherosclerosis 106:241–253, 1994.

51. Coresh J, Kwiterovich PJ, Smith H, Bachorik P. Association of plasma triglyceride concentration and LDL diameter density, and chemical composition with premature coronary artery disease in men and woman. J Lipid Res 34:1687–1697, 1993.

52. Stampfer MJ, Krauss RM, Ma J, Blanche PJ, Holl LG, Sacks FM, Hennekens CH. A prospective study of triglyceride level, low-density lipoprotein particle diameter, and risk of myocardial infarction. JAMA 276:882–888, 1996.

53. Gardner CD, Fortmann SP, Krauss RM. Association of small low-density lipoprotein particles with the incidence of coronary artery disease in men and women. JAMA 276:875–881, 1996.

54. Lamarche B, St-Pierre AC, Ruel IL, Cantin B, Dagenais GR, Despres JP. A prospective, population-based study of low density lipoprotein particle size as a risk factor for ischemic heart disease in men. Can J Cardiol 17:859–865, 2001.

55. St-Pierrc AC, Ruel IL, Cantin B, Dagenais GR, Bernard PM, Despres JP, Lamarche B. Comparison of various electrophoretic characteristics of LDL particles and their relationship to the risk of ischemic heart disease. Circulation 104:2295–2299, 2001.

56. Mykkanen L, Kuusisto J, Haffner SM, Laakso M, Austin MA. LDL size and risk of coronary heart disease in elderly men and women. Arterioscler Thromb Vasc Biol 19:2742–2748, 1999.

57. Campos H, Moye LA, Glasser SP, Stampfer SP, Sacks FM. Low-density lipoprotein size, pravastatin treatment, and coronary events. JAMA 286:1468–1474, 2001.

58. Barzilai N, Atzmon G, Schechter C, Schaefer EJ, Cupples AL, Lipton R, Cheng S, Shuldiner AR. Unique lipoprotein phenotype and genotype associated with exceptional longevity. JAMA 290:2030–2040, 2003.

59. Rizzo M, Barbagallo CM, Noto D, Pernice V, Polizzi F, Cefalu' AB, Notarbartolo A, Averna MR. Accumulation of ApoE-containing triglyceride-rich lipoproteins in normolipidemic men with premature coronary artery disease. Circulation Suppl 102:601–602, 2000 (abstract).

60. Rizzo M, Barbagallo CM, Pace A, Noto D, Maggiore M, Frasheri A, Pernice V, Barone G, Rubino A, Pieri D, Notarbartolo A, Averna M. LDL peak particle size and the extension of coronary atherosclerosis in 72 patients that underwent an angiographic exam. Nutr. Metabol. Cardiovasc. Dis. 14:316, 2004 (abstract).

61. Kingswood JC, Williams S, Owens DR. How soon after myocardial infarction should plasma lipid values be assessed? Br Med J 289:1651–1653, 1984.

62. Avogaro P, Bon GB, Cazzolato G et al. Variations in apolipoproteins B and A1 during the course of myocardial infarction. Eur J Clin Invest 8:121–129, 1978.

63. Rosenson RS. Myocardial injury: the acute phase response and lipoprotein metabolism. J Am Coll Cardiol 22:933–940, 1993.

64. Carlsson R, Lindberg G, Westin L, Israelsson B. Serum lipids four weeks after acute myocardial infarction are a valid basis for lipid lowering intervention in patients receiving thrombolysis. Br Heart J 74:18–20, 1995.

65. Pfohl M, Schreiber I, Liebich HM, Haring HU, Hoffmeister HM. Upregulation of cholesterol synthesis after acute myocardial infarction—is cholesterol a positive acute phase reactant? Atherosclerosis 142:389–393, 1999.

66. Lees RS, Fiser RH Jr, Beisel WR, Bartelloni PJ. Effects of an experimental viral infection on plasma lipid and lipoprotein metabolism. Metabolism 21:825–833, 1972.

67. Stubbe I, Gustafson A, Nilsson-Ehle P: Alterations in plasma proteins and lipoproteins in acute myocardial infarction: effects on activation of lipoprotein lipase. Scand J Clin Lab Invest 42:437–444, 1982.

68. Feingold KR, Pollock AS, Moser AH, Shigenaga JK, Grunfeld C. Discordant regulation of proteins of cholesterol metabolism during the acute phase response. J Lipid Res 36:1474–1482, 1995.

69. Mizock BA. Alterations in carbohydrate metabolism during stress: a review of the literature. Am J Med 98:75–84, 1995.

70. Barbagallo CM, Rizzo M, Cefalù AB et al. Changes in plasma lipids and low-density lipoprotein peak particle size during and after myocardial infarction. Am J Card 89:460–462, 2002.

71. O'Neal DN, Lewicki J, Ansari MZ et al. Lipid levels and peripheral vascular disease in diabetic and non-diabetic subjects. Atherosclerosis 136:1–8, 1998.

72. Landray MJ, Sagar G, Muskin J, Murray S, Holder RL, Lip GYH. Association of atherogenic low-density lipoprotein subfractions with carotid atherosclerosis. Q J Med 91:345–351, 1998.

73. Skoglund-Andersson C, Tang R, Bond MG, de Faire U, Hamsten A, Karpe F. LDL particle size distribution is associated with carotid intima-media thickness in healthy 50-year-old men. Arterioscler Thromb Vasc Biol 19:2422–2430, 1999.

74. Bokemark L, Wikstrand J, Attvall S, Hulthe J, Wedel H, Fagerberg B. Insulin resistance and intima-media thickness in the carotid and femoral arteries of clinically healthy 58-year-old men. The Atherosclerosis and Insulin Resistance Study (AIR). J Intern Med 249:59–67, 2001.

75. Hallman DM, Brown SA, Ballantyne CM, Sharrett AR, Boerwinkle E. Relationship between low-density lipoprotein subclasses and asymptomatic atherosclerosis in subjects from the Atherosclerosis Risk in Communities (ARIC) Study. Biomarkers 9:190–202, 2004.

76. Hulthe J, Bokemark L, Wikstrand J, Fagerberg B. The metabolic syndrome, LDL particle size, and atherosclerosis: the Atherosclerosis and Insulin Resistance (AIR) study. Arterioscler Thromb Vasc Biol 20:2140–2147, 2000.

77. Liu M-L, Ylitalo K, Nuotio I, Salonen R, Salonen JT, Taskinen MR. Association between carotid intima-media thickness and low-density lipoprotein size and susceptibility of low-density lipoprotein to oxidation in asymptomatic members of familial combined hyperlipidemia families. Stroke 33:1255–1260, 2002.

78. van Tits LJ, Smilde TJ, van Wissen S, de Graaf J, Kastelein JJ, Stalenhoef AF. Effects of atorvastatin and simvastatin on low-density lipoprotein subfraction profile, low-density lipoprotein oxidizability, and antibodies to oxidized low-density lipoprotein in relation to carotid intima media thickness in familial hypercholesterolemia. J Investig Med 52:177–184, 2004.

79. Watanabe T, Koba S, Kawamura M, Itokawa M, Idei T, Nakagawa Y, Iguchi T, Katagiri T. Small dense low-density lipoprotein and carotid atherosclerosis in relation to vascular dementia. Metabolism 53:476–482, 2004.

80. Hulthe J, Wiklund O, Olsson G, Fagerberg B, Bokemark L, Nivall S, Wikstrand J. Computerized measurement of LDL particle size in human serum. Reproducibility studies and evaluation of LDL particle size in relation to metabolic variables and the occurrence of atherosclerosis. Scand J Clin Lab Invest 59:649–661, 1999.

81. Berneis K, Jeanneret C, Muser J, Felix B, Miserez AR. Low-density lipoprotein size and subclasses are markers of clinically apparent and non-apparent atherosclerosis in type 2 diabetes. Metabolism 54:227–234, 2005.

82. Wallenfeldt K, Bokemark L, Wikstrand J, Hulthe J, Fagerberg B. Apolipoprotein B/apolipoprotein A-I in relation to the metabolic syndrome and change in carotid artery intima-media thickness during 3 years in middle-aged men. Stroke 35:2248–2252, 2004.

83. Gorter PM, Olijhoek JK, van der Graaf Y, Algra A, Rabelink TJ, Visseren FL; SMART Study Group. Prevalence of the metabolic syndrome in patients with coronary heart disease, cerebrovascular disease, peripheral arterial disease or abdominal aortic aneurysm. Atherosclerosis 173:363–369, 2004.

84. Grundy SM, Cleeman JI, Merz CNB, Brewer HB, Clark LT, Hunninghake DB et al. for the Coordinating Committee of the National Cholesterol Education Program. Implications of Recent Clinical Trials for the National Cholesterol Education Program Adult Treatment Panel III Guidelines. Circulation 110:227–239, 2004.

85. Rizzo M, Barbagallo CM, Noto D, Pace A, Cefalu' AB, Pernice V, Rubino A, Pinto V, Pieri D, Traina M, Frasheri A, Notarbartolo A, Averna MR. Diabetes, family history and extension of coronary atherosclerosis are strong predictors of

adverse events after PTCA: a one year follow-up study. Nutr Metabol Cardiovasc Dis, 15:361–367, 2005.

86. Reaven GM. Banting lecture. Role of insulin resistance in human disease. Diabetes 37:1595–1607, 1988.

87. Selby JV, Austin MA, Newman B et al. LDL subclass phenotypes and the insulin resistance syndrome in women. Circulation 88:381–387, 1993.

88. Syvanne M, Taskinen MR. Lipids and lipoproteins as coronary risk factors in non-insulin-dependent diabetes mellitus. Lancet 350 (Suppl 1):SI20–S123, 1997.

89. U.K. Prospective Diabetes Study 27. Plasma lipids and lipoproteins at diagnosis of NIDDM by age and sex. Diabetes Care 20:1683–1687, 1997.

90. Friedlander Y, Kidron M, Caslake M, Lamb T, McConnell M, Bar-On H. Low density lipoprotein particle size and risk factors of insulin resistance syndrome. Atherosclerosis 148:141–149, 2000.

91. Austin MA, Mykkanen L, Kuusisto J et al. Prospective study of small LDLs as a risk factor for non-insulin dependent diabetes mellitus in elderly men and women. Circulation 92:1770–1778, 1995.

92. Chambless LE, Heiss G, Folsom AR et al. Association of coronary heart disease incidence with carotid arterial wall thickness and major risk factors: the Atherosclerosis Risk in Communities (ARIC) Study, 1987–1993. Am J Epidemiol 146:483–494, 1997.

93. Craven TE, Ryu JE, Espeland MA et al. Evaluation of the associations between carotid artery atherosclerosis and coronary artery stenosis. A case–control study. Circulation 82:1230–1242, 1990.

94. Salonen JT, Salonen R. Ultrasound B-mode imaging in observational studies of atherosclerotic progression. Circulation 87:II56–II65, 1993.

95. Wofford JL, Kahl FR, Howard GR, McKinney WM, Toole JF, Crouse JR III. Relation of extent of extracranial carotid artery atherosclerosis as measured by B-mode ultrasound to the extent of coronary atherosclerosis. Arterioscler Thromb 11:1786–1794, 1991.

96. Goya K, Kitamura T, Inaba M et al. Risk factors for asymptomatic atherosclerosis in Japanese type 2 diabetic patients without diabetic microvascular complications. Metabolism 52:1302–1306, 2003.

97. Niskanen L, Rauramaa R, Miettinen H, Haffner SM, Mercuri M, Uusitupa M. Carotid artery intima-media thickness in elderly patients with NIDDM and in nondiabetic subjects. Stroke 27:1986–1992, 1996.

98. Rizzo M, Barbagallo CM, Pernice V, Rubino A, Noto D, Pace A, Amato S, Camarda C, Pinto V, Pieri D, Frasheri A, Notarbartolo A, Averna MR. Effects "beyond-cholesterol" by statins: a 3-months therapy with simvastatin reduces the atherogenic lipoproteins in patients with premature coronary artery disease and in absence of the traditional cardiovascular risk factors. Trends in Med 4:157–162, 2004 (in Italian).

99. Rizzo M, Berneis K. Lipid triad of atherogenic lipoprotein phenotype: a role in cardiovascular prevention? J Atheroscler Thromb, 12:237–239, 2005.

100. Superko RH, Berneis KK, William PT, Rizzo M, Wood PD. Gemfibrozil reduces small low-density lipoprotein more in normolipemic subjects classified as low-density lipoprotein pattern B compared with pattern A. Am J Card, 96:1266–1272, 2005.

101. Rizzo M, Berneis K. Low-density-lipoproteins size and cardiovascular prevention. Eur J Internal Med, 96:1266–1272; 2005.

102. Miwa K. Low density lipoprotein particles are small in patients with coronary vasospasm. Int J Cardiol 87:193–201, 2003.

10

Bile Acids: At the Crossroads of Sterol, Fat, and Carbohydrate Metabolism

LUIS B. AGELLON

Abstract

Bile acids are classically known as natural detergents that help keep cholesterol soluble in bile and aid in the absorption of lipids and lipid-soluble nutrients from the gut. Since bile acids are synthesized from cholesterol, they are also regarded as end products of cholesterol catabolism. It was recently known that bile acids can modulate cellular signaling cascades. Importantly, bile acids can regulate the expression of a variety of genes by serving as ligand and activator of a nuclear receptor transcription factor known as farnesoid X receptor (FXR). The identity of the genes that are responsive to regulation by bile acids via FXR suggests that bile acids can regulate, and possibly coordinate, the metabolism of sterols, fats, and carbohydrates.

Keywords: bile acids; cholesterol; enterohepatic circulation; fatty acids; gene regulation; glucose; liver; nuclear receptors; transgenic mice

Abbreviations: cyp7a, cholesterol 7α-hydroxylase; FXR, farnesoid X receptor; LXR, liver X receptor; LDL, low-density lipoproteins; LRH-1, liver receptor homolog protein-1; NR, nuclear receptor; PPAR, peroxisome proliferator-activated receptor; PXR, pregnane X receptor; SHP, short heterodimer partner; SREBP, sterol response element binding protein; VLDL, very low-density lipoproteins

Introduction

The metabolism of cholesterol is complicated and there are many factors involved in its synthesis, assimilation from dietary sources, transport, and disposal. It is generally accepted that the removal of excess cholesterol from peripheral tissues involves the "reverse cholesterol transport pathway" which operates to off-load cholesterol from peripheral cells and delivers it to the liver for disposal.

The liver is the only organ that is capable of eliminating meaningful amounts of cholesterol from the body. This is done directly by secretion of unesterified cholesterol into bile (biliary cholesterol) and indirectly by facilitating the conversion of cholesterol into bile acids for secretion into bile

(Fig. 10.1). However, biliary cholesterol can mix with diet-derived cholesterol and thus can reenter the body via absorption from the lumen of the intestine. Bile acids are also reabsorbed by the intestine but because cholesterol cannot be recovered from bile acids, the conversion of cholesterol into bile acids represents a terminal step in the catabolism of cholesterol. Bile acid synthesis can account for the elimination of 400–1000 mg of cholesterol per day compared to a combined total of about 100 mg via other pathways (steroid hormone synthesis and shedding of epithelial surfaces).

The synthesis of bile acids takes place only in the liver and the hepatic enzyme known as cholesterol 7α-hydroxylase (cyp7a) catalyzes the first and rate-controlling step in the classical bile acid biosynthetic pathway (Fig. 10.2). Peripheral cells can transform cholesterol into oxysterols, and upon their transport to the liver they complete their transformation into bile acids. This alternative pathway is distinguished from the classical pathway in that cholesterol is first transformed into an oxysterol other than 7α-hydroxycholesterol. The classical pathway is under complex regulatory control whereas the alternative pathway operates constitutively.

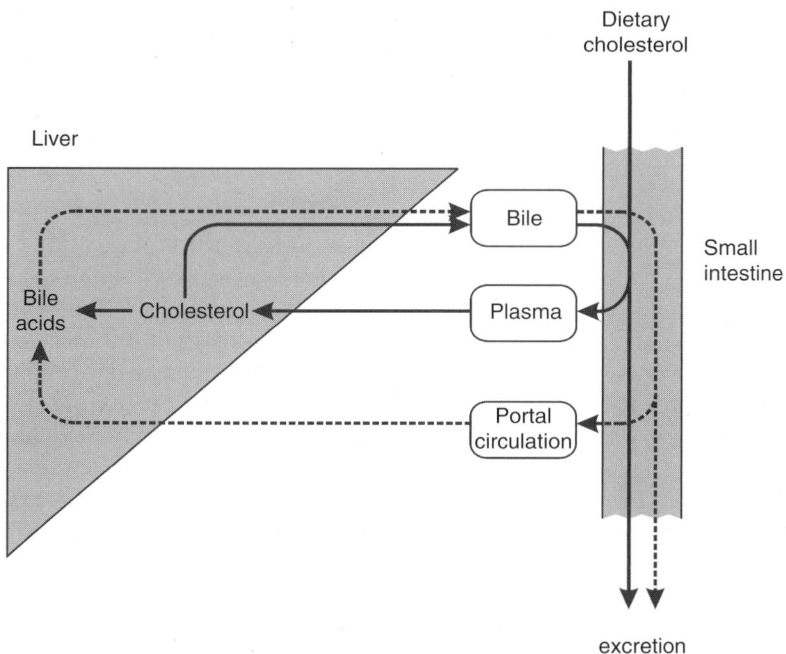

FIGURE 10.1. Transport of cholesterol and bile acids in the enterohepatic circulation. The path traced by cholesterol (solid lines) and bile acids (broken lines) in the enterohepatic circulation is shown.

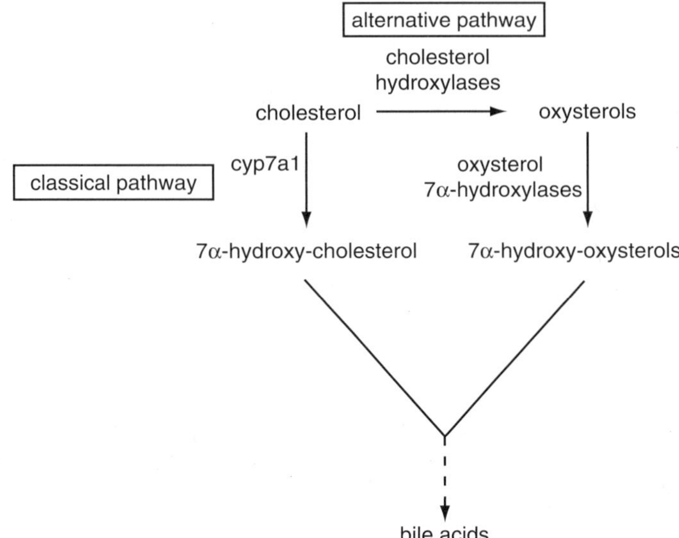

FIGURE 10.2. Bile acid biosynthetic pathways. The classical pathway operates entirely in the liver and is initiated by 7α-hydroxylation of cholesterol by cyp7a. The alternate pathway is initiated in various tissues by several cholesterol hydroxylases. The resulting oxysterols are transported to the liver for completion of bile acid synthesis. Enzymes distinct from cyp7a generate 7α-hydroxylated oxysterols.

Bile Acid Structure, Chemical Properties, and Biological Activities

The conversion of cholesterol into bile acids involves the modification of the steroid moiety (characterized by isomerization of the preexisting 3β-hydroxyl group, addition of one or two new hydroxyl groups, and saturation of all the carbon bonds) and the trimming of the side chain to yield a molecule containing 24 carbons. The result is a molecule that has a nearly flat shape with all the attached hydroxyl groups situated to one side of the steroid nucleus (Fig. 10.3). The products of the classical and alternative bile acid biosynthetic pathways are cholic and chenodeoxycholic acids, and referred to as primary bile acids. The formation of cholic acid requires sterol 12α-hydroxylase, the enzyme responsible for hydroxylating the C-12 carbon of the steroid nucleus. There is a wide assortment of bile acids that differ in the number and orientation of hydroxyl groups attached to the steroid nucleus in the bile of different species. The major bile acids found in human bile are chenodeoxycholic and cholic acids whereas muricholic and cholic acids are found in murine bile. Bile acids are conjugated to the amino acids taurine or glycine prior to their secretion into bile. This modification effectively decreases their

FIGURE 10.3. General structure of bile acids.

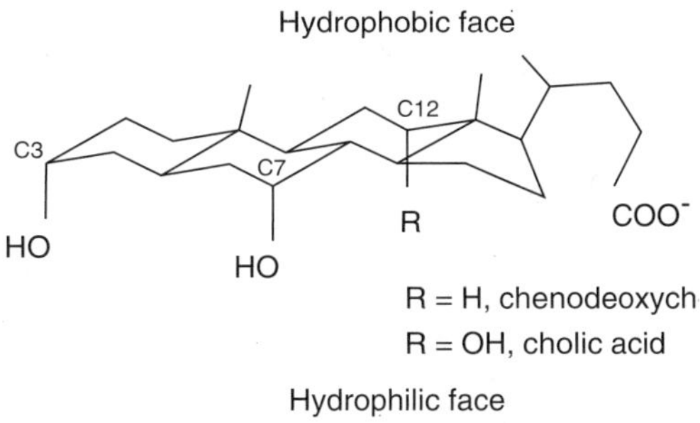

pK_a values, allowing them to exist in their ionized form even under acidic conditions. Primary bile acids are further modified as they circulate within the enterohepatic circulation (Fig. 10.1). The entry and exit of bile acids into and out of cells require specialized membrane transporters [1]. Dehydroxylation by enteric bacteria yields secondary bile acids, which are more hydrophobic than their parent primary bile acids. Of note is lithocholic acid, a secondary bile acid derived from chenodeoxycholic acid and the most hydrophobic of the bile acids. This bile acid is suspected of being involved in the pathogenesis of colon cancer and liver disease. Bile acids returning to the liver are further metabolized yielding glucuronidated and sulfated derivatives that eventually are excreted out of the body.

The unique orientation of the hydroxyl groups imparts an amphipathic character and underlies the most commonly regarded biological function of bile acids, as a detergent aiding in the absorption of lipids and lipid-soluble nutrients in the small intestine. Bile acids are also essential in maintaining the solubility of cholesterol in bile. More recently, bile acids have been shown to be capable of modulating a variety of cellular processes. One of these newly characterized functions is that of a potent activator of a transcription factor known as farnesoid X receptor (FXR, NR1H4) [2, 3]. FXR is a member of a superfamily of transcription factors referred to as nuclear receptors (NR) (see later), and is present in high abundance in the liver, intestine, kidneys, and adrenal glands. Lithocholic acid has been shown to activate two other NRs, pregnane X receptor (PXR; NR1I2) [4] and the vitamin D receptor (NR1I1) [5]. These observations imply that bile acids have the capacity to regulate the expression of many genes, and therefore the ability to influence the activity and efficiency of metabolic processes.

Studies have shown that a wide variety of bile acids species can activate FXR function albeit with varying degree of potency. This is partly explained by the fact that the ligand-binding pocket of FXR is organized in a way that can accommodate the uniquely flat and amphipathic structure of bile acids [6]. Only the steroid moiety of bile acids is bound by FXR and accounts for why both unconjugated and conjugated bile acids can serve as activators of FXR function. Interestingly, the 3α-hydroxyl group that is common to the majority of bile acids is not required for binding or activation of FXR function.

Importance of cyp7a in the Synthesis of Bile Acids from Cholesterol

The possibility that cyp7a activity can influence the concentration of cholesterol in the plasma has been extensively studied in various animal models. This idea stems from the fact that cyp7a catalyzes the key reaction that allows the entry of cholesterol into the classical bile acid biosynthetic pathway. Studies have shown that experimental alteration of hepatic cyp7a activity does have a significant impact on the concentration of cholesterol in the blood. The induction of cyp7a activity by experimentally increasing the cyp7a enzyme concentration in the liver cells *in vivo* significantly reduces the concentration of cholesterol not only in the low-density lipoprotein (LDL) fraction but also in other lipoprotein fractions. The elimination of cyp7a by targeted gene disruption was not initially seen to cause the elevation of plasma cholesterol in the original line of cyp7a-deficient mice [7]; however, a subline derived by selective breeding was subsequently shown to have increased concentration of cholesterol in the plasma [8]. In humans, documented cases of cyp7a deficiency are rare but the afflicted individuals do exhibit hypercholesterolemia and have increased susceptibility to atherosclerosis [9]. These findings demonstrate that cyp7a activity is an important factor in determining plasma cholesterol concentration. Indeed, transgenic C57BL/6J mice that constitutively express cyp7a in the liver are protected from diet-induced atherosclerosis [10]. Given its role in hepatic cholesterol metabolism, it is also likely that cyp7a is an important factor in determining susceptibility to gallstone formation.

The enzymes involved in the early steps of cholesterol into the alternative bile acid biosynthetic pathway (Fig. 10.2) have also been studied. In general, the elimination of these enzymes in the murine species does not result in drastic phenotypic changes (e.g., such as neonatal lethality) but aberrations in the metabolism of sterols are observed. Deficiency of sterol 27-hydroxylase, the enzyme that hydroxylates the cholesterol side chain, causes cerebrotendinous xanthomatosis in humans [11]. This syndrome is not evident in mice lacking the enzyme [12]. Loss of cyp7b1, the enzyme that catalyzes the 7α-hydroxylation of oxysterols, causes severe liver disease in human newborns but not in mice [13, 14]. Apparently, the requirement for the classical and alternative pathways for the synthesis of bile acids is different between humans and mice. This is probably also true for other species as well. On the other hand, the deficiency in cholesterol 24-hydroxylase, which is expressed at a much higher level in the brain than in the liver, appears to affect the metabolism of cholesterol only in the brain [15]. This enzyme may be of minor importance in the synthesis of bile acids in humans and mice.

NRs Play a Prominent Role in Regulating the Expression of the Gene Encoding cyp7a

The gene that encodes cyp7a (*CYP7A1* in humans; *Cyp7a1* in mice; Cyp7a1 for nonspecies-specific designation) is controlled by complex regulatory mechanisms that are sensitive to various physiological and pathological states. The mechanisms controlling cyp7a enzyme activity are exerted mainly at the level of Cyp7a1 gene transcription. The majority of transcription factors that interact with the Cyp7a1 gene promoter are NRs.

NRs comprise a family of transcription factors that are activated by a variety of small hydrophobic molecules [16]. Most NRs are bound to their cognate *cis*-elements, as homodimers or as heterodimers with retinoid X receptor (RXR). Some NRs bind to DNA as monomers and a few are tethered to other NRs. The DNA binding sites of NRs are composed of single or duplicates of a hexameric sequence termed hormone response element arranged in direct, inverted, or everted repeats with no or several nucleotides separating the repeated sequence. The identity of the NRs that bind to these elements is dictated by the exact sequence, orientation, and spacing of the hexamers. The activity of some of these transcription factors is activated upon binding of specific ligands; gene transcription is either stimulated or repressed (as compared to the basal unliganded state of the NR) depending on whether the bound ligand promotes the recruitment or release of coactivators or corepressors. The ligands are typically small hydrophobic molecules. Some NRs are activated by nutritional factors or derivatives of nutritional factors (e.g., oxysterols, fatty acids, and their metabolites), and regulate metabolic pathways responsible for the uptake, transport, utilization, and disposal of their own ligands. NRs whose ligands are still unidentified are referred to as orphan NRs.

It has long been known that bile acid production is stimulated by cholesterol feeding and inhibited by bile acid treatment. The molecular mechanism responsible for the induction by cholesterol and repression by bile acids of Cyp7a1 gene expression is now known (Fig. 10.4). The induction of Cyp7a1 gene transcription by cholesterol is mediated via liver X receptor (LXR)-α: RXR which binds to a direct repeat of hormone response elements separated by four nucleotides, termed a DR-4 element, within the murine *Cyp7a1* gene promoter [17] (Fig. 10.4b). The human *CYP7A1* gene promoter lacks the DR-4 element and is unresponsive to regulation by cholesterol [17]. The oxysterols that serve as LXR-α activators are generated by the bile acid biosynthetic pathways and by the cholesterol biosynthetic pathway resulting from the inhibition of de novo cholesterol synthesis. Targeted disruption of the gene for LXR-α in mice effectively abolishes the sensitivity of the murine *Cyp7a1* gene to stimulation by dietary cholesterol [18]. The discovery that bile acids are the natural ligands of FXR was the major development towards understanding the mechanism responsible for the inhibition of Cyp7a1 gene expression by bile acids [2]. This mechanism is indirect and involves FXR, the

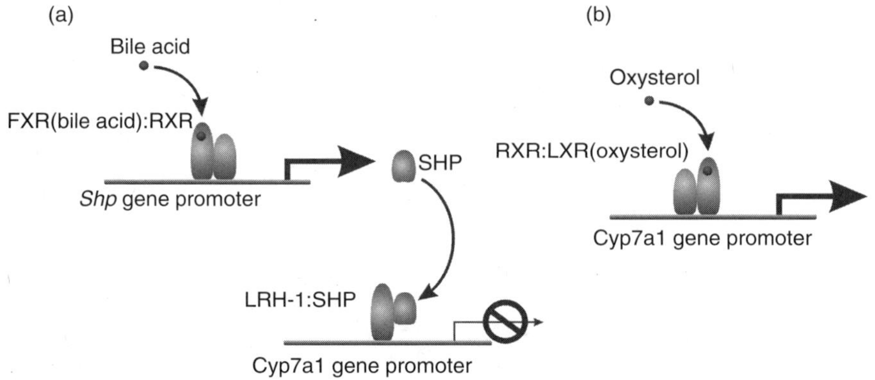

FIGURE 10.4. Regulation of the Cyp7a1 gene promoter by bile acids and oxysterols. Nuclear receptors play key roles in the induction and repression of Cyp7a1 gene expression. Induction by oxysterols is mediated by LXR-α. Repression by bile acids involves FXR, SHP, and LRH-1.

short heterodimer partner (SHP; NR0B2), and the liver receptor homolog protein-1 (LRH-1; NR5A2) [19, 20] (Fig. 10.4a). LRH-1 is normally bound to the Cyp7a1 gene promoter to enable expression in the liver. Bile acid-activated FXR stimulates the expression of the *Shp* gene; SHP binds to LRH-1 causing its inactivation, and rendering the Cyp7a1 gene promoter insensitive to stimulation by other transcription factors. Cyp7a1 gene expression in mice lacking FXR is no longer repressed by bile acid feeding [21]. Surprisingly, the elimination of SHP does not completely abolish the inhibition of Cyp7a1 gene expression by bile acids [22], and confirming the existence of FXR-independent pathways, involving PXR and c-Jun N-terminal kinase [23, 24], in the negative-feedback regulation of bile acid synthesis.

Other transcription factors that have been shown to interact and regulate the Cyp7a1 gene promoter include peroxisome proliferator-activated receptor α (PPAR-α; NR1C1), thyroid hormone receptors (NR1A1, NR1A2), HNF1, Foxo1 (HNF-3α) and HNF-4α (NR2A1). PPAR-α and thyroid hormone receptor-α also interact differentially with murine *Cyp7a1* and human *CYP7A1* gene promoters due to variations in promoter sequences between the two species [25, 26].

Role of Bile Acids in Metabolism

Under normal physiologic conditions, the synthesis of bile acids from cholesterol is a necessary process to replenish the bile acid pool in the entero-hepatic circulation as bile acids are required to keep cholesterol soluble in bile and for the proper absorption of dietary lipids and lipid-soluble nutrients from the lumen of intestine. It is now recognized that bile acids have a much wider repertoire of biological activities than originally thought. Bile acids can modulate cellular signaling cascades [27] that control a variety of ongoing cellular processes, such as cell viability and metabolic fate of hepatic cholesterol [28–30]. By serving as ligands for at least three NRs, bile acids also have the capacity to regulate the expression of many genes and thereby control metabolic pathways. It follows that aberrant metabolism of bile acids can have a major effect on metabolism.

The induction of cyp7a activity in experimental animals demonstrated that enhanced bile acid synthesis decreases the concentration of plasma lipids [31, 32]. Interestingly, the large stimulation of cyp7a activity does not appear to increase the fractional catabolic rate of LDL. Indeed, overproduction of the cyp7a enzyme can decrease plasma cholesterol concentration even in mice that do not have functional LDL receptors [33]. The reduction of plasma cholesterol concentration in cyp7a-overexpressing animals is accompanied by increase in the size of the bile acid pool in the enterohepatic circulation. Bile acids have been shown to decrease apolipoprotein B (the major structural protein of very low-density lipoproteins (VLDL)) secretion from cultured liver cells [30, 34]. Taken together, the findings support the concept that the reduction of cholesterol in cyp7a-overexpressing animals is not simply attributable to increased catabolism of LDL cholesterol. Rather, it is likely due to the decrease in the production of VLDL, which serve as the precursors of LDL.

Previous studies have noted that cholestyramine, a drug that sequesters bile acids in the intestine and thus prevent their return to the liver, increases the concentration of triacylglycerols in the plasma whereas bile acid feeding produces the opposite effect [35–39]. It appears therefore that bile acids themselves play an active role in decreasing the concentration of lipids in the plasma. Bile acids increase the expression of the genes encoding the VLDL receptor [40] and apolipoprotein CII [41], the cofactor of lipoprotein lipase. Complementary to these findings, bile acids have also been seen to suppress the expression of the gene for apolipoprotein CIII [42], an apolipoprotein that has been shown to interfere with apolipoprotein E-mediated clearance of triacylglycerol-rich lipoproteins from the plasma. Collectively, these findings point to a metabolic basis for the reduction of plasma triacylglycerol concentration in response to bile acid feeding.

Sterol response element binding proteins (SREBPs) are transcription factors that were originally identified as important regulators of genes involved in the de novo synthesis of cholesterol [43]. SREBPs are members of the basic helix-loop-helix leucine zipper family of transcription factors. These proteins are initially made as membrane-bound endoplasmic reticulum proteins. This precursor form of the SREBP is subsequently proteolytically processed by specific proteases in response to cellular cholesterol status to release the N-terminal domain that localizes to the nucleus. The nuclear form of SREBP is the active transcription factor that binds to gene promoters containing sterol response elements. Forced overexpression of the nuclear form of the SREBPs in transgenic mice revealed that this transcription factor also controls the expression of many genes involved in fat and phospholipid metabolism. In particular, the SREBP-1c isoform preferentially stimulates lipogenic genes, such as the fatty acid synthase gene (encodes the key enzyme complex for the synthesis of fatty acid) and glycerol-3-phosphate acyltransferase gene (encodes the enzyme that catalyzes the first committed step in biosynthesis of triacylglycerols and phospholipids). Recently, it was observed that bile acids can attenuate hepatic accumulation of triacylglycerols as well as decrease secretion of VLDL by hepatocytes [39]. Moreover, the feeding of bile acids or synthetic FXR agonist reduced the expression of SREBP-1c and lipogenic genes, as reflected by the decrease in the steady-state abundance of their respective mRNAs. This effect was demonstrated to be mediated via FXR through SHP using wild-type and SHP-deficient mouse strains as well as cell transfection studies. In humans, FXR stimulates the expression of the gene encoding PPAR-α [44], which in turn stimulates the expression of genes involved in the β-oxidation of fatty acids. This observation suggests an additional strategy for further reducing the accumulation of triacylglycerols.

Diabetes is a disorder characterized by inability to regulate properly the concentration of glucose in the blood, and is a risk factor for developing atherosclerosis. The activities of cyp7a and sterol 12α-hydroxylase enzymes are modestly elevated in insulin-dependent diabetes and likely explain the increase in bile acid pool size. Recently, it was found that glucose stimulates the expression of the gene for FXR in rats [45]. The diminished glucose uptake by diabetic hepatocytes likely results in decreased FXR gene expression thereby relieving SHP-mediated repression of gene expression. A role for bile acids in glucose metabolism is further supported by the finding that bile acids inhibit the expression of several genes encoding enzymes involved in the production of glucose [46, 47] (Fig. 10.5). Among these enzymes are phosphoenolpyruvate carboxykinase (PEPCK), the rate-limiting enzyme of the glucose biosynthetic pathway, and glucose-6-phosphatase (G6Pase), which dephosphorylates glucose-6-phosphate to yield free glucose. The transcription factors that are known to regulate the expression of these genes include HNF-4α, glucocorticoid receptor (GR, NR3C1) and Foxa1 (also known as HNF-3α, a member of the winged helix/forkhead family of transcription factors). Importantly, these transcription factors are known to regulate other genes involved in carbohydrate metabolism, and that their stimulatory effect on promoter activity is inhibited by SHP [46, 48–50]. It should be noted that bile acids have been found to stimulate glycogen synthase activity by enhancing insulin receptor activity and activation of the protein kinase Akt [51]. The stimulation of glycogen synthesis by a posttranscriptional mechanism may represent a strategy for the rapid clearance of free glucose. Thus bile acids can be seen as influencing carbohydrate metabolism by promoting the storage of glucose and by repressing the expression of genes for enzymes responsible for the synthesis of glucose.

Recent studies have clearly established that bile acids can modulate gene expression and cellular function. With regards to sterol, fat, and carbohydrate

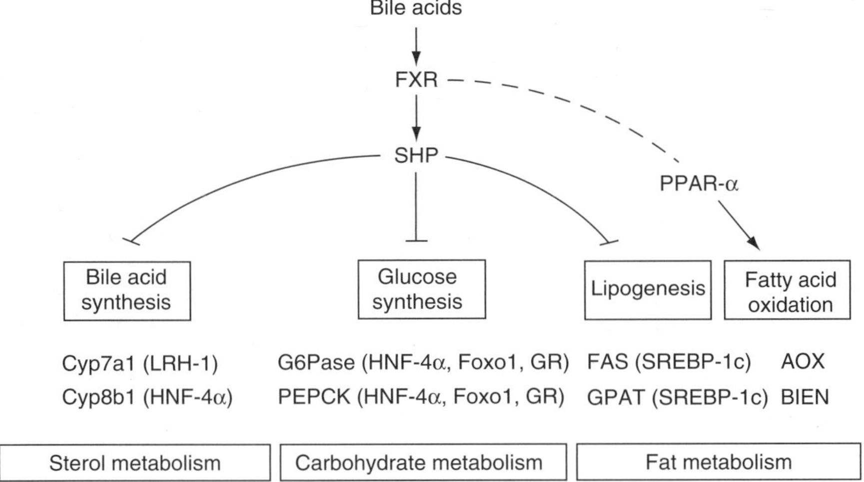

FIGURE 10.5. Role of bile acids in regulating sterol, fat, and carbohydrate metabolism. The scheme shows selected examples of enzymes involved in the metabolism of sterols, carbohydrates, and fats whose genes are affected by bile acids. The broken line denotes differential regulation in humans and mice. The regulation by bile acids is mediated directly through FXR, or indirectly via SHP which antagonizes the stimulatory action of other transcription factors (indicated in parenthesis). FXR, the bile acid-activated nuclear receptor NR1H4; PPAR-α, peroxisome proliferator-activated receptor-α NR1C1; SHP, short heterodimer partner NR0B2; LRH-1, liver receptor homolog protein-1 NR5A2; HNF-4α, hepatocyte nuclear factor-4α NR2A1; GR, glucocorticoid receptor NR3C1; Foxo1, winged helix/forkhead protein, also known as HNF3; SREBP-1c, sterol response element binding protein 1c; Cyp7a1, cholesterol 7α-hydroxylase, catalyzes the rate-limiting step in the classical bile acid biosynthetic pathway; Cyp7b1, oxysterol 7α-hydroxylase, catalyzes the formation of 7α-hydroxylation of oxysterols; PEPCK, phosphoenolpyruvate carboxykinase, catalyzes the rate-limiting step in glucose synthesis; G6Pase, glucose-6-phosphatase, catalyzes the dephosphorylation of glucose-6-phosphate; FAS, fatty acid synthase, the key enzyme complex responsible for the synthesis of fatty acids; GPAT, glycerol-3-phosphate acyltransferase, catalyzes the first committed step in biosynthesis of triacylglycerols and phospholipids; AOX, acyl-CoA oxidase, catalyzes the first and rate-limiting step in the peroxisomal β-oxidation of very long-chain fatty acids; BIEN, bifunctional enzyme, the second enzyme in the peroxisomal β-oxidation system that catalyzes the hydration and dehydrogenation of enoyl-CoA esters.

metabolism, bile acids can coordinately repress not only their own synthesis from cholesterol but also the synthesis of glucose, fatty acids, and triacylglycerols by controlling the expression of key genes involved in these processes (Fig. 10.5). This is achieved directly by activating promoter-bound FXR (e.g., SHP gene and the human PPAR-α gene) or indirectly by antagonizing the activity of other promoter-bound transcription factors (e.g., HNF-4α, Foxo1, GR, SREBP-1c) via the corepressor SHP.

Prospects

The murine species has become a popular model system to study metabolic disorders because transgenic technology has made it possible to study the effects of virtually any gene mutation. Much of our knowledge on bile acid metabolism has been compiled from studies using animals as experimental models. As our understanding of bile acid metabolism improved, it became apparent that significant differences exist in the way genes involved in the metabolism of bile acids are regulated in humans and mice. Nevertheless, the murine species has contributed much information regarding how mutations in genes involved in bile acid metabolism affects other metabolic pathways.

Primary sequence variations in gene promoter regions can lead to differences in the responses to specific modulators. In the case of the Cyp7a1 gene, sequence variations in the human *CYP7A1* and murine *Cyp7a1* gene promoters dictate distinct interactions with specific transcription factors and underlie the differential response of these genes to dietary fats and cholesterol [17, 25]. Primary sequence variations within regulatory regions may also manifest in the differential regulation and expression of genes encoding the transcription factors in different species, as observed for LXR-α, PPAR-α, and SHP [44, 52–55]. Some of these factors regulate the expression of their own genes as well as other transcription factors that regulate genes involved in the metabolism of sterols, fats, and carbohydrates. Thus, it will be important to determine how these genes are regulated in different species in order to gain insight into the metabolic strategies adopted by organisms to achieve similar physiological end points.

Another important factor that requires important consideration is gender. It is well established that gender is an independent risk factor for developing coronary artery disease in humans. In the murine species, male and female mice are also known to respond differently to environmental or experimental challenge. Specific examples of intergender differences relate to the metabolism of bile acids and other dietary lipids, susceptibility to diet-induced atherosclerosis, hypertriglyceridemia, and obesity [56–60]. More studies are beginning to analyze separately the responses of males and females to metabolic challenges. These studies should help uncover the mechanisms responsible for intergender metabolic differences that are not directly attributable to sex hormones.

Although most of the bile acids have a positive effect on FXR function, some species of bile acids have been seen to have little effect or a negative effect [2, 3, 61]. In fact, lithocholic acid, although an activator of PXR and vitamin D receptor, has been shown to behave as an antagonist of FXR [62]. It has also been suggested that bile acids interfere with the function of PPAR-α in mice, possibly by impeding coactivator recruitment [63]. This negative effect of bile acids on PPAR-α function in the murine species is in contrast with the observed positive effect of bile acids on PPAR-α action in humans

[44]. Similarly, guggulsterones have an antagonistic effect on FXR function [64, 65]. These compounds are found in the exudates of the guggul tree, and are purported to possess hypolipidemic, anti-obesity, and anti-inflammatory properties. The precise mechanisms underlying the therapeutic properties of guggulsterones remain unclear. Considering the antagonistic action of guggulsterones on FXR, their lipid-lowering effect might not be expected. However, guggulsterones appear to affect the expression of only some of the known FXR target genes suggesting that they may be able to regulate gene expression through FXR in a selective fashion. Additionally, these compounds can be bound by a number of different NRs, and can behave as agonists or antagonists depending on the identity of bound NR [66]. Given the desirable therapeutic effects of guggulsterones, it is likely that these compounds will receive continued attention.

In summary, ample evidence has been accumulated to demonstrate that bile acids can influence the expression of key genes involved in the metabolism of sterols, fats, and carbohydrates metabolism. The finding that bile acids inhibit the expression of genes involved in gluconeogenesis and lipogenesis suggest that de novo synthesis of high-energy fuel molecules such as glucose and fatty acids is not needed during times of increased bile acid flux within the enterohepatic circulation, when absorption of energy-dense dietary lipids occurs.

Acknowledgment: The work in the author's laboratory is supported by grants from the Canadian Institutes of Health Research and the Heart and Stroke Foundation of Alberta.

References

1. Agellon LB, Torchia EC: Intracellular transport of bile acids. Biochim Biophys Acta 1486: 198–209, 2000.
2. Makishima M, Okamoto AY, Repa JJ, Tu H, Learned RM, Luk A, Hull MV, Lustig KD, Mangelsdorf DJ, Shan B: Identification of a nuclear receptor for bile acids. Science 284: 1362–1365, 1999.
3. Parks DJ, Blanchard SG, Bledsoe RK, Chandra G, Consler TG, Kliewer SA, Stimmel JB, Willson TM, Zavacki AM, Moore DD, Lehman JM: Bile acids: natural ligands for an orphan nuclear receptor. Science 284: 1365–1368, 1999.
4. Staudinger JL, Goodwin B, Jones SA, Hawkins-Brown D, MacKenzie KI, LaTour A, Liu Y, Klaassen CD, Brown KK, Reinhard J, Willson TM, Koller BH, Kliewer SA: The nuclear receptor PXR is a lithocholic acid sensor that protects against liver toxicity. Proc Natl Acad Sci USA 98: 3369–3374, 2001.
5. Makishima M, Lu TT, Xie W, Whitfield GK, Domoto H, Evans RM, Haussler MR, Mangelsdorf DJ: Vitamin D receptor as an intestinal bile acid sensor. Science 296: 1313–1316, 2002.
6. Mi LZ, Devarakonda S, Harp JM, Han Q, Pellicciari R, Willson TM, Khorasanizadeh S, Rastinejad F: Structural basis for bile acid binding and activation of the nuclear receptor FXR. Mol Cell 11: 1093–1100, 2003.

7. Schwarz M, Lund EG, Setchell KDR, Kayden HJ, Zerwekh JE, Bjorkhem I, Herz J, Russell DW: Disruption of cholesterol 7α-hydroxylase gene in mice. II. Bile acid deficiency is overcome by induction of oxysterol 7α-hydroxylase. J Biol Chem 271: 18024–18031, 1996.

8. Erickson SK, Lear SR, Deane S, Dubrac S, Huling SL, Nguyen L, Bollineni JS, Shefer S, Hyogo H, Cohen DE, Shneider B, Sehayek E, Ananthanarayanan M, Balasubramaniyan N, Suchy FJ, Batta AK, Salen G: Hypercholesterolemia and changes in lipid and bile acid metabolism in male and female cyp7A1-deficient mice. J Lipid Res 44: 1001–1009, 2003.

9. Pullinger CR, Eng C, Salen G, Shefer S, Batta AK, Erickson SK, Verhagen A, Rivera CR, Mulvihill SJ, Malloy MJ, Kane JP: Human cholesterol 7α-hydroxylase (CYP7A1) deficiency has a hypercholesterolemic phenotype. J Clin Invest 110: 109–117, 2002.

10. Miyake JH, Duong-Polk XT, Taylor JM, Du EZ, Castellani LW, Lusis AJ, Davis RA: Transgenic expression of cholesterol 7α-hydroxylase prevents atherosclerosis in C57BL/6J mice. Arterioscler Thromb Vasc Biol 22: 121–126, 2002.

11. Bjorkhem I, Leitersdorf E: Sterol 27-hydroxylase deficiency: a rare cause of xanthomas in normocholesterolemic humans. Trends Endocrinol Metab 11: 180–183, 2000.

12. Rosen H, Reshef A, Maeda N, Lippoldt A, Shpizen S, Triger L, Eggertsen G, Bjorkhem I, Leitersdorf E: Markedly reduced bile acid synthesis but maintained levels of cholesterol and vitamin D metabolites in mice with disrupted sterol 27-hydroxylase gene. J Biol Chem 273: 14805–14812, 1998.

13. Setchell KD, Schwarz M, O'Connell NC, Lund EG, Davis DL, Lathe R, Thompson HR, Weslie Tyson R, Sokol RJ, Russell DW: Identification of a new inborn error in bile acid synthesis: mutation of the oxysterol 7α-hydroxylase gene causes severe neonatal liver disease. J Clin Invest 102: 1690–1703, 1998.

14. Li-Hawkins J, Lund EG, Turley SD, Russell DW: Disruption of the oxysterol 7α-hydroxylase gene in mice. J Biol Chem 275: 16536–16542, 2000.

15. Lund EG, Xie C, Kotti T, Turley SD, Dietschy JM, Russell DW: Knockout of the cholesterol 24-hydroxylase gene in mice reveals a brain-specific mechanism of cholesterol turnover. J Biol Chem 278: 22980–22988, 2003.

16. Gronemeyer H, Gustafsson JA, Laudet V: Principles for modulation of the nuclear receptor superfamily. Nat Rev Drug Discov 3: 950–964, 2004.

17. Agellon LB, Drover VA, Cheema SK, Gbaguidi GF, Walsh A: Dietary cholesterol fails to stimulate the human cholesterol 7α-hydroxylase gene (*CYP7A1*) in transgenic mice. J Biol Chem 277: 20131–20134, 2002.

18. Peet DJ, Turley SD, Ma W, Janowski BA, Lobaccaro J-MA, Hammer RE, Mangelsdorf DJ: Cholesterol and bile acid metabolism are impaired in mice lacking the nuclear oxysterol receptor LXRα. Cell 93: 693–704, 1998.

19. Lu TT, Makishima M, Repa JJ, Schoonjans K, Kerr TA, Auwerx J, Mangelsdorf DJ: Molecular basis for feedback regulation of bile acid synthesis by nuclear receptors. Mol Cell 6: 507–515, 2000.

20. Goodwin B, Jones SA, Price RR, Watson MA, McKee DD, Moore LB, Galardi C, Wilson JG, Lewis MC, Roth ME, Maloney PR, Willson TM, Kliewer SA: A regulatory cascade of the nuclear receptors FXR, SHP-1, and LRH-1 represses bile acid biosynthesis. Mol Cell 6: 517–526, 2000.

21. Sinal CJ, Tohkin M, Miyata M, Ward JM, Lambert G, Gonzales FJ: Targeted disruption of the nuclear receptor FXR/BAR impairs bile acid and lipid homeostasis. Cell 102: 731–744, 2000.

22. Kerr TA, Saeki S, Schneider M, Schaefer K, Berdy S, Redder T, Shan B, Russell DW, Schwarz M: Loss of nuclear receptor SHP impairs but does not eliminate negative feedback regulation of bile acid synthesis. Dev Cell 2: 713–720, 2002.

23. Gupta S, Stravitz RT, Dent P, Hylemon PB: Down-regulation of cholesterol 7α-hydroxylase (*CYP7A1*) gene expression by bile acids in primary rat hepatocytes is mediated by the c-Jun N-terminal kinase pathway. J Biol Chem 276: 15816–15822, 2001.

24. Wang L, Lee YK, Bundman D, Han Y, Thevananther S, Kim CS, Chua SS, Wei P, Heyman RA, Karin M, Moore DD: Redundant pathways for negative feedback regulation of bile acid production. Dev Cell 2: 721–731, 2002.

25. Cheema SK, Agellon LB: The murine and human cholesterol 7α-hydroxylase gene promoters are differentially responsive to regulation by fatty acids via peroxisome proliferator-activated receptor α. J Biol Chem 275: 12530–12536, 2000.

26. Drover VA, Wong NC, Agellon LB: A distinct thyroid hormone response element mediates repression of the human cholesterol 7α-hydroxylase (*CYP7A1*) gene promoter. Mol Endocrinol 16: 14–23, 2002.

27. Bouscarel B, Kroll SD, Fromm H: Signal transduction and hepatocellular bile acid transport: cross talk between acids and second messengers. Gastroenterology 117: 433–452, 1999.

28. Li D, Schuetz EG, Guzelian PS: Hepatocyte culture in study of P450 regulation. Methods Enzymol 206: 335–344, 1991.

29. Torchia EC, Stolz A, Agellon LB: Differential modulation of cellular death and survival pathways by conjugated bile acids. BMC Biochem 2: 11, 2001.

30. Elzinga BM, Baller JF, Mensenkamp AR, Yao Z, Agellon LB, Kuipers F, Verkade HJ: Inhibition of apolipoprotein B secretion by taurocholate is controlled by the N-terminal end of the protein in rat hepatoma McArdle-RH7777 cells. Biochim Biophys Acta 1635: 93–103, 2003.

31. Spady DK, Cuthbert JA, Willard MN, Meidell RS: Adenovirus-mediated transfer of a gene encoding cholesterol 7α-hydroxylase into hamsters increases hepatic enzyme activity and reduced plasma total and low density lipoprotein cholesterol. J Clin Invest 96: 700–709, 1995.

32. Agellon LB: Partial transfection of liver with a synthetic cholesterol 7α-hydroxylase transgene is sufficient to stimulate the reduction of cholesterol in the plasma of hypercholesterolemic mice. Biochem Cell Biol 75: 255–262, 1997.

33. Spady DK, Cuthbert JA, Willard MN, Meidell RS: Overexpression of cholesterol 7α-hydroxylase (CYP7A) in mice lacking the low density lipoprotein (LDL) receptor gene. LDL transport and plasma LDL concentrations are reduced. J Biol Chem 273: 126–132, 1998.

34. Li Q, Yokoyama S, Agellon LB: Active taurocholic acid flux through hepatoma cells increases the cellular pool of unesterified cholesterol derived from lipoproteins. Biochim Biophys Acta 1580: 22–30, 2002.

35. Angelin B, Einarsson K, Hellstrom K, Leijd B: Effects of cholestyramine and chenodeoxycholic acid on the metabolism of endogenous triglyceride in hyperlipoproteinemia. J Lipid Res 19: 1017–1024, 1978.

36. Carulli N, Ponz de Leon M, Podda M, Zuin M, Strata A, Frigerio G, Digrisolo A: Chenodeoxycholic acid and ursodeoxycholic acid effects in endogenous hypertriglyceridemias. A controlled double-blind trial. J Clin Pharmacol 21: 436–442, 1981.

37. Meijer GW, Beynen AC: Lipid and carbohydrate metabolism in cholestyramine-fed rats. Artery 16: 286–292, 1989.

38. Carrella M, Ericsson S, Del Piano C, Angelin B, Einarsson K: Effect of cholestyramine treatment on biliary lipid secretion rates in normolipidaemic men. J Intern Med 229: 241–246, 1991.

39. Watanabe M, Houten SM, Wang L, Moschetta A, Mangelsdorf DJ, Heyman RA, Moore DD, Auwerx J: Bile acids lower triglyceride levels via a pathway involving FXR, SHP, and SREBP-1c. J Clin Invest 113: 1408–1418, 2004.

40. Sirvent A, Claudel T, Martin G, Brozek J, Kosykh V, Darteil R, Hum DW, Fruchart JC, Staels B: The farnesoid X receptor induces very low density lipoprotein receptor gene expression. FEBS Lett 566: 173–177, 2004.

41. Kast HR, Nguyen CM, Sinal CJ, Jones SA, Laffitte BA, Reue K, Gonzalez FJ, Willson TM, Edwards PA: Farnesoid X-activated receptor induces apolipoprotein C-II transcription: a molecular mechanism linking plasma triglyceride levels to bile acids. Mol Endocrinol 15: 1720–1728, 2001.

42. Claudel T, Inoue Y, Barbier O, Duran-Sandoval D, Kosykh V, Fruchart J, Fruchart JC, Gonzalez FJ, Staels B: Farnesoid X receptor agonists suppress hepatic apolipoprotein CIII expression. Gastroenterology 125: 544–555, 2003.

43. Horton JD, Goldstein JL, Brown MS: SREBPs: activators of the complete program of cholesterol and fatty acid synthesis in the liver. J Clin Invest 109: 1125–1131, 2002.

44. Torra IP, Claudel T, Duval C, Kosykh V, Fruchart JC, Staels B: Bile acids induce the expression of the human peroxisome proliferator-activated receptor α gene via activation of the farnesoid X receptor. Mol Endocrinol 17: 259–272, 2003.

45. Duran-Sandoval D, Mautino G, Martin G, Percevault F, Barbier O, Fruchart JC, Kuipers F, Staels B: Glucose regulates the expression of the farnesoid X receptor in liver. Diabetes 53: 890–898, 2004.

46. Yamagata K, Daitoku H, Shimamoto Y, Matsuzaki H, Hirota K, Ishida J, Fukamizu A: Bile acids regulate gluconeogenic gene expression via small heterodimer partner-mediated repression of hepatocyte nuclear factor 4 and Foxo1. J Biol Chem 279: 23158–23165, 2004.

47. Stayrook KR, Bramlett KS, Savkur RS, Ficorilli J, Cook T, Christe ME, Michael LF, Burris TP: Regulation of carbohydrate metabolism by the farnesoid X receptor. Endocrinology 146: 984–991, 2005.

48. Lee YK, Dell H, Dowhan DH, Hadzopoulou-Cladaras M, Moore DD: The orphan nuclear receptor SHP inhibits hepatocyte nuclear factor 4 and retinoid X receptor transactivation: two mechanisms for repression. Mol Cell Biol 20: 187–195, 2000.

49. Borgius LJ, Steffensen KR, Gustafsson JA, Treuter E: Glucocorticoid signaling is perturbed by the atypical orphan receptor and corepressor SHP. J Biol Chem 277: 49761–49766, 2002.

50. Kim JY, Kim HJ, Kim KT, Park YY, Seong HA, Park KC, Lee IK, Ha H, Shong M, Park SC, Choi HS: Orphan nuclear receptor small heterodimer partner represses hepatocyte nuclear factor 3/Foxa transactivation via inhibition of its DNA binding. Mol Endocrinol 18: 2880–2894, 2004.

51. Han SI, Studer E, Gupta S, Fang Y, Qiao L, Li W, Grant S, Hylemon PB, Dent P: Bile acids enhance the activity of the insulin receptor and glycogen synthase in primary rodent hepatocytes. Hepatology 39: 456–463, 2004.

52. Tobin KA, Steineger HH, Alberti S, Spydevold O, Auwerx J, Gustafsson JA, Nebb HI: Cross-talk between fatty acid and cholesterol metabolism mediated by liver X receptor-α. Mol Endocrinol 14: 741–752, 2000.

53. Schultz JR, Tu H, Luk A, Repa JJ, Medina JC, Li L, Schwendner S, Wang S, Thoolen M, Mangelsdorf DJ, Lustig KD, Shan B: Role of LXRs in control of lipogenesis. Genes Dev 14: 2831–2838, 2000.

54. Laffitte BA, Joseph SB, Walczak R, Pei L, Wilpitz DC, Collins JL, Tontonoz P: Autoregulation of the human liver X receptor α promoter. Mol Cell Biol 21: 7558–7568, 2001.

55. Kim HJ, Kim JY, Park SK, Seo JH, Kim JB, Lee IK, Kim KS, Choi HS: Differential regulation of human and mouse orphan nuclear receptor small heterodimer partner promoter by sterol regulatory element binding protein-1. J Biol Chem 279: 28122–28131, 2004.

56. Paigen B, Morrow A, Holmes PA, Mitchell D, Williams RA: Quantitative assessment of atherosclerotic lesions in mice. Atherosclerosis 68: 231–240, 1987.

57. Costet P, Legendre C, More J, Edgar A, Galtier P, Pineau T: Peroxisome proliferator-activated receptor α-isoform deficiency leads to progressive dyslipidemia with sexually dimorphic obesity and steatosis. J Biol Chem 273: 29577–29585, 1998.

58. Djouadi F, Weinheimer CJ, Saffitz JE, Pitchford C, Bastin J, Gonzalez FJ, Kelly DP: A gender-related defect in lipid metabolism and glucose homeostasis in peroxisome proliferator-activated receptor α-deficient mice. J Clin Invest 102: 1083–1091, 1998.

59. Vassileva G, Huwyler L, Poirier K, Agellon LB, Toth MJ: The intestinal fatty acid binding protein is not essential for dietary fat absorption in mice. FASEB J 14: 2040–2046, 2000.

60. Schwarz M, Russell DW, Dietschy JM, Turley SD: Alternate pathways of bile acid synthesis in the cholesterol 7α-hydroxylase knockout mouse are not upregulated by either cholesterol or cholestyramine feeding. J Lipid Res 42: 1594–1603, 2001.

61. Lew JL, Zhao A, Yu J, Huang L, De Pedro N, Pelaez F, Wright SD, Cui J: The farnesoid X receptor controls gene expression in a ligand- and promoter-selective fashion. J Biol Chem 279: 8856–8861, 2004.

62. Yu J, Lo JL, Huang L, Zhao A, Metzger E, Adams A, Meinke PT, Wright SD, Cui J: Lithocholic acid decreases expression of bile salt export pump through farnesoid X receptor antagonist activity. J Biol Chem 277: 31441–31447, 2002.

63. Sinal CJ, Yoon M, Gonzalez FJ: Antagonism of the actions of peroxisome proliferator-activated receptor-α by bile acids. J Biol Chem 276: 47154–47162, 2001.

64. Cui J, Huang L, Zhao A, Lew JL, Yu J, Sahoo S, Meinke PT, Royo I, Pelaez F, Wright SD: Guggulsterone is a farnesoid X receptor antagonist in coactivator association assays but acts to enhance transcription of bile salt export pump. J Biol Chem 278: 10214–10220, 2003.

65. Urizar NL, Moore DD: Gugulipid: a natural cholesterol-lowering agent. Annu Rev Nutr 23: 303–313, 2003.

66. Burris TP, Montrose C, Houck KA, Osborne HE, Bocchinfuso WP, Yaden BC, Cheng CC, Zink RW, Barr RJ, Hepler CD, Krishnan V, Bullock HA, Burris LL, Galvin RJ, Bramlett K, Stayrook KR: The hypolipidemic natural product guggulsterone is a promiscuous steroid receptor ligand. Mol Pharmacol 67: 948–954, 2005.

Section II

Diabetes and Hypertension Induced Atherosclerosis

11

Lipoprotein Metabolism in Insulin-Resistant States

RITA KOHEN AVRAMOGLU, HEATHER BASCIANO, AND KHOSROW ADELI

Abstract

The incidence of insulin-resistant states, such as type 2 diabetes and obesity, has been rapidly increasing in both adult and pediatric populations worldwide. A major complication of insulin resistance is an atherogenic dyslipidemia that contributes to a significantly higher risk of atherosclerosis and cardiovascular disease. The most fundamental defect in these patients is resistance to cellular actions of insulin, particularly resistance to insulin-stimulated glucose uptake. Insulin insensitivity appears to cause hyperinsulinemia, enhanced hepatic gluconeogenesis and glucose output, reduced suppression of lipolysis in adipose tissue leading to a high free fatty acid (FFA) flux, and increased very low-density lipoprotein (VLDL) secretion causing hypertriglyceridemia and reduced plasma levels of high-density lipoprotein (HDL) cholesterol. Although the link between insulin resistance and dysregulation of lipoprotein metabolism is well established, a significant gap of knowledge exists regarding the underlying cellular and molecular mechanisms. Genetic and diet-induced animal models of insulin resistance have been recently employed to delineate the mechanistic link between perturbations in insulin-signaling pathways and dysregulation of hepatic lipid and lipoprotein metabolism in insulin-resistant states. A series of important and novel observations have been made and published in recent years that will be summarized in this chapter. The critical role of key phosphatases and protein kinases that mediate the signaling changes leading to dysregulation of hepatic lipogenesis and VLDL overproduction will be discussed. Emerging evidence suggests that insulin resistance and its associated metabolic dyslipidemia result from perturbations in key molecules of the insulin-signaling pathway, including overexpression of phosphatases, protein tyrosine phosphatase 1B (PTP-1B) and phosphatase and tensin homolog (PTEN), downregulation of the phosphatidylinositol-3-kinase (PI-3-K) pathway and basal activation of the mitogen-activated protein (MAP) kinase cascade, leading to a state of mixed hepatic insulin resistance and sensitivity. These signaling changes in turn cause an increased expression of sterol regulatory element-binding protein (SREBP) 1c, induction of de novo lipogenesis and higher activity of microsomal triglyceride transfer protein (MTP), which together with high exogenous FFA flux collectively stimulates the hepatic production of apolipoprotein B (apoB)-containing VLDL particles.

Keywords: apolipoprotein B; diabetes; free fatty acid; insulin resistance; insulin signaling; lipoprotein; metabolic syndrome

Abbreviations: ABC, ATP-binding cassette; ACC, acetyl-CoA carboxylase; apoB, apolipoprotein B; BAT, brown adipose tissue; BMI, body mass index; CHD, coronary heart disease; CPT, carnitine palmitoyl transferase; CR, chylomicron remnants; ER, endoplasmic reticulum; ERK, extracellular signal-regulated kinase; FAS, fatty acid synthase; FFA, free fatty acid; FXR, farnesoid X receptor; GLUT, glucose transporter; HDL, high-density lipoprotein; IDF, International Diabetes Federation; HMG-CoA reductase, 3-hydroxy-3-methylglutaryl coenzyme A reductase; HSL, hormone sensitive lipase; IR, insulin receptor; IRS, insulin receptor substrate; JAK2, Janus kinase 2; JNK, c-jun kinase; KO, knockout; LDL, low-density lipoprotein; LPL, lipoprotein lipase; LXR, liver X receptor; MAP, mitogen-activated protein; MEK, MAPK/ERK kinase; MTP, microsomal triglyceride transfer protein; NAFLD, nonalcoholic fatty liver disease; NASH, nonalcoholic steatohepatitis; PGC-1, PPARgamma coactivator-1; PI-3-K, phosphatidylinositol-3-kinase; PKB, protein kinase B; PKC, protein kinase C; PPAR, peroxisome proliferator-activated receptor; PTEN, phosphatase and tensin homolog; PTP-1B, phosphotyrosyl-protein phosphatase 1B; SCAP, SREBP cleavage-activating protein; SCD, stearoyl-CoA desaturase; SH, src homology; SRE, sterol regulatory element; SREBP, sterol regulatory element-binding protein; STZ, streptozotocin; TG, triglyceride; TNF, tumor necrosis factor; VLDL, very low-density lipoprotein; WHHL, Watanabe heritable hyperlipidemic; WHO, World Health Organization

The Rising Global Epidemic of Diabetes and Associated Dyslipidemic Complications

Social and economic changes within the last century have brought countless advances, and have vastly improved lifestyles in many parts of the world. Paradoxically, the global trend toward a more sedentary lifestyle, associated with a positive energy balance, has also led to an array of increased health risks in an aging population. Numerous sources including the World Health Organization (WHO) and International Diabetes Federation (IDF) report a worldwide diabetes epidemic, with a parallel rise in obesity and insulin resistance [1]. It is estimated that approximately 5% of the global population is diabetic, with 85–95% of this being attributed to noninsulin-dependent diabetes mellitus or type 2 diabetes [2]. The IDF currently estimates the number of individuals with type 2 diabetes at 125 million with a projected rise to between 200 and 300 million within the next decade [2]. In 2002, the American Diabetes Association estimated the total economic cost of diabetes in the United States alone at 132 billion dollars, with 18.2 million individuals (or 6.3% of the U.S. population) being affected [3]. Numerous ethnic populations appear to be particularly predisposed to the development of diabetes. Mexican–Americans have a 21% prevalence compared with 3% among non-Hispanic Americans [4]. Native Hawaiians are known to have a prevalence of type 2 diabetes that is four times higher than the general population of the

United States [5]. The potential for increased cases of type 2 diabetes is also very high in developing countries, including China and India [6–8].

Associated with diabetes is a wide array of very severe health risks, and long-term complications including eye, nerve, and kidney disease; as well as diseases of the circulatory system that may lead to amputation. Metabolic dyslipidemia is the most common complication of insulin resistance and type 2 diabetes, and is believed to be exacerbated by obesity, as well as numerous detrimental environmental factors such as a high-fat diet and sedentary lifestyle. The dyslipidemia accompanying insulin resistance is characterized by distinct changes from a normal plasma lipid and lipoprotein profile. Such changes include decreases in plasma levels of HDL with elevated plasma FFA and triglycerides (TG). These lipid changes are commonly associated with elevated VLDL production, a concomitant increase in apoB, an essential structural component of these atherogenic lipoproteins, and increased plasma levels of small dense low-density lipoprotein (LDL) [9, 10]. These parameters constitute a highly atherogenic dyslipidemic profile that significantly contributes to increased risk of cardiovascular disease, the most prevalent cause of death in industrialized countries and one which is rising at an alarming rate in less-developed countries.

Insulin Resistance and Metabolic Dyslipidemia: Components of a Complex Metabolic Syndrome

The current consensus is that obesity and insulin resistance may be part of a common pathologic state termed as "metabolic syndrome." The metabolic syndrome, formerly referred to as insulin resistance syndrome [11] or syndrome X [12], is characterized by a constellation of pathologies that include glucose intolerance, insulin resistance, obesity, dyslipidemia, and hypertension. Insulin resistance generally develops as the first indicator of type 2 diabetes and manifests as a decreased biological response to normal levels of circulating plasma insulin. Indicators of insulin resistance include impaired glucose tolerance, hyperglycemia, and elevated plasma insulin levels. As long as the pancreas can compensate for the decreased insulin response by increasing insulin secretion, the individual is able to control blood glucose level. Allowed to continue untreated, however, the pancreas eventually fails to produce sufficient insulin, and type 2 diabetes occurs. Although not formerly considered a disease of childhood, type 2 diabetes has begun to present with increasing frequency in the pediatric population [13, 14]. It is feared that the disease progression begins early in life, and persistence from childhood to adulthood produces type 2 diabetes and cardiovascular disease in early adulthood [15, 16].

The concept of the metabolic syndrome emerged progressively from clinical observations over many decades. The clustering of metabolic perturbations was first recognized in the early 1920s by the Swedish physician Eskil

Kylin who defined this multifactorial disease as including hypertension, gout, and hyperglycemia [17]. A quarter century later, Jean Vague a professor at the Université de Marseille, established a correlation between body fat distribution, in particular abdominal or android obesity, with the risk of diabetes and cardiovascular disease [18]. In 1965, Pietro Avogaro and Gaetano Crepaldi described the metabolic syndrome as being accompanied by hyperlipidemia due to elevated plasma TG levels, obesity, diabetes, hypertension, and a high risk of coronary artery disease [19]. In 1988, Gerald Reaven went on to further characterize the lipoprotein abnormalities associated with metabolic syndrome, and coined a new term, "syndrome X." This was followed by the terms "deadly quartet" [20] and "insulin resistance syndrome" [21]. In 1998, the WHO coined the term "metabolic syndrome" [22], which has since become the more popular designation of this metabolic disorder. The most current definition of metabolic syndrome according to the WHO includes insulin resistance as defined by glucose intolerance, impaired fasting glucose or type 2 diabetes accompanied by at least two of the following states; hypertension, elevated plasma TG, decreased plasma HDL, high body mass index (BMI), or elevated urinary albumin. Currently this definition is again shifting and the trend has been to diagnose metabolic syndrome based on individual clinical judgment, taking into account these previously described metabolic perturbations.

Link between Insulin Resistance, Metabolic Dyslipidemia, and Cardiovascular Disease

The single most common complication and the leading cause of mortality of the metabolic syndrome remains diabetic/metabolic dyslipidemia. The diabetic dyslipidemia that accompanies the metabolic syndrome is generally characterized by increased plasma TG, increased small dense LDL, and decreased HDL each of which is believed to be an independent risk factor for cardiovascular disease [23, 24]. Numerous clinical studies, as well as observations in animal models, have led to metabolic models of the possible causes of this common diabetic complication.

Although the molecular mechanisms underlying diabetic dyslipidemia are far from being completely understood, it is now generally believed that the elevation in plasma TG accompanying the disorder is central to development of this complex phenotype. Over the last several decades, a number of large-scale, long-term clinical studies testing several classes of lipid-lowering drugs have been conducted, and several have included diabetic individuals among their experimental subgroups. The general consensus among these studies has been that an increased risk of developing coronary heart disease (CHD) is associated with increased plasma lipids in insulin-resistant states. The larger of these studies include the Scandinavian Simvastatin Survival Study Group (4S), the Long-Term Intervention with Pravastatin in Ischemic Disease Study Group (LIPID), and the Cholesterol and Recurrent Events study (CARE)

[25–29]. Although these studies have focused on the cholesterol-lowering 3-hydroxy-3-methylglutaryl coenzyme A (HMG-CoA) reductase inhibitor drugs also known as statins, a general trend was found in lowering of CHD risk, both overall, as well as in a diabetes subgroup in the study. The Helsinki Heart Study was the first pioneering study that showed a small trend towards improvement of CHD risk in diabetic individuals using fibrates, a peroxisome proliferator-activated receptor (PPAR) alpha agonist class of compounds that reduce production and enhance clearance of VLDL [30]. These findings were reinforced recently with evidence from the Veterans Administration High-Density Lipoprotein Intervention Trial (VA-HIT) [31, 32] and the Diabetes Atherosclerosis Intervention Study (DAIS) [33–35], the latter showing a decrease in progression of atherosclerosis following 3 years of fenofibrate treatment. Numerous other studies such as the Lipids in Diabetes Study (LDS) are currently underway to evaluate the efficacy of fibrates or combination therapies in the treatment of diabetes-related CHD. The Fenofibrate Intervention and Event Lowering in Diabetes Study (FIELD) once completed will compile data from 12,000 individuals and will be one of the largest clinical trial studies performed to date. A newer class of compounds, the thiazolidinediones or glitazones, are PPAR gamma agonists that exert an insulin-sensitizing effect and improve glucose tolerance. As the thiazolidinediones have only been available for a few years, large-scale comparative clinical trials currently need to be undertaken in order to fully understand the effects of this class of drugs. In a smaller-scale clinical setting however, it has been repeatedly shown that use of these compounds result in normalization of glucose liver homeostasis accompanied by a decrease in plasma FFA, an increase in HDL, a decrease in small dense LDL, and a decrease in vascular inflammation [36–38].

Insulin Resistance and Metabolic Dyslipidemia: Lessons from Animal Models

In animal models, alterations in lipid and glucose metabolism may be induced through genetic manipulation, drug administration, or even relatively simple changes in diet. To date, a wide variety of models have been developed for the study of metabolic defects of insulin resistance or lipid metabolism [39, 40]. Use of animal models of insulin resistance has provided strong evidence for a link between insulin resistance and metabolic dyslipidemia (reviewed in Refs. [40, 41]). Genetic models such as the ob/ob mouse, the db/db mouse, the Zucker fa/fa (fatty) rat, and the ZDF/Drt-fa (diabetic/fatty) rat exhibit phenotypes of insulin resistance or diabetes, obesity, and dyslipidemia [39, 40-44]. The most widely recognized of these is perhaps the ob/ob mouse, which lacks the satiety factor leptin, a gene implicated in the development of obesity. The dramatic phenotype of the ob/ob mice includes morbid obesity, diabetes, infertility, as well as reduced activity, metabolism, and body temperature. The genetically diabetic or db/db mouse, a close relative of the ob/ob mouse, displays a similar phenotype although the genetic defect responsible appears to be a mutation in the leptin receptor gene, rather than in leptin itself. While the ob/ob mouse displays hyperplasia of pancreatic cells, the db/db mouse displays evidence of β-cell degranulation and apoptosis. Both the ob/ob and db/db mice display abnormal lipid profiles with hypertriglyceridemia, although in the ob/ob mouse, there also appears to be a defect in the catabolism of apoAI and apoAII leading to increased HDL levels [45]. It has recently been found that glucose-6-phosphate levels are elevated in both ob/ob and db/db mice [46]. This may partly account for the dysregulation of glucose metabolism and increased hepatic VLDL production in these animals. While treatment with thiazolidinediones improves the insulin sensitivity of muscle tissue and the fatty liver phenotype in ob/ob mice, hepatic VLDL production appears less sensitive to insulin treatment [47–49].

One of the earliest rat models used, the Zucker fatty rat (fa/fa), also carries a mutation in the leptin receptor gene [50]. This model exhibits hyperglycemia, hyperinsulinemia, type 2 diabetes, hypertriglyceridemia, hypercholesterolemia, in addition to increased appetite [51]. Early studies showed neointimal hyperplasia following injury in the diabetic animals compared with lean controls [52]. One of the striking features of Zucker rats is the accumulation of TG-rich lipoproteins such as chylomicrons in the plasma [53]. It has been shown that there is an elevation in MTP mRNA in Zucker rats compared with lean controls that may partly contribute to the VLDL overproduction [54]. Following insulin treatment, Zucker rats show resistance to completely normalize lipoprotein profiles [55]. Recently, treatment with thiazolidinediones showed an improvement in insulin-induced FFA clearance, improved FFA usage during fasting, improved glucose homeostasis, a decrease in hepatic TG production, and increased TG clearance [56, 57]. However, statin treatment resulted in marked decreases in plasma TG, total cholesterol, and total HDL. The currently available data suggests that in this model plasma TG but not cholesterol may be dependent on plasma insulin levels [58].

The homozygous JCR:LA corpulent rat (cp/cp) exhibits obesity, hypertriglyceridemia, hyperinsulinemia, and impaired glucose tolerance [59–61]. In these animals, insulin resistance appears to be more severe in homozygous males, while females and lean animals do not appear to develop atherosclerosis spontaneously. There is also enhanced secretion of VLDL particles that results from the conversion of dietary carbohydrate into TG [62].

Corroborating evidence from other animal models include the Watanabe heritable hyperlipidemic (WHHL) rabbit model of familial hypercholesterolemia. This animal model exhibits hyperinsulinemia and insulin resistance, and the addition of a small amount of cholesterol to the diet results in elevated plasma lipids and severe cardiovascular disease [63]. Administration of thiazolidinediones in combination with HMG-CoA reductase inhibitors lowers plasma lipid levels and ameliorates hypertension and insulin sensitivity [64–67]. Transgenic WHHL rabbits overexpressing human lipoprotein lipase (LPL) also showed improvements in hypertriglyceridemia, hypercholesterolemia, and obesity suggesting potential therapeutic benefits of LPL [68].

The transgenic mouse model using overexpression of the A1 adenosine receptor in adipose tissue has implicated this receptor in the metabolism of intracellular fat accumulation, FFA metabolism, and plasma glucose regulation [69–72]. Interestingly, although both control and transgenic animals were of the same size and body composition, the transgenic animals exhibited lower plasma FFA, and failed to develop glucose intolerance in oral glucose tolerance tests. Another genetically altered model, the ApoB/BATless mouse, is a cross between a human apoB transgenic mouse and a brown adipose tissue (BAT) knockout (KO) mouse. This model exhibits peripheral insulin resistance, resulting in obesity, hypertriglyceridemia, hypercholesterolemia, and hyperinsulinemia once placed on a high-fat diet [73]. Although increases in apoB levels were seen in the apoB/BATless mice, the mRNA levels of both MTP, as well as apoB, were similar between these and control mice expressing only the human apoB transgene. This suggests that VLDL assembly and secretion were instead regulated posttranslationally.

Several rodent models of diet-induced insulin resistance have also been developed and employed to enhance our understanding of the underlying mechanisms associated with complications due to dyslipidemia. The "sand rat" (*Psammomys obesus*), a gerbil native to the desert regions of the eastern Mediterranean and northern Africa, spontaneously develops obesity and insulin resistance when fed standard rodent chow instead of its customary diet of succulent plants [39, 41]. These animals exhibit signs of hyperglycemia, insulin resistance, dyslipidemia, obesity, and hyperphagia [74–77]. The defect is now believed to at least partly result from apoptosis of pancreatic β-cells. There is also evidence of insulin receptor (IR) signaling pathway dysfunction [44, 78], hepatic insulin resistance [79], and insensitivity to the satiety factor leptin [80]. Characterization of the lipid distribution within these animals showed significant increases in circulating VLDL and LDL in both hyperinsulinemic, as well as hyperinsulinemic and hyperglycemic animals [81]. Elevated levels of protein kinase C (PKC) epsilon in skeletal muscle are thought to contribute to the development of insulin resistance in these animals [82]. In addition, the novel protein factor, beacon, was also found to be differentially expressed in the brain of obese and control animals [83].

A fructose-fed Syrian golden hamster model has been developed in our laboratory and is increasingly used as a simple dietary model of insulin resistance and metabolic dyslipidemia. This diet-induced model of insulin resistance has been extensively characterized [84], and is of interest as its lipoprotein metabolism more closely resembles that of humans compared to other rodent models [85–88]. Hamsters develop hyperlipidemia and atherosclerosis in response to a modest increase in dietary cholesterol and saturated fat and can be made obese, hypertriglyceridemic, and insulin-resistant by fructose feeding [84, 89, 90]. Fructose-induced metabolic dyslipidemia is also usually accompanied by whole-body insulin resistance and reduced hepatic

212   Rita Kohen Avramoglu et al.

insulin sensitivity [84, 91]. Fructose feeding for a 2-week period induced significant increases in plasma TG, cholesterol, and FFA. Induction of insulin resistance was accompanied by a considerable rise in the *in vivo* production of hepatic VLDL-apoB and –TG [84]. These data suggest overall improved efficiency of VLDL assembly in animals fed a high-fructose diet. This may be due to the increased intracellular stability and availability of apoB, elevated levels of available neutral lipid, or increased MTP mass or activity [92, 93]. A more detailed discussion of mechanisms of metabolic dyslipidemia in the fructose-fed hamster model will appear later in this chapter.

Molecular Mechanisms of Insulin Resistance and Metabolic Dyslipidemia: Perturbations in Insulin Signaling Cascades

Under normal metabolic conditions, the binding of insulin to its plasma membrane receptor initiates a cascade of events, beginning with receptor autophosphorylation and activation of receptor tyrosine kinases (see Fig. 11.1) [94, 95]. Receptor autophosphorylation, in turn, results in tyrosine

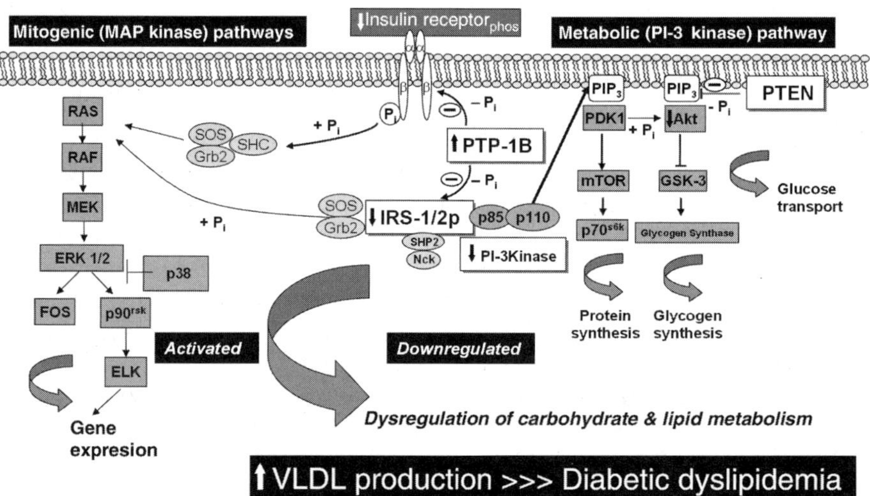

FIGURE 11.1. Postulated link between hepatic insulin signaling and downstream VLDL-apoB secretion. Insulin induces signal transduction via two major signaling pathways: the mitogenic, MAP-kinase pathway and the metabolic, PI 3-kinase pathway. Under normal conditions, insulin acutely reduces apoB secretion. Under conditions of insulin resistance, however, there is reduced sensitivity to inhibitory action of insulin on apoB resulting in enhanced VLDL secretion. Enhanced expression of PTP-1B, a key negative regulator of insulin signaling, may be a key initiating factor in inducing hepatic insulin resistance and consequently increased VLDL synthesis and secretion.

phosphorylation of adaptor proteins such as members of the insulin receptor substrate (IRS) family (IRS-1/2/3/4), and Shc [96–100]. Insulin signaling is regulated by the activity of phosphotyrosyl-protein phosphatases (PTPases) through dephosphorylation of the insulin receptor, IRS-1, IRS-2, and Shc leading to modulation of insulin action downstream of the receptor [101]. IRS-1 and IRS-2 are adaptor proteins for src homology 2 (SH2)-domain containing signaling proteins including the Grb-2–SOS complex, SHP2, Nck, and PI-3 kinase [102–105]. The production of 3′-phospholipids such as PI-3,4,5-P3 (PIP3) is dependent on activation of PI3-kinase. Binding of IRS to p85, the regulatory subunit of PI-3 kinase, activates the PI-3 kinase–PKB/Akt pathway, which is necessary for insulin action on glucose transport and glycogen synthesis [97, 106, 107].

Insulin resistance is thought to develop as a result of defects downstream of the IR activation. Because of the complexity of the numerous branches of the insulin signaling cascade, cross talk between signaling molecules, and the effects of detrimental environmental factors, attempts at understanding the precise mechanisms involved in the development of insulin-resistant states has proven challenging. Defects in insulin signaling appear to occur primarily in insulin sensitive tissues, namely liver, adipose, and muscle. Among obese and type 2 diabetic individuals, significant decreases in IRS-1-associated tyrosine phosphorylation and PI3-kinase activity have been observed in skeletal muscle and adipocytes; the two tissues predominantly targeted by insulin [107–111]. Obese and hyperinsulinemic individuals exhibited decreased IR expression level and activity, in addition to decreased tyrosine kinase activity, in both skeletal muscle and adipocytes [112, 113], while patients with type 2 diabetes exhibit reduced insulin signaling in skeletal muscle and the liver [112, 114]. In lean type 2 diabetic subjects, insulin-stimulated PI-3 kinase activity was also decreased suggesting that a defect in insulin signaling could contribute to insulin resistance [111]. Similar findings have been reported in numerous genetic [115–119] and induced [120, 121] rodent models of obesity. Targeted disruption of the IR and IRS-1 or IRS-2 [122–127] have all shown that an insensitivity to the effects of insulin in a combination of tissues including muscle, liver, and adipose tissue can lead to insulin resistance and type 2 diabetes [107].

Defects in molecules of the insulin-signaling pathway are now thought to be a major mechanism involved in the development of insulin resistance. Perturbations are documented in both arms of the insulin-signaling pathway, the PI-3 kinase and the mitogen-activated protein (MAP) kinase cascades. Activated PI3-kinase generates 3′-phospholipids, including PIP3, which mediates many cellular responses of insulin [128, 129]. PIP3 levels are attenuated by the action of a lipid phosphatase, PTEN [107]. Perturbations in PTEN can thus have important consequences on the PI-3 kinase signaling cascade.

Many gene regulatory events induced by receptor-associated kinases (such as IR) are mediated by the MAP kinase cascade (reviewed in Refs.

[130–132]). Several families of the MAP kinases exist including specifically extracellular signal-regulated kinase (ERK) 1/2, *c*-jun kinase (JNK), and p38. There is increasing evidence that members of the MAP kinase family of serine–threonine kinases may contribute to the development of insulin resistance. IRS-1 contains many potential MAP kinase phosphorylation sites [133], and it has been observed that certain MAP kinases are capable of phosphorylating and inactivating IRS-1 [134–136]. Furthermore, insulin stimulation is capable of activating members of the MAP kinase family, specifically ERK1/2, JNK, and p38 [137–139]. In human skeletal muscle, insulin signaling to ERK1/2 was normal from patients with type 2 diabetes compared with healthy control subjects, despite a diminished insulin responsiveness of components required for the metabolic effects of insulin such as components of the PI-3 kinase and AKT pathways [140, 141]. In other models an increased basal level of phosphorylation of ERK1/2, JNK, or p38 was observed in adipocytes from type 2 diabetic subjects. This was found to associate with a reduction in IRS-1 and glucose transporter 4 (GLUT4), suggesting that elevated MAP kinase activation may contribute to the type 2 diabetic condition [142, 143]. Interestingly, tumor necrosis factor (TNF) α-mediated insulin resistance appears to be mediated by the MAP/ERK kinase (MEK) 1/2-p42/44 MAP kinase pathway [144]. Overall, these very intriguing observations indicate a critical role for activated MAP kinase cascade in insulin resistance.

Recent evidence has also implicated alterations in expression of cellular phosphatases in the development of insulin resistance. Of particular interest is the PTP-1B, a member of the PTPase family of enzymes, which is widely expressed in insulin sensitive tissues [145]. PTP-1B dephosphorylates the IR *in vitro*, and as well inducing the downregulation of IRS-1 and insulin-stimulated PI-3 kinase activity [146–148]. Increased PTP-1B mass and activity have been shown to increase with carbohydrate-induced insulin resistance; while normalization of PTP-1B mass and activity result in the reversal of this type of insulin resistance [149, 150]. PTP-1B KO mice exhibit increased sensitivity toward insulin-induced IR and IRS-1 tyrosine phosphorylation, and are resistant to obesity [151]. Additional evidence of PTP-1B involvement in insulin resistance is seen in primary hepatocytes from an insulin resistant hamster where PTP-1B mass and activity were shown to increase, with a concomitant decrease in PI-3-K and PKB/Akt phosphorylation [93]. Interestingly, a significant increase in PTP-1B protein mass was seen following 2 days of insulin treatment in primary hepatocytes, which was then accompanied by a decrease in IR mass and phosphorylation. PTP-1B has also been shown to influence leptin, an important hormone known to be involved in the regulation of appetite and body mass [152]. In transfection studies, as well as in transgenic animals, PTP-1B has been shown to dephosphorylate the leptin receptor-associated kinase (Janus kinase 2, JAK2) [153–155]. Conversely, in PTP-1B deficient mice, there was an enhanced response toward leptin-mediated loss of body weight.

Adipose Tissue Insulin Resistance, Dysregulation of Fatty Acid Storage and Enhanced Free Fatty Acid Flux

Free fatty acid (FFA), whether obtained through dietary sources or produced via de novo lipogenesis, may generally undergo three fates: intracellular storage by adipose tissue in the form of TG [156]; export into the plasma in the form of VLDL-TG; or use as an energy source via FFA oxidation. Although FFA may also be stored in nonadipocyte cells as TG, this is usually a source of lipotoxicity [157]. Nonadipose TG storage usually occurs in situations where FFA export exceeds the catabolic capacity of metabolic tissues, or storage in adipose tissue is impaired [156]. Over 30% of energy expenditure occurs in skeletal muscle, which modulates the balance between stored lipids and energy utilization [158]. FFA oxidation is induced through both reductions in lipogenic gene expression and upregulation in genes controlling lipid oxidation [157]. These gene alterations are achieved through sophisticated nutrient sensors, and subsequent nuclear signaling cascades.

The net concentration of FFA in the plasma results from a delicate equilibrium between enzyme-regulated lipolysis of plasma triglyceride-rich lipoproteins such as VLDL, and FFA uptake by peripheral tissues. Hormonal regulation of extracellular and intracellular factors affecting FFA metabolism is currently not well understood. Under normal conditions, insulin stimulates postprandial uptake of glucose, as well as FFA esterification and storage, and can play a major role in the suppression of hormone sensitive lipase (HSL), the enzyme that is the major regulator of FFA release from adipose tissue [159]. Insulin and glucose are also believed to stimulate LPL activity, primarily from adipose tissue. LPL activity provides an essential first step in the delivery of FFA to adipose tissue for storage, as well as plasma TG removal.

Both fasting and postprandial levels of FFA are often elevated in obese and insulin-resistant individuals. In the insulin-resistant state, there is an increase in release of FFA from adipose tissue concomitant with a decrease in uptake by muscle tissue. This results in an increased influx of FFA into the liver. This "vicious cycle" may result in an attenuation in insulin signaling in these tissues and may exacerbate insulin resistance (see Fig. 11.2) [160, 161]. In muscle, insulin resistance may be mediated by soluble factors; likely derived from fat tissue [162]. Additional factors may influence FFA flux in insulin insensitive conditions. PPAR-gamma activation can lower circulating FFA and increase oxidation through stimulation of adiponectin production. Adiponectin stimulation results in decreased liver TG and increased insulin sensitivity [163]. Although these processes are by far exceeded by postprandial availability of FFA, insulin is known to stimulate enzymes involved in de novo lipogenesis. Generally under conditions where influx of fat to the liver exceeds efflux, there is excess hepatic fatty acid uptake, synthesis and secretion; and hepatic steatosis can ensue alongside insulin resistance [164]. In addition, overloading of white adipose tissue beyond its storage capacity can

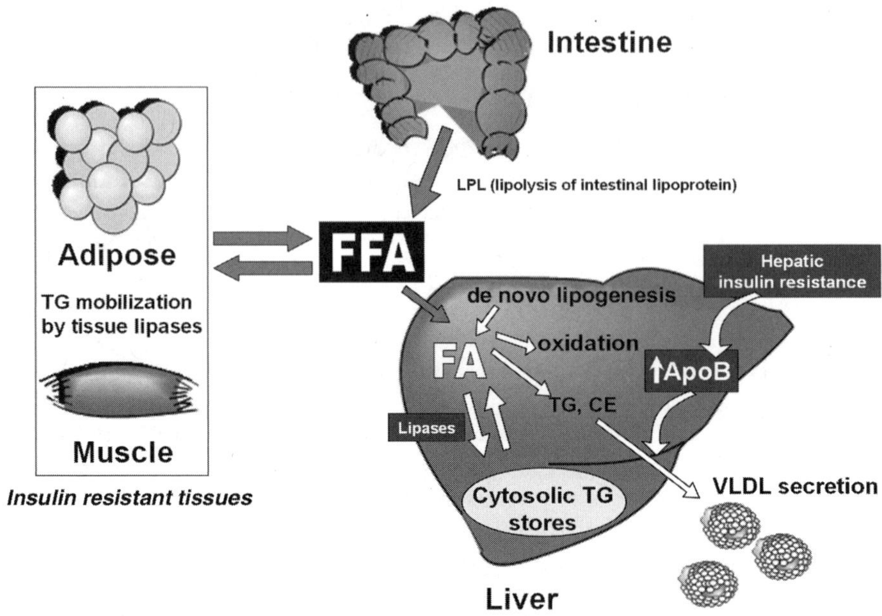

FIGURE 11.2. Mechanisms of diabetic dyslipidemia in insulin-resistant tissues. Insulin resistance results in enhanced TG lipolysis by adipocytes and decreased FFA absorption resulting in an enhanced FFA flux into peripheral tissues. Increased lipid availability resulting from elevated FFA flux combined with hepatic and intestinal insulin resistance appear to lead to a considerable overproduction of both hepatically- and intestinally derived apoB-containing lipoprotein particles. These events may, in turn, exacerbate fasting and postprandial metabolic dyslipidemia.

also lead to lipid disorders in skeletal and cardiac muscles, and pancreas, illustrating the importance of functional adipocytes [165].

The true underlying culprit to adipose dysfunction and metabolic syndrome may be a chronic positive energy balance—where dysregulation of adipose tissue metabolism may be a catalyst for the chain of events leading to metabolic dyslipidemia [166]. White adipose tissue is now well recognized as an endocrine tissue producing multiple hormones. Plasma levels of these adipocyte hormones, or adipokines, are altered in obese, insulin-resistant, and diabetic subjects [165]. In this way, adipose tissue participates in the regulation of glucose and insulin metabolism through the release of adipokines including leptin, adiponectin, and resistin [167].

The insulin-sensitizing hormone adiponectin is found in lower concentrations corresponding with greater adiposity [166, 168, 169], and is currently considered as an important link between obesity and insulin resistance. Decreased adipocyte differentiation, resulting in decreased adiponectin expression may be involved in the whole-body insulin resistance associated with obesity [170, 171]. Adiponectin is associated with inflammatory markers through underlying obesity, whereas association of resistin with inflammatory markers appear to be less clear [172]. This connection is further illustrated by significant weight loss with improvements in insulin resistance, atherosclerosis, and cardiovascular disease usually accompanied by increases in adiponectin levels [169, 173]. It has been further proposed that improvements in insulin resistance observed following bariatric surgery are modulated by changes in adipocytokine levels via loss of adipose mass, supporting a role for adiponectin and resistin in the development of insulin resistance and diabetes [162, 167].

Resistin is a protein secreted from adipose tissue that is also thought to provide a link between obesity, insulin sensitivity, and diabetes [174–176]. Although resistin may have an important involvement in insulin resistance, animal and human studies currently show conflicting results [167]. The regulation of resistin expression by agents known to modulate insulin sensitivity does not always support the association between resistin and obesity-induced insulin resistance [174]. It has been surprisingly found that high serum resistin levels can predict future increases in percent body fat, but do not predict insulin resistance in humans [177]. On the other hand, decreases in resistin levels may contribute to improving insulin action [175]. Consistent with this observation, high-fat feeding is known to induce the development of obesity and insulin resistance alongside increases in resistin expression. In addition, there are also positive correlations between glycemia and resistin gene expression [174]. Resistin protein level is elevated in obese mice and may be decreased by the use of insulin-sensitizing thiazolidinediones, while immunoneutralization of resistin improves insulin sensitivity in obese mice.

Dysregulation of Fatty Acid Oxidation in Insulin Resistance

Metabolic dyslipidemia of insulin-resistant states may arise from increased peripheral FFA exposure due to reduced FFA oxidation [178]. Impaired FFA oxidation may contribute to excess TG accumulation, and may mediate numerous adverse effects of metabolic insulin resistance syndrome [179]. The associated inflammation can lead to numerous dysfunctions in lipid and lipoprotein metabolism, alongside marked reductions in lipid oxidation [180]. In the insulin-resistant state, there is a failure of skeletal muscle to shift between glucose and lipid utilization in fasting and insulin-stimulated states. In addition, there is also a failure to suppress FFA efflux from adipose tissue, as well as an impaired FFA efflux and storage balance [181]. Thus TG accumulation in skeletal muscle results from a decrease in FFA oxidation, as well as increased FFA uptake [157].

Mechanisms of the reductions in FFA oxidation include reduced carnitine palmitoyl transferase (CPT) activity and the resultant increase in intracellular malonyl-CoA [157], chronic hyperglycemia and hyperinsulinemia [182], decreased expression of PPARgamma coactivator-1alpha (PGC-1alpha) [179], acquired or inherited defects in mitochondrial fatty acid oxidation [183], and decreased expression of mitochondria and mitochondrial enzymes in muscle [179]. Plasma TG levels increase following the suppression of FFA oxidation in patients with type 2 diabetes, where lipid oxidation and insulin-mediated glucose oxidation are reduced [178, 180].

Nuclear Signaling Cascades Regulating Fatty Acid Storage and Oxidation

The physiological regulation of lipogenic gene expression and metabolism is mediated by nuclear hormone receptors, which function as hormone and nutrient-regulated transcription factors that bind DNA [158]. Metabolic nuclear receptors such as receptors for oxysterols (LXRs), bile acids (FXR), and FFA (through PPAR) regulate key aspects of cellular and whole-body sterol homeostasis. These processes include lipoprotein synthesis, lipoprotein uptake by peripheral tissues, cholesterol absorption, bile acid synthesis, and reverse cholesterol transport [184]. The liver nuclear receptors also act as nutrient sensors of unsaturated FFA status to determine whether FFA are to be stored or oxidized. FFA may regulate lipid metabolism at the level of protein expression or degradation, or via gene transcription and mRNA decay [185]. Nuclear factors that regulate lipogenesis and the storage of FFA include the liver X receptor (LXR) and the sterol regulatory element-binding protein (SREBP) pathways. Transcriptional regulation through SREBP control de novo lipogenesis and lipoprotein uptake mediating cholesterol homeostasis [184]. The LXRs alpha and beta are also cholesterol metabolism and biosynthesis sensors, which result in induced expression of lipogenic enzymes upon stimulation [186]. LXRs, along with the farnesoid X receptor (FXR), also promote storage and catabolism of sterols and their metabolites to prevent excess intracellular cholesterol accumulation [184]. Upon stimulation, the LXRs, which are present in vascular smooth muscle tissue, may lead to foam cell formation, that, in turn, contributes to cardiovascular disease via upregulation of lipogenic genes such as SREBP and fatty acid synthase (FAS) [187].

Three PPAR isoforms exist namely, PPAR alpha, gamma, and delta [188]. These nuclear receptors are activated by natural FFA and lipid-derived ligands. PPAR gamma, which also activates lipogenic enzymes, may exacerbate steatosis and is transcriptionally upregulated in hepatic steatosis [189]. Activation of PPAR gamma induces TG storage in tissues in which it is expressed, including adipose tissue and the intestine [188]. Balancing out the lipogenic and TG storage capacities of PPAR gamma, LXR, and SREBP are the other members of the PPAR family of nuclear receptors, which are responsible for stimulating the oxidation of FFA. PPAR alpha is highly expressed in liver, heart, and to a lesser extent, in skeletal muscle, and can prevent steatosis upon activation through amplified FFA oxidation processes [188, 189]. In cellular responses to fasting and starvation, PPAR alpha-mediated FFA oxidation becomes important for energy metabolism [190, 191]. PPAR delta is ubiquitous, and may also augment FFA oxidation in tissues where PPAR alpha is less abundant [188]. The PPARs may also contribute to the amplification loop of the LXR-mediated cellular efflux of cholesterol, via the ATP-binding cassette (ABC) transporters, ABCA1 and ABCG1, further contributing to their antiatherogenic properties [186]. The balance and interactions of dietary factors via these nuclear receptors determines the partitioning of fatty acids between storage and oxidation.

Nuclear receptor signaling and lipid metabolism are tightly intertwined with insulin signaling and glucose/energy regulation. The IRS downstream of the IR have specific regulatory roles in liver function. IRS-1 is closely linked to glucose homeostasis, whereas IRS-2 is linked to lipid metabolism via downstream metabolic nuclear receptors [192]. Examples of these molecular relationships are illustrated by studies using PPAR alpha KO mice. As expected, PPAR alpha KO mice exhibit markedly reduced FFA oxidation, but are also hyperinsulinemic and hypoglycemic in the fasting state [191].

The pharmacologic ligands of PPAR alpha (fibrates) and PPAR gamma (thiazolidinediones) have been quite successfully developed [188, 193–195]. Fibrates have been found to play a role on insulin sensitivity, via decreases in circulating FFA and TG, and decreases in ectopic lipids deposited in nonadipose tissues [196]. In addition to insulin sensitization of muscle tissue, thiazolidinedione treatment is accompanied by enhanced total body fat oxidation, increases in insulin-stimulated glucose uptake, and decreased plasma FFAs [193]. Thiazolidinediones have also been used successfully in the treatment of nonalcoholic steatohepatitis (NASH) [197]. It appears that thiazolidinedione action on muscle insulin sensitivity may be the result of the secretion of insulin-sensitizing hormones such as adiponectin by adipocytes, and to the lowering of circulating lipids—all events that result in normalized glucose utilization [188, 189, 195]. In addition, another target gene of PPAR gamma is LXR alpha, which is also known for its lipid storage and glucose clearance capacity [186].

A new category of drug therapy for lipid disorders, insulin resistance, and NASH includes PPAR delta agonists. Functions of PPAR delta are less known, but these agonists may play a beneficial role in the treatment of lipid disorders, in particular obesity [194, 195]. PPAR delta can upregulate PGC-1alpha and regulate FFA oxidation in several tissues, which may prove beneficial in the treatment of insulin resistance and obesity [179, 194]. Combination products are also currently being developed for clinical use, targeting both PPAR alpha and PPAR gamma [198].

Aberrant De Novo Lipogenesis in Insulin Resistance: Critical Role of SREBP-1c

SREBP membrane proteins are members of the basic helix-loop-helix-leucine zipper family of transcription factors that act as dimers to activate genes involved in lipid metabolism, in addition to a range of enzymes required for endogenous cholesterol, FFA, TG, and phospholipid synthesis [199]. The SREBPs are synthesized as precursor proteins and require a sequential two-step proteolytic processing to release the amino-terminal active domain in the nucleus. SREBP precursor proteins in the endoplasmic reticulum (ER) form a complex with SREBP cleavage-activating protein (SCAP), which also contains a sterol sensory domain. In the presence of high cellular cholesterol concentrations, SCAP also binds the retention factor INSIG-1 and sequesters SREBP to the ER. Conversely, in presence of low cellular cholesterol concentrations SREBP is activated in the Golgi by a sequential two-step cleavage mediated by proteases S1P and S2P [200]. Transcriptionally active SREBP binds to sterol response elements (SRE) in the promoter/enhancer regions of target genes [201, 202]. Three SREBP isoforms, 1a, 1c, and 2, are expressed at varying levels in different tissues. SREBP-1a and -1c are isoforms produced from a single gene by alternate splicing, where SREBP-2 is encoded by a different gene and does not display any known isoforms [200, 201, 203]. *In vivo* studies using transgenic and KO mice suggest that SREBP-1c is involved in FFA synthesis and insulin-induced glucose metabolism, and importantly lipogenesis, whereas SREBP-2 is more specific to cholesterol synthesis and LDL receptor regulation. The SREBP-1a isoform appears to be implicated in both pathways [201, 204].

All SREBP isoforms are regulated by nutritional signals and intracellular cholesterol content, but the SREBP-1c isoform appears also to be primarily regulated at the transcriptional level by insulin [201]. Both insulin and LXR alpha induce SREBP-1c transcription, although full induction of the mature and transcriptionally active form of SREBP-1c protein does requires insulin. Activation of LXR alpha leads to the induction of SREBP-1c gene expression and its precursor protein, but it does not have a strong effect in inducing the mature nuclear form of the transcription factor until acute exposure to insulin via a phosphatidylinositol 3-kinase (PI-3-K)-dependent mechanism occurs [202, 205]. Insulin can enhance expression of SREBP in all three major target tissues: liver, fat, and skeletal muscle [206–209]. In the hyperinsulinemic state, levels of SREBP are similarly enhanced [210, 211]. There is evidence that the MAP kinase pathway [212] regulates the insulin-mediated stimulation of SREBP with ERK1/2 being shown to activate SREBP-1a by phosphorylating serine 117 [213]. Paradoxically, depletion of insulin and the resulting reduction in insulin signaling following streptozotocin (STZ) treatment induces SREBP-1c expression, despite the fact that normally, SREBP-1 is directly stimulated via insulin signaling.

SREBP-1a activation in the liver has a strong impact on plasma insulin levels, implicating the potential role of SREBPs in hepatic insulin metabolism relating to insulin resistance. Transgenic mice overexpressing nuclear SREBP-1a have enlarged pancreatic islets, a massively enlarged fatty liver, and a disappearance of peripheral white adipose tissue [214]. Excess glucose is also known to upregulate SREBP-1c precursor and nuclear proteins, and stimulate SREBP-1c maturation via a JAK/STAT dependent pathway [215]. Excess energy intakes can lead to the formation of remnant lipoproteins and hepatic insulin resistance through suppression of IRS-2, via activated hepatic SREBP-1c expression [204], and increased intramuscular lipid accumulation associated with muscle insulin resistance in obesity or type 2 diabetes may develop from de novo FFA synthesis in skeletal muscle. High-fructose diets induce insulin resistance through increased hepatic TG production following an upregulation of lipogenic pathways [216]. Following 7 days on a 60% fructose diet, there was an induction of lipogenic gene expression including FAS, acetyl-CoA carboxylase (ACC), and stearoyl-CoA desaturase (SCD), along with the hepatic isoform of SREBP-1 [217]. FAS has been extensively studied in rat livers, and is also an important downstream component of lipid synthesis machinery. In rats, fructose feeding created a 56% increase in TG secretion rate and an 86% increase in plasma TG, through the stimulation of FAS activity.

Recently, it has been found that overexpression of PTP-1B, which accompanies dysfunctional insulin signaling, leads to increases in the expression of FAS alongside increased mRNA and promoter activity of SREBP-1c. Insulin resistance may therefore be triggered by PTP-1B, contributing to the associated lipogenesis and hypertriglyceridemia observed [218]. In insulin-resistant rats fed a high-fructose diet, it has been reported that increased PTP-1B expression [218] occurs in correlation with the increases in hepatic SREBP-1 mRNA [219]. PTP-1B was shown to play a role in enhancing SREBP-1 gene expression via increases in protein phosphatase 2A activity and upregulation of Sp1 transcriptional activity [218].

Hepatic Steatosis: A Common Feature of Insulin Resistant Dyslipidemia

One of the most common liver diseases in the United States, nonalcoholic fatty liver disease (NAFLD), is found with high prevalence in the obese, type 2 diabetic population [220]. A second term, NASH, was conceived in the 1980s to describe this syndrome, whereby patients exhibit liver pathologies consistent with alcoholic hepatitis, despite the absence of excessive alcohol consumption [221]. NASH is characterized by inflammation, hepatocellular necrosis, and most notably an excessive deposition of fat in the liver [197]. The majority of patients showing characteristics of NASH tend to be overweight/obese males, exhibiting diabetes, dyslipidemia, and hyperinsulinemia [222, 223].

The general consensus is that steatohepatitis requires a secondary influence of inflammation, fibrosis, or necrosis for NASH to occur. Abnormalities in nutrition, oxidative stress, and lipid peroxidation may all interact, generating steatohepatitis [221]. Secondary oxidative stresses resulting from insulin resistance may also result in lipid peroxidation, metabolic stress, and NASH [220]. There is evidence of metabolic syndrome, with its intrinsic defect of insulin resistance, being a common link between NAFLD and the accompanying hypertriglyceridemia, obesity, and hypertension [221, 224, 225]. Further evidence of insulin resistance being a causative factor of NASH is illustrated by the fact that hepatic lipid accumulation, inflammation, and fibrosis are all improved if insulin sensitivity is regained [197].

In NAFLD patients, peripheral glucose removal is decreased, resulting from impaired glucose oxidation and stimulated glycogen synthesis, accompanied by increased lipid oxidation and the appearance of glycerol [225]. Lipid peroxidation is generally increased, and inversely correlated with fibrosis [223]. The degree of hepatic steatosis and associated LDL oxidizability is in turn also related to glucose disposal, and glucose production [225]. If fibrosis and inflammation of the liver related to these symptoms is evident, there is a higher likelihood of progression from NAFLD and NASH to cirrhosis. Complications of cirrhosis can include carcinoma and liver failure, requiring invasive therapy and liver transplantation [197, 220, 222, 224].

Overproduction of ApoB-Containing Lipoproteins in Insulin-Resistant States: A Major Feature of Metabolic Dylipidemia Observed in Insulin-Resistant States

Increased hepatic VLDL production may underlie many features of the dyslipidemia observed in insulin-resistant states. Early clinical evidence suggested increased production of TG-rich hepatic VLDL coupled with decreased VLDL clearance in individuals with metabolic syndrome [226–229]. There is also evidence for enhanced postprandial lipemia and the presence of small dense LDL particles (reviewed in Ref. [230]). All of these changes comprise an extremely atherogenic lipoprotein profile that contributes to the increased cardiovascular disease risk associated with the metabolic syndrome. The increase in VLDL production in insulin resistance appears to result from decreased sensitivity to the inhibitory effects of insulin on VLDL secretion [231, 232]. In humans, VLDL is assembled and secreted exclusively by the liver. The VLDL assembly process is complex, requiring the synthesis of the 550 kDa structural protein apoB100, followed by the assembly of lipoprotein particles with cholesterol, cholesterol ester, phospholipids, and triglyceride [233]. The availability of lipids within the lumen of ER appears to dictate the amount of apoB secreted extracellularly mostly via co- and posttranslational mechanisms [234, 235]. In lipid-poor states, a major proportion of newly synthesized apoB is degraded by proteasomal and non-proteasomal pathways [233, 236]. Thus, an important factor involved in the assembly of apoB-containing lipoproteins is MTP, which catalyzes the transfer of neutral lipids to the apoB molecule [237]. Chronic modulation of apoB and VLDL secretion can be achieved via changes in MTP expression and activity [237]. Insulin may control the rate of hepatic VLDL production directly by influencing the rate of apoB synthesis and degradation [229, 238], or modulation of MTP gene expression [239]. Insulin regulation of apoB appears to involve the activation of the PI-3 kinase pathway which may mediate the inhibitory effect of insulin on the VLDL assembly process [240, 241].

Links between Aberrant Free Fatty Acid Flux and Hepatic Lipoprotein Production

Increased hepatic VLDL-triglyceride production in insulin resistance appears to directly, or indirectly, result from decreased sensitivity to the inhibitory effects of insulin on VLDL secretion. Increased FFA can thus further attenuate insulin signaling and therefore exacerbate insulin resistance. Therefore, an increased FFA flux as is observed in insulin resistance may cause increased TG availability that may, in turn, stimulate assembly and secretion of VLDL in hepatocytes [242]. This has been observed in HepG2 cells [243–245] as well as some primary hepatocyte cell models [246–248]. However, studies in other primary systems including rat [249, 250], hamster [89, 92], and human [251] hepatocytes have failed to demonstrate FFA-mediated stimulation of apoB secretion, hence the matter of secretion remains controversial. Alternately, increased levels of apoB may instead be due to alterations in apoB stability as a result of changing FFA levels. Recently it has been shown that treatment of HepG2 cells with antiretroviral protease inhibitor compounds can prevent apoB degradation, despite increases in ubiquitinated apoB [252]. This resulted in a significant accumulation of intracellular apoB, but without any concurrent increase of secreted apoB. The impairment of apoB lipoprotein secretion was attributed to a sharp decrease in intracellular synthesis of neutral lipids. Secretion could only then be restored by the addition of exogenous fatty acid, suggesting that the intracellular pool of apoB could be secreted only upon lipid availability—as in the case of increased FFA flux to the liver in insulin resistance. It is thus reasonable to postulate that in insulin-resistant states, hepatocytes may respond to exogenous FFA by oversecreting VLDL.

De Novo Lipogenesis, SREBP-1c, MTP, and ER60 in Hepatic VLDL Overproduction

Studies in several models of insulin resistance, suggest a complex relationship between hepatic insulin resistance, hepatic FFA flux, SREBP-1c expression, de novo lipogenesis, MTP gene expression, and VLDL overproduction in insulin-resistant states (see Fig. 11.3) [253]. Development of insulin resistance in the

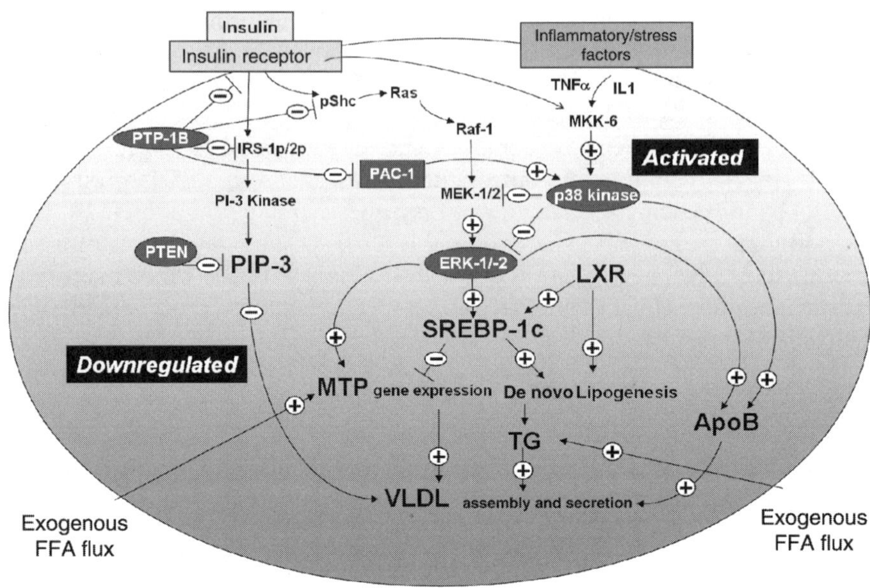

FIGURE 11.3. Molecular link between insulin resistance and metabolic dyslipidemia in the hepatocyte. Insulin binding to the insulin receptor triggers a signaling cascade that negatively regulate VLDL assembly and secretion principally via the PI-3 kinase pathway. In insulin-resistant states, this negative regulation appears to be lost, thus leading to stimulation of VLDL production. The upregulation of key phosphatases, including PTP-1B and PTEN, may play an important role in downregulation of PI-3 kinase signaling. Conversely, inflammatory and stress mechanisms result in stimulation of the MAP kinase pathway which in turn can upregulate MTP, SREBP-1c, and apoB, leading to increased de novo lipogenesis and increased VLDL secretion. Increased free fatty acid flux into the liver in the insulin-resistant state also plays an important role in activation of MTP and enhanced triglyceride synthesis. These factors collectively contribute to VLDL production and may exacerbate elevated circulating VLDL.

fructose-fed hamster model was found to result in high plasma levels of FFA and increased VLDL production, as well as higher MTP mass, mRNA, and activity [84]. As there is a tight correlation between MTP levels and VLDL production, increased MTP expression and activity would, therefore, be expected to stimulate the assembly and secretion of apoB-lipoproteins [237]. Higher MTP expression levels have also been observed in the liver of obese diabetic mice, as well as the Otsuka Long-Evans Toskushima fatty rat model of type 2 diabetes that presents with visceral fat obesity—confirming observations seen in the hamster model [254, 255]. The mechanisms stimulating MTP gene expression in insulin resistance and type 2 diabetes is currently unknown. Because insulin is a negative regulator of the MTP gene, MTP upregulation in insulin resistance is presumably due to the impaired insulin-regulatory system [256, 257]. This regulation appears to occur via activation of SREBPs, which, as mentioned previously, bind a putative SRE within the −124 to −116 of the 5′ MTP gene promoter [258]. Interestingly, the MTP promoter possesses overlapping insulin response elements and SREs [257, 259]. Emerging evidence also suggests that MTP gene expression is under the control of the MAP kinase cascade [239]. While ERK signaling inhibits MTP promoter activity, p38 kinase stimulates the promoter [239]. The signaling interactions are still unclear, but since MAPK is known to activate SREBP-1c, it may be expected that MAPK activation would also increase SREBP-1c.

SREBP-1c elevation would be expected to inhibit MTP gene expression, since the MTP promoter has been suggested to have negative SRE elements [258]. However, the opposite is observed; MTP gene expression is enhanced in the fructose-fed hamster liver in the setting of elevated SREBP-1c. This intriguing observation in the fructose-fed hamster compares well with observations in the obese diabetic mice, where MTP mRNA and activity were enhanced in the presence of elevated levels of SREBP-1a and SREBP-1c [254]. It has been suggested that increases in hepatic FFA flux and/or increased TG stores may stimulate MTP gene expression [239], and in this case, altered hepatic SREBP-1c may only play a minor role. Elevated SREBP-1c levels in transgenic mice that overexpress cholesterol 7α-hydroxylase (involved in bile acid metabolism) were also found to be associated with both accelerated lipogenesis, as well as increased MTP gene expression—leading to apoB oversecretion [260]. These findings suggest that MTP expression may be stimulated in insulin resistance by some unknown factors, which may block SREBP-mediated inhibition of the promoter. Recent studies using MTP gene promoter constructs have clearly shown a stimulatory effect of oleic acid on MTP promoter activity [261], despite earlier studies suggesting that FFA may not affect MTP expression [262]. This provides further evidence that external factors may generate stimulatory effects on MTP promoter activity. Interestingly, oleate-mediated stimulation of the MTP promoter was found to be independent of the SRE [261]. Since MAP kinases are now known to regulate both SREBP-1c and MTP gene expression, a potential link may exist between chronic basal activation of the MAP kinase cascade, the induction of de novo lipogenesis, and increased expression of MTP; which can together facilitate hepatic VLDL assembly and secretion.

Another contributing factor to insulin resistance includes the effects of lipid peroxidation on VLDL overproduction. High fructose-fed insulin-resistant rats have shown less protection from lipid peroxidation with a prooxidant effect, and have a hypertriglyceridemic phenotype [263]. It has been hypothesized that prooxidant stress response pathways may mediate delayed clearance and hepatic increases in VLDL secretion. Hypertriglyceridemic fructose-fed rats treated with lipoxygenase inhibitors showed a reversal of lipid dysregulations observed, and inflammatory protein activity responses [264]. Lipid peroxides, diene conjugates, and reactive substances are undeniably elevated in fructose-fed insulin-resistant animals, resulting in oxidative stress often being implicated in the pathology of insulin resistance—especially accompanying a deficient antioxidant system. Evidence of this is illustrated by the fact that administration of the antioxidant alpha-lipoic acid shows improvements in insulin sensitivity, and prevention of such negative oxidative changes [265]. Lipoic acid treatment can also prevent the increases in cholesterol, TG, activity of lipogenic enzymes, and VLDL secretion evident in insulin resistance due to fructose feeding; as well as prevent the reductions in LPL and HDL cholesterol. Lipoic acid may also even normalize the dyslipidemic cholesterol distribution of plasma lipoproteins [266]. Taken together, these studies show a clear role of insulin-resistant peroxidative stress pathways possibly involved in VLDL oversecretion.

In addition to the increased assembly and secretion of apoB, increased levels of small dense LDL particles have also been observed in insulin-resistant states [9]. Early evidence has shown that insulin resistance subsequent to fructose diets can alter the structure and function of VLDL particles causing an increased total cholesterol and phospholipid content, and an increase in the TG:protein ratio [267]. It has been found that higher TG levels can result in smaller, denser, more atherogenic LDL particles, where LDL particle size is inversely related to TG concentration [268]; this contributes to the morbidity of metabolic disorders associated with insulin resistance.

Emerging evidence suggests a potential role for other cellular factors in the fructose-induced dyslipidemia and enhanced hepatic VLDL production. Changes in the insulin-signaling pathway upon fructose feeding coincide with a drastic suppression of ER-60, a protein that potentially targets apoB to a nonproteasomal degradation pathway. As expected, this reduction in ER-60 is accompanied by an increase in the synthesis and secretion of apoB. ER-60 is directly associated with apoB [269]. Decreases in apoB secretion have been shown to occur via adenoviral-mediated overexpression of ER-60—a decrease that is unaffected by the addition of proteasomal inhibitors [270]. An important observation in the insulin resistant fructose-fed hamster model was that compared to chow-fed control animals, the hepatocytes of fructose-fed hamsters expressed a lower level of ER-60. Interestingly, when these fructose-fed hamsters were treated with the insulin sensitizing thiazolidinedione, rosiglitazone, normalization of the ER-60 protein in the liver occurred. In the same animal model, rosiglitazone significantly ameliorated VLDL secretion, both *in vivo* and ex vivo [256]. This suggests that ER-60 protein levels may be responsive to hepatic insulin signaling. The 5′ promoter of the ER-60 gene has been analyzed, and found to contain putative motifs that may mediate insulin and/or sterol regulation of ER-60, namely IREs, SRE, SRE3, and NF-Y. The enhanced stability of apoB in the livers of fructose-fed hamsters as a result of this downregulation of the ER-60 protease may thus contribute to, and result in, higher assembly and secretion of VLDL.

Intestinal ApoB48-Lipoprotein Overproduction in Insulin Resistance

The dyslipidemia that accompanies insulin-resistant states is increasingly being considered a postprandial phenomenon. Numerous reports have shown an elevation in postprandial TG-rich lipoproteins in insulin-resistant subjects, and that fasting hypertriglyceridemia may actually predict this abnormal postprandial response to a fat load [271–277]. Postprandial TG-rich lipoproteins and especially chylomicron remnants (CR) have been implicated as risk factors for atherosclerosis, based on both experimental work and clinical studies [278–282]. There is growing evidence that intestinal lipoprotein overproduction is a major contributor to the fasting and postprandial lipemia observed in insulin-resistant states [283]. Intestinal secretion of apolipoprotein B48-containing lipoprotein particles, accompanied by enhanced intestinal lipid synthesis in the form of free cholesterol, cholesterol ester, and triglyceride have been documented upon insulin resistance stimulated by chronic fructose feeding. Basal levels of lipoprotein secretion upon fructose feeding may enhance intestinal secretion through increased de novo lipogenesis, as well as increased MTP availability. Insulin-resistant animals fed a high-fructose diet, when compared to chow-fed animals, showed a significant shift toward secretion of larger, less dense chylomicrons [284]. Metabolic dyslipidemia is thus a common feature of insulin-resistant states and appears to arise from aberrant metabolism of apoB-containing lipoproteins produced not only in the liver, but also in the small intestine.

Concluding Remarks

Over the last decade, a great deal of progress has been made in understanding the molecular mechanisms of insulin resistance and its metabolic complications. The availability of both genetic and dietary animal models of this disorder has greatly facilitated this research effort. It is now clear that a strong link exists between perturbations in insulin signal transduction and alterations in lipid and lipoprotein metabolism. A central initiating factor appears to be defects in FFA storage and oxidation, leading to enhanced FA flux in key insulin sensitive tissues such as the liver and muscle. Emerging evidence also implicate a direct role for perturbations in key modulators of insulin signaling molecules and cascades leading to disordered lipoprotein metabolism, either directly or indirectly via changes in FFA flux. These initial abnormalities appear to lead to hepatic and intestinal overproduction of highly atherogenic lipoprotein particles that result in the commonly observed phenotype of hypertriglyceridemia, hyperapoB, and low HDL-cholesterol levels. Dysregulation of key nuclear transcription factors including LXR, PPARs, and SREBPs clearly play a central role in the transition from normolipidemia to a metabolic dyslipidemic state. It is also of particular interest that these major metabolic changes can be induced by relatively simple changes in dietary intake of certain nutrients, particularly fructose. Recent studies of the mechanisms by which fructose induces insulin resistance and metabolic dyslipidemia have been particularly fruitful in delineating the underlying molecular and cellular mechanisms of this increasingly common metabolic disorder. The recent explosion of interest in insulin resistance and its health consequences should lead to further understanding of the link between insulin insensitivity and lipoprotein dysregulation, and eventually to the development of novel therapeutic approaches in the prevention and treatment of the metabolic syndrome and its dyslipidemic complications.

References

1. Mackay J, Mensah GA, eds: The Atlas of Heart Disease and Stroke, 2004. World Health Organization, Geneva.
2. King H, Aubert RE, Herman WH: Global burden of diabetes, 1995–2025: prevalence, numerical estimates, and projections. Diabetes Care 21: 1414–1431, 1998.
3. Hogan P, Dall T, Nikolov P: Economic costs of diabetes in the US in 2002. Diabetes Care 26: 917–932, 2003.
4. Glaser NS, Jones KL: Non-insulin dependent diabetes mellitus in Mexican–American children. West J Med 168: 11–16, 1998.
5. Grandinetti A, Chang HK, Mau MK, Curb JD, Kinney EK, Sagum R, Arakaki RF: Prevalence of glucose intolerance among Native Hawaiians in two rural communities. Native Hawaiian Health Research (NHHR) Project. Diabetes Care 21: 549–554, 1998.
6. Amos AF, McCarty DJ, Zimmet P: The rising global burden of diabetes and its complications: estimates and projections to the year 2010. Diabet Med 14 (Suppl 5): S1–S85, 1997.
7. Pan XR, Yang WY, Li GW, Liu J: Prevalence of diabetes and its risk factors in China, 1994. National Diabetes Prevention and Control Cooperative Group. Diabetes Care 20: 1664–1669, 1997.
8. Ramachandran A, Snehalatha C, Latha E, Vijay V, Viswanathan M: Rising prevalence of NIDDM in an urban population in India. Diabetologia 40: 232–237, 1997.
9. Reaven GM, Chen YD, Jeppesen J, Maheux P, Krauss RM: Insulin resistance and hyperinsulinemia in individuals with small, dense low density lipoprotein particles. J Clin Invest 92: 141–146, 1993.
10. Taskinen MR: Insulin resistance and lipoprotein metabolism. Curr Opin Lipidol 6: 153–160, 1995.
11. Zimmet PZ: Kelly West Lecture 1991. Challenges in diabetes epidemiology—from West to the rest. Diabetes Care 15: 232–252, 1992.
12. Reaven GM: Banting lecture 1988. Role of insulin resistance in human disease. Diabetes 37: 1595–1607, 1988.
13. Pinhas-Hamiel O, Dolan LM, Daniels SR, Standiford D, Khoury PR, Zeitler P: Increased incidence of non-insulin-dependent diabetes mellitus among adolescents. J Pediatr 128: 608–615, 1996.

14. Sinha R, Fisch G, Teague B, Tamborlane WV, Banyas B, Allen K, Savoye M, Rieger V, Taksali S, Barbetta G, Sherwin RS, Caprio S: Prevalence of impaired glucose tolerance among children and adolescents with marked obesity. N Engl J Med 346: 802–810, 2002.

15. Bao W, Srinivasan SR, Berenson GS: Persistent elevation of plasma insulin levels is associated with increased cardiovascular risk in children and young adults. The Bogalusa Heart Study. Circulation 93: 54–59, 1996.

16. Freedman DS, Khan LK, Dietz WH, Srinivasan SR, Berenson GS: Relationship of childhood obesity to coronary heart disease risk factors in adulthood: the Bogalusa Heart Study. Pediatrics 108: 712–718, 2001.

17. Kylin E. Studien über das Hypertoni-Hyperglykemi-Hyperurikemi syndrom. Zentralblatt f Innere Medizin 44: 105–112, 1923.

18. Vague J. La differenciation sexuelle, facteur determinant des formes de l'obesite. Presse Med 30: 339–340, 1947.

19. Avogaro P, Crepaldi G, Enzi G, Tiengo A: [Metabolic aspects of essential obesity]. Epatologia 11: 226–238, 1965.

20. Kaplan NM: The deadly quartet. Upper-body obesity, glucose intolerance, hypertriglyceridemia, and hypertension. Arch Intern Med 149: 1514–1520, 1989.

21. Ferrannini E, Haffner SM, Mitchell BD, Stern MP: Hyperinsulinaemia: the key feature of a cardiovascular and metabolic syndrome. Diabetologia 34: 416–422, 1991.

22. Alberti KG: Impaired glucose tolerance: what are the clinical implications? Diabetes Res Clin Pract 40 (Suppl): S3–S8, 1998.

23. Krauss RM: Lipids and lipoproteins in patients with type 2 diabetes. Diabetes Care 27: 1496–1504, 2004.

24. Taskinen MR: Type 2 diabetes as a lipid disorder. Curr Mol Med 5: 297–308, 2005.

25. Pyorala K, Pedersen TR, Kjekshus J, Faergeman O, Olsson AG, Thorgeirsson G: Cholesterol lowering with simvastatin improves prognosis of diabetic patients with coronary heart disease. A subgroup analysis of the Scandinavian Simvastatin Survival Study (4S). Diabetes Care 20: 614–620, 1997.

26. Haffner SM, Alexander CM, Cook TJ, Boccuzzi SJ, Musliner TA, Pedersen TR, Kjekshus J, Pyorala K: Reduced coronary events in simvastatin-treated patients with coronary heart disease and diabetes or impaired fasting glucose levels: subgroup analyses in the Scandinavian Simvastatin Survival Study. Arch Intern Med 159: 2661–2667, 1999.

27. Goldberg RB, Mellies MJ, Sacks FM, Moye LA, Howard BV, Howard WJ, Davis BR, Cole TG, Pfeffer MA, Braunwald E: Cardiovascular events and their reduction with pravastatin in diabetic and glucose-intolerant myocardial infarction survivors with average cholesterol levels: subgroup analyses in the cholesterol and recurrent events (CARE) trial. The Care Investigators. Circulation 98: 2513–2519, 1998.

28. Downs JR, Clearfield M, Weis S, Whitney E, Shapiro DR, Beere PA, Langendorfer A, Stein EA, Kruyer W, Gotto AM Jr: Primary prevention of acute coronary events with lovastatin in men and women with average cholesterol levels: results of AFCAPS/TexCAPS. Air Force/Texas Coronary Atherosclerosis Prevention Study. JAMA 279: 1615–1622, 1998.

29. Ballantyne CM, Olsson AG, Cook TJ, Mercuri MF, Pedersen TR, Kjekshus J: Influence of low high-density lipoprotein cholesterol and elevated triglyceride on

coronary heart disease events and response to simvastatin therapy in 4S. Circulation 104: 3046–3051, 2001.

30. Koskinen P, Manttari M, Manninen V, Huttunen JK, Heinonen OP, Frick MH: Coronary heart disease incidence in NIDDM patients in the Helsinki Heart Study. Diabetes Care 15: 820–825, 1992.

31. Robins SJ: Targeting low high-density lipoprotein cholesterol for therapy: lessons from the Veterans Affairs High-density Lipoprotein Intervention Trial. Am J Cardiol 88: 19N–23N, 2001.

32. Robins SJ, Rubins HB, Faas FH, Schaefer EJ, Elam MB, Anderson JW, Collins D: Insulin resistance and cardiovascular events with low HDL cholesterol: the Veterans Affairs HDL Intervention Trial (VA-HIT). Diabetes Care 26: 1513–1517, 2003.

33. Diabetes Atherosclerosis Intervention Study Investigators. Effect of fenofibrate on progression of coronary-artery disease in type 2 diabetes: the Diabetes Atherosclerosis Intervention Study, a randomised study. Lancet 357: 905–910, 2001.

34. Forster T: [Milestone in the treatment of diabetic dyslipidemia: the DAIS Study]. Orv Hetil 142: 1588–1589, 2001.

35. Steiner G: The Diabetes Atherosclerosis Intervention Study (DAIS): a study conducted in cooperation with the World Health Organization. The DAIS Project Group. Diabetologia 39: 1655–1661, 1996.

36. Florkowski CM: Management of co-existing diabetes mellitus and dyslipidemia: defining the role of thiazolidinediones. Am J Cardiovasc Drugs 2: 15–21, 2002.

37. Olansky L, Marchetti A, Lau H: Multicenter retrospective assessment of thiazolidinedione monotherapy and combination therapy in patients with type 2 diabetes: comparative subgroup analyses of glycemic control and blood lipid levels. Clin Ther 25 (Suppl B): B64–B80, 2003.

38. Verges B: Clinical interest of PPARs ligands. Diabetes Metab 30: 7–12, 2004.

39. Shafrir E: Animal models of non-insulin-dependent diabetes. Diabetes Metab Rev 8: 179–208, 1992.

40. Shafrir E, Ziv E, Mosthaf L: Nutritionally induced insulin resistance and receptor defect leading to beta-cell failure in animal models. Ann N Y Acad Sci 892: 223–246, 1999.

41. Collier G, Walder K, De Silva A, Tenne-Brown J, Sanigorski A, Segal D, Kantham L, Augert G: New approaches to gene discovery with animal models of obesity and diabetes. Ann N Y Acad Sci 967: 403–413, 2002.

42. Kadowaki T: Insights into insulin resistance and type 2 diabetes from knockout mouse models. J Clin Invest 106: 459–465, 2000.

43. Kozak LP, Rossmeisl M: Adiposity and the development of diabetes in mouse genetic models. Ann N Y Acad Sci 967: 80–87, 2002.

44. Shafrir E, Ziv E: Cellular mechanism of nutritionally induced insulin resistance: the desert rodent *Psammomys obesus* and other animals in which insulin resistance leads to detrimental outcome. J Basic Clin Physiol Pharmacol 9: 347–385, 1998.

45. Silver DL, Jiang XC, Arai T, Bruce C, Tall AR: Receptors and lipid transfer proteins in HDL metabolism. Ann N Y Acad Sci 902: 103–111, 2000.

46. Park J, Rho HK, Kim KH, Choe SS, Lee YS, Kim JB: Overexpression of glucose-6-phosphate dehydrogenase is associated with lipid dysregulation and insulin resistance in obesity. Mol Cell Biol 25: 5146–5157, 2005.

47. Wiegman CH, Bandsma RH, Ouwens M, van der Sluijs FH, Havinga R, Boer T, Reijngoud DJ, Romijn JA, Kuipers F: Hepatic VLDL production in ob/ob mice is not stimulated by massive de novo lipogenesis but is less sensitive to the suppressive effects of insulin. Diabetes 52: 1081–1089, 2003.

48. Muurling M, Mensink RP, Pijl H, Romijn JA, Havekes LM, Voshol PJ: Rosiglitazone improves muscle insulin sensitivity, irrespective of increased triglyceride content, in ob/ob mice. Metabolism 52: 1078–1083, 2003.

49. Matsusue K, Haluzik M, Lambert G, Yim SH, Gavrilova O, Ward JM, Brewer B Jr, Reitman ML, Gonzalez FJ: Liver-specific disruption of PPARgamma in leptin-deficient mice improves fatty liver but aggravates diabetic phenotypes. J Clin Invest 111: 737–747, 2003.

50. Liao W, Angelin B, Rudling M: Lipoprotein metabolism in the fat Zucker rat: reduced basal expression but normal regulation of hepatic low density lipoprotein receptors. Endocrinology 138: 3276–3282, 1997.

51. Mathe D: Dyslipidemia and diabetes: animal models. Diabetes Metab 21: 106–111, 1995.

52. Park SH, Marso SP, Zhou Z, Foroudi F, Topol EJ, Lincoff AM: Neointimal hyperplasia after arterial injury is increased in a rat model of non-insulin-dependent diabetes mellitus. Circulation 104: 815–819, 2001.

53. Blay M, Peinado-Onsurbe J, Julve J, Rodriguez V, Fernandez-Lopez JA, Remesar X, Alemany M: Anomalous lipoproteins in obese Zucker rats. Diabetes Obes Metab 3: 259–270, 2001.

54. Phillips C, Owens D, Collins P, Tomkin GH: Microsomal triglyceride transfer protein: does insulin resistance play a role in the regulation of chylomicron assembly? Atherosclerosis 160: 355–360, 2002.

55. Sparks JD, Shaw WN, Corsetti JP, Bolognino M, Pesek JF, Sparks CE: Insulin-treated Zucker diabetic fatty rats retain the hypertriglyceridemia associated with obesity. Metabolism 49: 1424–1430, 2000.

56. Kalderon B, Mayorek N, Ben Yaacov L, Bar-Tana J: Adipose tissue sensitization to insulin induced by troglitazone and MEDICA 16 in obese Zucker rats *in vivo*. Am J Physiol Endocrinol Metab 284: E795–E803, 2003.

57. Oakes ND, Thalen PG, Jacinto SM, Ljung B: Thiazolidinediones increase plasma-adipose tissue FFA exchange capacity and enhance insulin-mediated control of systemic FFA availability. Diabetes 50: 1158–1165, 2001.

58. Noshiro O, Hirayama R, Shimaya A, Yoneta T, Niigata K, Shikama H: Role of plasma insulin concentration in regulating glucose and lipid metabolism in lean and obese Zucker rats. Int J Obes Relat Metab Disord 21: 115–121, 1997.

59. Russell JC, Koeslag DG, Amy RM, Dolphin PJ: Plasma lipid secretion and clearance in hyperlipidemic JCR:LA-corpulent rats. Arteriosclerosis 9: 869–876, 1989.

60. Russell JC, Graham S, Hameed M: Abnormal insulin and glucose metabolism in the JCR:LA-corpulent rat. Metabolism 43: 538–543, 1994.

61. Russell JC, Shillabeer G, Bar-Tana J, Lau DC, Richardson M, Wenzel LM, Graham SE, Dolphin PJ: Development of insulin resistance in the JCR:LA-cp rat: role of triacylglycerols and effects of MEDICA 16. Diabetes 47: 770–778, 1998.

62. Dolphin PJ, Stewart B, Amy RM, Russell JC: Serum lipids and lipoproteins in the atherosclerosis prone LA/N corpulent rat. Biochim Biophys Acta 919: 140–148, 1987.

63. Zhang B, Saku K, Hirata K, Liu R, Tateishi K, Yamamoto K, Arakawa K: Insulin resistance observed in WHHL rabbits. Atherosclerosis 91: 277–278, 1991.

64. Zhang B, Saku K, Arakawa K: Quantification of the effects of troglitazone on insulin sensitivity and beta-cell function in Watanabe heritable hyperlipidemic rabbits: a minimal model analysis. Metabolism 46: 273–281, 1997.

65. Zhang B, Shiomi M, Tanaka H, Mei J, Fan P, Tsujita Y, Horikoshi H, Saku K: Effects of high-dose troglitazone on insulin sensitivity and beta-cell function in Watanabe heritable hyperlipidemic rabbits. Eur J Drug Metab Pharmacokinet 26: 185–192, 2001.

66. Shiomi M, Ito T, Tsukada T, Tsujita Y, Horikoshi H: Combination treatment with troglitazone, an insulin action enhancer, and pravastatin, an inhibitor of HMG-CoA reductase, shows a synergistic effect on atherosclerosis of WHHL rabbits. Atherosclerosis 142: 345–353, 1999.

67. Saku K, Zhang B, Ohta T, Arakawa K: Troglitazone lowers blood pressure and enhances insulin sensitivity in Watanabe heritable hyperlipidemic rabbits. Am J Hypertens 10: 1027–1033, 1997.

68. Koike T, Liang J, Wang X, Ichikawa T, Shiomi M, Liu G, Sun H, Kitajima S, Morimoto M, Watanabe T, Yamada N, Fan J: Overexpression of lipoprotein lipase in transgenic Watanabe heritable hyperlipidemic rabbits improves hyperlipidemia and obesity. J Biol Chem 279: 7521–7529, 2004.

69. Dong Q, Ginsberg HN, Erlanger BF: Overexpression of the A1 adenosine receptor in adipose tissue protects mice from obesity-related insulin resistance. Diabetes Obes Metab 3: 360–366, 2001.

70. Green A: Catecholamines inhibit insulin-stimulated glucose transport in adipocytes, in the presence of adenosine deaminase. FEBS Lett 152: 261–264, 1983.

71. Schwabe U, Schonhofer PS, Ebert R: Facilitation by adenosine of the action of insulin on the accumulation of adenosine 3′:5′-monophosphate, lipolysis, and glucose oxidation in isolated fat cells. Eur J Biochem 46: 537–545, 1974.

72. Smith U, Kuroda M, Simpson IA: Counter-regulation of insulin-stimulated glucose transport by catecholamines in the isolated rat adipose cell. J Biol Chem 259: 8758–8763, 1984.

73. Siri P, Candela N, Zhang YL, Ko C, Eusufzai S, Ginsberg HN, Huang LS: Post-transcriptional stimulation of the assembly and secretion of triglyceride-rich apolipoprotein B lipoproteins in a mouse with selective deficiency of brown adipose tissue, obesity, and insulin resistance. J Biol Chem 276: 46064–46072, 2001.

74. Barnett M, Collier GR, Collier FM, Zimmet P, O'Dea K: A cross-sectional and short-term longitudinal characterisation of NIDDM in *Psammomys obesus*. Diabetologia 37: 671–676, 1994.

75. Collier GR, Walder K, Lewandowski P, Sanigorski A, Zimmet P: Leptin and the development of obesity and diabetes in *Psammomys obesus*. Obes Res 5: 455–458, 1997.

76. Habito RC, Barnett M, Yamamoto A, Cameron-Smith D, O'Dea K, Zimmet P, Collier GR: Basal glucose turnover in *Psammomys obesus*. An animal model of type 2 (non-insulin-dependent) diabetes mellitus. Acta Diabetol 32: 187–192, 1995.

77. Walder KR, Fahey RP, Morton GJ, Zimmet PZ, Collier GR: Characterization of obesity phenotypes in *Psammomys obesus* (Israeli sand rats). Int J Exp Diabetes Res 1: 177–184, 2000.

78. Kanety H, Moshe S, Shafrir E, Lunenfeld B, Karasik A: Hyperinsulinemia induces a reversible impairment in insulin receptor function leading to diabetes in the sand rat model of non-insulin-dependent diabetes mellitus. Proc Natl Acad Sci U S A 91: 1853–1857, 1994.

79. Ziv E, Kalman R, Hershkop K, Barash V, Shafrir E, Bar-On H: Insulin resistance in the NIDDM model *Psammomys obesus* in the normoglycaemic, normoinsulinaemic state. Diabetologia 39: 1269–1275, 1996.

80. Walder K, Lewandowski P, Morton G, Sanigorski A, De Silva A, Zimmet P, Collier GR: Leptin resistance in a polygenic, hyperleptinemic animal model of obesity and NIDDM: *Psammomys obesus.* Int J Obes Relat Metab Disord 23: 83–89, 1999.

81. Zoltowska M, Ziv E, Delvin E, Stan S, Bar-On H, Kalman R, Levy E: Circulating lipoproteins and hepatic sterol metabolism in *Psammomys obesus* prone to obesity, hyperglycemia and hyperinsulinemia. Atherosclerosis 157: 85–96, 2001.

82. Ikeda Y, Olsen GS, Ziv E, Hansen LL, Busch AK, Hansen BF, Shafrir E, Mosthaf-Seedorf L: Cellular mechanism of nutritionally induced insulin resistance in *Psammomys obesus*: overexpression of protein kinase Cepsilon in skeletal muscle precedes the onset of hyperinsulinemia and hyperglycemia. Diabetes 50: 584–592, 2001.

83. Collier GR, McMillan JS, Windmill K, Walder K, Tenne-Brown J, De Silva A, Trevaskis J, Jones S, Morton GJ, Lee S, Augert G, Civitarese A, Zimmet PZ: Beacon: a novel gene involved in the regulation of energy balance. Diabetes 49: 1766–1771, 2000.

84. Taghibiglou C, Carpentier A, Van Iderstine SC, Chen B, Rudy D, Aiton A, Lewis GF, Adeli K: Mechanisms of hepatic very low density lipoprotein overproduction in insulin resistance. Evidence for enhanced lipoprotein assembly, reduced intracellular ApoB degradation, and increased microsomal triglyceride transfer protein in a fructose-fed hamster model. J Biol Chem 275: 8416–8425, 2000.

85. Hoang VQ, Botham KM, Benson GM, Eldredge EE, Jackson B, Pearce N, Suckling KE: Bile acid synthesis in hamster hepatocytes in primary culture: sources of cholesterol and comparison with other species. Biochim Biophys Acta 1210: 73–80, 1993.

86. Jackson B, Gee AN, Martinez-Cayuela M, Suckling KE: The effects of feeding a saturated fat-rich diet on enzymes of cholesterol metabolism in the liver, intestine and aorta of the hamster. Biochim Biophys Acta 1045: 21–28, 1990.

87. Nistor A, Bulla A, Filip DA, Radu A: The hyperlipidemic hamster as a model of experimental atherosclerosis. Atherosclerosis 68: 159–173, 1987.

88. Sullivan MP, Cerda JJ, Robbins FL, Burgin CW, Beatty RJ: The gerbil, hamster, and guinea pig as rodent models for hyperlipidemia. Lab Anim Sci 43: 575–578, 1993.

89. Arbeeny CM, Meyers DS, Bergquist KE, Gregg RE: Inhibition of fatty acid synthesis decreases very low density lipoprotein secretion in the hamster. J Lipid Res 33: 843–851, 1992.

90. Liu GL, Fan LM, Redinger RN: The association of hepatic apoprotein and lipid metabolism in hamsters and rats. Comp Biochem Physiol A 99: 223–228, 1991.

91. Mangaloglu L, Cheung RC, Van Iderstine SC, Taghibiglou C, Pontrelli L, Adeli K: Treatment with atorvastatin ameliorates hepatic very-low-density lipoprotein

overproduction in an animal model of insulin resistance, the fructose-fed Syrian golden hamster: evidence that reduced hypertriglyceridemia is accompanied by improved hepatic insulin sensitivity. Metabolism 51: 409–418, 2002.

92. Taghibiglou C, Rudy D, Van Iderstine SC, Aiton A, Cavallo D, Cheung R, Adeli K: Intracellular mechanisms regulating apoB-containing lipoprotein assembly and secretion in primary hamster hepatocytes. J Lipid Res 41: 499–513, 2000.

93. Taghibiglou C, Rashid-Kolvear F, Van Iderstine SC, Le Tien H, Fantus IG, Lewis GF, Adeli K: Hepatic very low density lipoprotein-ApoB overproduction is associated with attenuated hepatic insulin signaling and overexpression of protein-tyrosine phosphatase 1B in a fructose-fed hamster model of insulin resistance. J Biol Chem 277: 793–803, 2002.

94. Kasuga M, Karlsson FA, Kahn CR: Insulin stimulates the phosphorylation of the 95,000-dalton subunit of its own receptor. Science 215: 185–187, 1982.

95. Olefsky JM: The insulin receptor. A multifunctional protein. Diabetes 39: 1009–1016, 1990.

96. Combettes-Souverain M, Issad T: Molecular basis of insulin action. Diabetes Metab 24: 477–489, 1998.

97. Fantin VR, Sparling JD, Slot JW, Keller SR, Lienhard GE, Lavan BE: Characterization of insulin receptor substrate 4 in human embryonic kidney 293 cells. J Biol Chem 273: 10726–10732, 1998.

98. Lavan BE, Lienhard GE: The insulin-elicited 60-kDa phosphotyrosine protein in rat adipocytes is associated with phosphatidylinositol 3-kinase. J Biol Chem 268: 5921–5928, 1993.

99. Ogawa W, Matozaki T, Kasuga M: Role of binding proteins to IRS-1 in insulin signalling. Mol Cell Biochem 182: 13–22, 1998.

100. White MF: The IRS-signaling system: a network of docking proteins that mediate insulin and cytokine action. Recent Prog Horm Res 53: 119–138, 1998.

101. Fischer EH, Charbonneau H, Tonks NK: Protein tyrosine phosphatases: a diverse family of intracellular and transmembrane enzymes. Science 253: 401–406, 1991.

102. Backer JM, Myers MG Jr, Shoelson SE, Chin DJ, Sun XJ, Miralpeix M, Hu P, Margolis B, Skolnik EY, Schlessinger J: Phosphatidylinositol 3′-kinase is activated by association with IRS-1 during insulin stimulation. EMBO J 11: 3469–3479, 1992.

103. Lavan BE, Kuhne MR, Garner CW, Anderson D, Reedijk M, Pawson T, Lienhard GE: The association of insulin-elicited phosphotyrosine proteins with src homology 2 domains. J Biol Chem 267: 11631–11636, 1992.

104. Lee CH, Li W, Nishimura R, Zhou M, Batzer AG, Myers MG Jr, White MF, Schlessinger J, Skolnik EY: Nck associates with the SH2 domain-docking protein IRS-1 in insulin-stimulated cells. Proc Natl Acad Sci U S A 90: 11713–11717, 1993.

105. Skolnik EY, Lee CH, Batzer A, Vicentini LM, Zhou M, Daly R, Myers MJ Jr, Backer JM, Ullrich A, White MF: The SH2/SH3 domain-containing protein GRB2 interacts with tyrosine-phosphorylated IRS1 and Shc: implications for insulin control of ras signalling. EMBO J 12: 1929–1936, 1993.

106. Czech MP, Corvera S: Signaling mechanisms that regulate glucose transport. J Biol Chem 274: 1865–1868, 1999.

107. Virkamaki A, Ueki K, Kahn CR: Protein–protein interaction in insulin signaling and the molecular mechanisms of insulin resistance. J Clin Invest 103: 931–943, 1999.

108. Bjornholm M, Kawano Y, Lehtihet M, Zierath JR: Insulin receptor substrate-1 phosphorylation and phosphatidylinositol 3-kinase activity in skeletal muscle from NIDDM subjects after *in vivo* insulin stimulation. Diabetes 46: 524–527, 1997.

109. Goodyear LJ, Giorgino F, Sherman LA, Carey J, Smith RJ, Dohm GL: Insulin receptor phosphorylation, insulin receptor substrate-1 phosphorylation, and phosphatidylinositol 3-kinase activity are decreased in intact skeletal muscle strips from obese subjects. J Clin Invest 95: 2195–2204, 1995.

110. Rondinone CM, Wang LM, Lonnroth P, Wesslau C, Pierce JH, Smith U: Insulin receptor substrate (IRS) 1 is reduced and IRS-2 is the main docking protein for phosphatidylinositol 3-kinase in adipocytes from subjects with non-insulin-dependent diabetes mellitus. Proc Natl Acad Sci U S A 94: 4171–4175, 1997.

111. Zierath JR, Krook A, Wallberg-Henriksson H: Insulin action in skeletal muscle from patients with NIDDM. Mol Cell Biochem 182: 153–160, 1998.

112. Caro JF, Ittoop O, Pories WJ, Meelheim D, Flickinger EG, Thomas F, Jenquin M, Silverman JF, Khazanie PG, Sinha MK: Studies on the mechanism of insulin resistance in the liver from humans with noninsulin-dependent diabetes. Insulin action and binding in isolated hepatocytes, insulin receptor structure, and kinase activity. J Clin Invest 78: 249–258, 1986.

113. Olefsky JM: Decreased insulin binding to adipocytes and circulating monocytes from obese subjects. J Clin Invest 57: 1165–1172, 1976.

114. Caro JF, Sinha MK, Raju SM, Ittoop O, Pories WJ, Flickinger EG, Meelheim D, Dohm GL: Insulin receptor kinase in human skeletal muscle from obese subjects with and without noninsulin dependent diabetes. J Clin Invest 79: 1330–1337, 1987.

115. Anai M, Funaki M, Ogihara T, Terasaki J, Inukai K, Katagiri H, Fukushima Y, Yazaki Y, Kikuchi M, Oka Y, Asano T: Altered expression levels and impaired steps in the pathway to phosphatidylinositol 3-kinase activation via insulin receptor substrates 1 and 2 in Zucker fatty rats. Diabetes 47: 13–23, 1998.

116. Friedman JE, Ishizuka T, Liu S, Farrell CJ, Bedol D, Koletsky RJ, Kaung HL, Ernsberger P: Reduced insulin receptor signaling in the obese spontaneously hypertensive Koletsky rat. Am J Physiol 273: E1014–E1023, 1997.

117. Hayakawa T, Shiraki T, Morimoto T, Shii K, Ikeda H: Pioglitazone improves insulin signaling defects in skeletal muscle from Wistar fatty (fa/fa) rats. Biochem Biophys Res Commun 223: 439–444, 1996.

118. Hurrell DG, Pedersen O, Kahn CR: Alterations in the hepatic insulin receptor kinase in genetic and acquired obesity in rats. Endocrinology 125: 2454–2462, 1989.

119. Kerouz NJ, Horsch D, Pons S, Kahn CR: Differential regulation of insulin receptor substrates-1 and -2 (IRS-1 and IRS-2) and phosphatidylinositol 3-kinase isoforms in liver and muscle of the obese diabetic (ob/ob) mouse. J Clin Invest 100: 3164–3172, 1997.

120. Folli F, Saad MJ, Backer JM, Kahn CR: Regulation of phosphatidylinositol 3-kinase activity in liver and muscle of animal models of insulin-resistant and insulin-deficient diabetes mellitus. J Clin Invest 92: 1787–1794, 1993.

121. Heydrick SJ, Gautier N, Olichon-Berthe C, Van Obberghen E, Marchand-Brustel Y: Early alteration of insulin stimulation of PI 3-kinase in muscle and adipocyte from gold thioglucose obese mice. Am J Physiol 268: E604–E612, 1995.

122. Accili D, Drago J, Lee EJ, Johnson MD, Cool MH, Salvatore P, Asico LD, Jose PA, Taylor SI, Westphal H: Early neonatal death in mice homozygous for a null allele of the insulin receptor gene. Nat Genet 12: 106–109, 1996.

123. Araki E, Lipes MA, Patti ME, Bruning JC, Haag B III, Johnson RS, Kahn CR: Alternative pathway of insulin signalling in mice with targeted disruption of the IRS-1 gene. Nature 372: 186–190, 1994.

124. Bruning JC, Winnay J, Bonner-Weir S, Taylor SI, Accili D, Kahn CR: Development of a novel polygenic model of NIDDM in mice heterozygous for IR and IRS-1 null alleles. Cell 88: 561–572, 1997.

125. Joshi RL, Lamothe B, Cordonnier N, Mesbah K, Monthioux E, Jami J, Bucchini D: Targeted disruption of the insulin receptor gene in the mouse results in neonatal lethality. EMBO J 15: 1542–1547, 1996.

126. Tamemoto H, Kadowaki T, Tobe K, Yagi T, Sakura H, Hayakawa T, Terauchi Y, Ueki K, Kaburagi Y, Satoh S: Insulin resistance and growth retardation in mice lacking insulin receptor substrate-1. Nature 372: 182–186, 1994.

127. Withers DJ, Gutierrez JS, Towery H, Burks DJ, Ren JM, Previs S, Zhang Y, Bernal D, Pons S, Shulman GI, Bonner-Weir S, White MF: Disruption of IRS-2 causes type 2 diabetes in mice. Nature 391: 900–904, 1998.

128. Fry MJ: Structure, regulation and function of phosphoinositide 3-kinases. Biochim Biophys Acta 1226: 237–268, 1994.

129. Srivastava AK, Pandey SK: Potential mechanism(s) involved in the regulation of glycogen synthesis by insulin. Mol Cell Biochem 182: 135–141, 1998.

130. Ballotti R, Tartare S, Van Obberghen E: [The insulin receptor: mechanism of activation and message transmission]. Pathol Biol (Paris) 40: 754–762, 1992.

131. Liu R, Bai H, Liu BW: [Signal transduction of insulin receptor]. Sheng Li Ke Xue Jin Zhan 32: 254–256, 2001.

132. Marchand-Brustel Y: Molecular mechanisms of insulin action in normal and insulin-resistant states. Exp Clin Endocrinol Diabetes 107: 126–132, 1999.

133. Sun XJ, Rothenberg P, Kahn CR, Backer JM, Araki E, Wilden PA, Cahill DA, Goldstein BJ, White MF: Structure of the insulin receptor substrate IRS-1 defines a unique signal transduction protein. Nature 352: 73–77, 1991.

134. Qiao LY, Goldberg JL, Russell JC, Sun XJ: Identification of enhanced serine kinase activity in insulin resistance. J Biol Chem 274: 10625–10632, 1999.

135. De Fea K, Roth RA: Modulation of insulin receptor substrate-1 tyrosine phosphorylation and function by mitogen-activated protein kinase. J Biol Chem 272: 31400–31406, 1997.

136. Aguirre V, Uchida T, Yenush L, Davis R, White MF: The c-Jun NH(2)-terminal kinase promotes insulin resistance during association with insulin receptor substrate-1 and phosphorylation of Ser(307). J Biol Chem 275: 9047–9054, 2000.

137. Desbois-Mouthon C, Blivet-Van Eggelpoel MJ, Auclair M, Cherqui G, Capeau J, Caron M: Insulin differentially regulates SAPKs/JNKs and ERKs in CHO cells overexpressing human insulin receptors. Biochem Biophys Res Commun 243: 765–770, 1998.

138. Kayali AG, Austin DA, Webster NJ: Stimulation of MAPK cascades by insulin and osmotic shock: lack of an involvement of p38 mitogen-activated protein kinase in glucose transport in 3T3-L1 adipocytes. Diabetes 49: 1783–1793, 2000.

139. Sweeney G, Somwar R, Ramlal T, Volchuk A, Ueyama A, Klip A: An inhibitor of p38 mitogen-activated protein kinase prevents insulin-stimulated glucose transport but not glucose transporter translocation in 3T3-L1 adipocytes and L6 myotubes. J Biol Chem 274: 10071–10078, 1999.

140. Cusi K, Maezono K, Osman A, Pendergrass M, Patti ME, Pratipanawatr T, DeFronzo RA, Kahn CR, Mandarino LJ: Insulin resistance differentially affects

the PI 3-kinase- and MAP kinase-mediated signaling in human muscle. J Clin Invest 105: 311–320, 2000.

141. Krook A, Bjornholm M, Galuska D, Jiang XJ, Fahlman R, Myers MG Jr, Wallberg-Henriksson H, Zierath JR: Characterization of signal transduction and glucose transport in skeletal muscle from type 2 diabetic patients. Diabetes 49: 284–292, 2000.

142. Fujishiro M, Gotoh Y, Katagiri H, Sakoda H, Ogihara T, Anai M, Onishi Y, Ono H, Abe M, Shojima N, Fukushima Y, Kikuchi M, Oka Y, Asano T: Three mitogen-activated protein kinases inhibit insulin signaling by different mechanisms in 3T3-L1 adipocytes. Mol Endocrinol 17: 487–497, 2003.

143. Yamamoto Y, Yoshimasa Y, Koh M, Suga J, Masuzaki H, Ogawa Y, Hosoda K, Nishimura H, Watanabe Y, Inoue G, Nakao K: Constitutively active mitogen-activated protein kinase increases GLUT1 expression and recruits both GLUT1 and GLUT4 at the cell surface in 3T3-L1 adipocytes. Diabetes 49: 332–339, 2000.

144. Engelman JA, Berg AH, Lewis RY, Lisanti MP, Scherer PE: Tumor necrosis factor alpha-mediated insulin resistance, but not dedifferentiation, is abrogated by MEK1/2 inhibitors in 3T3-L1 adipocytes. Mol Endocrinol 14: 1557–1569, 2000.

145. Goldstein BJ: Regulation of insulin receptor signaling by protein-tyrosine dephosphorylation. Receptor 3: 1–15, 1993.

146. Hashimoto N, Zhang WR, Goldstein BJ: Insulin receptor and epidermal growth factor receptor dephosphorylation by three major rat liver protein-tyrosine phosphatases expressed in a recombinant bacterial system. Biochem J 284 (Pt 2): 569–576, 1992.

147. Lammers R, Bossenmaier B, Cool DE, Tonks NK, Schlessinger J, Fischer EH, Ullrich A: Differential activities of protein tyrosine phosphatases in intact cells. J Biol Chem 268: 22456–22462, 1993.

148. Tonks NK, Diltz CD, Fischer EH: Characterization of the major protein-tyrosine-phosphatases of human placenta. J Biol Chem 263: 6731–6737, 1988.

149. Chen H, Wertheimer SJ, Lin CH, Katz SL, Amrein KE, Burn P, Quon MJ: Protein-tyrosine phosphatases PTP1B and syp are modulators of insulin-stimulated translocation of GLUT4 in transfected rat adipose cells. J Biol Chem 272: 8026–8031, 1997.

150. Maegawa H, Ide R, Hasegawa M, Ugi S, Egawa K, Iwanishi M, Kikkawa R, Shigeta Y, Kashiwagi A: Thiazolidine derivatives ameliorate high glucose-induced insulin resistance via the normalization of protein-tyrosine phosphatase activities. J Biol Chem 270: 7724–7730, 1995.

151. Elchebly M, Payette P, Michaliszyn E, Cromlish W, Collins S, Loy AL, Normandin D, Cheng A, Himms-Hagen J, Chan CC, Ramachandran C, Gresser MJ, Tremblay ML, Kennedy BP: Increased insulin sensitivity and obesity resistance in mice lacking the protein tyrosine phosphatase-1B gene. Science 283: 1544–1548, 1999.

152. Havel PJ: Control of energy homeostasis and insulin action by adipocyte hormones: leptin, acylation stimulating protein, and adiponectin. Curr Opin Lipidol 13: 51–59, 2002.

153. Cheng A, Uetani N, Simoncic PD, Chaubey VP, Lee-Loy A, McGlade CJ, Kennedy BP, Tremblay ML: Attenuation of leptin action and regulation of obesity by protein tyrosine phosphatase 1B. Dev Cell 2: 497–503, 2002.

154. Kaszubska W, Falls HD, Schaefer VG, Haasch D, Frost L, Hessler P, Kroeger PE, White DW, Jirousek MR, Trevillyan JM: Protein tyrosine phosphatase 1B

negatively regulates leptin signaling in a hypothalamic cell line. Mol Cell Endocrinol 195: 109–118, 2002.

155. Myers MP, Andersen JN, Cheng A, Tremblay ML, Horvath CM, Parisien JP, Salmeen A, Barford D, Tonks NK: TYK2 and JAK2 are substrates of protein-tyrosine phosphatase 1B. J Biol Chem 276: 47771–47774, 2001.

156. Yu YH, Ginsberg HN: Adipocyte signaling and lipid homeostasis: sequelae of insulin-resistant adipose tissue. Circ Res 96: 1042–1052, 2005.

157. Manco M, Calvani M, Mingrone G: Effects of dietary fatty acids on insulin sensitivity and secretion. Diabetes Obes Metab 6: 402–413, 2004.

158. Smith AG, Muscat GE: Skeletal muscle and nuclear hormone receptors: implications for cardiovascular and metabolic disease. Int J Biochem Cell Biol 37: 2047–2063, 2005.

159. Watt MJ, Holmes AG, Steinberg GR, Mesa JL, Kemp BE, Febbraio MA: Reduced plasma FFA availability increases net triacylglycerol degradation, but not GPAT or HSL activity, in human skeletal muscle. Am J Physiol Endocrinol Metab 287: E120–E127, 2004.

160. Dresner A, Laurent D, Marcucci M, Griffin ME, Dufour S, Cline GW, Slezak LA, Andersen DK, Hundal RS, Rothman DL, Petersen KF, Shulman GI: Effects of free fatty acids on glucose transport and IRS-1-associated phosphatidylinositol 3-kinase activity. J Clin Invest 103: 253–259, 1999.

161. Shulman GI: Cellular mechanisms of insulin resistance in humans. Am J Cardiol 84: 3J–10J, 1999.

162. Hazel M, Cooksey RC, Jones D, Parker G, Neidigh JL, Witherbee B, Gulve EA, McClain DA: Activation of the hexosamine signaling pathway in adipose tissue results in decreased serum adiponectin and skeletal muscle insulin resistance. Endocrinology 145: 2118–2128, 2004.

163. Yamauchi T, Kamon J, Minokoshi Y, Ito Y, Waki H, Uchida S, Yamashita S, Noda M, Kita S, Ueki K, Eto K, Akanuma Y, Froguel P, Foufelle F, Ferre P, Carling D, Kimura S, Nagai R, Kahn BB, Kadowaki T: Adiponectin stimulates glucose utilization and fatty-acid oxidation by activating AMP-activated protein kinase. Nat Med 8: 1288–1295, 2002.

164. Koteish A, Diehl AM: Animal models of steatosis. Semin Liver Dis 21: 89–104, 2001.

165. Faraj M, Lu HL, Cianflone K: Diabetes, lipids, and adipocyte secretagogues. Biochem Cell Biol 82: 170–190, 2004.

166. Frayn KN: Obesity and metabolic disease: is adipose tissue the culprit? Proc Nutr Soc 64: 7–13, 2005.

167. Ballantyne GH, Gumbs A, Modlin IM: Changes in insulin resistance following bariatric surgery and the adipoinsular axis: role of the adipocytokines, leptin, adiponectin and resistin. Obes Surg 15: 692–699, 2005.

168. Merl V, Peters A, Oltmanns KM, Kern W, Born J, Fehm HL, Schultes B: Serum adiponectin concentrations during a 72-hour fast in over- and normal-weight humans. Int J Obes Relat Metab Disord 29: 998–1001, 2005.

169. Singhal A, Jamieson N, Fewtrell M, Deanfield J, Lucas A, Sattar N: Adiponectin predicts insulin resistance but not endothelial function in young healthy adolescents. J Clin Endocrinol Metab 90: 4615–4621, 2005.

170. Jan V, Cervera P, Maachi M, Baudrimont M, Kim M, Vidal H, Girard PM, Levan P, Rozenbaum W, Lombes A, Capeau J, Bastard JP: Altered fat differentiation and adipocytokine expression are inter-related and linked to morphological

changes and insulin resistance in HIV-1-infected lipodystrophic patients. Antivir Ther 9: 555–564, 2004.

171. Kopp HP, Krzyzanowska K, Mohlig M, Spranger J, Pfeiffer AF, Schernthaner G: Effects of marked weight loss on plasma levels of adiponectin, markers of chronic subclinical inflammation and insulin resistance in morbidly obese women. Int J Obes Relat Metab Disord 29: 766–771, 2005.

172. Shetty GK, Economides PA, Horton ES, Mantzoros CS, Veves A: Circulating adiponectin and resistin levels in relation to metabolic factors, inflammatory markers, and vascular reactivity in diabetic patients and subjects at risk for diabetes. Diabetes Care 27: 2450–2457, 2004.

173. Reinehr T, Roth C, Menke T, Andler W: Adiponectin before and after weight loss in obese children. J Clin Endocrinol Metab 89: 3790–3794, 2004.

174. Asensio C, Cettour-Rose P, Theander-Carrillo C, Rohner-Jeanrenaud F, Muzzin P: Changes in glycemia by leptin administration or high-fat feeding in rodent models of obesity/type 2 diabetes suggest a link between resistin expression and control of glucose homeostasis. Endocrinology 145: 2206–2213, 2004.

175. Jung HS, Youn BS, Cho YM, Yu KY, Park HJ, Shin CS, Kim SY, Lee HK, Park KS: The effects of rosiglitazone and metformin on the plasma concentrations of resistin in patients with type 2 diabetes mellitus. Metabolism 54: 314–320, 2005.

176. Rajala MW, Qi Y, Patel HR, Takahashi N, Banerjee R, Pajvani UB, Sinha MK, Gingerich RL, Scherer PE, Ahima RS: Regulation of resistin expression and circulating levels in obesity, diabetes, and fasting. Diabetes 53: 1671–1679, 2004.

177. Vozarova de Court, Degawa-Yamauchi M, Considine RV, Tataranni PA: High serum resistin is associated with an increase in adiposity but not a worsening of insulin resistance in Pima Indians. Diabetes 53: 1279–1284, 2004.

178. Ortenblad N, Mogensen M, Petersen I, Hojlund K, Levin K, Sahlin K, Beck-Nielsen H, Gaster M: Reduced insulin-mediated citrate synthase activity in cultured skeletal muscle cells from patients with type 2 diabetes: evidence for an intrinsic oxidative enzyme defect. Biochim Biophys Acta 1741: 206–214, 2005.

179. McCarty MF: Up-regulation of PPARgamma coactivator-1alpha as a strategy for preventing and reversing insulin resistance and obesity. Med Hypotheses 64: 399–407, 2005.

180. Khovidhunkit W, Kim MS, Memon RA, Shigenaga JK, Moser AH, Feingold KR, Grunfeld C: Effects of infection and inflammation on lipid and lipoprotein metabolism: mechanisms and consequences to the host. J Lipid Res 45: 1169–1196, 2004.

181. Storlien L, Oakes ND, Kelley DE: Metabolic flexibility. Proc Nutr Soc 63: 363–368, 2004.

182. Aas V, Rokling-Andersen M, Wensaas AJ, Thoresen GH, Kase ET, Rustan AC: Lipid metabolism in human skeletal muscle cells: effects of palmitate and chronic hyperglycaemia. Acta Physiol Scand 183: 31–41, 2005.

183. Parish R, Petersen KF: Mitochondrial dysfunction and type 2 diabetes. Curr Diab Rep 5: 177–183, 2005.

184. Ory DS: Nuclear receptor signaling in the control of cholesterol homeostasis: have the orphans found a home? Circ Res 95: 660–670, 2004.

185. Clarke SD: The multi-dimensional regulation of gene expression by fatty acids: polyunsaturated fats as nutrient sensors. Curr Opin Lipidol 15: 13–18, 2004.

186. Schmitz G, Langmann T: Transcriptional regulatory networks in lipid metabolism control ABCA1 expression. Biochim Biophys Acta 1735: 1–19, 2005.

187. Davies JD, Carpenter KL, Challis IR, Figg NL, McNair R, Proudfoot D, Weissberg PL, Shanahan CM: Adipocytic differentiation and liver X receptor pathways regulate the accumulation of triacylglycerols in human vascular smooth muscle cells. J Biol Chem 280: 3911–3919, 2005.
188. Ferre P: The biology of peroxisome proliferator-activated receptors: relationship with lipid metabolism and insulin sensitivity. Diabetes 53 (Suppl 1): S43–S50, 2004.
189. Tanaka T, Masuzaki H, Nakao K: [Role of PPARs in the pathophysiology of nonalcoholic fatty liver disease]. Nippon Rinsho 63: 700–706, 2005.
190. Cheon Y, Nara TY, Band MR, Beever JE, Wallig MA, Nakamura MT: Induction of overlapping genes by fasting and a peroxisome proliferator in pigs: evidence of functional PPARalpha in nonproliferating species. Am J Physiol Regul Integr Comp Physiol 288: R1525–R1535, 2005.
191. Gremlich S, Nolan C, Roduit R, Burcelin R, Peyot ML, Delghingaro-Augusto V, Desvergne B, Michalik L, Prentki M, Wahli W: Pancreatic islet adaptation to fasting is dependent on peroxisome proliferator-activated receptor alpha transcriptional up-regulation of fatty acid oxidation. Endocrinology 146: 375–382, 2005.
192. Taniguchi CM, Ueki K, Kahn R: Complementary roles of IRS-1 and IRS-2 in the hepatic regulation of metabolism. J Clin Invest 115: 718–727, 2005.
193. Boden G, Homko C, Mozzoli M, Showe LC, Nichols C, Cheung P: Thiazolidinediones upregulate fatty acid uptake and oxidation in adipose tissue of diabetic patients. Diabetes 54: 880–885, 2005.
194. Fredenrich A, Grimaldi PA: PPAR delta: an uncompletely known nuclear receptor. Diabetes Metab 31: 23–27, 2005.
195. Zhang F, Lavan B, Gregoire FM: Peroxisome proliferator-activated receptors as attractive antiobesity targets. Drug News Perspect 17: 661–669, 2004.
196. Haluzik MM, Haluzik M: PPAR-alpha and insulin sensitivity. Physiol Res 2005.
197. Ratziu V, Tahiri M, Bonyhay L: Nonalcoholic steatohepatitis. Ann Endocrinol (Paris) 66: 1S71–1S80, 2005.
198. Blaschke F, Bruemmer D, Law RE: Will the potential of peroxisome proliferator-activated receptor agonists be realized in the clinical setting? Clin Cardiol 27: IV3–IV10, 2004.
199. Richardson MC, Cameron IT, Simonis CD, Das MC, Hodge TE, Zhang J, Byrne CD: Insulin and human chorionic gonadotropin cause a shift in the balance of sterol regulatory element-binding protein (SREBP) isoforms toward the SREBP-1c isoform in cultures of human granulosa cells. J Clin Endocrinol Metab 90: 3738–3746, 2005.
200. Weber LW, Boll M, Stampfl A: Maintaining cholesterol homeostasis: sterol regulatory element-binding proteins. World J Gastroenterol 10: 3081–3087, 2004.
201. Eberle D, Hegarty B, Bossard P, Ferre P, Foufelle F: SREBP transcription factors: master regulators of lipid homeostasis. Biochimie 86: 839–848, 2004.
202. Hegarty BD, Bobard A, Hainault I, Ferre P, Bossard P, Foufelle F: Distinct roles of insulin and liver X receptor in the induction and cleavage of sterol regulatory element-binding protein-1c. Proc Natl Acad Sci U S A 102: 791–796, 2005.
203. Toth JI, Datta S, Athanikar JN, Freedman LP, Osborne TF: Selective coactivator interactions in gene activation by SREBP-1a and -1c. Mol Cell Biol 24: 8288–8300, 2004.
204. Hitoshi S: [Lipid synthetic transcription factor, SREBP]. Nippon Rinsho 63: 897–907, 2005.

205. Datta S, Osborne TF: Activation domains from both monomers contribute to transcriptional stimulation by sterol regulatory element-binding protein dimers. J Biol Chem 280: 3338–3345, 2005.
206. Foretz M, Guichard C, Ferre P, Foufelle F: Sterol regulatory element binding protein-1c is a major mediator of insulin action on the hepatic expression of glucokinase and lipogenesis-related genes. Proc Natl Acad Sci U S A 96: 12737–12742, 1999.
207. Guillet-Deniau I, Mieulet V, Le Lay S, Achouri Y, Carre D, Girard J, Foufelle F, Ferre P: Sterol regulatory element binding protein-1c expression and action in rat muscles: insulin-like effects on the control of glycolytic and lipogenic enzymes and UCP3 gene expression. Diabetes 51: 1722–1728, 2002.
208. Kim JB, Sarraf P, Wright M, Yao KM, Mueller E, Solanes G, Lowell BB, Spiegelman BM: Nutritional and insulin regulation of fatty acid synthetase and leptin gene expression through ADD1/SREBP1. J Clin Invest 101: 1–9, 1998.
209. Sewter C, Berger D, Considine RV, Medina G, Rochford J, Ciaraldi T, Henry R, Dohm L, Flier JS, O'Rahilly S, Vidal-Puig AJ: Human obesity and type 2 diabetes are associated with alterations in SREBP1 isoform expression that are reproduced ex vivo by tumor necrosis factor-alpha. Diabetes 51: 1035–1041, 2002.
210. Boizard M, Le Liepvre X, Lemarchand P, Foufelle F, Ferre P, Dugail I: Obesity-related overexpression of fatty-acid synthase gene in adipose tissue involves sterol regulatory element-binding protein transcription factors. J Biol Chem 273: 29164–29171, 1998.
211. Shimomura I, Bashmakov Y, Horton JD: Increased levels of nuclear SREBP-1c associated with fatty livers in two mouse models of diabetes mellitus. J Biol Chem 274: 30028–30032, 1999.
212. Kotzka J, Lehr S, Roth G, Avci H, Knebel B, Muller-Wieland D: Insulin-activated Erk-mitogen-activated protein kinases phosphorylate sterol regulatory element-binding protein-2 at serine residues 432 and 455 *in vivo*. J Biol Chem 279: 22404–22411, 2004.
213. Roth G, Kotzka J, Kremer L, Lehr S, Lohaus C, Meyer HE, Krone W, Muller-Wieland D: MAP kinases Erk1/2 phosphorylate sterol regulatory element-binding protein (SREBP)-1a at serine 117 *in vitro*. J Biol Chem 275: 33302–33307, 2000.
214. Takahashi A, Shimano H, Nakagawa Y, Yamamoto T, Motomura K, Matsuzaka T, Sone H, Suzuki H, Toyoshima H, Yamada N: Transgenic mice overexpressing SREBP-1a under the control of the PEPCK promoter exhibit insulin resistance, but not diabetes. Biochim Biophys Acta 1740: 427–433, 2005.
215. Guillet-Deniau I, Pichard AL, Kone A, Esnous C, Nieruchalski M, Girard J, Prip-Buus C: Glucose induces de novo lipogenesis in rat muscle satellite cells through a sterol-regulatory-element-binding-protein-1c-dependent pathway. J Cell Sci 117: 1937–1944, 2004.
216. Kok N, Roberfroid M, Delzenne N: Dietary oligofructose modifies the impact of fructose on hepatic triacylglycerol metabolism. Metabolism 45: 1547–1550, 1996.
217. Miyazaki M, Dobrzyn A, Man WC, Chu K, Sampath H, Kim HJ, Ntambi JM: Stearoyl-CoA desaturase 1 gene expression is necessary for fructose-mediated induction of lipogenic gene expression by sterol regulatory element-binding protein-1c-dependent and -independent mechanisms. J Biol Chem 279: 25164–25171, 2004.
218. Shimizu S, Ugi S, Maegawa H, Egawa K, Nishio Y, Yoshizaki T, Shi K, Nagai Y, Morino K, Nemoto K, Nakamura T, Bryer-Ash M, Kashiwagi A: Protein-tyrosine phosphatase 1B as new activator for hepatic lipogenesis via sterol

regulatory element-binding protein-1 gene expression. J Biol Chem 278: 43095–43101, 2003.

219. Nagai Y, Nishio Y, Nakamura T, Maegawa H, Kikkawa R, Kashiwagi A: Amelioration of high fructose-induced metabolic derangements by activation of PPARalpha. Am J Physiol Endocrinol Metab 282: E1180–E1190, 2002.

220. Sass DA, Chang P, Chopra KB: Nonalcoholic fatty liver disease: a clinical review. Dig Dis Sci 50: 171–180, 2005.

221. McClain CJ, Mokshagundam SP, Barve SS, Song Z, Hill DB, Chen T, Deaciuc I: Mechanisms of non-alcoholic steatohepatitis. Alcohol 34: 67–79, 2004.

222. Bahrami H, Daryani NE, Mirmomen S, Kamangar F, Haghpanah B, Djalili M: Clinical and histological features of nonalcoholic steatohepatitis in Iranian patients. BMC Gastroenterol 3: 27–2003.

223. Loguercio C, De Simone T, D'Auria MV, de Sio I, Federico A, Tuccillo C, Abbatecola AM, Del Vecchio BC: Non-alcoholic fatty liver disease: a multicentre clinical study by the Italian Association for the Study of the Liver. Dig Liver Dis 36: 398–405, 2004.

224. Tagle AM: [Non-alcoholic fatty liver]. Rev Gastroenterol Peru 23: 49–57, 2003.

225. Bugianesi E, Gastaldelli A, Vanni E, Gambino R, Cassader M, Baldi S, Ponti V, Pagano G, Ferrannini E, Rizzetto M: Insulin resistance in non-diabetic patients with non-alcoholic fatty liver disease: sites and mechanisms. Diabetologia 48: 634–642, 2005.

226. Abrams JJ, Ginsberg H, Grundy SM: Metabolism of cholesterol and plasma triglycerides in nonketotic diabetes mellitus. Diabetes 31: 903–910, 1982.

227. Dunn FL, Grundy SM, Bilheimer DW, Havel RJ, Raskin P: Impaired catabolism of very low-density lipoprotein-triglyceride in a family with primary hyper-triglyceridemia. Metabolism 34: 316–324, 1985.

228. Howard BV, Abbott WG, Beltz WF, Harper IT, Fields RM, Grundy SM, Taskinen MR: Integrated study of low density lipoprotein metabolism and very low density lipoprotein metabolism in non-insulin-dependent diabetes. Metabolism 36: 870–877, 1987.

229. Sparks JD, Sparks CE: Insulin regulation of triacylglycerol-rich lipoprotein synthesis and secretion. Biochim Biophys Acta 1215: 9–32, 1994.

230. Hauner H: Insulin resistance and the metabolic syndrome—a challenge of the new millennium. Eur J Clin Nutr 56 (Suppl 1): S25–S29, 2002.

231. Howard BV: Lipoprotein metabolism in diabetes mellitus. J Lipid Res 28: 613–628, 1987.

232. Reaven GM, Chen YD: Role of insulin in regulation of lipoprotein metabolism in diabetes. Diabetes Metab Rev 4: 639–652, 1988.

233. Fisher EA, Ginsberg HN: Complexity in the secretory pathway: the assembly and secretion of apolipoprotein B-containing lipoproteins. J Biol Chem 277: 17377–17380, 2002.

234. Sniderman AD, Cianflone K: Substrate delivery as a determinant of hepatic apoB secretion. Arterioscler Thromb 13: 629–636, 1993.

235. Yao Z, McLeod RS: Synthesis and secretion of hepatic apolipoprotein B-containing lipoproteins. Biochim Biophys Acta 1212: 152–166, 1994.

236. Yao Z, Tran K, McLeod RS: Intracellular degradation of newly synthesized apolipoprotein B. J Lipid Res 38: 1937–1953, 1997.

237. Gordon DA, Jamil H: Progress towards understanding the role of microsomal triglyceride transfer protein in apolipoprotein-B lipoprotein assembly. Biochim Biophys Acta 1486: 72–83, 2000.

238. Adeli K, Theriault A: Insulin modulation of human apolipoprotein B mRNA translation: studies in an *in vitro* cell-free system from HepG2 cells. Biochem Cell Biol 70: 1301–1312, 1992.

239. Au WS, Kung HF, Lin MC: Regulation of microsomal triglyceride transfer protein gene by insulin in HepG2 cells: roles of MAPKerk and MAPKp38. Diabetes 52: 1073–1080, 2003.

240. Phung TL, Roncone A, Jensen KL, Sparks CE, Sparks JD: Phosphoinositide 3-kinase activity is necessary for insulin-dependent inhibition of apolipoprotein B secretion by rat hepatocytes and localizes to the endoplasmic reticulum. J Biol Chem 272: 30693–30702, 1997.

241. Brown AM, Gibbons GF: Insulin inhibits the maturation phase of VLDL assembly via a phosphoinositide 3-kinase-mediated event. Arterioscler Thromb Vasc Biol 21: 1656–1661, 2001.

242. Gibbons GF, Bartlett SM, Sparks CE, Sparks JD: Extracellular fatty acids are not utilized directly for the synthesis of very-low-density lipoprotein in primary cultures of rat hepatocytes. Biochem J 287 (Pt 3): 749–753, 1992.

243. Bostrom K, Boren J, Wettesten M, Sjoberg A, Bondjers G, Wiklund O, Carlsson P, Olofsson SO: Studies on the assembly of apo B-100-containing lipoproteins in HepG2 cells. J Biol Chem 263: 4434–4442, 1988.

244. Dixon JL, Furukawa S, Ginsberg HN: Oleate stimulates secretion of apolipoprotein B-containing lipoproteins from Hep G2 cells by inhibiting early intracellular degradation of apolipoprotein B. J Biol Chem 266: 5080–5086, 1991.

245. Pullinger CR, North JD, Teng BB, Rifici VA, Ronhild de Brito AE, Scott J: The apolipoprotein B gene is constitutively expressed in HepG2 cells: regulation of secretion by oleic acid, albumin, and insulin, and measurement of the mRNA half-life. J Lipid Res 30: 1065–1077, 1989.

246. Cartwright IJ, Higgins JA: Intracellular degradation in the regulation of secretion of apolipoprotein B-100 by rabbit hepatocytes. Biochem J 314 (Pt 3): 977–984, 1996.

247. White AL, Graham DL, LeGros J, Pease RJ, Scott J: Oleate-mediated stimulation of apolipoprotein B secretion from rat hepatoma cells. A function of the ability of apolipoprotein B to direct lipoprotein assembly and escape presecretory degradation. J Biol Chem 267: 15657–15664, 1992.

248. Zhang Z, Cianflone K, Sniderman AD: Role of cholesterol ester mass in regulation of secretion of ApoB100 lipoprotein particles by hamster hepatocytes and effects of statins on that relationship. Arterioscler Thromb Vasc Biol 19: 743–752, 1999.

249. Patsch W, Tamai T, Schonfeld G: Effect of fatty acids on lipid and apoprotein secretion and association in hepatocyte cultures. J Clin Invest 72: 371–378, 1983.

250. Sparks JD, Collins HL, Sabio I, Sowden MP, Smith HC, Cianci J, Sparks CE: Effects of fatty acids on apolipoprotein B secretion by McArdle RH-7777 rat hepatoma cells. Biochim Biophys Acta 1347: 51–61, 1997.

251. Lin Y, Smit MJ, Havinga R, Verkade HJ, Vonk RJ, Kuipers F: Differential effects of eicosapentaenoic acid on glycerolipid and apolipoprotein B metabolism in primary human hepatocytes compared to HepG2 cells and primary rat hepatocytes. Biochim Biophys Acta 1256: 88–96, 1995.

252. Liang JS, Distler O, Cooper DA, Jamil H, Deckelbaum RJ, Ginsberg HN, Sturley SL: HIV protease inhibitors protect apolipoprotein B from degradation by the proteasome: a potential mechanism for protease inhibitor-induced hyperlipidemia. Nat Med 7: 1327–1331, 2001.

253. Basciano H, Federico L, Adeli K: Fructose, insulin resistance, and metabolic dyslipidemia. Nutr Metab (Lond) 2: 5, 2005.
254. Bartels ED, Lauritsen M, Nielsen LB: Hepatic expression of microsomal triglyceride transfer protein and *in vivo* secretion of triglyceride-rich lipoproteins are increased in obese diabetic mice. Diabetes 51: 1233–1239, 2002.
255. Kuriyama H, Yamashita S, Shimomura I, Funahashi T, Ishigami M, Aragane K, Miyaoka K, Nakamura T, Takemura K, Man Z, Toide K, Nakayama N, Fukuda Y, Lin MC, Wetterau JR, Matsuzawa Y: Enhanced expression of hepatic acyl-coenzyme A synthetase and microsomal triglyceride transfer protein messenger RNAs in the obese and hypertriglyceridemic rat with visceral fat accumulation. Hepatology 27: 557–562, 1998.
256. Carpentier A, Taghibiglou C, Leung N, Szeto L, Van Iderstine SC, Uffelman KD, Buckingham R, Adeli K, Lewis GF: Ameliorated hepatic insulin resistance is associated with normalization of microsomal triglyceride transfer protein expression and reduction in very low density lipoprotein assembly and secretion in the fructose-fed hamster. J Biol Chem 277: 28795–28802, 2002.
257. Lin MC, Gordon D, Wetterau JR: Microsomal triglyceride transfer protein (MTP) regulation in HepG2 cells: insulin negatively regulates MTP gene expression. J Lipid Res 36: 1073–1081, 1995.
258. Sato R, Miyamoto W, Inoue J, Terada T, Imanaka T, Maeda M: Sterol regulatory element-binding protein negatively regulates microsomal triglyceride transfer protein gene transcription. J Biol Chem 274: 24714–24720, 1999.
259. Hagan DL, Kienzle B, Jamil H, Hariharan N: Transcriptional regulation of human and hamster microsomal triglyceride transfer protein genes. Cell type-specific expression and response to metabolic regulators. J Biol Chem 269: 28737–28744, 1994.
260. Miyake JH, Doung XD, Strauss W, Moore GL, Castellani LW, Curtiss LK, Taylor JM, Davis RA: Increased production of apolipoprotein B-containing lipoproteins in the absence of hyperlipidemia in transgenic mice expressing cholesterol 7alpha-hydroxylase. J Biol Chem 276: 23304–23311, 2001.
261. Qiu W, Taghibiglou C, Avramoglu RK, Van Iderstine SC, Naples M, Ashrafpour H, Mhapsekar S, Sato R, Adeli K: Oleate-mediated stimulation of microsomal triglyceride transfer protein (MTP) gene promoter: implications for hepatic MTP overexpression in insulin resistance. Biochemistry 44: 3041–3049, 2005.
262. Lin MC, Gordon D, Wetterau JR: Microsomal triglyceride transfer protein (MTP) regulation in HepG2 cells: insulin negatively regulates MTP gene expression. J Lipid Res 36: 1073–1081, 1995.
263. Busserolles J, Gueux E, Rock E, Mazur A, Rayssiguier Y: Substituting honey for refined carbohydrates protects rats from hypertriglyceridemic and prooxidative effects of fructose. J Nutr 132: 3379–3382, 2002.
264. Kelley GL, Allan G, Azhar S: High dietary fructose induces a hepatic stress response resulting in cholesterol and lipid dysregulation. Endocrinology 145: 548–555, 2004.
265. Thirunavukkarasu V, Anuradha CV: Influence of alpha-lipoic acid on lipid peroxidation and antioxidant defence system in blood of insulin-resistant rats. Diabetes Obes Metab 6: 200–207, 2004.
266. Thirunavukkarasu V, Anitha Nandhini AT, Anuradha CV: Effect of alpha-lipoic acid on lipid profile in rats fed a high-fructose diet. Exp Diabesity Res 5: 195–200, 2004.

267. Verschoor L, Chen YD, Reaven EP, Reaven GM: Glucose and fructose feeding lead to alterations in structure and function of very low density lipoproteins. Horm Metab Res 17: 285–288, 1985.
268. Rainwater DL, Mitchell BD, Comuzzie AG, Haffner SM: Relationship of low-density lipoprotein particle size and measures of adiposity. Int J Obes Relat Metab Disord 23: 180–189, 1999.
269. Adeli K, Macri J, Mohammadi A, Kito M, Urade R, Cavallo D: Apolipoprotein B is intracellularly associated with an ER-60 protease homologue in HepG2 cells. J Biol Chem 272: 22489–22494, 1997.
270. Qiu W, Kohen-Avramoglu R, Rashid-Kolvear F, Au CS, Chong TM, Lewis GF, Trinh DK, Austin RC, Urade R, Adeli K: Overexpression of the endoplasmic reticulum 60 protein ER-60 downregulates apoB100 secretion by inducing its intracellular degradation via a nonproteasomal pathway: evidence for an ER-60-mediated and pCMB-sensitive intracellular degradative pathway. Biochemistry 43: 4819–4831, 2004.
271. Battula SB, Fitzsimons O, Moreno S, Owens D, Collins P, Johnson A, Tomkin GH: Postprandial apolipoprotein B48- and B100-containing lipoproteins in type 2 diabetes: do statins have a specific effect on triglyceride metabolism? Metabolism 49: 1049–1054, 2000.
272. Chen YD, Swami S, Skowronski R, Coulston A, Reaven GM: Differences in postprandial lipemia between patients with normal glucose tolerance and non-insulin-dependent diabetes mellitus. J Clin Endocrinol Metab 76: 172–177, 1993.
273. Emken EA, Rohwedder WK, Adlof RO, DeJarlais WJ, Gulley RM: Absorption and distribution of deuterium-labeled trans- and cis-11-octadecenoic acid in human plasma and lipoprotein lipids. Lipids 21: 589–595, 1986.
274. Gangl A, Ockner RK: Intestinal metabolism of plasma free fatty acids. Intracellular compartmentation and mechanisms of control. J Clin Invest 55: 803–813, 1975.
275. Lewis GF, O'Meara NM, Soltys PA, Blackman JD, Iverius PH, Druetzler AF, Getz GS, Polonsky KS: Postprandial lipoprotein metabolism in normal and obese subjects: comparison after the vitamin A fat-loading test. J Clin Endocrinol Metab 71: 1041–1050, 1990.
276. Lewis GF, O'Meara NM, Soltys PA, Blackman JD, Iverius PH, Pugh WL, Getz GS, Polonsky KS: Fasting hypertriglyceridemia in noninsulin-dependent diabetes mellitus is an important predictor of postprandial lipid and lipoprotein abnormalities. J Clin Endocrinol Metab 72: 934–944, 1991.
277. O'Meara NM, Lewis GF, Cabana VG, Iverius PH, Getz GS, Polonsky KS: Role of basal triglyceride and high density lipoprotein in determination of postprandial lipid and lipoprotein responses. J Clin Endocrinol Metab 75: 465–471, 1992.
278. Boquist S, Ruotolo G, Tang R, Bjorkegren J, Bond MG, de Faire U, Karpe F, Hamsten A: Alimentary lipemia, postprandial triglyceride-rich lipoproteins, and common carotid intima-media thickness in healthy, middle-aged men. Circulation 100: 723–728, 1999.
279. Doi H, Kugiyama K, Oka H, Sugiyama S, Ogata N, Koide SI, Nakamura SI, Yasue H: Remnant lipoproteins induce proatherothrombogenic molecules in endothelial cells through a redox-sensitive mechanism. Circulation 102: 670–676, 2000.
280. Karpe F, Boquist S, Tang R, Bond GM, de Faire U, Hamsten A: Remnant lipoproteins are related to intima-media thickness of the carotid artery independently of LDL cholesterol and plasma triglycerides. J Lipid Res 42: 17–21, 2001.

281. McNamara JR, Shah PK, Nakajima K, Cupples LA, Wilson PW, Ordovas JM, Schaefer EJ: Remnant-like particle (RLP) cholesterol is an independent cardiovascular disease risk factor in women: results from the Framingham Heart Study. Atherosclerosis 154: 229–236, 2001.
282. Yu KC, Cooper AD: Postprandial lipoproteins and atherosclerosis. Front Biosci 6: D332–D354, 2001.
283. Taskinen MR: Diabetic dyslipidaemia: from basic research to clinical practice. Diabetologia 46: 733–749, 2003.
284. Guo Q, Avramoglu RK, Adeli K: Intestinal assembly and secretion of highly dense/lipid-poor apolipoprotein B48-containing lipoprotein particles in the fasting state: evidence for induction by insulin resistance and exogenous fatty acids. Metabolism 54: 689–697, 2005.

12

The Roles of Protein Glycation, Glycoxidation, and Advanced Glycation End-Product Formation in Diabetes-Induced Atherosclerosis

IMRAN RASHID, BRONWYN E. BROWN, DAVID M. VAN REYK, AND MICHAEL J. DAVIES

Abstract

Diabetes is known to induce a range of micro- and macrovascular complications, with the latter resulting in premature and accelerated atherosclerosis. Thus people with diabetes have a 2–4-fold increased risk of developing cardiovascular diseases which is responsible for ca. 50% of deaths amongst people with diabetes. The mechanisms behind this elevated risk are still not fully understood, though there is now increasing evidence for a role of glycation and glycoxidation reactions induced by hyperglycemia. This article reviews current knowledge of the role that these reactions play in diabetes-induced atherosclerosis with particular emphasis on the molecular reactions that result in the modification of lipoproteins, and the consequences of these reactions on cellular metabolism.

Keywords: advanced glycation end products; aldehydes; atherosclerosis; carbonyl stress; cholesterol; cholesterol esters; foam cell; glucose; glycation; glycoxidation; lipoprotein; macrophages; oxidation

Abbreviations: apoA-I, apolipoprotein A-I; apoB, apolipoprotein B-100; ACE, angiotensin-converting enzyme; AGE, advanced glycation end products; CAM, cellular adhesion molecule; CEL, carboxyethyl lysine; CML, carboxymethyl lysine; CVD, cardiovascular disease; DCCT, the Diabetes Control and Complications Trial; ECM, extracellular matrix; ERK, extracellular signal-regulated kinase; LDH, lactate dehydrogenase; GAPDH, glyceraldehyde-3-phosphate dehydrogenase; HDL, high-density lipoprotein; HMDM, human monocyte-derived macrophages; HUVEC, human umbilical vein endothelial cells; IL-1, interleukin-1; ICAM-1, intracellular adhesion molecule-1; LCAT, lecithin:cholesteryl acyl transferase; LDL, low-density lipoprotein; LOX-1, low-density lipoprotein receptor-1; MAK, mitogen-activated kinases; MCP-1, monocyte chemoattractant protein-1; M-CSF, macrophage colony-stimulating factor; MSR-A, macrophage scavenger receptor A; ox-LDL, oxidized LDL; PKC, protein kinase C; PTB, *N*-phenacylthiazolium bromide; RAGE, receptor for advanced glycation end products; sRAGE, soluble form of the receptor for AGE; SR-B1, scavenger receptor B1; STZ, streptozotocin; TNF-α, tumor necrosis factor-α; UKPDS, United Kingdom Prospective Diabetes Study; VCAM-1, vascular cellular adhesion molecule-1; VLDL, very low-density lipoprotein

Diabetes, Atherosclerosis, and Hyperglycemia

Cardiovascular disease (CVD) and diabetes represent two of the largest public health problems worldwide [1, 2]. More than 17 million people die from CVD every year and over 150 million have been diagnosed with diabetes [3, 4]. In Western countries and some non-Europid populations, such as Native American and Canadian communities, Pacific and Indian Ocean communities, and Australian Aboriginals [4], the prevalence of diabetes is higher than in the general population, with many more people in a "prediabetic" state [5]. Of further concern is that people with diabetes have an increased risk of CVD, leading to decreased life expectancy and quality of life [1, 4]. The World Health Organisation has reported that CVD is responsible for ca. 50% of deaths among people with diabetes [1]. Many studies have demonstrated that people with diabetes are at a 2–4-fold increased risk of CVD and mortality (reviewed in [6]). The risk of CVD always appears to be higher in people with diabetes than those without diabetes, regardless of geographical variation in risk [7].

People with diabetes manifest a number of clinical features considered to be CVD risk factors. Proatherogenic abnormalities in lipoprotein composition, subclass distribution, and lipoprotein metabolism have been identified in humans with Type 2 and Type 1 diabetes, particularly in the former where glycemic control is poor (reviewed in [8]). Notably, while the lipid profile of people with Type 1 diabetes with moderate to good glycemic control is relatively normal, the rate of CVD remains accelerated [8]. Hemostatic abnormalities indicative of hypercoagulability and increased platelet aggregation and adhesion are also demonstrable in people with diabetes (reviewed in [9, 10]). However, these abnormalities do not fully explain the increased risk of CVD.

It is widely held that the hyperglycemia in diabetes contributes to the increased incidence of atherosclerosis [10–12]. Large vessel disease appears at around the time of the first diagnosis of Type 2 diabetes, therefore hyperglycemia in the prediabetic state may be important in the development of atherosclerosis [10]. Tight hyperglycemic control has not been conclusively linked with improved macrovascular outcomes, although there are clear microvascular benefits. The Diabetes Control and Complications Trial (DCCT) demonstrated that in people with Type 1 diabetes good blood glucose control reduced the risk of retinopathy and cardiovascular events [13]. However, the numbers used in this study were small and a relatively young population was studied. A follow-up to the DCCT reported that intensive glycemic control inhibited the increases in carotid-intima thickness (a widely accepted predictor of atherosclerosis) detected in people with diabetes [14]. Glycated hemoglobin (HbA_{1c}) concentration is an indicator of average blood glucose over the previous 3 months and may be used as a monitoring tool for diabetes [15]. The United Kingdom Prospective Diabetes Study (UKPDS) provided clear evidence that decreasing HbA_{1c} reduced microvascular complications, but not macrovascular disease [16]. Indeed Brownlee [12] has put forward the concept of hyperglycemic memory in which there is a persistence or progression of hyperglycemic-induced vascular complications during periods of normoglycemia.

There are many possible mechanisms as to how hyperglycemia might induce atherosclerosis. These include increased polyol pathway flux, increased advanced glycation end-product (AGE) formation, activation of protein kinase C (PKC), increased oxidative stress, and increased hexosamine pathway flux. It has been suggested that these mechanisms may have a common element of overproduction of superoxide radicals by the mitochondrial electron transport chain (reviewed in [12]).

Glycation, Glycoxidation, and the Formation of Advanced Glycation End Products

Glycation of proteins involves reactions between a sugar, such as glucose or another reactive aldehyde, and a nucleophilic group on the protein. Thus, reaction occurs with the amine groups at the N-terminus and on Lys side chains, at the guanidine group of Arg side chains, and with the thiol group of Cys residues [17]. The extent of this spontaneous reaction is dependent on the duration of exposure to the modifying species and leads to the formation of a Schiff base (or thiohemiacetal in the case of Cys residues [18]). The Schiff bases formed with nitrogen nucleophiles undergo rearrangement to form an essentially irreversible ketoamine Amadori product [19]. Subsequent reactions of these species result in the generation of a heterogenous group of adducts collectively known as advanced glycation end products (AGE). Aldehydes such as glycolaldehyde and methyglyoxal, the levels of which are increased in diabetes (discussed later), can also glycate proteins, and the rates of reaction of these compounds are known to be significantly greater than that for glucose itself [17, 20]. Glucose and other aldehydes, whether free or protein-bound, can also undergo autoxidation reactions which yield radicals and other reactive intermediates (e.g., H_2O_2 and other peroxides) which can also contribute significantly to AGE generation. These latter processes are often termed glycoxidation [19, 21, 22].

Protein Glycation/Glycoxidation and AGE Accumulation in Diabetes

Modifications to proteins brought about by glycation and glycoxidation reactions include increases in chromophores and fluorophores, modification of Arg, Cys, and Lys side chains and the N-terminus, increases in net negative charge, unfolding, cross-linking, and fragmentation [18, 20, 21, 23–26]. It has been shown that the protein fragmentation and conformational changes brought about by the reaction of proteins with glucose are dependent upon hydroxylating/oxidizing agents (probably hydroxyl radicals) produced from H_2O_2, in the presence of metal ions, arising from glucose autoxidation and ketoaldehyde formation [21, 23]. The magnitude of these changes is dependent on the modifying agent used. Nagai et al. [26] have reported that the levels of the AGE-carboxymethyl lysine (CML) are high on glyoxal-modified BSA, low on glycolaldehyde-modified BSA, and undetectable on methylglyoxal-modified BSA. A similar pattern was also observed for antibodies against non-CML structures [26]. In contrast, the fluorescence spectra of the modified proteins were similar in all cases except for glyoxal-modified BSA, where they were weak, and all had similar changes in negative charge [26]. Of these modified species, only glucose- and glycolaldehyde-modified BSA were taken up by receptor-mediated endocytosis by a macrophage-type cell line [26]. Recent work has suggested that glycation often precedes oxidative modifications with the concentrations of metal ions and modifying aldehydes likely to be found *in vivo* [20].

These reactions are likely to be of relevance *in vivo* in people with diabetes and its associated complications. The most recent data from the EPIC-Norfolk study [27] indicate that in this cohort the risks of cardiovascular and coronary heart disease increased with increasing HbA_{1c} concentration and this was largely independent of a number of cardiovascular risks factors. HbA_{1c} concentrations correlated with the risk of disease in the cohort members with diagnosed diabetes. Previously, HbA_{1c} and other measures of hyperglycemia have been linked with increased mortality [28–33], cardiovascular disease/events [29–33], carotid artery stenosis [14, 34], fatty streaks [30], and endothelial dysfunction [35].

Levels of methylglyoxal [36–38] and 3-deoxyglucosone [39] have been reported to be elevated in the blood of people with diabetes. Significant increases in the levels of several AGE have been demonstrated in tissues, plasma, and urine from people with diabetes (reviewed in [19]), and in patients on hemodialysis [40].

The higher levels of AGE detected in urine, compared to plasma, suggest substantial turnover of AGE-modified material although this may be determined by the extent of modification. It has been shown that although moderately oxidized proteins are more sensitive to proteolytic attack, and are endocytosed more quickly than native proteins, the opposite is the case for heavily oxidized proteins [41, 42]. This may explain the observed accumulation of oxidized proteins over time in some tissues and disease states (age, atherosclerosis, diabetes) and contribute to these pathologies (reviewed in [43, 44]). Similar behavior may occur with glycated/glycoxidized proteins. Variation in the rates of protein turnover will also contribute to the level of tissue AGE accumulation; proteins with low turnover such as collagen and lens crystallins would therefore be expected to accumulate these materials and this has been detected experimentally [45–51].

In diabetes-associated atherosclerosis, both glycation and oxidative processes have been implicated in disease development and progression.

However, antioxidant therapies in the treatment of atherosclerosis have met with little success (reviewed in [52]). Baynes and Thorpe [53] and Knott et al. [20] have argued that when considering diabetic complications, damage to cell and tissue components via glycation reactions ("carbonyl stress") may precede and potentially lead to oxidative damage. These processes may also however be divorced from each other. This has potentially important therapeutic significance [53], as it raises the question as to whether there is increased benefit in using antioxidant therapies in the treatment of the vascular complications of diabetes, in addition to approaches designed at achieving glycemic control and prevention of glycation of cell and tissue components.

Immunohistochemical Localization of AGE Structures

Diabetes and increasing age are associated with a loss of elasticity of lung, skin, arterial, joint, and bone tissue. Collagen from the extracellular matrix (ECM) of these tissues is subjected to increased nonenzymatic glycation, and the extent of glycation correlates with the presence of late complications of diabetes [54, 55]. CML epitopes are not uniformly distributed, with the elastic fibres of blood vessels (and skin) staining more intensely than surrounding tissue [46]. CML formation is slow, and the accumulation of CML epitopes in ECM is probably exacerbated by the slow turnover of these proteins. Areas of intimal thickening and atherosclerotic plaques express more CML epitopes than areas with less severe morphologic alterations [46]. The cytoplasmic domains of foam cells also contain the CML epitope, which may be due to receptor-mediated uptake and degradation of AGE-modified proteins by macrophages within plaques [46, 56] (also discussed later). Serum proteins from people with diabetes also have an increased level of CML, which correlates with indices of glycemia such as HbA_{1c} [46]. Since CML can be formed from both glycation/glycoxidation and lipid oxidation reactions, the observed increase in CML may be accounted for by either an increase glycation or increased oxidative stress. However, the AGE product, pyrraline, has been reported to be formed by a nonoxidative mechanism, and both this material and CML levels are elevated in diabetic serum, supporting the notion that the nonoxidative glycation of proteins may contribute to AGE accumulation in diabetes [47, 57, 58].

Intracellular localization of glycolaldehyde-modified proteins and AGE in macrophage foam cells in atherosclerotic plaques has been demonstrated in several studies [56, 59, 60]. Protein modification by reactive carbonyl species has been detected in the sera and tissue of people with diabetes, with greater amounts of methylglyoxal-induced protein modification detected in people with, compared to without, diabetes [61]. Immunoelectron microscopic observation of atherosclerotic plaques obtained after autopsy has revealed that AGE-epitopes are located within the lysosomal lipid vacuoles or electron-dense granules of foam cells consistent with the impaired degradation of these proteins [59]. This localization might be indicative of two means by which protein glycation/glycoxidation may contribute to foam cell generation in diabetes-induced atherosclerosis (discussed in more detail later). Firstly, intracellular accumulation of lipoprotein-derived cholesterol may be due to resistance of the glycated (CML–protein adduct) or glycoxidation products on modified low-density lipoprotein (LDL) to lysosomal degradation. In our laboratory, glycated/glycoxidized proteins have been shown to inhibit a number of intracellular enzymes including lactate dehydrogenase (LDH), glyceraldehyde-3-phosphate dehydrogenase (GAPDH), glutathione reductase [62], and some of the lysosomal cathepsins (Zeng and Davies, unpublished results). Inhibition of the latter species may be of particular importance in modified protein turnover. Secondly, glycation/glycoxidation of LDL may promote particle uptake by macrophages. What follows is a consideration of mechanism of foam cell generation and evidence supporting a role for protein glycation/glycoxidation in this process.

The Macrophage Foam Cell in Atherosclerosis

Atherosclerosis is an inflammatory disease (reviewed [63]) and macrophages are present at all stages of the disease (reviewed [64]). Animal studies, using several different models of atherosclerosis, support a key role for macrophages in the development and progression of atherosclerosis. The *op* defect is a deficiency in macrophage colony-stimulating factor (M-CSF) that renders the host severely deficient in blood monocytes and tissue macrophages. Apolipoprotein E knockout (apoE$^{-/-}$) mice also carrying the *op* defect were found to have vessel lesions that were of smaller area with far fewer macrophages than those of littermates that were replete with M-CSF [65]. This was also the case for E^0 mice deficient in the monocyte chemoattractant protein-1 (MCP-1) receptor CCR2 [66]; mice transgenic for human apolipoprotein B deficient in MCP-1 [67]; and bone-marrow-irradiated LDL receptor knockout mice reconstituted with macrophages negative for the murine homolog of the interleukin-8 receptor, CXCR-2 [68].

Modifications to LDL such as oxidation, aggregation, incorporation into immune complexes and glycation, are strongly implicated in the pathogenesis and progression of atherosclerosis (reviewed [52, 69]). Modified LDL has been reported to be retained in the arterial wall (reviewed [70]). Macrophages, in their role of scavenging tissue and cellular debris, will take up the modified LDL. This is likely to involve several receptors, the expression of which are not regulated by cellular cholesterol levels. The excess cholesterol is esterified as a means of regulating cellular free cholesterol levels. Subsequent hydrolysis is required for export of the cholesterol from the cell to suitable transporters. Where export pathways are inhibited, deficient, or saturated, there would be an accumulation of cholesterol as cytosolic droplets leading to "foam cell" generation (reviewed [63, 71, 72]). Macrophages may also contribute to modification of LDL (reviewed [52, 63]).

Whether lipid accumulation in macrophages contributes principally to the development of a pro- or anti-inflammatory phenotype has not been fully elucidated. A major obstacle to achieving this is that existing *in vitro* models used to determine the pro- or anti-inflammatory effects of lipid loading often suffer limitations in terms of the biological relevance of both the lipid delivery systems used and the intracellular lipid composition achieved. However, other mechanisms may also come into play including matrix metalloproteinase activation and cytokine release. It has been reported that the accumulation of lipid destabilizes atherosclerotic lesions and increases the likelihood of rupture (e.g., [73]). Prolonged cholesterol accumulation is known to lead to cell death that is likely to involve both apoptosis and (proinflammatory) necrosis (reviewed [71]).

AGE-Modified Lipoproteins

The characteristic lipid profile of people with Type 2 diabetes consists of normal to high LDL, reduced high-density lipoprotein (HDL), and increased triglycerides. In contrast, the lipid profile of people with Type 1 diabetes with good glycemic control is typically near normal though vascular disease is still accelerated [8]. Glycation and AGE modification of LDL are increased in people with both Type 1 and Type 2 diabetes, and therefore may play an important role in the premature onset, and rapid progression, of atherosclerosis observed in both forms of the disease.

The reaction between glucose and the apolipoprotein B-100 (apoB) of LDL gives rise to similar modifications to those observed with other proteins. Thus increases in net negative charge, modification of Arg, Lys, and Trp residues, increases in fluorescence, enhanced cross-linking, and AGE generation have been reported. The formation of malondialdehyde, peroxides, and other lipid peroxidation products has also been observed [20, 74–81]. Specific AGE, such as CML, have also been identified in glucose-modified LDL, and in macrophage foam cells of atherosclerotic plaques [56, 82].

The rate, and extent, of LDL alteration is more marked on reaction with aldehydes, or mixtures of glucose plus Cu^{2+} ions, than with glucose alone [20]. A comparative study of the extent of alteration induced by aldehydes of different size and structure (glycolaldehyde, glyceraldehyde, erythrose, arabinose, and glucose; 2, 3, 4, 5, and 6 carbons, respectively) has shown that the lower the molecular mass of the aldehyde, the more rapid and more extensive were the changes in derivatization of free amino groups and increases in negative charge [74]. Modification of LDL by methylglyoxal and glycolaldehyde has also been shown to result in Arg, Trp, and Lys modifications, increases in net negative charge, and cross-linking [20]. Some studies have also demonstrated a stimulatory effect of glucose on Cu^{2+}-mediated LDL oxidation [77, 79], however this is dependent on the ratio of Cu^{2+} ions to LDL, and has not been observed with the low levels of Cu^{2+} that are likely to be present in the artery wall [20, 81, 83].

As with hemoglobin (HbA_{1c}), increased levels of glycated lipoproteins have been detected in the serum of people with diabetes [84–87]. AGE-modified LDL has been detected in the serum of people with diabetes, with the amount of glycated/glycoxidized lipoprotein proportional to the level of hyperglycemia as measured by HbA_{1c} [88, 89]. Increases of up to twofold have been detected [84, 85]. Steinbrecher and Witztum [90] reported that up to 5% of the Lys residues present on the apoB protein of LDL from people with diabetes was modified. Klein et al. [86] reported that the percentage of glycated/glycoxidized LDL is higher in people with diabetes than controls (21% and 5%, respectively). Interestingly, the levels of AGE-modified apoB were also observed to be greater in normolipidemic patients with occlusive atherosclerotic disease, but no diabetes, than in normal age-matched controls suggesting there may also be a causal link between AGE modification of lipoproteins and atherosclerosis in people without diabetes [91]. Bucala et al. [92] have identified AGE-epitopes near the LDL receptor-binding site as key sites of modification on apoB, though the exact nature of these modifications remains to be resolved. Glycation and/or glycoxidation of LDL has been reported to affect its catabolism and degradation [90, 93], and increase its proteoglycan binding properties [94]. Thus, glycated/glycoxidized LDL and/or AGE-modified LDL may contribute to diabetes-accelerated athero-sclerosis by exacerbating dyslipidemia, enhancing LDL–ECM affinity, immunogenicity, and macrophage endocytosis with resulting lipid loading of cells (i.e., foam cell formation).

Dyslipidemia

In vitro experiments have revealed that glucose preferentially targets Lys residues on the apoB protein of LDL and that this modification can interfere with the binding of LDL to the LDL receptor [84, 87]. Impaired binding of glucose- and AGE-modified LDL to the LDL receptor has been shown to reduce LDL receptor-mediated clearance mechanisms of LDL with the degree of impairment proportional to the amount of Lys derivatization [84, 87]. Glycation of LDL may therefore contribute to dyslipidemia in people with diabetes by inhibiting the hepatic clearance of LDL, effectively increas-ing the half-life of the lipoprotein. Administration of aminoguanidine, an inhibitor of AGE formation, to people with diabetes has been shown to decrease circulating LDL levels by up to 28% providing evidence for a con-tribution of LDL glycation to diabetic dyslipidemia [89].

AGE-LDL is Immunogenic

Glycation and glycoxidation of lipoproteins may render them immunogenic. Curtiss and Witztum [95] demonstrated that LDL glycated in the presence of cyanoborohydride, which was heavily modified (60% of Lys residues modi-fied), was a potent immunogen, stimulating antibodies that did not interact with unmodified LDL. When LDL was glycated with glucose alone (6% of

Lys residues modified), it was a less-potent immunogen. Differences in the antibody response and subsequent clearance were found to be dependent on the degree of modification.

Oxidized and glycated LDL are reported to elicit humoral immune responses in experimental animals and in humans leading to the formation of autoantibodies reactive to the modified forms of LDL [96]. The average avidity of AGE-LDL autoantibodies is higher than that of ox-LDL autoantibodies [97]. CML and carboxyethyl lysine (CEL) appear to be the major modifications recognized by autoantibodies, but antibody specificity is also dependent on the nature of the protein containing the AGE adducts [96]. It is unclear whether the net effect of the humoral immune response to glycated- or AGE-modified LDL is pro- or antiatherogenic. Autoantibodies to glycated LDL may form immune complexes with modified LDL, which could facilitate the uptake and removal of the complex constituting a possible antiatherogenic effect. However, the formation of immune complexes of LDL enhances cholesterol uptake by macrophages via Fc-γ receptors and thus the formation of foam cells [98]. In addition to cholesterol uptake, LDL-immune complexes stimulate the secretion of the proinflammatory cytokines interleukin-1 (IL-1) and tumor necrosis factor-α (TNF-α) from macrophages, that may contribute to the progression of atherosclerosis [97].

AGE-LDL: Enhanced Proteoglycan Binding

Glycation/glycoxidation of LDL has been shown to increase its proteoglycan binding properties [94]. Enhanced binding of LDL to ECM components such as proteoglycans may contribute to diabetes-induced atherosclerosis by promoting LDL retention in the artery wall. LDL retention may in turn enhance foam cell formation and the rate of atherosclerosis. Improved glycemic control in diabetic patients has been shown to reduce the extent of modification of LDL and decrease the binding capacity of LDL for proteoglycans [99]. In the same study, the extent of glycation was shown to be the only compositional variable that correlated with proteoglycan binding and removal of the glycated portion of LDL returned proteoglycan binding to normal. Greater subendothelial retention of lipoproteins has been reported in insulin-deficient rabbits [100], and thus perhaps contributes to atherosclerosis in people with diabetes. The "response to retention" hypothesis is probably the most widely accepted to explain lesion initiation [70, 101]. If LDL is retained in the intima, there is greater chance of uptake by macrophages, or other cells, to yield lipid-laden foam cells, and it has been demonstrated that macrophages *in vitro* can accumulate LDL isolated from human atherosclerotic lesions [102].

AGE-HDL

Glycation or glycoxidation of HDL may contribute to the enhanced levels of lipid present in the artery wall, by inhibiting reverse cholesterol transport (i.e., the process that removes excess cholesterol from lipid-laden cells). Glycated/glycoxidized HDL has been reported to remove cholesterol either to a similar extent [103], or less efficiently, from macrophages [104, 105] compared to nonmodified HDL. Cholesterol efflux has also been reported to be impaired in adipocytes [106] and fibroblasts [105] exposed to glycated (lipid free or lipid poor) apolipoprotein A-I (apoA-I, a major protein of HDL) or HDL itself. However the mechanism by which this occurs is unclear, and further studies are needed to clarify the effects of diabetes on other components of the reverse cholesterol pathway [107, 108]. HDL is known to have potent anti-inflammatory effects at the level of vascular endothelium and has been shown to reduce cellular adhesion molecule (CAM) expression and neutrophil infiltration into injured rabbit arteries [109]. In addition to glycation/glycoxidation potentially affecting the role of apoA-I and HDL in cholesterol efflux (though the data on this is contradictory), it may also have potential effects on the anti-inflammatory effects of HDL, and hence modulate the development of atherosclerosis via multiple pathways. It has recently been established that methylglyoxal-induced modification of HDL inhibits the ability of lecithin:cholesteryl acyl transferase (LCAT) to esterify free cholesterol associated with apoA-I (Nobecourt et al., unpublished data). It therefore remains to be established whether HDL retains its antiatherogenic effects in people with diabetes.

AGE-LDL and Foam Cell Formation

Modifications of LDL that result in the formation of AGE give rise to particles that promote lipid loading and foam cell formation. LDL isolated from people with Type 1 diabetes has been shown to induce significantly higher cholesterol ester accumulation in human macrophages than LDL isolated from control subjects, with the extent of cholesterol ester accumulation correlating with the extent of LDL glycation [85]. When the glycated/glycoxidized LDL was separated from the nonmodified LDL using affinity chromatography, the modified fraction induced significantly greater cholesterol and cholesteryl ester accumulation when incubated with human macrophages [86]. This modified LDL also significantly increased cholesteryl ester synthesis rates. Interestingly, there were no significant differences in the rates of cholesteryl ester synthesis for glycated/glycoxidized LDL obtained from people with or without diabetes [86]. LDL obtained from people with Type 2 diabetes has also been shown to induce greater macrophage uptake and cholesterol ester synthesis than LDL obtained from controls [110].

The modification of LDL with glucose *in vitro* has also been shown to induce cholesterol ester accumulation in human macrophages providing further evidence for the importance of LDL glycation/glycoxidation in foam cell formation in people with diabetes [85, 96, 110]. Using well-characterized, minimally oxidized LDL preincubated with glucose for 7 days, we were unable to induce *in vitro* foam cell formation in mouse [81] or human macrophages (Brown et al., unpublished data) whereas LDL modified by aldehydes under the same conditions resulted in significant lipid loading in both cell types ([81], Brown et al., unpublished results).

In vitro glycoxidation of LDL with glucose and metal ions can also affect macrophages. One study has shown using staining methods that LDL glycoxidized with glucose and iron can be taken up by mouse peritoneal macrophages, and that this also results in the production of superoxide radicals [80]. Another group has shown that human macrophages (that contain abundant scavenger receptors, but few LDL receptors) show increased rates of degradation of glycoxidized LDL when compared to control LDL, and that fibroblasts (that contain fewer scavenger receptors and more LDL receptors) show higher rates of degradation of native LDL than glycated/glycoxidized LDL. These human monocyte-derived macrophages (HMDM) also demonstrated enhanced uptake of glycoxidized LDL. Millican et al. [78] demonstrated that LDL glycoxidized with glucose and Cu^{2+} results in an increase in electrophoretic mobility that correlated with lipid oxidation, and that both of these correlated with the inhibition of DNA synthesis in macrophage-like cells. In recent work, LDL exposed to a combination of glucose and low (potentially more physiologically relevant) concentrations of Cu^{2+} was found not to induce *in vitro* foam cell formation in cultured murine macrophages [81].

AGE products are also known to be formed on the lipid components of LDL such as the amine groups of phospholipids. With glucose-modified LDL, modified phosphatidylethanolamines have been detected, and enrichment of LDL with these species has been shown to increase LDL uptake, and cholesteryl ester and triglyceride accumulation, in macrophages [111, 112].

Reactive aldehydes can be formed *in vivo* by a number of pathways including metabolic processes and the oxidation of glucose, lipids, and amino acids. The rate of AGE formation from the reaction of these aldehydes with proteins is orders of magnitude faster than glucose-induced AGE formation, and it is probable that reactive aldehydes play a key role in AGE formation *in vivo* [12]. Although many aldehydes have been identified that generate AGE, the ability of these aldehydes to induce the modifications of LDL that lead to LDL uptake and cholesterol accumulation in macrophages varies greatly. Modification of LDL by α-hydroxy aldehydes increases uptake by macrophages and glycolaldehyde modification appears to be particularly effective in inducing LDL endocytosis and degradation ([26, 74, 113], Brown et al., unpublished data). Endocytosis of α-hydroxy aldehyde-modified proteins has been shown to be inversely proportional to the mass of the modifying aldehyde [74]. Uptake of glycolaldehyde-modified LDL (and other proteins) is competitively inhibited by acetylated LDL, and substantially reduced in cells lacking the macrophage scavenger receptor A (MSR-A) suggesting that this receptor plays a key role in the binding and endocytosis of glycolaldehyde-modified proteins [26, 113].

Serum levels of glycolaldehyde have proven difficult to determine, probably due to the highly reactive nature of this aldehyde leading to low serum concentrations in its free form, though it is clear that this material must be present [60]. Methylglyoxal-modified LDL has been shown in one study to cause significantly less cholesteryl ester synthesis in murine macrophages when compared to native, oxidized, or acetylated LDL [75]. This is in contradiction to our recent studies where methylglyoxal-modified LDL was shown to increase the cellular cholesterol ester levels [81]. These differences may arise from the different conditions used to modify the LDL. The extent of cholesteryl ester accumulation detected with this methylglyoxal-modified LDL was less than that detected with glycolaldehyde-modified LDL [81]. Interestingly, modification of LDL with α-oxoaldehyde dicarbonyl compounds, such as methylglyoxal, gives rise to more modest increases in endocytosis and degradation of the apoB protein of LDL than glycolaldehyde, despite inducing a similar extent of protein charge modification (Brown et al., unpublished data). Modification of proteins by oxidation, glycation, or glycoxidation has been proposed to impair binding of LDL to the LDL receptor, and increase binding and endocytosis by unregulated scavenger receptor pathways. It has been proposed that an increase in net negative charge of a protein may be responsible for preferential binding to scavenger receptors [114]. Reaction with Lys and Arg residues, which are positively charged at physiological pH, appear to be a key determinant of the change in overall particle charge and are known to be key targets for modification. Glycolaldehyde has been shown to induce a greater degree of Lys derivatization than methylglyoxal, which preferentially targets Arg residues [17, 26, 75]. However, the large differences in uptake of glycolaldehyde- and methylglyoxal-modified LDL, despite similar levels of charge modification, suggest that overall charge modification is not the sole determinant for scavenger receptor binding, with this possibly related to the specific modifications induced on particular amino acids [81].

Our recent studies suggest that there is a degree of interdependence of proteolysis on lipolysis and vice versa, with perturbations in these processes, induced by the endocytosis of glycated LDL, potentially contributing to foam cell formation in diabetes-associated atherosclerosis. Thus a state of "carbonyl stress" in people with diabetes may contribute to the accelerated development of atherosclerosis. This has been further supported by studies that have demonstrated that carbonyl-scavenging compounds can inhibit glycation-induced changes and subsequent model foam cell formation using mouse (Brown et al., unpublished data) or human macrophages (Rashid et al., unpublished data).

Most studies that have tried to evaluate the extent of LDL glycation and glycoxidation in people with diabetes have used serum-derived LDL. Although lipoprotein glycation occurs within the vascular compartment, this may not represent the true extent of LDL modification, as extravasated LDL is likely to be subjected to further glycation and oxidation. Antioxidants such as α-tocopherol and carotene have been shown to protect, under certain circumstances, LDL (and other proteins) from oxidative modification [52].

Oxidized LDL (ox-LDL) levels in atheroma, however, have been shown to be 70-fold higher than levels in the circulation suggesting that ox-LDL is either formed, or preferentially accumulates, within the vessel wall [115]. Greater extents of LDL glycation/glycoxidation within the vascular wall are therefore a strong possibility. Intracellular hyperglycemia, particularly in the endothelium, has been proposed to be the primary initiating event in the formation of both intracellular and extracellular AGEs [45]. Glucose uptake by endothelial cells is predominantly mediated by GLUT1 transporters, which have a high affinity for glucose with K_m values around 2 mM [116]. Glucose uptake by such transporters is insulin-independent and the amount of glucose uptake is dependent on blood glucose levels. Hyperglycemia in diabetes thus directly increases endothelial intracellular glucose concentrations [116]. *In vitro* experiments have demonstrated that incubation of endothelial cells in the presence of glucose for 7 days results in an approximately 14-fold increase in intracellular AGE content [117]. Reactive aldehydes and AGE species released from endothelial cells may also react with extravasated LDL trapped in the subendothelial matrix leading to a greater extent of modification than observed in serum LDL. Furthermore, soluble AGE peptides released after macrophage-mediated degradation of AGE proteins can covalently bind other proteins forming "second generation" AGE [118]. AGE-peptides released from macrophages in the arterial wall may then bind and modify LDL trapped in the subendothelial matrix forming a cycle of AGE generation, endocytosis, and degradation. Alternatively, circulating glycated LDL may be subjected to oxidation on entering the vessel wall leading to elevated AGE modification of LDL within the arterial wall.

Receptors that Bind Advanced Glycation End Products

AGE are ligands for a number of pattern recognition receptors such as the receptor for advanced glycation end products (RAGE) and a number of so-called scavenger receptors [119, 120].

RAGE

RAGE is a 35-kDa protein that belongs to the immunoglobulin superfamily [121]. Low levels of RAGE expression are present in normal tissue and vasculature with higher levels of expression occurring in locations where RAGE ligands accumulate [122]. Similar to the pattern recognition receptors of the innate immune system, ligand binding to RAGE is dependent on the three-dimensional structure of the ligand rather than its specific amino acid sequence [123, 124]. RAGE has been shown to bind several peptides including AGE, S100 proteins, amyloid β-peptide, and the high-mobility group protein B1 (HMGB1). These ligands are characterized by β-sheets or fibril formation [122] and the AGE CML has been reported to bind RAGE and activate cell-signaling pathways [125]. The binding of RAGE by its ligands results in the activation of several signaling pathways including erk1/2 (p44/p42) MAP kinases, p21ras, p38, SAPK/JNK MAP kinases, rho GTPases, phosphoinositol-3-kinase and the JAK/STAT pathway (reviewed in [126]). These signaling pathways can lead to the activation of downstream effectors such as NF-κB and CREB, the former being able to stimulate adhesion molecule expression [127]. The ligation and activation of RAGE has been reported to lead to the formation of reactive oxygen species, and CML has been reported to stimulate superoxide radical and H_2O_2 generation via activation of NADPH oxidase in human endothelial cells [128]. Unlike scavenger receptors, the binding of AGE or other ligands to RAGE does not result in endocytosis of the ligand [125]. Recent studies have suggested that although AGE bind RAGE with high affinity, they do not uniformly induce CAM expression [129, 130]. Furthermore, S100b has been shown to induce CAM expression via RAGE with greater efficacy than AGE [130]. The ability of different AGE species to bind and activate RAGE and the role of S100 proteins in the pathogenesis of atherosclerosis requires further investigation. Studies with homozygous RAGE-null mice ($RAGE^{-/-}$) have provided insights into the function of RAGE *in vivo*. $RAGE^{-/-}$ mice are more resistant to the lethal effects of septic shock than wild-type mice suggesting that these receptors play a role in the initiation and propagation of the innate immune response in sepsis [131]. Deletion of the RAGE gene, however, does not compromise the adaptive immune response. RAGE itself has also been shown to function as an endothelial adhesion receptor for leukocytes, which may contribute to the increased inflammation observed in pathological conditions associated with increased RAGE expression such as diabetes, atherosclerosis, and chronic immune diseases [132].

Scavenger Receptors

Class A scavenger receptors (MSR-A) are expressed by macrophages and function as part of the host innate immune response. There are two subtypes of MSR-A, Type 1 and Type 2, which have a broad specificity for polyanionic ligands. Endogenous ligands for MSR-A include modified LDL, ox-LDL, and AGE-modified proteins, whereas exogenous ligands include lipid A, lipoteichoic acid, *Neisseria meningitidis*, and asbestos [133]. *In vivo*, MSR-A contributes to the innate immune response to gram-positive microbial infections and susceptibility to atherogenesis [133]. The uptake of acetylated LDL by macrophages via MSR-A and subsequent formation of model foam cells was first identified by Goldstein et al. [134]. Ox-LDL and AGE-modified proteins have also been shown to be endocytosed by macrophages via MSR-A [135]. The AGE-structure CML, however, is not recognized by the MSR-A [26] and uptake of proteins containing AGE adducts via this receptor is probably due to modification of the biophysical properties of proteins by glucose or aldehydes rather than the binding of a specific AGE structure. One identified determinant of MSR-A ligation is the modification of Lys residues, which play an important role in the maintenance of the tertiary structure of proteins [136]. Derivatization of protein Lys residues with oxidation products increases scavenger receptor binding, and blocking Lys residues by methylation prior to incubation with oxidation products prevents this effect [136].

CD36 and SR-B1 are class B scavenger receptors that bind AGE-modified proteins and unlike MSR-A, these receptors have two transmembrane domains [137]. CD36 is an 88-kDa glycosylated protein that binds ligands such as ox-LDL, fatty acids, collagen, anionic phospholipids, thrombospondin, *Plasmodium falciparum*, and some gram-positive/negative bacteria [133]. *In vivo*, macrophage CD36 appears to be responsible for the binding and uptake of ox-LDL [137]. CD36 deficiency has been shown to markedly reduce ox-LDL endocytosis in human macrophages and immunohistochemical studies have demonstrated CD36 expression in core regions of atherosclerotic plaques [138, 139]. Studies in mice lacking both MSR-A and CD36 have shown that these receptors account for 75–90% of degradation of acetylated or ox-LDL [140]. In the light of these findings, CD36 and MSR-A are regarded as the major pathways for ox-LDL-induced foam cell formation in atherosclerosis. In addition to ox-LDL, CD36 has demonstrated high affinity binding for AGE-BSA [141]. Scavenger receptor B1 (SR-B1) binds HDL and can mediate the transfer of cholesteryl esters without endocytic uptake of HDL. SR-B1 contributes to reverse cholesterol transport by promoting cholesterol efflux from peripheral cells such as macrophages and cholesterol uptake by hepatocytes [137]. In addition to its function in reverse cholesterol transport SR-B1 has also been shown to endocytose and degrade AGE-modified proteins [141]. Interestingly, AGE has been shown to inhibit SR-B1-dependent cholesterol efflux [141].

The lectin-like oxidized low-density lipoprotein receptor-1 (LOX-1) is a scavenger receptor for ox-LDL that is highly expressed on endothelial cells [142]. Upon binding to LOX-1, ox-LDL undergoes endocytosis and lysosomal degradation. Like other scavenger receptors, LOX-1 preferentially binds ligands that are negatively charged, and AGE-modified proteins have been shown to bind LOX-1 with high affinity [143]. Methylglyoxal- and glycolaldehyde-modified proteins in particular are effective ligands for LOX-1 [143]. In endothelial cells, the binding of ox-LDL to LOX-1 can induce increased expression of MCP-1, activation of NF-κB, and release of matrix metalloproteinases, and as such ligation of LOX-1 may contribute to the development of atherosclerosis [144–146].

Recent data suggests that LOX-1 may play a role in the development of atherosclerosis associated with diabetes. Expression of LOX-1 is significantly increased in the aorta of rats with diabetes compared to controls, with most prominent expression occurring in endothelial cells located at aortic bifurcations [147]. Interestingly, the very low-density lipoprotein (VLDL)/LDL fractions of diabetic serum were most effective at inducing LOX-1 expression, providing further evidence for the possible involvement of glycoxidized LDL. Furthermore, LOX-1 expression is increased in human macrophages exposed to hyperglycemic conditions [148]. This increase in macrophage LOX-1 expression is abrogated by antioxidants suggesting that LOX-1 expression may be increased by hyperglycemia-induced generation of reactive oxygen species. Increased macrophage LOX-1 expression may therefore contribute to cholesterol uptake and foam cell formation in people with diabetes.

Galectin-3

p60 and p90 are AGE receptors that have been purified from rat liver and a mouse macrophage-like cell line. Galectin-3 is a lectin-like protein that is a component of p90 that binds cell surface/matrix glycoprotein β-galactoside residues and is also capable of peptide–peptide associations [149]. Galectin-3 has been proposed to have a protective effect against diabetic glomerulopathy as galectin-3 knockout mice that have been rendered diabetic show increased renal AGE accumulation and accelerated glomerulopathy compared to wild-type mice [150, 151].

The Effects of AGE on the Endothelium

Endothelial cells may be exposed to AGE on both basal and apical surfaces. Furthermore, endothelial cells express RAGE and ligation of these receptors by AGE induces the expression of CAM, tissue factor, cytokines, and MCP-1 [126, 127]. Taken together these effects stimulate vascular inflammation by promoting inflammatory cell adhesion to the endothelium and migration into the artery wall. Endothelial dysfunction caused by increased cellular glucose or ligation of RAGE may contribute to diabetic atherosclerosis. RAGE may also provide a pathway for AGE delivery to the subendothelial space, other cells, and cross-linking with matrix components.

Elevated glucose has been shown to increase endothelial adhesion molecule expression *in vitro* [152, 153]. Interestingly, oscillating high/low-glucose concentrations simulating diabetic conditions have been shown to elicit more CAM expression than sustained high glucose [154]. The hyperglycemia-induced CAM expression coincides with increased concentrations of IκB-α suggesting that this expression is due to increased NF-κB activity [153]. Inhibition of PKC-β completely abrogated hyperglycemia-induced CAM expression, underscoring the importance of this enzyme [12].

Incubation of human umbilical vein endothelial cells (HUVEC) with AGE-modified bovine serum albumin (BSA) increases endothelial vascular cellular adhesion molecule-1 (VCAM-1) expression via activation of NF-κB [127]. This increase is completely attenuated by the administration of anti-RAGE IgG, a soluble form of the receptor for AGE (sRAGE) and the antioxidant *N*-acetylcysteine indicating that increased oxidative stress following AGE–RAGE interaction is essential for VCAM-1 induction. Furthermore, diabetic plasma has been shown to contain elevated levels of soluble VCAM-1 antigen consistent with increased production of VCAM-1 possibly due to increased activation of RAGE [127]. In support of this theory, the infusion of exogenous-modified AGE-albumin into rabbits causes focal intimal proliferation and increased expression of VCAM-1 and intracellular adhesion molecule-1 (ICAM-1) [155]. Concurrent administration of a cholesterol-rich diet with AGE infusion leads to an even greater increase in CAM expression and the development of lipid-rich atheromas containing foam cells. In this study, as observed in the murine models of diabetic atherosclerosis, the combination of hyperglycemia and hyperlipidemia markedly increased atherosclerosis highlighting the potential importance of the interaction of lipoproteins and elevated blood sugars to the progression of diabetic atherosclerosis.

In people with Type 2 diabetes, increased serum levels of AGE are associated with decreased endothelial-dependent brachial artery dilation, independent of other CVD risk factors [156]. Endothelium-dependent vasodilation is also abnormal in people with Type 1 diabetes with HbA_{1c} levels correlating with the level of impairment in response to acetylcholine [157]. These studies show that significant endothelial dysfunction exists in both types of diabetes with the level of impairment being proportional to indices of glycation such as HbA_{1c}/serum AGE concentration. Studies in our laboratory have demonstrated that glycated LDL does not induce endothelial cell death or proliferation changes [81]. However, glycoxidized LDL has been shown to impair endothelial function more than oxidized LDL, possibly due to enhanced nitric oxide inactivation by superoxide radicals [158], but another study using methylglyoxal-modified LDL demonstrated that this did not induce activation of human endothelial cells [75]. There are few other studies that have examined the effects of glycoxidized LDL on endothelial cells.

AGE and Smooth Muscle Cells

Vascular smooth muscle cells may be exposed to subendothelial and circulating AGE by an increase in endothelial permeability. Vascular smooth muscle cells show high levels of expression of receptors for AGE, which are unchanged by increasing age or the presence of vascular disease [159]. The functional significance of this is unclear, but the binding of AGE protein to smooth muscle cells can be associated with increased cellular proliferation, perhaps mediated by growth factors or cytokines [160, 161]. AGE-BSA has been shown to be endocytosed by smooth muscle cells, and also can induce chemotactic smooth muscle cell migration, but not their growth [162]. Glycoxidized BSA may also be cytotoxic to smooth muscle cells, possibly due to reactive oxygen species produced extracellularly by metal-ion-induced glycoxidation [82]. Exposure of vascular smooth cells to glycated albumin has also been reported to result in increased activation of NF-κB, increased activity of mitogen-activated kinases (MAK) and extracellular signal-regulated kinase (ERK), as well as increased expression of c-*fos*, c-*jun*, IL-6, and MCP-1 [163]. Glycated and glycoxidized LDL has also been shown to induce smooth muscle cell proliferation and migration [164, 165]. Studies in our laboratory however have shown no cytotoxic or proliferative effects of glycated, but minimally oxidized, LDL on smooth muscle cells [81].

Animal Models of Diabetic Atherosclerosis

The development of animal models of diabetes-accelerated atherosclerosis has lead to the elucidation of key processes in the pathogenesis of this disease. The first murine model to demonstrate hyperglycemia-accelerated atherosclerosis utilized BALB/c mice rendered diabetic with streptozotocin (STZ) [166] and fed an atherogenic diet [167]. The diabetic mice displayed a 17-fold increase in lesion area (fatty streaks) compared to nondiabetic mice fed the same atherogenic diet. However, C57BL/6 mice subjected to the same experimental conditions showed no differences in lesion size when compared to nondiabetic controls underscoring the importance of genetic background to the development of atherosclerotic lesions [167]. In another model of murine atherosclerosis, hyperlipidemia was induced by deletion of the apolipoprotein E gene (apoE$^{-/-}$) with consequent development of atherosclerotic plaques [168, 169]. The induction of hyperglycemia in these mice by STZ injection (STZ/apoE$^{-/-}$) lead to a fivefold increase in mean lesion area at the aortic sinus compared to nondiabetic apoE$^{-/-}$ mice, with lesions of STZ/apoE$^{-/-}$ mice being more advanced than controls, with fibrous cap formation frequently observed. STZ/apoE$^{-/-}$ mice also have increased levels of chylomicrons, LDL, intermediate density lipoprotein and enhanced formation of AGE [169]. Interestingly, aortic lysates from STZ/apoE$^{-/-}$ had increased levels of non-AGE–RAGE ligands, such as S100/calgranulin epitopes, compared with nondiabetic controls raising the possibility of a link between hyperglycemia and expression of these ligands [170]. Taken together, the increased plaque area observed in hyperlipidemic diabetic mice provides strong evidence for the involvement of hyperglycemia, and possibly the interaction between glucose and cholesterol, in the pathogenesis of atherosclerosis.

RAGE ligation and activation was implicated in the pathogenesis and progression of diabetic atherosclerosis when it was shown that infusion of sRAGE containing the RAGE extracellular binding domain decreased plaque area and number in the STZ/apoE$^{-/-}$ murine model of diabetic atherosclerosis [169]. The antiatherogenic effect of sRAGE is independent of glycemia, insulinemia, and lipid profile. Furthermore, the plaques of sRAGE-treated mice show less leukocyte infiltration and their plasma contains sRAGE–AGE complexes, which would be subjected to accelerated clearance [169]. Administration of sRAGE also inhibits the progression of established plaques of STZ/apoE$^{-/-}$ mice [169, 171]. Interestingly, sRAGE has been shown to decrease inflammation in RAGE knockout (RAGE$^{-/-}$) mice and as such sRAGE may inhibit the binding of AGE or other RAGE ligands to other receptors, such as the MSRs [131]. Inhibition of pathways distinct from RAGE may account for some of the antiatherogenic effects of sRAGE.

Aortic tissue levels of the proinflammatory RAGE ligand S100 are also decreased after administration of sRAGE to diabetic mice [170], suggesting that sRAGE may bind and inhibit non-AGE ligands of RAGE. Recent studies suggest that S100b is a more potent activator of RAGE than AGE-modified proteins and the plasma concentration of S100A12 has been shown to correlate with HbA_{1c} [129, 172]. The possible contribution of S100 proteins to the progression of diabetic atherosclerosis warrants further investigation. As AGE and S100 proteins have been shown to accumulate in the vessel wall in the absence of diabetes, it has been proposed that RAGE blockade may also be a potential therapeutic target in euglycemic atherosclerosis [173].

The inhibitor of AGE formation, aminoguanidine, and ALT-711, a compound suggested to be capable of cleaving AGE cross-links in proteins, have been shown to inhibit the progression of atherosclerosis in STZ/apoE$^{-/-}$ mice [174]. Both therapeutic interventions were shown to decrease total aortic collagen accumulation, profibrotic cytokine expression, and α-smooth muscle actin. The apparent efficacy of these interventions provides evidence for the involvement of AGE in the progression of diabetic atherosclerosis. The precise mechanisms by which AGE interventions inhibit the progression of diabetic atherosclerosis and the relative importance of RAGE to this process are still unknown.

Inhibition of angiotensin-converting enzyme (ACE) or blockade of type 1 angiotensin (AT_1) receptors attenuates the progression of atherosclerosis in STZ/apoE$^{-/-}$ mice [175, 176]. Induction of diabetes in the apoE$^{-/-}$ mice is associated with a significant increase in aortic ACE gene/protein expression and ACE binding density, linking diabetes to the activation of the vascular renin–angiotensin system [176]. Although ACE inhibition causes a small reduction in blood pressure, its antiatherogenic effects were shown to be independent of blood pressure as calcium-channel blockers produced a similar reduction in blood pressure but did not inhibit the progression of plaque formation [175]. In support of this, ACE inhibition and AT_1-receptor blockade but not calcium-channel blockade are associated with decreased formation of AGE *in vitro* [177]. Unlike inhibitors of AGE formation such as aminoguanidine, ACE inhibitors and AT_1 blockers have not been shown to scavenge reactive aldehydes but rather inhibit their formation by chelating transition metals and inhibiting oxidative stress induced by radical formation [177, 178].

The thiazolidinedione class of drugs, such as rosiglitazone, are PPAR-γ agonists that act to increase insulin sensitivity and are commonly used to improve glycemic control in people with Type 2 diabetes [179]. Rosiglitazone has also been shown to attenuate the progression of atherosclerosis in both apoE$^{-/-}$ and STZ/apoE$^{-/-}$ mice [180]. In STZ/apoE$^{-/-}$ mice, rosiglitazone has no effect on glycemia and reduces the correlation coefficient between plasma glucose levels and atherosclerosis. Although the precise mechanisms by which rosiglitazone exerts its antiatherogenic effects are unknown, PPAR-γ agonists have been shown to stimulate cholesterol removal via increased receptor (ABCA1)-mediated cholesterol efflux from macrophage foam cells [181, 182].

Administration of PPAR-α and PPAR-γ agonists has also been shown to inhibit endothelial cell adhesion molecule expression [183, 184]. Stimulation of reverse cholesterol transport and inhibition of endothelial dysfunction may contribute to the antiatherogenic effects of PPAR-modulating drugs.

Inhibitors of AGE Formation and Possible Therapeutic Interventions

The formation of reactive intermediate products, such as the α-oxoaldehydes and the α-hydroxyaldehydes, are important steps in the formation of AGE. Under normal conditions, α-oxoaldehydes are metabolized by the glutathione-dependent glyoxalase system or other detoxifying enzymes and very little proceeds to form AGE [185]. Increased formation of reactive carbonyl species and/or decreased metabolism can lead to increased concentrations of these aldehydes, referred to as "carbonyl stress," which in turn can lead to increased AGE formation.

AGE are a heterogeneous group of products that can be formed by a variety of pathways. Inhibition of AGE formation may be of therapeutic benefit to various AGE-related diseases such as diabetes. As both glycation and oxidation reactions contribute to AGE formation, inhibition of either process may decrease AGE formation. Chelation of metal ions has been shown to inhibit reactive oxygen species production, which in turn can inhibit the formation of reactive carbonyls and AGE [186, 187]. Studies conducted by Brownlee and coworkers [12, 188, 189] have shown that inhibition of mitochondrial-derived reactive oxygen species using uncoupling protein-1 or manganese superoxide dismutase inhibits glucose-induced activation of PKC, AGE formation, sorbitol accumulation, and NF-κB activation, identifying the critical importance of mitochondrial-derived oxidants in AGE formation. Recently benfotiamine, a lipid-soluble thiamine derivative that stimulates the pentose phosphate pathway enzyme transketolase, was shown to inhibit the hexosamine-, AGE- and diacylglycerol-PKC pathways by enhancing metabolism of glyceraldehyde-3-phosphate and fructose-6-phosphate [190]. Furthermore, benfotiamine was shown to inhibit experimental diabetic retinopathy and thus may be a potential therapy for vascular complications [190].

One potential mechanism to inhibit AGE formation involves the use of compounds that react rapidly with carbonyls, thereby preventing modification of endogenous targets by these compounds [191]. In order to effectively inhibit AGE formation *in vivo*, AGE inhibitors are required to stoichiometrically bind low molecular mass, soluble, reactive intermediates of AGE formation with greater affinity than natural targets of reactive aldehydes, such as lysine residues, which are present at much higher concentrations in plasma than achievable therapeutic concentrations of inhibitors [192]. Carbonyl-scavenging compounds provide a means to inhibit AGE formation by detoxifying the broad range of carbonyl-containing aldehydes that are produced by enzyme-independent pathways in the context of hyperglycemia and/or increased oxidative stress. Inhibiting the formation of reactive carbonyl species or the progression of glycated proteins to AGE are two strategies to reduce AGE formation. Furthermore, carbonyl-scavenging inhibitors of AGE formation need to bind carbonyl groups without interfering with the function of coenzymes or precursors that contain reactive aldehyde groups such as pyridoxal phosphate and retinal [193].

Aminoguanidine is a prototype therapeutic agent that has been shown to be a potent inhibitor of AGE formation *in vitro* and *in vivo* (reviewed in [193]). The two key reaction centers of aminoguanidine include the nucleophilic hydrazine group and the guanidino group, which react rapidly with α,β-dicarbonyl compounds such as glyoxal, methylglyoxal, and 3-deoxyglucosone, preventing them from binding protein Arg and Lys residues, thus retarding the formation of AGE [193]. We have shown that aminoguanidine can inhibit glycolaldehyde-induced modification of LDL and subsequent macrophage cholesterol ester accumulation (Rashid et al., unpublished data). Interestingly, aminoguanidine has no effects on hyperglycemia and seems to preferentially bind short-chain reactive aldehydes than glucose [191]. Aminoguanidine was initially shown to inhibit cross-linking and fluorescence in arterial collagen of rats with diabetes [194], and subsequent studies have shown that this compound retards the progression of nephropathy, neuropathy, vasculopathy, and atherosclerosis in animal models of diabetes (reviewed in [174, 193]). Two double blinded, placebo-controlled, and randomized clinical trials, ACTION I (in people with Type 1 diabetes) and ACTION II (in people with Type 2 diabetes), have been conducted to elucidate the therapeutic potential of aminoguanidine in human diabetic nephropathy [193, 195]. The primary end point for these two trials was the doubling of patient serum creatinine. In ACTION I, aminoguanidine showed a tendency to slow the doubling of serum creatinine but this did not reach significance. Aminoguanidine treatment did, however, decrease triglycerides, LDL cholesterol, and urinary protein. The ACTION II trial was prematurely halted due to a lack of efficacy and safety issues arising from side effects including pernicious-like anemia and the development of antinuclear antibodies [196]. In addition, aminoguanidine has been shown to inhibit nitric oxide synthases and semicarbazide-sensitive amine oxidase, and react with pyridoxal phosphate, thereby potentially interfering with vitamin B_6 metabolism [192]. These side effects, and its rapid plasma clearance, preclude aminoguanidine from being a suitable therapy for diabetic vascular disease. However, the inhibition of AGE accumulation, RAGE expression, and development of vascular complications in animal models provides supportive evidence for the involvement of reactive carbonyls in the development of diabetic vascular complications, and highlights the potential efficacy of therapies that scavenge such carbonyls. Thornalley [193] has suggested that hydrophobic derivatives of aminoguanidine would have a longer plasma half-life, and that a possible prodrug of aminoguanidine that becomes activated at vascular sites would increase its efficacy [193].

Pyridoxamine is an AGE inhibitor that is a structural analog of vitamin B_6 [197]. Vitamin B_6 is involved in the metabolism of carbonyl-containing compounds *in vivo* and pyridoxamine was developed in the search for novel inhibitors of AGE formation with fewer side effects than aminoguanidine [192]. Interestingly, pyridoxamine not only inhibits AGE formation by scavenging reactive carbonyls, but also modulates the progression of glycated proteins to AGE [192, 197]. It has therefore been described as a "post-Amadori inhibitor" of AGE formation. Administration of pyridoxamine to STZ-treated mice retards early nephropathy, inhibits dyslipidemia, and decreases the accumulation of CML and CEL [198]. Clinical trials testing the efficacy of pyridoxamine (under the trade name "Pyridorin") for diabetic nephropathy are currently being conducted. A number of other carbonyl-scavenging inhibitors of AGE formation are under development, which might circumvent the side effects observed with aminoguanidine (reviewed in [191]).

Recently compounds have been developed that have been reported to cleave AGE cross-links from proteins and thus potentially reverse AGE modifications. The first identified cross-link breaker, *N*-phenacylthiazolium bromide (PTB), was identified by Vasan et al. [199] as a breaker of glucose-derived protein cross-links. ALT-711 (3-phenacyl-4,5-dimethylthiazolium chloride) is a more stable cross-link breaker than PTB and has been shown to improve carotid artery cross-sectional compliance, decrease left ventricular mass, and increase collagen solubility in STZ-induced diabetic rats [200, 201]. It has also been shown to improve arterial and ventricular function in aging monkeys and enhanced cardiac diastolic compliance in aging dogs [202, 203]. Human trials have been conducted, with ALT-711 treatment resulting in improved total arterial compliance in aged humans with vascular stiffening and the drug was well tolerated [204]. Taken together, these *in vivo* experiments suggest that ALT-711 and AGE-breaking compounds may be of therapeutic value in diseases where AGE accumulates. However, the precise mechanism of action of ALT-711 in these animal and human studies remains uncertain. PTB has been shown to be unable to increase the solubility of cross-linked collagen derived from diabetic rats despite being able to cleave model AGE cross-links such as phenylpropanedione [205]. It is possible that, like inhibitors of AGE formation, cross-link breakers may scavenge reactive carbonyl species, which may contribute to observed effects *in vivo*.

Conclusions

Glycation and glycoxidation reactions appear to play a critical role in the development and progression of diabetes-associated atherosclerosis. A summary of these processes and potential protective strategies is given in Fig. 12.1. Such therapies, which have been designed to prevent or reverse

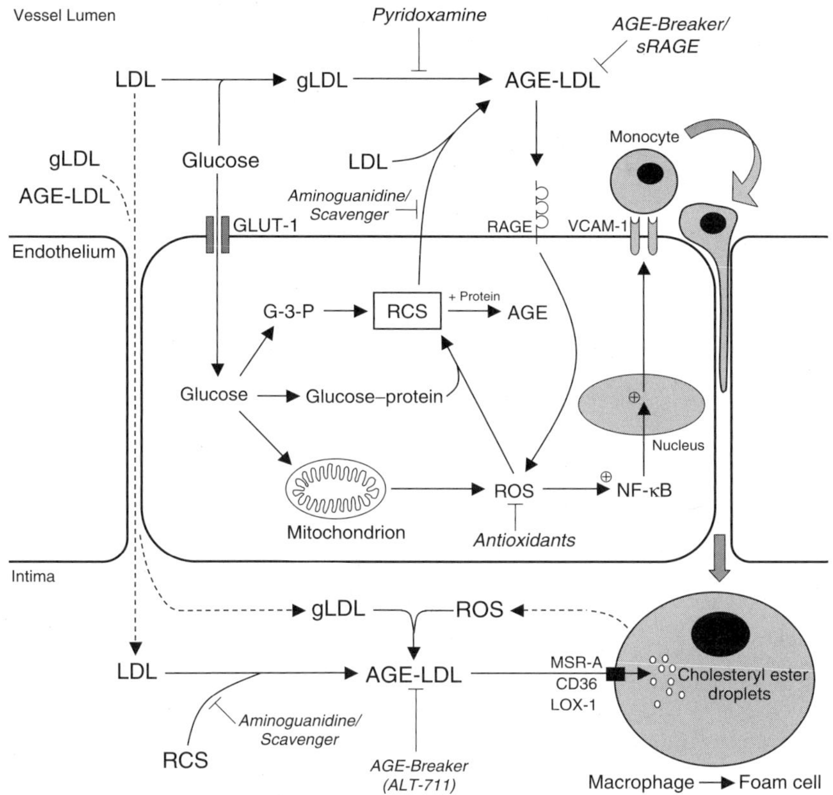

FIGURE 12.1. AGE formation and modification of LDL effects on endothelial cells and possible contribution to macrophage foam cell formation. *Abbreviations*: LDL, low-density lipoprotein; gLDL, glucose-bound LDL; G-3-P, glyceraldehyde-3-phosphate; RCS, reactive carbonyl species; ROS, reactive oxygen species; AGE, advanced glycation end product; NF-κB, nuclear factor-κB; RAGE, receptor for advanced glycation end products; VCAM-1, vascular cell adhesion molecule-1; sRAGE, soluble extracellular domain of RAGE; MSR-A, macrophage scavenger receptor-A; LOX-1, lectin-like oxidized low-density lipoprotein receptor-1. Glycation and AGE modification of LDL may occur within the bloodstream or in the subendothelial spaces. Hyperglycemia leads to glycation of serum proteins (including LDL) and an increase in endothelial glucose concentration via GLUT-1 transporters. Other cell types may also be subjected to increased intracellular glucose concentrations. Increased intracellular glucose leads to the formation of reactive carbonyl species such as methylglyoxal, glyoxal, and 3-deoxyglucosone, which may be released into the bloodstream or the subendothelial space and react with LDL to form AGE-LDL. RAGE ligation by AGE-modified proteins and intracellular hyperglycemia can lead to increased production of ROS, which induces NF-κB activation and increased expression of VCAM-1. Increased VCAM-1 promotes monocyte migration into the intima where these cells differentiate into macrophages. Uptake of AGE-LDL by macrophages via scavenger receptors may lead to the accumulation of cholesteryl ester droplets and the formation of foam cells. Scavengers of RCS, AGE-breakers, and antioxidants may inhibit the progression of hyperglycemia-accelerated atherosclerosis by decreasing AGE modification of proteins and subsequent endothelial dysfunction/foam cell formation.

these reactions, could play an adjunct role to existing approaches aimed at optimizing glycemic control and treating lipid abnormalities [6].

Acknowledgment: The authors gratefully acknowledge financial support from the Diabetes Australia Research Trust, the National Health and Medical Research Council, and the Australian Research Council.

References

1. WHO: The Cost of Diabetes. WHO, Geneva, 2003.
2. WHO: Cardiovascular Disease: Prevention and Control. WHO, Geneva, 2003.
3. Australian Institution of Health and Welfare: Australia's Health 2002. AIHW, Canberra, 2002.
4. Zimmet P, Alberti KGMM, Shaw J: Global and societal implications of the diabetes epidemic. Nature 414: 782–787, 2001.
5. Dunstan DW, Zimmet PZ, Welborn TA, de Courten MP, Cameron AJ, Sicree RA, Dwyer T, Colagiuri S, Jolley D, Knuiman M, Atkins R, Shaw JE: The rising prevalence of diabetes and impaired glucose tolerance. Diabetes Care 25: 829–834, 2002.
6. Sowers JR: Effects of statins on the vasculature: implications for aggressive lipid management in the cardiovascular metabolic syndrome. Am J Cardiol 91: 14–22, 2003.
7. Keen H, Clark C, Laakso M: Reducing the burden of diabetes: managing cardiovascular disease. Diabetes Metab Res Rev 15: 186–196, 1999.
8. Jenkins AJ, Best JD, Klein RL, Lyons TJ: Lipoproteins, glycoxidation and diabetic angiopathy. Diabetes Metab Res Rev 20: 349–368, 2004.
9. Sowers JR, Lester MA: Diabetes and cardiovascular disease. Diabetes Care 22: 14C–20C, 1999.
10. Eaton JW, Dean RT: Diabetes and atherosclerosis. In: Dean RT, Kelly DT (eds), Atherosclerosis—Gene Expression, Cell Interactions, and Oxidation. Oxford University Press, Oxford, 2000, pp 24–45.
11. Creager MA, Luscher TF, Cosentino F, Beckman JA: Diabetes and vascular disease. Pathophysiology, clinical consequences and medical therapy: Part I. Circulation 108: 1527–1532, 2003.
12. Brownlee M: Biochemistry and molecular cell biology of diabetic complications. Nature 414: 813–820, 2001.
13. The Diabetes Control and Complications Trial Research Group: The effect of intensive treatment of diabetes on the development and progression of long-term complications in insulin-dependant diabetes mellitus. N Engl J Med 329: 977–986, 1993.
14. Nathan DM, Lachin J, Cleary P, Orchard T, Brillon DJ, Backlund JY, O'Leary DH, Genuth S (DCCT Research Group): Intensive diabetes therapy and carotid intima-media thickness in type 1 diabetes mellitus. N Engl J Med 348: 2294–2303, 2003.
15. Marshall SM, Barth JH: Standardization of HbA_{1c} measurements—consensus statement. Diabetes Med 17: 5–6, 2000.

16. UK Prospective Studies Group: Intensive blood-glucose control with sulphonylureas or insulin compared with conventional treatment and risk of complications in patients with type 2 diabetes (UKPDS 33). Lancet 352: 837–883, 1998.

17. Lo TW, Westwood ME, McLellan AC, Selwood T, Thornalley PJ: Binding and modification of proteins by methylglyoxal under physiological conditions. A kinetic and mechanistic study with N-α-acetylarginine, N-α-acetylcysteine, and N-α-acetyllysine, and bovine serum albumin. J Biol Chem 269: 32299–32305, 1994.

18. Zeng J, Davies MJ: Evidence for the formation of adducts and S-(carboxymethyl)cysteine on reaction of alpha-dicarbonyl compounds with thiol groups on amino acids, peptides and proteins. Chem Res Toxicol 18: 1232–1242, 2005.

19. Ahmed N: Advanced glycation endproducts—role in pathology of diabetic complications. Diabetes Res Clin Pract 67: 3–12, 2005.

20. Knott HM, Brown BE, Davies MJ, Dean RT: Glycation and glycoxidation of low density lipoproteins by glucose and low molecular weight aldehydes: formation of modified and oxidised proteins. Eur J Biochem 270: 3572–3582, 2003.

21. Wolff SP, Dean RT: Glucose autoxidation and protein modification—the potential role of 'autoxidative glycosylation' in diabetes. Biochem J 245: 243–250, 1987.

22. Wolff SP, Dean RT: Monosaccharide autoxidation: a potential source of oxidative stress in diabetes? Formation and reactions of peroxides in biological systems. Bioelectrochem Bioenerg 18: 283–293, 1987.

23. Hunt JV, Dean RT, Wolff SP: Hydroxyl radical production and autoxidative glycosylation. Glucose autoxidation as the cause of protein damage in the experimental glycation model of diabetes mellitus and ageing. Biochem J 256: 205–212, 1988.

24. Hunt JV, Bottoms MA, Mitchinson MJ: Oxidative alterations in the experimental glycation model of diabetes mellitus are due to protein–glucose adduct oxidation. Biochem J 291: 529–535, 1993.

25. Fu M-X, Wells-Knecht KJ, Blackledge JA, Lyons TJ, Thorpe SR, Baynes JW: Glycation, glycoxidation, and cross-linking of late stages of the Maillard reaction. Diabetes 43: 676–683, 1994.

26. Nagai R, Matsumoto K, Ling X, Suzuki H, Araki T, Horiuchi S: Glycolaldehyde, a reactive intermediate for advanced glycation end products, plays an important role in the generation of an active ligand for the macrophage scavenger receptor. Diabetes 49: 1714–1723, 2000.

27. Khaw K-T, Wareham N, Bingham S, Luben R, Welch A, Day N: Association of hemoglobin A_{1c} with cardiovascular disease and mortality in adults: the European prospective investigation into cancer in Norfolk. Ann Intern Med 141: 413–420, 2004.

28. Lowe LP, Liu K, Greenland P, Metzger BE, Dyer AR, Stamler J: Diabetes, asymptomatic hyperglycaemia, and 22-year mortality in black and white men: the Chicago Heart Association Detection Project in Industry Study. Diabetes Care 20: 163–169, 1997.

29. Moss SE, Klein R, Klein BEK, Meuer SM: The association of glycemia and cause-specific mortality in a diabetic population. Arch Intern Med 154: 2473–2479, 1994.

30. Kuusisto J, Mykkanen L, Pyorala K, Laakso M: NIDDM and its metabolic control predict coronary heart disease in elderly subjects. Diabetes 43: 960–967, 1994.

31. Klein R: Hyperglycaemia and microvascular and macrovascular disease in diabetes. Diabetes Care 18: 258–268, 1995.

32. Standl E, Balletshofer B, Dahl B, Weichenhain B, Stiegler H, Hormann A, Holle R: Predictors of 10-year macrovascular and overall mortality in patients with NIDDM: the Munich General Practioner Project. Diabetologia 39: 1540–1545, 1996.
33. Khaw K-T, Wareham N, Luben R, Bingham S, Oakes S, Welch A, Day N: Glycated haemogloblin, diabetes, and mortality in men in the Norfolk cohort of European Prospective Investigation of Cancer and Nutrition (EPIC-Norfolk). Brit Med J 322: 1–6, 2001.
34. Beks PHJ, Mackaay AJC, de Vries H, de Neeling JND, Bouter LM: Carotid artery stenosis is related to blood glucose level in an elderly Caucasian population: the Hoorn study. Diabetologia 40: 290–298, 1997.
35. Jensen-Urstad KJ, Reichard PG, PRofors JS, Lindblad LE, Jensen-Urstad MT: Early atherosclerosis is retarded by improved long-term blood glucose control inpatients with IDDM. Diabetes 45: 1253–1258, 1996.
36. McLellan AC, Phillips SA, Thornalley PJ: The assay of methylglyoxal in biological systems by derivatization with 1,2-diamino-4,5-dimethoxybenzene. Anal Biochem 206: 17–23, 1992.
37. McLellan AC, Thornalley PJ, Benn J, Sonksen PH: Glyoxalase system in clinical diabetes mellitus and correlation with diabetic complications. Clin Sci 87: 21–29, 1994.
38. Thornalley PJ, Hooper NI, Jennings PE, Florkowski CM, Jones AF, Lunec J, Barnett AH: The human red blood cell glyoxalase system in diabetes mellitus. Diabetes Res Clin Pract 7: 115–120, 1989.
39. Niwa T: 3-Deoxyglucosone: metabolism, analysis, biological activity, and clinical implication. J Chromatogr B 731: 23–36, 1999.
40. Thornalley PJ, Battah S, Ahmed N, Karachalias N, Agalou S, Babaei-Jadidi R: Quantitative screening of advanced glycation endproducts in cellular and extracellular proteins by tandem mass spectrometry. Biochem J 375: 581–592, 2003.
41. Dean RT, Fu S, Stocker R, Davies MJ: Biochemistry and pathology of radical-mediated protein oxidation. Biochem J 324: 1–18, 1997.
42. Grant AJ, Jessup W, Dean RT: Accelerated endocytosis and incomplete catabolism of radical-damaged protein. Biochim Biophys Acta 1134: 203–209, 1992.
43. Linton S, Davies MJ, Dean RT: Protein oxidation and ageing. Exp Gerontol 36: 1503–1518, 2001.
44. Dunlop RA, Rodgers KJ, Dean RT: Recent developments in the intracellular degradation of oxidized proteins. Free Radic Biol Med 33: 894–906, 2002.
45. Degenhardt TP, Thorpe SR, Baynes JW: Chemical modification of proteins by methylglyoxal. Cell Mol Biol 44: 1139–1145, 1998.
46. Schleicher ED, Wagner E, Nerlich AG: Increased accumulation of the glycoxidation product N(epsilon)-(carboxymethyl)lysine in human tissues in diabetes and aging. J Clin Invest 99: 457–468, 1997.
47. Dunn JA, McCance DR, Thorpe SR, Lyons TJ, Baynes JW: AGE-dependent accumulation of N-epsilon-(carboxymethyl)lysine and N-epsilon-(carboxymethyl) hydroxylysine in human skin collagen. Biochemistry 30: 1205–1210, 1991.
48. Wells-Knecht MC, Lyons TJ, McCance DR, Thorpe SR, Baynes JW: AGE-dependent increase in ortho-tyrosine and methionine sulfoxide in human skin collagen is not accelerated in diabetes. Evidence against a generalized increase in oxidative stress in diabetes. J Clin Invest 100: 839–846, 1997.

49. Dyer DG, Dunn JA, Thorpe SR, Bailie KE, Lyons TJ, McCance DR, Baynes JW: Accumulation of Maillard reaction products in skin collagen in diabetes and aging. J Clin Invest 91: 2463–2469, 1993.

50. Meng J, Sakata N, Takebayashi S, Asano T, Futata T, Nagai R, Ikeda K, Horiuchi S, Myint T, Taniguchi N: Glycoxidation on aortic collagen from STZ-induced diabetic rats and its relevance to vascular damage. Atherosclerosis 136: 355–365, 1998.

51. Pokupec R, Kalauz M, Turk N, Turk Z: Advanced glycation endproducts in human diabetic and non-diabetic cataractous lenses. Graefes Archiv Clin Exp Ophthalmol 241: 378–384, 2003.

52. Stocker R, Keaney JR: Role of oxidative modifications in atherosclerosis. Physiol Rev: 1381–1478, 2004.

53. Baynes JW, Thorpe SR: Role of oxidative stress in diabetic complications—a new perspective on an old paradigm. Diabetes 48: 1–9, 1999.

54. Vogt BW, Schleicher ED, Wieland OH: Epsilon-Amino-lysine-bound glucose in human tissues obtained at autopsy. Increase in diabetes mellitus. Diabetes 31: 1123–1127, 1982.

55. Vishwanath V, Frank KE, Elmets CA, Dauchot PJ, Monnier VM: Glycation of skin collagen in type I diabetes mellitus. Correlation with long-term complications. Diabetes 35: 916–921, 1986.

56. Imanaga Y, Sakata N, Takebayashi S, Matsunaga A, Sasaki J, Arakawa K, Nagai R, Horiuchi S, Itabe H, Takano T: *In vivo* and *in vitro* evidence for the glycoxidation of low density lipoprotein in human atherosclerotic plaques. Atherosclerosis 150: 343–355, 2000.

57. Miyata S, Liu BF, Shoda H, Ohara T, Yamada H, Suzuki K, Kasuga M: Accumulation of pyrraline-modified albumin in phagocytes due to reduced degradation by lysosomal enzymes. J Biol Chem 272: 4037–4042, 1997.

58. Portero-Otin M, Pamplona R, Bellmunt MJ, Bergua M, Prat J: Glycaemic control and *in vivo* non-oxidative Maillard reaction: urinary excretion of pyrraline in diabetes patients. Eur J Clin Invest: 767–773, 1997.

59. Kume S, Takeya M, Mori T, Araki N, Suzuki H, Horiuchi S, Kodama T, Miyauchi Y, Takahashi K: Immunohistochemical and ultrastructural detection of advanced glycation end products in atherosclerotic lesions of human aorta with a novel specific monoclonal antibody. Am J Pathol 147: 654–667, 1995.

60. Nagai R, Hayashi CM, Xia L, Takeya M, Horiuchi S: Identification in human atherosclerotic lesions of GA-pyridine, a novel structure derived from glyco-laldehyde-modified proteins. J Biol Chem 277: 48905–48912, 2002.

61. Shamsi FA, Partal A, Sady C, Glomb MA, Nagaraj RH: Immunological evidence for methylglyoxal-derived modifications *in vivo*. Determination of antigenic epitopes. J Biol Chem 273: 6928–6936, 1998.

62. Morgan PE, Dean RT, Davies MJ: Inactivation of cellular enzymes by carbonyls and protein-bound glycation/glycoxidation products. Arch Biochem Biophys 403: 259–269, 2002.

63. Li AC, Glass CK: The macrophage foam cell as a target for therapeutic intervention. Nat Med 8: 1235–1242, 2002.

64. Takahashi K, Takeya M, Sakashita N: Multifunctional role of macrophages in the development and progression of atherosclerosis in humans and experimental animals. Med Electron Microsc 35: 179–203, 2002.

65. Smith JD, Trogan E, Ginsberg M, Grigaux C, Tian J, Miyata M: Decreased atherosclerosis in mice deficient in both macrophage colony-stimulating factor (*op*) and apolipoprotein E. Proc Natl Acad Sci USA 92: 8264–8268, 1995.

66. Boring L, Gosling J, Cleary M, Charo IF: Decreased lesion formation in CCR2–/– mice reveals a role for chemokines in the initiation of atherosclerosis. Nature 394: 894–897, 1998.

67. Gosling J, Slaymaker S, Gu L, Tseng S, Zlot CH, Young SG, Rollins BJ, Charo IF: MCP-1 deficiency reduces susceptibility to atherosclerosis in mice that overexpress human apolipoprotein B. J Clin Invest 103: 773–778, 1999.

68. Boisvert WA, Santiago R, Curtiss LK, Terkeltaub RA: A leukocyte homologue of the IL-8 receptor CXCR-2 mediates the accumulation of macrophages in atherosclerotic Lesions of LDL receptor-deficient mice. J Clin Invest 101: 353–363, 1998.

69. Tabas I: Nonoxidative modifications of lipoproteins in atherogenesis. Annu Rev Nutr 19: 123–139, 1999.

70. Williams KJ, Tabas I: The response-to-retention hypothesis of atherogenesis reinforced. Curr Opin Lipidol 9: 471–474, 1998.

71. Tabas I: Consequences of cellular cholesterol accumulation: basic concepts and physiological implications. J Clin Invest 110: 905–911, 2002.

72. Vaino S, Ikonen E: Macrophage cholesterol transport: a critical player in foam cell formation. Ann Med 35: 146–155, 2003.

73. Shiomi M, Yamada S, Ito T: Atheroma stabilizing effects of simvastatin due to depression of macrophages or lipid accumulation in the atheromatous plaques of coronary plaque-prone WHHL rabbits. Atherosclerosis 178: 287–294, 2005.

74. Kawamura M, Heinecke JW, Chait A: Increased uptake of alpha-hydroxy aldehyde-modified low density lipoprotein by macrophage scavenger receptors. J Lipid Res 41: 1054–1059, 2000.

75. Schalkwijk CG, Vermeer MA, Stehouwer CD, te Koppele J, Princen HM, van Hinsbergh VW: Effect of methylglyoxal on the physico-chemical and biological properties of low-density lipoprotein. Biochim Biophys Acta 1394: 187–198, 1998.

76. Hunt JV, Smith CC, Wolff SP: Autoxidative glycosylation and possible involvement of peroxides and free radicals in LDL modification by glucose. Diabetes 39: 1420–1424, 1990.

77. Mowri H-O, Frei B, Keaney JF: Glucose enhancement of LDL oxidation is strictly metal ion dependent. Free Radic Biol Med 29: 814–824, 2000.

78. Millican SA, Bagga M, Eddy R, Mitchinson MJ, Hunt JV: Effect of glucose-mediated LDL oxidation on the P388D1 macrophage-like cell line. Atherosclerosis 129: 17–25, 1997.

79. Kawamura M, Heinecke JW, Chait A: Pathophysiological concentrations of glucose promote oxidative modification of low density lipoprotein by a superoxide-dependant pathway. J Clin Invest 94: 771–778, 1994.

80. Kimura H, Minakami H, Kimura S, Sakurai T, Nakamura T, Kurashige S, Nakano M, Shoji A: Release of superoxide radicals by mouse macrophages stimulated by oxidative modification of glycated low density lipoproteins. Atherosclerosis 118: 1–8, 1995.

81. Brown BE, Dean RT, Davies MJ: Glycation of low-density lipoproteins by methylglyoxal and glycolaldehyde gives rise to the *in vitro* formation of lipid-laden foam cells. Diabetologia 48: 361–369, 2005.

82. Sakata N, Miyamoto K, Meng J, Tachikawa Y, Imanaga Y, Takebayashi S, Furukawa T: Oxidative damage of vascular smooth muscle cells by the glycated protein-cupric ion system. Atherosclerosis 136: 263–274, 1998.

83. Stadler N, Lindner RA, Davies MJ: Direct detection and quantification of transition metal ions in human atherosclerotic plaques: evidence for the presence of elevated levels of iron and copper. Arterioscler Thromb Vasc Biol 24: 949–954, 2004.
84. Schleicher E, Deufel T, Wieland OH: Non-enzymatic glycosylation of human serum lipoproteins. Elevated epsilon-lysine glycosylated low density lipoprotein in diabetic patients. FEBS Lett 129: 1–4, 1981.
85. Lyons TJ, Klein RL, Baynes JW, Stevenson HC, Lopes-Virella MF: Stimulation of cholesteryl ester synthesis in human monocyte-derived macrophages by low-density lipoproteins from type 1 (insulin-dependent) diabetic patients: the influence of non-enzymatic glycosylation of low-density lipoproteins. Diabetologia 30: 916–923, 1987.
86. Klein RL, Laimeins M, Lopes-Virella MF: Isolation, characterisation, and metabolism of the glycated and nonglycated subfractions of low-density lipoproteins isolated from Type 1 diabetic patients and nondiabetic patients. Diabetes 44: 1093–1098, 1995.
87. Sobenin IA, Tertov VV, Orekhov AN: Atherogenic modified LDL in diabetes. Diabetes 45(Suppl 3): S35–S39, 1996.
88. Bucala R, Makita Z, Koschinsky T, Cerami A, Vlassara H: Lipid advanced glycosylation: pathway for lipid oxidation in vivo. Proc Natl Acad Sci USA 90: 6434–6438, 1993.
89. Bucala R, Makita Z, Vega G, Grundy S, Koschinsky T, Cerami A, Vlassara H: Modification of low density lipoprotein by advanced glycation end products contributes to the dyslipidemia of diabetes and renal insufficiency. Proc Natl Acad Sci USA 91: 9441–9445, 1994.
90. Steinbrecher UP, Witztum JL: Glucosylation of low-density lipoproteins to an extent comparable to that seen in diabetes slows their catabolism. Diabetes 33: 130–134, 1984.
91. Stitt AW, He C, Friedman S, Scher L, Rossi P, Ong L, Founds H, Li YM, Bucala R, Vlassara H: Elevated AGE-modified apoB in sera of euglycemic, normolipidemic patients with atherosclerosis: relationship to tissue AGEs. Mol Med 3: 617–627, 1997.
92. Bucala R, Mitchell R, Arnolds K, Innerarity T, Vlassara H, Cerami A: Identification of the major site of apolipoprotein B modification by advanced glycosylation end products blocking uptake by the low density lipoprotein receptor. J Biol Chem 270: 10828–10832, 1995.
93. Witztum JL, Mahoney EM, Branks MJ, Fisher M, Elam R, Steinberg D: Nonenzymatic glucosylation of low-density lipoprotein alters its biologic activity. Diabetes 4: 283–291, 1982.
94. Edwards IJ, Wagner JD, Litwak KN, Rudel LL, Cefalu WT: Glycation of plasma low density lipoproteins increases interaction with arterial proteoglycans. Diabetes Res Clin Pract 46: 9–18, 1999.
95. Curtiss LK, Witztum JL: A novel method of generating region-specific monoclonal antibodies to modified proteins: application to the identification of human glucosylated low density lipoproteins. J Clin Invest 72: 1427–1438, 1983.
96. Virella G, Thorpe SR, Alderson NL, Stephan EM, Atchley D, Wagner F, Lopes-Virella MF (DCCT Group): Autoimmune response to advanced glycosylation end-products of human LDL. J Lipid Res 44: 487–493, 2003.
97. Lopes-Virella MF, Virella G: Cytokines, modified lipoproteins, and arteriosclerosis in diabetes. Diabetes 45(Suppl 3): S40–S44, 1996.

98. Lopes-Virella MF, Griffith RL, Shunk KA, Virella GT: Enhanced uptake and impaired intracellular metabolism of low density lipoprotein complexed with anti-low density lipoprotein antibodies. Arterioscler Thromb 11: 1356–1367, 1991.

99. Edwards IJ, Terry JG, Bell-Farrow AD, Cefalu WT: Improved glucose control decreases the interaction of plasma low-density lipoproteins with arterial proteoglycans. Metab Clin Exp 51: 1223–1229, 2002.

100. Proctor SD, Pabla CK, Mamo JCL: Arterial intimal retention of pro-atherogenic lipoproteins in insulin deficient rabbits and rats. Atherosclerosis 1449: 315–322, 2000.

101. Williams KJ: The response-to-retention hypothesis of early atherogenesis. Arterioscler Thromb Vasc Biol 15: 551–561, 1995.

102. Hoff HF, Morton RE: Lipoproteins containing apoB extracted from human aorta. Structure and function. Ann NY Acad Sci 454: 183–194, 1985.

103. Rashduni DL, Rifici VA, Schneider SH, Khachadurian AK: Glycation of high-density lipoprotein does not increase its susceptibility to oxidation or diminish its cholesterol efflux capacity. Metab Clin Exp 48: 139–143, 1999.

104. Passarelli M, Shimabukuro AFM, Catanozi S, Nakandakare ER, Rocha JC, Carrilho AJF, Quintao ECR: Diminished rate of mouse peritoneal macrophage cholesterol efflux is not related to the degree of HDL glycation in diabetes mellitus. Clin Chim Acta 301: 119–134, 2000.

105. Duell PB, Oram JF, Bierman EL: Nonenzymatic glycosylation of HDL and impaired HDL-receptor-mediated cholesterol efflux. Diabetes 40: 377–384, 1991.

106. Fievet C, Theret N, Shojaee N, Duchateau P, Castro G, Ailhaud G, Drouin P, Fruchart J-C: Apolipoprotein A-I-containing particles and reverse cholesterol transport in IDDM. Diabetes 41: 81–85, 1992.

107. Quintao ECR, Medina WL, Passarelli M: Reverse cholesterol transport in diabetes mellitus. Diabetes Metab Res Rev 16: 237–250, 2000.

108. Borggreve SE, de Vries R, Dullaart RPF: Alterations in high-density lipoprotein metabolism and reverse cholesterol transport in insulin resistance and type 2 diabetes mellitus: role of lipolytic enzymes, lecithin:cholesterol acyltransferase and lipid transfer proteins. Eur J Clin Invest 33: 1051–1069, 2003.

109. Nicholls SJ, Dusting GJ, Cutri B, Bao S, Drummond GR, Rye K-A, Barter PJ: Reconstituted high-density lipoproteins inhibit the acute pro-oxidant and proinflammatory vascular changes induced by a periarterial collar in normocholesterolemic rabbits. Circulation 111: 1543–1550, 2005.

110. Makita T, Tanaka A, Nakano T, Nakajima K, Numano F: Importance of glycation in the acceleration of low density lipoprotein (LDL) uptake into macrophages in patients with diabetes mellitus. Int Angiol 18: 149–153, 1999.

111. Ravandi A, Kuksis A, Shaikh NA: Glucosylated glycerophosphoethanolamines are the major LDL glycation products and increase LDL susceptibility to oxidation. Arterioscler Thromb Vasc Biol 20: 467–477, 2000.

112. Ravandi A, Kuksis A, Shaikh NA: Glycated phosphatidylethanolamine promotes macrophage uptake of low density lipoprotein and accumulation of cholesteryl esters and triacylglycerols. J Biol Chem 274: 16494–16500, 1999.

113. Jinnouchi Y, Sano H, Nagai R, Hakamata H, Kodama T, Suzuki H, Yoshida M, Ueda S, Horiuchi S: Glycolaldehyde-modified low density lipoprotein leads macrophages to foam cells via the macrophage scavenger receptor. J Biochem (Tokyo) 123: 1208–1217, 1998.

114. Haberland ME, Olch CL, Folgelman AM: Role of lysines in mediating interaction of modified low density lipoproteins with the scavenger receptor of human monocyte macrophages. J Biol Chem 259: 11305–11311, 1984.
115. Nishi K, Itabe H, Uno M, Kitazato KT, Horiguchi H, Shinno K, Nagahiro S: Oxidized LDL in carotid plaques and plasma associates with plaque instability. Arterioscler Thromb Vasc Biol 22: 1649–1654, 2002.
116. Mann GE, Yudilevich DL, Sobrevia L: Regulation of amino acid and glucose transporters in endothelial and smooth muscle cells. Physiol Rev: 183–252, 2003.
117. Giardino I, Edelstein D, Brownlee M: Nonenzymatic glycosylation *in vitro* and in bovine endothelial cells alters basic fibroblast growth factor activity. A model for intracellular glycosylation in diabetes. J Clin Invest 94: 110–117, 1994.
118. Bierhaus A, Hofmann MA, Ziegler R, Nawroth PP: AGEs and their interaction with AGE-receptors in vascular disease and diabetes mellitus. I. The AGE concept. Cardiovasc Res 37: 586–600, 1998.
119. Altannavch TS, Roubalova K, Kucera P, Andel M: Effect of high glucose concentrations on expression of ELAM-1, VCAM-1 and ICAM-1 in HUVEC with and without cytokine activation. Physiol Rev 53: 77–82, 2004.
120. Barter PJ, Baker PW, Rye KA: Effect of high-density lipoproteins on the expression of adhesion molecules in endothelial cells. Curr Opin Lipidol 13: 285–288, 2002.
121. Neeper M, Schmidt AM, Brett J, Yan SD, Wang F, Pan YC, Elliston K, Stern D, Shaw A: Cloning and expression of a cell surface receptor for advanced glycosylation end products of proteins. J Biol Chem 267: 14998–15004, 1992.
122. Schmidt AM, Yan SD, Yan SF, Stern DM: The multiligand receptor RAGE as a progression factor amplifying immune and inflammatory responses. J Clin Invest 108: 949–955, 2001.
123. Hoffmann JA, Kafatos FC, Janeway CA, Ezekowitz RA: Phylogenetic perspectives in innate immunity. Science 284: 1313–1318, 1999.
124. Ghosh S, May MJ, Kopp EB: NF-kappa B and Rel proteins: evolutionarily conserved mediators of immune responses. Ann Rev Immunol 16: 225–260, 1998.
125. Kislinger T, Fu C, Huber B, Qu W, Taguchi A, Du Yan S, Hofmann M, Yan SF, Pischetsrieder M, Stern D, Schmidt AM: N(epsilon)-(carboxymethyl)lysine adducts of proteins are ligands for receptor for advanced glycation end products that activate cell signalling pathways and modulate gene expression. J Biol Chem 274: 31740–31749, 1999.
126. Wautier JL, Schmidt AM: Protein glycation: a firm link to endothelial cell dysfunction. Circ Res 95: 233–238, 2004.
127. Schmidt AM, Hori O, Chen JX, Li JF, Crandall J, Zhang J, Cao R, Yan SD, Brett J, Stern D: Advanced glycation endproducts interacting with their endothelial receptor induce expression of vascular cell adhesion molecule-1 (VCAM-1) in cultured human endothelial cells and in mice. A potential mechanism for the accelerated vasculopathy of diabetes. J Clin Invest 96: 1395–1403, 1995.
128. Wautier MP, Chappey O, Corda S, Stern DM, Schmidt AM, Wautier JL: Activation of NADPH oxidase by AGE links oxidant stress to altered gene expression via RAGE. Am J Physiol Endocrinol Metab 280: E685–E694, 2001.
129. Valencia JV, Mone M, Zhang J, Weetall M, Buxton FP, Hughes TE: Divergent pathways of gene expression are activated by the RAGE ligands S100b and AGE-BSA. Diabetes 53: 743–751, 2004.
130. Valencia JV, Mone M, Koehne C, Rediske J, Hughes TE: Binding of receptor for advanced glycation end products (RAGE) ligands is not sufficient to induce

inflammatory signals: lack of activity of endotoxin-free albumin-derived advanced glycation end products. Diabetologia 47: 844–852, 2004.

131. Liliensiek B, Weigand MA, Bierhaus A, Nicklas W, Kasper M, Hofer S, Plachky J, Grone HJ, Kurschus FC, Schmidt AM, Yan SD, Martin E, Schleicher E, Stern DM, Hammerling GG, Nawroth PP, Arnold B: Receptor for advanced glycation end products (RAGE) regulates sepsis but not the adaptive immune response. J Clin Invest 113: 1641–1650, 2004.

132. Chavakis T, Bierhaus A, Al-Fakhri N, Schneider D, Witte S, Linn T, Nagashima M, Morser J, Arnold B, Preissner KT, Nawroth PP: The pattern recognition receptor (RAGE) is a counterreceptor for leukocyte integrins: a novel pathway for inflammatory cell recruitment. J Exp Med 198: 1507–1515, 2003.

133. Gordon S: Pattern recognition receptors: doubling up for the innate immune response. Cell 111: 927–930, 2002.

134. Goldstein JL, Ho YK, Basu SK, Brown MS: Binding site on macrophages that mediates uptake and degradation of acetylated low density lipoprotein, producing massive cholesterol deposition. Proc Natl Acad Sci USA 76: 333–337, 1979.

135. Araki N, Higashi T, Mori T, Shibayama R, Kawabe Y, Kodama T, Takahashi K, Shichiri M, Horiuchi S: Macrophage scavenger receptor mediates the endocytic uptake and degradation of advanced glycation end products of the Maillard reaction. Eur J Biochem 230: 408–415, 1995.

136. Zhang H, Yang Y, Steinbrecher UP: Structural requirements for the binding of modified proteins to the scavenger receptor of macrophages. J Biol Chem 268: 5535–5542, 1993.

137. Miyazaki A, Nakayama H, Horiuchi S: Scavenger receptors that recognize advanced glycation end products. Trends Cardiovasc Med 12: 258–262, 2002.

138. Nozaki S, Kashiwagi H, Yamashita S, Nakagawa T, Kostner B, Tomiyama Y, Nakata A, Ishigami M, Miyagawa J, Kameda-Takemura K: Reduced uptake of oxidized low density lipoproteins in monocyte-derived macrophages from CD36-deficient subjects. J Clin Invest 96: 1859–1865, 1995.

139. Nakata A, Nakagawa Y, Nishida M, Nozaki S, Miyagawa J, Nakagawa T, Tamura R, Matsumoto K, Kameda-Takemura K, Yamashita S, Matsuzawa Y: CD36, a novel receptor for oxidized low-density lipoproteins, is highly expressed on lipid-laden macrophages in human atherosclerotic aorta. Arterioscler Thromb Vasc Biol 19: 1333–1339, 1999.

140. Kunjathoor VV, Febbraio M, Podrez EA, Moore KJ, Andersson L, Koehn S, Rhee JS, Silverstein R, Hoff HF, Freeman MW: Scavenger receptors class A-I/II and CD36 are the principal receptors responsible for the uptake of modified low density lipoprotein leading to lipid loading in macrophages. J Biol Chem 277: 49982–49988, 2002.

141. Ohgami N, Nagai R, Miyazaki A, Ikemoto M, Arai H, Horiuchi S, Nakayama H: Scavenger receptor class B type I-mediated reverse cholesterol transport is inhibited by advanced glycation end products. J Biol Chem 276: 13348–13355, 2001.

142. Sawamura T, Kume N, Aoyama T, Moriwaki H, Hoshikawa H, Aiba Y, Tanaka T, Miwa S, Katsura Y, Kita T, Masaki T: An endothelial receptor for oxidized low-density lipoprotein. Nature 386: 73–77, 1997.

143. Jono T, Miyazaki A, Nagai R, Sawamura T, Kitamura T, Horiuchi S: Lectin-like oxidized low density lipoprotein receptor-1 (LOX-1) serves as an endothelial receptor for advanced glycation end products (AGE). FEBS Lett 511: 170–174, 2002.

144. Li D, Mehta JL: Antisense to LOX-1 inhibits oxidized LDL-mediated upregulation of monocyte chemoattractant protein-1 and monocyte adhesion to human coronary artery endothelial cells. Circulation 101: 2889–2895, 2000.
145. Cominacini L, Pasini AF, Garbin U, Davoli A, Tosetti ML, Campagnola M, Rigoni A, Pastorino AM, Lo Cascio V, Sawamura T: Oxidized low density lipoprotein (ox-LDL) binding to ox-LDL receptor-1 in endothelial cells induces the activation of NF-kappaB through an increased production of intracellular reactive oxygen species. J Biol Chem 275: 12633–12638, 2000.
146. Li D, Liu L, Chen H, Sawamura T, Ranganathan S, Mehta JL: LOX-1 mediates oxidized low-density lipoprotein-induced expression of matrix metalloproteinases in human coronary artery endothelial cells. Circulation 107: 612–617, 2003.
147. Chen M, Nagase M, Fujita T, Narumiya S, Masaki T, Sawamura T: Diabetes enhances lectin-like oxidized LDL receptor-1 (LOX-1) expression in the vascular endothelium: possible role of LOX-1 ligand and AGE. Biochem Biophys Res Commun 287: 962–968, 2001.
148. Li L, Sawamura T, Renier G: Glucose enhances human macrophage LOX-1 expression: role for LOX-1 in glucose-induced macrophage foam cell formation. Circ Res 94: 892–901, 2004.
149. Pricci F, Leto G, Amadio L, Iacobini C, Romeo G, Cordone S, Gradini R, Barsotti P, Liu FT, Di Mario U, Pugliese G: Role of galectin-3 as a receptor for advanced glycosylation end products. Kidney Int (Suppl 77): S31–S39, 2000.
150. Pugliese G, Pricci F, Leto G, Amadio L, Iacobini C, Romeo G, Lenti L, Sale P, Gradini R, Liu FT, Di Mario U: The diabetic milieu modulates the advanced glycation end product–receptor complex in the mesangium by inducing or upregulating galectin-3 expression. Diabetes 49: 1249–1257, 2000.
151. Pugliese G, Pricci F, Iacobini C, Leto G, Amadio L, Barsotti P, Frigeri L, Hsu DK, Vlassara H, Liu FT, Di Mario U: Accelerated diabetic glomerulopathy in galectin-3/AGE receptor 3 knockout mice. FASEB J 15: 2471–2479, 2001.
152. Esposito C, Gerlach H, Brett J, Stern D, Vlassara H: Endothelial receptor-mediated binding of glucose-modified albumin is associated with increased monolayer permeability and modulation of cell surface coagulant properties. J Exp Med 170: 1387–1407, 1989.
153. Kouroedov A, Eto M, Joch H, Volpe M, Luscher TF, Cosentino F: Selective inhibition of protein kinase C beta2 prevents acute effects of high glucose on vascular cell adhesion molecule-1 expression in human endothelial cells. Circulation 110: 91–96, 2004.
154. Piconi L, Quagliaro L, Da Ros R, Assaloni R, Giugliano D, Esposito K, Szabo C, Ceriello A: Intermittent high glucose enhances ICAM-1, VCAM-1, E-selectin and interleukin-6 expression in human umbilical endothelial cells in culture: the role of poly(ADP-ribose) polymerase. J Thromb Haemost: 1453–1459, 2004.
155. Vlassara H, Fuh H, Donnelly T, Cybulsky M: Advanced glycation endproducts promote adhesion molecule (VCAM-1, ICAM-1) expression and atheroma formation in normal rabbits. Mol Med 1: 447–456, 1995.
156. Tan KC, Chow WS, Ai VH, Metz C, Bucala R, Lam KS: Advanced glycation end products and endothelial dysfunction in type 2 diabetes. Diabetes Care 25: 1055–1059, 2002.
157. Chan NN, Vallance P, Colhoun HM: Endothelium-dependent and -independent vascular dysfunction in type 1 diabetes: role of conventional risk factors, sex, and glycemic control. Arterioscler Thromb Vasc Biol 23: 1048–1054, 2003.

158. Galle J, Schneider R, Winner B, Lehmann-Bodem C, Schinzel R, Munch G, Conzelmann E, Wanner C: Glyc-oxidized LDL impair endothelial function more potently than oxidised LDL: role of enhanced oxidative stress. Atherosclerosis 138: 65–77, 1998.
159. Ritthaler U, Deng Y, Zhang Y, Greten J, Abel M, Sido B, Allenberg J, Otto G, Roth H, Bierhaus A, Ziegler R, Schmidt A-M, Waldherr R, Wahl P, Stern DM, Nawroth PP: Expression of receptors or advanced glycation end products in peripheral occlusive vascular disease. Am J Pathol 146: 688–694, 1995.
160. Vlassara H, Bucala R: Recent progress in advanced glycation and diabetic vascular disease: role of advanced glycation end product receptors. Diabetes 45: S65–S66, 1996.
161. Chappey O, Dosquet C, Wautier M-P, Wautier J-L: Advanced glycation end products, oxidant stress and vascular lesions. Eur J Clin Invest 27: 97–108, 1997.
162. Higashi T, Sano H, Saihoji T, Ikeda K, Jinnouchi Y, Kanzaki T, Morisaki N, Rauvala H, Shichiri M, Horiuchi S: The receptor for advanced glycation end products mediates the chemotaxis of rabbit smooth muscle cells. Diabetes 46: 463–472, 1997.
163. Hattori Y, Suzuki M, Hattori S, Kasai K: Vascular smooth muscle cell activation by glycated albumin (Amadori adducts). Hypertension 39: 22–28, 2002.
164. Makita T, Tanaka A, Numano F: Effect of glycated low density lipoprotein on smooth muscle cell proliferation. Int Angiol 18: 331–334, 1999.
165. Taguchi S, Oinuma T, Yamada T: A comparative study of cultured smooth muscle cell proliferation and injury, utilizing glycated low density lipoproteins with slight oxidation, auto-oxidation, or extensive oxidation. J Atheroscler Thromb 7: 132–137, 2000.
166. Wilson GL, Leiter EH: Streptozotocin interactions with pancreatic b cells and the induction of insulin-dependent diabetes. Curr Top Microbiol Immunol 156: 27–54, 1990.
167. Kunjathoor VV, Wilson DL, LeBoeuf RC: Increased atherosclerosis in streptozotocin-induced diabetic mice. J Clin Invest 97: 1767–1773, 1996.
168. Plump AS, Smith JD, Hayek T, Aalto-Setala K, Walsh A, Verstuyft JG, Rubin EM, Breslow JL: Severe hypercholesterolemia and atherosclerosis in apolipoprotein E-deficient mice created by homologous recombination in ES cells. Cell 71: 343–353, 1992.
169. Park L, Raman KG, Lee KJ, Lu Y, Ferran LJ Jr, Chow WS, Stern D, Schmidt AM: Suppression of accelerated diabetic atherosclerosis by the soluble receptor for advanced glycation endproducts. Nat Med 4: 1025–1031, 1998.
170. Kislinger T, Tanji N, Wendt T, Qu W, Lu Y, Ferran LJ Jr, Taguchi A, Olson K, Bucciarelli L, Goova M, Hofmann MA, Cataldegirmen G, D'Agati V, Pischetsrieder M, Stern DM, Schmidt AM: Receptor for advanced glycation end products mediates inflammation and enhanced expression of tissue factor in vasculature of diabetic apolipoprotein E-null mice. Arterioscler Thromb Vasc Biol 21: 905–910, 2001.
171. Bucciarelli LG, Wendt T, Qu W, Lu Y, Lalla E, Rong LL, Goova MT, Moser B, Kislinger T, Lee DC, Kashyap Y, Stern DM, Schmidt AM: RAGE blockade stabilizes established atherosclerosis in diabetic apolipoprotein E-null mice. Circulation 106: 2827–2835, 2002.
172. Kosaki A, Hasegawa T, Kimura T, Iida K, Hitomi J, Matsubara H, Mori Y, Okigaki M, Toyoda N, Masaki H, Inoue-Shibata M, Nishikawa M, Iwasaka T:

Increased plasma S100A12 (EN-RAGE) levels in patients with type 2 diabetes. J Clin Endocrinol Metab: 5423–5428, 2004.

173. Hudson BI, Bucciarelli LG, Wendt T, Sakaguchi T, Lalla E, Qu W, Lu Y, Lee L, Stern DM, Naka Y, Ramasamy R, Yan SD, Yan SF, D'Agati V, Schmidt AM: Blockade of receptor for advanced glycation endproducts: a new target for therapeutic intervention in diabetic complications and inflammatory disorders. Arch Biochem Biophys 419: 80–88, 2003.

174. Forbes JM, Yee LT, Thallas V, Lassila M, Candido R, Jandeleit-Dahm KA, Thomas MC, Burns WC, Deemer EK, Thorpe SM, Cooper ME, Allen TJ: Advanced glycation end product interventions reduce diabetes-accelerated atherosclerosis. Diabetes 53: 1813–1823, 2004.

175. Candido R, Allen TJ, Lassila M, Cao Z, Thallas V, Cooper ME, Jandeleit-Dahm KA: Irbesartan but not amlodipine suppresses diabetes-associated atherosclerosis. Circulation 109: 1536–1542, 2004.

176. Candido R, Jandeleit-Dahm KA, Cao Z, Nesteroff SP, Burns WC, Twigg SM, Dilley RJ, Cooper ME, Allen TJ: Prevention of accelerated atherosclerosis by angiotensin-converting enzyme inhibition in diabetic apolipoprotein E-deficient mice. Circulation 106: 246–253, 2002.

177. Miyata T, van Ypersele de Strihou C, Ueda Y, Ichimori K, Inagi R, Onogi H, Ishikawa N, Nangaku M, Kurokawa K: Angiotensin II receptor antagonists and angiotensin-converting enzyme inhibitors lower *in vitro* the formation of advanced glycation end products: biochemical mechanisms. J Am Soc Nephrol 13: 2478–2487, 2002.

178. Miyata T, van Ypersele de Strihou C: Angiotensin II receptor blockers and angiotensin converting enzyme inhibitors: implication of radical scavenging and transition metal chelation in inhibition of advanced glycation end product formation. Arch Biochem Biophys 419: 50–54, 2003.

179. Li AC, Glass CK: PPAR- and LXR-dependent pathways controlling lipid metabolism and the development of atherosclerosis. J Lipid Res: 2161–2173, 2004.

180. Levi Z, Shaish A, Yacov N, Levkovitz H, Trestman S, Gerber Y, Cohen H, Dvir A, Rhachmani R, Ravid M, Harats D: Rosiglitazone (PPARgamma-agonist) attenuates atherogenesis with no effect on hyperglycaemia in a combined diabetes–atherosclerosis mouse model. Diabetes Obes Metab 5: 45–50, 2003.

181. Chinetti G, Lestavel S, Bocher V, Remaley AT, Neve B, Torra IP, Teissier E, Minnich A, Jaye M, Duverger N, Brewer HB, Fruchart JC, Clavey V, Staels B: PPAR-alpha and PPAR-gamma activators induce cholesterol removal from human macrophage foam cells through stimulation of the ABCA1 pathway. Nat Med 7: 53–58, 2001.

182. Chawla A, Boisvert WA, Lee CH, Laffitte BA, Barak Y, Joseph SB, Liao D, Nagy L, Edwards PA, Curtiss LK, Evans RM, Tontonoz P: A PPAR gamma-LXR-ABCA1 pathway in macrophages is involved in cholesterol efflux and atherogenesis. Mol Cell 7: 161–171, 2001.

183. Verrier E, Wang L, Wadham C, Albanese N, Hahn C, Gamble JR, Chatterjee VK, Vadas MA, Xia P: PPARgamma agonists ameliorate endothelial cell activation via inhibition of diacylglycerol–protein kinase C signalling pathway: role of diacylglycerol kinase. Circ Res 94: 1515–1522, 2004.

184. Rival Y, Beneteau N, Taillandier T, Pezet M, Dupont-Passelaigue E, Patoiseau JF, Junquero D, Colpaert FC, Delhon A: PPARalpha and PPARdelta activators

inhibit cytokine-induced nuclear translocation of NF-kappaB and expression of VCAM-1 in EAhy926 endothelial cells. Eur J Pharmacol 435: 143–151, 2002.

185. Thornalley PJ: The glyoxalase system: new developments towards functional characterization of a metabolic pathway fundamental to biological life. Biochem J 269: 1–11, 1990.

186. Jakus V, Hrnciarova M, Carsky J, Krahulec B, Rietbrock N: Inhibition of nonenzymatic protein glycation and lipid peroxidation by drugs with antioxidant activity. Life Sci 65: 1991–1993, 1999.

187. Price DL, Rhett PM, Thorpe SR, Baynes JW: Chelating activity of advanced glycation end-product inhibitors. J Biol Chem 276: 48967–48972, 2001.

188. Nishikawa T, Edelstein D, Du XL, Yamagishi S, Matsumura T, Kaneda Y, Yorek MA, Beebe D, Oates PJ, Hammes HP, Giardino I, Brownlee M: Normalizing mitochondrial superoxide production blocks three pathways of hyperglycaemic damage. Nature 404: 787–790, 2000.

189. Du X, Matsumura T, Edelstein D, Rossetti L, Zsengeller Z, Szabo C, Brownlee M: Inhibition of GAPDH activity by poly(ADP-ribose) polymerase activates three major pathways of hyperglycemic damage in endothelial cells. J Clin Invest 112: 1049–1057, 2003.

190. Hammes HP, Du X, Edelstein D, Taguchi T, Matsumura T, Ju Q, Lin J, Bierhaus A, Nawroth P, Hannak D, Neumaier M, Bergfeld R, Giardino I, Brownlee M: Benfotiamine blocks three major pathways of hyperglycemic damage and prevents experimental diabetic retinopathy. Nat Med 9: 294–299, 2003.

191. Rahbar S, Figarola JL: Novel inhibitors of advanced glycation endproducts. Arch Biochem Biophys 419: 63–79, 2003.

192. Metz TO, Alderson NL, Thorpe SR, Baynes JW: Pyridoxamine, an inhibitor of advanced glycation and lipoxidation reactions: a novel therapy for treatment of diabetic complications. Arch Biochem Biophys 419: 41–49, 2003.

193. Thornalley PJ: Use of aminoguanidine (pimagedine) to prevent the formation of advanced glycation endproducts. Arch Biochem Biophys 419: 31–40, 2003.

194. Brownlee M, Vlassara H, Kooney A, Ulrich P, Cerami A: Aminoguanidine prevents diabetes-induced arterial wall protein cross-linking. Science 232: 1629–1632, 1986.

195. Freedman BI, Wuerth JP, Cartwright K, Bain RP, Dippe S, Hershon K, Mooradian AD, Spinowitz BS: Design and baseline characteristics for the aminoguanidine Clinical Trial in Overt Type 2 Diabetic Nephropathy (ACTION II). Control Clin Trials 20: 493–510, 1999.

196. Nilsson BO: Biological effects of aminoguanidine: an update. Inflammation Res 48: 509–515, 1999.

197. Voziyan PA, Metz TO, Baynes JW, Hudson BG: A post-Amadori inhibitor pyridoxamine also inhibits chemical modification of proteins by scavenging carbonyl intermediates of carbohydrate and lipid degradation. J Biol Chem 277: 3397–3403, 2002.

198. Degenhardt TP, Alderson NL, Arrington DD, Beattie RJ, Basgen JM, Steffes MW, Thorpe SR, Baynes JW: Pyridoxamine inhibits early renal disease and dyslipidemia in the streptozotocin-diabetic rat. Kidney Int 61: 939–950, 2002.

199. Vasan S, Zhang X, Kapurniotu A, Bernhagen J, Teichberg S, Basgen J, Wagle D, Shih D, Terlecky I, Bucala R, Cerami A, Egan J, Ulrich P: An agent cleaving glucose-derived protein crosslinks *in vitro* and *in vivo*. Nature 382: 275–278, 1996.

200. Candido R, Forbes JM, Thomas MC, Thallas V, Dean RG, Burns WC, Tikellis C, Ritchie RH, Twigg SM, Cooper ME, Burrell LM: A breaker of advanced glycation end products attenuates diabetes-induced myocardial structural changes. Circ Res 92: 785–792, 2003.
201. Susic D, Varagic J, Ahn J, Frohlich ED: Crosslink breakers: a new approach to cardiovascular therapy. Curr Opin Cardiol 19: 336–340, 2004.
202. Asif M, Egan J, Vasan S, Jyothirmayi GN, Masurekar MR, Lopez S, Williams C, Torres RL, Wagle D, Ulrich P, Cerami A, Brines M, Regan TJ: An advanced glycation endproduct cross-link breaker can reverse age-related increases in myocardial stiffness. Proc Natl Acad Sci USA 97: 2809–2813, 2000.
203. Vaitkevicius PV, Lane M, Spurgeon H, Ingram DK, Roth GS, Egan JJ, Vasan S, Wagle DR, Ulrich P, Brines M, Wuerth JP, Cerami A, Lakatta EG: A cross-link breaker has sustained effects on arterial and ventricular properties in older rhesus monkeys. Proc Natl Acad Sci USA 98: 1171–1175, 2001.
204. Kass DA, Shapiro EP, Kawaguchi M, Capriotti AR, Scuteri A, deGroof RC, Lakatta EG: Improved arterial compliance by a novel advanced glycation end-product crosslink breaker. Circulation 104: 1464–1470, 2001.
205. Yang S, Litchfield JE, Baynes JW: AGE-breakers cleave model compounds, but do not break Maillard crosslinks in skin and tail collagen from diabetic rats. Arch Biochem Biophys 412: 42–46, 2003.

13

Molecular and Cellular Mechanisms by Which Diabetes Mellitus Promotes the Development of Atherosclerosis

GEOFF H. WERSTUCK

Abstract

Recent decades have witnessed a significant increase in the incidence of diabetes mellitus primarily driven by population trends toward weight gain and sedentary lifestyles. Diabetes mellitus is a leading cause of blindness, renal failure, lower limb amputation, and an independent risk factor for atherosclerotic cardiovascular disease (CVD) and stroke. In fact, individuals with diabetes have a two- to fourfold increased risk of myocardial infarction (MI) and stroke that accounts for over 65% of diabetic mortality. Despite a vast amount of research, the molecular and cellular mechanisms that predispose individuals with diabetes to the development and progression of atherosclerosis are not understood. This chapter summarizes the current state of our knowledge of these conditions as well as some of the animal models that are being used to further increase our understanding. In addition, pathways and mechanisms that may link diabetes, and hyperglycemia in particular, to the development and progression of atherosclerosis are discussed. The continued investigation of identified pathways, linking hyperglycemia and diabetes mellitus to atherosclerosis, and the discovery of novel mechanisms and targets will be important to the development of new and effective antiatherosclerotic therapies tailored to individuals with diabetes.

Keywords: accelerated atherosclerosis; animal models; diabetes mellitus; hyperglycemia; molecular mechanisms

Introduction

The last few decades have witnessed a dramatic, worldwide increase in the incidence of diabetes mellitus. Driven by changes in lifestyle and an escalating rate of obesity, the number of individuals with diabetes is expected to reach 300 million by the year 2025 [1, 2]. Therefore, the diabetes epidemic is, and will continue to be, a global health crisis. Complications associated with diabetes make it a leading cause of blindness, renal failure, and lower limb amputations as well as an important, independent risk factor for atherosclerotic cardiovascular disease (CVD). CVD accounts for over 65% of diabetic mortality. In fact, diabetes mellitus is "a coronary heart disease risk equivalent" meaning that individuals with diabetes have the same risk of myocardial infarction (MI) as nondiabetic persons who have had a previous MI [3–5]. Despite a tremendous amount of research, the treatment and prevention of diabetes-associated CVD is currently limited by our lack of understanding of the molecular and cellular mechanisms by which diabetes mellitus promotes atherosclerosis. This chapter will focus upon our current understanding of this problem, the model systems that are being used to further increase our knowledge and the potential future directions of this research.

Diabetes Mellitus

Diabetes mellitus is currently defined as a condition with casual blood glucose values are ≥11.1 mmol/L, or fasting plasma glucose ≥7.0 mmol/L, or glucose levels ≥11.1 mmol/L 2 h after an oral glucose tolerance test [6]. Pathologically, the term diabetes mellitus describes a heterogeneous group of metabolic disorders characterized by chronic hyperglycemia. These include conditions under which the body cannot produce insulin and/or cannot effectively respond to the insulin that is being produced. There are two major forms of diabetes; type 1 (T1D), formerly known as juvenile diabetes or insulin-dependent diabetes mellitus (IDDM) and type 2 (T2D), formerly known as adult onset or noninsulin-dependent diabetes mellitus (NIDDM). T1D and T2D are, in essence, two distinctly different diseases that present an overlapping array of signs and symptoms.

T1D is characterized by an absolute insulin deficiency that results from an autoimmune-mediated depletion of insulin secreting pancreatic β cells [7, 8]. Typically, T1D develops with a rapid onset in children, with a peak incidence at 14 years of age. T1D makes up approximately 10% of the total number of cases of diabetes mellitus. Recent breakthroughs in β cell transplantation may lead to an eventual cure for T1D. Until that time, the only treatment is the tight self-monitoring of plasma glucose and the administration of exogenous insulin according to carbohydrate intake.

T2D accounts for the remaining 90% of cases of diabetes mellitus. T2D is characterized by a relative insulin deficiency that arises from a combination of peripheral tissue insulin resistance and impaired β cell function [9, 10]. Tissues expressing the insulin responsive GLUT4 glucose transporter, including adipose, and skeletal muscle and liver are particularly relevant to insulin resistance. The onset of T2D is relatively gradual and, until recently, with rare exceptions, affected only adults. However with the dramatic increase in childhood obesity, there has been an alarming increase in the diagnosis of T2D in children. Depending on severity, T2D can be controlled by using a combination of lifestyle modifications (diet and exercise), medications that increase insulin sensitivity and exogenous insulin.

Despite their distinctly different etiologies, both T1D and T2D are characterized by chronic hyperglycemia. Furthermore, there is a strong correlation between hyperglycemia and both micro- and macrovascular disease [11–15].

The negative effects of elevated glucose levels on vascular function can include decreased proliferation of endothelial cells, the impairment of some parameters of vascular responsiveness, and increased endothelial programmed cell death [16–18]. It is well established that aggressive blood glucose lowering significantly decreases the incidence and severity of microvascular disease including retinopathy, renal failure, and peripheral nerve dysfunction [11, 12]. Recent evidence suggests that increased glycemic control also correlates with a reduction in macrovascular disease, however the relationship between glucose lowering and a decrease in CVD has been much more difficult to demonstrate [12, 19, 20]. Several explanations have been put forth to rationalize the inability to clinical trials to demonstrate a strong improvement in cardiovascular outcomes through glycemic control including the possibilities that the trials were underpowered, too short in duration or were too focused upon fasting glucose rather than postprandial glucose levels. Alternatively, these findings may indicate that the quality of glycemic control presently achievable is insufficient to be effective in protecting against macrovascular disease. Therefore, even short-term deviations in the control of blood glucose may promote vascular dysfunction.

The pathophysiology of T2D-associated CVD is further complicated by multiple risk factors, collectively known as the metabolic syndrome, that commonly accompany chronic hyperglycemia. The metabolic syndrome is clinically defined as a combination of abdominal obesity, insulin resistance (prediabetes), atherogenic dyslipidemia, and hypertension [21]. The metabolic syndrome is a major cause of morbidity and mortality with CVD being the primary clinical outcome [21]. Other complications can include respiratory difficulties, chronic skeletal muscle problems, gallbladder disease, infertility, hepatic steatosis, circulatory problems, and certain cancers [22, 23].

Therefore, while a role for hyperglycemia in the development and progression of atherosclerosis is supported by a great deal of basic research, the clinical role of elevated glucose levels in macrovascular disease is less clear. Furthermore, despite a great deal of research the mechanisms that may link high glucose concentrations to the molecular and cellular pathways of disease development are not fully understood. This review will focus on potential direct proatherogenic consequences of hyperglycemia.

Model Systems

A major challenge facing researchers, who study the role of diabetes mellitus in macrovascular disease, is the inherent resistance of most rodent strains to atherosclerotic plaque development. In commonly used rodent strains, hyperglycemia and diabetes alone are not sufficient to promote the development of lesions [24]. Even in strains that can develop vascular lesions, diabetes-accelerated atherosclerosis appears, in many cases, to depend upon hyperlipidemia, either resulting from a high fat diet or a genetic deficiency, as with the

ApoE-deficient or LDLR-deficient mouse [24, 25]. The general resistance of mice to atherogenesis may result, at least in part, from the observation that wild-type mice, in contrast with humans, have barely detectable low-density lipoprotein (LDL) and very low-density lipoprotein (VLDL) levels. The induction of hyperlipidemia, through dietary or genetic manipulation, "humanizes" the lipid profiles of mice by significantly elevating both VLDL and LDL [24–26].

Rodent models of diabetes mellitus and hyperglycemia can be generally categorized as; (i) models in which diabetes is chemically induced, (ii) strains that have been inbred and selected for the existence of chronic hyperglycemia, or (iii) genetically engineered mouse models of diabetes (reviewed in [27]). Each model will possess a number of advantages and disadvantages depending upon the particular study and the experimental design. In general, models of T1D are useful to study the effects of chronic hyperglycemia in isolation whereas T2D models may be more physiologically relevant, abet more complicated by additional CVD risk factors such as obesity, dyslipidemia, hypertension, and perhaps hyperinsulinemia [27].

Commonly used, inbred models of spontaneous, chronic hyperglycemia, including the nonobese diabetic (NOD) mouse, the KK mouse, the Goto Kakizaki rat, and the bio-breeding (BB) rat, which are very useful in the study of hyperglycemia and many aspects of diabetes mellitus, including microvascular complications, are resistant to the development of atherosclerosis.

Many researchers chemically induce hyperglycemia in existing models of atherosclerosis such as the LDLR or ApoE knockout mouse models that spontaneously develop atherosclerosis [28]. The most commonly used animal models involve the administration of compounds, such as alloxan or streptozotocin (STZ), which are selectively toxic to pancreatic β cells. Alloxan (2,4,5,6-tetraoxypyrimidine;5,6-dioxyuracil) can be administered parentally, intravenously intraperitoneally, or subcutaneously [29]. It is rapidly taken up not only by pancreatic β cells but also by hepatocytes. Beta cells are believed to be especially sensitive to the reactive oxygen species that are produced in the cyclical interconversion of alloxan to dialuric acid [30]. The effective range of diabetogenic dose is relatively narrow and great care must be taken to minimize the side effects of this compound [29]. STZ is usually administered as a single large dose that is directly toxic to β cells or by multiple low doses (e.g., five consecutive doses of 40 mg/kg body weight) that appear to induce β cell destruction by an immune-mediated process. STZ is a nitrosurea derivative that enters pancreatic β cell through the GLUT 2 transporter [31]. In low-to-moderate doses, STZ is selectively toxic to pancreatic β cells. STZ-induced diabetes accelerates atherosclerosis in ApoE-deficient mice and BALBc mice fed an atherogenic diet [24, 25]. An advantage of the alloxan and STZ-induced models of diabetes mellitus is that they permit the examination of the effects of hyperglycemia in the absence of other variables that are often associated with T2D including obesity and hyperinsulinemia. One complication of chemically induced hyperglycemia is that, over time, dyslipidemia develops thereby making it difficult to differentiate whether accelerated lesion development is a result of elevated plasma glucose or lipid levels [25, 32, 33]. In fact, in at least some models, the STZ-associated dyslipidemia appears to be essential for the acceleration of atherosclerosis and STZ-induced hyperglycemia alone fails to accelerate lesion development [33–35].

Transgenic models of T2D, such as the ob/ob mouse have been crossed with atherosclerosis-prone strains including the LDLR knockout mouse. Not surprisingly, these double knockout mice exhibit severe hypertriglyceridemia, hypercholesterolemia, and accelerated atherosclerosis beyond either model individually [36]. Recently a transgenic model of T1D has been developed in which β cell death can be induced, at will, by viral infection in an LDLR-deficient (proatherogenic) genetic background [32]. The advantage of this model is that β cell death can be specifically induced at any time and the uninfected mice act as a normoglycemic control. Hyperglycemia, in the absence of dyslipidemia, is sufficient to accelerate atherosclerosis in this model [32].

Potential Mechanisms by Which Hyperglycemia Accelerates Atherosclerosis

Atherosclerosis is a disease of the walls of major muscular arteries that involves inflammation and the accumulation of lipids (reviewed in [37, 38]). It is believed to originate as a response to insult or injury to the endothelial layer. The damaged endothelial cells increase the expression of adhesion proteins (β2 integrin, PCAM-1, VCAM, ICAM, P-, and E-selectins) and chemotactic factors (monocyte chemotactic protein, MCP-1) factors that actively recruit monocytes and T lymphocytes to the vessel wall. Increased endothelium permeability allows the infiltration of monocytes and T lymphocytes as well as of lipoprotein particles, especially LDL, into the intima. The activated monocytes differentiate into macrophages expressing scavenger receptors (SR-A, CD36) that recognize oxidized, and otherwise modified, LDL particles. Scavenging macrophages become engorged with LDL-derived cholesterol and take on the appearance of a foam cell. Early lesions consist almost solely of foam cells and are known as fatty streaks. As lesion development progresses, activated macrophages/foam cells and T lymphocytes further simulate the inflammatory process through the secretion of matrix metalloproteases (MMPs) and cytokines (IFN-γ) that stimulate vascular smooth muscle cell (VSMC) growth and migration into the intima. VSMCs produce and secrete extracellular matrix that becomes a fibrous cap that covers the lesion. Foam cell death leaves a growing mass of extracellular debris, known as the necrotic core that includes lipids and many procoagulant proteins (tissue factor). The action of MMPs on the components of the fibrous cap can result in cap thinning and plaque instability. The majority of MIs and strokes are thought to occur when plaques rupture and the blood comes into contact with the necrotic core, forming a thrombus or clot [37, 38].

In diabetic patients, the atherogenic process is thought to occur through an identical series of events. Atherosclerotic lesions isolated from diabetic patients are generally more advanced but otherwise indistinguishable from those isolated from nondiabetic patients [39, 40]. Individuals with diabetes have a two- to fourfold-increased risk of coronary heart disease and a fourfold-increased risk of mortality from heart disease [4]. The reasons for this are not clear. Diabetes/hyperglycemia may promote the development of CVD by increasing the incidence or extent of endothelial injury, by accelerating the development or growth of a plaque, and/or by decreasing plaque stability.

Several mechanisms by which hyperglycemia may promote atherosclerosis have been proposed. In general these involve the effects of elevated levels of intracellular glucose and the increased availability of glucose-derived molecules in cells of the vascular wall (Fig. 13.1).

The Polyol Pathway

The conversion of glucose to sorbitol occurs at a low rate under normoglycemic conditions due to the high K_{m} of aldose reductase for glucose. The flux through this pathway is significantly elevated under diabetic conditions as intracellular glucose concentrations rise [41]. High sorbitol concentration may be directly toxic to some cell types. In others, increased aldose reductase activity has been shown to lead the increased consumption of NADPH and

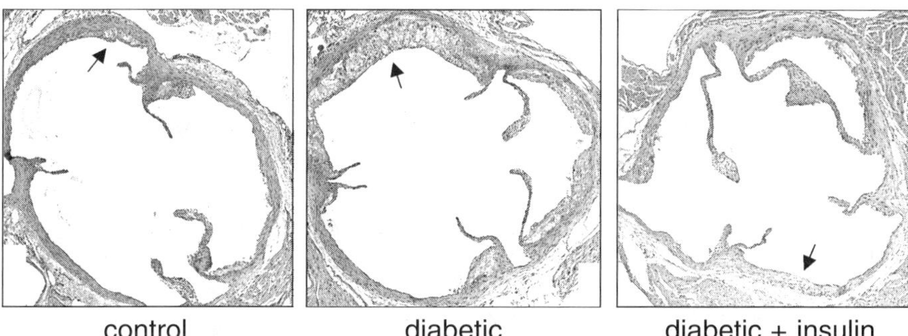

FIGURE 13.1. Streptozotocin (STZ)-induced hyperglycemia promotes accelerated atherosclerosis in ApoE-deficient mice. ApoE-deficient mice were give ten intraperitoneal injections of STZ (40 mg/kg body weight) or an equal volume of saline solution (control) before 8 weeks of age. Plasma glucose levels increase to approximately 20 to 25 mM in diabetic mice versus 7 to 9 mM in controls. Atherosclerotic lesions at the aortic sinus (indicated by arrows), examined at 15 weeks of age, are significantly larger and more advanced in the hyperglycemic mice than in nondiabetic mice or diabetic mice with a subcutaneous, slow release insulin pellet (diabetic + insulin).

depletion of GSH levels that can result in increased levels of reactive oxygen species (ROS) and subsequent cellular damage [42]. It has also been proposed that the conversion of sorbitol to 3-deoxyglucosone can feed into the production of advanced glycation endproducts (AGEs—see below).

Although the link between the polyol pathway and atherogenesis has been inferred, there is very little evidence to support a role in macrovascular disease. Studies focusing on the microvascular benefits of aldose reductase inhibition have, to date, produced mixed results [43–45]. Zenarestat, a potent aldose reductase inhibitor and potential treatment for diabetes-associated neuropathy, did reach phase III before the trial was suspended due to signs of renal toxicity in some patients [45, 46].

The Hexosamine Pathway

The shunting of excess intracellular glucose through the hexosamine pathway may also be relevant to the development of macrovascular complications of diabetes (Fig. 13.2). In a typical cell, under normoglycemic conditions, 1% to 3% of total glucose will be converted to glucosamine-6 phosphate by the enzyme glutamine:fructose-6 phosphate amidotransferase (GFAT) [47]. When intracellular glucose levels rise, flux through this pathway increases. In

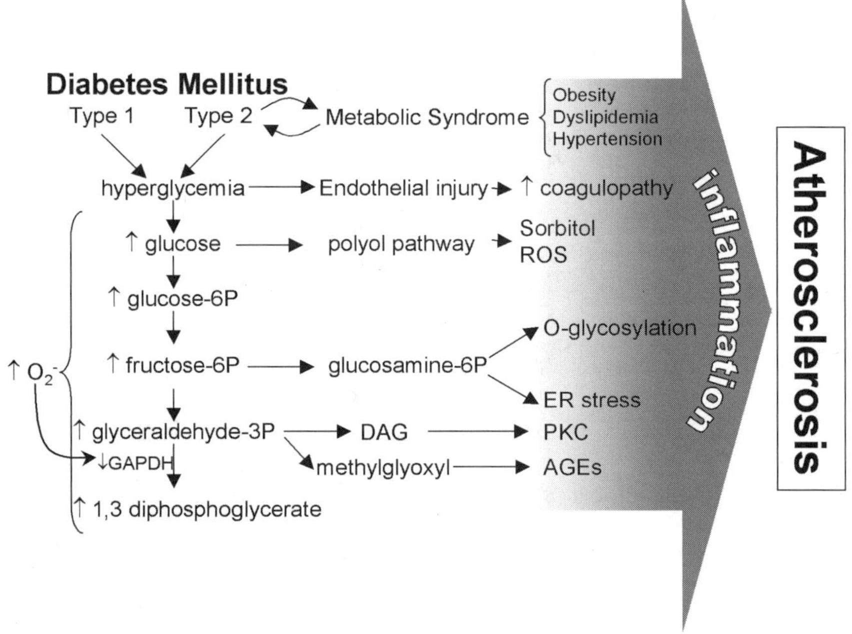

FIGURE 13.2. A summary of the potential links between diabetes mellitus, hyperglycemia, and accelerated atherosclerosis.

addition, increased GFAT expression and activity have been reported in tissues from humans with diabetes [48]. Increased hexosamine pathway flux has been implicated in several diabetes-associated complications including; insulin resistance [47, 49], pancreatic β cell death [50], as well as atherosclerosis [51]. While it is not entirely clear how hexosamine pathway byproducts promote cellular dysfunction most research has focused on the modification of proteins by O-linked glycosylation. The enzyme O-N-acetylglucosamine transferase (OGT) catalyses the addition of N-acetylglucosamine to serine and threonine residues of a variety of proteins. It is known that O-linked glycosylation can act in competition with phosphorylation of the same serine and threonine residues. It is well established that elevated glucosamine concentrations drive the O-linked glycosylation of proteins including transcription factors [52], nuclear pore proteins [53], as well as signaling factors [54] potentially altering their function, stability, and/or activity. Glucosamine has been shown to desensitize insulin-stimulated glucose uptake in both adipocytes [47] and skeletal muscle cells [55], probably by inhibiting the translocation of the glucose transporter, GLUT4, to the cell surface [56]. In addition, increased hexosamine pathway activity can promote the transcription of proinflammatory and prothrombotic factors including TGF-α, TGF-β, and PAI-1 [57, 58]. Therefore, the hyperglycemia-induced O-GlcNAc modification of proteins may change gene expression patterns and alter the function of specific factors that contribute to a proatherogenic, prothrombophilic vascular environment. More studies are required to test this theory and to precisely determine the factors and downstream pathways that may be involved in the acceleration of vascular disease.

Endoplasmic Reticulum Stress Pathways

Recently we have begun to investigate an additional intracellular effect of glucosamine that has not previously been examined in the context of diabetes and atherosclerosis—its ability to promote the accumulation of unfolded proteins in the endoplasmic reticulum (ER), thereby leading to ER dysfunction and cell injury [59–62]. In eukaryotic cells, the ER provides a contained environment for the synthesis and modification of membrane proteins and proteins destined to be secreted. These co- and posttranslational modifications, including disulfide bond formation and N-linked glycosylation, play an important role in the subsequent folding and oligomeric assembly of proteins [63]. Misfolded polypeptides are retained in the ER and subsequently targeted for degradation [64, 65]. Agents or conditions that disrupt ER homeostasis by interfering with glycosylation, disulfide bond formation, ER Ca^{2+} stores, and/or the general overloading of the ER with proteins will cause the accumulation of unfolded or misfolded proteins in the ER, a condition defined as ER stress (reviewed in [66, 67]).

In mammals, conditions of ER stress lead to the activation of the unfolded protein response (UPR) through activation of three distinct, integral ER membrane proteins designated PERK, Ire1, and ATF6 [67]. Together, these proteins signal (i) the general inhibition of protein expression, through the phosphorylation of the translation initiation factor eIF2α [68], and (ii) the specific induction of ER-resident chaperone expression including GRP78, GRP94, PDI, and calreticulin, that function to assist in the folding of the accumulated polypeptides [69–71]. The ER stress response also induces the expression of the transcription factor, GADD153, which is known to play a role in ER stress-induced growth arrest and programmed cell death [72]. The ultimate fate of a cell is determined by the balance of protective (i.e., ER chaperone expression) and proapoptotic signals triggered by ER dysfunction. We have shown that glucosamine can induce ER stress in cells that are relevant to atherogenesis—including human macrophages, aortic SMCs, endothelial cells as well as hepatocytes (Ref. [62] and unpublished data).

In recent years additional cellular responses to ER stress have been identified (Fig. 13.3). Our lab and others have shown that ER stress can activate

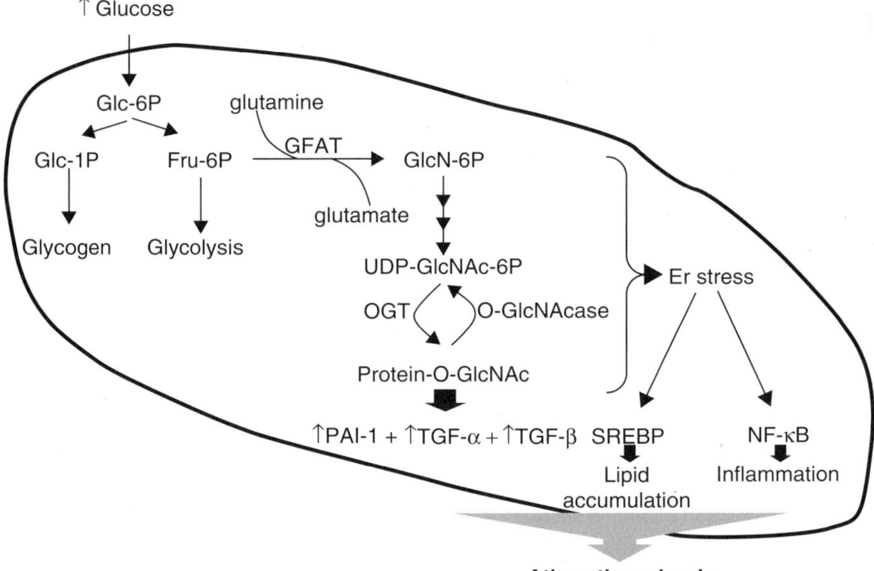

FIGURE 13.3. Links between the hexosamine pathway and atherothrombosis. The rate limiting enzyme of the hexosamine pathway, glutamine:fructose-6 phosphate amidotransferase (GFAT) converts fructose-6 phosphate to glucosamine 6-phosphate (GlcN-6P). Glycosylation of serine and threonine residues of intracellular proteins is catalyzed by OGT, a reaction that is reversed by O-GlcNAcase. The O-linked glycosylation of transcription factors including SP-1 have been shown to increase expression levels of PAI-1, TGF-α, and TGF-β. Glucosamine and/or downstream metabolites of glucosamine can also promote endoplasmic reticulum (ER) stress leading to lipid accumulation and activation of intracellular inflammatory pathways through the activation of the transcription factors SREBP and NF-κB, respectively.

and dysregulate the sterol regulatory element binding proteins (SREBPs), transcription factors that control lipid biosynthesis and uptake [62, 73, 74]. ER stress-inducing agents have been shown to promote the activation of NF-κB, the transcription factor that is responsible for promoting the expression of genes involved in inflammatory processes [75]. Finally, ER stress has been shown to induce the activation of caspases and promote apoptosis of human aortic endothelial cells and hepatocytes [62, 76]. Together, the dysregulation of lipid metabolism, activation of inflammatory pathways, and the induction of cell-specific apoptosis represent the hallmark features of atherosclerosis. In addition to glucosamine, other independent cardiovascular risk factors including homocysteine [73] and obesity [77] as well as the accumulation of intracellular free cholesterol [78] have been shown to induce ER stress. Thus ER stress pathway may be a common mechanism of general atherogenesis.

The potential role of ER stress-induced cellular dysfunction in the development and/or progression of human disease make this pathway a plausible therapeutic target. Studies are underway to develop and test potential antiatherogenic interventions that target ER stress pathways in models of diabetes mellitus.

Protein Kinase C Activation

The ubiquitous protein kinase C (PKC) family of signaling factors are known to play an important role in vascular function and in the pathogenesis of micro- and macrovascular disease. Specifically, glucose-induced PKC activation has been implicated in decreased endothelial vasodilation [79] and increased production of ROS [80] that could contribute to endothelial dysfunction. PKC inhibition blocks endothelial expression and activity of a number of adhesion molecules, including ICAM-1, P-, and E-selectins, under conditions of elevated glucose concentration [81]. AngII- and oxLDL-induced VSMC proliferation is dependent upon PKC activity [82, 83]. PKC may also be involved in foam cell formation through its involvement in the induction of receptors for oxidized LDL macrophages [84]. Most PKC isoforms can be activated by diacylglycerol (DAG), an intermediary of glucose and fat metabolism, and conditions of diabetes and high glucose concentration have been shown to increase membrane-associated PKC activity via the DAG activation pathway in cultured VSMCs and in animal models [85]. In addition, it has been proposed the PKC may also be activated by AGE–RAGE signaling pathways (see below). Selective inhibitors of PKC have shown promise in blocking nephropathy and retinopathy in animal models of diabetes [86]; however, there is no direct evidence that these inhibitors affect atherogenesis. The ubiquitous nature of expression, the large number of PKC isoforms and the lack of isoform-specific inhibitors are complicating factors in determining the potential role of this kinase in atherogenesis.

Advanced Glycation Endproducts

AGEs are formed through a nonenzymatic process, known as the Maillard reaction, involving the reaction of the aldehyde groups of reducing sugars with the amino groups of proteins [87, 88] (Fig. 13.4). Because the reaction is nonspecific, requiring only a free amine, AGEs are a very heterogeneous group of compounds. The final step in the reaction, the conversion of Amidori products into cross-linked advanced glycation products, occurs very slowly *in vivo*, and therefore AGEs are predominantly composed of stable proteins with long half-lives such as collagen, laminin, and lens crystallins.

AGEs gradually accumulate during the natural aging process and have been implicated in characteristic changes associated with aging including the loss of collagen elasticity and flexibility, the formation of cataracts and β amyloid plaques [89]. Formation of AGEs has been noted during the heating of food however, in healthy individuals, AGEs are poorly absorbed by the gut and rapidly cleared by the kidney [90]. Individuals with impaired renal function, a common complication of diabetes, are especially prone to the accumulation of dietary AGEs and therefore exogenous sources of AGEs may play a significant role in this population [91, 92].

In vivo, the Maillard reaction rate is directly proportional to the prevailing glucose concentration resulting in the enhanced levels of AGEs and AGE

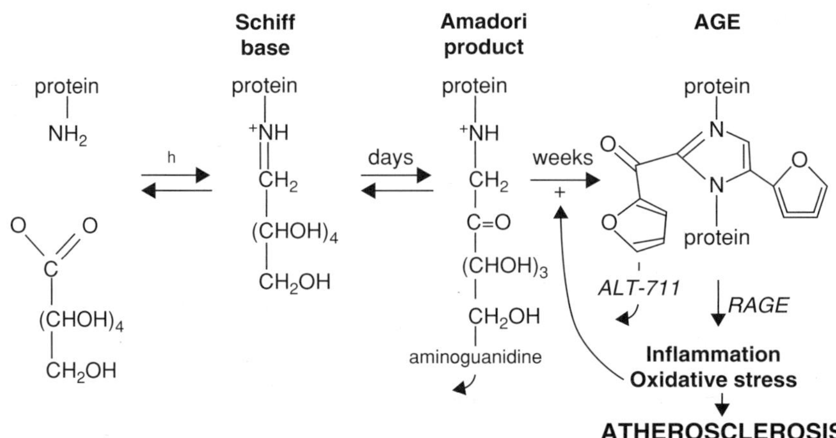

FIGURE 13.4. The formation of advanced glycation endproducts (AGE) by the Mallaird reaction. AGEs are produced by a multistep reaction involving the sequential formation of Schiff bases, Amidori products, and finally a heterogenous population of cross-linked advanced glycation endproducts. AGEs can interact with cellular receptors (RAGE) that activate intracellular inflammatory pathways and the production of ROS that may contribute to atherogenesis. Conditions of oxidative stress can also accelerate the formation of AGEs. The possible sites of post-Amidori-cross-link elimination by aminoguanidine and the cross-link breaker, ALT-711, are indicated.

intermediates in individuals with diabetes mellitus. This is clearly demonstrated by the routine clinical use of the Amadori-intermediate, glycated hemoglobin (HbA1c), as a diagnostic marker of chronic hyperglycemia.

The association of AGEs with chronic hyperglycemia has attracted a great deal of interest into the possible role of AGEs in diabetic complications. There are several potential ways that AGE-modified proteins could be damaging; the formation of AGEs may alter protein function [93], disrupt extracellular matrix [94, 95], and/or lead to the modification of lipoprotein particles thereby increasing their atherogenicity. However it now appears that the predominant vascular effect of AGEs occurs through their interaction with RAGE (receptor for AGE) found on macrophages, endothelial, and SMCs [96–98]. The AGE–RAGE interaction triggers a signal transduction cascade that results in the production of intracellular reactive oxidative species that can initiate an inflammatory response [99, 100]. The presence of ROS also feed back to accelerate the formation of AGEs.

There is a great deal of evidence supporting a role for AGEs in microvascular disease. AGEs accumulate in the kidney, retina, and peripheral nerve of diabetic rats [101]. Nephropathy can be induced in nondiabetic animals by the administration of AGEs [91, 92]. Neutralizing antibodies against RAGE also slow the progression of nephropathy in diabetic mice [102]. The administration of compounds that block AGE cross-linking (aminoguanidine and pyridoxamine) or interventions that break established cross-links (3-phenacyl-4,5-dimethylthiazolium chloride or ALT-711) have been shown to be protective against nephropathy and cardiovascular stiffness in diabetic models [93, 103].

There is a growing body of evidence supporting a role for AGEs in the development of atherosclerosis. AGEs have been localized to atherosclerotic plaques and RAGE is expressed by vascular endothelial cells, SMCs, and macrophages [104, 105]. Administration of both aminoguanidine and the AGE cross-link breaker, ALT-711, are associated with a significant decrease in plaque area and complexity in diabetic ApoE–/– mice [106]. The strongest evidence in support of the AGE–RAGE pathway comes from experiments that focus upon interfering with receptor function. This has been accomplished using a soluble form of RAGE (sRAGE) that retains its ligand binding ability [107]. Daily intraperitoneal injections of sRAGE into STZ-induced diabetic mice showed dose-dependent attenuation of early lesion development in terms of lesion area at the aortic root and lesion complexity relative to controls. Importantly, plasma glucose and lipid levels were not affected by sRAGE treatment suggesting the effect was glycemia and lipid independent [107]. It has further been demonstrated that sRAGE can block the progression and even induce the regression of established lesions in diabetic mice [108]. In support of these findings, in vascular damage assays RAGE null mice show [109] decreased intimal/medial thickening. As of yet there have been no published reports to show the effects of the RAGE knockout genotype on plaque development in mouse models of diabetes-accelerated atherosclerosis.

Interventions that interfere with AGE formation and action have shown some beneficial effects at the clinical level [110]. ALT-711 (under the commercial name, Alagebrium) has demonstrated safety and efficacy in a phase II clinical trial in the treatment of systolic hypertension and heart failure, conditions associated with vascular stiffening likely resulting from collagen cross-linking [111].

Some controversy still remains regarding the role of AGEs in atherosclerosis, in part due to the ability of other ligands, including cytokines of the S100/calgranulin family, to interact with and activate RAGE [112, 113]. The role of AGE–RAGE is further complicated by the fact that virtually every large, well-controlled clinical trial has shown that antioxidant supplementation has no positive impact on cardiovascular risk in diabetic patients [114–117]. This paradox supports the possibility that other, AGE-independent, mechanisms and pathways may play an important role in the atherogenic process.

It is important to note that the above mechanisms are not mutually exclusive. Atherosclerosis is a complex disease that involves the interaction of several different cell types and many different molecular factors. Therefore, it is very possible that atherogenesis is regulated and affected by several independent and interdependent pathways that are altered under conditions of hyperglycemia. Furthermore, diabetes mellitus, especially T2D, is often associated with a cluster of conditions including hypertension, obesity, and dyslipidemia—each a CVD risk factor in its own right. Thus the interplay between this combination of risk factors in the context of a particular genetic background and modulated by a myriad of environmental influences will all contribute to the development and progression of the disease.

Conclusions

Because of the cardiovascular risks of diabetes and the increasing prevalence of T2D, it is essential that we further increase our knowledge of how and why diabetes mellitus and hyperglycemia promote CVD. Currently, and for the near future, the primary strategy for managing CVD in the diabetic population will be through the control of hyperglycemia and through the treatment of associated complications such as hypertension and dyslipidemia using established medications such as ACE inhibitors, statins, and fibrates.

The continued identification and investigation of existing and novel pathways linking hyperglycemia and diabetes mellitus to atherosclerosis is important to the development of new and effective antiatherosclerotic therapies that are tailored to individuals with diabetes. A great deal of research has been focused upon the role of hyperglycemia in the development and progression of atherosclerosis in cell culture and animal model systems. Several mechanisms have been identified that appear to link glucose to proatherogenic processes. The most promising of these; the polyol pathway,

PKC activation, the hexosamine pathway, and the AGE–RAGE interaction show potential and are actively being evaluated as targets for putative antiatherogenic therapies. Thus far, however, all interventions targeting the effects of hyperglycemia, including direct glucose lowering, appear to show greater effect in the treatment of microvascular complications than in the control of macrovascular disease. This is likely due, at least in part, to the complexities of atherosclerosis and current limitations of the animal models available to researchers who study the development and progression of atherosclerosis. Additional studies are obviously required to further improve our understanding of this important disease.

Acknowledgments: G.H. Werstuck is supported by a New Investigator Award from the Heart and Stroke Foundation of Canada and research grants from the Canadian Institutes of Health Research (MOP-62910) and the Heart and Stroke Foundation of Ontario (NA5556).

References

1. King H, Aubert RE, Herman WH: Global burden of diabetes, 1995–2025: prevalence, numerical estimates, and projections. Diabetes Care 21: 1414–1431, 1998.
2. Zimmet P, Alberti KGMM, Shaw J: Global and societal implications of the diabetes epidemic. Nature 414: 782–788, 2001.
3. Kannel WB, McGee DL: Diabetes and cardiovascular disease. The Framingham study. JAMA 241: 2035–2038, 1979.
4. Haffner SM, Lehto S, Ronnemaa T, Pyorala K, Laakso M: Mortality from coronary heart disease in subjects with type2 diabetes and in nondiabetic subjects with and without prior myocardial infarction. N Engl J Med 339: 229–234, 1998.
5. Resnick HE, Howard BV: Diabetes and cardiovascular disease. Annu Rev Med 53: 345–367, 2002.
6. Canadian Diabetes Association Clinical Practice Guidelines Experts Committee: Definition, classification and diagnosis of diabetes and other dysglycemic categories. Canadian J Diabetes 27: S7–S9, 2003.
7. Tisch R, McDevitt H: Insulin dependent diabetes mellitus. Cell 85: 291–297, 1906.
8. Mathis D, Vence L, Benoist C: β cell death during progression to diabetes. Nature 414: 792–798, 1998.
9. Ginsberg HN: Insulin resistance and cardiovascular disease. J Clin Invest 106: 453–458, 2000.
10. Bell GI, Polonsky KS: Diabetes mellitus and genetically programmed defect in β-cell function. Nature 414: 788–791, 2001.
11. The Diabetes Control and Complications Trial/Epidemiology of Diabetes Interventions and Complications Research Group: Intensive diabetes therapy and carotid intima-medial thickness in type 1 diabetes mellitus. New Engl J Med 348: 2294–2303, 2003.
12. UK Prospective Diabetes Study (UKPDS) Group: Intensive blood-glucose control with sulphonylureas or insulin compared with conventional treatment and risk of complications in patients with type 2 diabetes (UKSPDS 33). Lancet 352: 837–853, 1998.

13. Haffner SM: The importance of hyperglycemia in the nonfasting state to the development of cardiovascular disease. Endocrine Rev 19: 583–592, 1998.
14. Standl E, Balletshofer B, Dahl B, Weichenhain B, Steigler H, Hormann A, Holl R: Predictors of 10-year macrovascular and overall mortality in patients with NIDDM: the Munich General Practitioner Project. Diabetologia 39: 1540–1545, 1996.
15. Lehto S, Ronnemaa T, Haffner SM, Pyorala K, Kallio V, Laakso M: Dyslipidemia and hyperglycemia predict coronary heart disease events in middle-aged patients with NIDDM. Diabetes 46: 1354–1359, 1997.
16. Santilli SM, Fiegel VD, Aldridge DE, Knighton DR: The effect of diabetes on the proliferation of aortic endothelial cells. Ann Vasc Surg 6: 503–510, 1992.
17. Huvers FC, De Leeuw PW, Houben AJ, De Haan CH, Hamulyak K, Schouten H, Wolffenbuttel BH, Schaper NC: Endothelium-dependent vasodilatation, plasma markers of endothelial function, and adrenergic vasoconstrictor responses in type 1 diabetes under near-normoglycemic conditions. Diabetes 48: 1300–1307, 1999.
18. Quagliaro L, Piconi L, Assaloni R, Martinelli L, Motz E, Ceriello A: Intermittent high glucose enhances apoptosis related to oxidative stress in human umbilical vein endothelial cells: the role of protein kinase C and NAD(P)H-oxidase activation. Diabetes 52: 2795–2804, 2003.
19. Nathan DM, Lachin J, Cleary P, Orchard T, Brillon DJ, Backlund JY, O'Leary DH, Genuth S; Diabetes Control and Complications Trial, Epidemiology of Diabetes Interventions and Complications Research Group: Intensive diabetes therapy and carotid intima-media thickness in type 1 diabetes mellitus. N Engl J Med 348: 2294–2303, 2003.
20. Lawes CM, Parag V, Bennett DA, Suh I, Lam TH, Whitlock G, Barzi F, Woodward M; Asia Pacific Cohort Studies Collaboration: Blood glucose and risk of cardiovascular disease in the Asia Pacific region. Diabetes Care 27: 2836–2842, 2004.
21. Grundy SM, Brewer HB Jr, Cleeman JI, Smith SC Jr, Lenfant C; American Heart Association; National Heart, Lung, and Blood Institute: Definition of metabolic syndrome: Report of the National Heart, Lung, and Blood Institute/American Heart Association conference on scientific issues related to definition. Circulation 109: 433–438, 2004.
22. World Health Organization: Obesity: preventing and managing the global epidemic. (WHO Technical Report Series, No. 894), 2003.
23. Grundy SM: Metabolic complications of obesity. Endocrine 13: 155–165, 2000.
24. Kunjathoor VV, Wilson D, LeBoeuf RC: Increased atherosclerosis in streptozotocin-induced diabetic mice. J Clin Invest 97: 1767–1773, 1996.
25. Park L, Raman KG, Lee KJ, Lu Y, Ferran LJ Jr, Chow WS, Stern D, Schmidt AM: Suppression of accelerated diabetic atherosclerosis by the soluble receptor for advanced glycation endproducts. Nat Med 4: 1025–1031, 1998.
26. Fazio S, Linton MF: Mouse models of hyperlipidemia and atherosclerosis. Front Biosci 6: D515–D525, 2001.
27. Rees DA, Alcolado JC: Animal models of diabetes mellitus. Diabet Med 22: 359–370, 2005.
28. Breslow JL: Mouse models of atherosclerosis. Science 272: 685–688, 1996.
29. Szkudelski T: The mechanism of alloxan and streptozotocin action in B cells of the rat pancreas. Physiol Res 50: 537–546, 2001.

30. Winterbourn CC, Munday R: Glutathione-mediated redox cycling of alloxan. Mechanisms of superoxide dismutase inhibition and of metal-catalyzed OH formation. Biochem Pharmacol 38: 271–277, 1989.
31. Schnedl WJ, Ferber S, Johnson JH, Newgard CB: STZ transport and cytotoxicity. Specific enhancement in GLUT2-expressing cells. Diabetes 43: 1326–1333, 1994.
32. Renard CB, Kramer F, Johansson F, Lamharzi N, Tannock LR, Herrath MG, Chait A, Bornfeldt KE: Diabetes and diabetes-associated lipid abnormalities have distinct effects on initiation and progression of atherosclerotic lesions. J Clin Invest 114: 659–668, 2004.
33. Reaven P, Merat S, Casandra F, Sutphin M, Palinski W: Effect of streptozotocin-induced hyperglycemia on lipid profiles, formation of advanced glycation end-products in lesions, and extent of atherosclerosis in LDL receptor-deficient mice. Arterioscler Thromb Vasc Biol 17: 2250–2256, 1997.
34. Kako Y, Huang L-S, Yang J, Katopodis T, Ramakrishnan R, Goldberg IJ: Streptozotocin-induced diabetes in human apolipoprotein B transgenic mice: effects on lipoproteins and atherosclerosis. J Lipid Res 40: 2185–2194, 1999.
35. Goldberg IJ, Isaacs A, Sehayek E, Breslow JL, Huang L-S: Effects of streptozotocin-induced diabetes in apolipoprotein AI deficient mice. Atherosclerosis 172: 47–53, 2004.
36. Hasty AH, Shimano H, Osuga J, Namatame I, Takahashi A, Yahagi N, Perrey S, Iizuka Y, Tamura Y, Amemiya-Kudo M, Yoshikawa T, Okazaki H, Ohashi K, Harada K, Matsuzaka T, Sone H, Gotoda T, Nagai R, Ishibashi S, Yamada N: Severe hypercholesterolemia, hypertriglyceridemia, and atherosclerosis in mice lacking both leptin and the low-density lipoprotein receptor. J Biol Chem 276: 37402–37408, 2001.
37. Ross R: Atherosclerosis—an inflammatory disease. New Engl J Med 340: 115–126, 1999.
38. Lusis AJ: Atherosclerosis. Nature 407: 233–241, 2000.
39. Drielsma RF, Burnett JR, Gray-Weale AC, Byrne K, Lusby RJ: Carotid artery disease: the influence of diabetes mellitus. J Cardiovasc Surg (Torino) 29: 692–696, 1988.
40. Silva JA, Escobar A, Collins TJ, Ramee SR, White CJ: Unstable angina. A comparison of angioscopic findings between diabetic and nondiabetic patients. Circulation 92: 1731–1736, 1995.
41. Srivastava SK, Ansari NH, Hair GA, Jaspan J, Rao MB, Das B: Hyperglycemia-induced activation of human erythrocyte aldose reductase and alterations in kinetic properties. Biochim Biophys Acta 870: 302–311, 1986.
42. Chung SS, Ho EC, Lam KS, Chung SK: Contribution of polyol pathway to diabetes-induced oxidative stress. J Am Soc Nephrol 14: S233–S236, 2003.
43. Greene DA, Arezzo JC, Brown MB, the Zenarestat Study Group: Effect of aldose reductase inhibition on nerve conduction and morphometry in diabetic neuropathy. Neurology 53: 580–591, 1999.
44. Sorbinil Retinopathy Trial Research Group: The Sorbinil Retinopathy Trial: Neuropathy results. Neurology 43: 1141–1149, 1993.
45. Engerman RL, Kern TS, Larson ME: Nerve conduction and aldose reductase inhibition during 5 years of diabetes or galactosaemia in dogs. Diabetologia 37: 141–144, 1994.

46. Brown MJ, Bird SJ, Watling S, Kaleta H, Hayes L, Eckert S, Foyt HL; Zenarest study: Natural progression of diabetic peripheral neuropathy in the Zenarestat study population. Diabetes Care 27: 1153–1159, 2004.

47. Marshall S, Bacote V, Traxinger RR: Discovery of a metabolic pathway mediating glucose-induced desensitization of the glucose transport system. J Biol Chem 266: 4706–4712, 1991.

48. Nerlich AG, Sauer U, Kolm-Litty V, Wagner E, Koch M, Schleicher ED: Expression of glutamine:fructose-6-phosphate amidotransferase in human tissues: evidence for high variability and distinct regulation in diabetes. Diabetes 47: 170–178, 1998.

49. Herbert LF, Daniels MC, Zhou J, Crook ED, Turner RL, Simmons ST, Neidigh JL, Zhu JS, Baron AD, McClain DA: Overexpression of glutamine: fructose-6-phosphate amidotransferase in transgenic mice leads to insulin resistance. J Clin Invest 98: 930–936, 1996.

50. Lui K, Paterson AJ, Chin E, Kudlow JE: Glucose stimulates protein modification by O-linked GlcNAc in pancreatic beta cells: linkage of O-linked GlcNAc to beta cell death. Proc Natl Acad Sci USA 14: 2820–2825, 2000.

51. Stender S, Astrup P: Glucosamine and experimental atherosclerosis. Increased wet weight and changed composition of cholesterol fatty acids in aorta of rabbits fed a cholesterol-enriched diet with added glucosamine. Atherosclerosis 26: 205–213, 1977.

52. Han I, Kudlow JE: Reduced O glycosylation of SP1 is associated with increased proteasome susceptibility. Mol Cell Biol 17: 2550–2558, 1997.

53. Han I, Oh E, Kudlow JE: Responsiveness of the state of O-linked N-acetyl-glucosamine modification of nuclear pore protein p62 to the extracellular glucose concentration. Biochem J 350: 109–114, 2000.

54. Du XL, Edelstein D, Dimmeler S, Ju Q, Sui C, Brownlee M: Hyperglycemia inhibits nitric oxide synthase activity by posttranslational modification at the Akt site. J Clin Invest 108: 1341–1348, 2001.

55. Hawkins M, Angelov I, Liu R, Barzilai N, Rossetti L: The tissue concentration of UDP-N-acetylglucosamine modulates the stimulatory effect of insulin on skeletal muscle glucose uptake. J Biol Chem 272: 4889–4895, 1997.

56. Ross SA, Chen X, Hope HR, Sun S, McMahon EG, Broschat K, Gulve EA: Development and comparison of two 3T3-L1 adipocyte models of insulin resistance: increased glucose flux vs glucosamine treatment. Biochem Biophys Res Commun 273: 1033–1041, 2000.

57. James LR, Fantus IG, Goldberg H, Ly H, Scholey JW: Overexpression of GFAT activates PAI-1 promoter in mesangial cells. Am J Physiol Renal Physiol 279: F718–F727, 2000.

58. Du X-L, Edelstein D, Rossetti L, Fantus IG, Goldberg H, Ziyadeh F, Wu J, Brownlee M: Hyperglycemia-induced mitochondrial superoxide overproduction activates the hexosamine pathway and induces plasminogen activator inhibitor-1 expression by increasing Sp1 glycosylation. Proc Natl Acad Sci USA 97: 12222–12226, 2000.

59. Morin MJ, Porter CW, McKernan P, Bernacki RJ: The biochemical and ultrastructural effects of tunicamycin and D-glucosamine in L1210 leukemic cells. J Cell Physiol 114: 162–172, 1983.

60. Lin H, Masso-Welsh P, Di Y, Cai J, Shen J, Subjeck JR: The 170-kDa glucose-regulated stress protein is an endoplasmic reticulum protein that binds immunoglobulin. Mol Biol Cell 4: 1109–1119, 1993.

61. Miskovic D, Salter-Cid L, Ohan N, Flajnik M, Heikkila JJ: Isolation and characterization of a cDNA encoding a Xenopus immunoglobulin binding protein, BiP (GRP78). Comp Biochem Physiol 116: 227–234, 1997.
62. Kim AJ, Shi YY, Austin RC, Werstuck GH: Valproate protects cells from endoplasmic reticulum stress-induced lipid accumulation and apoptosis by inhibiting glycogen synthase kinase 3. J Cell Sci 118: 89–99, 2005.
63. Helenius A: How-linked oligosaccharides effect protein folding in the endoplasmic reticulum. Mol Biol Cell 5: 253–265, 1994.
64. Travers KJ, Patil CK, Wodicka L, Lockhart DJ, Weissman JS, Walter P: Functional and genomic analyses reveal an essential coordination between the unfolded protein response and ER-associated degradation. Cell 101: 249–258, 2000.
65. Friedlander R, Jarosch E, Urban J, Volkwein C, Sommer T: A regulatory link between ER-associated protein degradation and the unfolded-protein response. Nat Cell Biol 2: 379–384, 2000.
66. Pahl HL: Signal transduction from the endoplasmic reticulum to the cell nucleus. Physiol Rev 79: 683–701, 1999.
67. Kaufman RJ: Stress signalling from the lumen of the endoplasmic reticulum: coordination of gene transcriptional and translational controls. Genes Dev 13: 1211–1233, 1999.
68. Harding HP, Zhang Y, Ron D: Protein translation and folding are coupled by an endoplasmic-reticulum-resident kinase. Nature 397: 271–274, 1999.
69. Cox JS, Shamu CE, Walter P: Translational induction of genes encoding endoplasmic reticulum resident proteins requires a transmembrane protein kinase. Cell 73: 1197–1206, 1993.
70. Wang X-Z, Harding HP, Zhang Y, Jolicoeur EM, Kuroda M, Ron D: Cloning of mammalian Ire1 reveals diversity in the ER stress response. EMBO J 17: 5708–5717, 1998.
71. Yoshida H, Haze K, Yanagi H, Yura T, Mori K: Identification of the *cis*-acting endoplasmic reticulum stress response element responsible for the transcriptional induction of mammalian glucose-regulated proteins. J Biol Chem 273: 33741–33749, 1998.
72. Zinszner H, Kuroda M, Wang X-Z, Batchvarova N, Lightfoot RT, Remotti H, Stevens JL, Ron D: CHOP is implicated in programmed cell death in response to impaired function of the endoplasmic reticulum. Genes Dev 12: 982–995, 1998.
73. Werstuck GH, Lentz SR, Dayal S, Shi Y, Hossain GS, Sood SK, Krisans SK, Austin RC: Homocysteine causes dysregulation of the cholesterol and fatty acid biosynthesis pathways. J Clin Invest 107: 1263–1273, 2001.
74. Lee JN, Ye J: Proteolytic activation of SREBP induced by cellular stress through the depletion of Insig-1. J Biol Chem 279: 45257–45265, 2004.
75. Jiang H-Y, Wek SA, McGrath BC, Scheuner D, Kaufman RJ, Cavener DR, Wek RC: Phosphorylation of the alpha subunit of eukaryotic initiation factor 2 is required for activation of NF-kappaB in response to diverse cellular stresses. Mol Cell Biol 23: 5651–5663, 2003.
76. Hossain GS, van Thienen JV, Werstuck GH, Zhou J, Sood SK, Dickhout JG, de Koning AB, Tang D, Wu D, Falk E, Poddar R, Jacobsen DW Zhang K, Kaufman RJ, Austin RC: TDAG51 is induced by homocysteine, promotes detachment-mediated programmed cell death and contributes to the development of atherosclerosis in hyperhomocysteinemia. J Biol Chem 278: 30317–30327, 2003.

77. Özcan U, Cao Q, Yilmaz E, Lee A-H, Iwakoshi NN, Özdelen E, Tuncman G, Görgun C, Glimcher LH, Hotamisligil GS: Endoplasmic reticulum stress links obesity, insulin action, and type 2 diabetes. Science 306: 457–461, 2004.

78. Li Y, Schwabe RF, Devries-Seimon T, Yao PM, Gerbod-Giannone MC, Tall AR, Davis RJ, Flavell R, Brenner DA, Tabas I: Free cholesterol-loaded macrophages are an abundant source of TNF-alpha and IL-6. Model of NF-kappa B- and MAP kinase-dependent inflammation in advanced atherosclerosis. J Biol Chem 280: 21763–21772, 2005.

79. Tesfamariam B, Brown ML, Cohen RA: Elevated glucose impairs endothelium-dependent relaxation by activating protein kinase C. J Clin Invest 87: 1643–1648, 1991.

80. Inoguchi T, Li P, Umeda F, Yu HY, Kakimoto M, Imamura M, Aoki T, Etoh T, Hashimoto T, Naruse M, Sano H, Utsumi H, Nawata H: High glucose level and free fatty acid stimulate reactive oxygen species production through protein kinase C—dependent activation of NAD(P)H oxidase in cultured vascular cells. Diabetes 49: 1939–1945, 2000.

81. Omi H, Okayama N, Shimizu M, Okouchi M, Ito S, Fukutomi T, Itoh M: Participation of high glucose concentrations in neutrophil adhesion and surface expression of adhesion molecules on cultured human endothelial cells: effect of antidiabetic medicines. J Diabetes Complications 16: 201–208, 2002.

82. Yang CM, Chien CS, Hsiao LD, Pan SL, Wang CC, Chiu CT, Lin CC: Mitogenic effect of oxidized low-density lipoprotein on vascular smooth muscle cells mediated by activation of Ras/Raf/MEK/MAPK pathway. Br J Pharmacol 132: 1531–1541, 2001.

83. Greene EL, Lu G, Zhang D, Egan BM: Signaling events mediating the additive effects of oleic acid and angiotensin II on vascular smooth muscle cell migration. Hypertension 37: 308–312, 2001.

84. Li L, Sawamura T, Renier G: Glucose enhances human macrophage LOX-1 expression: role for LOX-1 in glucose-induced macrophage foam cell formation. Circ Res 94: 892–901, 2004.

85. Inoguchi T, Xia P, Kunisaki M, Higashi S, Feener EP, King GL: Insulin's effect on protein kinase C and diacylglycerol induced by diabetes and glucose in vascular tissues. Am J Physiol 267: E369–E379, 1994.

86. Ishii H, Jirousek MR, Koya D, Takagi C, Xia P, Clermont A, Bursell SE, Kern TS, Ballas LM, Heath WF, Stramm LE, Feener EP, King GL: Amelioration of vascular dysfunctions in diabetic rats by an oral PKC beta inhibitor. Science 272: 728–731, 1996.

87. Wells-Knecht KJ, Zyzak DV, Litchfield JE, Thorpe SR, Baynes JW: Mechanism of autoxidative glycosylation: identification of glyoxal and arabinose as intermediates in the autoxidative modification of proteins by glucose. Biochemistry 34: 3702–3709, 1995.

88. Degenhardt TP, Thorpe SR, Baynes JW: Chemical modification of proteins by methylglyoxal. Cell Mol Biol 44: 1139–1145, 1998.

89. Ulrich P, Cerami A: Protein glycation, diabetes, and aging. Recent Prog Horm Res 56: 1–21, 2001.

90. Sgarbieri VC, Amaya J, Tanaka M, Chichester CO: Nutritional consequences of the Maillard reaction. Amino acid availability from fructose-leucine and fructose-tryptophan in the rat. J Nutr 103: 657–663, 1973.

91. Makita Z, Radoff S, Rayfield EJ, Yang Z, Skolnik E, Delaney V, Friedman EA, Cerami A, Vlassara H: Advanced glycosylation end products in patients with diabetic nephropathy. N Engl J Med 325: 836–842, 1991.

92. Koschinsky T, He C-J, Mitsuhashi T, Bucala R, Liu C, Buenting C, Heitmann K, Vlassara H: Orally absorbed reactive glycation products (glycotoxins): an environmental risk factor in diabetic nephropathy Proc Natl Acad Sci USA 94: 6474–6479, 1997.

93. Liu J, Masurekar MR, Vatner DE, Jyothirmayi GN, Regan TJ, Vatner SF, Meggs LG, Malhotra A: Glycation end-product cross-link breaker reduces collagen and improves cardiac function in aging diabetic heart. Am J Physiol Heart Circ Physiol 285:H2587–H2591, 2003.

94. Tanaka S, Avigad G, Brodsky B, Eikenberry EF: Glycation induces expansion of the molecular packing of collagen. J Mol Biol 203: 495–505, 1988.

95. Haitoglou CS, Tsilibary EC, Brownlee M, Charonis AS: Altered cellular interactions between endothelial cells and nonenzymatically glucosylated lamin/type IV collagen. J Biol Chem 267: 12404–12407, 1992.

96. Neeper M, Schmidt AM, Brett J, Yan SD, Wang F, Pan YC, Elliston K, Stern D, Shaw A: Cloning and expression of a cell surface receptor for advanced glycation end products of proteins. J Biol Chem 267: 14998–15004, 1992.

97. Li YM, Mitsuhashi T, Wojciechowicz D, Shimizu N, Li J, Stitt A, He C, Banerjee D, Vlassara H: Molecular identity and cellular distribution of advanced glycation endproduct receptors: relationship of p60 to OST-48 and p90 to 80K-H membrane proteins. Proc Natl Acad Sci USA 93: 11047–11052, 1997.

98. Smedsrod B, Melkko J, Araki N, Sano H, Horiuchi S: Advanced glycation end products are eliminated by scavenger-receptor-mediated endocytosis in hepatic sinusoidal Kupffer and endothelial cells. Biochem J 322: 567–573, 1997.

99. Hofmann MA, Lalla E, Lu Y, Gleason MR, Wolf BM, Tanji N, Ferran LJ Jr, Kohl B, Rao V, Kisiel W, Stern DM, Schmidt AM: Hyperhomocysteinemia enhances vascular inflammation and accelerates atherosclerosis in a murine model. J Clin Invest 107: 675–683, 2001.

100. Lander HM, Taurus JM, Ogiste JS, Hori O, Moss RA, Schmidt AM: Activation of the receptor for advanced glycation end products triggers a p21(ras)-dependent mitogen-activated protein kinase pathway regulated by oxidative stress. J Biol Chem 272: 17810–17814, 1997.

101. Karachalias N, Babaei-Jadidi R, Ahmed N, Thornalley PJ: Accumulation of fructosyl-lysine and advanced glycation end products in the kidney, retina and peripheral nerve of streptozotocin-induced diabetic rats. Biochem Soc Trans 31: 1423–1425, 2003.

102. Flyvbjerg A, Denner L, Schrijvers BF, Tilton RG, Mogensen TH, Paludan SR, Rasch R: Long-term renal effects of a neutralizing RAGE antibody in obese type 2 diabetic mice. Diabetes 53: 166–172, 2004.

103. Forbes JM, Thallas V, Thomas MC, Founds HW, Burns WC, Jerums G, Cooper ME: The breakdown of pre-existing advanced glycation end products is associated with reduced renal fibrosis in experimental diabetes. FASEB J 17: 1762–1764, 2003.

104. Meng J, Sakata N, Takebayashi S, Asano T, Futata T, Araki N, Horiuchi S: Advanced glycation end products of the Maillard reaction in aortic pepsin-insoluble and pepsin-soluble collagen from diabetic rats. Diabetes 45: 1037–1043, 1996.

105. Brett J, Schmidt AM, Yan SD, Zou YS, Weidman E, Pinsky D, Nowygrod R, Neeper M, Przysiecki C, Shaw A: Survey of the distribution of a newly characterized receptor for advanced glycation end products in tissues. Am J Pathol 143: 1699–1712, 1993.

106. Forbes JM, Yee LT, Thallas V, Lassila M, Candido R, Jandeleit-Dahm KA, Thomas MC, Burns WC, Deemer EK, Thorpe SM, Cooper ME, Allen TJ: Advanced glycation end product interventions reduce diabetes-accelerated atherosclerosis. Diabetes 53: 1813–1823, 2004.

107. Park L, Raman KG, Lee KJ, Lu Y, Ferran LJ, Chow WS, Stern D, Schmidt AM: Suppression of accelerated diabetic atherosclerosis by the soluble receptor for advanced glycation endproducts. Nat Med 4: 1025–1031, 1998.

108. Bucciarelli LG, Wendt T, Qu W, Lu Y, Lalla E, Rong LL, Goova MT, Moser B, Kislinger T, Lee DC, Kashyap Y, Stern DM, Schmidt AM: RAGE blockade stabilizes established atherosclerosis in diabetic apolipoprotein E-null mice. Circulation 106: 2827–2835, 2002.

109. Sakaguchi T, Yan SF, Yan SD, Belov D, Rong LL, Sousa M, Andrassy M, Marso SP, Duda S, Arnold B, Liliensiek B, Nawroth PP, Stern DM, Schmidt AM, Naka Y: Central role of RAGE-dependent neointimal expansion in arterial restenosis. J Clin Invest 111: 959–972, 2003.

110. Williams ME: Clinical studies of advanced glycation end product inhibitors and diabetic kidney disease. Curr Diab Rep 4: 441–446, 2004.

111. Little WC, Zile MR, Kitzman DW, Hundley WG, O'Brien TX, Degroof RC: The effect of alagebrium chloride (ALT-711), a novel glucose cross-link breaker, in the treatment of elderly patients with diastolic heart failure. J Card Fail 11: 191–195, 2005.

112. Valencia JV, Mone M, Koehne C, Rediske J, Hughes TE: Binding of receptor for advanced glycation end products (RAGE) ligands is not sufficient to induce inflammatory signals: lack of activity of endotoxin-free albumin-derived advanced glycation end products. Diabetologia 47: 844–852, 2004.

113. Valencia JV, Zhang MM, Weetall M, Buxton FP, Hughes TE: Divergent pathways of gene expression are activated by the RAGE ligands S100b and AGE-BSA. Diabetes 53: 743–751, 2004.

114. Heart Outcomes Prevention Evaluation Study Investigators: Effects of ramipril on cardiovascular and microvascular outcomes in people with diabetes mellitus: results of the HOPE study and MICRO-HOPE substudy. Lancet 355: 253–259, 2000.

115. Lonn EM, Yusuf S, Dzavik V, Doris CI, Yi Q, Smith S, Moore-Cox A, Bosch J, Riley WA, Teo KK, for the SECURE Investigators: Effects of Ramipril and vitamin E on atherosclerosis. The study to evaluate cartotid ultrasound changes in patients treated with ramipril and vitamin E (SECURE). Circulation 103: 919–925, 2001.

116. McQuillan BM, Hung J, Beilby JP, Nidorf M, Thompson PL: Antioxidant vitamins and the risk of carotid atherosclerosis. The Perth carotid ultrasound disease assessment study (CUDAS). J Am Coll Cardiol 38: 1788–1784, 2001.

117. Lonn E, Bosch J, Yusuf S, Sheridan P, Pogue J, Arnold JM, Ross C, Arnold A, Sleight P, Probstfield J, Dagenais GR, HOPE and HOPE-TOO Trial Investigators: Effects of long-term vitamin E supplementation on cardiovascular events and cancer: a randomized controlled trial. JAMA 293: 1338–1347, 2005.

14

Hypertension and Atherosclerosis: Advanced Glycation End Products—A Common Link

SUDESH VASDEV AND VICKI GILL

Abstract

The vascular diseases, hypertension and atherosclerosis, affect millions of individuals worldwide accounting for a large number of deaths globally. They share similar risk factors, and modify vascular structure and function. Although increased blood pressure itself can contribute to vascular injury, the relationship between hypertension and atherosclerosis likely has a biochemical component. We suggest that aldehyde conjugates, formed as a result of insulin resistance found in these two disorders, may be the common link. In insulin resistance, alterations in glucose metabolism lead to production of excess aldehydes including methylglyoxal. These aldehydes react nonenzymatically with free sulfhydryl (SH) and amino (NH_2) groups of proteins forming stable conjugates, also known as advanced glycation end products (AGEs). AGEs may act directly or via receptors to alter the function of cellular proteins including calcium channels, metabolic and antioxidant enzymes, receptors and structural proteins leading to endothelial dysfunction, inflammation, and increased oxidative stress. Alteration of vascular protein structure and function due to AGEs have been implicated in both hypertension and atherosclerosis. Therapies, which attenuate insulin resistance and lower AGEs, should be effective in the treatment of these conditions. Antioxidants including vitamin E and vitamin C, and thiols such as lipoic acid and cysteine, which improve glucose metabolism and lower AGEs, have been shown to prevent hypertension in animal models and humans with essential hypertension. Appropriate combinations of these antioxidants may also prevent atherosclerosis. A well-balanced diet containing these nutrients, such as one rich in fruits and vegetables, low in salt and sugar, with nuts and lean meats, and preventative measures such as not smoking, limiting alcohol, and participating in moderate exercise may also be effective in the treatment of these conditions.

Keywords: advanced glycation end products; aldehydes; atherosclerosis; hypertension; insulin resistance

Abbreviations: AGEs, advanced glycation end products; ALEs, advanced lipoxidation end products; BH4, tetrahydrobiopterin; $[Ca^{2+}]_i$, cytosolic free calcium; EGFR, endothelial growth factor receptor; eNOS, endothelial nitric oxide synthase; GAPDH, glyceraldehyde-3-phosphate dehydrogenase; G3P, glyceraldehyde-3-phosphate;

HSA, human serum albumin; H_2O_2, hydrogen peroxide; HOCl, hypochlorous acid; LDL, low-density lipoprotein; NAC, *N*-acetylcysteine; NAD+/NADH, nicotinamide adenine dinucleotide, oxidized/reduced; NADPH, nicotinamide adenine dinucleotide phosphate, reduced; NH_2, amino; NO, nitric oxide; O_2^-, superoxide radical; ·OH, hydroxyl radical; RAGEs, receptors of AGEs; ROO·/ROOH, free radical/nonradical; ROS, reactive oxygen species; SH, sulfhydryl; VSMC, vascular smooth muscle cell; WKY, Wistar-Kyoto

Introduction

Hypertension is a disease associated with the vascular system. It is generally symptomless and although the definition continues to evolve, the World Hypertension League characterizes hypertension as consistently elevated blood pressure greater than 140/90 [1]. Hypertension affects more than 600 million people worldwide and results in 13% of the total deaths globally [2]. Approximately 90% of all hypertension is classified as "essential" meaning that the cause is not known. Essential hypertension likely develops through a combination of genetic and lifestyle factors. Risk factors for hypertension include family history, diabetes, obesity, smoking, excessive alcohol intake, and a diet high in salt or low in antioxidant nutrients. Most of these risk factors are modifiable through lifestyle changes such as participating in moderate physical activity and eating a well-balanced diet. Healthy lifestyle choices also include not smoking and limiting alcohol intake [2, 3] (Table 14.1). Essential hypertension involves endothelial dysfunction with alterations in nitric oxide (NO)

TABLE 14.1. Risk factors common to hypertension and atherosclerosis, and interventions to reduce them.

	Risk factors	Interventions
Hypertension	Family history	Eat a well-balanced diet low in salt and
	Obesity	fat, rich in fruits and vegetables, with
	Diabetes	lean meats and low fat dairy (DASH diet)
	Smoking	Participate in regular moderate physical
	Excessive alcohol	activity to decrease obesity, improve
	consumption	carbohydrate metabolism and reduce
	Diet high in salt, low in	oxidative stress
	antioxidant nutrients	Stop smoking
		Limit alcohol intake to 1–2 drinks per day
Atherosclerosis	Family history	Be aware of family history and screen early
	Hypertension	for blood pressure, glucose, and lipid
	Dyslipidemias	abnormalities
	Obesity	Supplement with antioxidant combinations
	Diabetes	and if necessary, treat with
	Smoking	antihypertensive, antidiabetic, or lipid-
	Diet high in fat	lowering medications

bioavailability, alterations in calcium handling, smooth muscle cell proliferation, thickening of the vessel walls, and increased peripheral vascular resistance [4–7]. Individuals with hypertension are at increased risk for atherosclerotic diseases such as stroke, and heart and kidney disease. About 50% of people who have first myocardial infarction and two thirds with first stroke have blood pressures greater than 160/95 [8]. Long-term diastolic blood pressure decreases of 5–6 mm Hg are associated with 35–40% less incidence of stroke and 20–25% less coronary heart disease [9].

Atherosclerosis is a leading cause of death in the world. In 1996, it resulted in one third of total mortality in industrialized countries. Risk factors for atherosclerosis include hypertension, smoking, diabetes mellitus, obesity, dyslipidemias, and a diet high in saturated fat [2, 10] (Table 14.1). Atherosclerosis is an inflammatory condition of the blood vessels [11]. Damage to, or activation of the endothelium promotes low-density lipoprotein (LDL) entry into the intima, a process enhanced by an elevation in circulating levels of LDL. This alteration in endothelium also increases the expression of adhesion molecules on the cell surface resulting in recruitment of monocytes and platelet adhesion. The monocytes transmigrate to the subendothelial space where they differentiate into macrophages. Oxidized or modified LDL is scavenged by macrophages in the interstitial space transforming over time into foam cells. Accumulation of foam cells and other cellular debris evolve into atherosclerotic plaques [11–14]. As they grow, these plaques may cause narrowing and occlusion of the vessel, or may erupt into the vessel causing thrombosis. Atherosclerosis also involves increased smooth muscle cell migration and proliferation. Stiffening of the vessel walls may hinder elasticity exacerbating hypertension. Atherosclerotic lesions generally occur at junctions of large- and medium-size vessels but can arise throughout the vasculature [8]. Through stenosis or embolytic occlusion, lesions within the coronary, cerebral, or renal vessels result in the clinical manifestations of myocardial infarction, stroke, or renal failure [15]. Peripheral arterial disease is a condition where atherosclerotic lesions afflict the peripheral vessels, most commonly the superficial femoral and popliteal arteries, effecting widespread impairment in blood flow and oxygen deprivation in tissues. This condition manifests itself as muscle pain or weakness in the extremities including intermittent claudication and in extreme cases, ulceration, or gangrene [8, 16].

Hypertension and atherosclerosis share similar risk factors and both modify vascular structure and function [17] (Table 14.1). Increased blood pressure itself can contribute to vascular injury making vessels more susceptible to the inflammatory process leading to atherosclerosis. However, studies show that lowering blood pressure alone does not completely eliminate the risk of cardiovascular disease [18, 19] suggesting that the relationship of hypertension to atherosclerosis also has a biochemical component. Alteration of protein structure and function due to aldehyde conjugates/advanced glycation end products (AGEs) have been implicated in both hypertension and atherosclerosis [20–27]. In this chapter, we will focus on the origin of these aldehyde conjugates/AGEs and their role in the development and progression of hypertension and atherosclerosis. It has been proposed that insulin resistance with altered glucose metabolism may be the link between these two vascular disorders [28, 29] and we suggest that it may be the common source of aldehyde conjugates/AGEs (Fig. 14.1) .

Insulin Resistance

Insulin resistance is characterized by an inadequate glucose uptake in peripheral tissues at a given concentration of plasma insulin. It involves an impairment of the nonoxidative (glycolytic) pathways of intracellular glucose metabolism [30]. Humans with essential hypertension and normotensive offspring of essential hypertensives have insulin resistance [30–33]. Abnormalities in glucose metabolism exist in up to 80% of subjects with essential hypertension [30, 34]. Insulin resistance has also been documented in humans with atherosclerosis [28, 29, 35–37]. It has been suggested that hypertensives who are insulin resistant are at increased risk for cardiovascular disease [28, 38]. In metabolic syndrome, also known as syndrome X or insulin resistance syndrome, primary insulin resistance is linked to a group of coexisting conditions including hypertension, dyslipidemias, diabetes, and atherosclerotic cardiovascular disease [29].

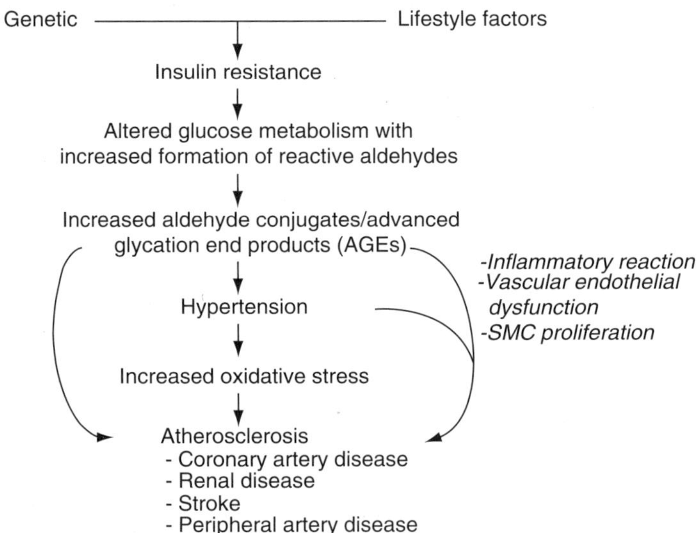

FIGURE 14.1. A schematic diagram of the common mechanism for hypertension and atherosclerosis showing the formation of advanced glycation end products (AGEs) from methylglyoxal.

Formation of Aldehydes due to Insulin Resistance

Under normal physiological conditions, glucose is metabolized via the glycolytic pathway to glyceraldehyde-3-phosphate (G3P), which is converted to 1,3-diphosphoglycerate by the enzyme glyceraldehyde-3-phosphate dehydrogenase (GAPDH), with further metabolism to pyruvate. Any factor which affects GAPDH, whether through inhibition or upregulation, has an impact on the rate of glucose metabolism. It has been shown that GAPDH is upregulated by insulin [39]. In insulin resistant states like essential hypertension, altered insulin function may downregulate GAPDH, slowing glucose metabolism through the glycolytic pathway, thus increasing metabolism via the polyol pathway. This may result in a buildup of G3P leading to an increase in the highly reactive aldehyde, methylglyoxal [40–43]. Methylglyoxal itself has been shown to inhibit GAPDH, likely via sulfhydryl (SH) groups, resulting in further abnormalities in glucose metabolism [44]. Methylglyoxal also induces aldose reductase in rat aortic vascular smooth muscle cells (VSMCs). Aldose reductase stimulates a further flux of glucose through the polyol pathway with increased formation of methylglyoxal [45]. Oxidative stress has also been shown to inhibit GAPDH and thus has the capacity to further exacerbate abnormalities in glucose metabolism [46].

Formation of Aldehyde Conjugates/Advance Glycation End Products (AGEs)

AGEs are formed when reactive aldehydes (e.g., methylglyoxal, glyoxal, glucose) react nonenzymatically with free SH and amino (NH_2) groups of amino acids including cysteine, arginine, or lysine, of proteins forming stable conjugates [47–51] (Fig. 14.2). This modification to protein structure results in

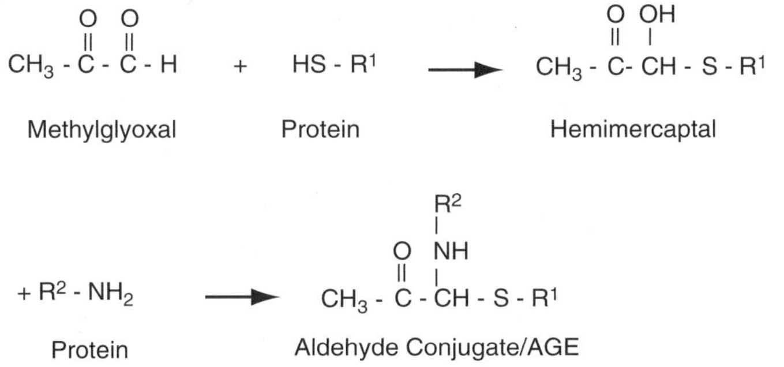

FIGURE 14.2. Methylglyoxal reacts nonenzymatically with free sulfhydryl (SH) and amino (NH_2) groups of amino acids of proteins forming irreversible compounds called adehyde conjugates or advanced glycation end products (AGEs). (R^1 and R^2 represent amino acids of protein)

functional changes [44, 52–55]. These conjugates are also known as AGEs and some have been identified and linked to complications of diabetes [52, 55, 56]. Aldehyde conjugates/ AGEs have also been shown to act on receptors of AGEs (RAGEs) and various scavenger receptors to influence protein function and expression [57–59]. Several aldehyde conjugates/AGEs including carboxymethyl-lysine, carboxyethyl-lysine, argpyrimidine, and glycoaldehyde-pyridine, have been identified, and have been implicated in hypertension and atherosclerosis [27, 60–63].

Under normal physiological conditions, methylglyoxal is formed but kept at a low level through catabolism via the glutathione-dependent glyoxalase enzyme system or by binding to cysteine and being excreted in bile and urine [49]. It has been suggested that aldehyde conjugates/AGEs formed under these conditions contribute to the regulation of normal tissue remodeling [64]. However, when there is an excess of methylglyoxal and other aldehydes, more conjugates/AGEs are formed leading to altered structure and function of tissue proteins. High levels of methylglyoxal also lead to a depletion of cysteine and glutathione resulting in increased oxidative stress.

Oxidative stress exists when there is an imbalance between the level of reactive oxygen species (ROS) and the body's antioxidant capacity. ROS include single electron oxidants such as superoxide radicals (O_2^-) and hydroxyl radical ($\cdot$OH), and the two-electron oxidants hydrogen peroxide (H_2O_2), hypochlorous acid (HOCl), and peroxynitrite [65]. Under normal conditions, ROS play a role in several processes; for example, they act as signaling molecules in regulating VSMC contraction–relaxation and VSMC growth [66]. However, excess ROS have the capacity to exert toxic effects. They can lead to a further formation of aldehydes including malondialdehyde and hydroxynonenal, through lipid peroxidation [49, 66–71]. These lipid-derived aldehydes have been shown to react with cysteine or lysine residues of proteins to form a type of AGE known as advanced lipoxidation end products (ALEs) [50, 72–74]. The myeloperoxidase system produces HOCl, which has been shown to form the AGE, glycoaldehyde pyridine, found in atherosclerotic lesions [27]. ROS also oxidize sugars, proteins, and nucleic acids leading to membrane dysfunction, tissue remodeling, enzyme inhibition, and alteration of gene expression [66, 75]. Levels of ROS are controlled by antioxidants or antioxidant enzymes, which neutralize or scavenge them. The antioxidant enzyme glutathione reductase acts to ensure that reduced glutathione is available to neutralize O_2^-. Another antioxidant enzyme, glutathione peroxidase, scavenges H_2O_2, preventing its reduction to $\cdot$OH [66]. Endogenous antioxidants such as glutathione, and antioxidant nutrients like vitamin C and vitamin E, lipoic acid, and coenzyme Q10, also function to control oxidative stress [21]. There is strong evidence that oxidative stress contributes to the progression of essential hypertension and the development of its complications including atherosclerosis [5, 66, 75–79]. For the purpose of uniformity, we will use the term AGEs throughout this chapter to denote aldehyde conjugates, AGEs, and ALEs.

Causative Role of AGEs in Hypertension and Atherosclerosis

We have shown that methylglyoxal given in the diet to Wistar-Kyoto (WKY) rats increased tissue AGEs and caused hypertension [80] (Fig. 14.3). Levels of tissue methylglyoxal and AGEs are higher in spontaneously hypertensive rats and sugar-induced hypertensive rats [22, 23, 62, 81, 82]. Although research on AGEs in human essential hypertension is scant, in preeclampsia, a hypertensive condition of pregnancy, RAGE expression was increased in vascular tissue [83]. There is strong evidence in human diabetes, another insulin resistance state, that AGEs are responsible for cellular protein modifications, which contribute to diabetic complications [52, 55, 56, 84–86]. AGE-mediated cross-links in collagen and elastin, also contribute to arterial stiffening, hindering vessel elasticity, and exacerbating hypertension [87]. Treatments which lower AGEs also lower the blood pressure [23, 81, 82, 88].

There is increasing evidence of a causative role for AGEs in atherosclerosis either through a hypertension-mediated increase in oxidative stress, and/or directly via stimulation of various inflammatory processes. AGEs have the capacity to adversely influence the function and expression of many of the body's proteins including calcium channels, metabolic, and antioxidant enzymes, receptors, and structural proteins. In the following section, we will discuss these adverse effects and their possible contribution to hypertension and atherosclerosis (Table 14.2).

AGEs and Increased Oxidative Stress

In addition to depleting glutathione, methylglyoxal, and AGEs may affect antioxidant enzyme activity. The antioxidant enzymes, glutathione peroxidase and glutathione reductase contain SH and NH_2 groups at their active sites [89, 90]. Formation of AGEs at these sites may result in inhibition of these enzymes. In rat VSMCs in culture, methylglyoxal inhibited these enzymes, leading to oxidative stress, low levels of reduced glutathione and increased levels of oxidized glutathione [22]. Vascular ROS are produced in endothelial, adventitial, and VSMCs by reduced nicotinamide adenine dinucleotide phosphate (NADPH) oxidase [66]. AGEs have been shown to activate RAGEs resulting in a NADPH oxidase-mediated increase in intracellular reactive oxygen intermediates and extracellular H_2O_2 [59]. AGE-modified bovine serum albumin increased O_2^- production in human platelets *in vitro* [91]. As discussed above, increased oxidative stress promotes further formation of aldehydes and thus AGEs. This creates a cycle where AGEs and ROS perpetuate a prooxidant state, which contributes to the development and progress of vascular disease.

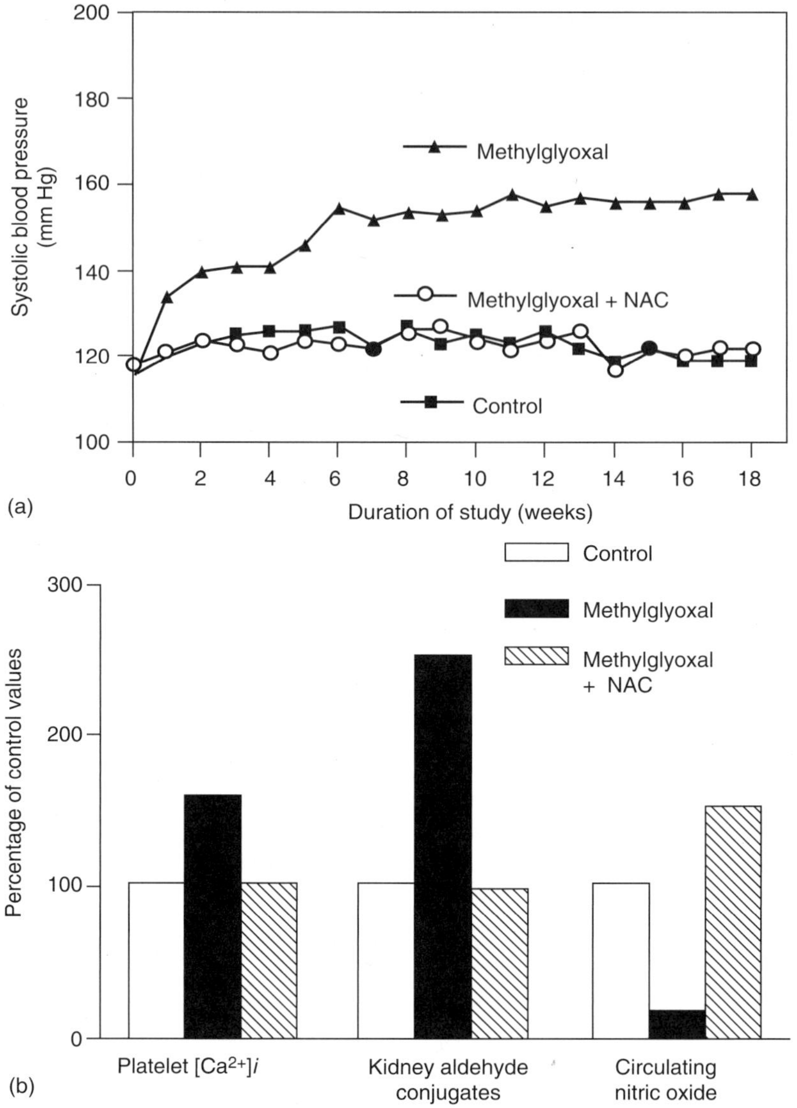

FIGURE 14.3. (a) The line graph shows the effect of *N*-acetylcysteine (NAC) on systolic blood pressure in methylglyoxal treated Wistar-Kyoto (WKY) rats. Starting at 7 weeks of age, WKY rats were divided into three groups of six animals each. Animals in the WKY-Control group were given normal diet and normal drinking water; WKY-Methylglyoxal, regular diet, and methylglyoxal in drinking water; WKY-Methylglyoxal+NAC, 1.5% NAC in diet and methylglyoxal in drinking water for the next 18 weeks. Methylglyoxal in drinking water was given at a concentration of 0.2% during weeks 0–5; 0.4%, weeks 6–10; and 0.8%, weeks 11–18. Values are mean of six animals in each group for each week. Standard deviation of mean was not more than 6 mm Hg in all cases. (b) The bar graph shows the effect of NAC on platelet $[Ca^{2+}]_i$, kidney aldehyde conjugates and circulating nitric oxide in methylglyoxal treated WKY rats. Experimental groups and treatment period were the same as in Figure 14.3a. Values are expressed as a percentage of the control group values at the completion of the study, age 25 weeks.

TABLE 14.2. Effect of AGEs and oxidative stress leading to hypertension and atherosclerosis.

Affected protein	Effect of alteration via AGEs or RAGEs	Effect of alteration via increased oxidative stress
Antioxidant enzymes (Glutathione peroxidase and glutathione reductase)	AGE formation with SH and NH_2 of enzymes resulting in an inhibition of activity causing a decrease in glutathione, and increase in oxidative stress with increased production of aldehydes leading to lipid peroxidation and formation of advanced lipoxidation end products	
Reduced nicotinamide adenine dinucleotide phosphate (NADPH)-oxidase	Activation of enzyme resulting in increased production of ROS (increased oxidative stress)	
Calcium channels	Alteration of SH group of calcium channel, with an increase in cytosolic free calcium, peripheral vascular resistance and blood pressure	Oxidation of calcium channel with an increase in cytosolic free calcium, peripheral vascular resistance and blood pressure
Nitric oxide synthase (NOS)	Alteration of SH group of NOS and decrease in arginine substrate with decreased NO production resulting in an increase in vasoconstriction, blood pressure, platelet aggregation, and vascular smooth muscle cell (VSMC) proliferation	Oxidation of BH4 with uncoupling of NOS resulting in a decreased production of NO and increased breakdown of NO to peroxynitrite, resulting in an increase in vasoconstriction, blood pressure, platelet aggregation, VSMC proliferation, and activation of monocyte adhesion molecules
Glyceraldehyde-3-phosphate dehydrogenase (GADPH)	AGE formation with SH group resulting in inhibition of GAPDH with exacerbation of insulin resistance, increase in aldehyde, AGE, and ROS formation	Oxidation of enzyme resulting in exacerbation of insulin resistance, increase in aldehyde, AGE, and ROS formation
Low-density lipoprotein (LDL)	Formation of LDL-AGE	Formation of oxidized LDL
Scavenger receptors	Activation of receptor by LDL-AGE promotes its endocytosis leading to formation of foam cells contributing to atherosclerotic plaques	Activation of receptor by oxidized LDL promotes its endocytosis leading to formation of foam cells contributing to atherosclerotic plaques
Receptors of AGEs (RAGE)	Activation of receptor influencing intracellular signals resulting in an increased formation of ROS, NF-κB, endothelin-1, adhesion molecule vascular cell adhesion molecule (VCAM)-1, insulin-like growth factors and interleukin-6 leading to monocyte recruitment and adhesion, platelet adhesion, cell proliferation, and vasoconstriction.	Activation of receptor influencing intracellular signals resulting in increased formation of NF-κB, endothelin-1, adhesion molecule VCAM-1, insulin-like growth factors and interleukin-6 leading to monocyte recruitment and adhesion, cell proliferation, and vasoconstriction.

AGEs and Increased Cytosolic Free Calcium

Vascular calcium channels are dependent on SH groups for normal function [49, 92] and their alteration can lead to increased cytosolic free calcium $[Ca^{2+}]_i$, abnormal contractile activity, and increased peripheral vascular resistance. AGEs impaired type 2 ryanodine receptor calcium release channels (calcium receptors which regulate cardiac contractility) during chronic diabetes [93]. ROS have been shown to increase intracellular calcium in VSMC and endothelial cells [66, 94]. Oxidized LDL was shown to enhance vasoconstriction likely via increased calcium influx into VSMCs [95]. AGE-modified human serum albumin (AGE-HSA) increased ROS and intracellular calcium in neonatal mesengial cells [96]. AGE-HSA also acts via RAGEs to produce ROS and increase the release of calcium from intracellular stores in human neutrophils [97]. In essential, human hypertensives and hypertensive animals, cytosolic $[Ca^{2+}]_i$ levels in VSMCs are elevated [98, 99]. We have shown elevated vascular tissue AGEs and increased platelet $[Ca^{2+}]_i$ in spontaneously hypertensive rats, a genetic model of hypertension, and in methylglyoxal and fructose-treated WKY rats, dietary models of hypertension [80–82] (Fig. 14.3). In addition to its role in vascular contractility and blood pressure homeostasis, calcium is a major determinant of platelet activation and aggregation, and increased levels may also contribute to enhanced atherogenic conditions [100].

AGEs and Endothelial Dysfunction

Maintaining normal endothelial function is essential to blood pressure homeostasis and vessel integrity. One of the major factors involved in regulation of endothelial function is NO. Endothelium-derived NO is not only a potent vasodilator but also inhibits platelet aggregation, VSMC migration and proliferation, monocyte adhesion, and adhesion molecule expression, thus regulating blood pressure and protecting vascular function [101]. Therefore, abnormalities in NO bioavailability impair endothelial function. This has been demonstrated in both hypertension and atherosclerosis [4, 102].

NO is synthesized from arginine by the thiol-containing endothelial nitric oxide synthase (eNOS) and cofactor tetrahydrobiopterin (BH4), and its production is regulated in part by insulin acting on specific receptors on the cell surface [40, 103–105]. Thus, the ability to form adequate amounts of NO depends on the availability of arginine, active eNOS, and BH4, and appropriate insulin response. Excess methylglyoxal and other aldehydes may limit substrate by forming AGEs with arginine [84]. Methylglyoxal has been shown to react with arginine residues to form several AGEs including argpyrimidine, which has been demonstrated in human serum and kidney [63, 106, 107]. The importance of SH groups in normal catalytic activity of eNOS has been demonstrated [104] and corroborates evidence that alteration of cysteine residues (c184) of human eNOS results in loss of activity [108]. AGEs inhibit eNOS activity and expression [109–111]. We have shown that methylglyoxal given in the diet to WKY rats increased tissue AGEs, decreased plasma NO, and caused hypertension [80] (Fig. 14.3). NOS activity requires the presence of BH4. It has been suggested that insulin stimulates BH4 synthesis and that this insulin action is impaired in insulin resistant states [102]. Also, oxidation of BH4 by ROS promotes uncoupling of eNOS therefore decreasing production of NO [102, 105]. Superoxide reacts with NO, also reducing bioavailability, forming peroxynitrite. Peroxynitrite and peroxynitrous acid are powerful and cytotoxic oxidants, also referred to as reactive nitrogen species, which may cause further damage to vascular tissue [112].

AGEs and Atherosclerotic Lesion Formation

Shear stress caused by elevated blood pressure results in endothelial injury enhancing formation of atherosclerotic plaques, particularly at junctions of vessels where turbulence is greatest. Subsequent oxidative stress will also contribute to this atherogenic environment. Oxidized LDL is recognized as a key component in atherosclerotic lesions and hypertensive subjects exhibit an enhanced susceptibility to LDL oxidation [13, 113]. AGEs may contribute to atherosclerosis in various ways. AGE-LDL caused cholesterol and cholesterol ester accumulation in macrophages *in vitro* [25] and has been identified in the cytoplasm of foam cells and extracellularly in the core of atherosclerotic lesions in humans and animals [14, 26, 27]. AGE-LDL and AGE-modified protein are ligands for scavenger receptors. Binding to these receptors leads to endocytic uptake of LDL and accumulation in human monocytes–macrophages [57, 114]. AGEs also bind to RAGEs generating a cascade of intracellular signals, which result in inflammatory responses including an increased expression of NF-κB, endothelin-1, vascular cell adhesion molecule-1, insulin-like growth factors, and interleukin-6. These changes result in monocyte recruitment and vascular adhesion, cell proliferation, and vasoconstriction [22, 58, 59, 64, 115, 116]. Endothelial growth factor receptor (EGFR) is involved in the regulation of multiple cellular processes such as cell growth, motility, differentiation, survival, and death. It is a common receptor shared by several growth factors. Methylglyoxal and glyoxal directly inhibit EGFR autophosphorylation and tyrosine kinase activity, by forming AGEs with the free NH_2 group of EGFR protein thus impairing downstream signaling [117]. ROS may also affect this signaling pathway by oxidation of the cysteine residue of tyrosine phosphatase, thus influencing receptor protein tyrosine kinases including EGFR [66]. This AGE-induced alteration in signaling may play a role in the cellular changes associated with hypertension and atherosclerosis.

AGEs may contribute to platelet aggregation and thrombus formation. AGEs increased O_2^- production and aggregation in human platelets *in vitro* [91]. Alterations due to AGEs reduce NO bioavailability resulting in an increase in platelet aggregation. It has been suggested that AGEs stimulate

externalization of phosphatidylserine, which activates clotting factors leading
to platelet adhesion [118].

Antioxidants in the Treatment of Hypertension and Atherosclerosis

We have presented evidence that insulin resistance leading to increased formation of AGEs and oxidative stress contributes to the etiology of essential hypertension and subsequent development of atherosclerosis. It is likely that by
attenuating insulin resistance and decreasing levels of AGEs and ROS it should
be possible to prevent or treat these diseases (Fig. 14.4). The protective effect of
individual antioxidants and antioxidant combinations has been investigated in
humans and animals. Supplementation with antioxidants including vitamin C,
E, or B6, lipoic acid and coenzyme Q10 has been shown to lower blood pressure in animal models and humans with essential hypertension [21]. However,
evidence supporting their efficacy in the treatment of atherosclerosis is somewhat less conclusive [119]. This may have less to do with the potential of these
compounds to achieve antiatherosclerotic effects and more to do with factors
such as the nature of the atherosclerotic lesion and degree of damage already
incurred, and which antioxidants were used and in what combinations.

Cardiovascular research usually entails monitoring the effect of dietary
intervention on the occurrence of hypertensive or atherosclerotic events such
as myocardial infarction, stroke, and kidney failure. If subjects have been

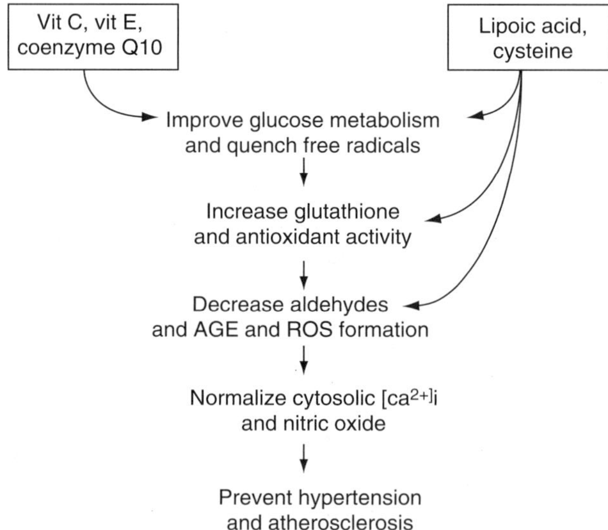

FIGURE 14.4. Mechanism of action of antioxidants for the prevention of hypertension
and atherosclerosis

hypertensive for a number of years and atherosclerosis is well established, antioxidant therapy may prevent further damage from AGEs and ROS but may not be able to completely restore vasculature to normal. The nature of atherosclerotic plaque itself may not be conducive to physical infiltration or repair by antioxidants. Tissue proteins already altered by AGEs will require time to regenerate and some tissues have long turnover times.

Typically, studies of cardiovascular disease use either high doses of single antioxidants, for example vitamin E, or combinations including β-carotene or vitamin C. Many of these studies have given variable results. Recent discussions, which suggest that vitamin E supplementation may be detrimental to cardiovascular health, may be an example of this [120]. Most of these studies do not consider that antioxidants in their oxidized form have the capacity to act as radicals and that it is necessary to maintain an adequate antioxidant balance such that these radicals are regenerated back into their reduced states. Antioxidants have been shown to regenerate each other from their oxidized to reduced forms [121–123] (Fig. 14.5). For example, vitamin E radical (oxidized) is regenerated to vitamin E (reduced) by vitamin C (ascorbate) or coenzyme Q10 (reduced). These antioxidants are in turn regenerated by dihydrolipoic acid. Thus, supplementation with vitamin E in high doses without the benefit of other regenerating antioxidants in the diet could conceivably result in damage from vitamin E radical. When choosing antioxidant

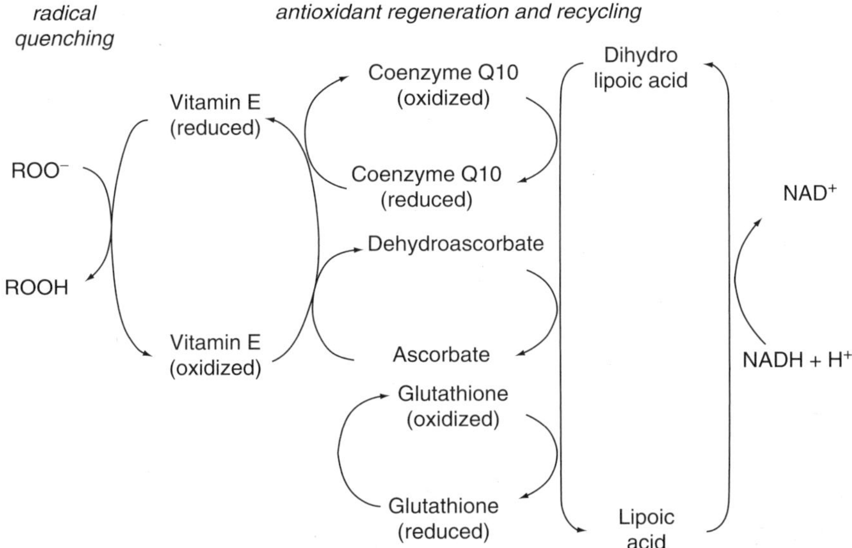

FIGURE 14.5. Regeneration of antioxidants in the body. Oxygen free radicals are quenched by antioxidants which themselves are oxidized. The oxidized antioxidant is regenerated to its original reduced form by another antioxidant of higher electronegativity. ROO⁻/ROOH (Free radical/non-radical); NAD⁺/NADH (Nicotinamide adenine dinucleotide oxidized/reduced).

combinations for cardiovascular studies, it may be as important to consider how these antioxidants work with each other, as it is to contemplate their mechanism of action with regards to the disease state. Studies using combinations of standard or moderate doses of combinations of antioxidants may show better efficacy in the treatment of hypertension and atherosclerosis. Choosing an antioxidant with high electronegativity (i.e., more reducing power) such as lipoic acid as one of the component antioxidants may ensure that radicals created from those antioxidants of lower electronegativity will be regenerated (Table 14.3). Using a combination of antioxidants, which are effective against both one- and two-electron radicals, should confer protection against a broader spectrum of radical agents. Lipoic acid, as well as cysteine, also improve glucose metabolism and in addition have the capacity to bind aldehydes directly preventing AGE formation and subsequent increase in ROS, making these antioxidants a more versatile treatment [124].

Therapies targeting insulin resistance, AGEs and oxidative stress should be effective in the treatment of hypertension and atherosclerosis (Fig. 14.4). A first line approach to these vascular disorders would include preventative measures such as participating in moderate physical activity, and eating a well-balanced diet rich in fruits and vegetables, low in salt and sugar, with nuts and lean meats. Healthy lifestyle choices include not smoking and limiting alcohol intake. Since antioxidants including vitamin E, vitamin C, and lipoic acid have been shown to improve glucose metabolism, lower AGEs and oxidative stress, reduce inflammatory response, and prevent hypertension, they will likely prevent subsequent damage due to atherosclerosis [21, 125, 126]. Considering the greater likelihood of benefit than harm, we suggest that an antioxidant combination supplement, as an additive to a well-balanced diet, is appropriate for most adults as a preventative measure. Early screening to evaluate blood pressure, glucose, and lipid abnormalities will also allow timely diagnosis and increase the opportunity to limit tissue damage and halt or slow disease progress (Table 14.1).

TABLE 14.3. Redox potentials of antioxidants in mammalian oxidation systems (Redox pairs given at increasing electronegativity and antioxidant activity).

Redox pair	Redox potential (E'_O Volts)
Oxygen/water	+0.82
Vitamin E oxidized/reduced	+0.37
Ubiquinone oxidized/reduced	+0.10
Vitamin C oxidized/reduced	+0.08
Cystine/cysteine	−0.22
Glutathione oxidized/reduced	−0.24
Lipoate oxidized/reduced	−0.29
Nicotinamide adenine dinucleotide oxidized/reduced (NAD$^+$/NADH)	−0.32
Nicotinamide adenine dinucleotide phosphate oxidized/ reduced (NADP$^+$/NADPH) H$^+$/H$_2$	−0.42

Conclusion

Hypertension and atherosclerosis share similar risk factor conditions and both involve structural and functional changes of the vasculature. Increased blood pressure contributes to atherosclerosis but it is likely that these two conditions also have a common biochemical causation. We suggest that AGEs, formed as a result of insulin resistance found in these two disorders, may be the common link. AGEs and subsequent oxidative stress have the capability to alter calcium channels, metabolic and antioxidant enzymes, receptors and structural proteins resulting in hypertension and atherosclerosis. We suggest that lifestyle choices and appropriate antioxidant supplements which attenuate insulin resistance, reduce the formation of aldehydes and AGEs and lower oxidative stress will prevent or attenuate these two vascular diseases.

Acknowledgments: We would like to thank the Canadian Institutes of Health Research Regional Partnership Program for their financial support.

References

1. World Hypertension League: Know your blood pressure. http://www. meduohio.edu/org/whl/know.html : 1, 2004.
2. World Health Organization: Global strategy on diet, physical activity and health: Chronic Disease. http:// www.who.int/dietphysicalactivity/publications/ facts/riskfactors/en/index/html :1–2, 2003.
3. Warnock DG, Textor SC: Core curriculum in Nephrology: hypertension. Am J Kidney Dis 44: 369–375, 2004.
4. Taddei S, Virdis D, Ghiadoni L, Salvetti G, Salvetti A: Endothelial dysfunction in hypertension. J Nephrol 13: 205–210, 2000.
5. Portaluppi F, Boari B, Manfredini R: Oxidative stress in essential hypertension. Curr Pharm Des 10: 1695–1698, 2004.
6. Resnick LM: Ionic basis of hypertension, insulin resistance, vascular disease, and related disorders. The mechanism of "syndrome X". Am J Hypertens 6: 123S–134S, 1993.
7. Oshima T, Young EW: Systemic and cellular calcium metabolism and hypertension. Semin Nephrol 15: 496–503, 1995.
8. American Stroke Association: Heart disease and stroke statistics – 2005 update, http://www.americanheart.org/downloadable/heart/1105390918119HDSStats 2005Update.pdf: 1–63, 2005.
9. Collins R, Peto R, MacMahon S, Hebert P, Fiebach NH, Eberland KA, Godwin J, Qizilbash N, Taylor JO, Hennekens CH: Blood pressure, stroke and coronary heart disease. Part 2, Short-term reductions in blood pressure:overview of randomised drug trials in their epidemiological context. Lancet 335: 827–838, 1990.
10. World Health Organization: World Health Report. Neglected Global Epidemics: Three growing threats. http://www.who.int/whr/2003/chapter6/en/ : 1–5, 2003.
11. Ross R: Atherosclerosis—an inflammatory disease. N Engl J Med 340: 115–126, 1999.

12. Hansson G: Inflammation, atherosclerosis, and coronary artery disease. N Engl J Med 352: 1685–1695, 2005.

13. Witzum JL, Steinberg D: Role of oxidized low density lipoprotein in atherogenesis. J Clin Invest 88: 1785–1792, 1991.

14. Sima A, Stancu C: Modified lipoproteins accumulate in human coronary atheroma. J Cell Mol Med 6: 110–111, 2002.

15. Tegos TJ, Kalodiki E, Sabetai MM, Nicolaides AN: The genesis of atherosclerosis and risk factors: a review. Angiology 52: 89–98, 2001.

16. Aronow WS: Management of peripheral arterial disease. Cardiol Rev 13: 61–68, 2005.

17. Pepine CJ, Handberg EM: The vascular biology of hypertension and atherosclerosis and intervention with calcium antagonists and angiotensin-converting enzyme inhibitors. Clin Cardiol 24: V1–V5, 2001.

18. Lembo G, Morisco C, Lanni F, Barbato E, Vecchione C, Fratta L, Trimarco B: Systemic hypertension and coronary artery disease: the link. Am J Cardiol 82: 2H–7H, 1998.

19. Neutel JM: Why lowering blood pressure is not enough: the hypertension syndrome and the clinical context of cardiovascular risk. Heart Dis 2: 370–374, 2000.

20. Vasdev S, Gill V, Longerich L: Role of methylglyoxal in essential hypertension. In: S.K. Gupta, et al. (eds), Pharmacotherapy of Heart Failure. Anamaya Publishers, New Delhi, India, 2004, pp 72–88.

21. Vasdev S, Gill V: Antioxidants in the treatment of hypertension. Internat J Ang 14: 60–73, 2005.

22. Wu L, Juurlink BHJ: Increased methylglyoxal and oxidative stress in hypertensive rat vascular smooth muscle cells. Hypertension 39: 809–814, 2002.

23. Midaoui AE, Elimadi A, Wu L, Haddad PS, de Champlain J: Lipoic acid prevents hypertension, hyperglycemia, and the increase in heart mitochondrial superoxide production. Am J Hypertens 16: 173–179, 2003.

24. Colaco CALS, Roser BJ: Atherosclerosis and glycation. Bioessays 16: 145–147, 1994.

25. Brown BE, Dean RT, Davies MJ: Glycation of low-density lipoproteins by methylglyoxal and glycoaldehyde gives rise to the *in vitro* formation of lipid-laden cells. Diabetologia 48: 361–369, 2005.

26. Palinski W, Koschinsky T, Butler SW, Miller E, Vlassara H, Cerami A, Witztum JL: Immunological evidence for the presence of advanced glycosylation end products in atherosclerotic lesions of euglycemic rabbits. Arterioscler Thromb Vasc Biol 15: 571–582, 1995.

27. Nagai R, Hayashi CM, Xia L, Takeya M, Horiuchi S: Identification in human atherosclerotic lesions of GA-pyridine, a novel structure derived from glycoaldehyde-modified proteins. J Biol Chem 277: 48905–48912, 2002.

28. Reaven GM: Insulin resistance/compensatory hyperinsulinemia, essential hypertension, and cardiovascular disease. J Clin Endocrinol Metab 88: 2399–2403, 2003.

29. DeFronzo R, Ferrannini E: Insulin resistance: a multifaceted syndrome responsible for NIDDM, obesity, hypertension and atherosclerotic cardiovascular disease. Diabetes Care 14: 173–194, 1991.

30. Ferrannini E, Buzzigoli G, Bonadonna R, Giorico MA, Oleggini M, Graziadei L, Pedrinelli R, Brandi L, Bevilacqua S: Insulin resistance in essential hypertension. N Engl J Med 313: 350–357, 1987.

31. Beatty OL, Harper R, Sheridan B, Atkinson AB, Bell PM: Insulin resistance in offspring of hypertensive parents. BMJ 307: 92–96.
32. Saad MF, Rewers M, Selby J, Howard G, Jinagouda S, Fahmi S, Zaccaro D, Bergman RN, Savage PJ, Haffner SM: Insulin resistance and hypertension. The insulin resistance atherosclerosis study. Hypertension 43: 1324–1331, 2004.
33. Vlasakova Z, Pelikanova T, Karasova L, Skibova J: Insulin secretion, sensitivity, and metabolic profile of young healthy offspring of hypertensive patients. Metabolism 53: 469–475, 2004.
34. Flack JM, Sowers JR: Epidemiologic and clinical aspects of insulin resistance and hyperinsulinemia. Am J Med 91:11S–21S, 1991.
35. Howard G, O'Leary DH, Rewers M, ammam R, Selby JV, Saad MF, Savage P, Bergman R: Insulin sensitivity and atherosclerosis. Circulation 93: 1809–1817, 1996.
36. Shinozaki K, Suzuki M, Ikebuchi M, Hara Y, Harano Y: Demonstration of insulin resistance in coronary artery disease documented with angiography. Diabetes Care 19: 1–7, 1996.
37. Harano Y, Suzuki M, Shinozaki K, Hara Y, Ryomoto K, Kanazawa A, Nishioheda Y, Tsushima M: Clinical impact of insulin resistance syndrome in cardiovascular diseases and its therapeutic approach. Hypertens Res 19: S81–S85, 1996.
38. Zavaroni I, Mazza S, Dall'Aglio E, Gasparini P, Passeri M, Reaven GM: Prevalence of hyperinsulinaemia in patients with high blood pressure. J Intern Med 231: 235–240, 1992.
39. Alexander MC, Lomanto M, Nasrin N, Ramaika C: Insulin stimulates glyceraldehyde-3-phosphate dehydrogenase gene expression through CIS-acting DNA sequences. Proc Natl Acad Sci 85: 5092–5096, 1988.
40. Mayes PA: Carbohydrate metabolism. In: Martin DW, Rodwell VW (eds), Harper's Review of Biochemistry, 19th edition. Lange Medical Publications, Los Altos, CA, USA, 1983, pp 161–187.
41. Thornalley PJ: The glyoxalase system in health and disease. Mol Aspects Med 14: 287–371, 1993.
42. Vander Jagt DL, Hunsaker LA: Methylglyoxal metabolism and diabetic complications: roles of aldose reductase, glyoxalase-I, betaine aldehyde dehydrogenase and 2-oxoaldehyde dehydrogenase. Chem Biol Interact 143–144: 341–351, 2003.
43. Beisswenger PJ, Howell SK, Smith K, Szwergold BS: Glyceraldehyde-3-phosphate dehydrogenase activity as an independent modifier of methylglyoxal levels in diabetes. Biochim Biophys Acta 1637: 98–106, 2003.
44. Morgan PE, Dean RT, Davies MJ: Inactivation of cellular enzymes by carbonyls and protein-bound glycation/ glycoxidation products. Arch Biochem Biophys 403: 259–269, 2002.
45. Chang KC, Paek KS, Kim HJ, Lee YS, Yabe-Nishimura C, Seo HG: Substrate-induced up-regulation of aldose reductase by methylglyoxal, a reactive oxoaldehyde elevated in diabetes. Mol Pharmacol 61: 1184–1191, 2002.
46. Morgan PE, Dean R, Davies MJ: Inhibition of glyceraldehyde-3-phosphate dehydrogenase by peptide and protein peroxides generated by singlet oxygen attack. Eur J Biochem 269: 1916–1925, 2002.
47. Randell ER, Vasdev S, Gill V: Measurement of methylglyoxal in rat tissues by electrospray ionization liquid chromatography and mass spectrophotometry. J Pharmacol Toxicol Methods 51: 153–157, 2004.

48. Thornalley PJ, Battah S, Ahmed N, Karachalias N, Agalou S, Babaei-Jadidi R, Dawnay A: Quantitative screening of advanced glycation end products in cellular and extracellular proteins by tandem mass spec. Biochem J 375: 581–592, 2003.

49. Schauenstein E, Esterbauer H, Zollner H: Aldehydes in biological systems. In: Lagnado JR (ed), Aldehydes in Biological Systems, Their Natural Occurrence and Biological Activities. Pion Limited, London, UK, 1977, pp 1–7.

50. Thorpe SR, Baynes JW: Maillard reaction products in tissue proteins: new products and new perspectives. Amino Acids 25: 275–281, 2003.

51. Ahmed N, Argirov OG, Minhas HS, Cordeiro CAA, Thornalley PJ: Assay of advanced glycation end products (AGEs): surveying AGEs by chromatographic assay with derivatization by 6-aminoquinolyl-N-hydroxysuccinimidyl-carbamate and application to Nε-(1-carboxyethyl)lysine-modified albumin. Biochem J 364: 1–14, 2002.

52. Nagaraj RH, Sarkar P, Mally A, Biemel KM, Lederer MO, Padayatti PS: Effect of pyridoxamine on chemical modification of protein by carbonyls in diabetic rats: characterization of a major product from the reaction of pyridoxamine and methylglyoxal. Arch Biochem Biophys 402: 110–119, 2002.

53. Frye EB, Degenhardt TP, Thorpe SR, Baynes JW: Role of Maillard reaction in aging of tissue proteins. J Biol Chem 273:18714–18719, 1998.

54. Odani H, Shinzato T, Usami J, Matsumoto Y, Fye EB, Baynes JW, Maeda K: Imidazolium crosslinks derived from reaction of lysine with glyoxal and methylglyoxal are increased in serum proteins of uremic patients: evidence for increased oxidative stress in uremia. FEBS Lett 427: 381–385, 1998.

55. Karachalias N, Babaei-Jadidi R, Ahmed N, Thornalley PJ: Accumulation of fructosyl-lysine and advanced glycation end products in the kidney, retina and peripheral nerve of streptozotocin-induced diabetic rat. Biochem Soc Trans 31:1423–1425, 2003.

56. Alt N, Carson JA, Alderson NL, Wang Y, Nagai R, Henle T, Thorpe SR, Baynes JW: Chemical modification of muscle protein in diabetes. Arch Biochem Biophys 425: 200–206, 2004.

57. Horiuchi S, Sakamoto Y, Sakai M: Scavenger receptors for oxidized and glycated proteins. Amino Acids 25: 283–292, 2003.

58. Schmidt AM, Yan SD, Yan SF, Stern DM: The multiligand receptor RAGE as a progression factor amplifying immune and inflammatory responses. J Clin Invest 108: 949–955, 2001.

59. Wautier, MP, Chappey O, Corda S, Stern DM, Schmidt AM, Wautier JL: Activation of NADPH oxidase by AGE links oxidant stress to altered gene expression via RAGE. Am J Physiol 280: E685–E694, 2001.

60. Alderson NL, Chachich ME, Youssef NN, Beattie J, Nachtigal M, Thorpe SR, Baynes JW: The AGE inhibitor pyridoxamine inhibits lipemia and development of renal and vascular disease. Kidney Int 63: 2123–2133, 2003.

61. Anderson MM, Requena JR, Crowley JR, Thorpe SR, Heinecke JW: The myeloperoxidase system in human phagocytes generates Nε-(carboxymethyl) lysine on proteins: a mechanism for producing advanced glycation end products at sites of inflammation. J Clin Invest 104: 103–113, 1999.

62. Wang X, Desai K, Clausen JT, Wu L: Increased methylglyoxal and advanced glycation end products in kidney from spontaneously hypertensive rats. Kidney Int 66: 2315–2321, 2004.

63. Oya T, Hattori N, Mizuno Y, Miyata S, Maeda S, Osawa T, Uchida K: Methylglyoxal modification of protein. J Biol Chem 274: 18492–18502, 1999.

64. Kirstein M, Aston C, Hintz R, Vlassara H: Receptor-specific induction of insulin-like growth factor I in human monocytes by advanced glycosylation end product-modified proteins. J Clin Invest 90: 439–446, 1992.

65. Stocker R, Keaney JF Jr: Role of oxidative modifications in atherosclerosis. Physiol Rev 84: 1381–1478, 2004.

66. Touyz RM, Schiffrin EL: Reactive oxygen species in vascular biology: implications for hypertension. Histochem Cell Biol 122: 339–352, 2004.

67. Brooks BR, Klamerth OJ: Interaction of DNA with bifunctional aldehydes. Eur J Biochem 5: 178–182, 1968.

68. Thornalley P, Wolff S, Crabbe J, Stern A: The autoxidation of glyceraldehyde and other simple monosaccharides under physiological conditions catalysed by buffer ions. Biochimica et Biophysica Acta 797: 276–287, 1984.

69. Poli G, Dianzani MU, Cheeseman KH, Slater TF, Lang J, Esterbauer H: Separation and characterization of the aldehydic products of lipid peroxidation stimulated by carbon tetrachloride or ADP-iron in isolated rat hepatocytes and rat liver microsomal suspensions. Biochem J 227: 629–638, 1985.

70. Benedetti A, Comporti M, Esterbauer H: Identification of 4-hydroxynonenal as a cytotoxic product originating from the peroxidation of liver microsomal lipids. Biochim Biophys Acta 620: 281–296, 1980.

71. Dargel R: Lipid peroxidation—a common pathogenetic mechanism? Exp Toxicol Pathol 44: 169–181, 1992.

72. Baynes JW, Thorpe SR: Glycoxidation and lipoxidation in atherogenesis. Free Rad Biol Med 28: 1708–1716, 2000.

73. Stitt AW, Frizzell N, Thorpe SR: Advanced glycation and advanced lipoxidation: possible role in initiation and progression of diabetic retinopathy. Curr Pharm Des 10: 3349–3360, 2004.

74. Voziyan PA, Metz TO, Baynes JW, Hudson BG: A post-amadori inhibitor pridoxamine also inhibits chemical modification of proteins by scavenging carbonyl intermediates of carbohydrate and lipid degradation. J Biol Chem 277: 3397–3403, 2002.

75. Kitiyakara C, Wilcox CS: Antioxidants for hypertension. Curr Opin Nephrol Hypertens 7: 531–538, 1998.

76. Kumar KV, Das UN: Are free radicals involved in the pathobiology of human essential hypertension? Free Rad Res Commun 19: 59–66, 1993.

77. Sagar S, Kallo U, Kaul N, Ganguly NK, Sharma BK: Oxygen free radicals in essential hypertension. Mol Cell Biochem 111: 103–108, 1992.

78. Koska J, Syrova D, Blazicek P, Marko M, Grna JD, Kvetnansky R, Vigas M: Malondialdehyde, lipofuscin and activity of antioxidant enzymes during physical exercise in patients with essential hypertension. J Hypertens 17: 529–535, 1999.

79. Chobanian AV, Alexander RW: Exacerbation of atherosclerosis by hypertension. Arch Intern Med 156: 1952–1956, 1996.

80. Vasdev S, Ford CA, Longerich L, Parai S, Gadag V, Wadhawan S: Aldehyde induced hypertension in rats: Prevention by N-acetylcysteine. Artery 23: 10–36, 1998.

81. Vasdev S, Ford CA, Parai S, Longerich L, Gadag V: Dietary alpha-lipoic acid supplementation lowers blood pressure in spontaneously hypertensive rats. J Hypertens 18(5): 567–573, 2000.

82. Vasdev S, Ford CA, Parai S, Longerich L, Gadag V: Dietary lipoic acid supplementation prevents fructose-induced hypertension in rats. Nutr Metab Cardiovasc Dis 10(6): 339–346, 2000.

83. Cooke CLM, Brockelsby JC, Baker PN, Davidge ST: The receptor for advanced glycation end products (RAGE) is elevated in women with preeclampsia. Hypertens Pregnancy 22: 173–184, 2003.

84. Degenhardt TP, Alderson NL, Arrington D, Beattie RJ, Basgen JM, Steffes MW, Thorpe SR, Baynes JW: Pyridoxamine inhibits early renal disease and dyslipidemia in streptozotocin-diabetic rat. Kidney Int 61: 939–950, 2002.

85. Babaei-Jadidi R, Karachalias N, Ahmed N, Battah S, Thornalley PJ: Prevention of incipient diabetic nephropathy by high-dose thiamine and benfotiamine. Diabetes 52: 2110–2120, 2003.

86. Wong, RKM, Pettit AI, Quinn PA, Jennings SC, Davies JE, Ng LL: Advanced glycation end products stimulate an enhanced respiratory burst mediated through the activation of cytosolic phospholipase A_2 and generation of arachidonic acid. Circulation 108: 1858–1864, 2003.

87. Aronson D: Cross-linking of glycated collagen in the pathogenesis of arterial and myocardial stiffening of aging and diabetes. J Hypertens 21: 3–12, 2003.

88. Vasdev S, Longerich L, Singal P: Nutrition and hypertension. Nutr Res 22: 111–123, 2002.

89. Stadtman TC: Selenium Biochemistry. Annu Rev Biochem 59: 111–127, 1990.

90. Park YS, Koh YH, Takahashi M, Miyamoto Y, Suzuki K, Dohmae N, Takio K, Honke K, Taniguchi N: Identification of the binding site of methylglyoxal on glutathione peroxidase: Methylglyoxal inhibits glutathione peroxidase activity via binding to glutathione binding sites Arg 184 and 185. Free Rad Res 37: 205–211, 2003.

91. Hangaishi M, Taguchi J, Miyata T, Ikari Y, Togo M, Hashimoto Y, Watanabe T, Kimura S, Kurokawa K, Ohno M: Increased aggregation of human platelets produced by advanced glycation end products *in vitro*. Biochem Biophys Res Commun 248: 285–292, 1998.

92. Zaidi NF, Lagenaur CF, Abramson JJ, Pessah I, Salama G: Reactive disulfides trigger Ca^{2+} release from sarcoplasmic reticulum via an oxidation reaction. J Biol Chem 264: 21725–21736, 1989.

93. Bidasee KR, Nallani K, Yu Y, Cocklin RR, Zhang Y, Wang M, Dincer UD, Besch HR Jr: Chronic diabetes increases advanced glycation end products on cardiac ryanodine receptors/ calcium-release channels. Diabetes 52: 1825–1836, 2003.

94. Hirosumi J, Ouchi Y, Watanabe M, Kusunoki J, Nakamura T, Orimo H: Effect of superoxide and lipid peroxide on cytosolic free calcium concentration in cultured pig aortic endothelial cells. Biochem Biophys Res Commun 152: 301–307, 1988.

95. Galle J, Bassenge E, Busse R: Oxidized low density liproteins potentiate vasoconstriction to various agonists by direct interaction with vascular smooth muscle. Circ Res 66: 1287–1293, 1990.

96. Scivattaro V, Ganz MB, Weiss MF. AGEs induce oxidative stress and activate protein kinase C-B_{II} in neonatal mesangial cells. Am J Physiol 278: F676–F683, 2000.

97. Collison KS, Parhar RS, Saleh SS, Meyer BF, Kwaasi AA, Hammami MM, Schmidt AM, Stern DM, Al-Mohanna F: RAGE-mediated neutrophil dysfunction is evoked by advanced glycation end products (AGEs). J Leukoc Biol 71: 433–444, 2002.

98. Aoki K, Kawaguchi Y, Sato K, Kondo S, Yamamoto M: Clinical and pharmacological properties of calcium antagonists in essential hypertension in humans and spontaneously hypertensive rats. J Cardiovasc Pharmacol 4: S298–S302, 1982.

99. Robinson BF: Altered calcium handling as a cause of primary hypertension. J Hypertens 2: 453–460, 1984.

100. Ding YA: Thrombogenic and lipid risk factors in hypertension and coronary artery disease. Jap Circ J 60: 75–84, 1996.

101. Taddei S, Ghiadoni L, Virdis D, Versari D, Salvetti A: Mechanisms of endothelial dysfunction: Clinical significance and preventative non-pharmacological therapeutic strategies. Curr Pharm Des 9: 2385–2402, 2003.

102. Shinozaki K, Kashiwagi A, Masada M, Okamura T: Molecular mechanism of impaired endothelial function associated with insulin resistance. Curr Drug Targets 4: 1–11, 2004.

103. Kahn NN, Acharya K, Bhattacharya S, Acharya R, Mazumder S, Bauman WA, Sinha AK: Nitric oxide: The "second messenger" of insulin. IUBMB Life 49: 441–450, 2000.

104. Patel JM, Block ER: Sulfhydryl-disulfide modulation and the role of disulfide oxidoreductases in regulation of the catalytic activity of nitric oxide synthase in pulmonary artery endothelial cells. Am J Respir Cell Mol Biol 13: 352–359, 1995.

105. Alp NJ, Channon KM: Regulation of endothelial nitric oxide synthase by tetrahydrobiopterin in vascular disease. Arterioscler Thromb Vasc Biol 24: 413–420, 2004.

106. Shipanova IN, Glomb MA, Nagaraja RH: Protein modification by methylglyoxal: chemical nature and synthetic mechanism of a major fluorescent adduct. Arch Biochem Biophys 344, 29–36, 1997.

107. Shamsi FA, Partal A, Sady C, Glomb MA, Nagaraj RH: Immunological evidence for methylglyoxal-derived modifications *in vivo*. J Biol Chem 273: 6928–6936, 1998.

108. Chen PF, Tsai AL, Wu KK: Cysteine 184 of endothelial nitric oxide synthase is involved in heme coordination and catalytic activity. J Biol Chem 269: 25062–25066, 1994.

109. Verbeke P, Perichon M, Friguet B, Bakala H: Inhibition if nitric oxide synthase activity by early and advanced glycation end products in cultured rabbit proximal tubular epithelial cells. Biochim Biophys Acta 1502: 481–494, 2000.

110. Chakravarthy U, Hayes RG, Stitt AW, McAuley E, Archer DB: Constitutive nitric oxide synthase expression in retinal vascular endothelial cells is suppressed by high glucose and advanced glycation end products. Diabetes 47: 945–952, 1998.

111. Rojas A, Romay S, Gonzalez D, Herrera B, Delgado R, Otero K: Regulation of endothelial nitric oxide synthase expression by albumin-derived advanced glycosylation end products. Circ Res 86: e50–e54, 2000.

112. Beckman JS, Koppenol WH: Nitric oxide, superoxide and peroxynitrite: the good, the bad and ugly. Am J Physiol 271: C1424–C1437, 1996.

113. Maggi E, Marcheso E, Ravetta V, Falaschi F, Finardi G, Bellomo G: Low density lipoprotein oxidation in essential hypertension. J Hypertens 11: 1103–1111, 1993.

114. Haberland ME, Fless GM, Scanu AM, Fogelman AM: Malondialdehyde modification of lipoprotein(a) produces avid uptake by human monocyte-macrophages. J Biol Chem 267: 4143–4151, 1992.

115. Quehenberger P, Bierhaus A, Fasching P, Muellner C, Klevesath M, Hong M, Stier G, Sattler M, Schleicher E, Speiser W, Nawroth PP: Endothelin 1 transcription is controlled by nuclear factor-kB in AGE-stimulated cultured endothelial cells. Diabetes 49: 1561–1570, 2000.

116. Kunt T, Forst T, Wilhelm A, Tritschler H, Pfuetzner A, Harzer O, Engelbach M, Zschaebitz A, Stofft E, Beyer J: α-Lipoic acid reduces expresson of vascular cell adhesion molecule-1 and endothelial adhesion of human monocytes after stimulation with advance glycation end products. Clin Sci 96: 75–82, 1999.
117. Portero-Otin M, Pamplona R, Bellmut MJ, Ruiz MC, Prat J, Salvayre R, Negre-Salvayre A: Advanced glycation end product precursors impair epidermal growth factor receptor signaling. Diabetes 51: 1535–1542, 2002.
118. Wang Y, Marshall SM, Thompson MG, Hoenich NA: Cardiovascular risk in patients with end-stage renal disease: a potential role for advanced glycation end products. Contrib Nephrol 149: 168–174, 2005.
119. Abudu N, Miller JJ, Attaelmannan M, Levinson SS: Vitamins in human arteriosclerosis with emphasis on vitamin C and vitamin E. Clin Chim Acta 339: 11–25, 2004.
120. Lonn E for the HOPE and HOPE-TOO Investigators: Effects of long-term vitamin E supplementation on cardiovascular events and cancer. A randomized controlled trial. JAMA 293: 1338–1347, 2005.
121. Packer L, Tritschler HJ, Wessel K: Neuroprotection by metabolic antioxidant α-lipoic acid. Free Rad Biol Med 22: 359–378, 1997.
122. Frei B, Kim MC, Ames BN: Ubiquinol-10 is an effective lipid-soluble antioxidant at physiological concentrations. Proc Natl Acad Sci USA 87: 4879–4883, 1990.
123. Kagan VE, Serbinova EA, Forte T, Scita G, Packer L: Recycling of vitamin E in human low density lipoproteins. J Lipid Res 33: 385–397, 1992.
124. Vasdev S, Gill V, Longerich L: Antihypertensive effect of alpha lipoic acid supplementation. Curr Top Nutr Res 2(3): 137–152, 2004.
125. Sola S, Mir MQS, Faiz AC, Khan-Merchant N, Menon RG, Parthasarthy S, Khan BV: Irbesartan and lipoic acid improve endothelial function and reduce markers of inflammation in the metabolic syndrome. Results of the Irbesartan and lipoic acid in endothelial dysfunction (ISLAND) study. Circulation 111: 343–348, 2005.
126. Osiecki H: The role of chronic inflammation in cardiovascular disease and its regulation by nutrients. Alter Med Rev 9: 32–53, 2004.

Section III

Hyperhomocysteinemia and Atherosclerosis

15

Homocysteine Metabolism

ENOKA P. WIJEKOON, MARGARET E. BROSNAN, AND JOHN T. BROSNAN

Abstract

Elevated plasma homocysteine is an independent risk factor for atherosclerotic vascular disease as well as for Alzheimer's disease and fractures. Since homocysteine arises as a result of *S*-adenosylmethionine-dependent methylation reactions, the major methyltransferases play an important role in determining the plasma level of this molecule; these include methyltransferases involved in the synthesis of phosphatidylcholine and creatine. Homocysteine removal is affected either by the remethylation pathway (which converts homocysteine back to methionine) or by the transsulfuration pathway, which produces cysteine. A number of B vitamins (in particular folic acid, pyridoxal, and vitamin B_{12}) are involved in homocysteine metabolism so that deficiencies of these vitamins are among the most common causes of hyperhomocysteinemia. Decreased renal function is also an important contributor to hyperhomocysteinemia. The effects of genetic polymorphisms, hormones, and drugs on plasma homocysteine are discussed. Plasma homocysteine encompasses a heterogeneous group of molecular species. The importance of these individual species to pathology remains to be determined.

Keywords: *S*-adenosylmethionine; end-stage renal disease; homocysteine thiolactone; methyltransferase; remethylation; transmethylation; transsulfuration; vitamins

Abbreviations: tHcy, total homocysteine; SAM, *S*-adenosylmethionine; MAT, methionine adenosyltransferase; SAH, *S*-adenosylhomocysteine; GNMT, glycine *N*-methyltransferase; MTHFR, methylenetetrahydrofolate reductase; BHMT, betaine:homocysteine methyltransferase; CBS, cystathionine β-synthase; CGL, cystathionine γ-lyase; L-DOPA, L-3,4-dihydroxyphenylalanine; COMT, catechol-*O*-methyltransferase; GAA, guanidinoacetic acid; AGAT, L-arginine:glycine amidinotransferase; GAMT, guanidinoacetate methyltransferase; PEMT, phosphatidylethanolamine *N*-methyltransferase; PE, phosphatidylethanolamine; PC, phosphatidylcholine

The original hypothesis, linking hyperhomocysteinemia with atherothrombotic vascular disease, was proposed after the observation of severe arteriosclerotic lesions in two children with elevated plasma homocysteine concentrations and homocysteinuria [1]. This led to the proposal that even less severe increases in plasma homocysteine may pose a risk. Since then, a plethora of prospective, retrospective, and cross-sectional studies have appeared that link moderate hyperhomocysteinemia to atherosclerotic disease. These studies show that patients with coronary, cerebrovascular, or peripheral arterial occlusive disease have mean plasma total homocysteine (tHcy) levels greater than control subjects. Kang et al. [2] reported mean plasma tHcy to be significantly higher both in male and female patients with coronary artery disease when compared to controls with angiographically normal coronary arteries. Since cross-sectional studies deal with patients who already exhibit disease characteristics, they do not answer the question whether hyperhomocysteinemia gives rise to the coronary artery disease or vice versa [3]. Two large prospective studies address this question. The Physicians' Health Study, which studied male physicians in the USA, showed that a plasma homocysteine concentration of only 1.7 µmol/L or 12% greater than the upper normal level was associated with a 3.4-fold increase in the risk of myocardial infarction [4]. The Tromso study, which investigated over 21,000 subjects between the ages of 12 and 61, showed an increase in the relative risk for coronary heart disease of 1.32 after adjusting for possible confounders [5]. Meta-analyses confirm homocysteine's status as an independent risk factor although they disagree on its strength. A meta-analysis of 27 studies relating homocysteine to coronary, cerebrovascular, and peripheral arterial vascular disease showed a very strong relationship between these diseases and tHcy [6]. It was concluded that a 5 µmol/L increment in tHcy was comparable to a 0.5 mmol/L increment in cholesterol in increasing the incidence of coronary artery disease. A more recent meta-analysis that examined 30 prospective and retrospective studies confirmed that increased plasma homocysteine is an independent predictor of ischemic heart disease but of only moderate strength [7]. The same group [8] has reported a meta-analysis of the MTHFR 677TT polymorphism, which tends to increase plasma homocysteine, particularly in the context of low folate. They found that individuals with the MTHFR 677TT genotype had a significantly higher risk of cardiovascular disease especially when found in conjunction with a low-folate status. The uncertainty about the degree of risk associated with elevated plasma homocysteine may be attributed to a number of factors, but Stamm and Reynolds [9] have emphasized one important factor: tHcy in plasma may not be the most appropriate marker of cardiovascular disease risk. In particular, risk may be more closely linked either to cellular homocysteine or to a particular component of plasma homocysteine.

There are certain situations in which the risk attributable to homocysteine may be elevated. Certainly, there is evidence that hyperhomocysteinemia is a stronger risk factor in patients with type 2 diabetes [10] and in patients with existing coronary disease. In this latter group, Nygard et al. [11] found a direct relationship between plasma homocysteine levels and overall mortality; at 4 years, Kaplan–Meier estimates of mortality were 3.8% for patients with tHcy levels less than 9 µmol/L, 8.6% for patients with homocysteine lcvel between 9 and 14.9 µmol/L, and 24.7% for those with levels 15 µmol/L or higher (Fig. 15.1).

There is considerable evidence that the cardiovascular risk posed by homocysteine is graded; there does not appear to be a threshold effect. A cross-sectional study of elderly subjects from the Framingham Heart Study clearly demonstrated a graded increase in the level of extracranial carotid artery stenosis with increasing plasma Hcy [12]. Similar observations were made in several other studies with regard to different cardiovascular disease states [5, 13].

Plasma Forms of Homocysteine

Although the term homocysteine is used generically, plasma contains several different forms of this amino acid. Plasma tHcy is made up of free and protein-bound homocysteine. The free homocysteine consists of homocysteine, homocystine, and cysteine–homocysteine-mixed disulfides [2] and a protein-bound fraction is linked to proteins by disulfide linkage, principally to cysteine 34 of albumin (Fig. 15.2). Typically, in humans, the protein-bound fraction makes up the bulk of plasma tHcy accounting for >70%, homocystine and cysteine–homocysteine-mixed disulfides make up 5–15% each, while only trace amounts (≤ 1%) are found as free reduced homocysteine [14]. These different forms of homocysteine along with reduced, free oxidized, and protein-bound forms of cysteine and cysteinylglycine comprise a dynamic system [15]. A change in any one of these species leads to alterations in the

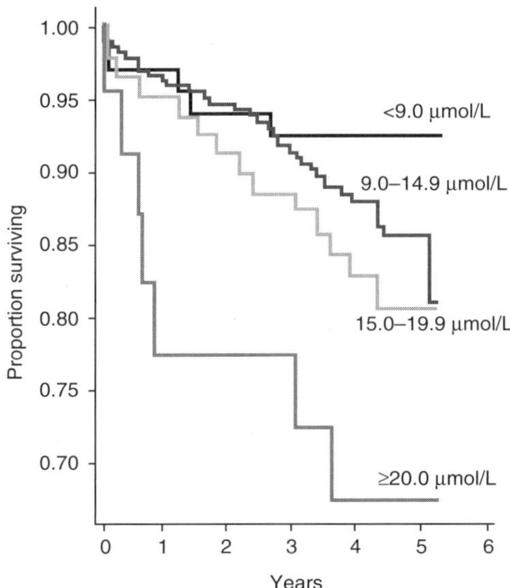

FIGURE 15.1. The relationship between plasma homocysteine level and the chance of survival in patients with coronary artery disease (reproduced from Ref. 11).

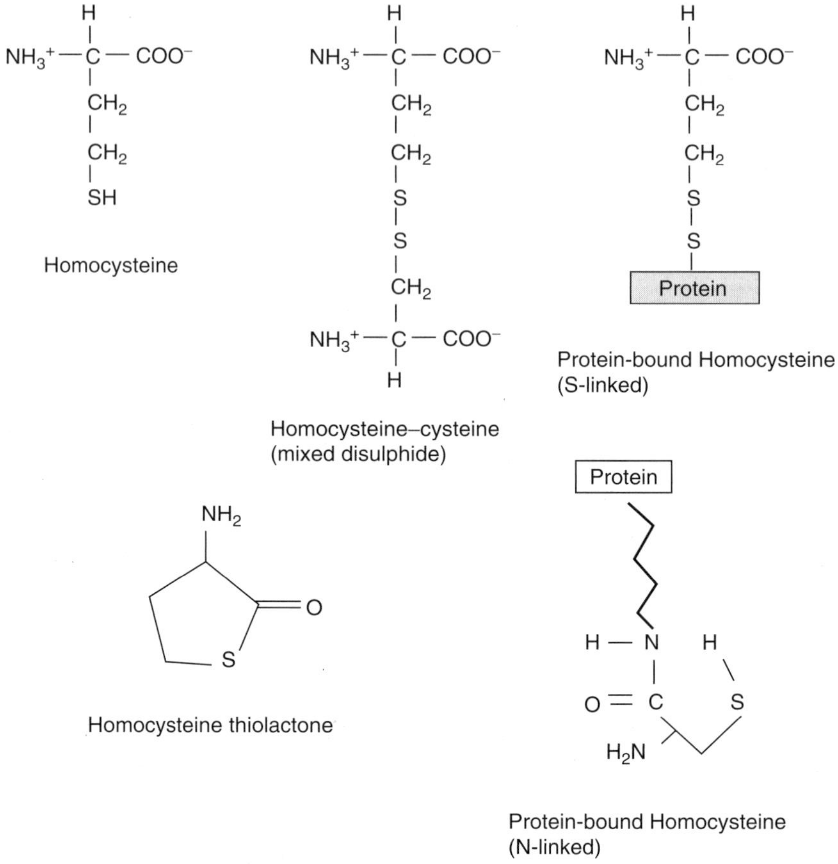

Homocysteine

Homocysteine–cysteine
(mixed disulphide)

Protein-bound Homocysteine
(S-linked)

Homocysteine thiolactone

Protein-bound Homocysteine
(N-linked)

FIGURE 15.2. Plasma forms of homocysteine.

thiol redox status [15]. Despite the existence of these different forms, typically the parameter measured is "total plasma homocysteine," since assays for the separate forms are not yet satisfactory for routine clinical measurement. Hyperhomocysteinemia is categorized as being moderate, intermediate, or severe with basal tHcy values of between 15 and 30, between 31 and 100, and greater than 100 nmol/mL, respectively [2].

"Total plasma homocysteine" does not encompass all forms of homocysteine in plasma. Homocysteine thiolactone (Fig. 15.2) can be formed in all human cell types; in general, its concentration is markedly increased under conditions of elevated plasma homocysteine. It is a cyclic thioester formed as a result of an error-correcting process in aminoacyl-tRNA synthetase [16]. Protein homocysteinylation, by means of the formation of an amide bond between the carboxyl group of homocysteine thiolactone and the ε-amino group of lysyl side chains in cellular and extracellular proteins, also occurs and can change the function of the protein so modified [17]. Such homocysteinylation of proteins, such as low-density lipoproteins, has been proposed as the basis for some of the pathological consequences of hyper-homocysteinemia. The N-linked homocysteine, which can account for up to 23% of plasma homocysteine, is not included in the calculation of plasma tHcy [18]. Homocysteine thiolactone is hydrolyzed to homocysteine by homocysteine thiolactonase, an enzyme associated with high-density lipoprotein [19].

Metabolism

The liver has been shown to play a significant role in the regulation of plasma homocysteine levels [20] because of its full complement of enzymes involved in methionine and homocysteine metabolism. Homocysteine is an intermediate in the pathway of methionine metabolism. The only dietary essential sulfur-containing amino acid in mammals, methionine, contributes to a variety of fundamental biological processes, including protein synthesis, methylation reactions involving S-adenosylmethionine (SAM), formation of the polyamines spermine and spermidine, and the synthesis of cysteine. Homocysteine is a key substrate in three essential biological processes: (1) recycling of intracellular folates; (2) catabolism of choline and betaine; and (3) formation of the nonessential amino acid cysteine and the antioxidant glutathione through the transsulfuration pathway [21].

The pathway of methionine metabolism (Fig. 15.3) consists of the ubiquitous methionine cycle and the transsulfuration pathway which has a more limited distribution. The methionine cycle is made up of the transmethylation and the remethylation pathways. In transmethylation, methionine is converted to the high-energy sulfonium compound, SAM in a reaction catalyzed by methionine adenosyltransferase (MAT) (EC 2.5.1.6). ATP provides the source of the adenosyl moiety. MAT exists in three isoenzymic forms MAT I, MAT II, and MAT III, which are products of two different genes *MAT 1A* and *MAT 2A* [22]. *MAT 1A* is expressed in mature liver while *MAT 2A* is expressed in all tissues and is induced during liver growth and dedifferentiation. The liver-specific isoenzymes, MAT I and MAT III are coded for by *MAT 1A*. MAT I, a tetramer, is also designated as an "intermediate K_m" isozyme (K_m for methionine ~40 μM) and is slightly inhibited by SAM. MAT III, a dimer, which appears to be derived from MAT I by a posttranslational modification [23] is designated as a "high K_m" isoform (K_m for methionine ~200 μM) and is activated by its product SAM, demonstrating a strong, positive cooperative modulation at physiological methionine and SAM concentrations. MAT III allows the liver to adjust immediately to an excess influx of methionine with increased SAM formation. MAT II, found in extrahepatic tissues and fetal liver and encoded by the gene *MAT 2A*, has a low K_m (~8 μM) for methionine and is inhibited by SAM [24].

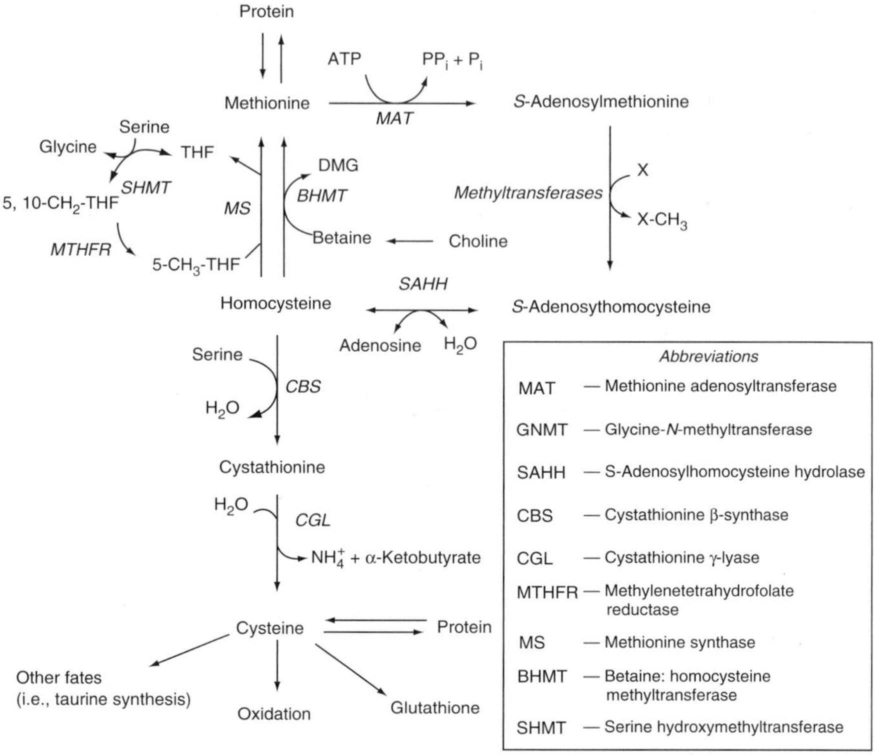

FIGURE 15.3. The pathway of methionine metabolism.

SAM, once formed, appears to be retained by the cell of origin and serves as the methyl donor for virtually all known biological methylation reactions. This transfer of the methyl group of SAM to a suitable methyl acceptor is catalyzed by any one of a large number of methyltransferases, leading to the formation of such diverse cellular components as creatine, epinephrine, carnitine, phospholipids as well as methylated proteins, DNA, and RNA [25]. Essentially all these methyltransferases are inhibited by the common product, S-adenosylhomocysteine (SAH). However, glycine N-methyltransferase (GNMT), which converts glycine to sarcosine, has unique kinetic properties. It is only weakly inhibited by SAH, has a relatively high K_m for SAM, and shows a sigmoidal dependence on SAM concentration [26]. GNMT thus functions as a benign high-capacity, SAM-dependent methyltransferase that is able to convert excess methionine/SAM to sarcosine, a nontoxic product. Sarcosine is then transported into mitochondria where its metabolism regenerates glycine.

The inhibition (albeit to different degrees) of essentially all of the different methyltransferases by SAH necessitates its rapid removal. SAH hydrolase, which occurs in all cells, converts SAH to homocysteine and adenosine [21]. Although the equilibrium of this reaction favors the formation of SAH, it is driven forward under physiological conditions due to the rapid removal of the products, homocysteine and adenosine. Adenosine can be removed by either of two enzymes, adenosine deaminase (EC 3.5.4.4) or adenosine kinase (EC 3.5.4.4). Homocysteine, on the other hand, occupies a central position in the pathway of methionine metabolism, lying at a crossroad between the transsulfuration and the remethylation pathways. SAH can also be removed by intracellular binding to specific, saturable protein sites and by export from the cell [21]. Once exported, the kidney appears to be active in the removal of SAH from the extracellular fluid [21]. Homocysteine can also bind to intracellular proteins and can be exported from cells [27]. However, unlike SAH, homocysteine can be taken up by many tissues including the liver [21]. Plasma homocysteine, therefore, represents homocysteine that has been exported from the cells of origin, to be transported to other cells/tissues that would catabolize it.

The remethylation of homocysteine to methionine completes the methionine cycle and in turn functions to conserve the carbon skeleton of methionine. Interestingly, the methylation of homocysteine to methionine is one exception where SAM does not provide the methyl group. Two different methyl donors, N^5-methyltetrahydrofolate and betaine, provide the methyl group necessary to convert homocysteine to methionine by two independent pathways. Methionine synthase, which transfers a methyl group from N^5-methyltetrahydrofolate to homocysteine, is widely distributed in mammalian tissue and contains methylcobalamin as an essential cofactor. The methyl group of N^5-methyltetrahydrofolate is synthesized by the folate coenzyme system using serine as the major donor of the one-carbon units [28]. N^5-methyltetrahydrofolate is formed by the irreversible reduction of $N^{5,10}$-methylenetetrahydrofolate, catalyzed by the flavoenzyme $N^{5,10}$-methylenetetrahydrofolate reductase (MTHFR) (EC 1.1.1.68) which uses NADPH as the reducing agent.

Betaine:homocysteine methyltransferase (BHMT) catalyzes a second route by which homocysteine is converted to methionine, using preformed methyl groups from betaine to methylate homocysteine [29]. BHMT, a zinc metalloenzyme [30], has a more limited distribution, being present only in liver, kidney, and lens in humans [29]. BHMT is feedback inhibited by its product N,N-dimethylglycine [31]. Dimethylglycine is converted to glycine in a series of reactions by which the methyl groups are oxidized and provide the methylene groups of $N^{5,10}$-methylenetetrahydrofolate.

If the carbon chain of methionine is not to be conserved, homocysteine can be irreversibly converted to cysteine through the transsulfuration pathway which is catalyzed by two pyridoxal 5′-phosphate (PLP)-containing enzymes, cystathionine β-synthase (CBS) (EC 4.2.1.22), and cystathionine γ-lyase (CGL) (EC 4.4.1.1). CBS catalyzes the condensation of homocysteine with serine to form cystathionine and commits the homocysteine moiety to the transsulfuration pathway [21]. Cystathionine is hydrolyzed by

CGL to form cysteine, α-ketobutyrate, and ammonia, thereby making cysteine a nonessential amino acid. Both the pyruvate dehydrogenase complex and the branched-chain α-keto acid dehydrogenase complex can catalyze the oxidative decarboxylation of α-ketobutyrate to propionyl CoA, which enters the Krebs' cycle at succinyl CoA [20]. Cysteine is a precursor of several essential molecules (e.g., glutathione, taurine, coenzyme A) as well as inorganic sulfate.

The transsulfuration pathway has a limited tissue distribution in rat tissues; only liver, kidney, pancreas, and small intestine are capable of synthesizing cysteine from methionine or homocysteine [32]. CBS is absent from heart, lung, testes, adrenal, and spleen in rats [21]. In humans, CBS was shown to be absent from heart muscle and primary aortic endothelial cells [33]. The highest levels of CBS mRNA expression in human tissues have been observed in adult liver and fetal liver and fetal brain [34]. Although CBS is present in the brain and adipose tissue of rats, CGL is absent from these tissues [21]. In agreement with this finding, very high levels of cystathionine have been found in human brains [35]. It may be that the function of CBS in the brain is to catalyze the production of H_2S, a neuromodulator and a smooth muscle relaxant [36].

Factors Influencing Methionine and Homocysteine Metabolism

Gender and Age

Plasma homocysteine has been shown to increase throughout life in both sexes [37]. In early childhood, both males and females have similar low plasma homocysteine levels, but sex differences become apparent at puberty [38]. The skewed distribution among the genders continues from puberty into adulthood with plasma homocysteine levels being 1–2 μmol/L higher in men than in women, and with a steady increase in both sexes with increasing age [38].

The gender differences could arise in a number of ways. It is possible that the differences are due to the higher formation of homocysteine in connection with increased creatine synthesis. Creatine synthesis is proportional to muscle mass and, therefore, is higher in men than in women [39]. A second possibility could be effects of sex steroids, as estrogens have been shown to lower plasma homocysteine levels [40]. This is also consistent with the observation that, after menopause, the gender difference in plasma homocysteine concentrations is attenuated. Postmenopausal women have been shown to have higher levels of plasma homocysteine than premenopausal women [41]. The changes in the plasma homocysteine levels that have been observed in the male-to-female and female-to-male transsexuals treated with sex steroids also show the effect of sex hormones on plasma homocysteine levels [42]. Plasma homocysteine was reduced in male–to-female transsexuals treated with ethinyl estradiol in combination with an antiandrogen while the female-to-male transsexuals treated with testosterone esters displayed an elevation in their plasma homocysteine [42]. Women who are pregnant have significantly reduced plasma homocysteine levels with tHcy decreasing between the first and the second trimesters and remaining low through the rest of the pregnancy [43]. Malinow et al. [44] suggested an increased uptake of maternal homocysteine by the developing fetus as the reason for this lowered plasma homocysteine in the mother.

A number of studies, conducted in different ethnic populations, have all demonstrated an age-related increase in plasma homocysteine, the increase being most marked in the oldest age group [45]. Plasma homocysteine is about 30% lower in children than in adults [38]. The age-dependent increase in plasma homocysteine could arise in a number of ways. Renal function is known to decline with age [46]. It is well established that renal disease is associated with increased plasma homocysteine [47]. This may be due to metabolism of homocysteine by the kidneys [48], although the clear-cut data obtained in rats are not evident in humans [49]. Suboptimal vitamin status, in particular impaired folate status [50] and cobalamin deficiency [51], may also play major roles.

Lifestyle

Several lifestyle factors are also known to affect plasma homocysteine levels. Chronic, high ethanol consumption [52], smoking [53], and consumption of caffeinated coffee [54] are all associated with an increase in the concentration of plasma homocysteine. Interestingly, a moderate consumption of alcohol appears to lower the plasma homocysteine concentration [55]. The consumption of beer has also been shown to lower plasma homocysteine, probably because beer contains folate [56]. Hyperhomocysteinemia associated with chronic alcoholism may arise due to impaired folate, vitamin B_{12}, and vitamin B_6 intake [51]. Acute alcohol intoxication in alcoholics leads to a transient increase in plasma homocysteine, which may be a direct result of inhibition of methionine synthase by acetaldehyde [57]. In the Hordaland homocysteine study, individuals who consumed more than six cups of coffee per day exhibited 2 to 3 μM higher level of homocysteine than individuals who did not consume coffee [54]. Decaffeinated coffee does not have the same effect on plasma homocysteine as does regular coffee [54].

Physical training has been shown to play a positive role in the reduction of plasma homocysteine [53]. The difference in homocysteine levels between subjects doing heavy training and those having a sedentary lifestyle was greater in the older age group where exercise was associated with an approximately 1 μmol/L reduction in homocysteine whereas those aged 40–42 years exhibited a reduction of 0.76 μmol/L. Acute exercise, however, does not have any effect on the plasma homocysteine levels [58].

Diet

SAM and SAH play a major role in the regulation of plasma homocysteine levels by affecting the distribution of homocysteine among competing pathways. Both these metabolites activate CBS [59, 60]. Finkelstein et al. [59] also showed that SAH is capable of inhibiting BHMT. Kutzbach and Stokstad [61] demonstrated that SAM is a strong allosteric inhibitor of MTHFR, while SAH functioned to partially reverse this inhibition. This inhibition will lead to reduced synthesis of N^5-methyltetrahydrofolate which itself is a modulator of methionine metabolism by its ability to bind and inhibit GNMT [62].

These mechanisms can combine to regulate methionine and homocysteine metabolism with regard to nutritional status, especially in response to protein and methionine intake, since increased methionine concentrations in the liver will lead to increased SAM levels. The fact that SAM activates MAT III [23] makes this system highly sensitive to methionine concentrations. The increased SAM will promote disposal via the transsulfuration pathway by activation of CBS and will inhibit remethylation by inhibiting both BHMT and MTHFR. A quantitative study of the methionine cycle *in vivo*, in healthy young men who were supplied with methionine at a rate of 198 µmol/kg/day, revealed a homocysteine synthesis through transmethylation of 238 µmol/kg; 38% of this homocysteine was channeled into remethylation and 62% underwent transsulfuration [63]. These data agreed with the labile methyl balances conducted by Mudd and Poole [39] who showed that remethylation could vary from 70% to 40% when changing from a restricted methionine intake to a diet containing increased amounts of methionine and choline.

Methylation Demand

The contribution of the different reactions involved in the formation of homocysteine must also be considered. SAM is the methyl donor of virtually all known biological methylation reactions, ranging from such diverse processes as metal detoxification, biosynthesis of many small molecules, and gene regulation via DNA methylation. Thirty nine different species of SAM-dependent methyltransferases have been well characterized in mammals [25], with the possibility that many more will be identified by means of genomic approaches which use conserved sequence motifs to detect methyltransferases in genomic open reading frames [64]. In fact, Katz et al. [64] calculated that class I methyltransferases make up ~0.6–1.6% of the genes in the yeast, human, mouse, *Drosophila melanogaster*, *Caenorhabditis elegans*, *Arabidopsis thaliana*, and *Escherichia coli* genomes.

The first demonstration of the importance of methylation to plasma homocysteine levels was provided by studies in patients with Parkinson's disease undergoing treatment with L-3,4-dihydroxyphenylalanine (L-DOPA). Parkinson's disease is characterized by a severe depletion of nigrostriatal dopaminergic neurons resulting in the deficiency of dopamine in the basal ganglia and melanin in the substantia nigra [65]. These patients are treated with L-DOPA alone or in combination with a peripheral decarboxylase inhibitor [66]. The dopamine deficiency is relieved by the decarboxylation of L-DOPA in the brain. Parkinson's patients undergoing treatment with L-DOPA have been shown to have plasma homocysteine levels that are about 50% higher than controls [67]. This increase in plasma homocysteine arises from the wasteful methylation of L-DOPA by catechol-*O*-methyltransferase (COMT), which catalyzes the SAM-dependent methylation of aromatic hydroxyl groups [66]. Due to the removal of the administered L-DOPA through this reaction, patients need to be given quite high doses (up to several grams per day) to achieve therapeutic levels of L-DOPA. This phenomenon was also evident in rats; the increased plasma tHcy was accompanied by decreased tissue levels of SAM [68]. Pretreatment with a COMT inhibitor was shown to alleviate or attenuate this response. These studies showed that plasma homocysteine levels were sensitive to the methylation of an exogenous substance.

Thereafter, studies conducted by the Brosnan group addressed the issue of the role of methylation demand imposed by physiological substrates on plasma homocysteine. Creatine synthesis, where SAM is used to methylate guanidinoacetic acid (GAA), has been considered to use more SAM than all of the other physiological methyltransferases combined [39] accounting for about 75% of homocysteine formation. Creatine synthesis is an interorgan pathway in which the two enzymes involved in creatine synthesis occur in two different organs: L-arginine:glycine amidinotransferase (AGAT) occurs in the kidney (EC 2.1.4.1) and guanidinoacetate methyltransferase (GAMT; EC 2.1.1.2) is localized to the liver. Stead et al. [69] studied the effect of changes in the methylation demand by modulating the rate of creatine synthesis by feeding rats GAA- or creatine-supplemented diets for 2 weeks. They found that homocysteine was significantly elevated in the plasma of rats fed GAA and that it was decreased by ~25% in the rats on a creatine-supplemented diet [69] (Fig. 15.4). Supplementation by either GAA or creatine resulted in a decrease in the activity of kidney AGAT. Incubation of hepatocytes with GAA resulted in the export of significantly higher levels of homocysteine, giving further evidence for the dependence of the plasma homocysteine level on the demand for methylation by physiological substrates. A separate series of studies showed that creatine supplementation was able to lower the plasma level of GAA as well as its arterio–venous difference across the kidney (E.E. Edison, M.E. Brosnan, and J.T. Brosnan, unpublished observations).

Noga et al. [70] demonstrated the importance of another physiological methylation reaction on the plasma homocysteine level. Phosphatidylethanolamine *N*-methyltransferase (PEMT), a liver-specific enzyme, is involved in the conversion of phosphatidylethanolamine (PE) to phosphatidylcholine

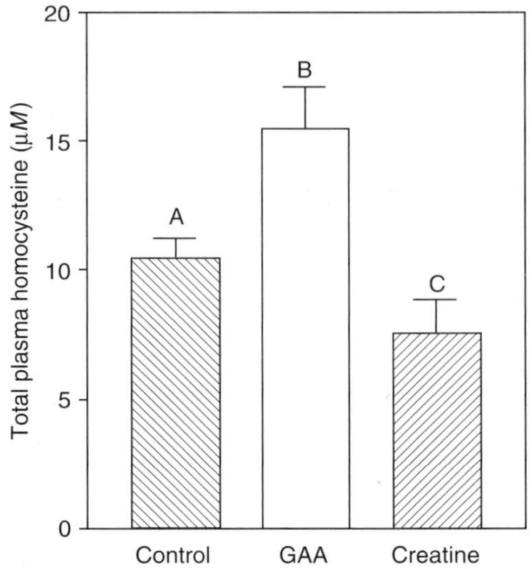

FIGURE 15.4. Total plasma homocysteine in rats fed guanidinoacetic acid or creatine [69].

(PC) [71]. 70% of hepatic PC is produced from choline by the activity of the enzymes of the Kennedy pathway; PEMT is responsible for the formation of the remaining 30% [72]. The synthesis of one molecule of PC through the PEMT pathway requires three successive SAM-dependent methylations. Plasma homocysteine in *pemt–/–* mice was decreased by about 50% [70] (Fig. 15.5). Hepatocytes isolated from the *pemt–/–* mice secreted ~50% less homocysteine

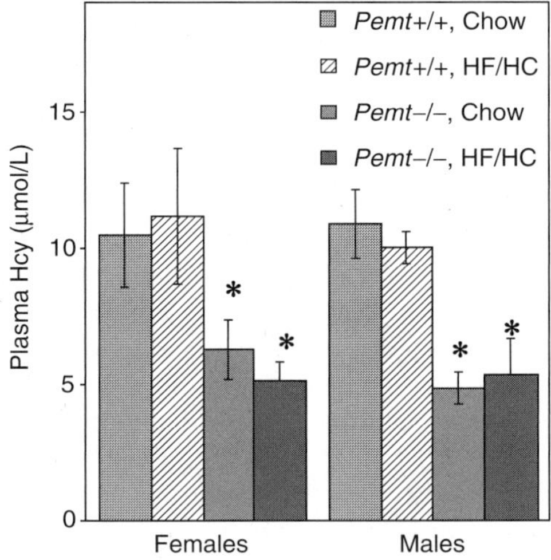

FIGURE 15.5. Total plasma homocysteine in *pemt–/–* mice [70].

than the cells isolated from the wild-type mice. McArdle RH7777 (rat hepatoma) cells, overexpressing PEMT, produced significantly more homocysteine than the wild-type cells. These results clearly demonstrated PC synthesis through the PEMT pathway to be a major contributor to the plasma homocysteine levels.

Hormonal Regulation of Plasma Homocysteine

Most of the early literature on homocysteine metabolism focused on traditional factors such as nutrition and genetics. However, studies conducted in many laboratories have now added to our understanding of the hormonal regulation of homocysteine metabolism. Diabetes mellitus, both type 1 and type 2, leads to increased plasma levels of homocysteine when accompanied by renal insufficiency [73, 74]. However, type 1 diabetic patients with normal plasma creatinine levels have been shown to have plasma tHcy levels significantly lower than healthy subjects [75]. Studies conducted in our laboratory confirmed this observation in streptozotocin-diabetic rats (type 1 model) [76]. In these rats, the decreased homocysteine was accompanied by concomitant increases in the activities of the hepatic transsulfuration enzymes, CBS and CGL, which were restored to normal by insulin treatment, suggesting regulation of plasma homocysteine by insulin. Administration of glucagon to rats was shown to reduce plasma homocysteine and was accompanied by increased flux of methionine through the transsulfuration pathway and by increases in the activities of the transsulfuration enzymes [77]. The concurrent increase in the activity and the mRNA of CBS suggested that glucagon may act at the level of gene transcription to alter homocysteine metabolism. This was directly confirmed by the experiments of Ratnam et al. [78]. Glucocorticoids or cAMP increased, and insulin attenuated, the CBS mRNA and protein levels in H4IIE cells, a rat hepatoma cell line, while insulin treatment of HepG2 cells, a human hepatoma cell line, led to a decreased level of CBS protein. The 70% reduction in CBS-1b promoter activity after insulin treatment further confirmed the effect of insulin, while the nuclear run-on experiments provided definitive evidence that both insulin and glucocorticoids act at the level of gene transcription (Fig. 15.6).

MAT is also under hormonal regulation. Adrenalectomy results in a three-fold decrease in its enzyme activity, immunoreactive protein, and mRNA levels in the liver; these effects could be reversed by triamcinolone treatment [79]. MAT mRNA content was increased, in a time- and dose-dependent manner, by both triamcinolone and dexamethasone, which effect was blocked by insulin. A direct effect of triamcinolone on the transcription of this gene was evident from experiments in which a luciferase reporter gene was driven by 1.4 kb of the 5'-flanking region of the hepatic MAT gene. Triamcinolone treatment resulted in a three-fold increase in the promoter activity [79].

342   Enoka P. Wijekoon et al.

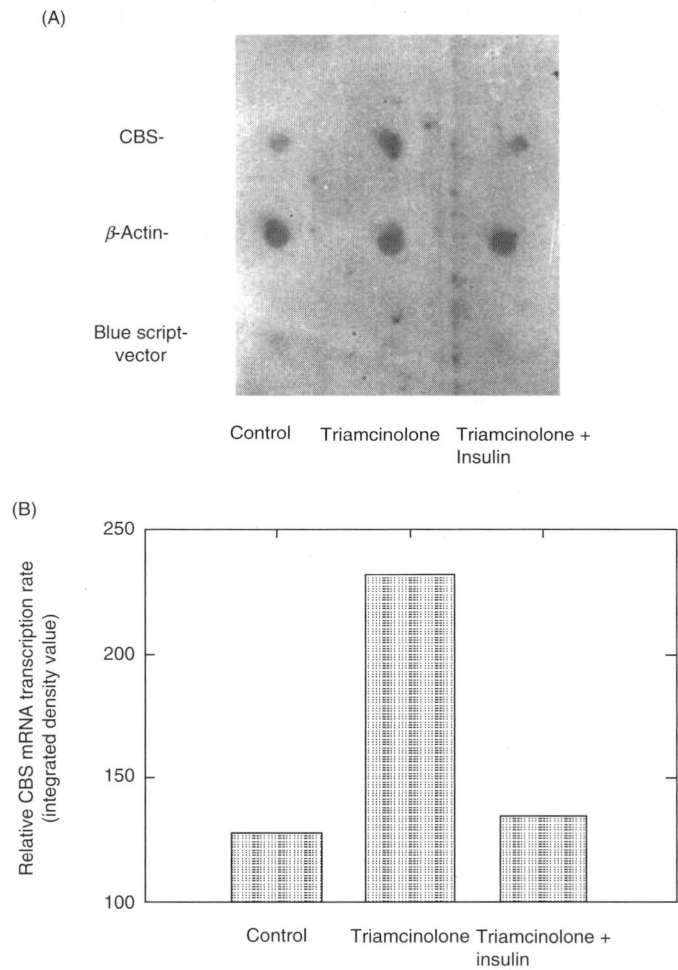

FIGURE 15.6. (A) Nuclear run on analysis of CBS transcription rate in H4IIE cells. (B) The relative transcription rate of CBS [78].

A recent study described the modulation of GNMT by insulin and glucocorticoids [80]. Pretreatment with insulin prevented the induction of GNMT by dexamethasone in rat pancreatic AR42J cells or in hepatoma H4IIE cells. It was also found that induction of diabetes by injection of streptozotocin leads to increased activity and abundance of GNMT.

Early studies conducted by Finkelstein et al. [81] showed that hydrocortisone increased the activity of BHMT threefold, whereas thyroxin treatment significantly reduced its activity. Methionine synthase was also shown to be responsive to hormones in tissues other than the liver; the kidney enzyme was increased by 151% by estrogen and decreased by 60% by growth hormone.

Hydrocortisone treatment led to a two-fold increase in the specific activity of methionine synthase in the pancreas [81].

It is evident therefore that glucagon and glucocorticoids function to increase the disposal of methionine and/or homocysteine, whereas insulin counteracts these effects.

Redox Regulation

CBS is a unique enzyme, in that it depends on both heme and PLP for its function [82]. The 63-kDa CBS subunit binds a molecule each of heme and PLP, and the presence of heme is required for PLP binding. Recently, Taoka et al. [83] found evidence for redox-linked regulation of CBS, dependent on heme. Under reducing conditions generated by the addition of titanium citrate, they observed a 41% decrease in the activity of CBS. Reoxidation of the ferrous enzyme with ferricyanide reversed the inhibition. Transsulfuration, catalyzed by the two enzymes, CBS and CGL, leads to the synthesis of cysteine, which is required for glutathione synthesis. The homocysteine-dependent transsulfuration pathway plays a major role in the maintenance of the intracellular glutathione pool under oxidative stress conditions and, indeed, the flux through the transsulfuration pathway responds to the oxidant load [84]. Approximately 50% of the intracellular glutathione pool in human liver cells is derived from cysteine produced by the transsulfuration pathway; thus the cellular redox environment may affect plasma homocysteine concentrations [84].

Two other enzymes in the homocysteine metabolic pathway have also been shown to be responsive to redox changes. S-Nitrosylation by nitric oxide, under conditions of septic shock or hypoxia, leads to the inactivation of both isoforms of MAT, MAT I and MAT III [85]; this inactivation can be reversed by millimolar concentrations of glutathione. H_2O_2 also inactivates MAT by reversibly and covalently oxidizing cysteine 121, located at a "flexible loop" over the active site cleft of MAT [86]. The GSH/GSSG ratio also modulates the redox state of MAT [87]. The second enzyme responsive to redox conditions, methionine synthase, is inhibited by oxidative conditions, potentially due to the oxidation of the cob(I)alamin form of the cobalamine cofactor [88] This curtailment of remethylation can be understood as a means of increasing the conversion of homocysteine to cysteine for GSH synthesis.

Nonphysiological Causes of Hyperhomocysteinemia

Vitamin Deficiency

The original report linking hyperhomocysteinemia to the occurrence of cardiovascular disease involved a patient who developed homocysteinuria due to a defect in cobalamin metabolism, now recognized as Cbl C deficiency [89].

Since then, many studies have confirmed the relationship between vitamins and plasma homocysteine levels. Inadequate plasma concentrations of one or more of the B vitamins are responsible for the hyperhomocysteinemia in about two thirds of hyperhomocysteinemic cases [90].

The ability of folate supplementation to decrease urinary homocysteine excretion was documented in patients with homocystinuria and mental retardation as early as 1968 [91]. Kang et al. [92] showed that depletion of tissue folate leads to hyperhomocysteinemia in nonhomocystinuric subjects. Isolated deficiency of vitamin B_{12} also leads to moderate to intermediate fasting hyperhomocysteinemia, with homocysteine levels exceeding that found in the obligate heterozygotes for CBS deficiency [93]. Vitamin B_6, the cofactor for both enzymes of the transsulfuration pathway, is important in determining post-methionine load plasma homocysteine levels [94]. Data from the Framingham Heart Study cohort showed that plasma homocysteine exhibits a strong inverse association with plasma folate and weaker associations with plasma vitamin B_{12} and PLP [90]. Vitamin intervention therapy with the three B vitamins alone and various combinations of them has been shown to reduce both basal plasma homocysteine as well as levels after a methionine load [95].

A few recent studies have tried to explain the mechanisms behind the homocysteine-lowering effects of these B vitamins. Supplementation with vitamin B_{12} was shown to increase the activity of methionine synthase, an effect ascribed by Gulati et al. [96] to a posttranslational regulation of methionine synthase by vitamin B_{12} since the induction of activity of this enzyme correlates with increased levels of protein. They were later able to show that B_{12} supplementation induces translational upregulation by shifting the mRNA from the ribonucleoprotein to the polysome pool. The B_{12} responsive element was localized to the 70-bp region located at the 3' end of the 5'-untranslated region of methionine synthase mRNA [97]. Miller et al. [98] postulated that deficiency of folate leads to increased plasma homocysteine not only through the impairment of homocysteine remethylation but also by impairing homocysteine removal via the transsulfuration pathway due to lowered tissue SAM levels which are inadequate to stimulate CBS.

Genetic Disorders

The most frequently described genetic defects of hyperhomocysteinemia are associated with the enzymes of transsulfuration. Plasma homocysteine levels are quite elevated in patients with homozygous CBS deficiency resulting in homocystinuria [99]. However, this is a rare disorder with a frequency estimated between 1:58,000 and 1:1,000,000 in newborns and a worldwide birth prevalence of 1:300,000. Heterozygosity for CBS in the general population is less than 1% [100]. The fasting homocysteine level in these individuals appears to be normal or only slightly elevated, although post-methionine load levels may be elevated [101]. Genetically determined defects have also been described in CGL, which leads to cystathioninuria [102].

On the other side of the spectrum are the patients with Down syndrome. Down syndrome or trisomy 21 is characterized by the failure of chromosome 21 to segregate normally during meiosis [103]. The gene for CBS is located on chromosome 21 and as a result is overexpressed in children with Down syndrome. A 157% increase in CBS enzyme activity in patients with Down syndrome is associated with reduced levels of plasma homocysteine [104]. However, the reduction in plasma homocysteine leads to a concurrent reduction in the folate-dependent resynthesis of methionine, creating a functional intracellular folate deficiency known as the "methyl trap" [105].

MTHFR mutations are also associated with hyperhomocysteinemia. Severe MTHFR deficiency (less than 2% of normal enzyme activity) is rare [106]. However, Kang et al. [107] identified a new variant of MTHFR, which displayed a distinctive thermolability with about 50% of the enzyme activity. This variant is associated with significantly higher levels of plasma homocysteine and is inherited as an autosomal recessive trait [107]. Patients who were postulated to be compound heterozygotes of the allele for the severe mutation and the allele for the thermolabile mutation have also been identified [108].

Two common functional polymorphisms of the MTHFR gene were later identified. C677T polymorphism in exon 4, resulting in an alanine to valine substitution at codon 222 [109], gives rise to the thermolability. Homozygotes for the C677T polymorphism (TT genotype) have about 30% of normal enzyme activity and frequently exhibit intermediate hyperhomocysteinemia [110]; heterozygotes have about 65% of normal activity [109]. The phenotypic expression of the TT genotype appears to relate to the folate status. In the Hordaland homocysteine study, a majority of subjects with intermediate hyperhomocysteinemia exhibited the TT genotype; 88% of these were folate deficient. [111]. The C677T mutation displays ethnic variability. Caucasians have a high (40%)-allele frequency [112], while it is almost absent in African Americans [113]. A second polymorphism, A1298C in exon 7, results in a substitution of glutamate with alanine at codon 429 [114]. Individuals with 1298CC genotype display 60% of the activity of those with AA genotype but it does not seem to give rise to hyperhomocysteinemia.

Cobalamin is bound to methionine synthase and acts as a methyl carrier between methyltetrahydrofolate and homocysteine. The methyl group of methyltetrahydrofolate is first transferred to the cobalamine cofactor to form methylcob(III)alamin, which in turn becomes cob(I)alamin after transfer of the methyl group [115]. Over time, cob(I)alamin may be oxidized to cob(II)alamin which renders the enzyme inactive. Cob(II)alamin needs to undergo a reductive methylation with SAM acting as the methyl donor to be converted back to its active state [116].

Two forms of methionine synthase deficiency are known which lead to the development of hyperhomocysteinemia [117]. Patients from the cblG complementation group of folate/cobalamin metabolism have mutations in the methionine synthase gene [118]. The second complementation group, cblE, shows reduced methionine synthase activity, due to a defect in the reducing system, which keeps the enzyme in the functional state [119]. cblE patients have been shown to have defective NADPH-dependent reducing activity. Leclerc et al. [120] isolated the cDNA corresponding to the reductive activation enzyme which they named methionine synthase reductase and also identified a number of mutations in three cblE patients, a 4-bp frame shift in two and a 3-bp deletion in one.

Renal Disorders

Patients with diabetes mellitus, either type 1 or type 2, have decreased plasma homocysteine levels when kidney function is normal [75, 76, 121]. However, with decreasing kidney function the concentration of plasma homocysteine changes and is elevated compared to controls. This has been observed in both type 1 and type 2 diabetes mellitus [73, 74]. Patients with end-stage renal disease with no diabetes also exhibit elevated plasma homocysteine [47]. These observations agree with the finding of Bostom et al. [48] that the rat kidney is a major organ involved in homocysteine metabolism. They showed a positive renal arteriovenous difference for homocysteine slightly greater than 20% of the mean arterial plasma homocysteine concentration, which suggested that the loss of the homocysteine metabolizing capacity of the kidneys may be a major cause of the increased plasma homocysteine in renal patients. The transsulfuration pathway was shown to be the major route of catabolism of homocysteine taken up by rat kidneys, accounting for 78% of the disappearance of homocysteine [122]. The important role played by the kidneys in maintaining plasma homocysteine homeostasis was shown by the ability of the kidneys to handle acute increases in plasma homocysteine [123]. Kidneys were able to significantly increase the uptake of homocysteine from the plasma and still manage to metabolize it with no change in urinary excretion, demonstrating the capacity of the kidneys for homocysteine catabolism.

Pharmacological Intervention

Intervention with a variety of pharmacological agents has been shown to disrupt plasma homocysteine metabolism. Many of these drugs act through their disruption of the absorption or the metabolism of the various vitamins of homocysteine metabolism. Methotrexate, used in a variety of diseases, interferes with folate metabolism by inhibiting dihydrofolate reductase [124], thereby reducing methyltetrahydrofolate levels in cells. The time course for the development of hyperhomocysteinemia depends on the dose of methotrexate used; thus it varies in patients with different diseases such as cancer [125], psoriasis [126], and rheumatoid arthritis [127]. Anticonvulsants also interfere with folate metabolism, thereby leading to hyperhomocysteinemia [128]. They are thought to deplete liver folate stores through inhibition of polyglutamation

[129]. Bile acid sequestrants interfering with the absorption of folate also cause hyperhomocysteinemia. This has been observed in coronary patients treated with niacin and colestipol as they exhibit significantly higher homocysteine levels when compared with similar patients receiving a placebo [130]. Cholestyramine [131], a bile acid sequestrant, and the antidiabetic drug metformin [132] interfere with the absorption of cobalamine and folate, leading to the development of hyperhomocysteinemia with long-term use of these drugs.

The anesthetic, nitrous oxide, is a well-known pharmacological agent that causes rapid (within 90 min) elevations in plasma homocysteine [133]. It is known to oxidize cob(I)alamin to cob(II)alamin, thereby inactivating methionine synthase irreversibly [134]. Azuridine, which is no longer in use, caused hyperhomocysteinemia in humans as well as animals probably through its interference with pyridoxal 5′-phosphate [135]. Isoniazid [136], niacin [137], and theophylline [138] also cause increases in plasma homocysteine by their interference with vitamin B_6 metabolism. The treatment of Parkinson's disease with L-DOPA leads to increased plasma homocysteine, the mechanism of which was discussed in the section on Methylation Demand.

Treatment with some other drugs has been shown to lead to reduction in plasma homocysteine. Oral penicillamine has been shown to reduce both free and plasma protein-bound homocysteine in homocystinuria patients [139]. Patients with acute lymphoblastic leukemia treated with 2-deoxycoformycin also have markedly reduced plasma homocysteine levels [140]. 2-Deoxycoformycin indirectly inhibits SAH hydrolase by blocking adenosine deaminase. The use of estrogen-containing oral contraceptives has also been shown to lower plasma homocysteine levels [141]. Cyclic variations in plasma homocysteine have been shown in oral contraceptive users with the variation in hormone levels in the contraceptives.

Perspective

The epidemiological evidence that links elevated plasma homocysteine to a number of chronic diseases, including atherosclerotic vascular disease, adequately justifies the many studies that seek to understand how plasma tHcy is determined. We have now reached the point where we have a good understanding of the determinants of plasma tHcy. As with the plasma level of any constituent, tHcy is determined both by its rate of production and its rate of removal. New information highlights the role of the major methyltransferases, in particular those involved in PC and creatine synthesis, in determining tHcy [142]. Classical work has always emphasized the role played by the homocysteine removal pathways—remethylation and transsulfuration. These studies emphasized the role of vitamin deficiencies, mutations, and polymorphisms. New data have also emphasized the effects of hormones and a number of drugs.

However, there are also significant areas of ignorance where work is urgently needed. What is the relation between renal function and tHcy? In most population studies, tHcy correlates quite well with plasma creatinine. It is clear that the kidney is a major organ for homocysteine removal in the rat but the balance of evidence on the human kidney does not support such a conclusion. However, it should be recognized that the high proportion of protein-bound (and, therefore, unfilterable) homocysteine in humans presents special problems to the *in vivo* investigation of renal homocysteine metabolism.

We know much too little about the membrane transport of homocysteine. We do know that the renal reabsorption of the homocysteine–cysteine mixed disulfide is via the cystine/dibasic amino acid transporter [143]. It has been suggested that there may be different mechanism for homocysteine ingress into and exit from cells. This is based on the fact that the bulk of the extracellular homocysteine is in the oxidized form but the more reduced intracellular environment ensures that the cellular homocysteine is largely reduced. We need to determine the mechanisms (and their control) responsible for homocysteine transport into and out of cells.

Which component of plasma homocysteine is most closely associated with chronic disease? Thus far, most studies have focused on tHcy. It is important to determine whether one of the components of tHcy is a better predictor of chronic disease. It is tempting to compare the present state of homocysteine epidemiology with that of cholesterol some decades ago. Then, it was apparent that elevated plasma cholesterol was a risk factor but the roles of the different lipoproteins were not appreciated. This analogy, however, may be misleading. LDL and HDL are quite different particles with different origins and metabolic fates. The components of tHcy, however, have a common origin and exist in dynamic equilibrium with each other. The importance of forms of plasma homocysteine other than the components of tHcy remains intriguing. Work from Jakubowski's laboratory has emphasized the role of both homocysteine thiolactone and of homocysteine bound either by amide or peptide linkages to human plasma proteins. Recent work has indicated that antibodies to N-homocysteinylated albumin can serve as an independent predictor of early coronary artery disease [144]. Future work should illuminate some of these novel aspects of homocysteine metabolism.

References

1. McCully KS: Vascular pathology of homocysteinemia: implications for the pathogenesis of arteriosclerosis. Am J Pathol 56: 111–128, 1969.
2. Kang SS, Wong PW, Malinow MR: Hyperhomocyst(e)inemia as a risk factor for occlusive vascular disease. Annu Rev Nutr 12: 279–298, 1992.
3. Stampfer MJ, Malinow MR: Can lowering homocysteine levels reduce cardiovascular risk? N Engl J Med 332: 328–932, 1995.
4. Stampfer MJ, Malinow MR, Willett WC, Newcomer LM, Upson B, Ullmann D, Tishler PV, Hennekens CH: A prospective study of plasma homocyst(e)ine and risk of myocardial infarction in US physicians. JAMA 268: 877–881, 1992.

5. Arnesen E, Refsum H, Bonaa KH, Ueland PM, Forde OH, Nordrehaug JE: Serum total homocysteine and coronary heart disease. Int J Epidemiol 24: 704–709, 1995.

6. Boushey CJ, Beresford SA, Omenn GS, Motulsky AG: A quantitative assessment of plasma homocysteine as a risk factor for vascular disease. Probable benefits of increasing folic acid intakes. JAMA 274: 1049–1057, 1995.

7. Homocysteine Studies Collaboration: Homocysteine and risk of ischemic heart disease and stroke: a meta-analysis. JAMA 288: 2015–2022, 2002.

8. Klerk M, Verhoef P, Clarke R, Blom HJ, Kok FJ, Schouten EG (MTHFR Studies Collaboration Group): MTHFR 677C—>T polymorphism and risk of coronary heart disease: a meta-analysis. JAMA 288: 2023–2031, 2002.

9. Stamm EB, Reynolds RD: Plasma total homocyst(e)ine may not be the most appropriate index for cardiovascular disease risk. J Nutr 129: 1927–1930, 1999.

10. Hoogeveen EK, Kostense PJ, Jakobs C, Dekker JM, Nijpels G, Heine RJ, Bouter LM, Stehouwer CD: Hyperhomocysteinemia increases risk of death, especially in type 2 diabetes: 5-year follow-up of the Hoorn Study. Circulation 101: 1506–1511, 2000.

11. Nygard O, Nordrehaug JE, Refsum H, Ueland PM, Farstad M, Vollset SE: Plasma homocysteine levels and mortality in patients with coronary artery disease. N Engl J Med 337: 230–236, 1997.

12. Selhub J, Jacques PF, Bostom AG, D'Agostino RB, Wilson PW, Belanger AJ, O'Leary DH, Wolf PA, Schaefer EJ, Rosenberg IH: Association between plasma homocysteine concentrations and extra-cranial carotid-artery stenosis. N Engl J Med 332: 286–291, 1995.

13. Genest JJ Jr, McNamara JR, Salem DN, Wilson PW, Schaefer EJ, Malinow MR: Plasma homocyst(e)ine levels in men with premature coronary artery disease. J Am Coll Cardiol 16: 1114–1119, 1990.

14. Jacobsen DW: Practical chemistry of homocysteine and other thiols. In: Carmel R, Jacobsen DW (eds), Homocysteine in Health and Disease. Cambridge University Press, Cambridge, UK, 2001, pp 9–20.

15. Ueland PM, Mansoor MA, Guttormsen AB, Muller F, Aukrust P, Refsum H, Svardal AM: Reduced, oxidized and protein-bound forms of homocysteine and other aminothiols in plasma comprise the redox thiol status—a possible element of the extracellular antioxidant defense system. J Nutr 126: 1281S–1284S, 1996.

16. Jakubowski H, Fersht AR: Alternative pathways for editing non-cognate amino acids by aminoacyl-tRNA synthetases. Nucleic Acids Res 9: 3105–3117, 1981.

17. Jakubowski H: Protein homocysteinylation: possible mechanism underlying pathological consequences of elevated homocysteine levels. FASEB J 13: 2277–2283, 1999.

18. Jakubowski H: Homocysteine is a protein amino acid in humans. Implications for homocysteine-linked disease. J Biol Chem 277: 30425–30428, 2002.

19. Jakubowski H: Calcium-dependent human serum homocysteine thiolactone hydrolase. A protective mechanism against protein N-homocysteinylation. J Biol Chem 275: 3957–3962, 2000.

20. Stead LM, Brosnan ME, Brosnan JT: Characterization of homocysteine metabolism in the rat liver. Biochem J 350: 685–692, 2000.

21. Finkelstein JD: The metabolism of homocysteine: pathways and regulation. Eur J Pediatr 157: S40–S44, 1998.

22. Chou JY: Molecular genetics of hepatic methionine adenosyltransferase deficiency. Pharmacol Ther 85: 1–9, 2000.
23. Cabrero C, Alemany S: Conversion of rat liver S-adenosyl-L-methionine synthetase from high-M_r form to low-M_r form by LiBr. Biochim Biophys Acta 952: 277–281, 1988.
24. Sullivan DM, Hoffman JL: Fractionation and kinetic properties of rat liver kidney MAT isozymes. Biochemistry 22: 1636–1641, 1983.
25. Clarke S, Banfield K: S-Adenosylmethionine-dependent methyltransferases. In: Carmel R, Jacobsen DW (eds), Homocysteine in Health and Disease, Cambridge University Press, Cambridge, UK, 2001, pp 9–20.
26. Takata Y, Huang Y, Komoto J, Yamada T, Konishi K, Ogawa H, Gomi T, Fujioka M, Takusagawa F: Catalytic mechanism of glycine N-methyltransferase. Biochemistry 42: 8394–8402, 2003.
27. Svardal A, Refsum H, Ueland PM: Determination of in $vivo$ protein binding of homocysteine and its relation to free homocysteine in the liver and other tissues of the rat. J Biol Chem 261: 3156–3163, 1986.
28. Davis SR, Stacpoole PW, Williamson J, Kick LS, Quinlivan EP, Coats BS, Shane B, Bailey LB, Gregory JF 3rd: Tracer-derived total and folate-dependent homocysteine remethylation and synthesis rates in humans indicate that serine is the main one-carbon donor. Am J Physiol Endocrinol Metab 286: E272–E279, 2004.
29. Finkelstein JD, Harris BJ, Kyle WE: Methionine metabolism in mammals: kinetic study of betaine-homocysteine methyltransferase. Arch Biochem Biophys 153: 320–324, 1972.
30. Breksa AP 3rd, Garrow TA: Recombinant human liver betaine-homocysteine S-methyltransferase: identification of three cysteine residues critical for zinc binding. Biochemistry 38: 13991–13998, 1999.
31. Finkelstein JD, Martin JJ, Harris BJ, Kyle WE: Regulation of hepatic betaine-homocysteine methyltransferase by dietary betaine. J Nutr 113: 519–521, 1983.
32. Mudd SH, Finkelstein JD, Irreverre F, Laster L: Transsulfuration in mammals. Microassays and tissue distributions of three enzymes of the pathway. J Biol Chem 240: 4382–4392, 1965.
33. Chen P, Poddar R, Tipa EV, Dibello PM, Moravec CD, Robinson K, Green R, Kruger WD, Garrow TA, Jacobsen DW: Homocysteine metabolism in cardiovascular cells and tissues: implications for hyperhomocysteinemia and cardiovascular disease. Adv Enzyme Regul 39: 93–109, 1999.
34. Quere I, Paul V, Rouillac C, Janbon C, London J, Demaille J, Kamoun P, Dufier JL, Abitbol M, Chasse JF: Spatial and temporal expression of the cystathionine beta-synthase gene during early human development. Biochem Biophys Res Commun 254: 127–137, 1999.
35. Tallan HH, Moore S, Stein WH: L-cystathionine in human brain. J Biol Chem 230: 707–716, 1958.
36. Chen X, Jhee KH, Kruger WD: Production of the neuromodulator H_2S by cystathionine beta-synthase via the condensation of cysteine and homocysteine. J Biol Chem 279: 52082–52086, 2004.
37. Refsum H, Ueland PM, Nygard O, Vollset SE: Homocysteine and cardiovascular disease. Annu Rev Med 49: 31–62, 1998.
38. Schneede J, Refsum H, Ueland PM: Biological and environmental determinants of plasma homocysteine. Semin Thromb Hemost 26: 263–279, 2000.

39. Mudd SH, Poole JR: Labile methyl balance for normal humans on various dietary regimens. Metabolism 24: 721–735, 1975.

40. Kang SS, Wong PW, Zhou JM, Cook HY: Total homocyst(e)ine in plasma and amniotic fluid of pregnant women. Metabolism 35: 889–891, 1986.

41. Boers GH, Smals AG, Trijbels FJ, Leermakers AI, Kloppenborg PW: Unique efficiency of methionine metabolism in premenopausal women may protect against vascular disease in the reproductive years. J Clin Invest 72: 1971–1976, 1983.

42. Giltay EJ, Hoogeveen EK, Elbers JM, Gooren LJ, Asscheman H, Stehouwer CD: Effects of sex steroids on plasma total homocysteine levels: a study in transsexual males and females. J Clin Endocrinol Metab 83: 550–553, 1998.

43. Andersson A, Hultberg B, Brattstrom L, Isaksson A: Decreased serum homocysteine in pregnancy. Eur J Clin Chem Clin Biochem 30: 377–379, 1992.

44. Malinow MR, Rajkovic A, Duell PB, Hess DL, Upson BM: The relationship between maternal and neonatal umbilical cord plasma homocyst(e)ine suggests a potential role for maternal homocyst(e)ine in fetal metabolism. Am J Obstet Gynecol 178: 228–233, 1998.

45. Selhub J, Jacques PF, Wilson PW, Rush D, Rosenberg IH: Vitamin status and intake as primary determinants of homocysteinemia in an elderly population. JAMA 270: 2693–2698, 1993.

46. Norlund L, Grubb A, Fex G, Leksell H, Nilsson JE, Schenck H, Hultberg B: The increase of plasma homocysteine concentrations with age is partly due to the deterioration of renal function as determined by plasma cystatin C. Clin Chem Lab Med 36: 175–178, 1998.

47. Bostom AG, Shemin D, Lapane KL, Miller JW, Sutherland P, Nadeau M, Seyoum E, Hartman W, Prior R, Wilson PW, Selhub J: Hyperhomocysteinemia and traditional cardiovascular disease risk factors in end-stage renal disease patients on dialysis: a case-control study. Atherosclerosis 114: 93–103, 1995.

48. Bostom A, Brosnan JT, Hall B, Nadeau MR, Selhub J: Net uptake of plasma homocysteine by the rat kidney *in vivo*. Atherosclerosis 116: 59–62, 1995.

49. van Guldener C, Donker AJ, Jakobs C, Teerlink T, de Meer K, Stehouwer CD: No net renal extraction of homocysteine in fasting humans. Kidney Int 54: 166–169, 1998.

50. Koehler KM, Pareo-Tubbeh SL, Romero LJ, Baumgartner RN, Garry PJ: Folate nutrition and older adults: challenges and opportunities. J Am Diet Assoc 97: 167–173, 1997.

51. Lindenbaum J, Rosenberg IH, Wilson PW, Stabler SP, Allen RH: Prevalence of cobalamin deficiency in the Framingham elderly population. Am J Clin Nutr 60: 2–11, 1994.

52. Cravo ML, Gloria LM, Selhub J, Nadeau MR, Camilo ME, Resende MP, Cardoso JN, Leitao CN, Mira FC: Hyperhomocysteinemia in chronic alcoholism: correlation with folate, vitamin B-12, and vitamin B-6 status. Am J Clin Nutr 63: 220–224, 1996.

53. Nygard O, Vollset SE, Refsum H, Stensvold I, Tverdal A, Nordrehaug JE, Ueland M, Kvale G: Total plasma homocysteine and cardiovascular risk profile. The Hordaland Homocysteine Study. JAMA 274: 1526–1533, 1995.

54. Nygard O, Refsum H, Ueland PM, Stensvold I, Nordrehaug JE, Kvale G, Vollset SE: Coffee consumption and plasma total homocysteine: the Hordaland Homocysteine Study. Am J Clin Nutr 65: 136–143, 1997.

55. Vollset SE, Nygård O, Kvåle G, Ueland PM, Refsum H: The Hordaland homocysteine study: lifestyle and plasma homocysteine in Western Norway. In: Graham IM, Refsum H, Rosenberg IH, Ueland PM (eds), Homocysteine Metabolism: From Basic Science to Clinical Medicine. Kluwer Academic Publishers, Boston, 1997, pp 177–182.

56. Mayer O Jr, Simon J, Rosolova H: A population study of the influence of beer consumption on folate and homocysteine concentrations. Eur J Clin Nutr 55: 605–609, 2001.

57. Kenyon SH, Nicolaou A, Gibbons WA: The effect of ethanol and its metabolites upon methionine synthase activity *in vitro*. Alcohol 15: 305–309, 1998.

58. Wright M, Francis K, Cornwell P: Effect of acute exercise on plasma homocysteine. J Sports Med Phys Fitness 38: 262–265, 1998.

59. Finkelstein JD, Kyle WE, Harris BJ: Methionine metabolism in mammals: regulatory effects of *S*-adenosylhomocysteine. Arch Biochem Biophys 165: 774–779, 1974.

60. Finkelstein JD, Kyle WE, Martin JL, Pick AM: Activation of cystathionine synthase by adenosylmethionine and adenosylethionine. Biochem Biophys Res Commun 66: 81–87, 1975.

61. Kutzbach C, Stokstad EL: Mammalian methylenetetrahydrofolate reductase. Partial purification, properties, and inhibition by *S*-adenosylmethionine. Biochim Biophys Acta 250: 459–477, 1971.

62. Wagner C, Briggs WT, Cook RJ: Inhibition of glycine *N*-methyltransferase activity by folate derivatives: implications for regulation of methyl group metabolism. Biochem Biophys Res Commun 127: 746–752, 1985.

63. Storch KJ, Wagner DA, Burke JF, Young VR: Quantitative study *in vivo* of methionine cycle in humans using [methyl-^{2}H$_3$]- and [1-^{13}C]methionine. Am J Physiol 255: E322–E331, 1988.

64. Katz JE, Dlakic M, Clarke S: Automated identification of putative methyltransferases from genomic open reading frames. Mol Cell Proteomics 2: 525–540, 2003.

65. Hague SM, Klaffke S, Bandmann O: Neurodegenerative disorders: Parkinson's disease and Huntington's disease. J Neurol Neurosurg Psychiatry 76: 1058–1063, 2005.

66. Cheng H, Gomes-Trolin C, Aquilonius SM, Steinberg A, Lofberg C, Ekblom J, Oreland L: Levels of L-methionine *S*-adenosyltransferase activity in erythrocytes and concentrations of *S*-adenosylmethionine and *S*-adenosylhomocysteine in whole blood of patients with Parkinson's disease. Exp Neurol 145: 580–585, 1997.

67. Muller T, Woitalla D, Hauptmann B, Fowler B, Kuhn W: Decrease of methionine and *S*-adenosylmethionine and increase of homocysteine in treated patients with Parkinson's disease. Neurosci Lett 308: 54–56, 2001.

68. Miller JW, Shukitt-Hale B, Villalobos-Molina R, Nadeau MR, Selhub J, Joseph JA: Effect of L-Dopa and the catechol-*O*-methyltransferase inhibitor Ro 41-0960 on sulfur amino acid metabolites in rats. Clin Neuropharmacol 20: 55–66, 1997.

69. Stead LM, Au KP, Jacobs RL, Brosnan ME, Brosnan JT: Methylation demand and homocysteine metabolism: effects of dietary provision of creatine and guanidinoacetate. Am J Physiol Endocrinol Metab 281: E1095–E1100, 2001.

70. Noga AA, Stead LM, Zhao Y, Brosnan ME, Brosnan JT, Vance DE: Plasma homocysteine is regulated by phospholipid methylation. J Biol Chem 278: 5952–5955, 2003.

71. Vance DE, Ridgway ND: The methylation of phosphatidylethanolamine. Prog Lipid Res 27: 61–79, 1988.

72. DeLong CJ, Shen YJ, Thomas MJ, Cui Z: Molecular distinction of phosphatidyl-choline synthesis between the CDP-choline pathway and phosphatidylethanola-mine methylation pathway. J Biol Chem 274: 29683–29688, 1999.

73. Hoogeveen EK, Kostense PJ, Beks PJ, Mackaay AJ, Jacobs C, Bouter LM, Heine RJ, Stehouwer CD: Diabetes mellitus: a population based study. Arterioscler Thromb Vasc Biol 18: 133–138, 1998.

74. Hultberg B, Agardh E, Andersson A, Brattstrom L, Isaksson A, Israelsson B, Agardh CD: Increased levels of plasma homocysteine are associated with nephropathy, but not severe retinopathy in type 1 diabetes mellitus. Scand J Clin Lab Invest 51: 277–282, 1991.

75. Robillon JF, Canivet B, Candito M, Sadoul JL, Jullien D, Morand P, Chambon P, Freychet P: Type 1 diabetes mellitus and homocyst(e)ine. Diabetes Metab 20: 494–496, 1994.

76. Jacobs RL, House JD, Brosnan ME, Brosnan JT: Effects of streptozotocin-induced diabetes and of insulin treatment on homocysteine metabolism in the rat. Diabetes 47: 1967–1970, 1998.

77. Jacobs RL, Stead LM, Brosnan ME, Brosnan JT: Hyperglucagonemia in rats results in decreased plasma homocysteine and increased flux through the transsulfuration pathway in liver. J Biol Chem 276: 43740–43747, 2001.

78. Ratnam S, Maclean KN, Jacobs RL, Brosnan ME, Kraus JP, Brosnan JT: Hormonal regulation of cystathionine β synthase expression in liver. J Biol Chem 277: 42912–42918, 2002.

79. Gil B, Pajares MA, Mato JM, Alvarez L: Glucocorticoid regulation of hepatic S-adenosylmethionine synthetase gene expression. Endocrinology 138: 1251–1258, 1997.

80. Nieman KM, Rowling MJ, Garrow TA, Schalinske KL: Modulation of methyl group metabolism by streptozotocin-induced diabetes and all-*trans*-retinoic acid. J Biol Chem 279: 45708–45712, 2004.

81. Finkelstein JD, Kyle W, Harris BJ: Methionine metabolism in mammals. Regulation of homocysteine methyltransferases in rat tissue. Arch Biochem Biophys 146: 84–92, 1971.

82. Kery V, Bukovska G, Kraus JP: Transsulfuration depends on heme in addition to pyridoxal 5′-phosphate. Cystathionine beta-synthase is a heme protein. J Biol Chem 269: 25283–25288, 1994.

83. Taoka S, Ohja S, Shan X, Kruger WD, Banerjee R: Evidence for heme-mediated redox regulation of human cystathionine beta-synthase activity. J Biol Chem 273: 25179–25184, 1998.

84. Mosharov E, Cranford MR, Banerjee R: The quantitatively important relation-ship between homocysteine metabolism and glutathione synthesis by the transsulfuration pathway and its regulation by redox changes. Biochemistry 39: 13005–13011, 2000.

85. Avila MA, Mingorance J, Martinez-Chantar ML, Casado M, Martin-Sanz P, Bosca L, Mato JM: Regulation of rat liver S-adenosylmethionine synthetase during septic shock: role of nitric oxide. Hepatology 25: 391–396, 1997.

86. Sanchez-Gongora E, Ruiz F, Mingorance J, An W, Corrales FJ, Mato JM: Interaction of liver methionine adenosyltransferase with hydroxyl radical. FASEB J 11: 1013–1019, 1997.

87. Pajares MA, Duran C, Corrales F, Pliego MM, Mato JM: Modulation of rat liver *S*-adenosylmethionine synthetase activity by glutathione. J Biol Chem 267: 17598–17605, 1992.

88. Chen Z, Chakraborty S, Banerjee R: Demonstration that mammalian methionine synthases are predominantly cobalamin-loaded. J Biol Chem 270: 19246–19249, 1995.

89. Gaull GE: Homocystinuria, vitamin B_6, and folate: metabolic interrelationships and clinical significance. J Pediatr 81: 1014–1018, 1972.

90. Selhub J, Jacques PF, Bostom AG, D'Agostino RB, Wilson PW, Belanger AJ, O'Leary DH, Wolf PA, Rush D, Schaefer EJ, Rosenberg IH: Relationship between plasma homocysteine, vitamin status and extracranial carotid-artery stenosis in the Framingham Study population. J Nutr 126: 1258S–1265S, 1996.

91. Carey MC, Fennelly JJ, FitzGerald O: Homocystinuria. II. Subnormal serum folate levels, increased folate clearance and effects of folic acid therapy. Am J Med 45: 26–31, 1968.

92. Kang SS, Wong PW, Norusis M: Homocysteinemia due to folate deficiency. Metabolism 36: 458–462, 1987.

93. Brattstrom L, Israelsson B, Lindgarde F, Hultberg B: Higher total plasma homocysteine in vitamin B_{12} deficiency than in heterozygosity for homocystinuria due to cystathionine beta-synthase deficiency. Metabolism 37: 175–178, 1988.

94. Brattstrom L, Israelsson B, Norrving B, Bergqvist D, Thorne J, Hultberg B, Hamfelt A: Impaired homocysteine metabolism in early-onset cerebral and peripheral occlusive arterial disease. Effects of pyridoxine and folic acid treatment. Atherosclerosis 81: 51–60, 1990.

95. Brattstrom L: Vitamins as homocysteine-lowering agents. J Nutr 126: 1276S–1280S, 1996.

96. Gulati S, Baker P, Li YN, Fowler B, Kruger W, Brody LC, Banerjee R: Defects in human methionine synthase in cblG patients. Hum Mol Genet 5: 1859–1865, 1996.

97. Oltean S, Banerjee R: Nutritional modulation of gene expression and homocysteine utilization by vitamin B_{12}. J Biol Chem 278: 20778–20784, 2003.

98. Miller JW, Nadeau MR, Smith J, Smith D, Selhub J: Folate-deficiency-induced homocysteinaemia in rats: disruption of *S*-adenosylmethionine's co-ordinate regulation of homocysteine metabolism. Biochem J 298: 415–419, 1994.

99. Mudd SH, Skovby F, Levy HL, Pettigrew KD, Wilcken B, Pyeritz RE, Andria G, Boers GH, Bromberg IL, Cerone R, Fowler B, Grobe H, Schmidt H, Schweitzer L: The natural history of homocystinuria due to cystathionine beta-synthase deficiency. Am J Hum Genet 37: 1–31, 1985.

100. Mudd SH, Levy HL, Skovby F: Disorders of transsulfuration. In: Scriver CR, Beaudet Al, Sly WS, Valle D (eds), The Metabolic and Molecular Basis of Inherited Disease. McGraw-Hill, New York, 1995, pp 1279–1327.

101. Tsai MY, Garg U, Key NS, Hanson NQ, Suh A, Schwichtenberg K: Molecular and biochemical approaches in the identification of heterozygotes for homocystinuria. Atherosclerosis 122: 69–77, 1996.

102. Frimpter GW: Cystathioninuria: nature of the defect. Science 149: 1095–1096, 1965.

103. Epstein CJ: Epilogue: toward the twenty-first century with Down syndrome—a personal view of how far we have come and of how far we can reasonably expect to go. Prog Clin Biol Res 393: 241–246, 1995.

104. Chadefaux B, Rethore MO, Raoul O, Ceballos I, Poissonnier M, Gilgenkranz S, Allard D: Cystathionine beta synthase: gene dosage effect in trisomy 21. Biochem Biophys Res Commun 128: 40–44, 1985.

105. Pogribna M, Melnyk S, Pogribny I, Chango A, Yi P, James SJ: Homocysteine metabolism in children with Down syndrome: *in vitro* modulation. Am J Hum Genet 69: 88–95, 2001.

106. Rozen R: Molecular genetics of methylenetetrahydrofolate reductase deficiency. J Inherit Metab Dis 19: 589–594, 1996.

107. Kang SS, Wong PW, Susmano A, Sora J, Norusis M, Ruggie N: Thermolabile methylenetetrahydrofolate reductase: an inherited risk factor for coronary artery disease. Am J Hum Genet 48: 536–545, 1991.

108. Kang SS, Wong PW, Bock HG, Horwitz A, Grix A: Intermediate hyperhomocysteinemia resulting from compound heterozygosity of methylenetetrahydrofolate reductase mutations. Am J Hum Genet 48: 546–551, 1991.

109. Frosst P, Blom HJ, Milos R, Goyette P, Sheppard CA, Matthews RG, Boers GJ, den Heijer M, Kluijtmans LA, van den Heuvel LP, Rozen R: A candidate genetic risk factor for vascular disease: a common mutation in methylenetetrahydrofolate reductase. Nat Genet 10: 111–113, 1995.

110. Kang SS, Zhou J, Wong PW, Kowalisyn J, Strokosch G: Intermediate homocysteinemia: a thermolabile variant of methylenetetrahydrofolate reductase. Am J Hum Genet 43: 414–421, 1988.

111. Guttormsen AB, Ueland PM, Nesthus I, Nygard O, Schneede J, Vollset SE, Refsum H: Determinants and vitamin responsiveness of intermediate hyperhomocysteinemia (> or = 40 micromol/liter). The Hordaland Homocysteine Study. J Clin Invest 98: 2174–2183, 1996.

112. van der Put NMJ, Eskes TKAB, Blom HJ: Is the common 677C-T mutation in the methylenetetrahydrofolate reductase gene a risk factor for neural tube defects? A meta-analysis. Q J Med 90: 111115, 1997.

113. McAndrew PE, Brandt JT, Pearl DK, Prior TW: The incidence of the gene for thermolabile methylene tetrahydrofolate reductase in African Americans. Thromb Res 83: 195–198, 1996.

114. van der Put NM, Gabreels F, Stevens EM, Smeitink JA, Trijbels FJ, Eskes TK, van den Heuvel LP, Blom HJ: A second common mutation in the methylenetetrahydrofolate reductase gene: an additional risk factor for neural-tube defects? Am J Hum Genet 62: 1044–1051, 1998.

115. Banerjee R: The Yin-Yang of cobalamin biochemistry. Chem Biol 4: 175–186, 1997.

116. Ludwig ML, Matthews RG: Structure-based perspectives on B_{12}-dependent enzymes. Annu Rev Biochem 66: 269–313, 1997.

117. Harding CO, Arnold G, Barness LA, Wolff JA, Rosenblatt DS: Functional methionine synthase deficiency due to cblG disorder: a report of two patients and a review. Am J Med Genet 71: 384–390, 1997.

118. Leclerc D, Campeau E, Goyette P, Adjalla CE, Christensen B, Ross M, Eydoux P, Rosenblatt DS, Rozen R, Gravel RA: Human methionine synthase: cDNA cloning and identification of mutations in patients of the cblG complementation group of folate/cobalamin disorders. Hum Mol Genet 5: 1867–1874, 1996.

119. Gulati S, Chen Z, Brody LC, Rosenblatt DS, Banerjee R: Defects in auxiliary redox proteins lead to functional methionine synthase deficiency. J Biol Chem 272: 19171–19175, 1997.

120. Leclerc D, Wilson A, Dumas R, Gafuik C, Song D, Watkins D, Heng HH, Rommens JM, Scherer SW, Rosenblatt DS, Gravel RA: Cloning and mapping of a cDNA for methionine synthase reductase, a flavoprotein defective in patients with homocystinuria. Proc Natl Acad Sci USA 95: 3059–3064, 1998.

121. Wijekoon EP, Hall B, Ratnam S, Brosnan ME, Brosnan JT: Homocysteine metabolism in ZDF (type 2) diabetic rats. Diabetes 54: 3245–3251, 2005.
122. House JD, Brosnan ME, Brosnan JT: Characterization of homocysteine metabolism in the rat kidney. Biochem J 328: 287–292, 1997.
123. House JD, Brosnan ME, Brosnan JT: Renal uptake and excretion of homocysteine in rats with acute hyperhomocysteinemia. Kidney Int 54: 1601–1607, 1998.
124. Bertino JR: Karnofsky memorial lecture. Ode to methotrexate. J Clin Oncol 11: 5–14, 1993.
125. Refsum H, Wesenberg F, Ueland PM: Plasma homocysteine in children with acute lymphoblastic leukemia: changes during a chemotherapeutic regimen including methotrexate. Cancer Res 51: 828–835, 1991.
126. Refsum H, Helland S, Ueland PM: Fasting plasma homocysteine as a sensitive parameter of antifolate effect: a study of psoriasis patients receiving low-dose methotrexate treatment. Clin Pharmacol Ther 46: 510–520, 1989.
127. Morgan SL, Baggott JE, Lee JY, Alarcon GS: Folic acid supplementation prevents deficient blood folate levels and hyperhomocysteinemia during long term, low dose methotrexate therapy for rheumatoid arthritis: implications for cardiovascular disease prevention. J Rheumatol 25: 441–446, 1998.
128. James GK, Jones MW, Pudek MR: Homocyst(e)ine levels in patients on phenytoin therapy. Clin Biochem 30: 647–649, 1997.
129. Carl GF, Smith ML, Furman GM, Eto I, Schatz RA, Krumdieck CL: Phenytoin treatment and folate supplementation affect folate concentrations and methylation capacity in rats. J Nutr 121: 1214–1221, 1991.
130. Blankenhorn DH, Azen SP, Crawford DW, Nessim SA, Sanmarco ME, Selzer RH, Shircore AM, Wickham EC: Effects of colestipol–niacin therapy on human femoral atherosclerosis. Circulation 83: 438–447, 1991.
131. Tonstad S, Refsum H, Ose L, Ueland PM: The C677T mutation in the methylenetetrahydrofolate reductase gene predisposes to hyperhomocysteinemia in children with familial hypercholesterolemia treated with cholestyramine. J Pediatr 132: 365–368, 1998.
132. Carlsen SM, Folling I, Grill V, Bjerve KS, Schneede J, Refsum H: Metformin increases total serum homocysteine levels in non-diabetic male patients with coronary heart disease. Scand J Clin Lab Invest 57: 521–527, 1997.
133. Ermens AA, Refsum H, Rupreht J, Spijkers LJ, Guttormsen AB, Lindemans J, Ueland PM, Abels J: Monitoring cobalamin inactivation during nitrous oxide anesthesia by determination of homocysteine and folate in plasma and urine. Clin Pharmacol Ther 49: 385–393, 1991.
134. Drummond JT, Matthews RG: Nitrous oxide inactivation of cobalamin-dependent methionine synthase from *Escherichia coli*: characterization of the damage to the enzyme and prosthetic group. Biochemistry 33: 3742–3750, 1994.
135. Drell W, Welch AD: Azaribine-homocystinemia-thrombosis in historical perspective. Pharmacol Ther 41: 195–206, 1989.
136. Krishnaswamy K: Isonicotinic acid hydrazide and pyridoxine deficiency. Int J Vitam Nutr Res 44: 457–465, 1974.
137. Basu TK, Mann S: Vitamin B-6 normalizes the altered sulfur amino acid status of rats fed diets containing pharmacological levels of niacin without reducing niacin's hypolipidemic effects. J Nutr 127: 117–121, 1997.
138. Ubbink JB, van der Merwe A, Delport R, Allen RH, Stabler SP, Riezler R, Vermaak WJ: The effect of a subnormal vitamin B-6 status on homocysteine metabolism. J Clin Invest 98: 177–184, 1996.

139. Kang SS, Wong PW, Curley K: The effect of D-penicillamine on protein-bound homocyst(e)ine in homocystinurics. Pediatr Res 16: 370–372, 1982.
140. Kredich NM, Hershfield MS, Falletta JM, Kinney TR, Mitchell B, Koller C: Effects of 2′-deoxycoformycin on homocysteine metabolism in acute lymphoblastic leukemia. Clin Res 29: 541A, 1981.
141. Steegers-Theunissen RP, Boers GH, Steegers EA, Trijbels FJ, Thomas CM, Eskes TK: Effects of sub-50 oral contraceptives on homocysteine metabolism: a preliminary study. Contraception 45: 129–139, 1992.
142. Stead LM, Brosnan JT, Brosnan ME, Vance DE, Jacobs RL: Is it time to re-evaluate methyl balance in humans? Am J Clin Nutr 83: 5–10, 2006.
143. Brosnan JT: Homocysteine and the kidney. In: Carmel R, Jacobsen DW (eds), Homocysteine in Health and Disease. Cambridge University Press, Cambridge, UK, 2001, pp 176–182.
144. Undas A, Jankowski M, Twardowska M, Padjas A, Jakubowski H, Szczeklik A: Antibodies to N-homocysteinylated albumin as a marker for early-onset coronary artery disease in men. Thromb Haemost 93: 346–350, 2005.

16

Role of Hyperhomocysteinemia in Atherosclerosis

STEPHEN M. COLGAN, DONALD W. JACOBSEN, AND RICHARD C. AUSTIN

Abstract

An elevated level of plasma total homocysteine is a pathological condition known as hyperhomocysteinemia. Studies have shown that homocysteine can induce apoptotic cell death and a relationship between hyperhomocysteinemia and atherosclerosis has been reported. This chapter will summarize the dietary and genetic factors that induce hyperhomocysteinemia. In addition, animal models of hyperhomocysteinemia will be discussed along with the potential cellular mechanisms that could cause hyperhomocysteinemia to induce cell death and accelerate atherosclerosis.

Keywords: apoptosis; atherosclerosis; endoplasmic reticulum stress; endothelial dysfunction; hyperhomocysteinemia

Abbreviations: apoE, apolipoprotein E; CBS, cystathionine β-synthase; GADD153, growth arrest- and DNA damage-inducible gene 153; HHcy, hyperhomocysteinemia; MCP-1, monocyte chemoattractant protein 1; MTHFR, 5,10-methylene-tetrahydrafolate reductase; PERK, PKR-like ER kinase; ROS, reactive oxygen species; SREBP, sterol regulatory element binding protein; TDAG51, T-cell death associated gene 51; TF, tissue factor; UPR, unfolded protein response; XBP-1, X-box binding protein-1

Introduction

Clinical manifestations of atherothrombotic disease, including peripheral vascular disease, myocardial infarction and stroke, account for the majority of deaths in North America [1–4]. Atherosclerosis is a complex, chronic process that is initiated at sites of endothelial cell injury and culminates in lesion disruption and thrombus formation [1–5]. The infiltration of monocytic cells, proliferation and migration of smooth muscle cells, cholesterol deposition, and elaboration of extracellular matrix are characteristic features of atherosclerotic lesions. Other characteristic features of lesion development are cholesterol-enriched macrophages, recognized as foam cells [6, 7]. It is well recognized that the acute clinical manifestations of atherothrombosis result from lesion rupture, thrombus formation, and vessel occlusion [2, 8].

A hallmark feature of animal and human atherosclerotic lesions is apoptotic cell death [9–11]. Apoptosis could lead to an increased risk of lesion rupture by diminishing the number of viable smooth muscle cells required for collagen synthesis and stabilization of the fibrous cap. Additionally, apoptosis enhances the thrombogenicity of the atherosclerotic lesion by increasing the number of tissue factor (TF)-rich apoptotic cells [12, 13]. The cellular pathways responsible for this effect and their significance to atherosclerosis are incompletely understood.

Hyperhomocysteinemia (HHcy) is a pathological condition distinguished by an elevated concentration of plasma total homocysteine [14–21]. Up to 40% of patients diagnosed with recurrent venous thrombosis, peripheral vascular disease or premature coronary artery disease have HHcy [14–19]. Previous studies have shown that homocysteine causes cell dysfunction and leads to apoptotic cell death in cell types relevant to atherosclerosis, including endothelial cells and smooth muscle cells [22–25]. A direct causal relationship between the induction of HHcy and atherosclerosis has previously been reported in animal models with diet-and/or genetic-induced HHcy [26–29]. Since previous studies have demonstrated that the administration of folic acid or a combination of B vitamins can decrease HHcy and attenuate atherogenesis in these animal models [27, 29], there is interest in vitamin supplementation as a strategy for prevention of atherosclerosis.

In this chapter, we will summarize the dietary and genetic factors that induce HHcy. In addition, the cellular mechanisms by which HHcy causes endothelial cell dysfunction and accelerates atherosclerosis will be discussed.

Genetic and Nutritional Factors that Induce Hyperhomocysteinemia

Homocysteine is a thiol-containing amino acid that is formed through the metabolic conversion of methionine to cysteine. Once produced, homocysteine can either be metabolized to cysteine via the transsulfuration pathway or converted to methionine via the remethylation pathway [17, 20, 21]. Severe forms of HHcy, termed homocystinuria, can be caused by mutations in genes responsible for the metabolism of homocysteine, including cystathionine β-synthase (CBS), methionine synthase (MS), 5,10-methylenetetrahydrofolate reductase (MTHFR), or betaine homocysteine methyltransferase (BHMT). Homozygous CBS deficiency, the most common genetic cause of homocystinuria, results in plasma total homocysteine concentrations of up to 500 μmol/L, compared to the normal range of 10–12 μmol/L [17, 21]. CBS deficiency is linked to various clinical manifestations, including skeletal abnormalities, osteoporosis, ectopia lentis, mental retardation, and hepatic steatosis [17]. In addition, patients are at high risk for premature atherosclerosis, which is the major cause of death associated with CBS deficiency [30–32]. Although homozygous CBS deficiency is uncommon, heterozygous CBS deficiency occurs in approximately 1% of the population and is associated with premature cardiovascular disease in phenotypically normal individuals [30–32]. A genetic mutation leading to MTHFR deficiency causes severe HHcy and can induce premature atherosclerosis and thrombotic disease [33–35]. Nutritional deficiencies of B vitamins required for the metabolism of homocysteine, such as folic acid, vitamin B6 (pyridoxine), and/or vitamin B12 (cyanocobalamin), can also lead to HHcy [36, 37]. It has been suggested that insufficient vitamin intake accounts for two thirds of all HHcy cases [37]. Although vitamin supplementation is effective in lowering plasma homocysteine levels, its significance to cardiovascular disease remains to be determined.

Many insights into the mechanism of homocysteine-induced cellular injury and atherosclerotic lesion development have been revealed in animal models of dietary and/or genetically induced HHcy. We will discuss the important findings from animal models that aid in the understanding of HHcy and atherosclerosis as well as recent studies from basic research that compliments the *in vivo* work.

Animal Models of Hyperhomocysteinemia-Induced Atherogenesis

Manipulation of plasma homocysteine concentrations can be accomplished by dietary and/or genetic approaches. Dietary supplementation of methionine, homocysteine, and/or depletion of B vitamins and folic acid can be used to induce mild to severe HHcy [26–29]. Animal models of HHcy include transgenic mice deficient in CBS or MTHFR. Homozygous CBS-deficient mice have 40-fold greater total plasma homocysteine levels than normal and suffer from hepatic steatosis, severe growth retardation, and dislocation of the lens [38]. These phenotypic changes are also characteristic in human patients with homocystinuria [16–21]. Although heterozygous CBS-deficient mice have twice the normal concentration of total plasma homocysteine, they do not have the same developmental defects observed in homozygotes, which make them ideal models to study mild HHcy. Heterozygous CBS-deficient mice present with endothelial dysfunction and impaired vasorelaxation through decreased vascular nitric oxide bioavailability [39, 40]. This relationship is not exclusive to CBS-deficient mice, as endothelial dysfunction has also been observed in rabbit and monkey models of HHcy [41–45]. Although endothelial dysfunction occurs in CBS-deficient mice and other animal models of HHcy, there is no evidence of atherosclerotic lesion development [45–47].

MTHFR-deficient mice have a comparable phenotype with CBS-deficient mice but are considered to be more susceptible to developing atherosclerotic lesions [46]. Adult heterozygous and homozygous MTHFR-deficient mice develop aortic lipid accumulation that is indicative of early atherosclerotic lesion development. Interestingly, the plasma lipid profile from these animals remains normal [46]. Although the aortic lipid accumulation is not as advanced as that seen in apolipoprotein E (apoE)- or LDL receptor-deficient mice, these observations provide the first evidence that mild HHcy contributes to the formation of athersclerotic lesions.

One of the earliest events in atherosclerotic lesion development is the binding of monocytic cells to the vascular endothelium. Recent studies in rats with diet-induced HHcy have demonstrated that the binding of monocytes to the endothelium is significantly elevated, although characteristic atherosclerotic lesions were not observed [47]. The increased expression of monocyte chemoattractant protein 1 (MCP-1), the proinflammatory adhesion molecule VCAM-1, and E-selectin in the aortic endothelium of hyperhomocysteinemic rats provides a potential cellular mechanism for enhanced monocyte binding. Further observations that dietary supplementation with folic acid prevented an increase in total plasma homocysteine levels, inhibited the expression of MCP-1, VCAM-1, and E-selectin, and decreased monocyte binding to the endothelium, provides further support of an antiatherogenic effect of lowering plasma homocysteine.

Given that atherosclerotic lesion development is limited in most animal models of HHcy, researchers have developed dietary HHcy in genetically altered mice that are prone to atherosclerotic lesion development. Studies have determined that diet-induced mild HHcy accelerates atherogenesis in apoE-deficient mice [26–29]. Zhou et al. [26, 27] demonstrated that apoE-deficient mice fed chow diets supplemented with methionine or homocysteine developed larger, more complex, and more numerous atherosclerotic lesions, compared to mice fed control diets (Fig. 16.1). Lesions in these mice were rich in smooth muscle cells and collagen, which is consistent with the ability of

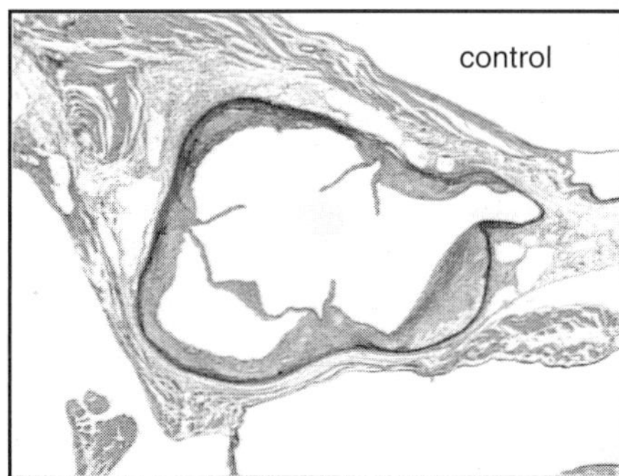

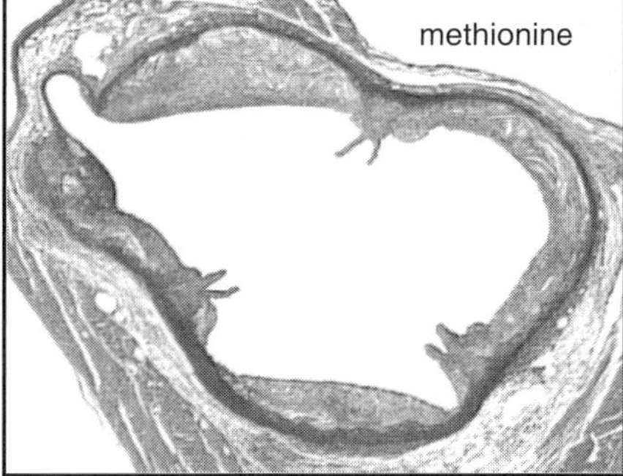

FIGURE 16.1. Atherosclerotic lesions in the aortic root of 25 week old apoE-deficient mice fed a control diet or a high methionine diet to induce hyperhomocysteinemia. Lesion sizes in mice fed high methionine diet were significantly larger, compared to mice fed control diet. Sections were stained with Orcein to reveal the elastic lamina (kindly provided by Dr. Ji Zhou).

homocysteine to induce proliferation of smooth muscle cells and deposition of the extracellular matrix [48]. Considering the effects of the hyperhomocysteinemic diets on lesion size and frequency were observed following 3 months, but not 12 months, of dietary supplementation, it suggests that HHcy mainly influences initial stages of atherosclerotic lesion development. Hofman et al. [29] also found that apoE-deficient mice supplemented with a high methionine diet, had significant increases in atherosclerotic lesion area compared to mice fed a control diet. Furthermore, HHcy was correlated with an increase in NF-κB activation, a transcription factor known to stimulate the production of cytokines, chemokines, leukocyte adhesion molecules, and haemopoetic atherogenesis [1–3]. The observed HHcy-induced NF-κB activation increased the expression of VCAM-1 and the pro-inflammatory mediators RAGE and EN-RAGE [29]. A recent study by Wang et al. [28] has also demonstrated that genetically induced HHcy in a hyperlipidemic background accelerates atherosclerosis. Using apoE/CBS-deficient double knockout mice, increased total plasma homocysteine concentrations correlated with increased atherosclerotic lesion area and lipid content, independent of diet [28].

In addition to its effects on lesion development, diet-induced HHcy in apoE-deficient mice increases the expression of the TF a critical mediator of thrombosis [29]. TF is an integral membrane glycoprotein essential for the initiation of blood coagulation and promotion of thrombosis [49]. Consistent with these findings, *in vitro* studies have demonstrated that homocysteine can induce TF procoagulant activity [50]. Evidence indicates that an increase in TF expression and/or activity can enhance thrombin generation, thereby increasing the risk of thrombotic complications [1, 51, 52].

Although these studies demonstrate that HHcy enhances atherosclerotic lesion development when combined with hyperlipidemia, additional work will be necessary to determine if HHcy accelerates atherogenesis in the presence of other cardiovascular risk factors, including diabetes and hypertension. The dietary and genetic models have supported many of the proposed *in vitro* mechanisms of the role of HHcy in atherosclerotic lesion development. We will further discuss the proposed *in vitro* mechanisms of HHcy and atherosclerosis that help to explain observations made *in vivo*.

Hyperhomocysteinemia and Atherosclerosis: Potential Cellular Mechanisms

Inflammation

Atherothrombotic disease has been suggested to be a form of chronic inflammation [1–3]. Evidence from cell culture studies has revealed that homocysteine induces the production of several proinflammatory cytokines. Treatment of monocytes, smooth muscle cells, and human vascular endothelial cells with homocysteine induces the expression of MCP-1 [53–55]. MCP-1 increases the adhesion of monocytes to the endothelium and their infiltration into the subendothelial cell space, which is an important step during atherogenesis. Homocysteine has also been shown to increase expression of IL-8, a T-lymphocyte, and neutrophil chemoattractant, in cultured endothelial cells [53] and human whole blood [56]. Homocysteine-induced expression of MCP-1 and IL-8 has been demonstrated to occur through the activation of the proinflammatory transcription factor NF-κB. Evidence that HHcy accelerates atherogenesis through vascular inflammation has been shown with the recent findings that NF-κB activation and downstream expression of proinflammatory mediators and cytokines are increased in atherosclerotic lesions from hyperhomocysteinemic apoE-deficient mice [29, 57].

Oxidative stress

The highly reactive thiol group of homocysteine is readily oxidized to form reactive oxygen species (ROS) [58], which suggests that homocysteine-induced cell injury and dysfunction occurs through a mechanism involving autooxidation and oxidative damage. This hypothesis fails to explain why cysteine, which is also readily autooxidized and present in plasma at much higher concentrations than homocysteine, does not cause endothelial cell injury and is not considered associated with cardiovascular disease [59]. Recent studies have also demonstrated that homocysteine does not significantly increase ROS through autooxidation and is involved in antioxidant and reductive cellular biochemistry [60]. In fact, the homocysteine-dependent transsulfuration pathway is critical for maintaining the intracellular levels of glutathione, and this pathway is sensitive to oxidative stress conditions [61, 62].

Homocysteine-induced oxidative stress could impact atherogenesis by mechanisms that are unrelated to autooxidation. Ex vivo studies using vascular tissues have revealed that HHcy causes abnormal vascular relaxation by inducing intracellular production of ROS such as superoxide [41, 63, 64]. Superoxide is believed to limit the normal vasodilation response by reacting with endothelial nitric oxide to generate peroxynitrite [65, 66]. Both superoxide and peroxynitrite result in the generation of lipid peroxides that contribute to the modification of tissues. Peroxynitrite also contributes to the modification of proteins through tyrosine nitration and the formation of 3-nitrotyrosine. Evidence that HHcy can inhibit the antioxidant potential of cells is shown by previous reports that homocysteine impairs heme oxygenase-1 (HO-1) and glutathione peroxidase (GPx) expression and activity in cultured vascular endothelial cells [67–69]. These findings are significant to atherothrombosis given that (i) HHcy enhances vascular dysfunction in GPx-deficient mice [67], (ii) GPx overexpression inhibits homocysteine-induced endothelial dysfunction [70], and (iii) HO-1 overexpression inhibits the atherosclerotic lesion development in apoE$^{-/-}$ mice [71].

Endoplasmic Reticulum Stress and the Unfolded Protein Response

In eukaryotic cells, the endoplasmic reticulum (ER) is the primary site for folding and maturation of secretory, transmembrane, and ER-resident proteins [72–76]. The ER contains a wide range of molecular chaperones such as GRP78, GRP94, calnexin, calreticulin, and protein disulphide isomerase to assist in the correct folding of newly synthesized proteins. These chaperones ensure that only correctly folded proteins are allowed to enter the Golgi for further processing and secretion. Pathological conditions and/or agents that interfere with protein folding activate the unfolded protein response (UPR). The UPR is an integrated intracellular signaling pathway that responds to ER stress by increasing the expression of UPR responsive genes (including the ER-resident chaperones), and attenuating global protein translation (Fig. 16.2).

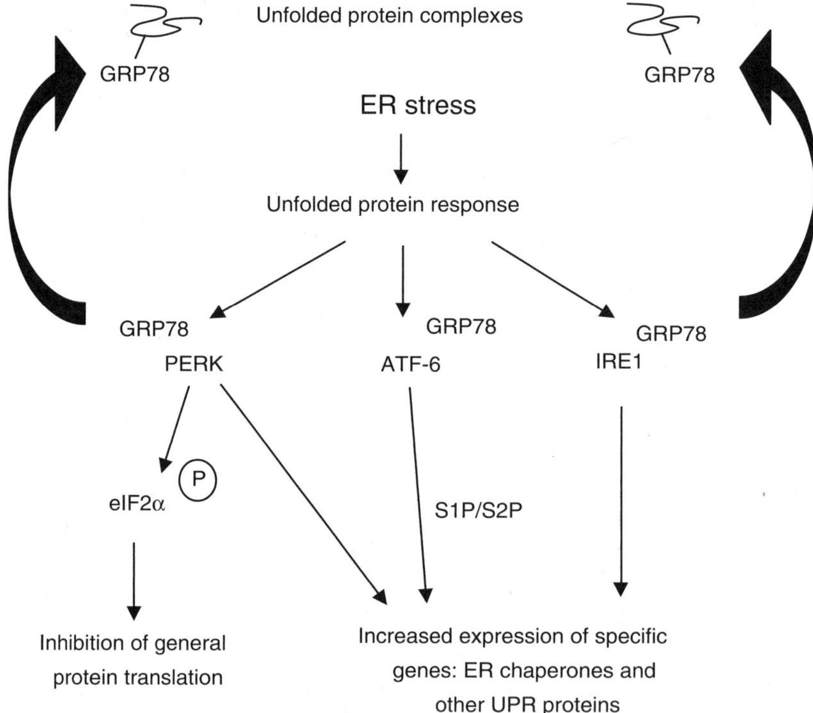

FIGURE 16.2. ER stress and the unfolded protein response (UPR). The UPR is regulated in eukaryotes by the proximal sensors IRE-1, ATF-6, and PERK. Activation of these sensors occurs following their dissociation from GRP78 in response to ER stress. Once activated the UPR functions as an integrated, intracellular signaling pathway to attenuate protein translation, and increase ER chaperone expression.

The UPR is mediated via three ER-resident sensors: a type-I ER transmembrane protein kinase (IRE-1), activating transcription factor 6 (ATF-6) and the PKR like ER kinase (PERK). Activation of these three sensors is mediated by the dissociation of GRP78 following ER stress [72–76]. During ER stress-induced accumulation of unfolded proteins, GRP78 dissociates from the ER-resident sensors to interact with unfolded proteins, thus causing the activation of IRE-1, ATF-6, and PERK [72–76]. In brief, the activation of both IRE-1 and ATF-6 increase the expression of ER-resident chaperones, and the activation of PERK leads to an inhibition of protein translation.

IRE-1 is a transmembrane protein kinase having endoribonuclease activity. Following ER stress and the dissociation of GRP78 from its ER luminal domain, IRE-1 dimerizes and is autophosphorylated, allowing IRE-1 to act as an endoribonuclease in the splicing of X-box binding protein-1 (XBP-1) mRNA. The removal of a 26 base pair intron results in a translation frameshift that permits XBP-1 to act as a transcription factor for genes containing ER stress response elements (ERSE).

ATF-6 is a transcription factor that is localized to the ER membrane in unstressed cells. ATF-6 translocates to the Golgi after release of GRP78. GRP78 regulates the activity of two Golgi localization sequences on ATF-6. In the absence of GRP78, the Golgi localization signal is dominant and results in constitutive translocation of ATF-6 to the Golgi and activation [77]. In the Golgi, site-1 protease (S1P) cleaves ATF-6 in the luminal domain and then the N-terminal membrane anchored half is cleaved by site-2 protease (S2P). The S1P and S2P also recognize and cleave sterol regulatory element binding proteins (SREBPs) under conditions of sterol depletion [78]. Following release, the transactivation domain of ATF-6 translocates to the nucleus where it binds to ERSE, activating transcription of numerous UPR-responsive genes including GADD153 and XBP-1 [79]. Associated with an increase in the transcriptional activation of UPR-responsive genes, ER stress also leads to a rapid, decrease in protein synthesis. This mechanism is mediated by the transmembrane protein kinase, PERK.

Similar to IRE1, PERK is activated by the release of GRP78 from its ER luminal domain. PERK then dimerizes, autophosphorylates, and phosphorylates substrate proteins such as eukaryotic initiation factor-2α (eIF-2α), which blocks mRNA translation to relieve the unfolded protein burden on the ER [72–76]. As a result, the UPR enhances cell survival by ensuring that the effects of ER stress are dealt with an efficient manner. Prolonged or severe ER stress can result in apoptotic cell death and contribute to a number of human diseases, including Alzheimer's disease, Parkinson's disease, and diabetes [74, 75].

One proposed mechanism of homocysteine-induced vascular injury involves ER stress and activation of the UPR [68, 76, 80–83]. It has been reported that elevated intracellular levels of homocysteine elicit ER stress and cause the activation of the UPR, leading to increased ER stress response gene expression, including GRP78, GRP94, homocysteine-responsive ER-resident protein (Herp), and reducing agents and tunicamycin-responsive protein (RTP)

[68, 80–83]. In addition, homocysteine has been shown to induce the expression of GADD153 [84] and T-cell death associated gene 51 (TDAG51) [22]. Both GADD153 and TDAG51 are proapoptotic factors that are induced by prolonged ER stress. The effects observed on gene expression directly involve the UPR because studies have shown that homocysteine treatment causes the activation of both ER-resident sensors PERK and IRE-1 [23, 85, 86].

Severe or prolonged ER stress caused by homocysteine has been shown to activate several cellular mechanisms involved in the development and progression of atherosclerosis, including inflammation, apoptotic cell death, and dysregulation of lipid metabolism. We as well as others have demonstrated that homocysteine-induced ER stress leads to dysregulation of lipid biosynthesis in hepatocytes by activating SREBPs [83, 87]. SREBPs are ER-resident transcription factors responsible for the expression of genes in the cholesterol/triglyceride biosynthesis and uptake pathways [88]. Overexpression of the ER chaperone GRP78, which protects cells from ER stress, inhibits homocysteine-induced SREBP controlled gene expression in cultured human cells [83]. This provides additional support for an association between ER stress and lipid metabolism. Consistent with these results, ER stress, increased expression of SREBP associated cholesterol biosynthetic genes and hepatic steatosis were observed in mice with diet-induced HHcy [83]. Recently, evidence that this mechanism may contribute to the atherogenic effect of HHcy was provided. HHcy-induced in a rat model through a high-methionine diet showed elevated hepatic lipid accumulation and an elevation of plasma cholesterol [89]. Three transcription factors, SREBP, cAMP response element-binding protein (CREB), and nuclear factor Y (NF-Y) were activated in livers of hyperhomocysteinemic rats. These transcription factors resulted in increased cholesterol biosynthesis by transcriptionally regulating HMG-CoA reductase expression causing hepatic lipid accumulation and subsequent hypercholesterolemia [89]. Further studies, however, are required to determine if this mechanism directly contributes to the effect of HHcy on atherosclerotic lesion development.

Molecular Targeting by Homocysteine: The Molecular Targeting Hypothesis

The molecular targeting hypothesis evolved from early reports on the formation of stable disulfide complexes between homocysteine and proteins in circulation [90–97]. Protein-bound homocysteine accounts for 70–80% of plasma total homocysteine in healthy individuals [93]. Mansoor and colleagues were the first to provide quantitative data on protein-bound homocysteine as well as protein-bound cysteine, cysteinylglycine, and glutathione in health and disease [98–101]. Current assays for plasma total homocysteine employ reducing agents to break protein-S–S-homocysteine disulfide bonds [102]. Although the goal of these early studies was to characterize the forms of circulating homocysteine, they also provided the ground work on which the molecular targeting hypothesis was based, namely, the ability of homocysteine to form stable disulfide complexes with specific protein targets and, in the process, alter or impair their function. A second type of molecular targeting is possible with the five-member ring compound homocysteine thiolactone, which contains an activated carbonyl group. The ε-amino group of protein lysine residues readily react with homocysteine thiolactone to form stable amide-bond linked homocysteine [103].

Extracellular Molecular Targeting

A number of plasma proteins form stable disulfide complexes with homocysteine including albumin [93, 104, 105], factor V [106], fibronectin [107], and transthyretin [108]. Although the mechanistic details for complex formation between homocysteine and factor V, fibronectin and transthyretin are incomplete, a fairly detailed picture on the formation of albumin-S–S-homocysteine has been obtained [104, 105].

The albumin molecule (molecular mass = 66.4 kDa) contains 35 cysteine residues, 34 of which form intrachain disulfide bonds and are buried within the globular structure of the molecule [109]. The cysteine at position 34 sits in a 10-Å hydrophobic crevice on the surface. Because the pKa of this sulfhydryl group is abnormally low (pKa ~ 5) [110], albumin-Cys^{34}SH exists as a thiolate anion at physiological pH, which allows it to actively participate in thiol/disulfide exchange reactions. Thus, Alb-Cys^{34}-S$^-$ will attack circulating cystine (Cys-S–S-Cys) to form albumin-Cys^{34}-S–S-Cys [104], which accounts for 60% of plasma total cysteine [90]. Alb-Cys^{34}-S–S-Cys is attacked by homocysteine thiolate anion (Hcy-S$^-$). The disulfide attack is directed towards the sulfur of liganded cysteine and not towards the sulfur of Alb-Cys^{34} [104]:

$$\text{Hcy-S}^- + \text{Alb-Cys}^{34}\text{-S–S-Cys} \rightarrow \text{Hcy-S–S-Cys} + \text{Alb-Cys}^{34}\text{-S}^-$$

It should be mentioned that less than 1% of the homocysteine entering circulation undergoes sulfhydryl group dissociation due to its high pKa (10.0). However, Hcy-S$^-$ is an extremely strong and reactive nucleophile. These reactions account for the formation of homocysteine–cysteine mixed disulfide (Hcy-S–S-Cys) in circulation, which makes up 10–15% of plasma total homocysteine. Approximately 20% of the homocysteine entering circulation undergoes autooxidation to homocystine (Hcy-S–S-Hcy) [105]. This reaction appears to be catalyzed by copper bound to albumin but not by copper bound to ceruloplasmin [105]. Having accounted for the formation of the "free" circulating disulfide forms of homocysteine, i.e., Hcy-S–S-Cys and Hcy-S–S-Hcy, we see that the final step leading to the formation albumin-bound homocysteine involves Alb-Cys^{34}-S$^-$ attack of these species:

$$\text{Alb-Cys}^{34}\text{-S}^- + \text{Hcy-S–S-Cys (or Hcy-S – S-Hcy)} \rightarrow$$

$$\text{Alb-Cys}^{34}\text{-S – S-Hcy} + \text{Cys-S}^- \text{ (or Hcy-S}^-)$$

The attack on the mixed disulfide is specific with 80% occurring at the sulfur of homocysteine. *in vitro* modeling of the biochemical events that occur in circulation with homocysteine have provided plausible reaction mechanisms, which may be useful in understanding other targeting processes [104, 105].

Human factor V (330 kDa) is activated by α-thrombin cleavages at Arg-709, Arg-1018, and Arg-1545 to produce factor Va. The clotting process is down-regulated by proteolytic cleavage of factor Va by activated protein C. Undas et al. [106] have shown that factor V is homocysteinylated *in vitro* and that the resulting homocysteinylated-factor Va is resistant to proteolytic inactivation by activated protein C. However, homocysteinylated-factor V or Va have yet to be identified in the plasma of patients with HHcy, and Lentz et al. [111] could find no evidence for the impairment of protein C activation by thrombin or the inactivation of factor Va by activated protein C in animal models of HHcy and in normal human subjects with acute HHcy after methionine loading.

Human plasma contains soluble fibronectin, a 440-kDa glycoprotein consisting of two similar but nonidentical subunits. Two disulfide bridges hold the subunits together, and each subunit contains >60 cysteine residues, most of which are in the form of intrachain disulfides. Fibronectin plays key roles in cell migration, cell adhesion, hemostasis, thrombosis, wound healing, and tissue remodeling. The fibronectin molecule has distinct domains for interactions with collagen, fibrin, heparin, and DNA. It was recently shown that L-^{35}S-homocysteine targets both fibronectin in human plasma and purified human fibronectin [107]. The mechanism of homocysteine targeting, be it by thiol/disulfide exchange, or by an oxidative process, has not been established. The targeting of fibronectin by homocysteine impairs its interaction with fibrin, which could lead to prolonged recovery from a thrombotic event and contribute to prolonged vascular occlusion [107].

In human plasma, transthyretin exists as a homotetramer and serves as a carrier protein for the hormone thyroxine and the retinol-binding protein-retinal complex [112]. Monomeric transthyretin (13.8 kDa) has a single cysteine residue at position 10, and in the normally folded tetrameric protein, the Cys^{10} residues are in exposed sites at the start of helical regions. Previous work had shown that the Cys^{10} residues formed covalent disulfide complexes with cysteine, glutathione, and cysteinylglycine [113, 114]. In a series of *in vitro* studies, it was demonstrated that transthyretin in normal human plasma and purified transthyretin could be homocysteinylated by L-^{35}S-homocysteine [108]. Plasma from normal donors, patients with end-stage renal disease and patients with homocystinuria contain Cys^{10}-homocysteinylated transthyretin as demonstrated by immunoprecipitation and high-performance liquid chromatography/electrospray mass spectrometry [108]. It is now clear that human transthyretin is targeted by homocysteine both *in vitro* and *in vivo* but the mechanism remains to be elucidated.

Intracellular Molecular Targeting

Although there is considerable evidence that extracellular proteins are targeted by homocysteine, this is not the case for intracellular proteins. It could be argued that the reducing redox potential within the cell would prevent or disrupt protein-S–S-homocysteine bond formation. However, there is now evidence that intracellular targets do exist and that the inactivation of these targets by homocysteine may be a major cause of cellular dysfunction.

Cellular glutathione peroxidase (GPx-1) activity is decreased in the presence of homocysteine. Homocysteine-induced inactivation of GPx-1 would result in higher levels of ROS, a decrease in the bioavailability of nitric oxide, and finally, endothelial dysfunction. Homocysteine, in a dose-dependent manner, inhibited GPx-1 activity in cultured rat aortic smooth muscle cells and bovine aortic endothelial cells [115, 116]. There are conflicting reports as to whether the observed decrease in cellular GPx-1 activity was due to direct inhibition of the enzyme by homocysteine. Upchurch et al. found no evidence for direct inhibition [116] but did find a homocysteine-induced decrease in the steady-state level of GPx-1 mRNA. In contrast, Nishio and Watanabe did observe direct inhibition of purified GPx-1 by homocysteine in a dose-dependent manner [115]. Recently, Handy et al. found that homocysteine inhibits the translation of GPx-1 by a mechanism that involves the selenocysteine incorporation sequence (SECIS) in the 3-untranslated region of the GPx-1 mRNA and other factors that allow for read-through at the UGA-stop codon and incorporation of selenocysteine into the protein [117]. Thus, homocysteine appears to target one or more factors essential for GPx-1 expression and the factor(s) involved could be at the level of transcription or translation, or the enzyme itself.

Arginine residues on proteins undergo S-adenosylmethionine-dependent methylation and when proteins are degraded, asymmetric and symmetric dimethylarginine (ADMA and SDMA) are released into the circulation. ADMA is a potent inhibitor of nitric oxide synthase but it can be eliminated by urinary excretion, or it can be hydrolyzed by the enzyme dimethylarginine dimethylaminohydrolase (DDAH). Stuhlinger et al. were the first to observe that homocysteine targets DDAH directly and inhibits its activity [118].

There is considerable evidence that elevated homocysteine induces ER stress [22, 23, 68, 80–83, 119–121] and that ER processing enzymes such as protein disulfide isomerase are targeted and inactivated by homocysteine. To our knowledge, entry of homocysteine into the ER has not been demonstrated and this, of course, would be a prerequisite before targeting could occur. It is also possible that the secretory proteins that are processed through the ER and Golgi apparatus are the targets themselves.

The chemical basis for molecular targeting by homocysteine is based on sulfhydryl group dissociation constants. The high pKa of the sulfhydryl group of homocysteine has several consequences: (i) the thiolate anion will be at relatively low concentration at physiological pH; (ii) although less than 1% of homocysteine will be in the thiolate anion form, it will be a highly reactive nucleophile as our studies with Alb-Cys^{34}-S–S-Cys have demonstrated [104, 105]; (iii) the protein-S–S-homocysteine disulfide bond will be stronger than those with other thiols having lower sulfhydryl group pKa's; (iv) as a consequence, displacement of homocysteine from protein-S–S-homocysteine by other thiols will be limited.

Hyperhomocysteinemia, ER Stress, and Apoptotic Cell Death

Studies have demonstrated that homocysteine-induced UPR activation causes apoptotic cell death in cultured human vascular endothelial cells [23]. Apoptotic cell death was dependent on IRE-1 activation and was also induced by other ER stress-inducing agents, including thapsigargin (affects ER calcium stores) and tunicamycin (affects N-linked glycosylation). Homocysteine-induced IRE-1 signaling leads to a rapid and sustained activation of c-Jun N-terminal protein kinases (JNK) [122]. This result is consistent with the finding that ER stress-induced JNK activation requires binding of IRE-1 to TRAF2 [123]. Since studies have shown that JNK activation is associated with apoptosis [124], it provides further evidence for a homocysteine-induced mechanism of apoptotic cell death. Furthermore, homocysteine-induced apoptosis requires caspase-3 activation, a result consistent with the ability of homocysteine thiolactone, a cyclic thioester derivative of homocysteine [125], to induce apoptotic cell death in HL-60 cells [126]. Although it has been suggested that caspase-7 and/or caspase-12 activation is involved in ER stress-induced apoptosis [127–129], the effect of homocysteine on the activation of caspase-7 or -12 has not been determined. Given that apoptosis has been reported to occur in both animal and human atherosclerotic lesions, and that mice fed hyperhomocysteinemic diets have increased ER stress and apoptotic cell death in atherosclerotic lesions [130], it is possible that homocysteine-induced ER stress and apoptosis could adversely influence the stability and/or thrombogenicity of atherosclerotic plaques.

Conclusions and Questions

HHcy is an independent risk factor for cardiovascular disease. Several cellular mechanisms have been proposed to elucidate the effects of HHcy on endothelial cell dysfunction and atherosclerosis. Some of the mechanisms include expression and/or activation of proinflammatory factors, oxidative stress, and ER stress. Animal models of HHcy have demonstrated a relationship between HHcy, endothelial cell dysfunction and atherosclerosis. These studies raise some interesting and relevant questions. Although HHcy has been observed to accelerate atherogenesis, what role does it play in plaque stability and/or thrombogenicity? In the absence of hypercholesterolemia or other cardiovascular risk factors, does HHcy enhance lesion development? Does dietary enrichment in B vitamins, essential for the metabolism of homocysteine, protect against cardiovascular disease? Do other risk factors such as hypercholesterolemia, diabetes, and/or hypertension contribute to atherosclerosis through a mechanism involving ER stress? This question is particularly interesting given that the accumulation of free cholesterol in the ER of cultured mouse peritoneal macrophages leads to the depletion of ER calcium stores and leads to UPR activation and apoptotic cell death [131]. Furthermore, it was recently shown that there is an association between free cholesterol accumulation, apoptotic cell death and ER stress in advanced atherosclerotic lesions from apoE-deficient mice [132]. Answers to these and other important questions will enhance our understanding of the mechanisms by which HHcy atherothrombotic disease and its clinical outcomes.

Acknowledgments: D. W. Jacobsen is a recipient of a grant to support homocysteine research (HL52234) from the National Institutes of Health. R.C. Austin is a Career Investigator of the Heart and Stroke Foundation of Ontario and is supported by Research Grants from the Heart and Stroke Foundation of Ontario (T5385) and the Canadian Institutes of Health Research (MOP-67116, MOP-74477). We apologize to many of our colleagues whose work could not be directly acknowledged due to space limitations.

References

1. Ross R: Atherosclerosis-an inflammatory disease. N Engl J Med 340: 115–126, 1999.
2. Fuster V: Mechanisms leading to myocardial infarction: insight from studies of vascular biology. Circulation 90: 2126–2146, 1994.
3. Berliner JA, Navab M, Fogelman AM, Frank JS, Demer LL, Edwards PA, Watson AD, Lusis AJ: Atherosclerosis: basic mechanisms. Oxidation, inflammation, and genetics. Circulation 91: 2488–2496, 1995.
4. Navab M, Berliner JA, Watson AD, Hama SY, Territo MC, Lusis AJ, Shih DM, Van Lenten BJ, Frank JS, Demer LL, Edwards PA, Fogelman AM: The Yin and Yang of oxidation in the development of the fatty streak. Arterioscler Thromb Vasc Biol 16: 831–842, 1996.
5. Spady DK: Reverse cholesterol transport and atherosclerosis regression. Circulation 100: 576–578, 1999.
6. Stary HC, Chandler AB, Dinsmore RE, Fuster V, Glagov S, Insull W Jr, Rosenfeld ME, Schwartz CJ, Wagner WD, Wissler RW: A definition of advanced types of atherosclerotic lesions and a histological classification of atherosclerosis. A report from the Committee on Vascular Lesions of the Council on Arteriosclerosis, American Heart Association. Circulation 92: 1355–1374, 1995.
7. Stary HC, Chandler AB, Glagov S, Guyton JR, Insull W Jr, Rosenfeld ME, Schaffer SA, Schwartz CJ, Wagner WD, Wissler RW: A definition of initial, fatty streak, and intermediate lesions of atherosclerosis. A report from the Committee on Vascular Lesions of the Council on Arteriosclerosis, American Heart Association. Circulation 89: 2462–2478, 1994.
8. Lee RT, Libby P: The unstable atheroma. Arterioscler Thromb Vasc Biol 17: 1859–1867, 1997.
9. Isner JM, Kearney M, Bortman S, Passeri J: Apoptosis in human atherosclerosis and restenosis. Circulation 91: 2703–2711, 1995.
10. Kockx MM, De Meyer GR, Muhring J, Bult H, Bultinck J, Herman AG: Distribution of cell replication and apoptosis in atherosclerotic plaques of cholesterol-fed rabbits. Atherosclerosis 120: 115–124, 1996..

11. Hegyi L, Skepper JN, Cary NR, Mitchinson MJ: Foam cell apoptosis and the development of the lipid core of human atherosclerosis. J Pathol 180: 423–442, 1996.
12. Tedgui A, Mallat Z: Apoptosis as a determinant of atherosclerosis. Thromb Haemost 86: 420–426, 2001.
13. Bennet MR: Apoptosis of vascular smooth muscle cells in vascular remodeling and atherosclerotic plaque rupture. Cardiovascular Res 41: 361–368, 1999.
14. Clarke R, Daly L, Robinson K, Naughten E, Cahalane S, Fowler B, Graham I: Hyperhomocysteinemia: an independent risk factor for vascular disease. N Engl J Med 324: 1149–1155, 1991.
15. den Heijer M, Koster T, Blom HJ, Bos GM, Briet E, Reitsma PH, Vandenbroucke JP, Rosendaal FR: Hyperhomocysteinemia as a risk factor for deep-vein thrombosis. N Engl J Med 334: 759–762, 1996.
16. McCully KS: Homocysteine and vascular disease. Nat Med 2: 386–389, 1996.
17. Mudd SH, Levy HL, Kraus JP: Disorders of transsulfation. In: The Metabolic and Molecular Basis of Inherited Disease. Scriver CR, Beadet AL, Sly WS, Vallee D, Childs B, Kinzler KW, Vogelstein B (eds), McGraw-Hill, New York, NY, 2001.
18. Selhub J, Jacques PF, Bostom AG, D'Agostino RB, Wilson PW, Belanger AJ, O'Leary DH, Wolf PA, Schaefer EJ, Rosenberg IH: Association between plasma homocysteine concentrations and extracranial carotid-artery stenosis. N Engl J Med 332: 286–291, 1995.
19. Ueland PM, Refsum H, Beresford SAA, Vollset SE: The controversy over homocysteine and cardiovascular risk. Am J Clin Nutr 72: 324–332, 2000.
20. Welch GN, Loscalzo JN: Homocysteine and atherothrombosis. N Engl J Med 338: 1042–1050, 1998.
21. Wilcken DEL, Dudman NPB: Homocystinuria and atherosclerosis. In: Lusis AJ, Rotter JI, Sparkes RS (eds), Molecular Genetics of Coronary Artery Disease; Candidate Genes and Process in Atherosclerosis. Monograms in Human Genetics. Karger, New York, NY, 1992.
22. Hossain GS, van Thienen JV, Werstuck GH, Zhou J, Sood SK, Dickhout JG, de Koning AB, Tang D, Wu D, Falk E, Poddar R, Jacobsen DW, Zhang K, Kaufman RJ, Austin RC: TDAG51 is induced by homocysteine, promotes detachment-mediated programmed cell death, and contributes to the development of atherosclerosis in hyperhomocysteinemia. J Biol Chem 278: 30317–30327, 2003.
23. Zhang C, Cai Y, Adachi MT, Oshiro S, Aso T, Kaufman RJ, Kitajama S: Homocysteine induces programmed cell death in human vascular endothelial cells through activation of the unfolded protein response. J Biol Chem 276: 35867–35874, 2001.
24. Suhara T, Fukuo K, Yasuda O, Tsubakimoto M, Takemura Y, Kawamoto H, Yokoi T, Mogi M, Kaimoto T, Ogihara T: Homocysteine enhances endothelial apoptosis via upregulation of Fas-mediated pathways. Hypertension 43: 1208–1213, 2004.
25. Buemi M, Marino D, Di Pasquale G, Floccari F, Ruello A, Aloisi C, Corica F, Senatore M, Romea A, Frisina N: Effects of homocysteine on proliferation, necrosis and apoptosis of vascular smooth muscle cells in culture and influence of folic acid. Thromb Res 104: 207–213, 2001.
26. Zhou J, Møller J, Danielsen CC, Bentzon J, Ravn HB, Austin RC, Falk E: Hyperhomocysteinemia promotes the development of collagen-rich and stable

plaques in apoE-deficient mice. Arterioscler Thromb Vasc Biol 21: 1470–1476, 2001.

27. Zhou J, Møller J, Ritskes-Hoitinga M, Larsen ML, Austin RC, Falk E: Effects of vitamin supplementation and hyperhomocysteinemia on atherosclerosis in apoE-deficient mice. Atherosclerosis 168: 255–262, 2003.

28. Wang H, Jiang X, Yang F, Gaubatz JW, Ma L, Magera MJ, Yang X, Berger PB, Durante W, Pownall HJ, Schafer AI: Hyperhomocysteinemia accelerates atherosclerosis in cystathionine β-synthase and apolipoprotein E double knock-out mice with and without dietary perturbation. Blood 101: 3901–3907, 2003.

29. Hofmann MA, Lalla E, Lu Y, Gleason MR, Wolf BM, Tanji N, Ferran LJ Jr, Kohl B, Rao V, Kisiel W, Stern DM, Schmidt AM: Hyperhomocysteinemia enhances vascular inflammation and accelerates atherosclerosis in a murine model. J Clin Invest 107: 675–683, 2001.

30. Mudd SH, Havlik1 R, Levy HL, McKusick VA, Feinleib M: Cardiovascular risk in heterozygotes for homocystinuria. Am J Hum Genet 34: 1018–1021, 1982.

31. Ueland PM, Refsum H: Plasma homocysteine, a risk factor for vascular disease: plasma levels in health, disease, and drug therapy. J Lab Clin Med 114: 473–501, 1989.

32. McCully KS: Vascular pathology of homocysteinemia: implications for the pathogenesis of arteriosclerosis. Am J Pathol 56: 111–128, 1969.

33. Frosst P, Blom HJ, Milos R, Goyette P, Sheppard CA, Matthews RG, Boers GJ, den Heijer M, Kluijtmans LA, van den Heuvel LP, Rozen R: A candidate genetic risk factor for vascular disease: a common mutation in methylenetetrahydrofolate reductase. Nat Genet 10: 111–113, 1995.

34. Goyette P, Sumner JS, Milos R, Duncan AM, Rosenblatt DS, Matthews RG, Rozen R: Human methylenetetrahydrofolate reductase: isolation of cDNA, mapping and mutation identification. Nat Genet 7: 195–200, 1994.

35. Rosenblatt DS, Fenton WA: Inherited disorders of folate and cobalamin transport and metabolism. In: The Metabolic and Molecular Basis of Inherited Disease. Scriver CR, Beaudet AL, Sly WS, Valle D, Childs B, Kinzler KW, Vogelstein B (eds), 6th edition. McGraw-Hill, New York, NY, 2001.

36. Franken DG, Boers GHJ, Blom HJ, Trijibels FJM, Kloppenborg PW: Treatment of mild hyperhomocysteinemia in vascular disease patients. Arterioscler Thromb 14: 465–470, 1994.

37. Selhub J, Jacques PF, Wilson PWF, Rush D, Rosenberg IH: Vitamin status and intake as primary determinants of homocysteinemia in an elderly population. J Am Med Assoc 270: 2693–2698, 2001.

38. Watanabe M, Osada J, Aratani Y, Kluckman K, Reddick R, Malinow MR, Maeda N: Mice deficient in cystathionine β-synthase: animal models for mild and severe homocyst(e)inemia. Proc Natl Acad Sci USA 92: 1585–1589, 1995.

39. Eberhardt RT, Forgione MA, Cap A, Leopold JA, Rudd MA, Tolliet M, Heyrick S, Stark R, Klings ES, Moldovan NI, Yaghoubi M, Goldschmidt-Clermont PJ, Farber HW, Cohen R, Loscalzo J: Endothelial dysfunction in a murine model of mild hyperhomocyst(e)inemia. J Clin Invest 106: 483–491, 2000.

40. Weiss N, Heydrick S, Zhang YY, Bierl C, Cap A, Loscalzo J: Cellular redox state and endothelial dysfunction in mildly hyperhomocysteinemic cystathionine β-synthase-deficient mice. Arterioscler Thromb Vasc Biol 22: 34–41, 2002.

41. Lang D, Kredan MB, Moat SJ, Hussain SA, Powell CA, Bellamy MF, Powers HJ, Lewis MJ: Homocysteine-induced inhibition of endothelium-dependent

relaxation in rabbit aorta: role for superoxide anions. Arterioscler Thromb Vasc Biol 20: 422–427, 2000.

42. Dayal S, Bottiglieri T, Arning E, Maeda N, Malinow MR, Sigmund CD, Heistad DD, Faraci FM, Lentz SR: Endothelial dysfunction and elevation of S-adenosylhomocysteine in cystathionine β-synthase-deficient mice. Circ Res 88: 1203–1209, 2001.

43. Lentz SR, Erger RA, Dayal S, Maeda N, Malinow MR, Heistad DD, Faraci FM: Folate dependence of hyperhomocysteinemia and vascular dysfunction in cystathionine β-synthase-deficient mice. Am J Physiol Heart Circ Physiol 279: H970–H975, 2000.

44. Lentz SR, Malinow MR, Piegors DJ, Bhopatkar-Teredesai M, Faraci FM, Heistad DD: Consequences of hyperhomocyst(e)inemia on vascular function in atherosclerotic monkeys. Arterioscler Thromb Vasc Biol 17: 2930–2934, 1997.

45. Lentz SR, Piegors DJ, Malinow RM, Heistad DD: Supplementation of atherogenic diet with B vitamins does not prevent atherosclerosis or vascular dysfunction in monkeys. Circulation 103: 1006–1011, 2001.

46. Chen Z, Karaplis AC, Ackerman SL, Pogribny IP, Melnyk S, Lussier-Cacan S, Chen MF, Pai A, John SW, Smith RS, Bottiglieri T, Bagley P, Selhub J, Rudnicki MA, James SJ, Rozen R: Mice deficient in methylenetetrahydrofolate reductase exhibit hyperhomocysteinemia and decreased methylation capacity, with neuropathology and aortic lipid deposition. Hum Mol Genet 10: 433–443, 2001.

47. Wang G, Woo CWH, Sung FL, Siow YL, O K: Increased monocyte adhesion to aortic endothelium in rats with hyperhomocysteinemia: role of chemokine and adhesion molecules. Arterioscler Thromb Vasc Biol 22: 1777–1783, 2002.

48. Majors A, Ehrhart LA, Pezacka EH: Homocysteine as a risk factor for vascular disease. Enhanced collagen production and accumulation by smooth muscle cells. Arterioscler Thromb Vasc Biol 17: 2074–2081, 1997.

49. Nemerson Y: The tissue factor pathway of blood coagulation. Semin Hematol 29: 170–176, 1992.

50. Fryer RH, Wilson BD, Gubler DB, Fitzgerald LA, Rodgers GM: Homocysteine, a risk factor for premature vascular disease and thrombosis, induces tissue factor activity in endothelial cells. Arterioscler Thromb Vasc Biol 13: 1327–1333, 1993.

51. Toschi V, Gallo R, Lettino M, Fallon JT, Gertz SD, Fernandez-Ortiz A, Chesebro JH, Badimon L, Nemerson Y, Fuster V, Badimon JJ: Tissue factor modulates thrombogenecity of human atherosclerotic plaques. Circulation 95: 594–599, 1997.

52. Ardissino D, Merlini PA, Ariens R, Coppola R, Bramucci E, Mannucci PM: Tissue-factor antigen and activity in human coronary atherosclerotic plaques. Lancet 349: 769–771, 1997.

53. Poddar R, Sivasubramanian N, Dibello PM, Robinson K, Jacobsen D: Homocysteine induces expression and secretion of monocyte chemoattractant protein-1 and interleukin-8 in human aortic endothelial cells: implications for vascular disease. Circulation 103: 2717–2723, 2001.

54. Wang G, O K: Homocysteine stimulates the expression of monocyte chemoattractant protein-1 receptor (CCR2) in human monocytes: possible involvement of oxygen free radicals. Biochem J 357: 233–240, 2001.

55. Wang G, Siow YL, O K: Homocysteine induces monocyte chemoattractant protein-1 expression by activating NF-kappa B in THP-1 macrophages. Am J Physiol Heart Circ Physiol 280: H2840–H2847, 2001.

56. Zeng XK, Remick DG, Wang X: Homocysteine induces production of mono-cyte chemoattractant protein-1 and interleukin-8 in cultured human whole blood. Acta Pharmacol Sin 25: 1419–1425, 2004.

57. Zhou J, Werstuck GH, Lhotak S, de Koning ABL, Sood SK, Hossain GS, Moller J, Ritskes-Hoitinga M, Falk E, Dayal S, Lentz SR, Austin RC: Association of multiple cellular stress pathways with accelerated atherosclerosis in hyperhomocysteinemic ApoE-deficient mice. Circulation 110: 207–213, 2004.

58. Starkebaum G, Harlan JM: Endothelial cell injury due to copper-catalyzed hydrogen peroxide generation from homocysteine. J Clin Invest 77: 1370–1376, 1986.

59. Jacobsen DW: Hyperhomocysteinemia and oxidative stress: time for a reality check? Arterioscler Thromb Vasc Biol 20: 1182–1184, 2000.

60. Zappacosta B, Mordente A, Persichilli S, Minucci A, Carlino P, Martorana GE, Giardina B, De Sole P: Is homocysteine a pro-oxidant? Free Rad Res 35: 499–505, 2001.

61. Vitvitsky V, Mosharov E, Tritt M, Ataullakhanov F, Banerjee R: Redox regulation of homocysteine-dependent glutathione synthesis. Redox Rep 8: 57–63, 2003.

62. Mosharov E, Cranford MR, Banerjee R: The quantitatively important relationship between homocysteine metabolism and glutathione synthesis by the transsulfuration pathway and its regulation by redox changes. Biochemistry 39: 13005–13011, 2000.

63. Duan J, Murohara T, Ikeda H, Sasaki K, Shintani S, Akita T, Shimada T, Imaizumi T: Hyperhomocysteinemia impairs angiogenesis in response to hindlimb ischemia. Arterioscler Thromb Vasc Biol 20: 2579–2585, 2000.

64. Franken DG, Boers GHJ, Blom HJ, Trijibels FJM, Kloppenborg PW: Treatment of mild hyperhomocysteinemia in vascular disease patients. Arterioscler Thromb Vasc Biol 14: 465–470, 1994.

65. Gryglewski RJ, Palmer RM, Moncada S: Superoxide anion is involved in the breakdown of endothelium-derived vascular relaxing factor. Nature 320: 454–456, 1986.

66. Heinecke JW, Rosen H, Suzuki LA, Chait A: The role of sulfur-containing amino acids in superoxide production and modification of low-density lipoprotein by arterial smooth muscle cells. J Biol Chem 262: 10098–10103, 1987.

67. Dayal S, Brown KL, Weydert CJ, Oberley LW, Arning E, Bottiglieri T, Faraci FM, Lentz SR: Deficiency of glutathione peroxidase-1 sensitizes hyperhomocysteinemic mice to endothelial dysfunction. Arterioscler Thromb Vasc Biol 22: 1996–2002, 2002.

68. Outinen PA, Sood SK, Pfeifer SI, Pamidi S, Podor TJ, Li J, Weitz JI, Austin RC: Homocysteine-induced endoplasmic reticulum stress and growth arrest leads to specific changes in gene expression in human vascular endothelial cells. Blood 94: 959–967, 1999.

69. Handy DE, Zhang Y, Loscalzo J: Homocysteine down regulates cellular glutathione peroxidase (GPx1) by decreasing translation. J Biol Chem 208: 15518–15525, 2005.

70. Weiss N, Zhang YY, Heydrick S, Bierl C, Loscalzo J: Overexpression of cellular glutathione peroxidase rescues homocyst(e)ine-induced endothelial dysfunction. Proc Natl Acad Sci USA 98: 12503–12508, 2001.

71. Juan SH, Lee TS, Tseng KW, Liou JY, Shyue SK, Wu KK, Chau LY: Adenovirus-mediated heme oxygenase-1 gene transfer inhibits the development

of atherosclerosis in apolipoprotein E-deficient mice. Circulation 104: 1519–1525, 2001.

72. Ron D: Translational control in the endoplasmic reticulum stress response. J Clin Invest 110: 1383–1388, 2002.

73. Lee AS: The glucose-regulated proteins: stress induction and clinical applications. Trends Biochem Sci 26: 504–510, 2001.

74. Kaufman RJ: Orchestrating the unfolded protein response in health and disease. J Clin Invest 110: 1389–1398, 2002.

75. Rutkowski DT, Kaufman RJ: A trip to the ER: coping with stress. Trends Cell Biol 14: 20–28, 2004.

76. de Koning ABL, Werstuck GH, Zhou J, Austin RC: Hyperhomocysteinemia and its role in the development of atherosclerosis. Clin Biochem 36: 431–41, 2003.

77. Shen J, Chen X, Hendershot L, Prywes R: ER stress regulation of ATF6 localization by dissociation of BiP/GRP78 binding and unmasking of Golgi localization signals. Dev Cell 3: 99–111, 2002.

78. Ye J, Rawson RB, Komuro R, Chen X, Dave UP, Prywes R, Brown MS, Goldstein JL: ER stress induces cleavage of membrane-bound ATF6 by the same proteases that process SREBPs. Mol Cell 6: 1355–1364, 2000.

79. Yoshida H, Okada T, Haze K, Yanagi H, Yura T, Negishi M, Mori K: ATF6 activated by proteolysis binds in the presence of NF-Y (CBF) directly to the cis-acting element responsible for the mammalian unfolded protein response. Mol Cell Biol 20: 6755–6767, 2000.

80. Outinen PA, Sood SK, Liaw PC, Sarge KD, Maeda N, Hirsh J, Ribau J, Podor TJ, Weitz JI, Austin RC: Characterization of the stress-inducing effects of homocysteine. Biochem J 332: 213–221, 1998.

81. Kokame K, Agarwala KL, Kato H, Miyata T: Herp, a new ubiquitin-like membrane protein induced by endoplasmic reticulum stress. J Biol Chem 275: 32846–32853, 2000.

82. Kokame K, Kato H, Miyata T: Homocysteine-respondent genes in vascular endothelial cells identified by differential display analysis. GRP78/BiP and novel genes. J Biol Chem 271: 29659–29665, 1996.

83. Werstuck GH, Lentz SR, Dayal S, Hossain GS, Sood SK, Shi YY, Zhou J, Maeda N, Krisans SK, Malinow MR, Austin RC: Homocysteine-induced endoplasmic reticulum stress causes dysregulation of the cholesterol and fatty acid biosynthetic pathways. J Clin Invest 107: 1263–1273, 2001.

84. Zinszner H, Kuroda M, Wang XZ, Batchvarova N, Lightfoot RT, Remotti H, Stevens JL, Ron D: CHOP is implicated in programmed cell death in response to impaired function of the endoplasmic reticulum. Genes Dev 12: 982–995, 1998.

85. Ma K, Vattem KM, Wek RC: Dimerization and release of molecular chaperone inhibition facilitate activation of eukaryotic initiation factor-2 kinase in response to endoplasmic reticulum stress. J Biol Chem 277: 18728–18735, 2002.

86. Nonaka K, Tsujino T, Watari Y, Emoto N, Yokoyama M: Taurine prevents the decrease in expression and secretion of extracellular superoxide dismutase induced by homocysteine: amelioration of homocysteine-induced endoplasmic reticulum stress by taurine. Circulation 104: 1165–1170, 2001.

87. Ji C, Kaplowitz N: Betaine decreases hyperhomocysteinemia, endoplasmic reticulum stress, and liver injury in alcohol-fed mice. Gastroenterology 124: 1488–1499, 2003.

88. Horton JD, Goldstein JL, Brown MS: SREBPs: activators of the complete program of cholesterol and fatty acid synthesis in the liver. J Clin Invest 109: 1125–1131, 2002.

89. Woo CW, Siow YL, Pierce GN, Choy PC, Minuk GY, Mymin D, O K: Hyperhomocysteinemia induces hepatic cholesterol biosynthesis and lipid accumulation via activation of transcription factors. Am J Physiol Endocrinol Metab 288: E1002–E1010, 2005.

90. Kang S-S, Wong PWK, Becker N: Protein-bound homocyst(e)ine in normal subjects and in patients with homocystinuria. Pediatr Res 13: 1141–1143, 1979.

91. Malloy MH, Rassin DK, Gaull GE: Plasma cysteine in homocysteinemia. Am J Clin Nutr 34: 2619–2621, 1981.

92. Smolin LA, Benevenga NJ: Accumulation of homocyst(e)ine in vitamin B6 deficiency: a model for the study of cystathionine β-synthase deficiency. J Nutr 112: 1264–1272, 1982.

93. Refsum H, Helland S, Ueland PM: Radioenzymic determination of homocysteine in plasma and urine. Clin Chem 31: 624–628, 1985.

94. Kang S-S, Wong PWK, Cook HY, Norusis M, Messer JV: Protein-bound homocysteine: A possible risk factor for coronary artery disease. J Clin Invest 77: 1482–1486, 1986.

95. Sartorio R, Carrozzo R, Corbo L, Andria G: Protein-bound plasma homocyt(e)ine and identification of heterozygotes for cystathionine-synthase deficiency. J Inherit Metab Dis 9: 25–29, 1986.

96. Wiley VC, Dudman NPB, Wilcken DEL: Interrelations between plasma free and protein-bound homocysteine and cysteine in homocystinuria. Metabolism 37: 191–195, 1988.

97. Wiley VC, Dudman NPB, Wilcken DEL: Free and protein-bound homocysteine and cysteine in cystathionine β-synthase deficiency: Interrelations during short- and long-term changes in plasma concentrations. Metabolism 38: 734–739, 1989.

98. Mansoor MA, Svardal AM, Ueland PM: Determination of the *in vivo* redox status of cysteine, cysteinylglycine, homocysteine, and glutathione in human plasma. Anal Biochem 200: 218–229, 1992.

99. Mansoor MA, Ueland PM, Aarsland A, Svardal AM: Redox status and protein binding of plasma homocysteine and other aminothiols in patients with homocystinuria. Metabolism 42: 1481–1485, 1993.

100. Mansoor MA, Ueland PM, Svardal AM: Redox status and protein binding of plasma homocysteine and other aminothiols in patients with hyperhomocysteinemia due to cobalamin deficiency. Am J Clin Nutr 59: 631–635, 1994.

101. Mansoor MA, Bergmark C, Svardal AM, Lonning PE, Ueland PM: Redox status and protein binding of plasma homocysteine and other aminothiols in patients with early-onset peripheral vascular disease. Homocysteine and peripheral vascular disease. Arterioscler Thromb Vasc Biol 15: 232–240, 1995.

102. Refsum H, Smith AD, Ueland PM, Nexo E, Clarke R, McPartlin J, Johnston C, Engbaek F, Schneede J, McPartlin C, Scott JM: Facts and recommendations about total homocysteine determinations: an expert opinion. Clin Chem 50: 3–32, 2004.

103. Jakubowski H: Molecular basis of homocysteine toxicity in humans. Cell Mol Life Sci 61: 470–87, 2004.

104. Sengupta S, Chen H, Togawa T, DiBello PM, Majors AK, Büdy B, Ketterer ME, Jacobsen DW: Albumin thiolate anion is an intermediate in the formation of albumin-SS-homocysteine. J Biol Chem 276: 30111–30117, 2001.
105. Sengupta S, Wehbe C, Majors AK, Ketterer ME, DiBello PM, Jacobsen DW: Relative roles of albumin and ceruloplasmin in the formation of homocysteine, homocysteine-cysteine mixed disulfide, and cystine in circulation. J Biol Chem 276: 46896–46904, 2001.
106. Undas A, Williams EB, Butenas S, Orfeo T, Mann KG: Homocysteine inhibits inactivation of Factor Va by activated protein C. J Biol Chem 276: 4389–4397, 2001.
107. Majors AK, Sengupta S, Willard B, Kinter MT, Pyeritz RE, Jacobsen DW: Homocysteine binds to human plasma fibronectin and inhibits its interaction with fibrin. Arterioscler Thromb Vasc Biol 22: 1354–1359, 2002.
108. Lim A, Sengupta S, McComb ME, Theberge R, Wilson WG, Costello CE, Jacobsen DW: in vitro and in vivo interactions of homocysteine with human plasma transthyretin. J Biol Chem 278: 49707–49713, 2003.
109. He XM, Carter DC: Atomic structure and chemistry of human serum albumin. Nature 358: 209–215, 1992.
110. Peters T Jr: All About Albumin: Biochemistry, Genetics, and Medical Applications. Academic Press, San Diego, 1996.
111. Lentz SR, Piegors DJ, Fernández JA, Erger RA, Arning E, Malinow MR: Effects of hyperhomocysteinemia on protein C activation and activity. Blood 100: 2108–2112, 2002.
112. Benson MD: Amyloidosis. In: Scriver CR, Beaudet AL, Sly WS, Valle D (eds), The Metabolic and Molecular Bases of Inherited Disease, 8th edition. McGraw-Hill Medical Publishing Division, New York, 5345–5378, 2001.
113. Théberge R, Conners L, Skinner M, Costello CE: Characterization of transthyretin from serum using immunoprecipitation, HPLC/electrospray ionization and matrix-assisted laser desorption/ionization mass spectrometry. Anal Chem 71: 452–459, 1999.
114. Kishikawa M, Nakanishi T, Miyazaki A, Shimizu A: A simple and reliable method of detecting variant transthyretins by multidimensional liquid chromatography coupled to electrospray ionization mass spectrometry. Amyloid: Int J Exp Clin Invest 6: 48–53, 1999.
115. Nishio E, Watanabe Y: Homocysteine as a modulator of platelet-derived growth factor action in vascular smooth muscle cells: a possible role for hydrogen peroxide. Br J Pharmacol 122: 269–274, 1997.
116. Upchurch GR Jr, Welch GN, Fabian AJ, Freedman JE, Johnson JL, Keaney JF Jr, Loscalzo J: Homocyst(e)ine decreases bioavailable nitric oxide by a mechanism involving glutathione peroxidase. J Biol Chem 272: 17012–17017, 1997.
117. Handy DE, Zhang Y, Loscalzo J: Homocysteine down-regulates cellular glutathione peroxidase (GPx1) by decreasing translation. J Biol Chem 280: 15518–15525, 2005.
118. Stuhlinger MK, Tsao PS, Her J-H, Kimoto M, Balint RF, Cooke JP: Homocysteine impairs the nitric oxide synthase pathway role of asymmetric dimethylarginine. Circulation 104: 2569–2575, 2001.
119. Lentz SR, Sadler JE: Inhibition of thrombomodulin surface expression and protein C activation by the thrombogenic agent homocysteine. J Clin Invest 88: 1906–1914, 1991.

120. Hayashi T, Honda G, Suzuki K: An atherogenic stimulus homocysteine inhibits cofactor activity of thrombomodulin and enhances thrombomodulin expression in human umbilical vein endothelial cells. Blood 79: 2930–2936, 1992.
121. Lentz SR, Sadler JE: Homocysteine inhibits von Willebrand factor processing and secretion by preventing transport from the endoplasmic reticulum. Blood 81: 683–689, 1993.
122. Cai Y, Zhang C, Nawa T, Aso T, Tanaka M, Oshiro S, Ichijo H, Kitajima S: Homocysteine-responsive ATF3 gene expression in human vascular endothelial cells: activation of c-Jun NH(2)-terminal kinase and promoter response element. Blood 96: 2140–2148, 2000.
123. Urano F, Wang X, Bertolotti A, Zhang Y, Chung P, Harding HP, Ron D: Coupling of stress in the ER to activation of JNK protein kinases by transmembrane protein kinase IRE1. Science 287: 664–666, 2000.
124. Chen Y-T, Meyer CF, Tan TH: Persistent activation of c-Jun N-terminal kinase 1 (JNK1) in gamma radiation-induced apoptosis. J Biol Chem 271: 631–634, 1996.
125. Jakubowski H: Metabolism of homocysteine thiolactone in human cell cultures. Possible mechanism for pathological consequences of elevated homocysteine levels. J Biol Chem 272: 1935–1942, 1997.
126. Huang RFS, Huang SM, Lin BS, Wei JS, Liu T-Z: Homocysteine thiolactone induces apoptotic DNA damage mediated by increased intracellular hydrogen peroxide and caspase 3 activation in HL-60 cells. Life Sci 68: 2799–2811, 2001.
127. Reddy RK, Mao C, Baumeister P, Austin RC, Kaufman RJ, Lee AS: Endoplasmic reticulum chaperone protein GRP78 protects cells from apoptosis induced by topoisomerase inhibitors: role of ATP binding site in suppression of caspase-7 activation. J Biol Chem 278: 20915–20924, 2003.
128. Rao RV, Hermel E, Castro-Obregon S, del Rio G, Ellerby LM, Ellerby HM, Bredesen DE: Coupling endoplasmic reticulum stress to the cell death program. Mechanism of caspase activation. J Biol Chem 276: 33869–33874, 2001.
129. Nakagawa T, Zhu H, Morishima N, Li E, Xu J, Yankner BA, Yuan J: Caspase-12 mediates endoplasmic-reticulum-specific apoptosis and cytotoxicity by amyloid-beta. Nature 403: 98–103, 2000.
130. Zhou J, Lhotak S, Hilditch BA, Austin RC: Activation of the unfolded protein response occurs at all stages of atherosclerotic lesion development in apolipoprotein E-deficient mice. Circulation 111: 1814–1821, 2005.
131. Feng B, Yao PM, Li Y, Devlin CM, Zhang D, Harding HP, Sweeney M, Rong JX, Kuriakose G, Fisher EA, Marks AR, Ron D, Tabas I: The endoplasmic reticulum is the site of cholesterol-induced cytotoxicity in macrophages. Nat Cell Biol 5: 781–792, 2003.

17

Molecular and Biochemical Mechanisms of Hyperhomocysteinemia-Induced Cardiovascular Disorders

KARMIN O, YAW L. SIOW, PATRICK C. CHOY, AND GRANT M. HATCH

Abstract

Hyperhomocysteinemia, an elevation of homocysteine (Hcy) levels in the blood, is regarded as an independent risk factor for cardiovascular disorders due to atherosclerosis. It is estimated that up to 40% of patients diagnosed with premature coronary artery disease, peripheral vascular disease, or recurrent venous thrombosis are present with hyperhomocysteinemia. Monocyte infiltration into the subendothelial space in the arterial wall and later differentiation into macrophages are important initial steps in the development of atherosclerotic lesions. Monocyte chemoattractant protein-1 (MCP-1) is a potent chemokine that stimulates monocyte migration into the intima of arterial walls. MCP-1 exerts its action mainly through the interaction with the C–C chemokine receptor (CCR2) on the surface of monocytes. The expression of MCP-1 and other inflammatory factors in atherosclerotic lesions can be upregulated by the transcription factor, nuclear factor-κB (NF-κB). Results from *in vivo* and *in vitro* experiments suggest that Hcy-induced MCP-1 expression in vascular cells together with increased CCR2 expression in monocytes may represent a mechanism for Hcy-enhanced monocyte infiltration into the arterial wall during atherogenesis. Hcy-induced oxidative stress appears to contribute to increased inflammatory reactions in vascular cells. Epidemiological and laboratory studies also indicate that hyperhomocysteinemia is a risk factor for disorders that involve not only cardiovascular system but also other organs. Here we review recent studies on the Hcy-induced endothelial dysfunction, chemokine expression in vascular cells, the involvement of oxidative stress, and the role of NF-κB activation in gene expression. The impact of hyperhomocysteinemia on liver and kidney is also discussed.

Keywords: atherosclerosis; chemokine; cholesterol; homocysteine; oxidative stress

Abbreviations: Hcy, homocysteine; VCAM-1, vascular cell adhesion molecule-1; TNF-α, tumor necrosis factor-α; ROS, reactive oxygen species; NF-κB, nuclear factor-κB; CBS, cystathionine β-synthase; IκB, inhibitory protein; MCP-1, monocyte chemoattractant protein-1; CCR2, C–C chemokine receptor; iNOS, inducible nitric oxide synthase; HMG-CoA, 3-hydroxy-3-methylglutaryl coenzyme A; SREBP, sterol regulatory element binding protein

Introduction

Hyperhomocysteinemia, a condition of elevated blood levels of homocysteine (Hcy), is regarded as one of the common and independent risk factors for atherosclerosis [1–12]. Atherosclerosis is the principal contributor to the pathogenesis of myocardial and cerebral infarctions, which are the leading causes of mortality and morbidity in many countries. Abnormal elevations of Hcy levels, up to 100–250 μM in the blood, have been reported in patients with hyperhomocysteinemia [2, 5, 11]. Many factors may regulate plasma levels of Hcy. For example, severe hyperhomocysteinemia seen in children is usually the result of rare homozygous deficiency of enzymes necessary for Hcy metabolism [11]. Moderate increases in blood Hcy levels occur more frequently and are often found in patients with heterozygous enzyme deficiency, folate or pyridoxine (vitamin B_6) deficiency, impaired renal function, as well as in elderly people and in postmenopausal women [6, 7, 11].

Metabolism of Homocysteine

Hcy is a sulfur-containing amino acid formed during the conversion of methionine to cysteine. Plasma Hcy is found primarily in three molecular forms, namely Hcy, disulfide Hcy, and the mixed disulfide Hcy–cysteine [7, 11]. Majority of Hcy molecules in the circulation are in the oxidized and the protein-bound form. Reduced or free Hcy (nonprotein-bound form) constitutes about 1% of the total Hcy level in the blood. The normal range of total Hcy (sum of all forms of Hcy) in adults is 5–15 μM, with a mean level of 10 μM. Hyperhomocysteinemia refers to the total plasma Hcy level above 15 μM nuclear factor-κB (NF-κB) [6, 7, 11]. The cellular homeostasis of Hcy is tightly regulated under normal conditions [1]. Hcy can be metabolized by two major pathways: (1) the transulfuration pathway to form cysteine, which requires vitamin B_6 as a cofactor and (2) the remethylation pathway to form methionine, which requires methyltetrahydrofolate as a cosubstrate and vitamin B_{12} as a cofactor (Fig. 17.1). Factors that perturb steps in Hcy metabolic pathways can cause an increase in its cellular levels and lead to hyperhomocysteinemia (plasma Hcy level higher than 15 μM). Abnormal elevations of Hcy levels up to 100–250 μM in the blood have been reported in patients with hyperhomocysteinemia [8]. Severe hyperhomocysteinemia seen in children is usually the result of a rare homozygous deficiency of enzymes necessary for Hcy catabolism. Moderate increases in blood Hcy levels occur frequently and are often seen in patients with heterozygous enzyme deficiency, impaired kidney function, deficiency of folate, vitamin B_6, or vitamin B_{12}, in elderly people, etc. [2, 5, 11, 13]. There is an inverse correlation between plasma levels of Hcy and kidney function [13].

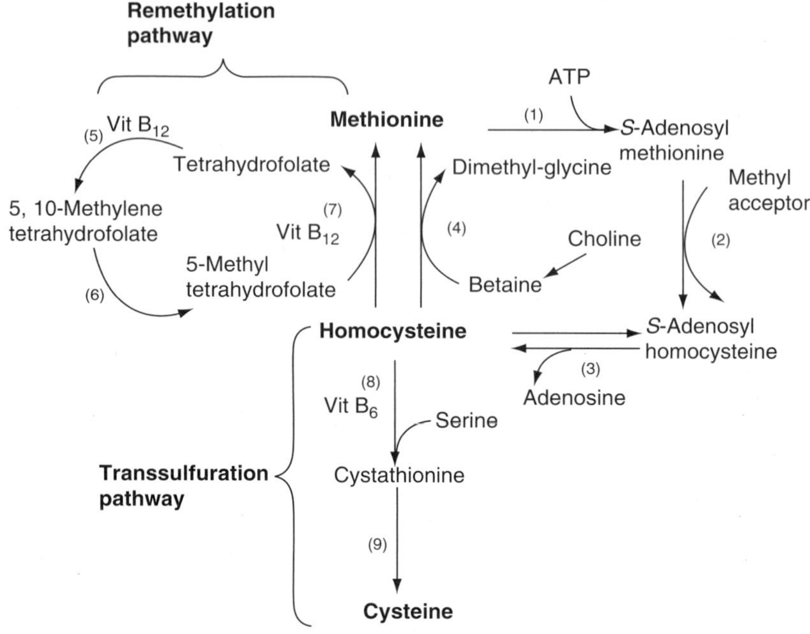

FIGURE 17.1. Homocysteine metabolic pathways. The remethylation pathway and the transsulfuration pathway are the two major pathways for homocysteine metabolism. Reactions involved in these two pathways are catalyzed by the following enzymes: (1) methionine adenosyltransferase; (2) *S*-adenosylmethionine-dependent methyltransferase; (3) *S*-adenosyl homocysteine hydrolase; (4) betaine-homocysteine methyltransferase; (5) serine hydroxymethyltransferase; (6) 5,10-methylenetetrahydrofolate reductase; (7) 5-methyltetrahydrofolate-homocysteine methyltransferase; (8) cystathionine β-synthase; and (9) γ-cystathionase.

Hyperhomocysteinemia and Cardiovascular Disorders

Although the precise molecular mechanisms responsible for the pathogenicity of hyperhomocysteinemia remain uncertain, several potential mechanisms have been proposed. These include endothelial dysfunction [10, 14, 15], increased proliferation of smooth muscle cells [16–18], enhanced coagulability [18], and increased cholesterol biosynthesis in hepatocytes [19–21]. Endothelial injury and dysfunction are considered to be one of the leading mechanisms contributing to atherogenesis. Upon injury, endothelial cells are capable of producing various cytokines and growth factors that in turn participate in the development of atherosclerotic lesions.

Hyperhomocysteinemia and Endothelial Function

It has been proposed that Hcy-caused endothelial injury may be due to oxidative stress, attenuation of nitric oxide-mediated vasodilatation, and disturbances in the antithrombotic activities of the endothelium [20–28]. This topic was recently reviewed by Austin et al. [10]. Several animal models with hyperhomocysteinemia have been developed in monkeys [26], apoE-null mice [27], cystathionine β-synthase (CBS)-deficient mice [23], and high-methionine fed rats [28]. In diet-induced moderate hyperhomocysteinemic monkeys, Lentz et al. [26] noticed increased platelet-mediated vasoconstriction, impaired endothelium-dependent vasodilatation, and decreased thrombomodulin-dependent activation of protein C, when compared to that of monkeys fed a normal diet. In the apoE-null mice with dietary-induced hyperhomocysteinemia, Hofmann et al. [27] reported a twofold increase in the aortic root lesion size. These mice also had significantly elevated levels of vascular cell adhesion molecule-1 (VCAM-1) and tumor necrosis factor-α (TNF-α). An impaired endothelium-dependent vasodilatation function, likely due to diminished nitric oxide bioactivity was observed in the CBS-deficient mice, which had impaired Hcy metabolism [23]. These studies indicate that Hcy may enhance vascular inflammation and endothelial dysfunction in animals that are prone to the development of atherosclerosis. In our recent study, hyperhomocysteinemia was induced in rats that were fed a high-methionine diet for 4 weeks [28]. In this animal model, serum Hcy level was 4–5-fold higher than that of control rats. In aortic rings isolated from hyperhomocysteinemic rats, endothelium-dependent vessel relaxation was impaired while endothelium-independent vessel relaxation was not affected. This study clearly demonstrates that in the absence of other risk factors, hyperhomocysteinemia alone is able to cause endothelial dysfunction. Taken together, results obtained from various animal models indicate that Hcy, at pathological concentrations, can act synergistically with other risk factors as well as act independently causing endothelial dysfunction.

Oxidative Stress and Activation of NF-κB

Many risk factors causing atherosclerosis share a common feature of generating intracellular oxidative stress [15, 29–34]. It has been suggested that generation of reactive oxygen species (ROS) is responsible for Hcy-induced cell injury [15, 29, 33, 34]. ROS are often generated in cells as by-products in many metabolic and signal transduction pathways [32]. Under normal conditions, small amount of ROS generated inside cells can be scavenged by cellular antioxidant defense mechanism. For example, superoxide anions generated can be scavenged by superoxide dismutase to form oxygen and hydrogen peroxide, the latter is decomposed by catalase and glutathione peroxidase [22, 35–37]. However, ROS scavengers would no longer be sufficient for their removal when ROS are above a certain level. As a result, ROS accumulate inside cells and oxidative stress occurs. The major ROS include superoxide anion, hydrogen peroxide, hydroxyl radical, and peroxynitrite, which are formed during the sequential reduction of oxygen. Overproduction of these free radicals can induce oxidation of DNA, proteins, and lipids resulting in mutations, loss of enzymatic activities, and alteration of lipid functions. The initial product of ROS is superoxide anion. Many *in vitro* studies demonstrated that Hcy was able to stimulate intracellular superoxide anion generations in vascular cells as well as in other types of cells [10, 38–40]. We observed that Hcy, at higher concentrations (50–200 μM), caused an increase in intracellular superoxide anion levels in vascular endothelial cells [38], in vascular smooth muscle cells [41], and in monocytes/macrophages [39, 40]. As a consequence, there was a marked increase in the expression of inflammatory genes such as monocyte chemoattractant protein-1 (MCP-1), chemokine receptor (CCR2), and inducible nitric oxide synthase (iNOS) in these cells [39–41]. Pretreatment of cells with cell-permeable polyethylene glycol-bound superoxide dismutase, a known superoxide scavenger, reversed Hcy-induced elevation of superoxide anion levels in endothelial cells, vascular smooth muscle cells, and monocytes/macrophages. The polyethylene glycol-bound superoxide dismutase treatment also abolished Hcy-induced mRNA expression of MCP-1, CCR2, and iNOS. These results suggest that oxidative stress contributes to inflammatory responses elicited by Hcy in vascular cells.

The expression of inflammatory factors is mainly controlled by transcription factors such as NF-κB. NF-κB plays an important role in the expression of inflammatory factors in the vasculature [42–44]. Oxidative stress can activate this transcription factor leading to upregulating the expression of its target genes [39, 41–45]. Our laboratory reported that Hcy treatment could activate NF-κB in vascular cells as well as in macrophages via superoxide anion generation [39]. In resting cells, NF-κB is normally present in the cytoplasm in an inactive form associated with an inhibitory protein (IκB) [45–49]. IκB-α is one of the best-characterized forms of IκB proteins. Upon stimulation, there is a rapid phosphorylation of IκB-α and subsequent degradation of IκB-α by the proteasome, leading to the release of NF-κB. After dissociation from IκB, the activated NF-κB is translocated into the nucleus where it binds to the κB binding motifs in the promoters or enhancers of the genes encoding cytokines. Activated NF-κB (found in the nucleus) has been detected in macrophages, endothelial cells, and smooth muscle cells of human atherosclerotic lesions [43, 44, 46]. In contrast, little or no activation of NF-κB is found in normal arterial walls. In diet-induced hyperhomocysteinemic rat, we observed a marked elevation of superoxide anion level and the presence of activated NF-κB in the aorta while little superoxide anion and activated NF-κB were detected in the control aorta [38]. Several lines of evidence indicated that NF-κB was activated in endothelial cells upon Hcy treatment. First, nuclear translocation of NF-κB occurred in endothelial cells after incubation with Hcy for 15–30 min. Second, the results from electrophoretic mobility shift assay demonstrated that Hcy treatment caused a significant increase in the NF-κB/DNA binding activity. Third, results from transient transfection demonstrated an enhanced NF-κB-regulated transcriptional activity in Hcy-treated cells. Further investigation revealed that oxidative stress and subsequent activation of IκB kinases (IKK-α and IKK-β) are essential for Hcy-induced activation of NF-κB in endothelial cells [38]. Such a mechanism may regulate the inflammatory response in the vascular wall during the early stage of atherosclerosis in hyperhomocysteinemia. ROS have been implicated to stimulate IκB-α degradation and NF-κB activation in vascular cells [50]. It was reported that hydrogen peroxide stimulated NF-κB activity via activation of IKK-α and IKK-β, which were kinases that phosphorylate IκB protein, in HeLa cells [51]. Antioxidants were shown to be able to block IκB-α degradation and NF-κB activation [52, 53].

Homocysteine and Chemokine Expression in Vascular Cells

One of the important features during the early stage of atherogenesis is monocyte adhesion to the injured arterial endothelium followed by their differentiation into macrophages. Macrophages are able to uptake large amounts of lipids, particularly cholesterol from oxidized lipoproteins contributing to lipid accumulation in the atherosclerotic lesion [54–57]. MCP-1 is a potent chemokine that stimulates monocyte migration into the intima of arterial walls [53–57]. The amount of this chemokine appears to be increased in atherosclerotic lesions in human and in experimental animals [28, 54–57]. The expression of MCP-1 and other inflammatory factors (i.e., adhesion molecules) in atherosclerotic lesions can be upregulated by NF-κB. Our recent studies demonstrated that Hcy, at pathological concentrations, stimulated the expression of MCP-1 mRNA in cultured endothelial cells [58, 59], in vascular smooth muscle cells [41], and in macrophages [60]. The experimental results suggest that NF-κB activation and protein kinase signaling pathways may play important roles in Hcy-induced MCP-1 expression. Elevated MCP-1 production by these cells, in turn, stimulated monocyte chemotaxis *in vitro* [41, 58–60]. Pretreatment of cells with NF-κB inhibitors could alleviate the stimulatory effect of Hcy on MCP-1 expression, supporting the notion that Hcy-stimulated chemokine expression was mediated via NF-κB activation [41, 60]. It is generally believed that endothelial expression of MCP-1 initiates the migration of monocytes into the arterial wall. Based on results obtained from our laboratory [41, 60, 61] as well as from other investigators, we speculate that Hcy-induced endothelial MCP-1 expression may be associated with early development of atherosclerosis by stimulating monocyte migration into the subendothelial space and differentiation into macrophages. On the other hand, MCP-1 produced in smooth muscle cells as

well as in macrophages may facilitate the recruitment of additional monocytes into the lesion at later stages of atherosclerosis in patients with hyperhomocysteinemia. MCP-1 exerts its action mainly through the interaction
with the CCR2 on the surface of monocytes. We observed that Hcy was able
to stimulate CCR2 expression in human peripheral blood monocytes as well
as in THP-1 cells (derived from human monocytic cell line) [40]. We hypothesize that Hcy-induced MCP-1 expression in vascular cells, together with
enhanced CCR2 expression in peripheral blood monocytes, may represent a
mechanism for monocyte/macrophage accumulation in the arterial wall during atherogenesis. Indeed, we observed that Hcy-treated monocytes/
macrophages as well as endothelial cells were able to take up oxidized LDL
causing intracellular lipid accumulation [61]. We also examined the *in vivo*
effect of Hcy on MCP-1 expression leading to monocyte adhesion to the
endothelium [28]. Male Sprague–Dawley rats developed hyperhomocysteinemia after being fed a high-methionine diet (regular chow plus 1.7% methionine, wt/wt) for 4 weeks. In diet-induced hyperhomocysteinemic rats, the
number of monocytes present on the surface of the aortic endothelium was
significantly elevated when compared to the control rats. There was a significant increase in the expression of MCP-1 protein in the endothelium.
Further analysis revealed that the expression of adhesion molecules such as
VCAM-1 and E-selectin was also significantly elevated in the aortic endothelium of hyperhomocysteinemic rats. Pretreatment with specific antibodies
against MCP-1, VCAM, or E-selectin could block monocyte binding to the
aortic endothelium of hyperhomocysteinemic rats. These results indicated
that elevation of MCP-1 and adhesion molecules in vascular cells was
responsible for enhanced monocyte adhesion to the endothelium. Increased
monocyte/macrophage binding and adhesion to the vascular endothelium
may represent an early feature of atherosclerotic development in hyperhomocysteinemia [12, 28]. In the same animal model, we also observed that
hyperhomocysteinemia was associated with reduced endothelium-dependent
vessel relaxation [28]. These findings suggest that hyperhomocysteinemia can
act independently in the development of vascular dysfunction. Folic acid supplementation to rats fed a high-methionine diet prevented an elevation of
MCP-1, VCAM-1, and E-selectin expression in rat aortic endothelium.

Hyperhomocysteinemia as a Risk Factor for Other Disorders

Although hyperhomocysteinemia is regarded as an independent risk factor
for atherosclerosis, elevation of plasma Hcy levels is often associated with
diseases that involve other organs. Hyperhomocysteinemia has been indicated as a potential risk factor for Alzheimer's disease [62–65], osteoporo-
sis/hip fracture [66, 67], liver dysfunction [11, 14, 21, 68, 69], and kidney
injury [11, 14, 70–73].

Homocysteine and Hepatic Lipid Metabolism

A positive correlation between the plasma concentrations of Hcy and cholesterol was observed in patients with hyperhomocysteinemia [11, 68]. Over three decades ago, McCully [14] reported postmortem observation of extensive arteriosclerosis in two children with severe hyperhomocysteinemia and proposed a pathogenic link between elevated blood Hcy levels and atherogenesis. In subsequent studies, a correlation between the plasma levels of Hcy and cholesterol was found in patients [68]. Abnormal lipid metabolism was also found in hyperhomocysteinemic animal models [20, 21]. In hyperhomocysteinemic mice caused by CBS deficiency, there was excessive accumulation of lipid droplets in hepatocytes due to increased endoplasmic reticulum stress [21]. In a recent study, ethanol was shown to induce fatty liver and apoptosis in mice through Hcy-induced endoplasmic reticulum stress [74]. We previously reported that Hcy stimulated the production and secretion of cholesterol in human hepatoma cells (HepG2) via activation of 3-hydroxy-3- methylglutaryl coenzyme A (HMG-CoA) reductase [19]. Subsequent study revealed an activation of HMG-CoA reductase in the liver of hyperhomocysteinemic rat [20]. In diet-induced hyperhomocysteinemic rats, there were accumulation of hepatic lipids and elevation of serum cholesterol contents due to increased HMG-CoA reductase gene expression in the liver. The HMG-CoA reductase is the rate-limiting enzyme, catalyzing the conversion of HMG-CoA to mevalonate, in cholesterol biosynthesis [75, 76]. The activity of this enzyme is regulated by several mechanisms including transcriptional regulation, posttranslational modification, allosteric regulation, and levels of endogenous and exogenous cholesterol [77–79]. Sterol regulatory element binding proteins (SREBPs), a subfamily of basic helix-loop-helix zipper proteins, are important transcription factors regulating lipid homeostasis [80–82]. The SREPB-2 preferentially activates enzymes involved in cholesterol production such as HMG-CoA reductase [80, 81]. Two other transcription factors, namely, cAMP response element binding protein and nuclear factor Y, have been identified as important SREBP-2 coregulators and act synergistically for upregulation of HMG-CoA reductase gene expression [82–85]. In the rat model, further investigation revealed that hyperhomocysteinemia could upregulate HMG-CoA reductase gene expression in the liver via the activation of transcription factors (SREBP-2, cAMP response element binding protein, nuclear factor Y) in hepatocytes [20]. Abnormal cholesterol metabolism can cause serious complications including hepatic fatty infiltration, which can progress to fibrosis and cirrhosis leading to liver failure [19–21, 86, 87]. There were small lipid droplets accumulated in the liver of hyperhomocysteinemic rat, a condition assembled as fatty liver [20, 21]. We postulated that the ability of Hcy to promote cholesterol biosynthesis in hepatocytes is a mechanism of hyperhomocysteinemia-associated liver pathology. Although hyperhomocysteinemia is regarded as an independent risk factor for the development of atherosclerosis, a high blood level of Hcy is often associated with other risk factors. In the Hordaland Study [68], it was found that elevated plasma Hcy concentration was associated with other cardiovascular risk factors including elevated cholesterol level, male sex, old age, smoking, high blood pressure, and lack of exercise. Another study demonstrated that hyperhomocysteinemia was associated with sudden death resulting from coronary atherosclerosis with fibrous plaques [88]. However, the causality between various risk factors remains to be investigated. Nevertheless, chronic exposure of the vessel wall to a moderate elevation of plasma cholesterol concentration induced by Hcy together with hyperhomocysteinemia may lead to vessel injury over a prolonged period of time. The long-term effect of Hcy-induced hypercholesterolemia on the cardiovascular system remains to be investigated.

Homocysteine and Kidney Injury

Chronic kidney disease is one of the risk factors causing hyperhomocysteinemia possibly due to impaired Hcy metabolism in the kidney [70–73]. In patients with chronic renal failure, there is a marked elevation of blood Hcy concentrations and a striking increase in the risk for vascular diseases [70]. Apart from cardiovascular findings, the postmortem examination also revealed a moderate increase in mesangial matrix as well as in the number of mesangial cells and endothelial cells in the kidney indicating pathological changes observed [14]. Although kidney dysfunction is one of the risk factors causing hyperhomocysteinemia, the impact of elevated blood Hcy levels on renal function is largely unknown. One study revealed that diet-induced hyperhomocysteinemia caused vascular remodeling and tubulointerstitial injury in the kidney [71]. Results from another study suggested that hyperhomocysteinemia might be an important pathogenic factor for glomerular damage in hypertensive animals independent of blood pressure [72]. A recent epidemiological investigation in the general population revealed a positive association between an elevation of serum Hcy levels and the development of chronic kidney disease [70]. The data suggests that hyperhomocysteinemia may be a risk factor for kidney disease [70]. We observed that diet-induced hyperhomocysteinemia could activate NF-κB and induce iNOS expression leading to increased peroxynitrite formation in the kidney [73]. Increased iNOS expression, in turn, caused increased production of nitric oxide in the tissue. Nitric oxide can rapidly interact with superoxide anion to form highly reactive peroxynitrite. Peroxynitrite is a potent oxygen free radical that is capable of causing extensive protein tyrosine nitration and mediates iron-catalyzed lipid peroxidation. Increased production of free radicals may represent one of the important mechanisms underlying Hcy-induced kidney injury [73].

Conclusion Remarks

Results from epidemiological and laboratory studies indicate that elevated blood Hcy levels is not only a risk factor for cardiovascular disorders due to atherosclerosis but also a potential risk factor for disorders that involve other organs. Biochemical and molecular mechanisms by which Hcy, at an elevated level, affects various organs remains to be further investigated. The knowledge obtained will provide a scientific basis for prevention and therapeutic intervention of patients with cardiovascular disorders associated with hyperhomocysteinemia or with other risk factors.

Acknowledgment: The work was supported, in part, by Heart and Stroke Foundation of Manitoba, Canadian Institute of Health Research, and Natural Sciences and Engineering Research Council. Grant.M.Hatch is a Canada Research Chair in Molecular Cardiolipin Metabolism.

References

1. Clarke R, Daly L, Robinson K, Naughten E, Cahalane S, Fowler B, Graham I: Hyperhomocysteinemia: an independent risk factor for vascular disease. N Engl J Med 324: 1149–1155, 1991.
2. McCully KS: Homocysteine and vascular disease. Nat Med 2: 386–389, 1996.
3. Malinow MR: Hyperhomocyst(e)inemia. A common and easily reversible risk factor for occlusive atherosclerosis. Circulation 81: 2004–2006, 1990.
4. Nygard O, Nordrehaug JE, Refsum H, Ueland PM, Farstad M, Vollset SE: Plasma homocysteine levels and mortality in patients with coronary artery disease. N Engl J Med 337: 230–236, 1997.
5. Booth GL, Wang EE: Preventive health care, 2000 update: screening and management of hyperhomocysteinemia for the prevention of coronary artery disease events. The Canadian Task Force on Preventive Health Care. Can Med Assoc J 163: 21–29, 2000.
6. Refsum H, Ueland PM, Nygard O, Vollset SE: Homocysteine and cardiovascular disease. Annu Rev Med 49: 31–62, 1998.
7. Duell PB, Malinow MR: Homocyst(e)ine: an important risk factor for atherosclerotic vascular disease. Curr Opin Lipidol 8: 28–34, 1997.
8. Harker LA, Slichter SJ, Scott CR, Ross R: Homocystinemia. Vascular injury and arterial thrombosis. N Engl J Med 291: 537–543, 1974.
9. Welch GN, Loscalzo J: Homocysteine and atherothrombosis. N Engl J Med 338: 1042–1050, 1998.
10. Austin RC, Lentz SR, Werstuck GH: Role of hyperhomocysteinemia in endothelial dysfunction and atherothrombotic disease. Cell Death Differ 11(Suppl 1): S56–S64, 2004.
11. Mudd SH, Levy HL, Skovby F: Disorders of transsulfuration. In: Scriver CR, Beaudet AL, Sly WS, Valle D (eds), The Metabolic Basis of Inherited Disease. McGraw-Hill, New York, NY, 1995, pp 1279–1327.
12. Choy PC, Siow YL, Mymin D, O K: Lipids and atherosclerosis. Biochem Cell Biol 82: 212–224, 2004.
13. Wheeler DC: Cardiovascular disease in patients with chronic renal failure. Lancet 348: 1673–1674, 1996.
14. McCully KS: Vascular pathology of homocysteinemia: implications for the pathogenesis of arteriosclerosis. Am J Pathol 56: 111–128, 1969.
15. Weiss N, Heydrick S, Zhang YY, Bierl C, Cap A, Loscalzo J: Cellular redox state and endothelial dysfunction in mildly hyperhomocysteinemic cystathionine beta-synthase-deficient mice. Arterioscler Thromb Vasc Biol 22: 34–41, 2002.

16. Tsai JC, Perrella MA, Yoshizumi M, Hsieh CM, Haber E, Schlegel R, Lee ME: Promotion of vascular smooth muscle cell growth by homocysteine: a link to atherosclerosis. Proc Natl Acad Sci USA 91: 6369–6373, 1994.

17. Chen C, Halkos ME, Surowiec SM, Conklin BS, Lin PH, Lumsden AB: Effects of homocysteine on smooth muscle cell proliferation in both cell culture and artery perfusion culture models. J Surg Res 88: 26–33, 2000.

18. Durand P, Lussier-Cacan S, Blache D: Acute methionine load-induced hyperhomocysteinemia enhances platelet aggregation, thromboxane biosynthesis, and macrophage-derived tissue factor activity in rats. FASEB J 11: 1157–1168, 1997.

19. O K, Lynn EG, Chung YH, Siow YL, Man RY, Choy PC: Homocysteine stimulates the production and secretion of cholesterol in hepatic cells. Biochim Biophys Acta 1393: 317–324, 1998.

20. Woo CW, Siow YL, Pierce GN, Choy PC, Minuk GY, Mymin D, O K: Hyperhomocysteinemia induces hepatic cholesterol biosynthesis and lipid accumulation via activation of transcription factors. Am J Physiol Endocrinol Metab 288: E1002–E1010, 2005.

21. Werstuck GH, Lentz SR, Dayal S, Hossain GS, Sood SK, Shi YY, Zhou J, Maeda N, Krisans SK, Malinow MR, Austin RC: Homocysteine-induced endoplasmic reticulum stress causes dysregulation of the cholesterol and triglyceride biosynthetic pathways. J Clin Invest 107: 1263–1273, 2001.

22. Schlaich MP, John S, Jacobi J, Lackner KJ, Schmieder RE: Mildly elevated homocysteine concentrations impair endothelium dependent vasodilation in hypercholesterolemic patients. Atherosclerosis 153: 383–389, 2000.

23. Eberhardt RT, Forgione MA, Cap A, Leopold JA, Rudd MA, Trolliet M, Heydrick S, Stark R, Klings ES, Moldovan NI, Yaghoubi M, Goldschmidt-Clermont PJ, Farber HW, Cohen R, Loscalzo J: Endothelial dysfunction in a murine model of mild hyperhomocyst(e)inemia. J Clin Invest 106: 483–491, 2000.

24. Morita H, Saito Y, Kurabayashi M, Nagai R: Diet-induced mild hyperhomocysteinemia and increased salt intake diminish vascular endothelial function in a synergistic manner. J Hypertens 20: 55–62, 2002.

25. Mujumdar VS, Aru GM, Tyagi SC: Induction of oxidative stress by homocyst(e)ine impairs endothelial function. J Cell Biochem 82: 491–500, 2001.

26. Lentz SR, Sobey CG, Piegors DJ, Bhopatkar MY, Faraci FM, Malinow MR, Heistad DD: Vascular dysfunction in monkeys with diet-induced hyperhomocyst(e)inemia. J Clin Invest 98: 24–29, 1996.

27. Hofmann MA, Lalla E, Lu Y, Gleason MR, Wolf BM, Tanji N, Ferran LJ Jr, Kohl B, Rao V, Kisiel W, Stern DM, Schmidt AM: Hyperhomocysteinemia enhances vascular inflammation and accelerates atherosclerosis in a murine model. J Clin Invest 107: 675–683, 2001.

28. Wang G, Woo CW, Sung FL, Siow YL, O K: Increased monocyte adhesion to aortic endothelium in rats with hyperhomocysteinemia: role of chemokine and adhesion molecules. Arterioscler Thromb Vasc Biol 22: 1777–1783, 2002.

29. Alvarez-Maqueda M, El Bekay R, Monteseirin J, Alba G, Chacon P, Vega A, Santa Maria C, Tejedo JR, Martin-Nieto J, Bedoya FJ, Pintado E, Sobrino F: Homocysteine enhances superoxide anion release and NADPH oxidase assembly by human neutrophils. Effects on MAPK activation and neutrophil migration. Atherosclerosis 172: 229–238, 2004.

30. Miller FJ Jr, Gutterman DD, Rios CD, Heistad DD, Davidson BL: Superoxide production in vascular smooth muscle contributes to oxidative stress and impaired relaxation in atherosclerosis. Circ Res 82: 1298–1305, 1998.

31. Hathaway CA, Heistad DD, Piegors DJ, Miller FJ Jr: Regression of atherosclerosis in monkeys reduces vascular superoxide levels. Circ Res 90: 277–283, 2002.
32. Martindale JL, Holbrook NJ: Cellular response to oxidative stress: signaling for suicide and survival. J Cell Physiol 192: 1–15, 2002.
33. Kanani PM, Sinkey CA, Browning RL, Allaman M, Knapp HR, Haynes WG: Role of oxidant stress in endothelial dysfunction produced by experimental hyperhomocyst(e)inemia in humans. Circulation 100: 1161–1168, 1999.
34. Dayal S, Brown KL, Weydert CJ, Oberley LW, Arning E, Bottiglieri T, Faraci FM, Lentz SR: Deficiency of glutathione peroxidase-1 sensitizes hyperhomocysteinemic mice to endothelial dysfunction. Arterioscler Thromb Vasc Biol 22: 1996–2002, 2002.
35. Lassegue B, Clempus RE: Vascular NAD(P)H oxidases: specific features, expression, and regulation. Am J Physiol Regul Integr Comp Physiol 285: R277–R297, 2003.
36. Cai H, Griendling KK, Harrison DG: The vascular NAD(P)H oxidases as therapeutic targets in cardiovascular diseases. Trends Pharmacol Sci 24: 471–478, 2003.
37. Cathcart MK: Regulation of superoxide anion production by NADPH oxidase in monocytes/macrophages: contributions to atherosclerosis. Arterioscler Thromb Vasc Biol 24: 23–28, 2004.
38. Au-Yeung KK, Woo CW, Sung FL, Yip JC, Siow YL, O K: Hyperhomocysteinemia activates nuclear factor-kappaB in endothelial cells via oxidative stress. Circ Res 94: 28–36, 2004.
39. Woo CW, Cheung F, Chan VW, Siow YL, O K: Homocysteine stimulates inducible nitric oxide synthase expression in macrophages: antagonizing effect of ginkgolides and bilobalide. Mol Cell Biochem 243: 37–47, 2003.
40. Wang G, O K: Homocysteine stimulates the expression of monocyte chemoattractant protein-1 receptor (CCR2) in human monocytes: possible involvement of oxygen free radicals. Biochem J 357: 233–240, 2001.
41. Wang G, Siow YL, O K: Homocysteine stimulates nuclear factor kappaB activity and monocyte chemoattractant protein-1 expression in vascular smooth-muscle cells: a possible role for protein kinase C. Biochem J 352(Pt 3): 817–826, 2000.
42. Marumo T, Schini-Kerth VB, Fisslthaler B, Busse R: Platelet-derived growth factor-stimulated superoxide anion production modulates activation of transcription factor NF-kappaB and expression of monocyte chemoattractant protein 1 in human aortic smooth muscle cells. Circulation 96: 2361–2367, 1997.
43. Brand K, Page S, Rogler G, Bartsch A, Brandl R, Knuechel R, Page M, Kaltschmidt C, Baeuerle PA, Neumeier D: Activated transcription factor nuclear factor-kappa B is present in the atherosclerotic lesion. J Clin Invest 97: 1715–1722, 1996.
44. Brand K, Page S, Walli AK, Neumeier D, Baeuerle PA: Role of nuclear factor-kappa B in atherogenesis. Exp Physiol 82: 297–304, 1997.
45. Baldwin AS Jr: The NF-kappa B and I kappa B proteins: new discoveries and insights. Annu Rev Immunol 14: 649–683, 1996.
46. Thurberg BL, Collins T: The nuclear factor-kappa B/inhibitor of kappa B autoregulatory system and atherosclerosis. Curr Opin Lipidol 9: 387–396, 1998.
47. Simeonidis S, Stauber D, Chen G, Hendrickson WA, Thanos D: Mechanisms by which IkappaB proteins control NF-kappaB activity. Proc Natl Acad Sci USA 96: 49–54, 1999.
48. Karin M: The beginning of the end: IkappaB kinase (IKK) and NF-kappaB activation. J Biol Chem 274: 27339–27342, 1999.

49. Mercurio F, Zhu H, Murray BW, Shevchenko A, Bennett BL, Li J, Young DB, Barbosa M, Mann M, Manning A, Rao A: IKK-1 and IKK-2: cytokine-activated IkappaB kinases essential for NF-kappaB activation. Science 278: 860–866, 1997.

50. Bowie A, O'Neill LA: Oxidative stress and nuclear factor-kappaB activation: a reassessment of the evidence in the light of recent discoveries. Biochem Pharmacol 59: 13–23, 2000.

51. Kamata H, Manabe T, Oka S, Kamata K, Hirata H: Hydrogen peroxide activates IkappaB kinases through phosphorylation of serine residues in the activation loops. FEBS Lett 519: 231–237, 2002.

52. Kretz-Remy C, Mehlen P, Mirault ME, Arrigo AP: Inhibition of I kappa B-alpha phosphorylation and degradation and subsequent NF-kappa B activation by glutathione peroxidase overexpression. J Cell Biol 133: 1083–1093, 1996.

53. Manna SK, Zhang HJ, Yan T, Oberley LW, Aggarwal BB: Overexpression of manganese superoxide dismutase suppresses tumor necrosis factor-induced apoptosis and activation of nuclear transcription factor-kappaB and activated protein-1. J Biol Chem 273: 13245–13254, 1998.

54. Takahashi M, Ikeda U, Masuyama J, Kitagawa S, Kasahara T, Saito M, Kano S, Shimada K: Involvement of adhesion molecules in human monocyte adhesion to and transmigration through endothelial cells in vitro. Atherosclerosis 108: 73–81, 1994.

55. Yla-Herttuala S, Lipton BA, Rosenfeld ME, Sarkioja T, Yoshimura T, Leonard EJ, Witztum JL, Steinberg D: Expression of monocyte chemoattractant protein 1 in macrophage-rich areas of human and rabbit atherosclerotic lesions. Proc Natl Acad Sci USA 88: 5252–5256, 1991.

56. Nelken NA, Coughlin SR, Gordon D, Wilcox JN: Monocyte chemoattractant protein-1 in human atheromatous plaques. J Clin Invest 88: 1121–1127, 1991.

57. Valente AJ, Rozek MM, Sprague EA, Schwartz CJ: Mechanisms in intimal monocyte-macrophage recruitment. A special role for monocyte chemotactic protein-1. Circulation 86: III20–III25, 1992.

58. Sung FL, Slow YL, Wang G, Lynn EG, O K: Homocysteine stimulates the expression of monocyte chemoattractant protein-1 in endothelial cells leading to enhanced monocyte chemotaxis. Mol Cell Biochem 216: 121–128, 2001.

59. Sung FL, O K: Homocysteine stimulates monocyte adhesion to endothelial cells. FASEB J 13: A1359, 1999.

60. Wang G, Siow YL, O K: Homocysteine induces monocyte chemoattractant protein-1 expression by activating NF-kappaB in THP-1 macrophages. Am J Physiol Heart Circ Physiol 280: H2840–H2847, 2001.

61. O K, Siow YL: Biochemical mechanisms of hyperhomocysteinemia in atherosclerosis: role of chemokine expression. In: Pierce GN, Nagano M, Zahradka P, Dhalla NS (eds), Atherosclerosis, Hypertension and Diabetes. Kluwer Academic, Boston, 2003, pp 53–62.

62. Diaz-Arrastia R: Homocysteine and neurologic disease. Arch Neurol 57: 1422–1427, 2000.

63. Seshadri S, Beiser A, Selhub J, Jacques PF, Rosenberg IH, D'Agostino RB, Wilson PW, Wolf PA: Plasma homocysteine as a risk factor for dementia and Alzheimer's disease. N Engl J Med 346: 476–483, 2002.

64. Irizarry MC: Biomarkers of Alzheimer disease in plasma. NeuroRx 1: 226–234, 2004.

65. Morris MS: Homocysteine and Alzheimer's disease. Lancet Neurol 2: 425–428, 2003.

66. van Meurs JB, Dhonukshe-Rutten RA, Pluijm SM, van der Klift M, de Jonge R, Lindemans J, de Groot LC, Hofman A, Witteman JC, van Leeuwen JP, Breteler MM, Lips P, Pols HA, Uitterlinden AG: Homocysteine levels and the risk of osteoporotic fracture. N Engl J Med 350: 2033–2041, 2004.

67. McLean RR, Jacques PF, Selhub J, Tucker KL, Samelson EJ, Broe KE, Hannan MT, Cupples LA, Kiel DP: Homocysteine as a predictive factor for hip fracture in older persons. N Engl J Med 350: 2042–2049, 2004.

68. Nygard O, Vollset SE, Refsum H, Stensvold I, Tverdal A, Nordrehaug JE, Ueland M, Kvale G: Total plasma homocysteine and cardiovascular risk profile. The Hordaland Homocysteine Study. JAMA 274: 1526–1533, 1995.

69. Watanabe M, Osada J, Aratani Y, Kluckman K, Reddick R, Malinow MR, Maeda N: Mice deficient in cystathionine beta-synthase: animal models for mild and severe homocyst(e)inemia. Proc Natl Acad Sci USA 92: 1585–1589, 1995.

70. Ninomiya T, Kiyohara Y, Kubo M, Tanizaki Y, Tanaka K, Okubo K, Nakamura H, Hata J, Oishi Y, Kato I, Hirakata H, Iida M: Hyperhomocysteinemia and the development of chronic kidney disease in a general population: the Hisayama study. Am J Kidney Dis 44: 437–445, 2004.

71. Kumagai H, Katoh S, Hirosawa K, Kimura M, Hishida A, Ikegaya N: Renal tubulointerstitial injury in weanling rats with hyperhomocysteinemia. Kidney Int 62: 1219–1228, 2002.

72. Li N, Chen YF, Zou AP: Implications of hyperhomocysteinemia in glomerular sclerosis in hypertension. Hypertension 39: 443–448, 2002.

73. Zhang F, Siow YL, O K: Hyperhomocysteinemia activates NF-kappaB and inducible nitric oxide synthase in the kidney. Kidney Int 65: 1327–1338, 2004.

74. Ji C, Kaplowitz N: Betaine decreases hyperhomcysteinemia, endoplasmic reticulum stress, and liver injury in alcohol-fed mice. Gastroenterology 124: 1488–1499, 2003.

75. Hylemon PB, Pandak WM, Vlahcevic ZR: Regulation of hepatic cholesterol homeostasis. In: Arias IM (ed), The Liver Biology and Pathobiology. Lippincott Williams & Wilkins, Philadelphia, 2001.

76. Qureshi N, Dugan RE, Cleland WW, Porter JW: Kinetic analysis of the individual reductive steps catalyzed by beta-hydroxy-beta-methylglutaryl-coenzyme A reductase obtained from yeast. Biochemistry 15: 4191–4207, 1976.

77. Beg ZH, Stonik JA, Brewer HB Jr: Phosphorylation of hepatic 3-hydroxy-3-methylglutaryl coenzyme A reductase and modulation of its enzymic activity by calcium-activated and phospholipid-dependent protein kinase. J Biol Chem 260: 1682–1687, 1985.

78. Jurevics H, Hostettler J, Barrett C, Morell P, Toews AD: Diurnal and dietary-induced changes in cholesterol synthesis correlate with levels of mRNA for HMG-CoA reductase. J Lipid Res 41: 1048–1054, 2000.

79. Nakanishi M, Goldstein JL, Brown MS: Multivalent control of 3-hydroxy-3-methylglutaryl coenzyme A reductase. Mevalonate-derived product inhibits translation of mRNA and accelerates degradation of enzyme. J Biol Chem 263: 8929–8937, 1988.

80. Horton JD, Goldstein JL, Brown MS: SREBPs: activators of the complete program of cholesterol and fatty acid synthesis in the liver. J Clin Invest 109: 1125–1131, 2002.

81. Horton JD, Shimomura I, Brown MS, Hammer RE, Goldstein JL, Shimano H: Activation of cholesterol synthesis in preference to fatty acid synthesis in liver

and adipose tissue of transgenic mice overproducing sterol regulatory element-binding protein-2. J Clin Invest 101: 2331–2339, 1998.

82. Vallett SM, Sanchez HB, Rosenfeld JM, Osborne TF: A direct role for sterol regulatory element binding protein in activation of 3-hydroxy-3-methylglutaryl coenzyme A reductase gene. J Biol Chem 271: 12247–12253, 1996.

83. Jackson SM, Ericsson J, Mantovani R, Edwards PA: Synergistic activation of transcription by nuclear factor Y and sterol regulatory element binding protein. J Lipid Res 39: 767–776, 1998.

84. Ngo TT, Bennett MK, Bourgeois AL, Toth JI, Osborne TF: A role for cyclic AMP response element-binding protein (CREB) but not the highly similar ATF-2 protein in sterol regulation of the promoter for 3-hydroxy-3-methylglutaryl coenzyme A reductase. J Biol Chem 277: 33901–33905, 2002.

85. Bennett MK, Osborne TF: Nutrient regulation of gene expression by the sterol regulatory element binding proteins: increased recruitment of gene-specific coregulatory factors and selective hyperacetylation of histone H3 *in vivo*. Proc Natl Acad Sci USA 97: 6340–6344, 2000.

86. Durrington P: Dyslipidaemia. Lancet 362: 717–731, 2003.

87. Harrison SA, Di Bisceglie AM: Advances in the understanding and treatment of nonalcoholic fatty liver disease. Drugs 63: 2379–2394, 2003.

88. Burke AP, Fonseca V, Kolodgie F, Zieske A, Fink L, Virmani R: Increased serum homocysteine and sudden death resulting from coronary atherosclerosis with fibrous plaques. Arterioscler Thromb Vasc Biol 22: 1936–1941, 2002.

Section IV

Other Factors in Atherosclerosis and the Associated Complications

18

The Role of the Immune System in Atherosclerosis: Lessons Learned from Using Mouse Models of the Disease

STEWART C. WHITMAN AND TANYA A. RAMSAMY

Abstract

Atherosclerosis is a multifactor, highly complex disease with numerous etiologies that work synergistically to promote lesion development. One of the emerging components that drive the development of both early- and late-stage atherosclerotic lesions has been shown to be the participation of both the innate and the acquired immune systems. In both humans and animal models of atherosclerosis, the most prominent cells that infiltrate evolving lesions are macrophages and T lymphocytes. The ablation of either of these cell types reduces the extent of atherosclerosis in mice that were rendered susceptible to the disease by deficiency of either apolipoprotein E (*Apoe*) or the LDL receptor (*Ldl-r*). In addition to these major immune cell participants, a number of less prominent leukocyte populations that can modulate the atherogenic process are also involved. This chapter will focus on the participatory role of two "less prominent" immune components, namely natural killer (NK) cells and natural killer T (NKT) cells. Although this chapter will highlight the fact that both NK and NKT cells are not sufficient for causing the disease, the roles played by both these cells are becoming increasingly important in understanding the complexity of this disease process.

Keywords: apolipoprotein E null mice; atherosclerosis; cytokines; innate immunity; LDL receptor null mice; NK cells; NKT cells; review

Abbreviations: *Apoe*, apolipoprotein E; *Ldl-r*, LDL receptor; NK, natural killer; NKT, natural killer T; IFN; interferon; MHC, major histocompatibility complex; TCR, T cell receptor; α-GalCer, α-galactosylceramide; DN, double negatives; Th, T helper; IL, Interleukin

Introduction

Atherosclerosis and its complications lead to half of all adult deaths in Canada, the United States, and other Western societies [1, 2]. Atherosclerosis is a multifactorial, highly complex disease with numerous etiologies that work synergistically to promote lesion development. One of the emerging components that drive the development of both early- and late-stage atherosclerotic lesions has been shown to be the participation of both the innate and acquired immune systems [3, 4]. With this said, atherosclerotic lesions are characterized by a pronounced infiltration of leukocytes at all stages of disease progression [2]. In both humans and animal models of atherosclerosis, the most prominent cells that infiltrate evolving lesions are macrophages and T lymphocytes [5–7]. The ablation of either of these cell types reduces the extent of atherosclerosis in mice that were rendered susceptible to the disease by deficiency of either apolipoprotein E (*Apoe*$^{-/-}$) or the LDL receptor (*Ldl-r*$^{-/-}$) [8–12]. In addition to these major immune cell participants, a number of less prominent leukocyte populations that can modulate the atherogenic process are also involved. Although this chapter will focus on the participatory role of two "less prominent" immune components, namely natural killer (NK) cells [13, 14] and natural killer T (NKT) cells [15–19], it should be pointed out that the process of atherogenesis has been shown to be modulated by small numbers of other types of immune cells, such as B lymphocytes [20–22], mast cells [23], and dendritic cells [24], as reviewed in greater detail by VanderLaan and Reardon [4].

The Contributions of the Immune System to the Various Stages of Atherosclerosis

In the arteries of healthy children, preexisting mononuclear cell infiltrations have been identified at regions known to have a greater likelihood for later developing atherosclerosis [25, 26]. These sites are speculated to function as local immunosurveillance systems that monitor the bloodstream for potentially harmful endogenous and exogenous antigens [26, 27]. The morphological feature of early-stage lesions in both *Apoe*$^{-/-}$ [28–30] and *Ldl-r*$^{-/-}$ mice [31] are very similar to those found in humans [32–34]. These lesions consist of an abnormal accumulation of lipoproteins and an assembly of immune cells, consisting mainly of T lymphocytes and macrophages [7], and to a lesser extent, NK cells, B cells, mast cells, and dendritic cells [13, 20, 24, 35].

The accumulation of plasma-derived lipoproteins is considered a necessary event in initiating atherosclerosis [36]. *In vitro* studies have shown that lipoproteins are trapped by matrix components [37–39] and modified by oxidation [40] within the tunica media to a form that is chemotactic for monocytes [41]. Oxidized lipoproteins may also be chemotactic for other immune cells such as NK cells, NKT cells, T- and B lymphocytes, facilitating their recruitment to the vessel wall [42]. Within the intima, monocytes differentiate to macrophages and begin to clear the modified lipoproteins via their scavenger receptors [43]. This culminates in the generation of numerous cholesterol ester-enriched "foam" cells and form what is commonly termed a *fatty streak*.

Advanced-stage atherosclerotic lesions have an accumulation of extracellular lipid, known as the lipid core. Macrophages, macrophage-derived foam cells and lymphocytes are found densely concentrated along the periphery of the lipid core of these lesions [33, 44]. In advance-stage human atherosclerotic lesions, lymphocytes [44], mast cells [45], NK cells [46], and NKT cells [47] have all been identified in the regions bordering the shoulder of the lipid core and the fibrous cap, with NK and NKT cells representing 0.1% and 2%, respectively, of the total lymphocyte population in these regions [46, 47]. Fibrous plaques that contain a hematoma and/or thrombotic deposits are termed *complicated lesions*, and are well documented in *Apoe$^{-/-}$* mice [31, 48–51], but less so in *Ldl-r$^{-/-}$* mice [31].

NK Cells and their Role in Innate Immunity

NK cells represent a subset of bone marrow-derived lymphocytes distinguishable from T- and B lymphocytes by their morphology, phenotype, and their ability to kill aberrant cells without prior sensitization [52]. In mice, as in humans, NK cells constitute only 10–15% of peripheral blood lymphocytes, but this distribution increases to 45% in certain areas such as the liver, peritoneal cavity, and placenta [53–55]. NK cells play an important role in maintaining the integrity of the innate immune defenses, as their primary role is believed to be one of providing early defense against pathogens during the initial response period while the adaptive immune system is being activated [56]. Functionally, NK cells act as effectors, either directly through the process of cell-mediated cytotoxicity upon degranulation with subsequent perforin and granzyme release, or through cytokine production, the most prominent being interferon (IFN)-γ, which in turn can mediate the activation of other effector cells that are in close apposition.

NK Cells and Experimental Atherosclerosis

Although a direct participatory role of NK cells in the process of human atherogenesis has not yet been shown, detailed immunohistochemical analysis of human autopsy specimens has shown the presence of NK cells at all stages of atherosclerotic lesion development [26, 46]. NK cells have also been detected in the *Ldl-r$^{-/-}$* mouse model of atherosclerosis, yet unlike human lesions, only early-stage lesions in these mice were found to stain positive for NK cells [13, 57]. Interestingly, in both human and mouse atherosclerotic lesions, NK cells were found to make up only a small fraction of the lymphocyte population present in these lesions; approximately 0.1–0.5% of the total lymphocytes.

Animal models that combine genetic risks for atherosclerosis with an altered immune system have been invaluable in demonstrating a link between atherosclerosis and immunity [3, 58]. The identification of the beige mutation mouse and the creation of the Ly49A transgenic mouse, two mouse strains that exhibit partial and complete NK cell deficiency, respectively, has allowed for the creation of a similar animal model aimed at defining the true role of NK cells in atherosclerosis.

Beige Mutation Mice

NK cell function is decreased in mice having the beige mutation [59–61] and these mice have been used in two separate atherosclerosis studies, yet these have yielded different results. Beige mice fed a diet enriched in saturated fat, cholesterol, and cholate did not exhibit any change in atherosclerotic lesion formation [62]. However, when the beige defect was bred onto the $Ldl\text{-}r^{-/-}$ background, there was a modest, but statistically significant increase in lesion size [57]. The beige mouse has a very complex phenotype, and while NK cell activity is decreased in these mice, the defect is not complete [59–61] allowing for residual NK cell activity to persist. Furthermore, given the nature of the mutation in the beige mice, which involves a poorly characterized protein required for proper lysosomal trafficking [59], disturbances in cell populations that are distinct from that of NK cells may ultimately have been responsible for the antiatherogenic effect noted [57].

Ly49A Transgenic Mice

Recently, transgenic mice have been developed that have defective natural cytotoxicity and a selective deficiency in functional $NK1.1^+ CD3^-$ cells, while maintaining functionally normal T- and B lymphocytes [63]. This phenotype was achieved by expressing the inhibitory major histocompatibility complex (MHC) class I specific receptor, Ly49A, under the control of the granzyme A promoter. Ly49A is present on all NK cells and is a C-type lectin-like receptor that recognizes the MHC class I ligands, H-2D(d) and D(k). Interactions of these ligands with Ly49A inhibits activation of NK cells, which provides the rationale for the absence of the functional cells in these transgenic mice [63].

The development of transgenic mice with selective deficiency in NK activity affords the ability to define the specific role of NK cells in the development of atherosclerosis. Using the Ly49A transgenic mouse, Whitman and colleagues have shown that the deficiency of functional NK cells in both $Ldlr^{-/-}$ [13] and $Apoe^{-/-}$ [14] mice results in a significant reduction in the development of early-stage atherosclerotic lesions. Interestingly, lesions in $Apoe^{-/-}$ mice that carry the Ly49A transgene were found to advance in both size and complexity, such that after these mice had been fed an atherosclerosis-promoting diet for 12 weeks, the protective effect of NK cell deficiency was greatly diminished in female mice and completely lost in male mice [14].

NKT Cells and their Role in Linking the Innate and Acquired Immune Systems

Almost 20 years ago, three independent laboratories published studies identifying a previously unknown subset of T lymphocytes [64–66]. Nevertheless, it was not until 1995 that Makino et al. [67] coined the term NKT cell to describe a heterogeneous subset of mouse T lymphocytes that share some characteristics with NK cells [68] and appear to provide a link between the adaptive and innate immune systems. Despite this, it quickly became apparent that this broad definition of an NKT cell was inadequate, since some NKT cells lack an NK cell receptor. Furthermore, most murine strains, with the exception of the commonly used C57BL/6 strain in atherosclerosis studies, completely lack the expression of the classical NK cell receptor, NK1.1 [69]. In the absence of a truly definitive marker ubiquitously expressed on all NKT cells, a number of alternative criteria have been used to define this class of lymphocyte as unique from that of NK cells. The four most often used criteria are: (i) the ability of NKT cells to show autoreactivity to the nonclassical MHC molecule CD1d; (ii) the expression of a specific T cell receptor (TCR) reservoir of the NKT cell; (iii) the presence of NK cells receptors; and (iv) the responsiveness of the cell to the synthetic CD1d ligand, α-galactosylceramide

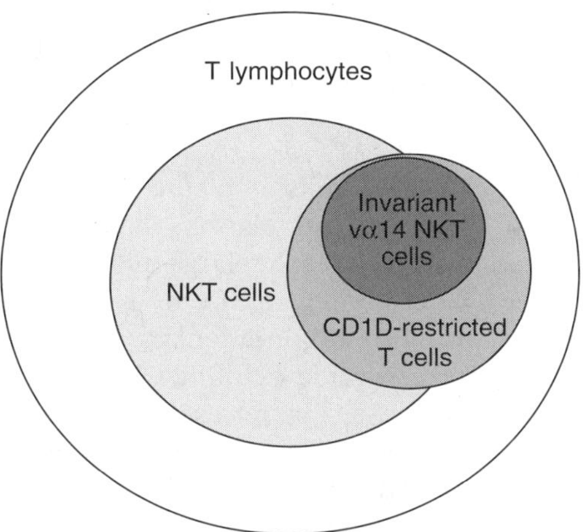

FIGURE 18.1. NKT cells are a subset of T lymphocytes that share many characteristics with both NK cell and T lymphocytes. NKT cells are best described as a heterogeneous cell population that can be partitioned based on their reactivity to CD1d. The CD1d-dependent NKT cells can further be segregated based on their TCR reservoir. Vα14 NKT cells, the most abundant form of NKT cells in the mouse, are composed of an invariant α chain and one of three β chains. CD1d-dependent T lymphocytes, including Vα14 T cells, can either be T lymphocytes or NKT cells depending on their receptor repertoire. Adapted from Wilson and Delovitch [71].

(α-GalCer), which is derived from marine sponges. In addition to being used as a way to distinguish NKT cells from NK cells, these four selection criteria have allowed NKT cells to be themselves partitioned into four broad categories as outlined in Fig. 18.1.

The ability of one unique subset of T lymphocytes to interact with CD1d results from the expression of a highly biased, evolutionarily conserved, TCR consisting of an invariant α-chain (a Vα14 segment joined to a Jα18 segment) that preferentially binds to one of three β-chains (Vβ8.2, Vβ-7, and Vβ-2) [70, 71]. As such, these T lymphocytes are known as invariant NKT cells [72] or Vα14 NKT cells [73]. In C57BL/6 mice, this unique subset of T lymphocytes generally, but not consistently, expresses the NK1.1 receptor to varying degrees [69, 74–76]. Although the physiological substrate for this subset of T lymphocytes is still unclear, invariant NKT cells have a strong response to and are selectively activated by the exogenous substrate, α-GalCer, a marine sponge-derived glycosphingolipid [77], which binds specifically to CD1d [74, 78]. Furthermore, invariant Vα14 T cells do not express CD8, a marker characteristic of cytotoxic T lymphocytes, but instead are either CD4$^+$, a marker characteristic helper T lymphocytes, or double negatives (DN) [75, 76]. Although Vα14 NKT cells are by far the more abundant NKT cells present, other, less common NKT cells have been identified.

Like the Vα14 NKT cells, the non-Vα14 NKT cells are also CD1d-dependent and can either be CD4$^+$ or DN [79–81]. This additional class of NKT cells, which expresses either the Vα3.2-Jα9 or Vα8 α-chain and preferentially binds the Vβ8 β-chain, is less common than the Vα14 NKT cells in the murine model [81]. Yet another class of NKT cells are those that are CD1d-independent and include NKT cells that express the NK1.1 receptor, have diverse TCR can be CD8$^+$, CD4$^+$, or DN [82, 83] or those that express CD49B, an antigen present on a majority of the murine NK cells [84].

Once activated, all NKT cell populations have the capacity to exert immunoregulatory functions by releasing large quantities of T helper (Th) 1 or Th2 cytokines. The Th1 cytokines release by NKT cells upon activation include the proatherogenic cytokines interleukin (IL)-12 [85, 86] and IFN-γ [87–90], whereas the antiatherogenic cytokines [86] and IL-10 [91–94] are associated with the Th2 response. The release of cytokines from NKT cells initiates a cascade of events and causes the bystander activation of adjacent NK cells, B lymphocytes, and dendritic cells [95], as well as the activation of conventional CD4$^+$ and CD8$^+$ T lymphocytes [71, 96]. Although the natural ligands for CD1d-dependent NKT cells, which includes the Vα14 NKT cells, still remain to be identified, α-GalCer has been shown to be a potent and specific activator of NKT cells [74]. As such, evidence exists to suggest that NKT cells can modulate antitumoral [97–99] and antimicrobial immunity [100, 101] as well as a number of autoimmune diseases, including experimental autoimmune encephalomyelitis [102] (a model for multiple sclerosis) and type 1 diabetes [103, 104], and as reviewed by van der Vliet et al. [105].

Atherosclerosis is a chronic inflammatory disorder that is known to involve components of both the innate and adaptive immune systems [3, 5, 106, 107]. Given that NKT cells link the two arms of the immune system, it seems likely that they would also play a role in the pathogenesis of atherosclerosis. In fact, several studies have shown, using lipopolysaccharide or α-GalCer, that NKT cells are present in atherosclerotic lesions [15, 18, 108]. Furthermore, α-GalCer administration has recently been shown to specifically exacerbate atherosclerosis in *Apoe*$^{-/-}$ mice, a model predisposed to developing atherosclerotic lesions spontaneously, when compared to vehicle treated animals [16, 18], and those also deficient in CD1d [15]. Incidentally, CD1d has been detected in human atherosclerotic lesions [109] underscoring the probable involvement of NKT cells in the human disease process. The administration of α-GalCer to *Apoe*$^{-/-}$ mice has been shown to have dramatic effects on the local cytokine environment of the aortas with established atherosclerotic lesions by inducing the production of IFN-γ, IL-4, and IL-10 [16]. Others have observed an early burst of inflammatory cytokines, both Th1 (IFN-γ, tumor necrosis factor-α, IL-2) and Th2 (IL-4, IL-5) cytokines, in the sera of *Apoe*$^{-/-}$ [16]. The effect of α-GalCer on the development of atherosclerosis and the production of proatherogenic cytokines by NKT cells suggests that CD1d-dependent NKT cells might play a participatory role in the atherogenic process.

CD1d Null Mice

Animal models that combine genetic risks for atherosclerosis, such as the *Apoe*$^{-/-}$ or the *Ldl-r*$^{-/-}$ [90] mouse, with an altered immune system have been invaluable in demonstrating a link between atherosclerosis and immunity [3, 5]. Using these models, several groups have provided evidence to suggest that NKT cells are potentially proatherogenic [15–19]. The most commonly used model of NKT cells deficiency is the CD1d$^{-/-}$ mouse, which was first created by Mendiratta et al. [110]. In this model, the CD1d$^{-/-}$ mice lack the molecule necessary for normal development of the CD1d-dependent NKT cell and are therefore deficient in functional CD1d-dependent NKT cells. Recent studies have shown that deficiency of CD1d significantly slows the development of atherosclerosis in wild-type mice on an atherogenic diet [18] or in *Apoe*$^{-/-}$ mice whether they were fed a chow diet [15] or an atherosclerotic diet, which is a diet enriched in both cholesterol and fat [16]. The reduction in atherosclerotic lesion size ranged from 25% in double knockouts on a chow diet to 70% on an atherosclerotic diet. Similar observations were shown when CD1d$^{-/-}$ was put onto an *Ldl-r*$^{-/-}$ genetic background. These mice had 50% less lesions than the CD1d$^{+/+}$ controls when on an atherosclerotic diet for 4 weeks [17]. However, the authors did note that this difference was lost when both groups were compared at 8 and 12 weeks.

Using the technique of bone marrow transplantation following lethal irradiation, Nakai et al. [18] showed that *Ldl-r*$^{-/-}$ mice reconstituted with CD1d$^{-/-}$

bone marrow had significantly less atherosclerosis compared to lethally irradiated $Ldl\text{-}r^{-/-}$ mice reconstituted with wild-type bone marrow. These data suggest that replenishing the hematopoietic stem cells with bone marrow from mice that have CD1d-dependent NKT cells is sufficient to stimulate the development of early stage atherosclerotic lesions. These results, combined with those obtained using CD1d null mice, suggest that the absence of CD1d-reactive NKT cells attenuates lesion formation in mice during the early fatty streak formation.

Jα18 Null Mice

The exclusive expression of the invariant Vα14/Vβ receptor on Vα14 NKT cells and the essential requirement of Vα14 expression for the development of Vα14 NKT cells was demonstrated in Vα14 NKT cell-deficient (*Jα18*$^{-/-}$) mice [111]. Here, Cui et al. [111] showed that targeted deletion of the *Jα18* gene caused a complete failure of the mice to develop Vα14 NKT cells, leaving other lymphoid lineages intact. This observation strongly suggests that the invariant Vα14 segment is indispensable for the generation of Vα14 NKT cells. Targeted deletion of the *Jα18* gene resulting in the selective depletion of CD1d-dependent, Vα14 NKT cells without affecting the population of other NKT (including the other CD1d-dependent NKT cells), NK, and conventional T lymphocytes [73, 111]. By superimposing the *Jα18* deficiency onto mice that are already susceptible to atherosclerosis by virtue of being $Ldl\text{-}r^{-/-}$, Whitman et al. have shown that the loss of functionally active Vα14 NKT cells significantly slows the formation of lesions by 20% in both genders of mice [19]. As such, this study shows that a single population of NKT cells participates in the process of early-stage atherosclerotic lesion formation. Upon stimulation, Vα14 NKT cells have the capacity to exert immunoregulatory functions by releasing large amounts of inflammatory cytokines, including the proatherogenic cytokine IFN-γ [87–90]. This release of cytokines will, in turn, cause the activation of adjacent NK cells, B cells, CD4^{+} and CD8^{+} T lymphocytes [71, 112], as well as adjacent antigen-presenting cells [95]. This hierarchical form of activation may be one of the key mechanisms by which NKT cells promote atherosclerosis at the level of the vessel wall.

Concluding Remarks

Despite the ever-growing list of immune cells now identified within the developing atherosclerotic lesion of both humans and mice, this chapter has only focused on two unique populations of lymphocytes, namely NK cells and NKT cells. As noted in the introduction, macrophages and conventional T lymphocytes constitute the major cell types involved in the early- and latter stages of this disease. However, despite the apparent underrepresentation of both cell types, NK cells and NKT cells have now been shown to have a significant participatory role alongside these other more "prominent" classes of lymphocytes. As highlighted above, recent experimental evidence from our laboratory [13, 14, 19] and others [15–18] suggest that neither NK cells nor NKT cells cause the disease, but rather these immune cells play an important role in accelerating lesion development by modulating, via the elaboration of IFN-γ, the function of other more prominent immune cells (conventional T lymphocytes and macrophages) found within the developing atherosclerotic lesion. In this context, research from a number of laboratories strongly suggests that the proinflammatory cytokines IL-12 [85, 86] and IL-18 [113, 114] play a significant role in promoting atherosclerosis via the activation of NKT and NK cells, respectively.

Current *in vivo* models that combine genetic risks for atherosclerosis with an altered immune system have revealed to us the important modulatory role played by both NK and NKT cells. These studies have also highlighted the fact that the transgenic mouse models of atherosclerosis, although not a perfect substitute for human lesion pathology, are currently our best all-around model that affords us a significant degree of flexibility by allowing us access to a vast array of different transgenic and gene-targeted mice. As new advancements occur in mouse gene manipulation techniques, the use of the mouse will increase our ability to define these complicated mechanistic pathways, leaving us with the easier task of having to only confirm such pathways in our human atherosclerotic specimens.

Acknowledgments: S. C. Whitman is the recipient of a Great-West Life & London Life New Investigator Award from the Heart and Stroke Foundation of Canada and is supported by the Heart and Stroke Foundation of Ontario Grant NA-5086 and Canadian Institutes of Health Research Grants MOP-53344 and GHS-60663. T. A. Ramsamy is the recipient of a Heart and Stroke Foundation of Canada Post-Doctoral Fellowship awarded to the Coronary Disease and Heart Failure Group at the University of Ottawa Heart Institute. The authors apologize to those investigators whose excellent contributions to the field of atherosclerosis and immunity research, which could not be included in this review due to space limitations.

References

1. McMillan GC: Historical review of research on atherosclerosis. Adv Exp Med Biol 369: 1–6, 1995.
2. Ross R: Atherosclerosis—an inflammatory disease. N Engl J Med 340: 115–126, 1999.
3. Hansson GK, Libby P, Schonbeck U, Yan ZQ: Innate and adaptive immunity in the pathogenesis of atherosclerosis. Circ Res 91: 281–291, 2002.
4. VanderLaan PA, Reardon CA: Thematic review series: the immune system and atherogenesis. The unusual suspects: an overview of the minor leukocyte populations in atherosclerosis. J Lipid Res 46: 829–838, 2005.

5. Hansson GK: Immune mechanisms in atherosclerosis. Arterioscler Thromb Vasc Biol 21: 1876–1890, 2001.

6. Daugherty A, Rateri DL: T lymphocytes in atherosclerosis: the yin-yang of Th1 and Th2 influence on lesion formation. Circ Res 90: 1039–1040, 2002.

7. Roselaar SE, Kakkanathu PX, Daugherty A: Lymphocyte populations in atherosclerotic lesions of apoE –/– and LDL receptor –/– mice. Decreasing density with disease progression. Arterioscler Thromb Vasc Biol 16: 1013–1018, 1996.

8. Dansky HM, Charlton SA, Harper MM, Smith JD: T and B lymphocytes play a minor role in atherosclerotic plaque formation in the apolipoprotein E-deficient mouse. Proc Natl Acad Sci U S A 94: 4642–4646, 1997.

9. Daugherty A, Pure E, Delfel-Butteiger D, Chen S, Leferovich J, Roselaar SE, Rader DJ: The effects of total lymphocyte deficiency on the extent of atherosclerosis in apolipoprotein E–/– mice. J Clin Invest 100: 1575–1580, 1997.

10. Song L, Leung C, Schindler C: Lymphocytes are important in early atherosclerosis. J Clin Invest 108: 251–259, 2001.

11. Smith JD, Trogan E, Ginsberg M, Grigaux C, Tian J, Miyata M: Decreased atherosclerosis in mice deficient in both macrophage colony-stimulating factor (op) and apolipoprotein E. Proc Natl Acad Sci U S A 92: 8264–8268, 1995.

12. de Villiers WJ, Smith JD, Miyata M, Dansky HM, Darley E, Gordon S: Macrophage phenotype in mice deficient in both macrophage-colony-stimulating factor (op) and apolipoprotein E. Arterioscler Thromb Vasc Biol 18: 631–640, 1998.

13. Whitman SC, Rateri DL, Szilvassy SJ, Yokoyama W, Daugherty A: Depletion of natural killer cell function decreases atherosclerosis in low-density lipoprotein receptor null mice. Arterioscler Thromb Vasc Biol 24: 992–994, 2004.

14. Ramsamy TA, Rogers L, Hasu M, Siddiqui MM, Huynh T, Tam NL, Yokoyama WM, Whitman SC: Natural killer cell deficiency impairs early-stage lesion development in apolipoprotein E null mice. Arterioscler Thromb Vasc Biol 25: E-82, 2005.

15. Tupin E, Nicoletti A, Elhage R, Rudling M, Ljunggren HG, Hansson GK, Berne GP: CD1d-dependent activation of NKT cells aggravates atherosclerosis. J Exp Med 199: 417–422, 2004.

16. Major AS, Wilson MT, McCaleb JL, Ru SY, Stanic AK, Joyce S, Van Kaer L, Fazio S, Linton MF: Quantitative and qualitative differences in proatherogenic NKT cells in apolipoprotein E-deficient mice. Arterioscler Thromb Vasc Biol 24: 2351–2357, 2004.

17. Aslanian AM, Chapman HA, Charo IF: Transient role for CD1d-restricted natural killer T cells in the formation of atherosclerotic lesions. Arterioscler Thromb Vasc Biol 25: 628–632, 2005.

18. Nakai Y, Iwabuchi K, Fujii S, Ishimori N, Dashtsoodol N, Watano K, Mishima T, Iwabuchi C, Tanaka S, Bezbradica JS, Nakayama T, Taniguchi M, Miyake S, Yamamura T, Kitabatake A, Joyce S, Van Kaer L, Onoe K: Natural killer T cells accelerate atherogenesis in mice. Blood 104: 2051–2059, 2004.

19. Purdy L, Hasu M, Whitman SC: Depletion of invariant natural killer T cells decreases atherosclerosis in LDL receptor null mice. Arterioscler Thromb Vasc Biol 24: E57, 2004.

20. Sohma Y, Sasano H, Shiga R, Saeki S, Suzuki T, Nagura H, Nose M, Yamamoto T: Accumulation of plasma cells in atherosclerotic lesions of Watanabe heritable hyperlipidemic rabbits. Proc Natl Acad Sci U S A 92: 4937–4941, 1995.

21. Zhou X, Hansson GK: Detection of B cells and proinflammatory cytokines in atherosclerotic plaques of hypercholesterolaemic apolipoprotein E knockout mice. Scand J Immunol 50: 25–30, 1999.
22. Major AS, Fazio S, Linton MF: B-lymphocyte deficiency increases atherosclerosis in LDL receptor-null mice. Arterioscler Thromb Vasc Biol 22: 1892–1898, 2002.
23. Kaartinen M, Penttila A, Kovanen PT: Accumulation of activated mast cells in the shoulder region of human coronary atheroma, the predilection site of atheromatous rupture. Circulation 90: 1669–1678, 1994.
24. Bobryshev YV, Taksir T, Lord RS, Freeman MW: Evidence that dendritic cells infiltrate atherosclerotic lesions in apolipoprotein E-deficient mice. Histol Histopathol 16: 801–808, 2001.
25. Strong JP, Malcom GT, Oalmann MC, Wissler RW: The PDAY study: natural history, risk factors, and pathobiology. Pathobiological determinants of atherosclerosis in youth. Ann N Y Acad Sci 811: 226–235, 1997.
26. Millonig G, Malcom GT, Wick G: Early inflammatory–immunological lesions in juvenile atherosclerosis from the Pathobiological Determinants of Atherosclerosis in Youth (PDAY)-study. Atherosclerosis 160: 441–448, 2002.
27. Millonig G, Schwentner C, Mueller P, Mayerl C, Wick G: The vascular-associated lymphoid tissue: a new site of local immunity. Curr Opin Lipidol 12: 547–553, 2001.
28. Piedrahita JA, Zhang SH, Hagaman JR, Oliver PM, Maeda N: Generation of mice carrying a mutant apolipoprotein E gene inactivated by gene targeting in embryonic stem cells. Proc Natl Acad Sci U S A 89: 4471–4475, 1992.
29. Plump AS, Smith JD, Hayek T, Aalto-Setala K, Walsh A, Verstuyft JG, Rubin EM, Breslow JL: Severe hypercholesterolemia and atherosclerosis in apolipoprotein E-deficient mice created by homologous recombination in ES cells. Cell 71: 343–353, 1992.
30. Nakashima Y, Plump AS, Raines EW, Breslow JL, Ross R: ApoE-deficient mice develop lesions of all phases of atherosclerosis throughout the arterial tree. Arterioscler Thromb 14: 133–140, 1994.
31. Ishibashi S, Goldstein JL, Brown MS, Herz J, Burns DK: Massive xanthomatosis and atherosclerosis in cholesterol-fed low density lipoprotein receptor-negative mice. J Clin Invest 93: 1885–1893, 1994.
32. Stary HC: Macrophages, macrophage foam cells, and eccentric intimal thickening in the coronary arteries of young children. Atherosclerosis 64: 91–108, 1987.
33. Stary HC: The sequence of cell and matrix changes in atherosclerotic lesions of coronary arteries in the first forty years of life. Eur Heart J 11 (Suppl E): 3–19, 1990.
34. Stary HC: Composition and classification of human atherosclerotic lesions. Virchows Arch A Pathol Anat Histopathol 421: 277–290, 1992.
35. Kaartinen M, Penttila A, Kovanen PT: Mast cells in rupture-prone areas of human coronary atheromas produce and store TNF-alpha. Circulation 94: 2787–2792, 1996.
36. Getz GS: The involvement of lipoproteins in atherogenesis. Evolving concepts. Ann N Y Acad Sci 598: 17–28, 1990.
37. Schwenke DC, Carew TE: Initiation of atherosclerotic lesions in cholesterol-fed rabbits. I. Focal increases in arterial LDL concentration precede development of fatty streak lesions. Arteriosclerosis 9: 895–907, 1989.

38. Schwenke DC, Carew TE: Initiation of atherosclerotic lesions in cholesterol-fed rabbits. II. Selective retention of LDL vs. selective increases in LDL permeability in susceptible sites of arteries. Arteriosclerosis 9: 908–918, 1989.
39. Fry DL: Mass transport, atherogenesis, and risk. Arteriosclerosis 7: 88–100, 1987.
40. Steinberg D, Witztum JL: Lipoproteins and atherogenesis. Current concepts. JAMA 264: 3047–3052, 1990.
41. Quinn MT, Parthasarathy S, Fong LG, Steinberg D: Oxidatively modified low density lipoproteins: a potential role in recruitment and retention of monocyte/macrophages during atherogenesis. Proc Natl Acad Sci U S A 84: 2995–2998, 1987.
42. Cybulsky MI, Iiyama K, Li H, Zhu S, Chen M, Iiyama M, Davis V, Gutierrez-Ramos JC, Connelly PW, Milstone DS: A major role for VCAM-1, but not ICAM-1, in early atherosclerosis. J Clin Invest 107: 1255–1262, 2001.
43. Boullier A, Bird DA, Chang MK, Dennis EA, Friedman P, Gillotre-Taylor K, Horkko S, Palinski W, Quehenberger O, Shaw P, Steinberg D, Terpstra V, Witztum JL: Scavenger receptors, oxidized LDL, and atherosclerosis. Ann N Y Acad Sci 947: 214–222, 2001.
44. Jonasson L, Holm J, Skalli O, Bondjers G, Hansson GK: Regional accumulations of T cells, macrophages, and smooth muscle cells in the human atherosclerotic plaque. Arteriosclerosis 6: 131–138, 1986.
45. Jeziorska M, McCollum C, Woolley DE: Mast cell distribution, activation, and phenotype in atherosclerotic lesions of human carotid arteries. J Pathol 182: 115–122, 1997.
46. Bobryshev YV, Lord RS: Identification of natural killer cells in human atherosclerotic plaque. Atherosclerosis 180: 423–427, 2005.
47. Bobryshev YV, Lord RS: Co-accumulation of dendritic cells and natural killer T cells within rupture-prone regions in human atherosclerotic plaques. J Histochem Cytochem 53: 781–785, 2005.
48. Rosenfeld ME, Polinsky P, Virmani R, Kauser K, Rubanyi G, Schwartz SM: Advanced atherosclerotic lesions in the innominate artery of the ApoE knockout mouse. Arterioscler Thromb Vasc Biol 20: 2587–2592, 2000.
49. Calara F, Silvestre M, Casanada F, Yuan N, Napoli C, Palinski W: Spontaneous plaque rupture and secondary thrombosis in apolipoprotein E-deficient and LDL receptor-deficient mice. J Pathol 195: 257–263, 2001.
50. Williams H, Johnson JL, Carson KG, Jackson CL: Characteristics of intact and ruptured atherosclerotic plaques in brachiocephalic arteries of apolipoprotein E knockout mice. Arterioscler Thromb Vasc Biol 22: 788–792, 2002.
51. Zhou J, Moller J, Danielsen CC, Bentzon J, Ravn HB, Austin RC, Falk E: Dietary supplementation with methionine and homocysteine promotes early atherosclerosis but not plaque rupture in ApoE-deficient mice. Arterioscler Thromb Vasc Biol 21: 1470–1476, 2001.
52. Lanier LL: NK cell receptors. Annu Rev Immunol 16: 359–393, 1998.
53. Trinchieri G: Biology of natural killer cells. Adv Immunol 47: 187–376, 1989.
54. Whiteside TL, Herberman RB: Role of human natural killer cells in health and disease. Clin Diagn Lab Immunol 1: 125–133, 1994.
55. Cerwenka A, Lanier LL: Natural killer cells, viruses and cancer. Nat Rev Immunol 1: 41–49, 2001.
56. Yokoyama WM, Kim S, French AR: The dynamic life of natural killer cells. Annu Rev Immunol 22: 405–429, 2004.

57. Schiller NK, Boisvert WA, Curtiss LK: Inflammation in atherosclerosis: lesion formation in LDL receptor-deficient mice with perforin and Lyst(beige) mutations. Arterioscler Thromb Vasc Biol 22: 1341–1346, 2002.

58. Hansson GK: Immune mechanisms in atherosclerosis. Arterioscler Thromb Vasc Biol 21: 1876–1890, 2001.

59. Roder J, Duwe A: The beige mutation in the mouse selectively impairs natural killer cell function. Nature 278: 451–453, 1979.

60. Roder JC: The beige mutation in the mouse. I. A stem cell predetermined impairment in natural killer cell function. J Immunol 123: 2168–2173, 1979.

61. Roder JC, Lohmann-Matthes ML, Domzig W, Wigzell H: The beige mutation in the mouse. II. Selectivity of the natural killer (NK) cell defect. J Immunol 123: 2174–2181, 1979.

62. Paigen B, Holmes PA, Novak EK, Swank RT: Analysis of atherosclerosis susceptibility in mice with genetic defects in platelet function. Arteriosclerosis 10: 648–652, 1990.

63. Kim S, Iizuka K, Aguila HL, Weissman IL, Yokoyama WM: *In vivo* natural killer cell activities revealed by natural killer cell-deficient mice. Proc Natl Acad Sci U S A 97: 2731–2736, 2000.

64. Budd RC, Miescher GC, Howe RC, Lees RK, Bron C, MacDonald HR: Developmentally regulated expression of T cell receptor beta chain variable domains in immature thymocytes. J Exp Med 166: 577–582, 1987.

65. Fowlkes BJ, Kruisbeek AM, Ton-That H, Weston MA, Coligan JE, Schwartz RH, Pardoll DM: A novel population of T-cell receptor alpha beta-bearing thymocytes which predominantly expresses a single V beta gene family. Nature 329: 251–254, 1987.

66. Ceredig R, Lynch F, Newman P: Phenotypic properties, interleukin 2 production, and developmental origin of a "mature" subpopulation of Lyt-2-. Proc Natl Acad Sci U S A 84: 8578–8582, 1987.

67. Makino Y, Kanno R, Ito T, Higashino K, Taniguchi M: Predominant expression of invariant V alpha 14+ TCR alpha chain in NK1.1+ T cell populations. Int Immunol 7: 1157–1161, 1995.

68. Reichlin A, Yokoyama WM: Natural killer cell proliferation induced by anti-NK1.1 and IL-2. Immunol Cell Biol 76: 143–152, 1998.

69. Hammond KJ, Pellicci DG, Poulton LD, Naidenko OV, Scalzo AA, Baxter AG, Godfrey DI: CD1d-restricted NKT cells: an interstrain comparison. J Immunol 167: 1164–1173, 2001.

70. Bendelac A, Bonneville M, Kearney JF: Autoreactivity by design: innate B and T lymphocytes. Nat Rev Immunol 1: 177–186, 2001.

71. Wilson SB, Delovitch TL: Janus-like role of regulatory iNKT cells in autoimmune disease and tumour immunity. Nat Rev Immunol 3: 211–222, 2003.

72. Kronenberg M, Gapin L: The unconventional lifestyle of NKT cells. Nat Rev Immunol 2: 557–568, 2002.

73. Taniguchi M, Harada M, Kojo S, Nakayama T, Wakao H: The regulatory role of Valpha14 NKT cells in innate and acquired immune response. Annu Rev Immunol 21: 483–513, 2003.

74. Kawano T, Cui J, Koezuka Y, Toura I, Kaneko Y, Motoki K, Ueno H, Nakagawa R, Sato H, Kondo E, Koseki H, Taniguchi M: CD1d-restricted and TCR-mediated activation of valpha14 NKT cells by glycosylceramides. Science 278: 1626–1629, 1997.

75. Benlagha K, Weiss A, Beavis A, Teyton L, Bendelac A: *In vivo* identification of glycolipid antigen-specific T cells using fluorescent CD1d tetramers. J Exp Med 191: 1895–1903, 2000.

76. Matsuda JL, Naidenko OV, Gapin L, Nakayama T, Taniguchi M, Wang CR, Koezuka Y, Kronenberg M: Tracking the response of natural killer T cells to a glycolipid antigen using CD1d tetramers. J Exp Med 192: 741–754, 2000.

77. Wilson MT, Singh AK, Van Kaer L: Immunotherapy with ligands of natural killer T cells. Trends Mol Med 8: 225–231, 2002.

78. Singh N, Hong S, Scherer DC, Serizawa I, Burdin N, Kronenberg M, Koezuka Y, Van Kaer L: Cutting edge: activation of NK T cells by CD1d and alpha-galactosylceramide directs conventional T cells to the acquisition of a Th2 phenotype. J Immunol 163: 2373–2377, 1999.

79. Cardell S, Tangri S, Chan S, Kronenberg M, Benoist C, Mathis D: CD1-restricted CD4+ T cells in major histocompatibility complex class II-deficient mice. J Exp Med 182: 993–1004, 1995.

80. Behar SM, Dascher CC, Grusby MJ, Wang CR, Brenner MB: Susceptibility of mice deficient in CD1D or TAP1 to infection with *Mycobacterium tuberculosis*. J Exp Med 189: 1973–1980, 1999.

81. Park SH, Weiss A, Benlagha K, Kyin T, Teyton L, Bendelac A: The mouse CD1d-restricted repertoire is dominated by a few autoreactive T cell receptor families. J Exp Med 193: 893–904, 2001.

82. Eberl G, Lees R, Smiley ST, Taniguchi M, Grusby MJ, MacDonald HR: Tissue-specific segregation of CD1d-dependent and CD1d-independent NK T cells. J Immunol 162: 6410–6419, 1999.

83. Hammond KJ, Pelikan SB, Crowe NY, Randle-Barrett E, Nakayama T, Taniguchi M, Smyth MJ, van Driel IR, Scollay R, Baxter AG, Godfrey DI: NKT cells are phenotypically and functionally diverse. Eur J Immunol 29: 3768–3781, 1999.

84. Arase H, Saito T, Phillips JH, Lanier LL: Cutting edge: the mouse NK cell-associated antigen recognized by DX5 monoclonal antibody is CD49b (alpha 2 integrin, very late antigen-2). J Immunol 167: 1141–1144, 2001.

85. Lee TS, Yen HC, Pan CC, Chau LY: The role of interleukin 12 in the development of atherosclerosis in ApoE-deficient mice. Arterioscler Thromb Vasc Biol 19: 734–742, 1999.

86. Davenport P, Tipping PG: The role of interleukin-4 and interleukin-12 in the progression of atherosclerosis in apolipoprotein E-deficient mice. Am J Pathol 163: 1117–1125, 2003.

87. Gupta S, Pablo AM, Jiang X, Wang N, Tall AR, Schindler C: IFN-gamma potentiates atherosclerosis in ApoE knock-out mice. J Clin Invest 99: 2752–2761, 1997.

88. Whitman SC, Ravisankar P, Daugherty A: IFN-gamma deficiency exerts gender-specific effects on atherogenesis in apolipoprotein E–/– mice. J Interferon Cytokine Res 22: 661–670, 2002.

89. Whitman SC, Ravisankar P, Elam H, Daugherty A: Exogenous interferon-gamma enhances atherosclerosis in apolipoprotein E–/– mice. Am J Pathol 157: 1819–1824, 2000.

90. Buono C, Come CE, Stavrakis G, Maguire GF, Connelly PW, Lichtman AH: Influence of interferon-gamma on the extent and phenotype of diet-induced atherosclerosis in the LDLR-deficient mouse. Arterioscler Thromb Vasc Biol 23: 454–460, 2003.

91. Caligiuri G, Rudling M, Ollivier V, Jacob MP, Michel JB, Hansson GK, Nicoletti A: Interleukin-10 deficiency increases atherosclerosis, thrombosis, and low-density lipoproteins in apolipoprotein E knockout mice. Mol Med 9: 10–17, 2003.

92. Mallat Z, Besnard S, Duriez M, Deleuze V, Emmanuel F, Bureau MF, Soubrier F, Esposito B, Duez H, Fievet C, Staels B, Duverger N, Scherman D, Tedgui A: Protective role of interleukin-10 in atherosclerosis. Circ Res 85: e17–e24, 1999.

93. Pinderski LJ, Fischbein MP, Subbanagounder G, Fishbein MC, Kubo N, Cheroutre H, Curtiss LK, Berliner JA, Boisvert WA: Overexpression of interleukin-10 by activated T lymphocytes inhibits atherosclerosis in LDL receptor-deficient mice by altering lymphocyte and macrophage phenotypes. Circ Res 90: 1064–1071, 2002.

94. Pinderski Oslund LJ, Hedrick CC, Olvera T, Hagenbaugh A, Territo M, Berliner JA, Fyfe AI: Interleukin-10 blocks atherosclerotic events *in vitro* and *in vivo*. Arterioscler Thromb Vasc Biol 19: 2847–2853, 1999.

95. Fujii S, Shimizu K, Smith C, Bonifaz L, Steinman RM: Activation of natural killer T cells by alpha-galactosylceramide rapidly induces the full maturation of dendritic cells *in vivo* and thereby acts as an adjuvant for combined CD4 and CD8 T cell immunity to a coadministered protein. J Exp Med 198: 267–279, 2003.

96. Godfrey DI, Hammond KJ, Poulton LD, Smyth MJ, Baxter AG: NKT cells: facts, functions and fallacies. Immunol Today 21: 573–583, 2000.

97. Kobayashi E, Motoki K, Uchida T, Fukushima H, Koezuka Y: KRN7000, a novel immunomodulator, and its antitumor activities. Oncol Res 7: 529–534, 1995.

98. Kawano T, Cui J, Koezuka Y, Toura I, Kaneko Y, Sato H, Kondo E, Harada M, Koseki H, Nakayama T, Tanaka Y, Taniguchi M: Natural killer-like nonspecific tumor cell lysis mediated by specific ligand-activated Valpha14 NKT cells. Proc Natl Acad Sci U S A 95: 5690–5693, 1998.

99. Nakagawa R, Motoki K, Ueno H, Iijima R, Nakamura H, Kobayashi E, Shimosaka A, Koezuka Y: Treatment of hepatic metastasis of the colon26 adenocarcinoma with an alpha-galactosylceramide, KRN7000. Cancer Res 58: 1202–1207, 1998.

100. Exley MA, Bigley NJ, Cheng O, Tahir SM, Smiley ST, Carter QL, Stills HF, Grusby MJ, Koezuka Y, Taniguchi M, Balk SP: CD1d-reactive T-cell activation leads to amelioration of disease caused by diabetogenic encephalomyocarditis virus. J Leukoc Biol 69: 713–718, 2001.

101. Gonzalez-Aseguinolaza G, de Oliveira C, Tomaska M, Hong S, Bruna-Romero O, Nakayama T, Taniguchi M, Bendelac A, Van Kaer L, Koezuka Y, Tsuji M: Alpha-galactosylceramide-activated Valpha 14 natural killer T cells mediate protection against murine malaria. Proc Natl Acad Sci U S A 97: 8461–8466, 2000.

102. Miyamoto K, Miyake S, Yamamura T: A synthetic glycolipid prevents autoimmune encephalomyelitis by inducing TH2 bias of natural killer T cells. Nature 413: 531–534, 2001.

103. Sharif S, Arreaza GA, Zucker P, Mi QS, Sondhi J, Naidenko OV, Kronenberg M, Koezuka Y, Delovitch TL, Gombert JM, Leite-De-Moraes M, Gouarin C, Zhu R, Hameg A, Nakayama T, Taniguchi M, Lepault F, Lehuen A, Bach JF, Herbelin A: Activation of natural killer T cells by alpha-galactosylceramide treatment prevents the onset and recurrence of autoimmune Type 1 diabetes. Nat Med 7: 1057–1062, 2001.

104. Hong S, Wilson MT, Serizawa I, Wu L, Singh N, Naidenko OV, Miura T, Haba T, Scherer DC, Wei J, Kronenberg M, Koezuka Y, Van Kaer L: The natural killer T-cell ligand alpha-galactosylceramide prevents autoimmune diabetes in nonobese diabetic mice. Nat Med 7: 1052–1056, 2001.

105. van der Vliet HJ, Molling JW, von Blomberg BM, Nishi N, Kolgen W, van den Eertwegh AJ, Pinedo HM, Giaccone G, Scheper RJ: The immunoregulatory role of CD1d-restricted natural killer T cells in disease. Clin Immunol 112: 8–23, 2004.

106. Binder CJ, Chang MK, Shaw PX, Miller YI, Hartvigsen K, Dewan A, Witztum JL: Innate and acquired immunity in atherogenesis. Nat Med 8: 1218–1226, 2002.

107. Linton MF, Major AS, Fazio S: Proatherogenic role for NK cells revealed. Arterioscler Thromb Vasc Biol 24: 992–994, 2004.

108. Ostos MA, Recalde D, Zakin MM, Scott-Algara D: Implication of natural killer T cells in atherosclerosis development during a LPS-induced chronic inflammation. FEBS Lett 519: 23–29, 2002.

109. Melian A, Geng YJ, Sukhova GK, Libby P, Porcelli SA: CD1 expression in human atherosclerosis. A potential mechanism for T cell activation by foam cells. Am J Pathol 155: 775–786, 1999.

110. Mendiratta SK, Martin WD, Hong S, Boesteanu A, Joyce S, Van Kaer L: CD1d1 mutant mice are deficient in natural T cells that promptly produce IL-4. Immunity 6: 469–477, 1997.

111. Cui J, Shin T, Kawano T, Sato H, Kondo E, Toura I, Kaneko Y, Koseki H, Kanno M, Taniguchi M: Requirement for Valpha14 NKT cells in IL-12-mediated rejection of tumors. Science 278: 1623–1626, 1997.

112. Godfrey DI, Hammond KJ, Poulton LD, Smyth MJ, Baxter AG: NKT cells: facts, functions and fallacies. Immunol Today 21: 573–583, 2000.

113. Whitman SC, Ravisankar P, Daugherty A: Interleukin-18 enhances atherosclerosis in apolipoprotein E(–/–) mice through release of interferon-gamma. Circ Res 90: E34–E38, 2002.

114. Elhage R, Jawien J, Rudling M, Ljunggren HG, Takeda K, Akira S, Bayard F, Hansson K: Reduced atherosclerosis in interleukin-18 deficient apolipoprotein E-knockout mice. Cardiovasc Res 59: 234–240, 2003.

Biochemistry of Atherosclerosis edited by S.K. Cheema, Springer, New York, 2006

19

The Role of Infectious Agents in Atherogenesis

PAUL K.M. CHEUNG AND GRANT N. PIERCE

Abstract

Atherosclerosis has been widely accepted and investigated as an autoimmune disease of the blood vessel. While the contribution of the inflammatory components is significant in plaque development, there are emerging evidences that various human infectious pathogens can also be involved in the processes of endothelial injury, leading to smooth muscle cell migration and proliferative responses. These events are also major mechanisms in the formation of pathogenic lesions. Several of these pathogens, including herpesvirus, cytomegalovirus, *Helicobacter pylori, Porphyromonas gingivalis*, and *Chlamydia pneumonia* have generated major clinical interest. When in combination with other cardiovascular risk factors such as lipidemia and hypertension, these infectious agents can stimulate coronary heart disease and ischemic stroke. In this review article, we summarize the findings that implicate these viral and bacterial organisms as important contributors to the process of atherogenesis. In particular, we provide an in-depth review on the broad spectrum of statistical and experimental data regarding the roles of *P. gingivalis* and *C. pneumonia* as major causative agents in atherosclerotic diseases. Based on our review, we have suggested the use of ex vivo techniques to further investigate the atherosclerotic properties of the bacteria. We believe understanding the mechanism and contribution of these organisms in atherosclerosis can help the development of specific treatments and preventive measures.

Keywords: atherosclerosis; cytomegalovirus; endothelial injury; herpesvirus; *Helicobacter pylori*; oxidized LDL; *Porphyromonas gingivalis* and *Chlamydia pneumonia*; vascular smooth muscle cells

Infections and Atherosclerosis

Atherosclerosis is a chronic disease that results in lesion development in arteries. The arterial lesion reduces blood flow in the vessel and, with increasing severity, causes coronary heart disease and ischemic stroke [1, 2]. At present, atherosclerosis is also commonly accepted as an inflammatory disease. Much of the scientific investigation in the past three decades has been championed by Dr. Russell Ross [3, 4]. His hypothesis of atherosclerosis as a response to endothelial injury remains a guiding framework within which most studies have been conducted. Specifically, the hypothesis can be summarized as:

Injury to the lining of the artery (endothelium) results in protective inflammatory and fibroproliferative responses which, when in excess, becomes the disease process of atherogenesis.

There are several well-studied risk factors of endothelial injury in arteries, such as hypertension [5–7], diabetes [8, 9], obesity [10], high-plasma homocysteine [11–13], and hyperlipidemia [14, 15]. Several comprehensive review articles on these subjects are available [16, 17]. These factors, however, may account for only 50% to 70% of atherosclerotic events in a general population [18]. One less recognized risk factor is the role of infection in stimulating atherogenic events. In recent years, through statistical associations and investigation in animal models, several viral and bacterial pathogens have been identified as potentially atherogenic. Current opinions remains divided over the significance of these infectious agents in the overall, long-term process of lesion and plaque development. This skepticism largely arises from conflicting reports of clinical correlation and the lack of success in treating atherosclerotic patients with the antibiotics. In this article, we shall focus our review on the *in vitro* and *in vivo* investigations that link several viral and bacterial pathogens to lesion formation. In particular, we shall discuss the extensive evidence that suggests the mechanism through which *Chlamydia pneumoniae*, a pathogen that is widely present in the human population, may cause atherosclerosis.

The general development of an atherosclerotic lesion can be summarized as follows. Early stage lesions can occur in young children as an accumulation of foam cells in the subendothelial layer (and above the smooth muscle cell intimal layer) of the arterial environment [3]. These foam cells typically originate from circulating monocytes and T lymphocytes which adhere to the endothelial layer, migrate into the intimal space, and absorb excess lipoproteins [19]. A typical early lesion like this is termed a "fatty streak" and is considered insignificant in affecting blood flow or endothelial function. However, over decades of development, continued foam cell accumulation causes significant narrowing of the arterial lumen. Biochemical interactions among cells in the lesion milieu lead to the heightened expression of extracellular proteases (such as matrix metalloproteinases), chemokines (such as IL-8), and cytokines (such as TNF-α) that results in fibrosis, cell proliferation, cell migration, and accelerated plaque formation. At this stage, the lesion is termed a "plaque." The development of a plaque has two important outcomes: (i) reduction in the lumen area (flow volume) with dysfunction of the intimal layer including a loss in arterial contractility and (ii) an increased likelihood of autoimmune responses against the lesion and hence the evolution of a lesion into an unstable "fibrotic plaque." These late stage fibrotic plaques are characterized by an increased proportion of calcified structures and necrotic cells. Fibrotic plaques can "grow" to occlude the entire blood vessel, resulting in acute myocardial infarction if it happens inside the coronary arteries, or ischemic stroke if it occurs inside cerebral arteries. Destabilization and rupturing of the fibrotic lesions can also occur. Plaque material is released into the lumen of the artery and elicits an immediate thrombotic response, which can result in a blockage of the artery. Figure 19.1 summarizes this process of atherosclerosis.

Infectious Agents and Atherosclerosis

Several viral and bacterial agents are strongly associated with atherogenesis. These include herpesviruses, cytomegalovirus, *Helicobacter pylori*, *Porphyromonas gingivalis*, and *C. pneumoniae*. In the following sections, we shall review current findings that suggest the potential roles of these infectious agents in causing atheroclerosis.

Herpesvirus and Cytomegalovirus

Cytomegalovirus (CMV) and herpesviruses (HPV) have been detected in atherosclerotic arteries [20–25]. Herpes simplex virus type 2 is more frequently observed in coronary artery biopsy specimens than the type 1 virus [26]. A small HIV study involving eight young patients revealed unusually severe coronary lesions that may be associated with the HIV or coexisting CMV and HPV [27]. The CMV genome has also been detected by PCR in arterial samples taken from patients undergoing vascular surgery [28]. CMV-seropositive patients are also associated with more advanced atherosclerotic plaques [28]. Other clinical studies of allograft transplant patients have demonstrated associations between CMV infection and thickening of the intima/endothelial layers [29]. Patients suffering acute myocardial infarction demonstrate a significantly higher prevalence of serum CMV DNA [30].

The important limitation in all of the above studies, however, is that they cannot distinguish the viral infections as causative agents of atherosclerosis from infections that are enhanced by preexisting factors in the lesion milieu. To support the assertion that CMV and HPV can cause atherosclerosis, studies in animal models have been carried out. The Marek's disease virus (herpes-type DNA virus) is reported to cause atherosclerosis in chicken, leading to a pathology that is analogous to those observed in human arteries [31]. Interestingly, this atherogenic condition can be prevented by vaccination, supporting a causative role of the virus in atherogenesis. A more definitive test of a causative agent, however, would be to verify that the human virus can also result in similar pathologies in animal models. Data for CMV infection in rats appears to satisfy this criterion. CMV can increase atherosclerotic lesions in the presence of hypercholesterolemia, resembling what is known about low-density lipoprotein (LDL) with atherosclerosis in human [32]. At

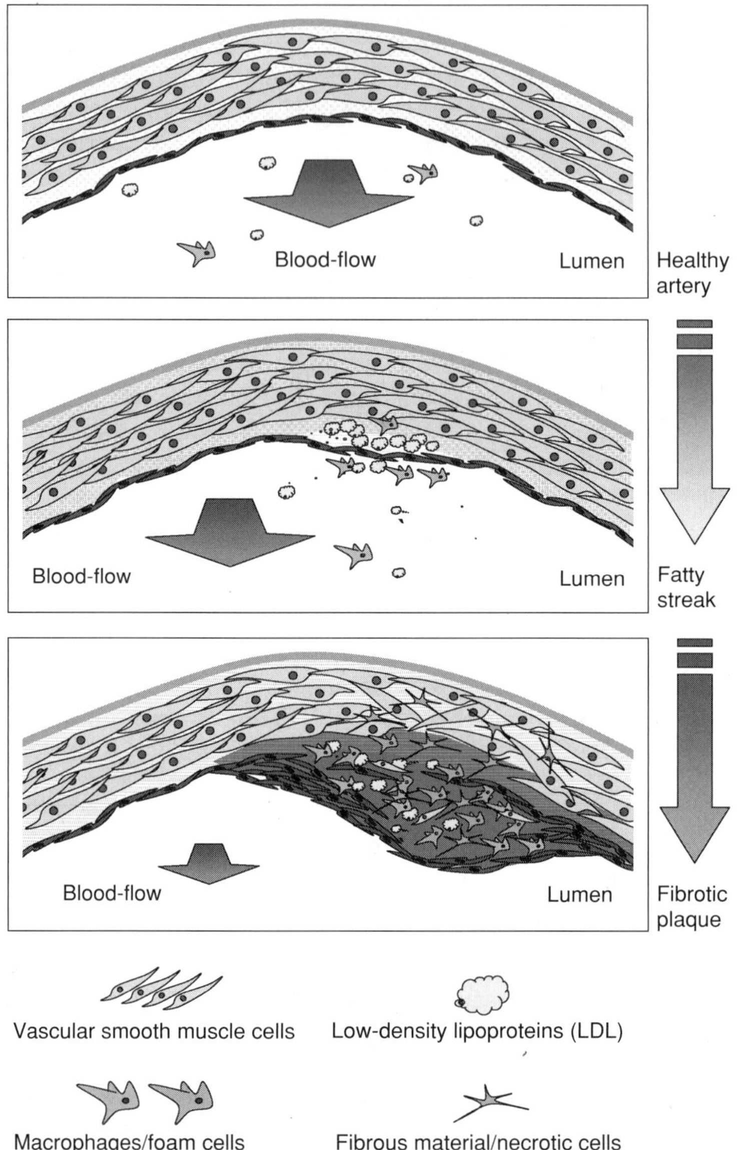

FIGURE 19.1. Development of an atherosclerotic lesion. Accumulation of low-density lipoprotein and foam cells in the subendothelial space results in the formation of a fatty streak. Continued endothelial injury and inflammatory insult leads to fibrosis and cell death within the lesion. The endothelium is thickened in the fibrotic plaque. Blood flow is reduced as the lesion grows into the lumen.

the same time, there are *in vitro* studies that demonstrate the ability of the CMV infection in upregulating the oxidized LDL scavenger receptor CD 36 [33], ICAM-1 [34], VCAM-1 [35], PDGF-β receptor [36], IL-6, and MCP-1 [37]. These molecules are important players in stimulating atherosclerotic lesions and hence this provides important biochemical evidence for the mechanism by which CMV may cause lesions *in vivo*. Recently, Hillebrands et al. [38] have demonstrated that CMV-infected rats with transplanted aortic allografts demonstrate enhanced atherogenesis if the infection occurs early post-transplantation (1–5 days but not later). This may be relevant to coronary transplant patients who already suffer chronic CMV infections. While this finding supports a role of CMV in stimulating atherogenesis, the authors also cautioned that the conclusion should not be generalized because the effect appeared to be dependent on the combination of host and viral strain [38].

The main weakness in the postulate that CMV and HPV are causative agents in atherosclerosis is the lack of data on treatments that are effective concurrently against both atherosclerosis and the viruses. The presence of studies that fail to find significant correlations between viral replication (HPV and CMV) and plaque development further adds to the controversy. For instance, in the ARIC study which examined 340 patients, it was concluded that vessel wall thickening may not have a significant association with the detection of CMV [39]. In summary, the cases for CMV and HPV as significant atherosclerotic agents are not definitive but do warrant further investigation. Current data seems to suggest that the contribution of these viral infections may be more prominent in certain immune-compromised subjects (e.g., HIV-infected and/or transplant patients). The presence of these viruses in the arterial and atherosclerotic environment of susceptible individuals may further aggravate the already injured endothelial layer with heightened immune responses. This also implies that CMV and HPV infections may not be important atherosclerotic risk factors in the general population. Thus the role of CMV and HPV may resemble more of "environmental" factors that enhance the severity of lesion formation.

Helicobacter pylori

H. pylori (Hpy) is a well-studied gram-negative bacteria. It is now understood that Hpy infection is a major cause of persistent gastritis in adults and children [40, 41]. Given its prevalence and its ability to stimulate inflammatory responses, the possible role of this bacterium in causing atherosclerosis has received significant attention. Several observational studies have demonstrated strong correlations between Hpy infection and coronary heart disease in humans [42, 43]. Hpy infection may enhance atherogenesis in patients with hyperhomocysteinemia [44]. There is an indication that infection by Hpy specifically (and not by CMV nor *C. pneumoniae* (Cpn)) attenuates coronary vasodilation and induces atherosclerosis [45]. This suggests that Hpy infection may enhance atherogenesis by mechanisms similar to those induced by hypertension and is acting somewhat differently from CMV or Cpn. Also, there is a recent report that suggests Hpy does not enhance myocardial infarct in lieu with oxidized LDL [46], an observation that again sets it apart from other potential atherogenic agents such as Cpn. A recent clinical investigation of 81 male patients who had an average age 40 years old concluded that Hpy infection is associated with endothelial dysfunction and with elevated levels of C-reactive protein, an inflammatory marker of atherosclerotic disease [47]. There are other similar clinical reports which conclude positive correlations of Hpy with cardiovascular events such as ischemic stroke and coronary heart disease [48–50]. These data provide strong support to the assertion that Hpy is a proatherogenic agent. However, at present, there is a lack of understanding of the biochemical events that demonstrate the causative mechanism between Hpy infection and plaque formation. Various strains of Hpy have been demonstrated to cause endothelial dysfunction and proliferation [51]. Current opinion [50] seems to agree that while this pathogen can elicit significant inflammatory diseases, further studies and *in vitro* experiments on the molecular effects of the pathogen may help evaluate the value of controlling this pathogen in coronary heart disease patients.

Porphyromonas gingivalis

P. gingivalis (Pgn) and Cpn are probably the most important infectious proatherogenic pathogens known to date. A strong correlation between periodontal disease and atherosclerosis have been reported since 1989 [52–56]. Gingivitis and periodontal diseases are common in the human population and these infections can lead to dental plaque development and destruction of tooth supporting tissues. The usual periodontal pathogens are Pgn, *Actinobacillus actinomycetemcomitans* and *Bacteroides forsythus*. The rDNA of these pathogens have been detected in 18–30% of carotid atheromas [57, 58]. In the National Health and Nutrition Examination Survey (NHANES III, a cross-sectional study) that involves 5564 patients of 40 years and older, after adjustment for age, sex, race, social status, smoking, and diabetes, there was a fourfold increase in the incidence of myocardial infarction in patients with periodontal disease [56]. Further adding weight to this association is the report that cardiovascular risk is related to the severity of periodontal infection [59].

It is particularly noteworthy that longitudinal studies have generated very different conclusions regarding the role of periodontal infections in atherosclerosis. In longitudinal studies, periodontal infection appears weaker (between 1 and 2.5-fold increase in risk) in predicting coronary diseases [60, 61]. Furthermore, two other large longitudinal studies (involving 8000+ patients) failed to find an association, after adjustments for other risk factors, between periodontal disease and atherosclerosis [62, 63]. The apparent discrepancies between cross-sectional and longitudinal studies may reflect the complex multifactorial nature of atherogenesis. The adjustment for other risk factors that utilize similar inflammatory pathways as do periodontal infections could attenuate, especially over long-term progression of atherosclerosis, the observed significance of periodontal disease. At the same time, it is plausible that the differences between longitudinal and cross-sectional studies may indeed reflect the acute nature of periodontal disease in stimulating cardiovascular incidences. In myocardial infarct and periodontal disease patients, significant inflammatory responses are often observed with high levels of C-reactive proteins and bacteremia [64, 65]. Thus, it is possible that periodontal disease results in atherosclerotic outcomes by causing acute severe immune responses in patients who are already predisposed to atherogenesis in the coronary arteries.

To understand more about the mechanism by which periodontal infection can induce atherosclerotic plaques, two recent investigations studied the molecular mechanisms of lesion aggravation by Pgn. Pgn is receiving special attention over other periodontal disease agents due to a recent report that has identified a serum antibody against Pgn, and not those against *A. actinomycetemcomitans*, was a strong indicator for coronary heart disease [66]. Lourbakos et al. [67] demonstrated that gingipains (cysteine proteinases) produced by this pathogen are potent coagulation factors. These enzymes increase intracellular calcium and stimulate platelet aggregation at efficiencies similar to that of thrombin. Thus, the presence of Pgn in atheromas such as carotid lesions [57, 58] may increase the likelihood of an arterial blockade. In apo-E null mice, Lalla et al. [68] have demonstrated the acceleration of atherosclerotic plaque formation by Pgn. The infection resulted in an upregulation of IL-6, VCAM-1, and tissue factor (TF) in the animals while a minority of mice (two out of nine) demonstrated the presence of Pgn in the aortic tissue. These findings, combined with other *in vitro* data that demonstrates the proatherogenic effects of Pgn [69, 70], makes a strong argument for Pgn periodontal disease as a direct inducer of atherosclerotic diseases. However, despite these indirect pieces of evidence implicating the agent in atherogenesis, the causative role of Pgn remains to be proven by one crucial observation: anti-Pgn treatments can reduce atherogenic and cardiovascular events. To this end, therefore, the outcome of the ongoing clinical NIH approved trial of periodontal treatments in several thousand patients with atherosclerotic disease will be an important piece to complete the puzzle that defines the causal relationship between the periodontal infection and atherogenesis (http://www.cscc.unc.edu/pave/).

Chlamydia pneumoniae

In the past 15 years, infection by Cpn and its role in atherosclerosis has been more intensively studied than other infectious agents. This is largely due to findings from various clinical and laboratory studies that strongly implicate its contribution to inflammation, cell proliferation, and plaque formation. However, conflicting reports from animal models and the failure of antibiotics to reduce atherosclerosis or clinical events associated with the infection have fueled the debate over the atherogenic nature of Cpn. Thus, current opinions remain divided regarding the role of Cpn as a significant agent in human atherosclerosis. In the following sections, we shall provide a concise summary of what is known about the bacteria, its atherogenic effects, and current data that argues for and against its role in plaque formation.

A general serological test for *Chlamydia* has been available since 1960s (the microimmunofluorescence serologic test) [71]. However, it was not until 1985 that Saikku et al. [72] investigated and distinguished Cpn from *Chlamydia psittaci*. Subsequent retrospective serology work by Grayston and coworkers [73–78] revealed the presence of this previously undetected species in various outbreaks of pneumonia and acute respiratory disease epidemics from 1950s to 1990s. Thus, Cpn has been an important pathogen in the human population for a considerable time. Cpn is distinguished from the other *Chlamydia* species in four aspects: (i) Cpn exhibits a unique morphology of a pear-shaped inclusion body in a cellular environment [79, 80]; (ii) the Cpn DNA sequence is homologous but discrete from other *Chlamydia* species [81, 82]; (iii) Cpn serology is highly conserved and only a single serovar has been observed [83–85]; and (iv) Cpn infections cause symptoms in the respiratory system and secondary clinical outcomes are different from other *Chlamydia* species [86, 87].

During active replication, Cpn resides in membrane-bound vesicles (reticulate bodies) that are metabolically active and susceptible to antibiotic treatments. During unfavorable conditions, the reticulate bodies can relapse into elementary bodies that are resistant to antibiotics but are metabolically inactive. The precise biochemical mechanism through which Cpn persists as a chronic infection in a host cell environment remains unknown. Latency of the bacteria has been observed in both animal and cell culture models [88, 89]. A typical life cycle of Cpn is summarized in Fig. 19.2.

In terms of epidemiology, serological studies have suggested that at age 20, 50% of adults have been infected. Up to 80% of men and 70% of women late in life have been exposed to the pathogen [90, 91]. Considering the nature of a persistent infection and the high frequency of reinfection, it has been suggested that all individuals in a population have been exposed to the pathogen at some point in their life [92].

Symptoms of a Cpn infection are mild and continued coughing and malaise [93]. Treatments with the macrolide class antibiotics (azithromycin, erythromycin, and doxycycline) are effective only if they are administered early postinfection or over a long period of time [94]. Chronic infection is a possible outcome and the pathogen can disseminate to various vital organs via infections of monocytes and macrophages [95, 96]. Secondary complications include sinusitis, pneumonia, myocarditis, and endocarditis [87, 97–99].

The postulate that Cpn infection can cause atherosclerosis is supported by findings in three major areas: Clinical associations, *in vitro* molecular studies and *in vivo* infection of animal models. In 1988, Saikku et al. [100] identified

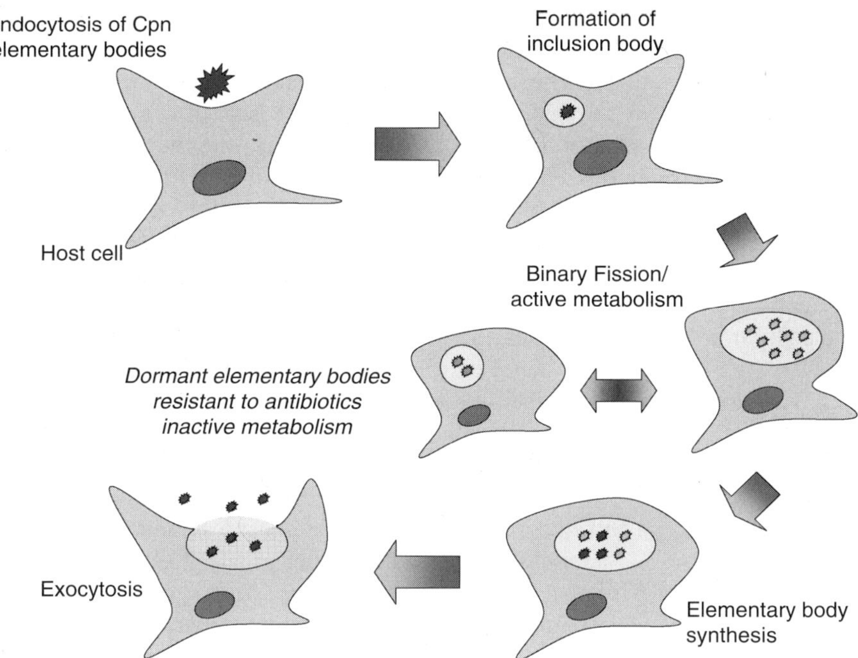

FIGURE19.2. The life cycle of a *Chlamydia pneumonia* infection. Elementary bodies enter a host cell by endocytosis. The bacteria multiply by binary fission and new reticulate bodies are formed. Reticulate bodies are active, and rely on nutrients and ATP extracted from the host cytoplasm for survival. Under unfavorable conditions, the bacteria may relapse into inactive elementary bodies, which can be resistant to antibiotics. Infection spreads by exocytosis of progeny elementary bodies to neighboring cells.

a significantly higher amount of Cpn antibody in the serum of patients with coronary artery disease. Subsequent work by Thom and coworkers [101–106] have demonstrated strong correlations between the detection of Cpn and coronary artery disease, with a 2–7-fold increase in risk. The presence of Cpn in coronary arterial tissues has been confirmed by electron microscopy, immunocytochemistry, and PCR techniques. Of note, Cpn has been found not only in fatty streak tissues (early stage atherosclerosis) and fibrous plaques (late stage), but also in macrophages and smooth muscle cells [107–109], both of which are important cell types in the composition of plaques. Recovery of cultivable live pathogen from atherosclerotic lesion was performed in 1996 [110]. Cpn infection is more likely to occur in plaques of patients who have undergone balloon angioplasty or stent operations than in native arterial lesions [111]. Unlike findings in the CMV-related atherosclerosis, Cpn is rarely found in nonatherosclerotic tissues [112–114]. Furthermore, Cpn antibody titers from patients strongly correlate with the occurrence of coronary heart disease [115] and Cpn can be found in 52% of atheromatous lesions but in only 5% of control samples. Thus, the available observations suggest a statistically significant role for Cpn infection in atherosclerosis. To further support the importance of Cpn in atherosclerotic lesions, recent *in vivo* and *in vitro* studies have provided insights into the biochemical pathways by which the pathogen may directly stimulate plaque formation.

Findings from *in vitro* studies have confirmed the ability of Cpn to infect vascular endothelial cells, smooth muscle cells, and macrophages [116, 117]. Persistent infection of human endothelial cells have been reported, and the replication of Cpn in endothelial cells can induce upregulation of vascular and cellular adhesion molecules, including ICAM-1, VCAM-1, and ELAM-1, all of which signify endothelial cell injury [118, 119]. In addition, inflammatory markers such as IL-6, IL-8, MCP-1, NF-κB, and plasminogen-activator–inhibitor-1 are upregulated in Cpn-infected vascular endothelial and smooth muscle cells [120, 121]. The presence of Cpn lipopolysaccharide (LPS) can be a strong stimulator of inflammatory responses in vascular endothelial cells via toll-like receptor-4 (TLR-4). At the same time, LPS can stimulate TNF-α, IL-1, and TF secretion, leading to a 2–20-fold increase in adhesion of monocytes and polymorphonuclear leukocytes to the endothelial surface [122, 123]. These molecular events are accompanied by an apparent increased efficiency at which neutrophils and monocytes can transmigrate through the Cpn-infected endothelial layer [124] into subendothelial space where monocytes can transform to foam cells and take up residence. *in vitro* experiments have verified that Cpn can transform the macrophage into a foam cell phenotype [125–127]. At the same time, the proximity of monocytes at the endothelial surface can in turn enhance the susceptibility of endothelial cells to Cpn infection [128].

Cpn also affect vascular smooth muscle function. Besides the proadhesive responses, Cpn infection of endothelial cells can also increase the proliferation of smooth muscle cells in the intimal layer [129]. Cpn can stimulate this proliferation via upregulation of endogenous heat shock protein 60 [96]. Of note, heat shock protein 60 is an important target of the autoimmune response that leads to enhanced atherosclerosis [130]. It has also been suggested that the Cpn-hsp60 can upregulate the expression of macrophage TNF-α and matrix metalloproteinases [131]. These data provide important cellular biochemical evidence of the pathways involved in the proatherogenic effects of Cpn. This lends further strength to the argument that Cpn is an important atherogenic agent. A schematic representation of the pathways that can be employed by Cpn infection to induce atherogenesis is presented in Fig. 19.3.

One of the limitations of these *in vitro* cellular studies is that the significance of these infection-mediated events in the context of the overall long-term *in vivo* atherogenesis remains unknown. To address this limitation, investigations using animal models of Cpn infection have provided some important data. The causative role of Cpn in human atherosclerosis is often

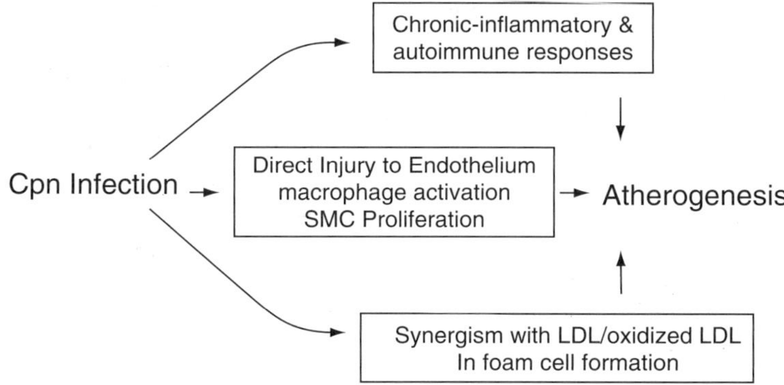

FIGURE 19.3. The three major pathways that *Chlamydia pneumonia* can contribute to atherogenesis: stimulation of immune and autoimmune responses (against molecules such as hsp60 and α-myosin), direct infection of the endothelial and smooth muscle cells at the lesion, and synergism with LDL and oxidized LDL in the formation of foam cells and the migration of monocytes into the lesion.

studied in tandem with low-density lipoprotein (LDL). LDL is a primary determinant in the atherogenic process. The development of fatty streaks and plaques involves the uptake of LDL by smooth muscle cells and monocytes. It is well known that hyperlipidemia is an important risk factor for human atherosclerosis [106]. In transgenic mouse models that are defective in LDL receptor expression and generate a high level of circulating LDL, Cpn infection can significantly enlarge the lesion area in the presence of a high-cholesterol diet [132]. The atherogenic effect appears to be Cpn dependent as heat-inactivated Cpn does not produce the same effect [133]. In addition, infection with Cpn induced significant atherosclerotic lesions in mice only when they ingested a cholesterol-enriched diet [134]. However, in rabbits, the Cpn infection alone was sufficient to stimulate significant atherosclerosis [135]. In those experiments, Fong et al. [135] noted that lesions in animals fed with low-cholesterol diets were not prominent to the naked eye but microscopically as significant as in animals fed with much higher cholesterol diet. They also demonstrated that atherosclerosis is amplified in the rabbit aorta with Cpn inoculations but not with *Mycoplasma pneumoniae*. This is consistent with other experiments that have shown *Chlamydia trachomatis* does not induce atherosclerotic lesions in the same setting that Cpn does [134]. All in all, these findings suggest that Cpn functions strongly as an atherogenic agent when in synergism with LDL but it may also induce initial atherosclerotic lesions on its own. Currently, there is no definitive biochemical knowledge regarding the molecular pathways that LDL and Cpn can interact to enhance plaque formation in human. While Cpn infection does enhance LDL uptake in macrophages *in vitro*, smooth muscle cells are also capable of transforming into foam cells. To date, there is no data that indicates Cpn infection can stimulate accumulation of LDL molecules within the subendothelial space.

Not all *in vivo* investigations identify Cpn as a causative agent in atherosclerosis. There are reports that fail to find any atherogenic effects of Cpn infection in mouse models [136, 137]. The presence of conflicting observations in animal models is likely due to different conditions in animal handling, efficiency of infection, and genetic factors that affect the overall process of lesion development. This brings up the possibility that components of the immune system from animal models can react differently to the LDL/Cpn infection (e.g., mouse and rabbits) and thus the development of plaques (or the lack of) in certain animals does not necessarily fully resemble atherogenesis in humans. Several thorough reviews on the strengths and weaknesses of various animal models used in the study of atherosclerosis and Cpn infection have been published recently [96, 138, 139]. In summary, although some differences have been identified in the literature, the *in vivo* data generally provides strong support for the role of Cpn and circulating lipids as strong risk factors for atherosclerosis.

The amount of *in vitro* and *in vivo* data on Cpn infection has provided significant impetus for the development of several clinical trials that investigated the potential for using antibiotics to treat preexisting atherosclerotic diseases. In the late 1990s, the effectiveness of azithromycin and roxithromycin in reducing cardiovascular events of Cpn-induced patients [140–142] was successfully demonstrated. Cardiovascular events were defined as nonfatal myocardial infarction, cardiovascular death, and recurrent angina. However, recent reports from trials involving larger patient population do not demonstrate observable differences between treated and nontreated groups over coronary events and cerebral events [143, 144]. These reports have cast some doubts on the role of Cpn as an important atherogenic agent. On the other hand, the effectiveness of administering antibiotics in eradicating persistent Cpn from patients may be questionable. Frequent reinfection of the pathogen from other humans is likely considering the presence of this pathogen in the general population. There have been reports that Cpn can be found in lesions of patients who are serologically negative of infection [145, 146], and yet serology is the main end point utilized in these clinical trials. In the mouse model, Cpn antibodies can recur within 3 weeks of apparent eradication [147]. Another trial identified protective effects of antibiotic therapy against clinical cardiovascular disease but only in subgroups with one or more additional risk factor present [148]. This would agree well with the Cpn/LDL synergism observations demonstrated in animal work [132, 134].

In summary, there are legitimate questions concerning the association of Cpn infection with atherosclerosis. Is it possible that the *in vitro* proliferative and proinflammatory events mediated by Cpn infection may not be significant compared to other factors during the clinical development of *in vivo* atherosclerosis over long period of time? Is it also possible that the immune responses in animals (whether transgenic or not) to Cpn infection can be different from those in the human arterial environment, leading to potentially misleading observations of Cpn-induced atherosclerosis? Is it possible or justified to develop a novel Cpn-specific treatment for atherosclerotic patients and coronary heart disease?

Thus, at the present time, a causative role for Cpn in atherosclerosis remains highly suggestive but not ultimately proven. As in the case with other potential atherogenic pathogens, more experiments are needed to generate a definitive verdict. To begin to settle the controversy, it may be necessary to employ an ex vivo organ vessel culture environment to study the biochemistry of Cpn infection under controlled conditions to more effectively understand the complex interactions among different cell types and the consequences in structural and morphological changes. Further, it may also be critical to isolate the host immune components (T cell, monocytes, and their cytokine secretions), all of which have significant but host-dependent atherogenic effects on their own. These components contribute to atherosclerosis in response to a wide range of stimulants and thus, their presence can mask the investigation of *in vivo* Cpn-specific biochemical pathways that lead to cell proliferation, vascular remodeling, and lesion formation. Understanding the biochemistry of Cpn "exclusive" pathways and effects may allow for the quantification and to determine the significance of Cpn in plaque formation in a host environment. Ultimately, it is only with knowledge such as this that we can develop and design new vaccines or more potent treatments against the persistent infection.

Acknowledgments: This work was supported by a grant from the Canadian Institutes for Health Research (CIHR). P. Cheung was a CIHR/Heart and Stroke Foundation of Canada IMPACT postdoctoral fellow. G. Pierce was a CIHR senior scientist.

References

1. Breslow JL: Cardiovascular disease burden increases, NIH funding decreases. Nat Med 3: 600–601, 1997.
2. Braunwald E: Shattuck lecture—cardiovascular medicine at the turn of the millennium: triumphs, concerns, and opportunities. N Engl J Med 337: 1360–1369, 1997.
3. Ross R: The pathogenesis of atherosclerosis: a perspective for the 1990s. Nature 362: 801–809, 1993.
4. Ross R, Glomset JA: Atherosclerosis and the arterial smooth muscle cell: proliferation of smooth muscle is a key event in the genesis of the lesions of atherosclerosis. Science 180: 1332–1339, 1973.
5. Griendling KK, Ushio-Fukai M, Lassegue B, Alexander RW: Angiotensin II signaling in vascular smooth muscle. New concepts. Hypertension 29: 366–373, 1997.
6. Kranzhofer R, Schmidt J, Pfeiffer CA, Hagl S, Libby P, Kubler W: Angiotensin induces inflammatory activation of human vascular smooth muscle cells. Arterioscler Thromb Vasc Biol 19: 1623–1629, 1999.

7. Hernandez-Presa M, Bustos C, Ortego M, Tunon J, Renedo G, Ruiz-Ortega M, Egido J: Angiotensin-converting enzyme inhibition prevents arterial nuclear factor-kappa B activation, monocyte chemoattractant protein-1 expression, and macrophage infiltration in a rabbit model of early accelerated atherosclerosis. Circulation 95: 1532–1541, 1997.
8. Schmidt AM, Yan SD, Wautier JL, Stern D: Activation of receptor for advanced glycation end products: a mechanism for chronic vascular dysfunction in diabetic vasculopathy and atherosclerosis. Circ Res 84: 489–497, 1999.
9. Baynes JW, Thorpe SR: Role of oxidative stress in diabetic complications: a new perspective on an old paradigm. Diabetes 48: 1–9, 1999.
10. Yudkin JS, Stehouwer CD, Emeis JJ, Coppack SW: C-reactive protein in healthy subjects: associations with obesity, insulin resistance, and endothelial dysfunction: a potential role for cytokines originating from adipose tissue? Arterioscler Thromb Vasc Biol 19: 972–978, 1999.
11. McCully KS: Vascular pathology of homocysteinemia: implications for the pathogenesis of arteriosclerosis. Am J Pathol 56: 111–128, 1969.
12. Nygard O, Nordrehaug JE, Refsum H, Ueland PM, Farstad M, Vollset SE: Plasma homocysteine levels and mortality in patients with coronary artery disease. N Engl J Med 337: 230–236, 1997.
13. Malinow MR: Plasma homocyst(e)ine and arterial occlusive diseases: a mini-review. Clin Chem 41: 173–176, 1995.
14. Berliner J, Leitinger N, Watson A, Huber J, Fogelman A, Navab M: Oxidized lipids in atherogenesis: formation, destruction and action. Thromb Haemost 78: 195–199, 1997.
15. Stemme S, Faber B, Holm J, Wiklund O, Witztum JL, Hansson GK: T lymphocytes from human atherosclerotic plaques recognize oxidized low density lipoprotein. Proc Natl Acad Sci USA 92: 3893–3897, 1995.
16. Libby P, Ridker PM, Maseri A: Inflammation and atherosclerosis. Circulation 105: 1135–1143, 2002.
17. Ross R: Atherosclerosis—an inflammatory disease. N Engl J Med 340: 115–126, 1999.
18. Gorelick PB: Stroke prevention therapy beyond antithrombotics: unifying mechanisms in ischemic stroke pathogenesis and implications for therapy: an invited review. Stroke 33: 862–875, 2002.
19. Crossman D: The future of the management of ischaemic heart disease. BMJ 314: 356–359, 1997.
20. Gyorkey F, Melnick JL, Guinn GA, Gyorkey P, DeBakey ME: Herpesviridae in the endothelial and smooth muscle cells of the proximal aorta in arteriosclerotic patients. Exp Mol Pathol 40: 328–339, 1984.
21. Yamashiroya HM, Ghosh L, Yang R, Robertson AL Jr: Herpesviridae in the coronary arteries and aorta of young trauma victims. Am J Pathol 130: 71–79, 1988.
22. Benditt EP, Barrett T, McDougall JK: Viruses in the etiology of atherosclerosis. Proc Natl Acad Sci USA 80: 6386–6389, 1983.
23. Melnick JL, Petrie BL, Dreesman GR, Burek J, McCollum CH, DeBakey ME: Cytomegalovirus antigen within human arterial smooth muscle cells. Lancet 2: 644–647, 1983.
24. Latsios G, Saetta A, Michalopoulos NV, Agapitos E, Patsouris E: Detection of cytomegalovirus, *Helicobacter pylori* and *Chlamydia pneumoniae* DNA in carotid atherosclerotic plaques by the polymerase chain reaction. Acta Cardiol 59: 652–657, 2004.

25. Nerheim PL, Meier JL, Vasef MA, Li WG, Hu L, Rice JB, Gavrila D, Richenbacher WE, Weintraub NL: Enhanced cytomegalovirus infection in atherosclerotic human blood vessels. Am J Pathol 164: 589–600, 2004.

26. Raza-Ahmad A, Klassen GA, Murphy DA, Sullivan JA, Kinley CE, Landymore RW, Wood JR: Evidence of type 2 herpes simplex infection in human coronary arteries at the time of coronary artery bypass surgery. Can J Cardiol 11: 1025–1029, 1995.

27. Paton P, Tabib A, Loire R, Tete R: Coronary artery lesions and human immunodeficiency virus infection. Res Virol 144: 225–231, 1993.

28. Hendrix MG, Daemen M, Bruggeman CA: Cytomegalovirus nucleic acid distribution within the human vascular tree. Am J Pathol 138: 563–567, 1991.

29. Koskinen PK, Nieminen MS, Krogerus LA, Lemstrom KB, Mattila SP, Hayry PJ, Lautenschlager IT: Cytomegalovirus infection and accelerated cardiac allograft vasculopathy in human cardiac allografts. J Heart Lung Transplant 12: 724–729, 1993.

30. Gabrylewicz B, Mazurek U, Ochala A, Sliupkas-Dyrda E, Garbocz P, Pyrlik A, Mroz I, Wilczok T, Tendera M: Cytomegalovirus infection in acute myocardial infarction. Is there a causative relationship? Kardiol Pol 59: 283–292, 2003.

31. Fabricant CG, Fabricant J, Minick CR, Litrenta MM: Herpesvirus-induced atherosclerosis in chickens. Fed Proc 42: 2476–2479, 1983.

32. Span AH, Grauls G, Bosman F, van Boven CP, Bruggeman CA: Cytomegalovirus infection induces vascular injury in the rat. Atherosclerosis 93: 41–52, 1992.

33. Carlquist JF, Muhlestein JB, Horne BD, Hart NI, Lim T, Habashi J, Anderson JG, Anderson JL: Cytomegalovirus stimulated mRNA accumulation and cell surface expression of the oxidized LDL scavenger receptor, CD36. Atherosclerosis 177: 53–59, 2004.

34. Kloover JS, Soots AP, Krogerus LA, Kauppinen HO, Loginov RJ, Holma KL, Bruggeman CA, Ahonen PJ, Lautenschlager IT: Rat cytomegalovirus infection in kidney allograft recipients is associated with increased expression of intracellular adhesion molecule-1 vascular adhesion molecule-1, and their ligands leukocyte function antigen-1 and very late antigen-4 in the graft. Transplantation 69: 2641–2647, 2000.

35. Martelius T, Salmi M, Wu H, Bruggeman C, Hockerstedt K, Jalkanen S, Lautenschlager I: Induction of vascular adhesion protein-1 during liver allograft rejection and concomitant cytomegalovirus infection in rats. Am J Pathol 157: 1229–1237, 2000.

36. Zhou YF, Yu ZX, Wanishsawad C, Shou M, Epstein SE: The immediate early gene products of human cytomegalovirus increase vascular smooth muscle cell migration, proliferation, and expression of PDGF beta-receptor. Biochem Biophys Res Commun 256: 608–613, 1999.

37. Rott D, Zhu J, Zhou YF, Burnett MS, Zalles-Ganley A, Epstein SE: IL-6 is produced by splenocytes derived from CMV-infected mice in response to CMV antigens, and induces MCP-1 production by endothelial cells: a new mechanistic paradigm for infection-induced atherogenesis. Atherosclerosis 170: 223–228, 2003.

38. Hillebrands JL, Dam JG, Onuta G, Klatter FA, Grauls G, Bruggeman CA, Rozing J: Cytomegalovirus-enhanced development of transplant arteriosclerosis in the rat; effect of timing of infection and recipient responsiveness. Transpl Int 18: 735–742, 2005.

39. Sorlie PD, Adam E, Melnick SL, Folsom A, Skelton T, Chambless LE, Barnes R, Melnick JL: Cytomegalovirus/herpesvirus and carotid atherosclerosis: the ARIC Study. J Med Virol 42: 33–37, 1994.

40. Hunt R, Thomson AB: Canadian *Helicobacter pylori* consensus conference. Canadian Association of Gastroenterology. Can J Gastroenterol 12: 31–41, 1998.

41. NIH Consensus Conference: *Helicobacter pylori* in peptic ulcer disease. NIH Consensus Development Panel on *Helicobacter pylori* in Peptic Ulcer Disease. JAMA 272: 65–69, 1994.

42. Mendall MA, Goggin PM, Molineaux N, Levy J, Toosy T, Strachan D, Camm AJ, Northfield TC: Relation of *Helicobacter pylori* infection and coronary heart disease. Br Heart J 71: 437–439, 1994.

43. Strachan DP, Mendall MA, Carrington D, Butland BK, Yarnell JW, Sweetnam PM, Elwood PC: Relation of *Helicobacter pylori* infection to 13-year mortality and incident ischemic heart disease in the caerphilly prospective heart disease study. Circulation 98: 1286–1290, 1998.

44. Sung JJ, Sanderson JE: Hyperhomocysteinaemia, *Helicobacter pylori*, and coronary heart disease. Heart 76: 305–307, 1996.

45. Prasad A, Zhu J, Halcox JP, Waclawiw MA, Epstein SE, Quyyumi AA: Predisposition to atherosclerosis by infections: role of endothelial dysfunction. Circulation 106: 184–190, 2002.

46. Kayo S, Ohsawa M, Ehara S, Naruko T, Ikura Y, Hai E, Yoshimi N, Shirai N, Tsukamoto Y, Itabe H, Higuchi K, Arakawa T, Ueda M: Oxidized low-density lipoprotein levels circulating in plasma and deposited in the tissues: comparison between *Helicobacter pylori*-associated gastritis and acute myocardial infarction. Am Heart J 148: 818–825, 2004.

47. Oshima T, Ozono R, Yano Y, Oishi Y, Teragawa H, Higashi Y, Yoshizumi M, Kambe M: Association of *Helicobacter pylori* infection with systemic inflammation and endothelial dysfunction in healthy male subjects. J Am Coll Cardiol 45: 1219–1222, 2005.

48. Sawayama Y, Ariyama I, Hamada M, Otaguro S, Machi T, Taira Y, Hayashi J: Association between chronic *Helicobacter pylori* infection and acute ischemic stroke: Fukuoka Harasanshin Atherosclerosis Trial (FHAT). Atherosclerosis 178: 303–309, 2005.

49. Shmuely H, Passaro DJ, Vaturi M, Sagie A, Pitlik S, Samra Z, Niv Y, Koren R, Harell D, Yahav J: Association of CagA+ *Helicobacter pylori* infection with aortic atheroma. Atherosclerosis 179: 127–132, 2005.

50. Franceschi F, Leo D, Fini L, Santoliquido A, Flore R, Tondi P, Roccarina D, Nista EC, Cazzato AI, Lupascu A, Pola P, Silveri NG, Gasbarrini G, Gasbarrini A: *Helicobacter pylori* infection and ischaemic heart disease: an overview of the general literature. Dig Liver Dis 37: 301–308, 2005.

51. Kalia N, Jones C, Bardhan DK, Reed MW, Atherton JC, Brown NJ: Effects of genotypically different strains of *Helicobacter pylori* on human microvascular endothelial cells *in vitro*. Dig Dis Sci 46: 54–61, 2001.

52. Position paper: epidemiology of periodontal diseases. American Academy of Periodontology. J Periodontol 67: 935–945, 1996.

53. Mattila KJ, Nieminen MS, Valtonen VV, Rasi VP, Kesaniemi YA, Syrjala SL, Jungell PS, Isoluoma M, Hietaniemi K, Jokinen MJ: Association between dental health and acute myocardial infarction. BMJ 298: 779–781, 1989.

54. Grau AJ, Buggle F, Ziegler C, Schwarz W, Meuser J, Tasman AJ, Buhler A, Benesch C, Becher H, Hacke W: Association between acute cerebrovascular ischemia and chronic and recurrent infection. Stroke 28: 1724–1729, 1997.

55. Syrjanen J, Peltola J, Valtonen V, Iivanainen M, Kaste M, Huttunen JK: Dental infections in association with cerebral infarction in young and middle-aged men. J Intern Med 225: 179–184, 1989.

56. Arbes SJ Jr, Slade GD, Beck JD: Association between extent of periodontal attachment loss and self-reported history of heart attack: an analysis of NHANES III data. J Dent Res 78: 1777–1782, 1999.

57. Haraszthy VI, Zambon JJ, Trevisan M, Zeid M, Genco RJ: Identification of periodontal pathogens in atheromatous plaques. J Periodontol 71: 1554–1560, 2000.

58. Chiu B: Multiple infections in carotid atherosclerotic plaques. Am Heart J 138: S534–S536, 1999.

59. Armitage GC: Periodontal infections and cardiovascular disease—how strong is the association? Oral Dis 6: 335–350, 2000.

60. Morrison HI, Ellison LF, Taylor GW: Periodontal disease and risk of fatal coronary heart and cerebrovascular diseases. J Cardiovasc Risk 6: 7–11, 1999.

61. DeStefano F, Anda RF, Kahn HS, Williamson DF, Russell CM: Dental disease and risk of coronary heart disease and mortality. BMJ 306: 688–691, 1993.

62. Joshipura KJ, Rimm EB, Douglass CW, Trichopoulos D, Ascherio A, Willett WC: Poor oral health and coronary heart disease. J Dent Res 75: 1631–1636, 1996.

63. Hujoel PP, Drangsholt M, Spiekerman C, DeRouen TA: Periodontal disease and coronary heart disease risk. JAMA 284: 1406–1410, 2000.

64. Wu T, Trevisan M, Genco RJ, Falkner KL, Dorn JP, Sempos CT: Examination of the relation between periodontal health status and cardiovascular risk factors: serum total and high density lipoprotein cholesterol, C-reactive protein, and plasma fibrinogen. Am J Epidemiol 151: 273–282, 2000.

65. Slade GD, Offenbacher S, Beck JD, Heiss G, Pankow JS: Acute-phase inflammatory response to periodontal disease in the US population. J Dent Res 79: 49–57, 2000.

66. Pussinen PJ, Jousilahti P, Alfthan G, Palosuo T, Asikainen S, Salomaa V: Antibodies to periodontal pathogens are associated with coronary heart disease. Arterioscler Thromb Vasc Biol 23: 1250–1254, 2003.

67. Lourbakos A, Yuan YP, Jenkins AL, Travis J, Andrade-Gordon P, Santulli R, Potempa J, Pike RN: Activation of protease-activated receptors by gingipains from *Porphyromonas gingivalis* leads to platelet aggregation: a new trait in microbial pathogenicity. Blood 97: 3790–3797, 2001.

68. Lalla E, Lamster IB, Hofmann MA, Bucciarelli L, Jerud AP, Tucker S, Lu Y, Papapanou PN, Schmidt AM: Oral infection with a periodontal pathogen accelerates early atherosclerosis in apolipoprotein E-null mice. Arterioscler Thromb Vasc Biol 23: 1405–1411, 2003.

69. Deshpande RG, Khan MB, Genco CA: Invasion of aortic and heart endothelial cells by *Porphyromonas gingivalis*. Infect Immun 66: 5337–5343, 1998.

70. Khlgatian M, Nassar H, Chou HH, Gibson FC, III, Genco CA: Fimbria-dependent activation of cell adhesion molecule expression in *Porphyromonas gingivalis*-infected endothelial cells. Infect Immun 70: 257–267, 2002.

71. Wang SP, Grayston JT: Immunologic relationship between genital TRIC, lymphogranuloma venereum, and related organisms in a new microtiter indirect immunofluorescence test. Am J Ophthalmol 70: 367–374, 1970.

72. Saikku P, Wang SP, Kleemola M, Brander E, Rusanen E, Grayston JT: An epidemic of mild pneumonia due to an unusual strain of *Chlamydia psittaci*. J Infect Dis 151: 832–839, 1985.

73. Grayston JT: *Chlamydia pneumoniae*, strain TWAR pneumonia. Annu Rev Med 43: 317–323, 1992.

74. Aldous MB, Grayston JT, Wang SP, Foy HM: Seroepidemiology of *Chlamydia pneumoniae* TWAR infection in Seattle families, 1966–1979. J Infect Dis 166: 646–649, 1992.

75. Mordhorst CH: Chlamydial Infections. Cambridge University Press, Cambridge, UK, 1986.

76. Mordhorst CH: Chlamydial Infections. Cambridge University Press, Cambridge, UK, 1990.

77. Kleemola M, Saikku P, Visakorpi R, Wang SP, Grayston JT: Epidemics of pneumonia caused by TWAR, a new *Chlamydia* organism, in military trainees in Finland. J Infect Dis 157: 230–236, 1988.

78. Pether JV, Wang SP, Grayston JT: *Chlamydia pneumoniae*, strain TWAR, as the cause of an outbreak in a boys' school previously called psittacosis. *Epidemiol Infect* 103: 395–400, 1989.

79. Chi EY, Kuo CC, Grayston JT: Unique ultrastructure in the elementary body of *Chlamydia* sp. strain TWAR. J Bacteriol 169: 3757–3763, 1987.

80. Kuo CC, Chi EY, Grayston JT: Ultrastructural study of entry of *Chlamydia* strain TWAR into HeLa cells. Infect Immun 56: 1668–1672, 1988.

81. Rockey DD, Lenart J, Stephens RS: Genome sequencing and our understanding of *Chlamydiae*. Infect Immun 68: 5473–5479, 2000.

82. Kalman S, Mitchell W, Marathe R, Lammel C, Fan J, Hyman RW, Olinger L, Grimwood J, Davis RW, Stephens RS: Comparative genomes of *Chlamydia pneumoniae* and *C. trachomatis*. Nat Genet 21: 385–389, 1999.

83. Gaydos CA, Quinn TC, Bobo LD, Eiden JJ: Similarity of *Chlamydia pneumoniae* strains in the variable domain IV region of the major outer membrane protein gene. Infect Immun 60: 5319–5323, 1992.

84. Molestina RE, Dean D, Miller RD, Ramirez JA, Summersgill JT: Characterization of a strain of *Chlamydia pneumoniae* isolated from a coronary atheroma by analysis of the *omp1* gene and biological activity in human endothelial cells. Infect Immun 66: 1370–1376, 1998.

85. Jackson LA, Campbell LA, Kuo CC, Rodriguez DI, Lee A, Grayston JT: Isolation of *Chlamydia pneumoniae* from a carotid endarterectomy specimen. J Infect Dis 176: 292–295, 1997.

86. Grayston JT, Wang SP, Kuo CC, Campbell LA: Current knowledge on *Chlamydia pneumoniae*, strain TWAR, an important cause of pneumonia and other acute respiratory diseases. Eur J Clin Microbiol Infect Dis 8: 191–202, 1989.

87. Grayston JT, Aldous MB, Easton A, Wang SP, Kuo CC, Campbell LA, Altman J: Evidence that *Chlamydia pneumoniae* causes pneumonia and bronchitis. J Infect Dis 168: 1231–1235, 1993.

88. Schachter J, Dawson C: Human Chlamydial Infections. PSG Publishing, Littleton, MA, 1978.

89. Beatty WL, Morrison RP, Byrne GI: Persistent *Chlamydiae*: from cell culture to a paradigm for chlamydial pathogenesis. Microbiol Rev 58: 686–699, 1994.

90. Wang S, Grayston JT: *Chlamydia pneumoniae* (TWAR) microimmunofluorescence antibody studies—1998 update. In: Stephens R, Christiansen G, Byrne B

et al. (eds), Chlamydial Infections. University of California, Berkley, CA, 1998, pp 155–158.

91. Wang S, Grayston J: Population prevalence antibody to *Chlamydia pneumoniae*, strain TWAR. In: Bowie W, Caldwell H, Jones R et al. (eds), Chlamydial Infections. Cambridge University Press, Cambridge, UK, 1990, pp 402–405.

92. Grayston JT: Background and current knowledge of *Chlamydia pneumoniae* and atherosclerosis. J Infect Dis 181(Suppl 3): S402–S410, 2000.

93. Thom DH, Grayston JT: Infections with *Chlamydia pneumoniae* strain TWAR. Clin Chest Med 12: 245–256, 1991.

94. Schachter J, Dawson C: Human Chlamydial Infections. PSG Publishing, Littleton, MA, 1978.

95. Balin BJ, Gerard HC, Arking EJ, Appelt DM, Branigan PJ, Abrams JT, Whittum-Hudson JA, Hudson AP: Identification and localization of *Chlamydia pneumoniae* in the Alzheimer's brain. Med Microbiol Immunol (Berl) 187: 23–42, 1998.

96. Hirono S, Pierce GN: Dissemination of *Chlamydia pneumoniae* to the vessel wall in atherosclerosis. Mol Cell Biochem 246: 91–95, 2003.

97. Odeh M, Oliven A: Chlamydial infections of the heart. Eur J Clin Microbiol Infect Dis 11: 885–893, 1992.

98. Marrie TJ, Harczy M, Mann OE, Landymore RW, Raza A, Wang SP, Grayston JT: Culture-negative endocarditis probably due to *Chlamydia pneumoniae*. J Infect Dis 161: 127–129, 1990.

99. Jones RB, Priest JB, Kuo C: Subacute chlamydial endocarditis. JAMA 247: 655–658, 1982.

100. Saikku P, Leinonen M, Mattila K, Ekman MR, Nieminen MS, Makela PH, Huttunen JK, Valtonen V: Serological evidence of an association of a novel *Chlamydia*, TWAR, with chronic coronary heart disease and acute myocardial infarction. Lancet 2: 983–986, 1988.

101. Thom DH, Grayston JT, Campbell LA, Kuo CC, Diwan VK, Wang SP: Respiratory infection with *Chlamydia pneumoniae* in middle-aged and older adult outpatients. Eur J Clin Microbiol Infect Dis 13: 785–792, 1994.

102. Thom DH, Wang SP, Grayston JT, Siscovick DS, Stewart DK, Kronmal RA, Weiss NS: *Chlamydia pneumoniae* strain TWAR antibody and angiographically demonstrated coronary artery disease. Arterioscler Thromb 11: 547–551, 1991.

103. Thom DH, Grayston JT, Wang SP, Kuo CC, Altman J: *Chlamydia pneumoniae* strain TWAR, *Mycoplasma pneumoniae*, and viral infections in acute respiratory disease in a university student health clinic population. Am J Epidemiol 132: 248–256, 1990.

104. Melnick SL, Shahar E, Folsom AR, Grayston JT, Sorlie PD, Wang SP, Szklo M: Past infection by *Chlamydia pneumoniae* strain TWAR and asymptomatic carotid atherosclerosis. Atherosclerosis Risk in Communities (ARIC) Study Investigators. Am J Med 95: 499–504, 1993.

105. Patel P, Mendall MA, Carrington D, Strachan DP, Leatham E, Molineaux N, Levy J, Blakeston C, Seymour CA, Camm AJ et al.: Association of *Helicobacter pylori* and *Chlamydia pneumoniae* infections with coronary heart disease and cardiovascular risk factors. BMJ 311: 711–714, 1995.

106. Dahlen GH, Boman J, Birgander LS, Lindblom B: Lp(a) lipoprotein, IgG, IgA and IgM antibodies to *Chlamydia pneumoniae* and HLA class II genotype in early coronary artery disease. Atherosclerosis 114: 165–174, 1995.

107. Kuo CC, Shor A, Campbell LA, Fukushi H, Patton DL, Grayston JT: Demonstration of *Chlamydia pneumoniae* in atherosclerotic lesions of coronary arteries. J Infect Dis 167: 841–849, 1993.

108. Kuo CC, Gown AM, Benditt EP, Grayston JT: Detection of *Chlamydia pneumoniae* in aortic lesions of atherosclerosis by immunocytochemical stain. Arterioscler Thromb 13: 1501–1504, 1993.

109. Shor A, Kuo CC, Patton DL: Detection of *Chlamydia pneumoniae* in coronary arterial fatty streaks and atheromatous plaques. S Afr Med J 82: 158–161, 1992.

110. Ramirez JA: Isolation of *Chlamydia pneumoniae* from the coronary artery of a patient with coronary atherosclerosis. The *Chlamydia pneumoniae*/ Atherosclerosis Study Group. Ann Intern Med 125: 979–982, 1996.

111. Campbell LA, O'Brien ER, Cappuccio AL, Kuo CC, Wang SP, Stewart D, Patton DL, Cummings PK, Grayston JT: Detection of *Chlamydia pneumoniae* TWAR in human coronary atherectomy tissues. J Infect Dis 172: 585–588, 1995.

112. Kuo CC, Grayston JT, Campbell LA, Goo YA, Wissler RW, Benditt EP: *Chlamydia pneumoniae* (TWAR) in coronary arteries of young adults (15–34 years old). Proc Natl Acad Sci USA 92: 6911–6914, 1995.

113. Muhlestein JB, Hammond EH, Carlquist JF, Radicke E, Thomson MJ, Karagounis LA, Woods ML, Anderson JL: Increased incidence of *Chlamydia* species within the coronary arteries of patients with symptomatic atherosclerotic versus other forms of cardiovascular disease. J Am Coll Cardiol 27: 1555–1561, 1996.

114. Thom DH, Grayston JT, Siscovick DS, Wang SP, Weiss NS, Daling JR: Association of prior infection with *Chlamydia pneumoniae* and angiographically demonstrated coronary artery disease. JAMA 268: 68–72, 1992.

115. Danesh J, Collins R, Peto R: Chronic infections and coronary heart disease: is there a link? Lancet 350: 430–436, 1997.

116. Quinn TC, Gaydos CA: *in vitro* infection and pathogenesis of *Chlamydia pneumoniae* in endovascular cells. Am Heart J 138: S507–S511, 1999.

117. Godzik KL, O'Brien ER, Wang SK, Kuo CC: *in vitro* susceptibility of human vascular wall cells to infection with *Chlamydia pneumoniae*. J Clin Microbiol 33: 2411–2414, 1995.

118. Kaukoranta-Tolvanen SS, Laitinen K, Saikku P, Leinonen M: *Chlamydia pneumoniae* multiplies in human endothelial cells *in vitro*. Microb Pathog 16: 313–319, 1994.

119. Kaukoranta-Tolvanen SS, Ronni T, Leinonen M, Saikku P, Laitinen K: Expression of adhesion molecules on endothelial cells stimulated by *Chlamydia pneumoniae*. Microb Pathog 21: 407–411, 1996.

120. Dechend R, Maass M, Gieffers J, Dietz R, Scheidereit C, Leutz A, Gulba DC: *Chlamydia pneumoniae* infection of vascular smooth muscle and endothelial cells activates NF-kappaB and induces tissue factor and PAI-1 expression: a potential link to accelerated arteriosclerosis. Circulation 100: 1369–1373, 1999.

121. Molestina RE, Miller RD, Ramirez JA, Summersgill JT: Infection of human endothelial cells with *Chlamydia pneumoniae* stimulates transendothelial migration of neutrophils and monocytes. Infect Immun 67: 1323–1330, 1999.

122. Bevilacqua MP, Pober JS, Wheeler ME, Cotran RS, Gimbrone MA Jr: Interleukin-1 activation of vascular endothelium. Effects on procoagulant activity and leukocyte adhesion. Am J Pathol 121: 394–403, 1985.

123. Bevilacqua MP, Pober JS, Wheeler ME, Cotran RS, Gimbrone MA Jr: Interleukin 1 acts on cultured human vascular endothelium to increase the

adhesion of polymorphonuclear leukocytes, monocytes, and related leukocyte cell lines. J Clin Invest 76: 2003–2011, 1985.

124. Molestina RE, Miller RD, Ramirez JA, Summersgill JT: Infection of human endothelial cells with *Chlamydia pneumoniae* stimulates transendothelial migration of neutrophils and monocytes. Infect Immun 67: 1323–1330, 1999.

125. Rutherford JD: *Chlamydia pneumoniae* and atherosclerosis. Curr Atheroscler Rep 2: 218–225, 2000.

126. Rosenfeld ME, Blessing E, Lin TM, Moazed TC, Campbell LA, Kuo C: *Chlamydia*, inflammation, and atherogenesis. J Infect Dis 181(Suppl 3): S492–S497, 2000.

127. Kalayoglu MV, Byrne GI: Induction of macrophage foam cell formation by *Chlamydia pneumoniae*. J Infect Dis 177: 725–729, 1998.

128. Lin TM, Campbell LA, Rosenfeld ME, Kuo CC: Monocyte-endothelial cell coculture enhances infection of endothelial cells with *Chlamydia pneumoniae*. J Infect Dis 181: 1096–1100, 2000.

129. Coombes BK, Mahony JB: *Chlamydia pneumoniae* infection of human endothelial cells induces proliferation of smooth muscle cells via an endothelial cell-derived soluble factor(s). Infect Immun 67: 2909–2915, 1999.

130. Xu Q, Wick G: The role of heat shock proteins in protection and pathophysiology of the arterial wall. Mol Med Today 2: 372–379, 1996.

131. Bachmaier K, Neu N, de la Maza LM, Pal S, Hessel A, Penninger JM: *Chlamydia* infections and heart disease linked through antigenic mimicry. Science 283: 1335–1339, 1999.

132. Liu L, Hu H, Ji H, Murdin AD, Pierce GN, Zhong G: *Chlamydia pneumoniae* infection significantly exacerbates aortic atherosclerosis in an LDLR–/– mouse model within six months. Mol Cell Biochem 215: 123–128, 2000.

133. Sharma J, Niu Y, Ge J, Pierce GN, Zhong G: Heat-inactivated *C. pneumoniae* organisms are not atherogenic. Mol Cell Biochem 260: 147–152, 2004.

134. Hu H, Pierce GN, Zhong G: The atherogenic effects of *Chlamydia* are dependent on serum cholesterol and specific to *Chlamydia pneumoniae*. J Clin Invest 103: 747–753, 1999.

135. Fong IW, Chiu B, Viira E, Jang D, Mahony JB: De Novo induction of atherosclerosis by *Chlamydia pneumoniae* in a rabbit model. Infect Immun 67: 6048–6055, 1999.

136. Caligiuri G, Rottenberg M, Nicoletti A, Wigzell H, Hansson GK: *Chlamydia pneumoniae* infection does not induce or modify atherosclerosis in mice. Circulation 103: 2834–2838, 2001.

137. Aalto-Setala K, Laitinen K, Erkkila L, Leinonen M, Jauhiainen M, Ehnholm C, Tamminen M, Puolakkainen M, Penttila I, Saikku P: *Chlamydia pneumoniae* does not increase atherosclerosis in the aortic root of apolipoprotein E-deficient mice. Arterioscler Thromb Vasc Biol 21: 578–584, 2001.

138. Saikku P, Laitinen K, Leinonen M: Animal models for *Chlamydia pneumoniae* infection. Atherosclerosis 140(Suppl 1): S17–S19, 1998.

139. Moghadasian MH, Frohlich JJ, McManus BM: Advances in experimental dyslipidemia and atherosclerosis. Lab Invest 81: 1173–1183, 2001.

140. Gupta S, Leatham EW, Carrington D, Mendall MA, Kaski JC, Camm AJ: Elevated *Chlamydia pneumoniae* antibodies, cardiovascular events, and azithromycin in male survivors of myocardial infarction. Circulation 96: 404–407, 1997.

141. Meier CR, Derby LE, Jick SS, Vasilakis C, Jick H: Antibiotics and risk of subsequent first-time acute myocardial infarction. JAMA 281: 427–431, 1999.
142. Gurfinkel E, Bozovich G, Daroca A, Beck E, Mautner B: Randomised trial of roxithromycin in non-Q-wave coronary syndromes: ROXIS Pilot Study. ROXIS Study Group. Lancet 350: 404–407, 1997.
143. Anderson JL, Muhlestein JB, Carlquist J, Allen A, Trehan S, Nielson C, Hall S, Brady J, Egger M, Horne B, Lim T: Randomized secondary prevention trial of azithromycin in patients with coronary artery disease and serological evidence for *Chlamydia pneumoniae* infection: The Azithromycin in Coronary Artery Disease: Elimination of Myocardial Infection with *Chlamydia* (ACADEMIC) study. Circulation 99: 1540–1547, 1999.
144. Vainas T, Stassen FR, Schurink GW, Tordoir JH, Welten RJ, van den Akker LH, Kurvers HA, Bruggeman CA, Kitslaar PJ: Secondary prevention of atherosclerosis through *Chlamydia pneumoniae* eradication (SPACE Trial): a randomised clinical trial in patients with peripheral arterial disease. Eur J Vasc Endovasc Surg 29: 403–411, 2005.
145. Campbell LA, O'Brien ER, Cappuccio AL et al.: Detection of *Chlamydia penumoniae* in atherectomy tissue from patients with symptomatic coronary artery disease. In: Orfila J, Byrne GL, Cherneskey MA et al. (eds), Chlamydial Infections. Societa Editrice Esculapio, Bologna, Italy, 1994, pp 212–215.
146. Maass M, Gieffers J, Krause E, Engel PM, Bartels C, Solbach W: Poor correlation between microimmunofluorescence serology and polymerase chain reaction for detection of vascular *Chlamydia pneumoniae* infection in coronary artery disease patients. Med Microbiol Immunol (Berl) 187: 103–106, 1998.
147. Malinverni R, Kuo CC, Campbell LA, Grayston JT: Reactivation of *Chlamydia pneumoniae* lung infection in mice by cortisone. J Infect Dis 172: 593–594, 1995.
148. Kinjo K, Sato H, Sato H, Ohnishi Y, Hishida E, Nakatani D, Mizuno H, Ohgitani N, Kubo M, Shimazu T, Akehi N, Takeda H, Hori M: Joint effects of *Chlamydia pneumoniae* infection and classic coronary risk factors on risk of acute myocardial infarction. Am Heart J 146: 324–330, 2003.

20

Transplant Arteriopathy: Role of Nitric Oxide Synthase

NANDINI NAIR, HANNAH VALANTINE, AND JOHN P. COOKE

Abstract

Endothelial vasodilator dysfunction is a major manifestation of transplant arteriopathy. This impairment reflects abnormalities in the production or activity of several endothelial vasoactive substances. Most notably, there is a significant deficit in the nitric oxide synthase (NOS) pathway. The impairment of the NOS pathway contributes to alterations in vascular reactivity, structure, and interaction with circulating factors. Since endothelium-derived NO suppresses vascular cell proliferation and vascular inflammation, a deficit in vascular NO facilitates the initiation and progression of transplant arteriopathy. The allograft endothelium is made dysfunctional by a number of factors including ischemia–reperfusion during transplantation; metabolic abnormalities posttransplantation including dyslipidemia, insulin resistance, and hypertension due to the use of immunosuppressive agents; the direct effects of some immunosuppressive agents on endothelial function; and infectious agents, most notably cytomegalovirus (CMV). This chapter focuses on factors adversely influencing endothelial dysfunction in transplant arteriopathy. The further delineation of the mechanisms by which the NOS pathway becomes dysregulated in transplant arteriopathy will be useful in the pursuit of new diagnostic and therapeutic modalities.

Keywords: asymmetric dimethylarginine (ADMA); cardiac transplant; cytomegalovirus (CMV); dimethylarginine dimethylaminohydrolase (DDAH); nitric oxide synthase (NOS); transplant arteriopathy

Abbreviations: ADMA, asymmetric dimethylarginine; DDAH, dimethylarginine dimethylaminohydrolase; FMVD, flow-mediated vasodilation; NO, nitric oxide; cGMP, cyclic GMP; eNOS, endothelial nitric oxide synthase; iNOS, inducible nitric oxide synthase; VSMC, vascular smooth muscle cell; CAT, cationic amino acid transporter; VCAM-1, vascular cell adhesion molecule; MCP-1, monocyte chemoattractant protein-1; L-NA, L-nitroarginine; CMV, cytomegalovirus

Introduction

Transplant arteriopathy is an accelerated form of arterial occlusive disease, and is the major cause of death in long-term survivors after heart transplantation [1]. Endothelial vasodilator dysfunction is prominently

435

observed in patients with transplant arteriopathy. This impairment reflects abnormalities in the production or activity of several endothelial vasoactive substances. Most notably, there is a significant deficit in the nitric oxide synthase (NOS) pathway. The impairment of the NOS pathway contributes to alterations in vascular reactivity, structure, and interaction with circulating blood elements. Since endothelium-derived NO suppresses vascular cell proliferation and vascular inflammation, a deficit in vascular NO facilitates the initiation and progression of transplant arteriopathy. This chapter focuses on identification of factors adversely influencing endothelial dysfunction in transplant arteriopathy. Understanding the mechanisms of endothelial dysfunction may lead to endothelial-targeted therapies for prevention of transplant arteriopathy.

Vascular Alterations in Transplant Arteriopathy

Transplant arteriopathy is characterized by vascular inflammation and intimal proliferation, and it ultimately results in luminal stenosis of epicardial branches, occlusion of smaller vessels, and myocardial infarction. Histopathological analysis has revealed that morphologic manifestation of transplant arteriopathy may range from concentric, diffuse intimal hyperplasia to fibrofatty plaques indistinguishable from spontaneously occurring atherosclerosis [2].

Immunologic and nonimmunologic factors likely influence the evolution and progression of transplant arteriopathy. Allograft coronary endothelial cells can serve as potent stimulators (antigen-presenting cells) as well as targets of allogeneic lymphocyte reactivity. T-cell-interaction with graft endothelial cells, initiates and sustains the chronic immune response injury [3–5]. Furthermore, a number of conditions occurring in the context of transplantation may cause alterations in endothelial function, and the expression of adhesion molecules and chemokines, which participate in the inflammatory process leading to a prothrombogenic state. These predisposing conditions include preservation injury, ischemia–reperfusion, acute rejection, T cell activation, antibody deposition and complement fixation, and viral infection. The ongoing inflammation is thought to accelerate the development of transplant arteriopathy.

A major contributor to the inflammation and vascular cell proliferation that characterizes transplant arteriopathy is endothelial dysfunction. The allograft endothelium is made dysfunctional by a host of factors including ischemia–reperfusion during transplantation; metabolic abnormalities posttransplantation including dyslipidemia, insulin resistance, and hypertension due to the use of immunosuppressive agents; the direct effects of some immunosuppressive agents on endothelial function; and infectious agents, most notably cytomegalovirus (CMV).

Evaluation of Endothelial Vasodilator Function of the Coronary Arteries

One of the important endothelial functions is its ability to modulate vessel tone. The healthy endothelium exerts a vasodilator influence by releasing a panoply of paracrine factors that relax vascular smooth muscle cell (VSMC), such as prostacyclin, endothelium-derived hyperpolarizing factor, and nitric oxide (NO). Endothelial vasodilator function of the epicardial coronary arteries is assessed by coronary angiography after intra-arterial infusions of acetylcholine or substance P. In addition, the endothelium can be stimulated to release vasodilator factors by increases in flow. The effect of increased flow can be assessed by infusing an endothelium-independent vasodilator (such as adenosine) downstream of the proximal epicardial coronary artery. During these interventions, coronary angiography permits measurement of the diameter of the epicardial coronary arteries. An increase in diameter of these conduit vessels is expected in response to stimulation of the endothelium.

The endothelial vasodilator function of the resistance vessels in the heart can be assessed by measuring coronary blood flow before and after intracoronary artery infusion of endothelium-dependent vasodilators such as substance P or acetylcholine. During these infusions, a Doppler flow wire is used to measure flow velocity, and thereby calculate coronary blood flow.

These approaches have been used to assess the endothelial function of the allograft coronary arteries. Endothelial dysfunction is commonly observed in the transplanted heart [6–8]. The etiology of endothelial dysfunction is multifactorial (Fig. 20.1) and time dependent. In the early months after transplant, epicardial vasodilatation is relatively preserved in response to tachycardia. During follow-up, exercise-induced flow-mediated endothelium-dependent vasodilation becomes impaired [9]. Some have estimated the prevalence of endothelial vasodilator dysfunction in the epicardial coronary arteries to be 20% to 30% of the patients during the first year and 30% to 40% at long-term follow-up [10]. This is almost certainly an underestimate, as endothelial dysfunction of the epicardial arteries was defined as a paradoxical vasoconstriction of >10% in response to acetylcholine [10]. The Stanford experience in recent years is that about 90% of cardiac transplant patients show endothelial vasodilator dysfunction, as defined by the absence of vasodilation or active vasoconstriction [11]. Epicardial endothelial dysfunction may be segmental in nature [12], which could cause a sampling error, and thereby account for some of the variability in reports of endothelial dysfunction. Furthermore, endothelium-dependent flow responses declined significantly (approximately 50%) in a 3-year follow-up period [13].

The severity of microvascular endothelial dysfunction does not correlate with vasodilator responses of the epicardial arteries suggesting independent determinants of the two processes [14, 15].

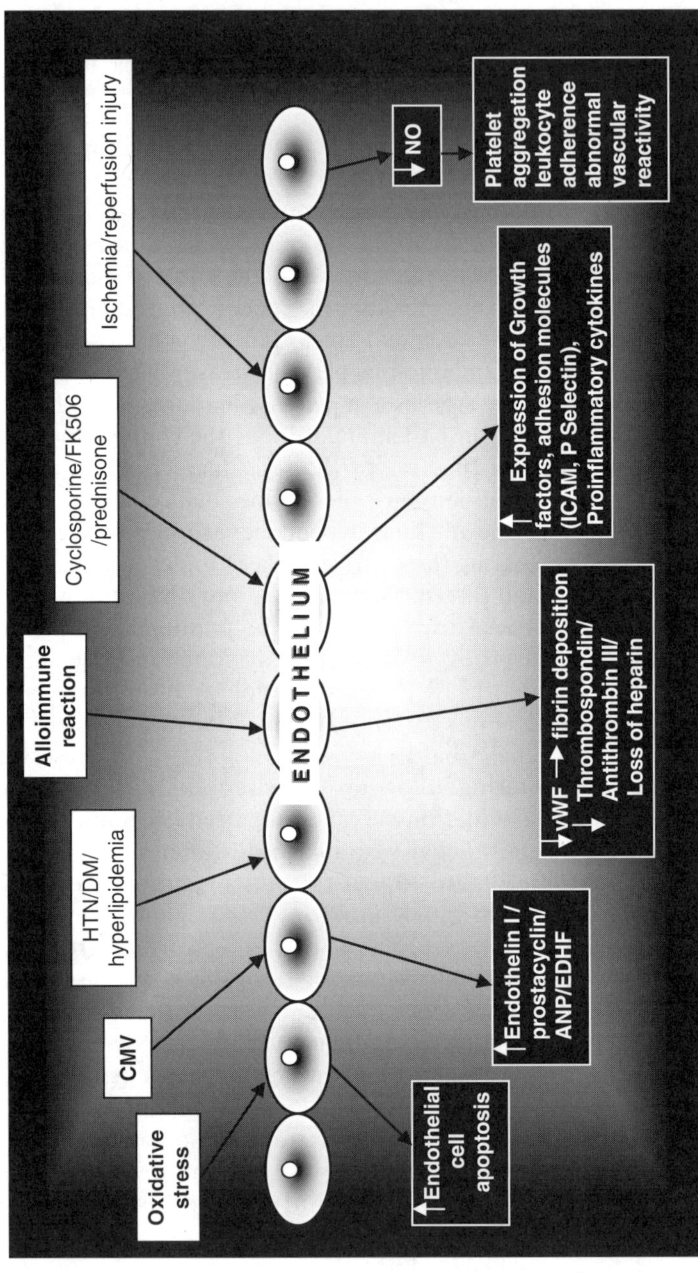

FIGURE 20.1. The multifactorial etiology of transplant arteriosclerosis: role of immunological and nonimmunological factors.

Endothelial Dysfunction and Loss of Vascular Homeostasis

This endothelial dysfunction observed in the cardiac catheterization laboratory may contribute to alterations in vascular reactivity and coronary blood flow that adversely affect the allograft heart. The endothelium is a major determinant of vascular tone and blood flow. It synthesizes a variety of vasodilator substances, such as NO, prostacyclin, atrial natriuretic peptide, endothelium-derived hyperpolarizing factor, and adrenomedullin. The shear stress of coronary blood flow is a primary stimulus for the endothelial release of these substances. Endothelial vasodilator dysfunction may be associated with reduced release of vasodilators and increased endothelial production of vasoconstrictors such as endothelin and angiotensin II. Notably, a loss of the homeostatic balance between endothelial vasodilators (which predominate in health) and endothelial vasoconstrictors (which are increased in a number of vascular disorders, most notably arteriosclerosis) may contribute to adverse changes in vascular structure as well as vascular reactivity. Endothelium-derived NO is paradigmatic of an endothelial factor that is vasoprotective as well as vasodilatory. We and others have provided evidence that a reduction in the synthesis or bioactivity of endothelium-derived NO promotes processes favoring coronary vascular disease [16–19].

The Vasoprotective Effects of NO

Our studies have focused on endothelium-derived NO as a paradigm of an endothelial factor that regulates coronary vessel tone, vessel structure, and interaction of the vessel with circulating blood elements. Endothelium-derived NO is the most potent endogenous vasodilator known NO induces vasodilation by stimulating soluble guanylate cyclase to produce cyclic GMP (cGMP). NO has a short half-life, and avidly interacts with sulfhydryl-containing proteins, heme proteins, and oxygen-derived free radicals. By virtue of its ability to nitrosylate proteins, it may change their activity or behavior [20]. The physiological importance of this endothelium-derived vasodilator is reflected by the significant increase in vascular resistance induced in animals and humans exposed to pharmacological antagonists of NOS [21].

Endothelium-derived NO also inhibits platelet adherence to the vessel wall. NO released into the lumen affects the behavior of circulating platelets. As platelets traverse the healthy myocardial microvasculature, they exhibit an elevation of cGMP, and a suppression of their aggregability. This effect of the healthy cardiac microvasculature can be suppressed by pharmacological antagonists of the NOS pathway. Furthermore, endothelium-derived NO inhibits leukocyte adherence to the vessel wall. An acute effect of NO to inhibit leukocyte adhesion is likely mediated by effects of NO on intracellular signaling of adhesion molecules [22]. A more chronic effect of NO is mediated by its suppression of specific adhesion molecules and chemokines. Finally, endothelium-derived NO also inhibits VSMC proliferation [23–25]. This is partly mediated by an effect of NO to increase VSMC apoptosis [26]. These observations indicate that NO is an endogenous antiatherogenic molecule.

Impairment of the NOS Pathway in Atherosclerosis

Atherosclerosis and transplant arteriopathy share some common pathophysiological processes, hence it is instructive to review the evidence for a role of NOS impairment in atherosclerosis. In animal models and in patients, endothelium-mediated vasodilation is impaired [27, 28]. The mechanism of impairment may include endothelial generation of superoxide anion and increased degradation of NO; elaboration of vasoconstrictor prostanoids and endothelin; reduced elaboration of prostacyclin; and/or impaired biosynthesis of NO [29–31]. With respect to the latter defect, impaired biosynthesis of NO may be due to alterations in NOS affinity for L-arginine; lipid-induced impairment of the high-affinity cationic amino acid transporter (CAT); reduced availability of the cofactor tetrahydrobiopterin; or increased levels of asymmetric dimethylarginine (ADMA), the competitive inhibitor of NOS. Our group and others have accumulated extensive data to indicate that O_2^- and ADMA are major determinants of endothelial vasodilator dysfunction induced by cardiovascular risk factors [32–35]. Whereas O_2^- degrades NO to reduce its bioactivity, ADMA inhibits NO synthesis. Moreover, recent data from our laboratory indicates that the elevations in O_2^- and ADMA are inextricably linked (Fig. 20.2).

Multiple lines of evidence point to a pathophysiological role for ADMA. We were the first to demonstrate that endothelial vasodilator dysfunction in hypercholesterolemic animals or humans could be reversed by administration of the NO precursor L-arginine [35, 36]. In patients with atherosclerotic or transplant coronary artery disease, the impairment of acetylcholine-induced

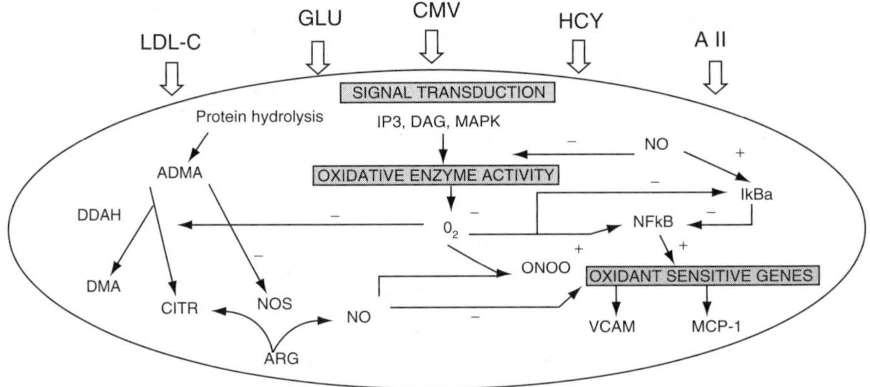

FIGURE 20.2. The effect of oxidative stress on intracellular signaling: the inextricable link between ADMA and oxidative stress.

vasodilation of the epicardial and coronary resistance vessels is reversed by an intravenous infusion of L-arginine. Several groups of investigators have observed an arginine-induced improvement in endothelium-dependent NO-mediated vasodilation in patients with native or transplant coronary artery disease [37–40]. Since the K_m of NOS for L-arginine is in a micromolar range [40], with plasma L-arginine levels at 50–100 µm, L-arginine should not be rate limiting. There are several possible explanations for this "arginine paradox," including effects of hypercholesterolemia and/or atherosclerosis upon NOS affinity; the effect of glutamine to inhibit activation of NOS; a reduction in the availability of the cofactor tetrahydrobiopterin; the colocalization to the caveolar membrane of NOS and CAT (the cationic amino acid transporter that mediates L-arginine influx) and reduced arginine transport due to lipid-induced impairment of the high-affinity CAT. Localization of NOS and CAT to the caveolar membrane may explain the sensitivity of the NOS pathway to reductions in extracellular L-arginine levels [41, 42]. However, there is accumulating data in preclinical and clinical studies, that ADMA is the major determinant of the "arginine paradox."

ADMA: An Endogenous Inhibitor of NO synthesis

ADMA is a competitive inhibitor of NOS. This modified amino acid is derived from the methylation of internal arginine residues in proteins and the subsequent hydrolysis of these proteins [34]. ADMA is not produced by methylation of free arginine, and is not affected by L-arginine intake [43, 44]. ADMA is excreted by the kidney or metabolized by dimethylarginine dimethylaminohydrolase (DDAH) to citrulline and dimethylamine [45]. Normal plasma levels are 0.5 to 1 µm [34]. Plasma levels of ADMA are elevated in a number of conditions associated with endothelial vasodilator dysfunction including renal failure, hypercholesterolemia, hyperhomocysteinemia, hypertension, diabetes mellitus, and heart failure [46–48]. In young hypercholesterolemic subjects, we have shown that plasma ADMA levels are doubled in association with an arginine-reversible impairment of flow-mediated vasodilation (FMVD) of the brachial artery.

It appears that the levels of ADMA observed in these conditions are sufficient to explain the impairment of endothelial function, and to explain the observations made repeatedly by multiple investigators that the endothelial dysfunction is reversible by administration of exogenous L-arginine [35–39, 49, 50]. Faraci et al. [51] found that the IC50 for ADMA was 2 µM. Added to crude purified preparations of eNOS, ADMA in a physiological/pathophysiological range (1 to 10 µm) induces a dose-dependent inhibition of NO synthesis. At the concentrations observed in hypercholesterolemic patients, ADMA inhibits NO biosynthesis. It should also be noted that the plasma level of ADMA is only an indirect reflection of tissue levels. In endothelial cells regenerating after vascular injury the intracellular levels of ADMA are three- to eightfold higher than plasma levels [52]. Furthermore, one must take into consideration the recent observation by Ignarro's group that in the setting of inflammation and atherogenesis, arginase II is induced in the vessel wall (unpublished communication). The same cytokines that cause the accumulation of ADMA also increase the expression of arginase II. Accordingly, vascular levels of L-arginine may be reduced under conditions where vascular levels of ADMA are increased. It is very likely that the arginine/ADMA ratio in the vessel wall is a regulator of NOS activity that becomes disordered in atherosclerosis, or with risk factors for atherosclerosis. In hypercholesterolemic rabbits, the plasma arginine/ADMA ratio is a better predictor of endothelial vasodilator dysfunction than is LDL cholesterol [44].

In patients with peripheral arterial disease, plasma ADMA levels are elevated two to fivefold and are correlated with clinical severity [53]. In a randomized placebo-controlled trial, intravenous administration of L-arginine (8 g twice daily) improved walking distance by 150%, significantly better than vehicle and the active control PGE1 [50]. In patients with critical limb ischemia the observed increase in limb blood flow is due to the conversion of the exogenous L-arginine to NO as reflected by parallel increases in urinary nitrogen oxides and urinary cGMP [53].

Oxidative Stress and Atherosclerosis

A balance exists between O_2^- and NO production in the endothelial cell. Risk factors, such as hypercholesterolemia, impair this balance, favoring superoxide anion generation, which activates oxidant-sensitive transcriptional pathways that induce genes (e.g., MCP-1 and VCAM) that are involved in atherogenesis (Fig. 20.2). NO inhibits the generation of superoxide anion by activated neutrophils by nitrosylation or nitrosation of NADPH oxygenase [54]. We have observed that flow-stimulated endothelial cells produced more NO, which was associated with reduced superoxide anion elaboration and suppressed expression of vascular cell adhesion molecule (VCAM-1). These effects of shear stress were blocked by inhibitors of NO synthesis [55]. We have shown that NO regulates oxidant-sensitive gene expression *in vivo*. Hypercholesterolemia increases monocyte chemoattractant protein-1 (MCP-1) expression in the thoracic aorta of the NZW rabbit, an effect that is attenuated by increasing vascular NO synthesis by administration of L-arginine. By contrast, inhibition of NO generation by the NOS antagonist L-nitroarginine (L-NA) markedly increases expression of MCP-1 in the rabbit aorta. Chronic dietary manipulation of NO synthesis affects the progression of lesion formation in hypercholesterolemic animals. Administration of L-arginine to fat-fed NZW rabbits for 10 weeks markedly inhibits plaque progression. Our preclinical studies suggest that NO is a potent endogenous antiatherogenic molecule, suppressing key processes in atherosclerosis.

Alterations in gene expression as well as NOS activity may reduce the vaso-protective influence of this pathway. For example, certain endothelial nitric oxide synthase (eNOS) gene polymorphisms appear to be predictive of coronary artery disease. The eNOS gene Glu298Asp polymorphism has been implicated in the occurrence of variant angina, essential hypertension, and acute myocardial infarction [56, 57].

ADMA and Vascular Disease

Hyperlipidemia, hyperhomocysteinemia, hypertension, and hyperglycemia are all conditions that are associated with the posttransplant state, due to the use of cyclosporine and steroids for immunosuppression. We have studied the effect of each of these conditions on the NOS pathway of humans, with particular reference to the role of ADMA. Studies in individuals who only had one of these risk factors have revealed that, in each of these conditions, the observed endothelial vasodilator dysfunction is associated with elevations in ADMA [58]. These observations yield the provocative hypothesis that ADMA may be a common pathophysiological pathway by which these risk factors cause endothelial dysfunction, and initiate atherogenesis.

DDAH is the enzyme that degrades ADMA to citrulline and dimethylamine [59]. The increase in ADMA accumulation induced by metabolic perturbations is temporally related to a decline in DDAH activity [58]. The decline in DDAH activity appears to be due to an increase in endothelial oxidative stress induced by cardiovascular risk factors. The sensitivity of DDAH to oxidative stress is conferred by a sulfhydryl group in its active site. Oxidation of this sulfhydryl impairs activity of the enzyme [60]. We have shown that homocysteine mounts an oxidative attack on DDAH, forming a mixed disulfide and inactivating the enzyme [61]. This effect of homocysteine may contribute to its impairment of endothelial vasodilator function, and promotion of vascular disease. Indeed, we have shown that an oral methionine challenge increases plasma ADMA and plasma homocysteine levels in human subjects, an effect temporally related to a decline in endothelial vasodilator function [62]. A recent study also revealed that tissue ADMA is elevated in human atheroma [63].

In another study of Japanese patients undergoing duplex ultrasonography of the carotid artery, plasma levels of ADMA were positively correlated to age ($p<0.0001$), mean arterial pressure ($p<0.0001$), and glucose ($p<0.0006$). ADMA levels were better correlated to IMT of the carotid artery ($r = 0.51$, $p<0.0001$) than all traditional risk factors except age in a multivariate analysis. The correlation between ADMA and IMT remained significant after adjusting for age ($r = 0.33$, $p = 0.0003$) [64]. This finding has recently been confirmed and extended by Zoccali et al. [65] who studied 225 individuals with end-stage renal disease. In a mean follow up of 33.4 months of these patients, they found that an elevation in ADMA level was the strongest predictor of vascular events, with those in the upper quintile of plasma ADMA level having an odds ratio greater than 10.

In animal models of "response to vascular injury," balloon angioplasty denudes the overlying endothelium, and causes vascular smooth muscle migration and proliferation resulting in myointimal hyperplasia. Intriguingly, under these circumstances, regenerating endothelial cells have higher intracellular levels of ADMA. The regenerating endothelium manifests reduced vasodilatory capacity [27]. Notably, the level of intracellular endothelial ADMA is directly related to the resulting intimal thickening of the injured vessel [64].

In summary, these studies implicate ADMA in endothelial vasodilator dysfunction and the pathophysiology of atherosclerosis. The emergence of ADMA as a global cardiovascular risk marker has been summarized recently [58, 66].

Derangement of the NOS Pathway in Transplant Arteriopathy

The beneficial effect on transplant arteriosclerosis of vascular NO is not limited to that derived from eNOS. It has been shown that inhibition of iNOS activity in the aortic allograft significantly increases intimal hyperplasia at 4 weeks [67]. The protective role for iNOS was confirmed by Koglin and coworkers [68]. The role of inducible nitric oxide synthase (iNOS) in vascular pathology is controversial. Because iNOS has a rate constant 1000-fold greater than eNOS, arginine may become depleted, particularly in states of inflammation where arginase is expressed. Under such circumstances, iNOS produces superoxide anion as well as NO. NO and superoxide anion rapidly react to form peroxynitrite anion, which is a highly reactive free radical that can be cytotoxic. However, when arginine is not rate limiting, the product of iNOS is likely to be predominantly NO. Under these circumstances, iNOS could be protective, explaining the results obtained with the iNOS deficient animals. Endothelial NOS has a well-established protective role in the endothelium. In a murine chronic rejection model, eNOS-deficient aortic allografts developed significantly worse arteriosclerosis compared with controls. Aortic allografts from eNOS knockout had a significant increase in intima/media ratios compared to those obtained from wild-type and iNOS knockout mice [69].

eNOS Dysfunction in Cardiac Transplant Recipients

Coronary endothelial vasodilator dysfunction is a common finding in cardiac transplant recipients that represents an early marker for the development of intimal thickening and graft atherosclerosis. Our group and others indicate that dysfunction of the NOS pathway contributes to transplant arteriopathy.

We tested the hypothesis that endothelial dysfunction precedes intimal thickening and that administration of L-arginine, the precursor of endothelium-derived relaxing factor, improves endothelial vasodilator function of coronary conduit and resistance vessels if given at an early stage of graft atherosclerosis [38].

In our studies in 18 cardiac transplant patients acetylcholine tended to elicit vasoconstriction in epicardial coronary arteries. Epicardial coronary vasoconstriction elicited by acetylcholine was attenuated by infusion of L-arginine and was observed predominantly in patients with normal intravascular ultrasound characteristics. Blood flow was significantly enhanced with L-arginine.

Early epicardial endothelial dysfunction (vasoconstriction to acetylcholine) predicted the development of intimal thickening as assessed by intravascular ultrasound in human heart transplant recipients at one-year posttransplantation [70]. Conversely, enhanced myocardial endothelin expression has been associated with coronary endothelial dysfunction in transplant patients [71]. Therefore, an imbalance between NO and endothelin bioactivity in the allograft may contribute to development of transplant arteriopathy.

The coronary vasculature of cardiac transplant recipients therefore exhibits a generalized endothelial dysfunction of conduit and resistance vessels which is improved by L-arginine consistent with the hypothesis that ADMA, the endogenous competitive inhibitor of NOS, plays a role in the derangement of the NOS pathway observed in cardiac transplant recipients.

Role of CMV in Transplant Endothelial Dysfunction

Human CMV, a member of the herpesviruses, can infect human vascular endothelial cells and induces changes relevant to atherogenesis [72]. Human CMV has been shown to be associated with transplant arteriopathy [73–75, 80]. There is increasing evidence that endothelial dysfunction plays a major role in CMV-induced transplant arteriosclerosis [81–83]. CMV infection increases expression of endothelial surface adhesion molecules, which upregulate the recruitment of leukocytes. CMV infection promotes mononuclear adhesion, activation, and transendothelial migration within the allograft vasculature and shifts the balance between endothelial factors mediating blood fluidity to a procoagulant state. The most direct evidence for a link between CMV and transplant arteriosclerosis was recently produced by our group. In this study, prophylactic treatment of cardiac transplant recipients with ganciclovir reduced the incidence of transplant arteriosclerosis [76]. Thus, a therapy directed towards CMV infection dramatically improved the outcome of patients after transplantation.

The mechanisms by which CMV may contribute to atherogenesis are incompletely defined. The immediate early gene of human CMV can code for a protein that has sequence homology and immunologic cross-reactivity with a domain of human leukocyte antigen-DR [74]. Additionally, CMV interferes with the action of p53, a protein that inhibits proliferation and induces apoptosis of VSMC. We propose that a major mechanism by which CMV could initiate and/or accelerate arteriosclerosis is by impairing the NOS pathway. Viral infections are known to impair endothelium-dependent vasodilation in humans contributing to atherosclerosis [77]. In the hypercholesterolemic mouse, infection with murine forms of chlamydia or herpesvirus accelerates plaque growth [78].

Studies by Weis et al. [79] suggest that CMV may impair endothelial function in part by elevating plasma levels of ADMA, the endogenous inhibitor of NOS. Weis et al. observed a doubling of serum ADMA levels in cardiac transplant patients as compared to control subjects. Higher ADMA levels were associated with greater likelihood of CMV infection (as detected by the presence of CMV DNA in leukocytes). *in vitro* experiments by Weis et al. [79] showed human microvascular endothelial cells infected with CMV produced increased levels of ADMA. Elevations in production of the superoxide anion were associated with reductions in the activity of DDAH, the enzyme that degrades ADMA. Reduced intracellular levels of cGMP in CMV infected cells reflected a reduction in the bioactivity of NO. Such an impairment of the function of the endothelial cells secondary to CMV infection could predispose patients to vascular disease. This hypothesis is consistent with the observations in experimental models that pharmacological or genetic inhibition of the NOS pathway accelerates atherogenesis, whereas increased levels of NO synthesis reduce vascular lesion formation [84–87].

Therapeutic Reduction of ADMA in Transplant Arteriopathy

Most recently, we have tested the role of ADMA in a murine model of transplant arteriopathy. We hypothesized that overexpression of DDAH would reduce plasma and tissue ADMA concentrations and thereby increase NO production [88]. Indeed, DDAH transgenic animals manifested an increase in plasma NOx levels and a reduction in systemic vascular resistance [88]. We further hypothesized that the increase in vascular NO production would have a vasoprotective effect. Specifically, we hypothesized that the transgenic mice would develop less transplant arteriopathy after cardiac transplantation.

To test this hypothesis, donor hearts of C-H-2^{bm12}KhEg (H-2^{bm12}) wild-type mice were heterotopically transplanted into C57BL/6 (H-2^{b}) transgenic mice overexpressing human DDAH-I, or transplanted into wild-type (WT) littermates. In some studies, the allografts were procured after 4 h of reperfusion (WT and DDAH-I recipients). In a second series, donor hearts were transplanted into DDAH-I transgenic or WT mice and procured 30 days after transplantation.

In DDAH-I recipients plasma ADMA concentrations were lower, in association with reduced myocardial generation of superoxide anion (WT vs. DDAH: 465.7±79.8 vs. 173.4±32.3 µM/mg/h; p = 0.02). Overexpression of DDAH was also associated with a reduction in several inflammatory cytokines, adhesion molecules, and chemokines in the cardiac tissue. In the allografts harvested after 30 days, transplant arteriopathy was markedly reduced in cardiac allografts of DDAH-I transgenic recipients (Fig. 20.3) as assessed by luminal narrowing (WT vs. DDAH: 79±2% vs. 33±7%; $p<0.01$), intima/media ratio (WT vs. DDAH: 1.1±0.1 vs. 0.5±0.1; $p<0.01$), and the percentage of diseased vessels (WT vs. DDAH: 100±0% vs. 62±10%; $p<0.01$). To conclude, overexpression of DDAH-I attenuated oxidative stress, inflammatory cytokines, and transplant arteriopathy in murine cardiac allografts. These murine studies show that ADMA is an important regulator of NO synthesis. Furthermore, a genetic reduction in plasma or tissue ADMA levels has a vasculoprotective effect, as manifested in this model by a reduction in transplant arteriopathy.

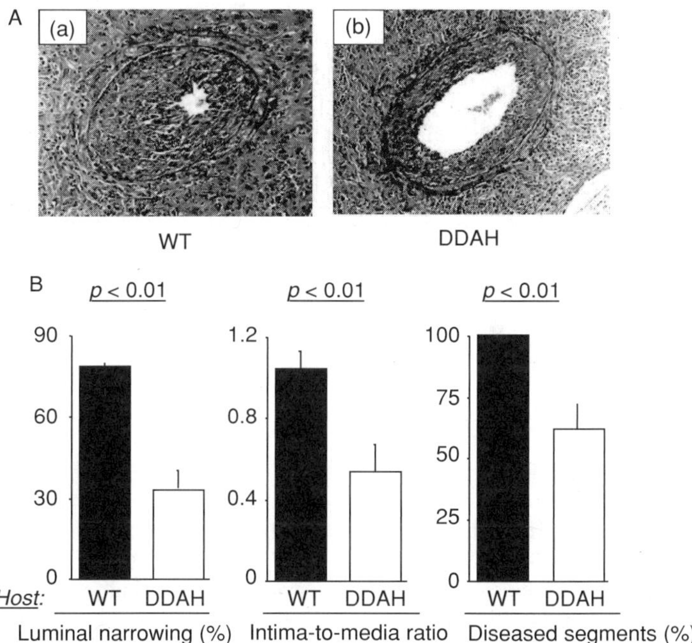

FIGURE 20.3. Representative sections of cardiac allografts stained with Elastica van Gieson for evaluation of coronary arteries in donor hearts transplanted into wild-type recipients (a) and into DDAH-I transgenic recipients (b). Note the marked fibrointimal thickening and luminal narrowing, morphologically resembling typical human GCAD, in donor hearts in wild-type recipients (a). In contrast, preserved vessel lumen and decreased luminal narrowing is observed in donor hearts in DDAH-I transgenic recipients (b). From Ref. [89] with permission.

Conclusion

In organ transplantation, allograft eNOS expression and activity can be impaired by preexisting arteriosclerotic disease in the graft, graft ischemia before transplantation, immunosuppressive agents such as cyclosporin A and tacrolimus, classical risk factors (hyperlipidemia, hypertension, diabetes, hyperhomocysteinemia) and possibly, infectious diseases such as CMV. We have substantial preliminary data indicating that there is an impairment of the NOS pathway in cardiac transplant recipients, and that this impairment is mediated by ADMA, the endogenous inhibitor of NOS. We furthermore have evidence that a number of metabolic disturbances common to transplant recipients may disrupt the NOS pathway by increasing the production of ADMA and superoxide anion. The increase in plasma ADMA observed in patients with cardiac allografts may contribute to transplant arteriopathy. The further delineation of the mechanisms by which the NOS pathway becomes dysregulated in transplant arteriopathy will be useful in the pursuit of new diagnostic and therapeutic modalities.

Acknowledgments and Disclosures: This study was supported by grants from the National Heart, Lung, and Blood Institute (R01 HL-63685; RO1 HL-75774; R01 CA098303; and PO1AI50153), the Tobacco Related Disease Research Program (11RT-0147), Philip Morris USA Inc. Dr. Cooke is the inventor of patents owned by Stanford University for diagnostic and therapeutic applications of the NOS pathway from which he receives royalties.

References

1. Hosenpud JD, Benell LE, Keck BM, Fiol B, Boucek MM, Novick RJ: The Registry of the International Society for Heart and Lung Transplantation: Sixteenth Official Report—1999. J Heart Lung Transplant 18: 611–626, 1999.
2. Billingham ME: Histopathology of graft coronary disease. J Heart Lung Transplant 11: S38–S44, 1992.
3. Weis M, VonScheidt W: Coronary atherosclerosis in the transplanted heart. Ann Rev Med 51: 81–100, 2000.
4. Hancock WW: Basic science aspects of chronic rejection: induction of protective genes to prevent development of transplant arteriosclerosis. Transplant Proc 30: 1585–1589, 1998.
5. Andersen H: Heart allograft vascular disease. An obliterative vascular disease in transplanted hearts. Atherosclerosis 142: 243–263, 1999.
6. Fish RD, Nabel EG, Selwyn AP, Ludmer PL, Mudge GH, Kirshenbaum JM, Schoen FJ, Alexander RW, Ganz P: Response of coronary arteries of transplant patients to acetylcholine. J Clin Invest 81: 21–31, 1988.
7. Mügge A, Heublein B, Kuhn B, Nolte C, Haverich A, Warnecke J, Forrsmann WG, Lichtlen PR: Impaired coronary dilator response to substance P and

impaired flow-dependent dilator responses in heart transplant patients with graft vasculopathy. J Am Coll Cardiol 21: 163–171, 1993.

8. Hartmann A, Weis M, Olbrich HG, Cieslinski G, Schacherer C, Burger W, Beyersdorf F, Schrader R: Endothelium-dependent and endothelium-independent vasomotion in large coronary arteries and in the microcirculation after cardiac transplantation. Eur Heart J 15: 1486–1493, 1994.

9. Aptecar E, Dupouy P, Benvenuti C, Mazzucotelli JP, Teiger E, Geschwind H, CastaigneA, Loisance D, Dubois-Rande JL: Sympathetic stimulation overrides flow-mediated endothelium-dependent epicardial coronary vasodilation in transplant patients. Circulation 94: 2542–2550, 1996.

10. Weis M, Pehlivali S, von Scheidt W: Heart allograft endothelial cell dysfunction. Cause, course, and consequences. Z Kardiol 89: IX/58–62, 2000.

11. Fearon WF, Nakamura M, Lee DP, Rezaee M, Vagelos RH, Hunt SA, Fitzgerald PJ, Yock PG, Yeung AC: Simultaneous assessment of fractional and coronary flow reserves in cardiac transplant recipients: Physiologic Investigation for Transplant Arteriopathy (PITA Study). Circulation 108: 1605–1610, 2003.

12. Weis M, Wolf WP, Mazilli N, Olbrich HG, Schacherer C, Weimer J, Burger W, Hartmann A: Variations of segmental endothelium dependent and endothelium independent vasomotor tone in the long term follow up after cardiac transplantation (qualitative changes in endothelial function). Am Heart J 134: 306–315, 1997.

13. Treasure CB, Vita JA, Ganz P, Ryan TJ Jr, Schoen FJ, Vekshtein VI, Yeung AC, Mudge GH, Alexander RW, Selwyn AP: Loss of the coronary microvascular response to acetylcholine in cardiac transplant patients. Circulation 86: 1156–1164, 1992.

14. Hollenberg SM, Tamburro P, Klein LW, Burns D, Easington C, Costanzo MR, Parrillo JE, Johnson MR: Discordant epicardial and microvascular endothelial responses in heart transplant recipients early after transplantation. J Heart Lung Transplant 17: 487–494, 1998.

15. Clausell N, Butany J, Molossi S, Lonn E, Gladstone P, Rabinovitch M, Daly PA: Abnormalities in intramyocardial arteries detected in cardiac transplant biopsy specimens and lack of correlation with abnormal intracoronary ultrasound or endothelial dysfunction in large epicardial coronary arteries. J Am Coll Cardiol 26: 110–119, 1995.

16. Maxwell AJ, Cooke JP: The role of nitric oxide in atherosclerosis. Coron Art Dis 10: 277–286, 1999.

17. Ignarro LJ, Napoli C: Novel features of nitric oxide, endothelial nitric oxide synthase, and atherosclerosis Curr Diab Rep 5: 17–23, 2005.

18. Cooke JP: Flow, NO, and atherogenesis. Proc Natl Acad Sci U S A 100: 768–770, 2003.

19. Davignon J, Ganz P: Role of endothelial dysfunction in atherosclerosis. Circulation 109 (23 Suppl 1): III27–III32, 2004.

20. Ignarro LJ, Burke TM, Wood KS, Wolin MS, Kadowitz PJ: Association between cyclic GMP accumulation and acetylcholine-elicited relaxation of bovine intra-pulmonary artery. J Pharmacol Exp Ther 228: 682–690, 1984.

21. Vallance P, Collier J, Moncada S: Effects of endothelium-derived nitric oxide on peripheral arteriolar tone in man. Lancet 2: 997–1000, 1989.

22. Tsao P, McEnvoy LM, Drexler H: Enhanced endothelial adhesiveness in hypercholesterolemia is attenuated by L-arginine. Circulation 89: 2176–2182, 1994.

23. Kubes P, Suzuki M, Granger DN: Nitric oxide: an endogenous modulator of leukocyte adhesion. Proc Natl Acad Sci U S A 88 (11): 4651–4655, 1991.

24. Tsao PS, Buitrago R, Chan JR, Cooke JP: Fluid flow inhibits endothelial adhesiveness. Nitric oxide and transcriptional regulation of VCAM-1. Circulation 94: 1682–1689, 1996.

25. Garg UC, Hassid A: Nitric oxide-generating vasodilators and 8-bromo-cyclic guanosine monophosphate inhibit mitogenesis and proliferation of cultured rat vascular smooth muscle cells. J Clin Invest 83: 1774–1777, 1989.

26. Tanner FC, Meier P, Greutert H, Champion C, Nabel EG, Luscher TF: Nitric oxide modulates expression of cell cycle regulatory proteins: a cytostatic strategy for inhibition of human vascular smooth muscle cell proliferation. Circulation 101: 1982–1989, 2000.

27. Weidinger FF, McLenachan JM, Cybulski MI, Gordon JB, Rennke HG, Hollenberg NK, Fallon JT, Ganz P, Cooke JP: Persistent dysfunction of regenerated endothelium after balloon angioplasty of rabbit iliac artery. Circulation 81: 1667–1679, 1990.

28. Ganz P, Vita JA: Testing endothelial vasomotor function: nitric oxide, a multipotent molecule. Circulation 108: 2049–2053, 2003.

29. Verbeuren TJ, Coene MC, Jordaens FH, VanHove CE, Zonnekeyan LL, Herman AG: Effect of hypercholesterolemia on vascular reactivity in the rabbit. II. Influence of treatment with dipyridamole on endothelium-dependent and endothelium-independent responses in isolated aortas of control and hypercholesterolemic rabbits. Circ Res 59: 496–504, 1986.

30. Ohara Y, Petersen TE, Harrison DG: Hypercholesterolemia increases endothelial superoxide anion production. J Clin Invest 91: 2546–2551, 1993.

31. Tesfamariam B, Cohen R: Free radicals mediate endothelial cell dysfunction caused by elevated glucose. Am J Physiol 263: H321–H326, 1992.

32. Lerman A, Holmes DR Jr, Bell MR, Garratt KN, Nishimura RA, Burnett JC Jr: Endothelin in coronary endothelial dysfunction and early atherosclerosis in humans. Circulation 92: 2426–2431, 1995.

33. Cosentino F, Katusic ZS: Tetrahydrobiopterin and dysfunction of endothelial nitric oxide synthase in coronary arteries. Circulation 91: 139–144, 1995.

34. Vallance P, Leone A, Calver A, Collier J, Moncada S: Endogenous dimethlarginine as an inhibitor of nitric oxide synthesis. J Cardiovasc Pharmacol 20: S60–S62, 1992.

35. Cooke JP, Andon NA, Girerd XJ, Hirsch AT, Creager MA: Arginine restores cholinergic relaxation of hypercholesterolemic rabbit thoracic aorta. Circulation 83: 1057–1062, 1991.

36. Creager MA, Gallagher SJ, Girerd XJ, Coleman SM, Dzau VJ, Cooke JP: L-Arginine improves endothelium-dependent vasodilation in hypercholesterolemic humans. J Clin Invest 80: 1248–1253, 1992.

37. Drexler H, Zeiher AM, Meinzer K, Just H: Correction of endothelial dysfunction in coronary microcirculation of hypercholesterolemic patients by L-arginine. Lancet 338: 1546–1550, 1991.

38. Drexler H, Fischell TA, Pinto FJ: Effect of L-arginine on coronary endothelial function in cardiac transplant recipients. Relation to vessel wall morphology. Circulation 89: 1615–1623, 1994.

39. Lerman A, Burnett JC Jr, Higano ST, McKinley LJ, Holmes DR Jr: Long-term L-arginine supplementation improves small-vessel coronary endothelial function in humans. Circulation 97: 2123–2128, 1998.

40. Pollock JS, Forstermann U, Mitchell JA, Warner TD, Schmidt HH, Nakane M, Murad F: Purification and characterization of particulate endothelium-derived relaxing factor synthase from cultured and native bovine aortic endothelial cells. Proc Natl Acad Sci U S A 88: 10480–10484, 1991.

41. Arnal JF, Munzel T, Venema RC, James NL, Bai CL, Mitch WE, Harrison DG: Interactions between L-arginine and L-glutamine change endothelial NO production. An effect independent of NO synthase substrate availability. J Clin Invest 95: 2565–2572, 1995.

42. McDonald KK, Zharikov S, Block ER, Kilberg MS: A caveolar complex between the cationic amino acid transporter 1 and endothelial nitric-oxide synthase may explain the "arginine paradox". J Biol Chem 272: 31213–31216, 1997.

43. Najbauer J, Johnson BA, Young AL, Aswad DW: Peptides with sequences similar to glycine, arginine-rich motifs in proteins interacting with RNA are efficiently recognized by methyltransferase(s) modifying arginine in numerous proteins. J Biol Chem 268: 10501–10509, 1993.

44. Bode-Boger S, Boger RH, Kienke S, Junker W, Frolich JC: Elevated L-arginine/dimethylarginine ratio contributes to enhanced systemic NO production by dietary L-arginine in hypercholesterolemic rabbits. Biochem Biophys Res Commun 219: 598–603, 1996.

45. MacAllister R, Parry H, Kimoto M: Regulation of nitric oxide synthesis by dimethylargnine dimethylaminohydrolase. Br J Pharmacol 119: 1533–1540, 1997.

46. Usui M, Matsuoka H, Miyazaki H, Ueda S, Okuda S, Imaizumi T: Increased endogenous nitric oxide synthase inhibitor in patients with congestive heart failure. Life Sci 62: 2425–2430, 1998.

47. Matsuoka H, Itoh S, Kimoto M, Kohno K, Tamai O, Wada Y, Yasukawa H, Iwami G, Okuda S, Imaizumi T: Asymmetrical dimethylarginine, an endogenous nitric oxide synthase inhibitor, in experimental hypertension. Hypertension 29: 242–247, 1997.

48. Vallance P, Leone A, Calver A, Collier J, Moncada S: Accumulation of an endogenous inhibitor of nitric oxide synthesis in chronic renal failure. Lancet 339: 572–575, 1992.

49. Clarkson P, Adams MR, Powe AJ, Donald AE, McCredie R, Robinson J, McCarthy SN, Keech A, Celermajer DS, Deanfield JE: Oral L-arginine improves endothelium-dependent dilation in hypercholesterolemic young adults. J Clin Invest 97: 1989–1994, 1996.

50. Boger R, Bode-Boger S, Thiele W: Restoring vascular nitric oxide formation by L-arginine improves the symptoms of intermittent claudication in patients with peripheral arterial occlusive disease. J Am Coll Cardiol 32: 1336–1344, 1998.

51. Faraci FM, Brian JE Jr, Heistad DD: Response of cerebral blood vessels to an endogenous inhibitor of nitric oxide synthase. Am J Physiol 269: H1522–H1527, 1995.

52. Azuma H, Sato J, Hamasaki H, Sugimoto A, Isotani E, Obayashi S: Accumulation of endogenous inhibitors for nitric oxide synthesis and decreased content of L-arginine in regenerated endothelial cells. Br J Pharmacol 116: 1001–1004, 1995.

53. Boger RH, Bode-Boger SM, Thiele W, Junker W, Alexander K, Frolich JC: Biochemical evidence for impaired nitric oxide synthesis in patients with peripheral arterial occlusive disease. Circulation 95: 2068–2074, 1997.

54. Clancy RM, Leszczynska-Piziak J, Abramson SB: Nitric oxide, an endothelial cell relaxation factor, inhibits neutrophil superoxide anion production via a direct action on NADPH oxidase. J Clin Invest 90: 1116–1121,1992.

55. Tsao PS, Buitrago R, Chan JR, Cooke JP: Fluid flow inhibits endothelial adhesiveness. Nitric oxide and transcriptional regulation of VCAM-1. Circulation 94: 1682–1689, 1996.

56. Shimasaki Y, Yasue H, Yoshimura M, Nakayama M, Kamitani S, Harada M, Ishikawa M, Kuwahara K, Ogawa E, Hamanaka I, Takahashi N, Kaneshige T, Teraoka H, Akamizu T, Azuma N, Yoshimasa Y, Yoshimasa T, Itoh H, Masuda Y, Yasue H, Nakao K: Association of the missense Glu298Asp variant of the endothelial nitric oxide synthase gene with myocardial infarction. J Am Coll Cardiol 31: 1506–1510, 1998.

57. Lacolley P, Gautier S, Poirier O, Pannier B, Cambien F, Benetos A: Nitric oxide synthase gene polymorphisms, blood pressure and aortic stiffness in normotensive and hypertensive subjects. J Hypertens 16: 31–35, 1998.

58. Cooke JP: Asymmetrical dimethylarginine: the Uber marker? Circulation 109: 1813–1818, 2004.

59. MacAllister RJ, Parry H, Kimoto M, Ogawa T, Russell RJ, Hodson H, Whitley GS, Vallance P: Regulation of nitric oxide synthesis by dimethylarginine dimethylaminohydrolase. Br J Pharmacol 119: 1533–1540, 1996.

60. Murray-Rust J, Leiper J, McAlister M, Phelan J, Tilley S, Santa Maria J, Vallance P, McDonald N: Structural insights into the hydrolysis of cellular nitric oxide synthase inhibitors by dimethylarginine dimethylaminohydrolase. Nat Struct Biol 8: 679–683, 2001.

61. Stuhlinger MC, Tsao PS, Her JH, Kimoto M, Balint RF, Cooke JP: Homocysteine impairs the nitric oxide synthase pathway: role of asymmetric dimethylarginine. Circulation 104: 2569–2575, 2001.

62. Stuhlinger MC, Oka RK, Graf EE, Schmolzer I, Upson BM, Kapoor O, Szuba A, Malinow MR, Wascher TC, Pachinger O, Cooke JP: Endothelial dysfunction induced by hyperhomocyst(e)inemia: role of asymmetric dimethylarginine. Circulation 108: 933–938, 2003.

63. Matsuoka H, Miyazaki H, Miyazaki H: Endogenous nitric oxide synthase inhibitor is elevated in human atheromatous lesion. Circulation 98: 1-116, 1998.

64. Miyazaki H, Matsuoka H, Cooke JP: Endogenous nitric oxide synthase inhibitor: a novel marker of atherosclerosis. Circulation 99: 1141–1146, 1999.

65. Zoccali C, Bode-Boger S, Mallamaci F, Benedetto F, Tripepi G, Malatino L, Cataliotti A, Bellanuova I, Fermo I, Frolich J, Boger R: Plasma concentration of asymmetrical dimethylarginine and mortality in patients with end-stage renal disease: a prospective study. Lancet 358: 2113–2117, 2001.

66. Fliser D: Asymmetric dimethylarginine (ADMA): the silent transition from an 'uraemic toxin' to a global cardiovascular risk molecule. Eur J Clin Invest 35: 71–79, 2005.

67. Shears LL, Kawaharada N, Tzeng E, Billiar TR, Watkins SC, Kovesdi I, Lizonova A, Pham SM: Inducible nitric oxide synthase suppresses the development of allograft arteriosclerosis. J Clin Invest 100: 2035–2042, 1997.

68. Koglin J, Glysing-Jensen T, Mudgett JS, Russell, MS: Exacerbated transplant arteriosclerosis in inducible nitric oxide-deficient mice. Circulation 97: 2059–2065, 1998.

69. Lee PC, Wang ZL, Qian S, Watkins SC, Lizonova A, Kovesdi I, Tzeng E, Simmons RL, Billiar TR, Shears LL II: Endothelial nitric oxide synthase protects aortic allografts from the development of transplant arteriosclerosis. Transplantation 69: 1186–1192, 2000.

70. Davis SF, Yeung AC, Meredith IT, Charbonneau F, Ganz P, Selwyn AP, Anderson TJ: Early endothelial dysfunction predicts the development of transplant coronary artery disease at 1 year posttransplant. Circulation 93: 457–462, 1996.

71. Weis M, Wildhirt SM, Schulze C, Rieder G, Wilbert-Lampen U, Wolf WP, Arendt RM, Enders G, Meiser BM, von Scheidt W: Endothelin in coronary endothelial dysfunction early after human heart transplantation. J Heart Lung Transplant 18: 1071–1079, 1999.

72. Epstein SE, Zhou YF, Zhu J: Infection and atherosclerosis: emerging mechanistic paradigms. Circulation 100: e20–e28, 1999.

73. Knight DA, Briggs BR, Bennett CF, Harindranath N, Waldman WJ, Sedmak DD: Attenuation of cytomegalovirus-induced endothelial intercellular adhesion molecule-1 mRNA/protein expression and T lymphocyte adhesion by a 2'-O-methoxyethyl antisense oligonucleotide. Transplantation 69:417–426, 2000.

74. Fujinami RS, Nelson JA, Walker L, Oldstone MB: Sequence homology and immunologic cross-reactivity of human cytomegalovirus with HLA-DR beta chain: a means for graft rejection and immunosuppression. J Virol 62: 100–105, 1988.

75. Valantine HA: The role of viruses in cardiac allograft vasculopathy. Am. J Trans 4: 169–177, 2004.

76. Valantine HA, Gao SZ, Menon SG, Renlund DG, Hunt SA, Oyer P, Stinson EB, Brown BW Jr, Merigan TC, Schroeder JS: Impact of prophylactic immediate post transplant ganciclovir on development of transplant atherosclerosis: a post hoc analysis of a randomized, placebo-controlled study. Circulation 100: 61–66, 1999.

77. Hingorani AD, Cross J, Kharbanda RK, Mullen MJ, Bhagat K, Taylor M, Donald AE, Palacios M, Griffin GE, Deanfield JE, MacAllister RJ, Vallance P: Acute systemic inflammation impairs endothelium-dependent dilatation in humans. Circulation 102: 994–999, 2000.

78. Alber DG, Powell KL, Vallance P, Goodwin DA, Grahme-Clarke C: Herpesvirus infection accelerates atherosclerosis in the apolipoprotein E-deficient mouse. Circulation 102: 779–785, 2000.

79. Weis M, Kledal TN, Lin KY, Panchal SN, Gao SZ, Valantine HA, Mocarski ES, Cooke JP: Cytomegalovirus infection impairs the nitric oxide synthase pathway: role of asymmetric dimethylarginine in transplant arteriosclerosis. Circulation 109: 500–505, 2004.

80. Grattan MT, Moreno-Cabral CE, Starnes VA, Oyer PE, Stinson EB, Shumway NE: Cytomegalovirus infection is associated with cardiac allograft rejection and atherosclerosis. JAMA 261: 3561–3566, 1989.

81. van Dam-Mieras MC, Muller AD, van Hinsbergh VW, Mullers WJ, Bomans PH, Bruggeman CA: The procoagulant response of cytomegalovirus infected endothelial cells. Thromb Haemost 68: 364–370, 1992.

82. Billstrom Schroeder M, Worthen GS: Viral regulation of RANTES expression during human cytomegalovirus infection of endothelial cells. J Virol 75: 3383–3390, 2001.

83. Levi M: CMV endothelitis as a factor in the pathogenesis of atherosclerosis. Cardiovasc Res 50: 432–433, 2001.

84. Cayatte AJ, Palacino JJ, Horten K, Cohen RA: Chronic inhibition of nitric oxide production accelerates neointima formation and impairs endothelial function in hypercholesterolemic rabbits. Arterioscler Thromb 14: 753–759, 1994.

85. Huang PL: Disruption of the endothelial nitric oxide synthase gene: effect on vascular response to injury. Am J Cardiol 82: 57S–59S, 1998.

86. Cooke J, Singer A, Tsao P: Anti-atherogenic effects of L-arginine in the hypercholesterolemic rabbit. J Clin Invest 90: 1168–1172, 1992.

87. von der Leyen HE, Gibbons GH, Morishita R, Lewis NP, Zhang L, Nakajima M, Kaeda Y, Cooke JP, Dzau VJ: Gene therapy inhibiting neointimal vascular lesion: *in vivo* transfer of endothelial cell nitric oxide synthase gene. Proc Natl Acad Sci U S A 92: 1137–1141, 1995.

88. Dayoub H, Achan V, Adimoolam S, Jacobi J, Stuehlinger M, Wang B, Tsao PS, Kimoto M, Vallance P, Patterson AJ, Cooke JP: DDAH regulates NO synthesis: genetic and physiological evidence. Circulation 108: 1043–1048, 2003.

89. Tanaka M, Sydow K, Gunawan F, Jacobi J, Tsao PS, Robbins RC, Cooke JP: DDAH overexpression suppresses graft coronary artery disease in murine cardiac allografts. Circulation 112: 1549–1556, 2005.

21

Insights into the Molecular Mechanisms of Plaque Rupture and Thrombosis

P.K. SHAH AND BEHROOZ SHARIFI

Abstract

Understanding cellular and molecular mechanisms that contribute to plaque instability and plaque rupture will continue to provide novel insights into prevention of atherosclerosis, plaque rupture, and thrombosis.

Ruptured plaques and, by inference, plaques at risk for rupture (vulnerable or unstable plaques) generally contain a large acellular lipid core with a reduced collagen content and fewer collagen synthesizing smooth muscle cells, a thinned out fibrous cap, intimal and adventitial inflammation with destruction of internal and external elastic membranes, enhanced plaque neovascularity, and outward remodeling of the vessel wall. Coronary culprit lesions responsible for 60–70% of acute coronary syndromes evolve from only mildly or moderately stenotic lesions suggesting that such less obstructive lesions may be more prone to plaque rupture and thrombosis. Inflammation is implicated not only in the initiation and progression of atherosclerosis but also appears to play a critical role in promoting plaque vulnerability, plaque rupture, and eventual thrombosis through its effect on matrix turnover and release of thrombogenic mediators.

Keywords: atherosclerosis; inflammation; plaque erosion; plaque rupture; thrombosis

Introduction

Cardiovascular disease is the leading cause of death for both men and women in the United States and much of the western world and is predicted to be the leading global killer by 2020 [1]. Coronary heart disease, most strokes, and limb ischemia are all caused by atherosclerosis. Some of its clinical manifestations result from progressive luminal narrowing by an atherosclerotic plaque with inward or constrictive remodeling and exaggerated or paradoxical vasoconstriction; however, it is the development of a thrombus over an underlying plaque that causes the most acute and serious clinical manifestations of atherosclerotic vascular disease. Coronary thrombosis, therefore, is responsible for the vast majority of cases of unstable angina, acute myocardial infarction, and ischemic sudden death [2–6].

Rupture of Atherosclerotic Plaque

A large number of studies have revealed that coronary thrombosis is the proximate cause for abrupt coronary occlusion leading to acute myocardial infarction, unstable angina, and many cases of sudden cardiac death [7–18]. Rupture of the fibrous cap is the underlying basis for 70–80% of coronary thrombi with extension of the thrombus into the plaque as well as into the lumen, and with propagation of the thrombus upstream from the site of cap rupture [14, 18–20]. Coronary stenoses produced by plaques with a ruptured fibrous cap and superimposed thrombus often produce a distinctive pattern on contrast angiography characterized as a "complex lesion." These lesions have eccentric stenoses bearing irregular or overhanging margins and lucencies or filling defects [9, 21].

Relationship to Plaque Size and Stenosis Severity

Plaques producing nonflow-limiting and less than severe stenoses have been shown to account for more cases of plaque rupture and thrombosis than plaques producing a more severe luminal diameter stenosis. In some cases a coronary angiogram performed weeks or months before the acute event had shown the culprit site to have <70% (often <50%) diameter narrowing [2, 15, 16, 21–27]. In line with the angiographic data, stress testing in stable coronary disease patients has shown that the site of ischemia on stress myocardial perfusion scintigraphy does not accurately predict the future site of acute myocardial infarction [28]. This apparent clinical and angiographic paradox may be attributed to several factors [29]: less stenotic plaques outnumber the more severely stenotic plaques, more stenotic plaques are likely to promote collaterals which protect from clinically overt manifestations of coronary occlusion, angiography underestimates stenosis severity, and finally less stenotic plaques may be more vulnerable to plaque rupture. In addition to plaque size, positive remodeling (outward expansion) versus negative remodeling (vessel shrinkage or contraction) can play an important role in determining the net effect of a plaque on lumen size. Outward remodeling of unstable or vulnerable plaques may minimize luminal encroachment despite large plaque size. Human studies using intravascular ultrasound have, in fact, shown that outward arterial expansion due to positive remodeling is more common at culprit lesion sites in unstable angina, whereas inward or negative remodeling is more common in stable angina [30–32]. Similarly, computer models show that larger lumens create greater circumferential stress on the fibrous caps, thereby increasing their likelihood of rupture [33]. Finally, recent histomorphometric data suggest that plaques with prominent outward remodeling, on an average, contain a larger lipid core and more inflammatory cells than plaques without outward remodeling [34]. These histological attributes are known to be more prevalent in ruptured plaques and by inference in plaques at risk for rupture (vulnerable plaques).

Relationship to Plaque Composition to Plaque Rupture

The concept of "plaque vulnerability" assumes that certain plaques have distinctive histological features present before the plaque ruptures (Table 21.1). Plaques that rupture tend to be large, to demonstrate outward or positive remodeling, have a large lipid core often occupying $\geq 40\%$ plaque volume, show inflammatory cell infiltration of the fibrous cap and adventitia, possess a thin fibrous cap depleted of collagen, glycoseaminoglycans, and smooth muscle cells and have increased adventitial and plaque neovascularity [2, 4, 30, 35–44].

The acellular lipid core is composed of free cholesterol, cholesterol crystals, and cholesterol esters derived from lipids that have infiltrated the arterial wall and also lipids derived from the death, by apoptosis or necrosis, of foam cells, mostly macrophages. Accumulation of large quantities of free cholesterol has been shown to induce macrophage apoptosis through the activation of endoplasmic reticulum mediated apoptotic gene program involving Caspases, which can be abrogated by partial deficiency of the Niemen–Pick disease type C gene; this enhanced apoptosis of macrophage–foam cells may thus contribute to expansion of the acellular lipid core [35, 45]. Recently it has been suggested that red cell membranes may contribute to expansion of the lipid core when intraplaque hemorrhage brings red cells into the plaque [46]. Such hemorrhage may occur from rupture of increased number of neovessels that are abundant in atherosclerotic plaques. A large eccentric lipid core could contribute to plaque instability by conferring a mechanical disadvantage to the plaque through redistribution of circumferential stress to the shoulder regions of the plaque where nearly 60% of plaque ruptures tend to occur [33, 36, 47–49]. Recent studies using genetic profiling have shown selective expression of a novel gene, perilipin, in ruptured human plaques. This is of considerable interest since perilipin inhibits lipid hydrolysis and could contribute to an accumulation of lipids in the core, thereby contributing to

TABLE 21.1. Phenotype of a plaque at risk for rupture.

- Large acellular lipid core ($\geq 40\%$ plaque volume)
- Thin fibrous cap depleted of collagen and smooth muscle cells
- Adventitial expansion with outward (positive) remodeling
- Inflammatory cell infiltration (monocyte–macrophages, activated T cells, dendritic cells, and degranulating mast cells)
    Of fibrous cap and around lipid core
    Adventitia
- Increased plaque neovascularity and hemorrhage

plaque vulnerability [50]. In addition, the lipid core contains prothrombotic, oxidized lipids and is impregnated with procoagulant tissue factor derived from apoptotic macrophages. These make the lipid core highly thrombogenic when exposed to circulating blood [51–55].

A number of studies have shown that ruptured plaques contain more inflammatory cells compared to intact plaques. These cells are mostly monocyte–macrophages, but also include activated T cells, dendritic cells, and activated degranulating mast cells expressing proteolytic enzymes tryptase and chymase. Inflammatory cells are often found adjacent to the sites of fibrous cap rupture and around the lipid core as well as in the adventitia around areas of neovascularization [38, 56–60]. Inflammatory cells are probably recruited into the atherosclerotic plaques by cell adhesion molecules, such as vascular cell adhesion molecule (VCAM)-1, and chemokines, such as monocyte chemoattractant protein (MCP)-1, interleukin-8, and eotaxin [61]. They are then retained and activated in the vessel wall through the activity of other cytokines such as the macrophage colony-stimulating factor (M-CSF). Another potential avenue for the entry and recruitment of inflammatory cells inside the atherosclerotic lesion may be through the adventitial neovasculature, which is enhanced in atherosclerosis [62–67]. It has also been suggested that inflammatory cells may be derived from either resident arterial or bone marrow-derived progenitor cells. Factors that contribute to recruitment and activation of inflammatory cells and the inflammatory response in atherosclerosis include oxidized lipids, cytokines, such as M-CSF, increased angiotensin II activity, elevated arterial pressure, diabetes, obesity, insulin resistance, smoking, chronic infections remote from the arterial wall, possible infectious organisms in the vessel wall (*Chlamydia pneumoniae*, cytomegalovirus, etc.) and activation of the immune system with release of proinflammatory mediators, such as interferon-γ (IFN-γ), CD40-ligand, etc., in response to antigens such as oxidized low-density lipoprotein (LDL), heat shock proteins (HSP), beta glycoprotein, and possibly others [4, 61]. In addition, a deficiency of natural anti-inflammatory molecules such as IL-10 and transforming growth factor beta (TGF-β) may also promote plaque inflammation and an unstable plaque phenotype [68–70]. Our laboratory has also identified a role for toll-like receptor 4 (TLR-4) and myeloid differentiation factor (MyD88) mediated innate immune signaling in the pathogenesis of experimental atherosclerosis [71, 72].

The Relationship between Plaque Inflammation and Plaque Rupture

Thinning of the fibrous cap is generally considered to be a prelude to rupture and is a sign of vulnerability. The cap protects the deeper components of the plaque from contact with circulating blood, but thins out in the vicinity of rupture. The fibrous cap is composed of matrix molecules, such as collagen, elastin, and proteoglycans, derived from smooth muscle cells. Fibrous caps from ruptured plaques contain less extracellular matrix (collagen and proteoglycans) and fewer smooth muscle cells than caps from intact plaques [39].

The depletion of matrix components has been shown by us and others to lead to cap thinning. Specifically fibrillar collagens from the fibrous cap are depleted due to an imbalance between synthesis and breakdown. This predisposes the fibrous cap to rupture, either spontaneously or in response to hemodynamic or other triggers [2, 4]. Enhanced matrix breakdown has been primarily attributed to a host of matrix-degrading metalloproteinases (MMPs) and other proteases, such as cathepsins and tryptase/chymase, which are expressed in atherosclerotic plaques by inflammatory cells (macrophages, foam cells, and mast cells), and to a lesser extent, by smooth muscle cells and endothelial cells [73–82]. These enzymes can degrade all components of the extracellular matrix and have been shown to be catalytically active both *in vitro* as well as *in vivo* [74, 83–85]. The activity of MMPs is tightly regulated at the level of gene transcription, and also by their secretion in an inactive zymogen form that requires extracellular activation and cosecretion of the tissue inhibitors of metalloproteinases (TIMPs) [84]. Thus, increased gene transcription, enhanced activation, and reduced activity of TIMPs can individually or together create an environment for increased matrix degradation.

All the components necessary for the activation of the MMP pathway exist in atherosclerotic plaques. Latent forms of MMPs can be activated by plasmin (produced by the plasminogen activator (uPA) from plasminogen by macrophages), trypsin, and chymase (derived from degranulating mast cells). Increased MMP production can be induced by oxidized lipids, reactive oxygen species, chlamydial HSP, CD40 ligation, inflammatory cytokines, tenascin-C derived from macrophages and hemodynamic stress [59, 79–80, 86–92]. In addition to MMPs, increased expression of cysteine and aspartate proteases of the cathepsin family, as well as reduced expression of their inhibitor, cystatin-c, in human atherosclerotic lesions, may also contribute to increased matrix breakdown in atherosclerosis [93, 94].

In addition to matrix degradation, matrix depletion may also result from reduced synthesis due to a decrease in the number of smooth muscle cells or a reduction in their synthetic function [38, 39]. The activated T cell-derived cytokine, IFN-γ, inhibits collagen gene expression in smooth muscle cells *in vitro*. This suggests that activated T cells in the plaque may inhibit matrix synthesis by producing IFN-γ. Several investigators have demonstrated increased smooth muscle cell death by apoptosis in human plaques, and several key players of the death-signaling pathway have been identified in atherosclerotic lesions [95–106]. Other stimuli that may induce smooth muscle cell death in atherosclerosis include oxidized lipids and the epidermal growth factor (EGF)-like domain of macrophage-derived tenascin-C (normally cryptic, but exposed when MMPs cleave the intact tenascin-C molecule) and apo C1-enriched high-density lipoprotein (HDL) [107, 108].

Role of Plaque Inflammation in Thrombosis

The thrombogenic components of the plaque include collagen and the lipid core and following rupture they are exposed to the circulation. The lipid core tends to be the most thrombogenic part of the plaque in part due to direct platelet activating effects of oxidized lipids but in large measure due to lipid core being impregnated with catalytically active tissue factor that activates the extrinsic clotting cascade leading to thrombin generation and thrombus formation at sites of plaque rupture [52, 54, 55]. The major source of tissue factor in the lipid core appears to be the apoptotic macrophage [106]. Thus inflammatory cells primarily contribute to plaque thrombogenicity by providing a source of tissue factor. Lipid ingestion, exposure to oxidized lipids, cytokines such as CD40-ligand, and other proinflammatory stimuli activate macrophages to produce tissue factor [4]. Apoptosis of endothelial cells, which has been shown to occur in response to hypochlorous acid produced through inflammatory cell-derived myeloperoxidase enzyme, may also contribute to the thrombogenicity of atherosclerotic lesions [109, 110].

Insights from Experimental Models of Plaque Rupture

Difficulties caused by a lack of resemblance to human disease have rendered the data generated from animal models difficult to interpret and despite numerous attempts no convincing and consistently reproducible animal model of spontaneous atherosclerotic plaque rupture and thrombosis is currently available. In the past, investigators have injected catecholamines, lipopolysaccharide (LPS), and Russel's viper venom to trigger thrombosis in rabbits with atherosclerosis [111], Rekhter and colleagues used a balloon incorporated in the arterial wall to study the role of lipid accumulation and macrophage infiltration on vulnerability to rupture [112]. However, these models bear little resemblance to human disease. Mouse models provide a similar story. Endothelin injections in apo E null mice have been shown to trigger acute myocardial necrosis, but coronary plaque rupture and thrombosis were not the underlying mechanism [113]. Recent research with apo E null mice revealed that there were frequent atherosclerotic lesions resembling vulnerable plaques in the innominate artery. Although intraplaque hemorrhage was observed, frank rupture and thrombosis were not demonstrated [114]. Other investigators have described findings suggestive of plaque rupture and thrombosis in the innominate artery of genetic variations of apo E null mice that were fed a lard-based high-fat diet [115].

Overexpression of MMP-1 has failed to produce plaque rupture in mice, and paradoxical reduction in atherosclerosis was actually observed with MMP-1 overexpression, raising some questions about the role of MMPs as the critical mediator of plaque rupture [116]. Recently, Calara et al. [117] reported findings suggestive of plaque rupture and thrombosis in apo E and LDL-receptor null mice, but the overall frequency was quite low. Von der

Thusen and colleagues reported evidence of plaque rupture in murine models of atherosclerosis in response to vasopressor infusion when the proapoptotic gene p53 was overexpressed locally in carotid plaques [118]. However, this model again suffers from the drawback that both p53 overexpression and pressor stimuli were required making it less of a model of spontaneous plaque rupture and thrombosis. Despite these limitations, search for a model of spontaneous plaque rupture and thrombosis continues.

Potential Role of Extrinsic Triggers in Plaque Rupture

Plaque rupture may follow events, such as extreme physical activity (especially in someone unaccustomed to regular exercise), severe emotional trauma, sexual activity, exposure to illicit drugs (cocaine, marijuana, amphetamines), exposure to cold, or acute infection [119–127]. However it may also occur spontaneously without obvious triggers.

While plaque rupture often leads to thrombosis with the clinical manifestations of an acute coronary syndrome, it may also occur without clinical manifestations (silent plaque rupture). In approximately 40–80% of cases of acute coronary syndrome, multiple plaque ruptures have been demonstrated in arterial segments remote from the acute culprit site [128]. The thrombotic response to a plaque rupture is probably regulated by the thrombogenicity of the exposed plaque constituents, the local hemorrheology (determined by the severity of underlying stenosis), shear-induced platelet activation, and also by systemic thrombogenicity and fibrinolytic activity [2, 4]. Lipid-rich plaques may be more thrombogenic than fibrous plaques, probably because of the high content of tissue factor in the lipid core [55]. The major source of tissue factor appears to be the macrophage. Apoptosis of macrophages may impregnate the lipid core with tissue factor-laden microparticles, making the lipid core highly thrombogenic [53]. Inflammatory cells, therefore, may be critical in influencing plaque thrombogenicity.

Recent studies in our laboratory have shown that plaques of smokers contain more tissue factor and inflammatory cells (macrophages) compared to nonsmokers, perhaps contributing to the high thrombotic risk in smokers [129]. Furthermore, coronary collaterals may also influence the clinical consequences of acute coronary occlusion. Several investigators have suggested that organization and healing at the site of plaque rupture and thrombosis may eventually lead to rapid progression of plaque and worsening of stenosis, thereby providing a mechanism for atherosclerosis progression [130].

Plaque Erosion and Calcified Nodules

Coronary thrombi have been observed overlying atherosclerotic plaques in 20–40% of cases, without rupture of the fibrous cap [2, 4, 13, 131, 132]. Such thrombi occur over plaques with superficial endothelial erosion. These erosions are particularly common in young victims of sudden death, in smokers, and in women. Plaques under such thrombi do not have a large lipid core, but rather a proteoglycan-rich matrix. The prevalence of inflammation is also lower than that in plaque rupture. The precise mechanisms of thrombosis in this scenario are unknown. It is conceivable that thrombosis in such cases is triggered by an enhanced systemic thrombogenic state (enhanced platelet aggregability, increased circulating tissue factor levels, depressed fibrinolytic state) [2, 4]. Activated circulating leukocytes may transfer active tissue factor by shedding microparticles and transferring them onto adherent platelets [133, 134]. It is possible that these circulating sources of tissue factor (rather than plaque-derived tissue factor) contribute to thrombosis at sites of superficial endothelial denudation such as in plaque erosion. In addition, endothelial cell apoptosis may also increase local thrombogenicity accounting for both endothelial denudation and thrombosis in plaque erosion [109, 110]. Furthermore, severe deficiencies of antithrombotic molecules, thrombomodulin, and protein-C receptor, on advanced atherosclerotic lesions may also contribute to thrombosis [135]. Erosion of a calcified nodule within an atherosclerotic plaque has also been reported as an uncommon cause of coronary thrombosis.

Implications for Plaque Stabilization through Change in Plaque Phenotype

Several angiographic studies have shown that risk factor modification leads to reduced new lesion formation, less lesion progression, and in some cases, actual regression. However, these studies have also shown that the magnitude of vaso-occlusive clinical event reduction is far greater than that accounted for by the relatively small changes in stenosis severity. This apparent discrepancy has led to the following hypotheses: (a) risk factor modification may induce plaque regression and reverse remodeling with little net change in stenosis severity; or (b) risk factor modification may not change plaque mass or stenosis severity, but might reduce the propensity for plaque rupture and thrombosis by changing the composition of the plaque. The latter possibility is referred to as "plaque stabilization" [2, 4, 136–141]. Studies in animals have in fact shown that lowering lipids through diet, statin therapy, or direct administration of apo A-I and HDL-like particles, can deplete lipids, reduce inflammation, sometimes reduce MMP and tissue factor levels, and increase the collagen content of atherosclerotic lesions [142–148]. Thus, plaque composition change can be achieved in animal models.

Similarly our laboratory has recently demonstrated that 3 months of therapy with pravastatin also favorably modifies human carotid plaque composition to a more stable phenotype, providing the first human data paralleling the results from animal models [149]. It can be postulated, therefore, that reducing lipids and inflammation in atherosclerotic plaques may help to lower the risk of plaque rupture and subsequent thrombosis. Also, such a plaque-stabilizing effect may account for the clinical benefits of risk factor modification by lifestyle changes and drug therapy (lipid-modifying drugs, angiotensin-converting enzyme, and angiotensin-II receptor blockers) [2, 4]. Future additional approaches may include direct administration of HDL and its apolipoproteins, and novel HDL-boosting compounds, such as inhibitors of cholesterol ester transfer protein (CETP), orally effective apo A-I mimetic peptides, PPAR agonists, and inhibitors of the endocannabinoid system such as rimonabant [150–155].

Summary

Atherosclerosis is a chronic immunoinflammatory disease characterized by lipid and matrix deposition, neoangiogenesis, inflammation and immune activation, vessel wall remodeling, and abnormal vasomotor regulation. Inflammatory/immune gene activation appears to be a common pathophysiologic underpinning in the evolution and progression of atherosclerosis. The natural history of atherosclerotic vascular disease is characterized by episodes of plaque rupture and superficial endothelial erosion leading to thrombus formation, which is the proximate event responsible for acute ischemic syndromes. Considerable evidence implicates inflammation in the process of plaque rupture and subsequent thrombosis with multiple risk factors serving as potential proinflammatory triggers. Risk factor modification appears to reduce acute vaso-occlusive events primarily by changing the plaque phenotype from one that is vulnerable to rupture into the one that is less prone to rupture. This process of plaque stabilizations represents a novel paradigm in atherosclerosis management.

Acknowledgment: Contributions of many colleagues and collaborators in the Atherosclerosis Research Center at Cedars Sinai Medical Center are also gratefully acknowledged.

References

1. Murray CJ, Lopez AD: Global mortality, disability and contribution of risk factors: Global Bureau of Disease Study. Lancet 349: 1436–1442, 1997.
2. Falk E, Shah PK, Fuster V: Coronary plaque disruption. Circulation 92: 657–671, 1995.
3. Virmani R, Burke AP, Farb A: Plaque rupture and plaque erosion. Thromb Haemat 82 (Suppl 1): 1–3, 1999.
4. Shah PK: Pathophysiology of coronary thrombosis: role of plaque rupture and plaque erosion. Prog Cardiovasc Dis 44: 357–368, 2002.
5. Libby P: Current concepts of the pathogenesis of the acute coronary syndromes. Circulation 104: 365–372, 2001.

6. DeWood MA, Spores J, Notske R, et al: Prevalence of total coronary occlusion during the early hours of transmural myocardial infarction. N Engl J Med 303: 897–902, 1980.

7. DeWood MA, Spores J, Hensley GR, et al: Coronary arteriographic findings in acute transmural myocardial infarction. Circulation 68: I39–I49, 1983.

8. DeWood MA, Stifter WF, Simpson CS, et al: Coronary arteriographic findings soon after non-Q-wave myocardial infarction. N Engl J Med 315: 417–423, 1986.

9. Levin DC, Fallon JT: Significance of the angiographic morphology of localized coronary stenoses: histopathologic correlations. Circulation 66: 316–320, 1982.

10. Sherman CT, Litvack F, Grundfest W, et al: Coronary angioscopy in patients with unstable angina pectoris. N Engl J Med 315: 913–919, 1986.

11. Kruskal JB, Commerford PJ, Franks JJ, Kirsch RE: Fibrin and fibrinogen-related antigens in patients with stable and unstable coronary artery disease. N Engl J Med 317: 1361–1365, 1987.

12. Folts JD: Platelet aggregation in stenosed coronary or cerebral arteries: a mechanism for sudden death? Wis Med J 79: 24–26, 1980.

13. Davies MJ, Thomas A: Thrombosis and acute coronary–artery lesions in sudden cardiac ischemic death. N Engl J Med 310: 1137–1140, 1984.

14. Falk E: Plaque rupture with severe pre-existing stenosis precipitating coronary thrombosis. Characteristics of coronary atherosclerotic plaques underlying fatal occlusive thrombi. Br Heart J 50: 127–134, 1983.

15. Ambrose JA, Winters SL, Stern A, et al: Angiographic morphology and the pathogenesis of unstable angina pectoris. J Am Coll Cardiol 5: 609–616, 1985.

16. Ambrose JA, Winters SL, Arora RR, et al: Coronary angiographic morphology in myocardial infarction: a link between the pathogenesis of unstable angina and myocardial infarction. J Am Coll Cardiol 6: 1233–1238, 1985.

17. Gorlin R, Fuster V, Ambrose JA: Anatomic–physiologic links between acute coronary syndromes. Circulation 74: 6–9, 1986.

18. Friedman M: Pathogenesis of coronary thrombosis, intramural and intraluminal hemorrhage. Adv Cardiol 4: 20–46, 1970.

19. Friedman M: The pathogenesis of coronary plaques, thromboses, and hemorrhages: an evaluative review. Circulation 52: III34–III40, 1975.

20. Constantinides P: Atherosclerosis—a general survey and synthesis. Surv Synth Pathol Res 3: 477–498, 1984.

21. Ambrose JA, Hjemdahl-Monsen C, Borrico S, et al: Quantitative and qualitative effects of intracoronary streptokinase in unstable angina and non-Q wave infarction. J Am Coll Cardiol 9: 1156–1165, 1987.

22. Ambrose JA, Winters SL, Arora RR, et al: Angiographic evolution of coronary artery morphology in unstable angina. J Am Coll Cardiol 7: 472–478, 1986.

23. Ambrose JA, Monsen C: Significance of intraluminal filling defects in unstable angina. Am J Cardiol 57: 1003–1004, 1986.

24. Little WC, Constantinescu M, Applegate RJ, et al: Can coronary angiography predict the site of a subsequent myocardial infarction in patients with mild-to-moderate coronary artery disease? Circulation 78: 1157–1166, 1988.

25. Hackett D, Davies G, Maseri A: Pre-existing coronary stenoses in patients with first myocardial infarction are not necessarily severe. Eur Heart J 9: 1317–1323, 1988.

26. Giroud D, Li JM, Urban P, Meier B, Rutishauer W: Relation of the site of acute myocardial infarction to the most severe coronary arterial stenosis at prior angiography. Am J Cardiol 69: 729–732, 1992.

27. Brown G, Albers JJ, Fisher LD, et al: Regression of coronary artery disease as a result of intensive lipid-lowering therapy in men with high levels of apolipoprotein B. N Engl J Med 323: 1289–1298, 1990.
28. Naqvi TZ, Hachamovitch R, Berman D, Buchbinder N, Kiat H, Shah PK: Does the presence and site of myocardial ischemia on perfusion scintigraphy predict the occurrence and site of future myocardial infarction in patients with stable coronary artery disease? Am J Cardiol 79: 1521–1524, 1997.
29. Shah PK: Plaque size, vessel size and plaque vulnerability: bigger may not be better. J Am Coll Cardiol 32: 663–664, 1998.
30. Schoenhagen P, Ziada KM, Kapadia SR, Crowe TD, Nissen SE, Tuzcu EM: Extent and direction of arterial remodeling in stable versus unstable coronary syndromes: an intravascular ultrasound study. Circulation 101: 598–603, 2000.
31. von Birgelen C, Klinkhart W, Mintz GS, et al: Plaque distribution and vascular remodeling of ruptured and nonruptured coronary plaques in the same vessel: an intravascular ultrasound study *in vivo*. J Am Coll Cardiol 37: 1864–1870, 2001.
32. Takano M, Mizuno K, Okamatsu K, Yokoyama S, Ohba T, Sakai S: Mechanical and structural characteristics of vulnerable plaques: analysis by coronary angioscopy and intravascular ultrasound. J Am Coll Cardiol 38: 99–104, 2001.
33. Loree HM, Kamm RD, Stringfellow RG, Lee RT: Effects of fibrous cap thickness on peak circumferential stress in model atherosclerotic vessels. Circ Res 71: 850–858, 1992.
34. Varnava AM, Mills PG, Davies MJ: Relationship between coronary artery remodeling and plaque vulnerability. Circulation 105: 939–943, 2002.
35. Guyton JR, Klemp KF: Development of the lipid-rich core in human atherosclerosis. Arterioscler Thromb Vasc Biol 16: 4–11, 1996.
36. Richardson PD, Davies MJ, Born GV: Influence of plaque configuration and stress distribution on fissuring of coronary atherosclerotic plaques. Lancet 2: 941–944, 1989.
37. Davies MJ, Richardson PD, Woolf N, Katz DR, Mann J: Risk of thrombosis in human atherosclerotic plaques: role of extracellular lipid, macrophage, and smooth muscle cell content. Br Heart J 69: 377–381, 1993.
38. Felton CV, Crook D, Davies MJ, Oliver MF: Relation of plaque lipid composition and morphology to the stability of human aortic plaques. Arterioscler Thromb Vasc Biol 17: 1337–1345, 1997.
39. Burleigh MC, Briggs AD, Lendon CL, Davies MJ, Born GV, Richardson PD: Collagen types I and III, collagen content, GAGs and mechanical strength of human atherosclerotic plaque caps: span-wise variations. Atherosclerosis 96: 71–81, 1992.
40. Moreno PR, Falk E, Palacios IF, Newell JB, Fuster V, Fallon JT: Macrophage infiltration in acute coronary syndromes. Implications for plaque rupture. Circulation 90: 775–778, 1994.
41. Moreno PR, Bernardi VH, Lopez-Cuellar J, et al: Macrophages, smooth muscle cells, and tissue factor in unstable angina. Implications for cell-mediated thrombogenicity in acute coronary syndromes. Circulation 94: 3090–3097, 1996.
42. Lendon CL, Davies MJ, Born GV, Richardson PD: Atherosclerotic plaque caps are locally weakened when macrophages density is increased. Atherosclerosis 87: 87–90, 1991.
43. Depre C, Havaux X, Wijns W. Neovascularization in human coronary atherosclerotic lesions. Cathet Cardiovasc Diagn 39: 215–220, 1996.

44. Tenaglia AN, Peters KG, Sketch MH Jr, Annex BH: Neovascularization in atherectomy specimens from patients with unstable angina: implications for pathogenesis of unstable angina. Am Heart J 135: 10–14, 1998.

45. Feng B, Yao PM, Li Y, Devlin CM, Zhang D, Harding HP, Sweeney M, Rong JX, Kuriakes G, Fisher EA, Marks AR, Ron D, Tabas I: The endoplasmic reticulum is the site of cholesterol-induced cytotoxicity in macrophages. Nat Cell Biol 5: 781–792, 2003.

46. Kolodgie FD, Gold HK, Burke AP, Fowler DR, Kruth HS, Weber DK, Farb A, Guerrero LJ, Hayase M, Kutys R, Narula J, Finn AV, Virmani R: Intraplaque hemorrhage and progression of atheroma. N Engl J Med 349: 2316–2325, 2003.

47. Cheng GC, Loree HM, Kamm RD, Fishbein MC, Lee RT: Distribution of circumferential stress in ruptured and stable atherosclerotic lesions. A structural analysis with histopathological correlation. Circulation 87: 1179–1187, 1993.

48. Loree HM, Tobias BJ, Gibson LJ, Kamm RD, Small DM, Lee RT: Mechanical properties of model atherosclerotic lesion lipid pools. Arterioscler Thromb 14: 230–234, 1994.

49. Huang H, Virmani R, Younis H, Burke AP, Kamm RD, Lee RT: The impact of calcification on the biomechanical stability of atherosclerotic plaques. Circulation 103: 1051–1056, 2001.

50. Faber BC, Cleutjens KB, Niessen RL, et al: Identification of genes potentially involved in rupture of human atherosclerotic plaques. Circ Res 89: 547–554, 2001.

51. Essler M, Retzer M, Bauer M, Zangl KJ, Tigyi G, Siess W: Stimulation of platelets and endothelial cells by mildly oxidized LDL proceeds through activation of lysophosphatidic acid receptors and the Rho/Rho-kinase pathway. Inhibition by lovastatin. Ann N Y Acad Sci 905: 282–286, 2000.

52. Toschi V, Gallo R, Lettino M, et al: Tissue factor modulates the thrombogenicity of human atherosclerotic plaques. Circulation 95: 594–599, 1997.

53. Mallat Z, Hugel B, Ohan J, Leseche G, Freyssinet JM, Tedgui A: Shed membrane microparticles with procoagulant potential in human atherosclerotic plaques: a role for apoptosis in plaque thrombogenicity. Circulation 99: 348–353, 1999.

54. Badimon JJ, Lettino M, Toschi V, et al: Local inhibition of tissue factor reduces the thrombogenicity of disrupted human atherosclerotic plaques: effects of tissue factor pathway inhibitor on plaque thrombogenicity under flow conditions. Circulation 99: 1780–1787, 1999.

55. Fernandez-Ortiz A, Badimon JJ, Falk E, et al: Characterization of the relative thrombogenicity of atherosclerotic plaque components: implications for consequences of plaque rupture. J Am Coll Cardiol 23: 1562–1569, 1994.

56. van der Wal AC, Becker AE, van der Loos CM, Das PK: Site of intimal rupture or erosion of thrombosed coronary atherosclerotic plaques is characterized by an inflammatory process irrespective of the dominant plaque morphology. Circulation 89: 36–44, 1994.

57. Kovanen PT: The mast cell—a potential link between inflammation and cellular cholesterol deposition in atherogenesis. Eur Heart J 14 (Suppl K): 105–117, 1993.

58. Kovanen PT, Kaartinen M, Paavonen T: Infiltrates of activated mast cells at the site of coronary atheromatous erosion or rupture in myocardial infarction. Circulation 92: 1084–1088, 1995.

59. Kaartinen M, van der Wal AC, van der Loos CM, et al: Mast cell infiltration in acute coronary syndromes: implications for plaque rupture. J Am Coll Cardiol 32: 606–612, 1998.

60. Laine P, Kaartinen M, Penttila A, Panula P, Paavonen T, Kovanen PT: Association between myocardial infarction and the mast cells in the adventitia of the infarct-related coronary artery. Circulation 99: 361–369, 1999.

61. Libby P: Inflammation in atherosclerosis. Nature 420: 868–874, 2002.

62. Barger AC, Beeuwkes R III, Lainey LL, Silverman KJ. Hypothesis: vasa vasorum and neovascularization of human coronary arteries. A possible role in the pathophysiology of atherosclerosis. N Engl J Med 310: 175–177, 1984.

63. Kamat BR, Galli SJ, Barger AC, Lainey LL, Silverman KJ: Neovascularization and coronary atherosclerotic plaque: cinematographic localization and quantitative histologic analysis. Hum Pathol 18: 1036–1042, 1987.

64. Barger AC, Beeuwkes R III: Rupture of coronary vasa vasorum as a trigger of acute myocardial infarction. Am J Cardiol 66: 41G–43G, 1990.

65. Heistad DD, Armstrong ML: Blood flow through vasa vasorum of coronary arteries in atherosclerotic monkeys. Arteriosclerosis 6: 326–331, 1986.

66. Williams JK, Heistad DD: The vasa vasorum of the arteries. J Mal Vasc 21: 266–269, 1996.

67. Kwon HM, Sangiorgi G, Ritman EL, et al: Enhanced coronary vasa vasorum neovascularization in experimental hypercholesterolemia. J Clin Invest 101: 1551–1556, 1998.

68. Gojova A, Brun V, Bruno E, Cottrez F, Gourdy P, Ardouin P, Tedgui A, Mallat Z, Groux H: Specific abrogation of transforming growth factor beta signaling in T cells alters atherosclerolic lesion size and composition in mice. Blood 102: 4052–4058, 2003.

69. Robertson AL, Rudling M, Zhore X, Gorelik L, Flavell RA, Hansson GK: Disruption of TGF-beta signaling in T-cells accelerates atherosclerosis. J Clin Invest 112: 1342–1350, 2003.

70. Tedgui A, Mallat Z: Anti-inflammatory mechanisms in the vascular wall. Circ Res 88: 877–887, 2001.

71. Xu XH, Shah PK, Faure E, Equils O, Thomas L, Fishbein MC, Luthringer D, Xu XP, Rajavashisth TB, Yano J, Kaul S, Arditi M: Toll-like receptor-4 is expressed by macrophages in murine and human lipid-rich atherosclerotic plaques and upregulated by oxidized LDL. Circulation. 104(25): 3103–3108, 2001.

72. Michelsen KS, Wong MH, Shah PK, Zhang W, Yano J, Doherty TM, Akira S, Rajavashisth TB, Arditi M: Lack of toll-like receptor 4 or myeloid differentiation factor 88 reduces atherosclerosis and alters plaque phenotype in mice deficient in apolipoprotein E. Proc Natl Acad Sci U S A 101(29): 10679–10684, 2004.

73. Henney AM, Wakeley PR, Davies MJ, et al: Localization of stromelysin gene expression in atherosclerotic plaques by in situ hybridization. Proc Natl Acad Sci U S A 88: 8154–8158, 1991.

74. Galis ZS, Sukhova GK, Lark MW, Libby P: Increased expression of matrix metalloproteinases and matrix degrading activity in vulnerable regions of human atherosclerotic plaques. J Clin Invest 94: 2493–2503, 1994.

75. Brown DL, Hibbs MS, Kearney M, Loushin C, Isner JM: Identification of 92-kD gelatinase in human coronary atherosclerotic lesions. Association of active enzyme synthesis with unstable angina. Circulation 91: 2125–2131, 1995.

76. Nikkari ST, O'Brien KD, Ferguson M, et al: Interstitial collagenase [MMP-1] expression in human carotid atherosclerosis. Circulation 92: 1393–1398, 1995.

77. Li Z, Li L, Zielke HR, et al: Increased expression of 72-kD type IV collagenase [MMP-2] in human aortic atherosclerotic lesions. Am J Pathol 148: 121–128, 1996.

78. Galis ZS, Sukhova GK, Libby P: Microscopic localization of active proteases by in situ zymography: detection of matrix metalloproteinase activity in vascular tissue. FASEB J 9: 974–980, 1995.

79. Xu XP, Meisel SR, Ong JM, et al: Oxidized low-density lipoprotein regulates matrix metalloproteinase-9 and its tissue inhibitor in human monocyte-derived macrophages. Circulation 99: 993–998, 1999.

80. Rajavashisth TB, Xu XP, Jovinge S, et al: Membrane type 1 matrix metalloproteinase expression in human atherosclerotic plaques: evidence for activation by proinflammatory mediators. Circulation 99: 3103–3109, 1999.

81. Rajavashisth TB, Liao JK, Galis ZS, et al: Inflammatory cytokines and oxidized low density lipoproteins increase endothelial cell expression of membrane type 1-matrix metalloproteinase. J Biol Chem 274: 11924–11929, 1999.

82. Herman MP, Sukhova GK, Libby P, et al: Expression of neutrophil collagenase [matrix metalloproteinase-8] in human atheroma: a novel collagenolytic pathway suggested by transcriptional profiling. Circulation 104: 1899–1904, 2001.

83. Shah PK, Falk E, Badimon JJ, et al: Human monocyte-derived macrophages induce collagen breakdown in fibrous caps of atherosclerotic plaques. Potential role of matrix-degrading metalloproteinases and implications for plaque rupture. Circulation 92: 1565–1569, 1995.

84. Shah PK: Role of inflammation and metalloproteinases in plaque disruption and thrombosis. Vasc Med 3: 199–206, 1998.

85. Sukhova GK, Schonbeck U, Rabkin E, et al: Evidence for increased collagenolysis by interstitial collagenases-1 and -3 in vulnerable human atheromatous plaques. Circulation 99: 2503–2509, 1999.

86. Lee RT, Schoen FJ, Loree HM, Lark MW, Libby P: Circumferential stress and matrix metalloproteinase 1 in human coronary atherosclerosis. Implications for plaque rupture. Arterioscler Thromb Vasc Biol 16: 1070–1073, 1996.

87. Galis ZS, Muszynski M, Sukhova GK, Simon-Morrissey E, Libby P: Enhanced expression of vascular matrix metalloproteinases induced *in vitro* by cytokines and in regions of human atherosclerotic lesions. Ann N Y Acad Sci 748: 501–507, 1995.

88. Galis ZS, Sukhova GK, Kranzhofer R, Clark S, Libby P: Macrophage foam cells from experimental atheroma constitutively produce matrix-degrading proteinases. Proc Natl Acad Sci U S A 92: 402–406, 1995.

89. Kol A, Sukhova GK, Lichtman AH, Libby P. Chlamydial heat shock protein 60 localizes in human atheroma and regulates macrophage tumor necrosis factor-alpha and matrix metalloproteinase expression. Circulation 98: 300–307, 1998.

90. Mach F, Schonbeck U, Fabunmi RP, et al: T lymphocytes induce endothelial cell matrix metalloproteinase expression by a CD40L-dependent mechanism: implications for tubule formation. Am J Pathol 154: 229–238, 1999.

91. Schonbeck U, Mach F, Sukhova GK, et al: Regulation of matrix metalloproteinase expression in human vascular smooth muscle cells by T lymphocytes: a role for CD40 signaling in plaque rupture? Circ Res 81: 448–454, 1997.

92. Wallner K, Li C, Shah PK, et al: Tenascin-C is expressed in macrophage-rich human coronary atherosclerotic plaque. Circulation 99: 1284–1289, 1999.

93. Sukhova GK, Shi GP, Simon DI, Chapman HA, Libby P: Expression of the elastolytic cathepsins S and K in human atheroma and regulation of their production in smooth muscle cells. J Clin Invest 102: 576–583, 1998.

94. Shi GP, Sukhova GK, Grubb A, et al: Cystatin C deficiency in human atherosclerosis and aortic aneurysms. J Clin Invest 104: 1191–1197, 1999.

95. Bennett MR, Evan GI, Schwartz SM: Apoptosis of human vascular smooth muscle cells derived from normal vessels and coronary atherosclerotic plaques. J Clin Invest 95: 2266–2274, 1995.

96. Bjorkerud S, Bjorkerud B: Apoptosis is abundant in human atherosclerotic lesions, especially in inflammatory cells [macrophages and T cells], and may contribute to the accumulation of gruel and plaque instability. Am J Pathol 149: 367–380, 1996.

97. Kockx MM, Knaapen MW: The role of apoptosis in vascular disease. J Pathol 190: 267–280, 2000.

98. Ihling C, Haendeler J, Menzel G, et al: Co-expression of p53 and MDM2 in human atherosclerosis: implications for the regulation of cellularity of atherosclerotic lesions. J Pathol 185: 303–312, 1998.

99. Crisby M, Kallin B, Thyberg J, et al: Cell death in human atherosclerotic plaques involves both oncosis and apoptosis. Atherosclerosis 130: 17–27, 1997.

100. Bennett MR: Apoptosis of vascular smooth muscle cells in vascular remodelling and atherosclerotic plaque rupture. Cardiovasc Res 41: 361–368, 1999.

101. Galle J, Heermeier K, Wanner C: Atherogenic lipoproteins, oxidative stress, and cell death. Kidney Int Suppl 71: S62–S65, 1999.

102. Vieira O, Escargueil-Blanc I, Jurgens G, et al: Oxidized LDLs alter the activity of the ubiquitin-proteasome pathway: potential role in oxidized LDL-induced apoptosis. FASEB J 14: 532–542, 2000.

103. Rossig L, Dimmeler S, Zeiher AM: Apoptosis in the vascular wall and atherosclerosis. Basic Res Cardiol 96: 11–22, 2001.

104. Geng YJ, Wu Q, Muszynski M, Hansson GK, Libby P: Apoptosis of vascular smooth muscle cells induced by *in vitro* stimulation with interferon-gamma, tumor necrosis factor-alpha, and interleukin-1 beta. Arterioscler Thromb Vasc Biol 16: 19–27, 1996.

105. Geng YJ, Henderson LE, Levesque EB, Muszynski M, Libby P: Fas is expressed in human atherosclerotic intima and promotes apoptosis of cytokine-primed human vascular smooth muscle cells. Arterioscler Thromb Vasc Biol 17: 2200–2208, 1997.

106. Mallat Z, Tedgui A: Apoptosis in the vasculature: mechanisms and functional importance. Br J Pharmacol 130: 947–962, 2000.

107. Wallner K, Li C, Shah PK, Wu KJ, Schwartz SM, Sharifi BG: EGF-like domain of tenascin-C is proapoptotic for cultured smooth muscle cells. Arterioscler Thromb Vasc Biol 24(8): 1416–1421, 2004.

108. Kolmakova A, Kwiterovich P, Virgil D, Alaupovic P, Knight-Gibson C, Martin SF, Chatterjee S: Apolipoprotein C-I induces apoptosis in human aortic smooth muscle cells via recruiting neutral sphingomyelinase. Arterioscler Thromb Vasc Biol 24(2): 264–269, 2004.

109. Durand E, Scoazec A, Lafont A, Boddaert J, Al Hajzen A, Addad F, Mirshahi M, Desnos M, Tedgui A, Mallat Z: *In vivo* induction of endothelial apoptosis leads to vessel thrombosis and endothelial denudation: a clue to the understanding of the mechanisms of thrombotic plaque erosion. Circulation 109(21): 2503–2506, 2004.

110. Sugiyama S, Kugiyama K, Aikawa M, Nakamura S, Ogawa H, Libby P: Hypochlorous acid, a macrophage product, induces endothelial apoptosis and tissue factor expression: involvement of myeloperoxidase-mediated oxidant in plaque erosion and thrombogenesis. Arterioscler Thromb Vasc Biol 24(7): 1309–1314, 2004.

111. Abela GS, Picon PD, Friedl SE, et al: Triggering of plaque disruption and arterial thrombosis in an atherosclerotic rabbit model. Circulation 91: 776–784, 1995.

112. Rekhter MD, Hicks GW, Brammer DW, et al: Animal model that mimics atherosclerotic plaque rupture. Circ Res 83: 705–713, 1998.

113. Caligiuri G, Levy B, Pernow J, Thoren P, Hansson GK: Myocardial infarction mediated by endothelin receptor signaling in hypercholesterolemic mice. Proc Natl Acad Sci U S A 96: 6920–6924, 1999.

114. Rosenfeld ME, Polinsky P, Virmani R, Kauser K, Rubanyi G, Schwartz SM: Advanced atherosclerotic lesions in the innominate artery of the ApoE knockout mouse. Arterioscler Thromb Vasc Biol 20: 2587–2592, 2000.

115. Johnson JL, Jackson CL:. Atherosclerotic plaque rupture in the apolipoprotein E knockout mouse. Atherosclerosis 154: 399–406, 2001.

116. Lemaitre V, O'Byrne TK, Borczuk AC, Okada Y, Tall AR, D'Armiento J: ApoE knockout mice expressing human matrix metalloproteinase-1 in macrophages have less advanced atherosclerosis. J Clin Invest 107: 1227–1234, 2001.

117. Calara F, Silvestre M, Casanada F, Yuan N, Napoli C, Palinski W: Spontaneous plaque rupture and secondary thrombosis in apolipoprotein E-deficient and LDL receptor-deficient mice. J Pathol 195: 257–263, 2001.

118. von der Thusen JH, van Vlijmen BJ, Hoeben RC, Kockx MM, Havekes LM, van Berkel TJ, Biessen EA: Induction of atherosclerotic plaque rupture in apolipoprotein E-/- mice after adenovirus-mediated transfer of p53. Circulation 105: 2064–2070, 2002.

119. Muller JE, Tofler GH, Stone PH: Circadian variation and triggers of onset of acute cardiovascular disease. Circulation 79: 733–743, 1989.

120. Muller JE: Morning increase of onset of myocardial infarction. Implications concerning triggering events. Cardiology 76: 96–104, 1989.

121. Muller JE, Tofler GH: Triggering and hourly variation of onset of arterial thrombosis. Ann Epidemiol 2: 393–405, 1992.

122. Willich SN, Jimenez AH, Tofler GH, DeSilva RA, Muller JE: Pathophysiology and triggers of acute myocardial infarction: clinical implications. Clin Invest 70: S73–S78, 1992.

123. Willich SN, Maclure M, Mittleman M, Arntz HR, Muller JE: Sudden cardiac death. Support for a role of triggering in causation. Circulation 87: 1442–1450, 1993.

124. Peters A, Dockery DW, Muller JE, Mittleman MA: Increased particulate air pollution and the triggering of myocardial infarction. Circulation 103: 2810–2815, 2001.

125. Mittleman MA, Lewis RA, Maclure M, Sherwood JB, Muller JE: Triggering myocardial infarction by marijuana. Circulation 103: 2805–2809, 2001.

126. Muller JE: Circadian variation and triggering of acute coronary events. Am Heart J 137: S1–S8, 1999.

127. Muller JE: Triggering of cardiac events by sexual activity: findings from a case-crossover analysis. Am J Cardiol 86: 14F–18F, 2000.

128. Goldstein JA, Demetriou D, Grines CL, Pica M, Shoukfeh M, O'Neill WW: Multiple complex coronary plaques in patients with acute myocardial infarction. N Engl J Med 343: 915–922, 2000.

129. Matetzky S, Tani S, Kangavari S, et al: Smoking increases tissue factor expression in atherosclerotic plaques: implications for plaque thrombogenicity. Circulation 102: 602–604, 2000.

130. Burke AP, Kolodgie FD, Farb A, et al: Healed plaque ruptures and sudden coronary death: evidence that subclinical rupture has a role in plaque progression. Circulation 103: 934–940, 2001.
131. Burke AP, Farb A, Malcom GT, Liang YH, Smialek J, Virmani R: Coronary risk factors and plaque morphology in men with coronary disease who died suddenly. N Engl J Med 336: 1276–1282, 1997.
132. Farb A, Burke AP, Tang AL, et al: Coronary plaque erosion without rupture into a lipid core. A frequent cause of coronary thrombosis in sudden coronary death. Circulation 93: 1354–1363, 1996.
133. Rauch U, Bonderman D, Bohrmann B, et al: Transfer of tissue factor from leukocytes to platelets is mediated by CD15 and tissue factor. Blood 96: 170–175, 2000.
134. Mallat Z, Benamer H, Hugel B, et al: Elevated levels of shed membrane microparticles with procoagulant potential in the peripheral circulating blood of patients with acute coronary syndromes. Circulation 101: 841–843, 2000.
135. Laszik ZG, Zhou XJ, Ferrell GL, Silva FG, Esmon CT: Down-regulation of endothelial expression of endothelial cell protein C receptor and thrombomodulin in coronary atherosclerosis. Am J Pathol 159: 797–802, 2001.
136. Schwartz CJ, Valente AJ, Sprague EA, Kelley JL, Cayatte AJ, Mowery J: Atherosclerosis. Potential targets for stabilization and regression. Circulation 86: III117–III123, 1992.
137. Waters D: Plaque stabilization: a mechanism for the beneficial effect of lipid-lowering therapies in angiography studies. Prog Cardiovasc Dis 37: 107–120, 1994.
138. Shah PK: Pathophysiology of plaque rupture and the concept of plaque stabilization. Cardiol Clin 21: 303–314, 2003.
139. Shah PK: Mechanisms of plaque vulnerability and rupture. J Am Coll Cardiol 41: 15S–22S, 2003.
140. Kullo IJ, Edwards WD, Schwartz RS: Vulnerable plaque: pathobiology and clinical implications. Ann Intern Med 129: 1050–1060, 1998.
141. Lee RT: Plaque stabilization: the role of lipid lowering. Int J Cardiol 74 (Suppl 1): S11–S15, 2000.
142. Aikawa M, Rabkin E, Okada Y, et al: Lipid lowering by diet reduces matrix metalloproteinase activity and increases collagen content of rabbit atheroma: a potential mechanism of lesion stabilization. Circulation 97: 2433–2444, 1998.
143. Aikawa M, Libby P: Lipid lowering reduces proteolytic and prothrombotic potential in rabbit atheroma. Ann N Y Acad Sci 902: 140–152, 2000.
144. Aikawa M, Rabkin E, Sugiyama S, et al: An HMG-CoA reductase inhibitor, cerivastatin, suppresses growth of macrophages expressing matrix metalloproteinases and tissue factor *in vivo* and *in vitro*. Circulation 103: 276–283, 2001.
145. Fukumoto Y, Libby P, Rabkin E, et al: Statins alter smooth muscle cell accumulation and collagen content in established atheroma of Watanabe heritable hyperlipidemic rabbits. Circulation 103: 993–999, 2001.
146. Ameli S, Hultgardh-Nilsson A, Cercek B, et al: Recombinant apolipoprotein A-I Milano reduces intimal thickening after balloon injury in hypercholesterolemic rabbits. Circulation 90: 1935–1941, 1994.
147. Shah PK, Nilsson J, Kaul S, et al: Effects of recombinant apolipoprotein A-I [Milano] on aortic atherosclerosis in apolipoprotein E-deficient mice. Circulation 97: 780–785, 1998.

148. Shah PK, Yano J, Reyes O, et al: High-dose recombinant apolipoprotein A-I [Milano] mobilizes tissue cholesterol and rapidly reduces plaque lipid and macrophage content in apolipoprotein E-deficient mice: potential implications for acute plaque stabilization. Circulation 103: 3047–3050, 2001.
149. Crisby M, Nordin-Fredriksson G, Shah PK, Yano J, Zhu J, Nilsson J: Pravastatin treatment increases collagen content and decreases lipid content, inflammation, metalloproteinases, and cell death in human carotid plaques: implications for plaque stabilization. Circulation 103: 926–933, 2001.
150. Shah PK, Kaul S, Nilsson J, Cercek B: Exploiting the vascular protective effects of high-density lipoprotein and its apolipoproteins: an idea whose time for testing is coming, part I. Circulation 104: 2376–2383, 2001.
151. Shah PK, Kaul S, Nilsson J, Cercek B: Exploiting the vascular protective effects of high-density lipoprotein and its apolipoproteins: an idea whose time for testing is coming, part II. Circulation 104: 2498–2502, 2001.
152. Forrester J, Makkar R, Shah PK: Increasing HDL cholesterol in dyslipidemia by cholesterol ester transfer protein inhibition: an update for clinicians. Circulation 111: 1847–1854, 2005.
153. Li X, Chyu KY, Neto JR, Yano J, Nathwani N, Ferreira C, Dimayuga PC, Cercek B, Kaul S, Shah PK: Differential effects of apolipoprotein A-I-mimetic peptide on evolving and established atherosclerosis in apolipoprotein E-null mice. Circulation 110(12): 1701–1705, 2004.
154. Claudel T, Leibowitz MD, Fievet C, et al: Reduction of atherosclerosis in apolipoprotein E knockout mice by activation of the retinoid X receptor. Proc Natl Acad Sci U S A 98: 2610–2615, 2001.
155. Kirkhamm TC, Williams CM: Endocannabinoid receptor antagonists: potential for obesity treatment. Treat Endocrinol 3(6): 345–360, 2004.

Section V

Management of Atherosclerosis

22

Modification of Biochemical and Cellular Processes in the Development of Atherosclerosis by Red Wine

HARJOT K. SAINI, PARAMBIR DHAMI, YAN-JUN XU, SUKHINDER KAUR CHEEMA, AMARJIT S. ARNEJA, AND NARANJAN S. DHALLA

Abstract

Atherosclerosis is commonly associated with unstable angina and acute myocardial infarction. It occurs as a result of a cascade of events caused by various environmental, dietary, genetic, and inflammatory factors. Different epidemiological studies have suggested that moderate amounts of red wine consumption reduce the risk of complications associated with atherosclerosis. This contention is further supported by a variety of experimental investigations demonstrating that both alcoholic and phenolic components are responsible for the apparent protective effects of red wine. The maintenance of endothelial function, augmentation in the levels of high-density lipoproteins (HDLs), prevention of low-density lipoprotein (LDL) oxidation, attenuation of smooth muscle proliferation and migration, inhibition of platelet aggregation and adhesion, as well as reduction in inflammatory mediators are major mechanisms linked with the protective effects of red wine. Despite these beneficial effects, insufficient information is available to recommend red wine as a therapeutic strategy to prevent atherosclerosis. Particularly, in view of high alcoholic content, excessive consumption of red wine can be seen to produce harmful effects. Therefore, a large-scale clinical trial is needed to determine the exact amount of red wine required for the beneficial effects and to categorize it as a future antiatherosclerotic agent.

Keywords: endothelial function; inflammation; LDL oxidation; lipoproteins; nitric oxide; red wine; thrombosis; vascular smooth muscle cells

Abbreviations: CAD, coronary artery disease; eNOS, endothelial NO synthase; ET-1, endothelin-1; HDL, high-density lipoprotein; hs-CRP, high-sensitivity C-reactive protein; HUVEC, human umbilical vein endothelial cells; ICAM, intercellular adhesion molecule; IL, interleukin; iNOS, inducible NO synthase; LDL, low-density lipoprotein; MCP, monocyte chemoattractant protein; NFkB, nuclear factor kappa B; NO, nitric oxide; Ox-LDL, oxidized LDL; PDGF, platelet-derived growth factor; SMC, smooth muscle cell; TNF, tumor necrosis factor; VCAM, vascular cell adhesion molecule; VSMC, vascular smooth muscle cell

Introduction

Atherosclerosis, manifested by coronary artery disease (CAD), peripheral arterial disease, and cerebrovascular disease, is the major problem of global concern [1]. Multiple Risk Factor Intervention Trial and Framingham Studies have identified that hypercholesterolemia, hypertension, diabetes, as well as lifestyle factors such as stress, physical inactivity, and cigarette smoking are the major risk factors associated with the development of atherosclerosis [2, 3]. Over the past decades, primary and secondary prevention trials with hydroxyl-methyl glutaryl coenzyme A (HMG-CoA) inhibitors/statins have supported the contention that lowering low-density lipoprotein (LDL) cholesterol is the primary target for the prevention of atherosclerotic disease development [4]. However, a significant gap exists between the current statin therapy and the third report of the National Cholesterol Education Program (NCEP) Expert Panel on Detection, Evaluation, and Treatment of High Blood Cholesterol in Adults (Adult Treatment Panel III) guidelines [5]. It has been suggested that treatment goal for aggressive lipid lowering for LDL for high-risk patients of CAD should be <100 mg/dL [5]. To achieve this goal, higher doses of statins are needed; however, such higher doses have greater risk of adverse effects and each two-, three-, and fourfold increase causes 6%, 12%, and 18% decrease (only a minor change) in LDL concentration, respectively [6]. In addition, for every 1% fall in serum LDL level only 2% reduction in the incidence of CAD has been reported [7]. Because of the pleiotropic effects of statins (beyond LDL lowering) including the maintenance of endothelial function, attenuation of smooth muscle cell (SMC) proliferation, reduction of inflammatory response, stabilization of plaque and prevention of oxidative stress [8], the relationship between reducing cholesterol and atherosclerotic plaque development remains questionable. Therefore, new treatment strategies, especially the modification of lifestyle factors are considered necessary, which can be opted in combination with the moderate doses of statins to prevent the development and progression of the disease.

Epidemiological studies have demonstrated that despite the consumption of high saturated fat intake, the prevalence of death due to CAD in France is relatively less as compared to USA and other Western countries [9]. An inverse relationship between intake of red wine and mortality due to CAD has been revealed as the basis for the French paradox [9]. In addition, a significant reduction (70%) in the risk of newer coronary events has been reported by the intake of Mediterranean diet, which includes moderate consumption of red wine [10]. Since red wine is a combination of ethanol and other polyphenolic substances (Table 22.1) [11, 12], the participation of these individual components in the prevention of atherosclerotic plaque development is not completely elucidated. Accordingly, this review has attempted to describe the effects of red wine on the modification of biochemical and cellular events occurring during the disease process. In addition, the pathophysiology of atherosclerosis has been discussed to have a

TABLE 22.1. Different components of red wine.

Components
Ethanol
Polyphenolic substances
Flavonoids
Flavonols (Myricetin, Quercetin, Kaempferol)
Flavan-3-ols (Tannins, Proanthocyanidines, Gallocatechin, Catechin, Epicatechin)
Anthocyanins (Cyanidin, Malvidin)
Nonflavonoids
Hydroxycinnamic acid derivatives (Caffeic acid, Ferulic acid, p-Coumaric acid)
Hydroxy benzoic acid derivatives (Gallic acid, Vanillic acid)
Stilbenes and Stilbene glycosides (Resveratrol)

better understanding of its therapeutic targets in general and red wine in particular.

Biochemical and Cellular Events in the Development of Atherosclerosis

Atherosclerosis is multifactorial and polygenic in origin and develops decades earlier than its clinical manifestation [13]. The clinical silent phase of the disease initiates in the form of an inflammatory response, associated with fatty streak formation subsequent to intimal thickening, leading to the development of atherosclerotic plaque, which consists of cholesterol crystals, lymphocytes, proliferating SMCs, foam cells, proteoglycans, and cell debris [13, 14]. The later apparent stage of the disease is manifested by arterial calcification, thinning of fibrous cap, plaque rupture producing microemboli, hemorrhage, and intravascular thrombosis [14, 15]. Different biochemical and cellular events underlying the disease processes due to various pathophysiological factors including diabetes, hypertension, and hypercholesterolemia are shown in Fig. 22.1.

Intimal Thickening and Plaque Formation

The classic response-to-injury hypothesis has postulated that alteration in endothelial function is the initial step of disease process [16]. The potential causes of endothelial dysfunction include oxidative stress, mechanical stress, genetic alterations, elevated plasma homocysteine levels, and infectious microorganisms [17]. In spite of the different means of endothelial injury, the end result occurs as an increase in endothelial permeability, imbalance between endothelium-derived relaxing factors and endothelium-derived contracting factors, expression of adhesion molecules, release of growth factors and chemotactic factors leading to intimal thickening and plaque formation [18]. It is important to note that under physiological conditions, the secretion

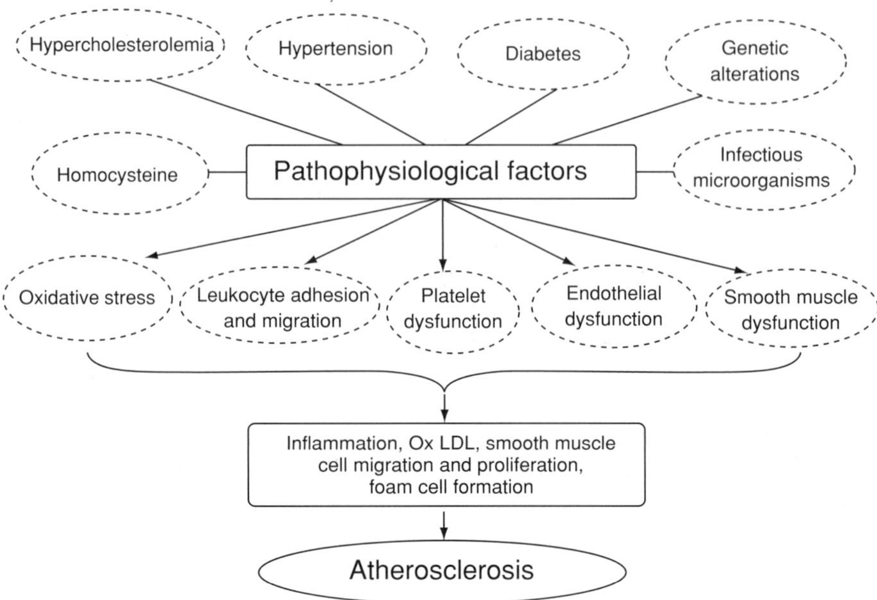

FIGURE 22.1. Different biochemical and cellular processes participating in the development of atherosclerosis.

of adhesion molecules such as vascular cell adhesion molecule-1 (VCAM-1), intercellular adhesion molecule-1 (ICAM-1), E-selectin, and endothelial leukocyte adhesion molecule is regulated by proinflammatory cytokines such as tumor necrosis factor-α (TNF-α), interleukin (IL)-1, IL-4, IL-6, and interferon-γ [19]. However, in the presence of endothelial dysfunction, the levels of these cytokines have been shown to be elevated, which in turn cause more production of adhesion molecules and thus favor the monocyte recruitment and adhesion to the endothelium [20]. The recruited monocytes differentiate to macrophages in the subendothelial space and further aggravate the development of atherosclerosis [21].

Formation of Fatty Streak

The alteration in endothelial permeability triggers the transmigration of LDL particles into the intima through the endothelial layer, where it is modified by oxidative stress to oxidized LDL (Ox-LDL) [22]. Macrophage lipoxygenase, myeloperoxidase, and NADPH oxidase are the major sources of reactive oxygen species production [23]. Oxidation of polyunsaturated fatty acids within the LDL causes the release of aldehyde and ketones such as malondialdehyde and 4-hydroxynonenal, which can alter the lysine residues on apolipoproteins B-100 (apoB-100), the major protein of LDL [24]. The modified apoB-100 is no longer recognized by the macrophage apoB receptor, but instead is taken by the macrophage scavenger receptor leading to the formation of foam cells and results in fatty streak [24].

Formation of Lipid Core

The uptake of Ox-LDL by macrophages is the major stimulus for the production and release of various cytokines such as TNF-α, IL-1β, and IL-8 as well as cytotoxic substances leading to an inflammatory response [25]. These cytokines cause the recruitment of macrophages, T cells, and SMCs, as well as upregulate the expression of endothelial adhesion molecules [17]. The cytotoxic substances released by macrophages further oxidize LDL particles in the intima and promote their uptake by macrophages. The continuation of this process converts the foam cells into lipid pools and ultimately forms the lipid core of the atherosclerotic plaque [17].

Stabilization of the Plaque

The proliferation and migration of SMCs from media to intima form the fibrous cap, which causes stabilization of the plaque [15]. Once in the intima, SMCs cause the production of cytokines, growth factors, and extracellular matrix consisting of collagen and proteoglycans [15]. In addition, lysophosphatidylcholine, a product formed by LDL oxidation and a potent chemoattractant for monocytes and T-lymphocytes, induces the expression of VCAM-1 and ICAM-1, increases the level of platelet-derived growth factor (PDGF) and heparin-binding epidermal growth factor in endothelial cells and SMCs [15]. Furthermore, lysophosphatidic acid is released from Ox-LDL, which stimulates vascular smooth muscle cell (VSMC) proliferation as well as platelet aggregation, increases intracellular Ca^{2+} concentration and thus promotes the formation of atherosclerotic plaque [26, 27]. Ox-LDL itself is toxic to macrophages and therefore contributes to the amplification of inflammatory process and triggers the formation of necrotic core in the advanced atherosclerotic lesions [23]. The lipid droplets released from the dead macrophages are phagocytized by SMCs leading to the production of SMC foam cells [28]. It is emphasized that the degree of SMC proliferation, migration, extracellular matrix formation, and extent of mature plaque formation is determined by the mutual interaction between platelets, endothelium, SMCs, and macrophages.

Progression of the Disease and Destabilization of Atherosclerotic Plaque

The slow progression of disease leads to ischemia by reduction in the lumen size; this condition is clinically manifested as angina [29]. Chronic inflammatory reaction and induction of calcification-mediated changes in the mechanistic characteristics of arteries predisposes the plaque to rupture [29]. The increased expression of matrix metalloproteases is mainly responsible for atherosclerotic plaque rupture [15]. In addition, thinning of the fibrous cap triggers the plaque rupturing process and exposes the thrombogenic contents of plaque to blood stream leading to the formation of thrombus [17]. Furthermore, production of Ox-LDL, platelet accumulation, local increase in tissue factor, thromboxane A_2, serotonin, ADP, and platelet-activating factor aggravate the formation of thrombosis [17], which is manifested by acute coronary syndromes such as acute myocardial infarction and unstable angina [29]. Because of the complex acquaintance between environmental, genetic, cellular, and biochemical factors, the sequential characterization of atherosclerotic lesions in individual patients is far from clear. Therefore, evidence-based therapies as a result of epidemiological studies through risk factor modification are considered to provide significant clinical benefits.

Antiatherosclerotic Effects of Red Wine

Various experimental studies have demonstrated that both the alcoholic and polyphenolic components of red wine prevent the development and progression of atherosclerosis by modifying the following biochemical and cellular events during the disease process.

Modification of Nitric Oxide (NO)-Mediated Vasoprotection by Red Wine

NO plays an important role in maintaining the vascular tone because of its strong vasorelaxing ability [30]. NO-induced protection in the early phase of atherosclerosis is mainly mediated by prevention of leukocyte migration and adhesion to the vascular endothelium by a decrease in monocyte chemoattractant protein-1 (MCP-1) [31], surface adhesion molecules such as CD11/CD18, P selectin, VCAM-1, and ICAM-1 [32, 33]. The protection by NO in the latter stages of atherosclerosis is manifested by inhibition of DNA synthesis, mitogenesis, VSMC proliferation, and migration [34]. In addition, NO is a potent inhibitor of platelet aggregation and adhesion to the vascular wall [35] and prevents the release of PDGFs, which are known to stimulate SMC proliferation [36]. Furthermore, the antiatherosclerotic properties of NO are associated with its ability to decrease endothelial permeability, reduce influx of LDL into the intima and inhibit LDL oxidation [37]. Red wine has been shown to cause the endothelium-dependent relaxation of blood vessels via enhanced generation and increased biological activity of NO [38]. Fitzpatrick et al. [39] have demonstrated that this effect of red wine is due to its polyphenolic components such as quercitin and tannic acid, unlike resveratrol and malvidin, which have the ability to generate NO from the vascular endothelium as seen in phenyle-phrine-contracted rat aortic rings. Similarly, NO-mediated vasodilation after incubation with red wine polyphenols has been observed in human coronary arteries [40].

Andriambeloson et al. [41] have demonstrated that in rat aortic rings pre-constricted with norepinephrine, red wine polyphenols-induced endothelium-dependent relaxation to acetylcholine is linked with enhanced NO synthesis instead of increasing the biological effectiveness of NO or via protecting its breakdown by superoxide anions. On the other hand, Zenebe et al. [42] have reported that polyphenols in red wine prolong the half-life and increase the bioavailability of NO by preventing its degradation by reactive oxygen species. Although the molecular mechanisms associated with red wine polyphenols-induced vasoprotection are not completely elucidated, recent studies have shown the involvement of endothelial NO synthase (eNOS) activation in the presence of polyphenols [43]. An increase in the activity of eNOS promoter with eNOS mRNA stabilization has been demonstrated in human endothelial cells by red wine polyphenol treatment [43]; resveratrol has been reported to be the major red wine polyphenol responsible for such an effect [44]. In contrast, the ethanol present in red wine has not been associated with upregulation of eNOS expression under similar experimental conditions [43]. However, the alterations in eNOS expression in the presence of red wine are still controver-sial as results vary with the variety of red wine. Exposure of French red wine independent of their maturation in oak barrels or steel tanks, significantly enhances eNOS activity, mRNA, and protein content in cultured human umbilical vein endothelial cells (HUVEC) both in a time and concentration-dependent manner [43]. On the other hand, no such effect on eNOS expression was observed in the presence of German red wines [43]. The difference in the polyphenol content may be the reason for these controversial findings because the French red wine has higher contents of polyphenols than other wines [45].

Additional mechanisms underlying NO-dependent vasoprotection have been demonstrated in different cell types after red wine polyphenol treat-ment. An increase in the intracellular Ca^{2+} concentration, which causes aug-mentation of NO biosynthesis, has been reported in rat thoracic aorta and bovine aortic endothelial cells upon red wine polyphenol treatment [41, 46]. In addition, an increase in the expression of cyclooxgenase-2, release of endothelial thromboxane A_2 and the expression of inducible NO synthase (iNOS) have been observed in rat aorta after treatment with dry powder of red wine [47]. These investigators have suggested that increased expression of iNOS may compensate the extra endothelial NO-induced hyporeactivity in response to norepinephrine in order to maintain the agonist-induced con-traction [47]. Besides NO production, the formation of other mediators of vascular tone such as prostacyclin (PGI_2), a potent vasodilator, and antiplatelet agent [48] as well as endothelium-derived hyperpolarizing factor, a redox-sensitive vasodilator [49], are also increased by red wine polyphenols. In addition, the synthesis of endothelin-1 (ET-1), a potent vasoconstrictor substance, is reduced by polyphenol treatment [50]. Furthermore, red wine polyphenols decrease the levels of high-sensitivity C-reactive protein (hs-CRP), which is a marker of early inflammatory disease and a potential risk predictor for future atherosclerosis [51].

Modification of High-Density Lipoprotein (HDL) Content by Red Wine

The antiatherosclerotic properties of HDL are related with the role of HDL in reverse cholesterol transport as HDL removes excess cholesterol from peripheral tissues and transports it to the liver [52]. In addition, HDL-induced prevention of LDL oxidation plays an important role in the prevention of atherosclerotic plaque development [53]. Conversion of bioactive lipid peroxidation products into inactive compounds by HDL associated enzymes such as paraoxonase [54], platelet activating factor acetylhydrolase [55] and lecithin-cholesterol acyltransferase [56] appear to be the major mechanisms for such an effect. Furthermore, shift of lipid peroxidation products from LDL to HDL followed by conversion of cholesteryl ester hydroperoxides to stable cholesteryl ester hydroxides are implicated in the protective effects of HDL [57]. Red wine consumption increases the level of HDL [58], which is linked with an increase in HDL cholesterol and apolipoprotein A-I [59]. Enrichment of HDL particles with polyunsaturated phospholipids such as arachidonic acid and eicosapentaenoic acid, especially those containing omega-3 (C20:5) are associated with the antiatherosclerotic effects of red wine [59]. Furthermore, recent studies have shown that HDL causes an increase in the production of NO [60], a known vasoprotective agent [30–37]. Red wine-induced increase in the plasma HDL content along with an increase in NO and attenuation of ET-1 as well as hs-CRP play an important role in maintenance of endothelial function in the initial stages of atherosclerosis as shown in Fig. 22.2.

Modification of LDL Oxidation by Red Wine

Polyphenols from red wine such as catechin and quercetin have the capability to protect the LDL from oxidation [61]. It has been demonstrated that polyphenols bind to LDL and polyphenol-enriched LDL is resistant to oxidation in contrast to native LDL [62]. Reduction in LDL oxidation followed by a significant decrease in atherosclerotic lesion area was also observed in apolipoprotein E (apoE) deficient mice in which accelerated atherosclerosis is associated with increased lipid peroxidation of plasma, LDL and very low-density lipoproteins (VLDL), as well as increased susceptibility of these lipoproteins to undergo lipid peroxidation under oxidative stress [63, 64]. In addition, LDL aggregation which is directly related with LDL oxidation, and shown to be taken up by macrophages at an enhanced rate leading to foam cell

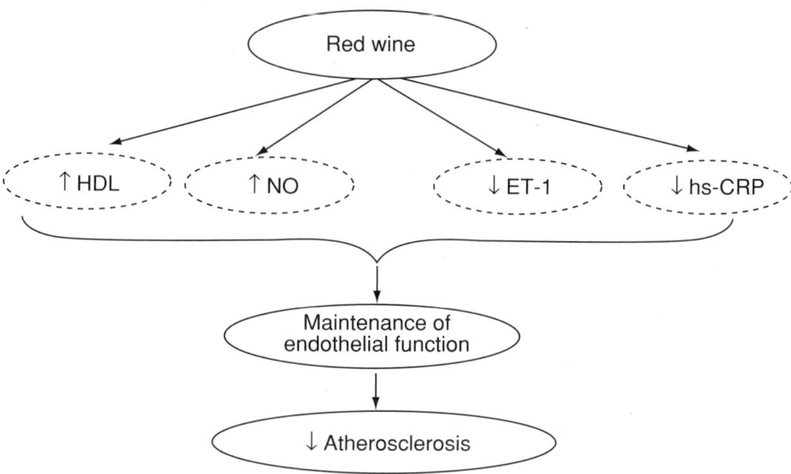

FIGURE 22.2. Effect of red wine on maintenance of endothelial function. HDL, high-density lipoproteins; NO, nitric oxide; ET-1, endothelin-1; hs-CRP, high sensitivity C-reactive protein; ↑, increase; ↓, decrease.

formation, is also reduced by red wine, quercetin, and catechin in atherosclerotic apoE knockout mice [64]. These observations were further verified by treating LDL, isolated from human volunteers, who consumed wine for a period of 2 weeks. Cu^{2+}-induced lipid peroxidation manifested by the formation of lipid peroxides, thiobarbituric acid reaction substances, and conjugated dienes was reduced 72%, 54%, and 46%, respectively in these subjects after red wine consumption [65]. Two possible mechanisms have been proposed; firstly, the phenolic compounds complex with Cu^{2+} reductase and covert Cu^{2+} to Cu^{+}, which in turn reduces hydroperoxides; secondly, phenols in wine may act as self-regenerating reducing components [61]. Although grape juice has been reported to inhibit the Cu^{2+} catalyzed human LDL oxidation *in vitro*, red wine consumption-induced enhancement of LDL resistance to oxidation *in vivo* was not mimicked by grape juice consumption [66]. Increased intestinal absorption of flavonoids due to alcohol content of red wine may be the reason for such an effect [67]. In addition, it has been demonstrated that alcohol is a natural stabilizing agent for polyphenolic components in red wine [68] and thus the effectiveness of polyphenols in red wine and grape juice appears to be different. Red wine polyphenols have also been reported to prevent LDL oxidation by reducing oxidative stress in macrophages through inhibition of oxygenases such as NADPH oxidase, 15-lipoxygenase, cytochrome P450, myeloperoxidase as well as by increasing the cellular antioxidants such as glutathione system [69].

Modification of VSMC Proliferation by Red Wine

VSMC proliferation and migration are the critical events associated with the progressive intimal thickening and arterial wall sclerosis [16]. Red wine polyphenols have been shown to inhibit the SMC proliferation and DNA synthesis in rat aortic SMCs [38]. Different mechanisms have been proposed; the potential pathway involves the downregulation of activating transcriptional factor-1 and cAMP response element leading to the decreased expression of cyclin A, a cell cycle regulator in DNA replication at G1/S transition and in the S and G2/M phases [38]. Red wine polyphenol, resveratrol, has been shown to be the major component linked with inhibition of the cell cycle progression at S/G2 phase transition and prevention of VSMC proliferation [70]. Resveratrol-mediated production of reactive oxygen species by binding with DNA and subsequent reduction of Cu^{2+} to Cu^+ has also been suggested as the DNA cleaving effect of resveratrol [71]. Another mechanism involves the downregulation of phosphatidylinositol 3′-kinase and p38 mitogen-activated protein kinase activity [38]. In addition, red wine polyphenols has been shown to be associated with inhibition of VSMC-mediated migration during atherosclerosis [72]. Inhibition of PDGFβ has also been linked with red wine polyphenol-mediated VSMC proliferation and migration [72]. Furthermore, PDGF-induced overexpression of vascular endothelial growth factor and other growth factors such as α-thrombonin and transforming growth factor-β, which are mainly involved in VSMC proliferation, have been shown to be inhibited by red wine polyphenols [73].

Modification of Inflammation by Red Wine

Moderate amount of red wine has been reported to inhibit the expression of MCP-1, a potent chemoattractant for circulating monocytes, and to cause reduction in neointimal hyperplasia after balloon injury in cholesterol fed rabbits [74]. In addition, red wine prevents the activation of nuclear factor kappa B (NFκB), a transcriptional factor involved in immune and inflammatory response, in peripheral blood mononuclear cells [75]. Furthermore, a significant reduction in the inflammatory mediators of atherosclerosis such as lymphocyte function associated antigen-1, Mac-1, very late activation antigen-4, VCAM-1, ICAM-1, IL-1α, hs-CRP, and fibrogen content was observed in healthy individuals by red wine consumption [51]. Although no effect on serum TNF-α content was observed after red wine polyphenols, TNF-α induced VCAM-1 expression was reduced by polyphenols especially proanthocyanidines in human umbilical endothelial cells leading to reduced adhesion with leukocytes and T-cells [76, 77]. Recent studies have demonstrated that resveratrol in red wine has the capability to reduce TNF-α and lipopolysaccharide stimulated expression of VCAM-1 in HUVEC [78]. This effect of red wine was simulated by *N*-acetyl cysteine, a known antioxidant, suggesting the involvement of reactive oxygen species in this process [78].

Since VCAM-1 promoter has various binding regions for transcriptional factors such as NFκB and activator protein-1, resveratrol has also been shown to inhibit the activation of these transcriptional factors in addition to VCAM-1 [79]. Previous studies have shown that quercetin and resveratrol prove to be effective inhibitors of 5-lipoxygenase pathways, which are known to be involved in the synthesis of leukotriens, powerful mediators of inflammation [80]. Resveratrol also inhibits the release of elastase and β-glucuronidase from neutrophil, which inhibits the expression of β_2-integrin as well as Mac-1, and therefore it downregulates the adhesion-dependent thrombogenic functions [81]. From these observations, it emerges that red wine consumption causes the inhibition of LDL oxidation, VSMC proliferation and migration, as well as various inflammatory mediators, which are important targets for impairment of the plaque formation and progression as shown in Figs. 22.3 and 22.4.

Modification of Platelet Aggregation and Thrombosis by Red Wine

Red wine supplementation has been shown to prevent experimental thrombosis in mechanically stenosed canine coronary arteries independent of its alcohol content [82]. Xia et al. [83] have demonstrated that polyphenols are responsible for the antiplatelet aggregation effect of red wine. Keevil et al. [84] have further confirmed the involvement of polyphenols in the inhibition of platelet aggregation as drinking approximately two cups of purple grape juice for 1 week inhibits the platelet aggregation in healthy humans as

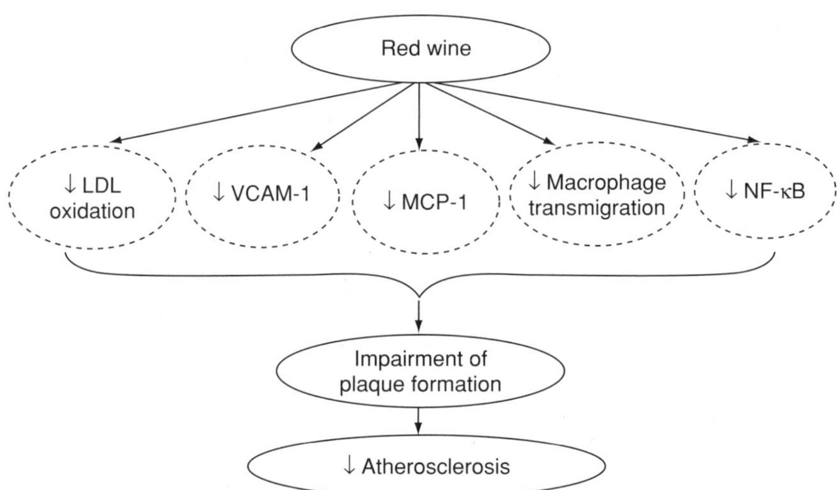

FIGURE 22.3. Effect of red wine on impairment of plaque formation. LDL, low-density lipoprotein; VCAM-1, vascular cell adhesion molecule-1; MCP-1, monocyte chemoattractant protein-1; NFκB, nuclear factor kappa B; ↓, decrease.

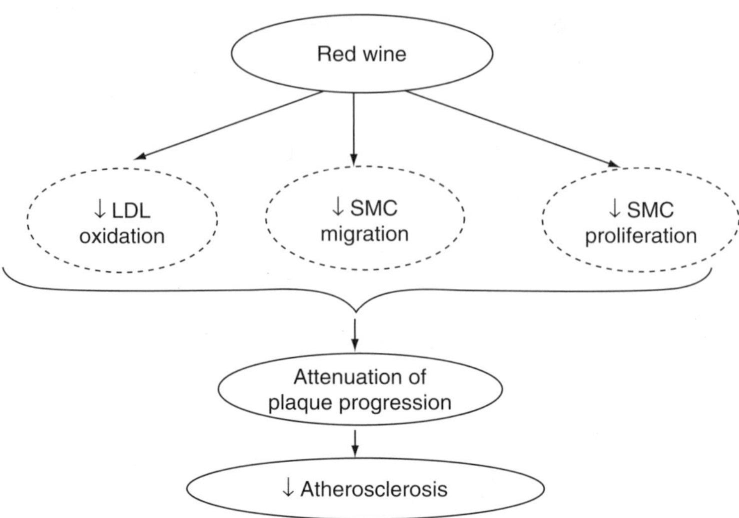

FIGURE 22.4. Effect of red wine on attenuation of plaque progression. LDL, low-density lipoprotein; SMC, smooth muscle cell; ↓, decrease.

determined by whole blood aggregometry with collagen, ADP or thrombin. To further establish whether ethanol or phenolic components of red wine exert inhibitory effect on hemostasis and thrombosis, Wollny et al. [85] treated rats with both of these components. The authors concluded that red wine causes attenuation of hemostatic parameters, such as template bleeding time and platelet adhesion to fibrillar collagen as well as prevents the experimental thrombosis regardless of its alcohol content [85]. In addition, alcohol-free red wine has been shown to be as effective as original wine suggesting that red wine components other than alcohol may be responsible for the observed effects [85]. *In vitro* experiments have demonstrated that polyphenols such as resveratrol and quercitin have the maximum antiplatelet aggregating effect [86]. Quercitin and resveratrol have also been found to potentiate PGI_2 levels by increasing cAMP in platelets, which cause a decrease in cytosolic Ca^{2+} leading to prevention of platelet aggregation [86]. Resveratrol has also been reported as an inhibitor of cyclooxygenase pathway as it causes a decrease in thromboxane A_2 production [87]. In addition, quercetin inhibits lipoxygenase system by impairing 12-hydroxyeicosatetraenoic production. Furthermore, it has been observed that red wine polyphenols inhibit phosphodiesterase, phospholipase A_2 and reduce oxidative stress on thrombocytes [87]. Wolley et al. [85] have demonstrated the involvement of NO in the prevention of experimental thrombosis by red wine polyphenol treatment. It is pointed out that ethanol has no effect on hemostatic parameters, whereas it causes a reduction in experimental thrombosis [85]. These studies indicated that alcohol and non-alcoholic components of red wine have different effects on atherosclerosis plaque development and progression [85]. On the other hand, Pellegrini et al. [88] in a randomized crossover study have reported that alcoholic content of red wine, unlike nonalcoholic components, causes a decrease in collagen-induced platelet aggregation and fibrinogen content suggesting that beneficial effects of red wine on platelet aggregation and haemostatic variables are associated with the alcohol content. Nonetheless, both alcohol and polyphenolic compounds in red wine have antithrombotic properties as the reduction in the levels of tissue factor, von Willebrand factor and factor VII was observed after moderate amounts of alcohol consumption [89] with an increase in tissue plasminogen activator and reduction in plasminogen activator inhibitor antigen-1 after red wine intake [90, 91]. The schematic representation of antithrombotic effect of red wine is given in Fig. 22.5.

Conclusions

From the forgoing discussion, it is evident that we have described the sequence of events associated with the biochemical and cellular processes for the development and progression of atherosclerosis. Various factors such as hypercholesterolemia, oxidative stress, inflammation, endothelial dysfunction, SMC proliferation and migration, platelet activation, and adhesion of monocytes seem to play a critical role in the genesis and progression of atherosclerosis. Although a variety of possible mechanisms have been proposed for the antiatherosclerotic effects of alcoholic and polyphenolic

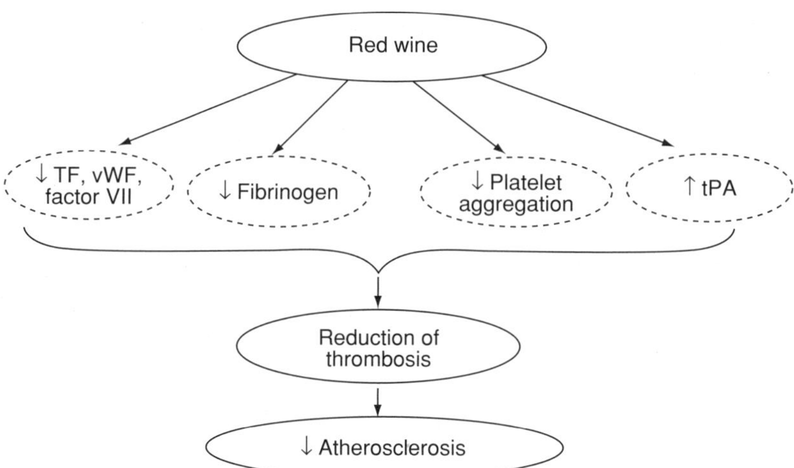

FIGURE 22.5. Effect of red wine on reduction of thrombus formation. TF, tissue factor; vWF, von Willebrand factor; tPA, tissue plasminogen activator; ↑, increase; ↓, decrease.

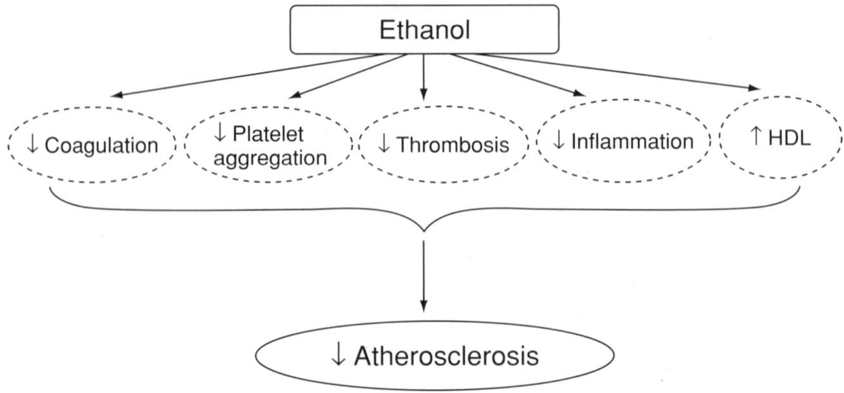

FIGURE 22.6. Effect of red wine ethanol on different mediators of atherosclerosis. ↑, increase; ↓, decrease.

components of red wine [92, 93] as shown in Figs. 22.6 and 22.7, the conclusions drawn from different studies vary from each other. The differences in the amount of wine consumption, experimental conditions, and varieties of wines in terms of the type of grapes, the region in which grown or the method of cultivation may be the reason for such discrepancies [94]. Because of the multifactorial nature of the disease, the results are also influenced by the socioeconomic, genetic, dietary, and environmental factors. Furthermore, American Heart Association Science Advisory Committee has stated that red wine consumption is no more beneficial than moderate amounts of other alcoholic beverages [95]. In fact, the additional beneficial effects due to red wine polyphenolic components can be achieved by grape juice consumption [95]. Therefore, the question related to the therapeutic recommendation of red wine consumption remains unanswered and still a

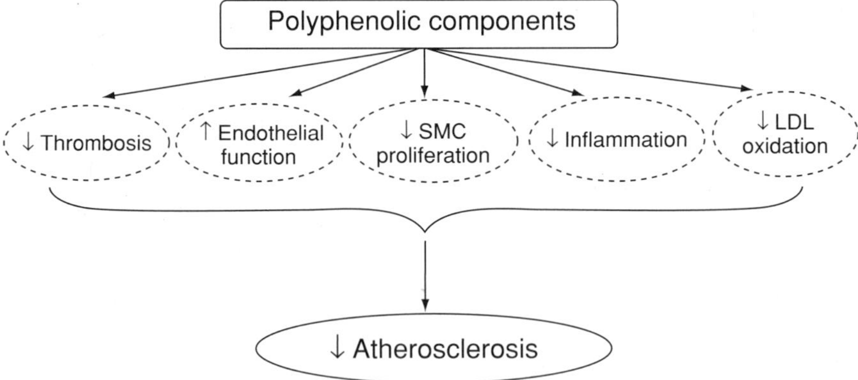

FIGURE 22.7. Effect of red wine polyphenolic components on different mediators of atherosclerosis. ↑, increase; ↓, decrease.

matter of discussion between the patient and the physician. Further studies focusing on differentiating the effects of types of alcoholic beverages, identifying and adjustment of the confounding factors, as well as assessing the pattern of consumption, may provide definitive evidence of the antiatherosclerotic potential of red wine.

Acknowledgments: The work reported in the article was supported by a grant from the Canadian Institutes of Health Research. HKS is a predoctoral fellow of the Heart and Stroke Foundation of Canada. SKC holds a New Investigator Award from the Canadian Institutes of Health Research.

References

1. Murray CJ, Lopez AD: Alternative projections of mortality and disability by cause 1990–2020: Global burden of disease Study. Lancet 349:1498–1504, 1997.
2. Neaton JD, Wentworth D: Serum cholesterol, blood pressure, cigarette smoking, and death from coronary heart disease. Overall findings and differences by age for 316,099 white men. Multiple Risk Factor Intervention Trial Research Group. Arch Intern Med 152: 56–64, 1992.
3. Kannel WB, Wilson PW: An update on coronary risk factors. Med Clin North Am 79: 951–971, 1995.
4. LaRosa JC, He J, Vupputuri S: Effect of statins on risk of coronary disease: a meta-analysis of randomized controlled trials. JAMA 282: 2340–2346, 1999.
5. National Cholesterol Education Program (NCEP) Expert Panel on Detection, Evaluation, and Treatment of High Blood Cholesterol in Adults (Adult Treatment Panel III). Third Report of the National Cholesterol Education Program (NCEP) Expert Panel on Detection, Evaluation, and Treatment of High Blood Cholesterol in Adults (Adult Treatment Panel III) final report. Circulation 106: 3143–3421, 2002.
6. Grundy SM: Alternative approaches to cholesterol-lowering therapy. Am J Cardiol 90: 1135–1138, 2002.
7. Lipid Research Clinics Program. JAMA 252: 2545–2448, 1984.
8. Davignon J: Beneficial cardiovascular pleiotropic effects of statins. Circulation 109: 39–43, 2004.
9. Renaud S, de Lorgeril M: Wine, alcohol, platelets, and the French paradox for coronary heart disease. Lancet 339: 1523–1526, 1992.
10. de Lorgeril M, Salen P, Martin JL, Monjaud I, Delaye J, Mamelle N: Mediterranean diet, traditional risk factors, and the rate of cardiovascular complications after myocardial infarction: final report of the Lyon Diet Heart Study. Circulation 99: 779–785, 1999.
11. van de Wiel A, van Golde PH, Hart HC: Blessings of the grape. Eur J Intern Med 12: 484–489, 2001.
12. Waterhouse AL: Wine phenolics. Ann NY Acad Sci 957: 21–36, 2002.
13. Stary HC, Chandler AB, Dinsmore RE, Fuster V, Glagov S, Insull W Jr, Rosenfeld ME, Schwartz CJ, Wagner WD, Wissler RW: A definition of advanced types of atherosclerotic lesions and a histological classification of atherosclerosis. A report from the Committee on Vascular Lesions of the Council on Arteriosclerosis, American Heart Association. Circulation 92: 1355–1374, 1995.

14. Virmani R, Kolodgie FD, Burke AP, Farb A, Schwartz SM: Lessons from sudden coronary death: a comprehensive morphological classification scheme for atherosclerotic lesions. Arterioscler Thromb Vasc Biol 20: 1262–1275, 2000.

15. Berliner JA, Navab M, Fogelman AM, Frank JS, Demer LL, Edwards PA, Watson AD, Lusis AJ: Atherosclerosis: basic mechanisms. Oxidation, inflammation, and genetics. Circulation 91: 2488–2496, 1995.

16. Ross R, Glomset JA: Atherosclerosis and the arterial smooth muscle cell: Proliferation of smooth muscle is a key event in the genesis of the lesions of atherosclerosis. Science 180: 1332–1339, 1973.

17. Ross R: Atherosclerosis—an inflammatory disease. N Engl J Med 340: 115–126, 1999.

18. Matsuoka H: Endothelial dysfunction associated with oxidative stress in human. Diabetes Res Clin Pract 54: 65–72, 2001.

19. da Luz PL, Coimbra SR: Wine, alcohol and atherosclerosis: clinical evidences and mechanisms. Braz J Med Biol Res 37: 1275–1295, 2004.

20. Eriksson EE: Mechanisms of leukocyte recruitment to atherosclerotic lesions: future prospects. Curr Opin Lipidol 15: 553–558, 2004.

21. Osterud B, Bjorklid E: Role of monocytes in atherogenesis. Physiol Rev 83: 1069–1112, 2003.

22. Holvoet P, Collen D: Oxidation of low density lipoproteins in the pathogenesis of atherosclerosis. Atherosclerosis 137: 33–38, 1998.

23. Chisolm GM, Steinberg D: The oxidative modification hypothesis of atherogenesis: an overview. Free Radic Biol Med 28: 1815–1826, 2000.

24. Lindner A, Charra B, Sherrard DJ, Scribner BH: Accelerated atherosclerosis in prolonged maintenance hemodialysis. N Engl J Med 290: 697–701, 1974.

25. Saini HK, Arneja AS, Dhalla NS: Role of cholesterol in cardiovascular dysfunction. Can J Cardiol 20: 333–346, 2004.

26. Xu YJ, Aziz OA, Bhugra P, Arneja AS, Mendis MR, Dhalla NS: Potential role of lysophosphatidic acid in hypertension and atherosclerosis. Can J Cardiol 19: 1525–1536, 2003.

27. Xu YJ, Rathi SS, Chapman DC, Arneja AS, Dhalla NS: Mechanisms of lysophosphatidic acid-induced DNA synthesis in vascular smooth muscle cells. J Cardiovasc Pharmacol 41: 381–387, 2003.

28. Wolfbauer G, Glick JM, Minor LK, Rothblat GH: Development of the smooth muscle foam cell: uptake of macrophage lipid inclusions. Proc Natl Acad Sci USA 83: 7760–7764, 1986.

29. Shah PK: Pathophysiology of plaque rupture and the concept of plaque stabilization. Cardiol Clin 21: 303–314, 2003.

30. Moncada S, Palmer RM, Higgs EA: Nitric oxide: physiology, pathophysiology, and pharmacology. Pharmacol Rev 43: 109–142, 1991.

31. Zeiher AM, Fisslthaler B, Schray-Utz B, Busse R: Nitric oxide modulates the expression of monocyte chemoattractant protein 1 in cultured human endothelial cells. Circ Res 76: 980–986, 1995.

32. Gauthier TW, Scalia R, Murohara T, Guo JP, Lefer AM: Nitric oxide protects against leukocyte–endothelium interactions in the early stages of hypercholesterolemia. Arterioscler Thromb Vasc Biol 15: 1652–1659, 1995.

33. Forstermann U, Nakane M, Tracey WR, Pollock JS: Isoforms of nitric oxide synthase: functions in the cardiovascular system. Eur Heart J 14: 10–15, 1993.

34. Garg UC, Hassid A: Nitric oxide-generating vasodilators and 8-bromo-cyclic guanosine monophosphate inhibit mitogenesis and proliferation of cultured rat vascular smooth muscle cells. J Clin Invest 83: 1774–1777, 1989.

35. Radomski MW, Palmer RM, Moncada S: The role of nitric oxide and cGMP in platelet adhesion to vascular endothelium. Biochem Biophys Res Commun 148:1482–1489, 1987.

36. Hirafuji M, Nezu A, Shinoda H, Minami M: Involvement of platelet cyclic GMP but not cyclic AMP suppression in leukocyte-dependent platelet adhesion to endothelial cells induced by platelet-activating factor *in vitro*. Br J Pharmacol 117: 299–304, 1996.

37. Goss SP, Kalyanaraman B, Hogg N: Antioxidant effects of nitric oxide and nitric oxide donor compounds on low-density lipoprotein oxidation. Methods Enzymol 301: 444–453, 1999.

38. Dell'Agli M, Busciala A, Bosisio E: Vascular effects of wine polyphenols. Cardiovasc Res 63: 593–602, 2004.

39. Fitzpatrick DF, Hirschfield SL, Coffey RG: Endothelium-dependent vasorelaxing activity of wine and other grape products. Am J Physiol Heart Circ Physiol 265: H774–H778, 1993.

40. Flesch M, Schwarz A, Bohm M: Effects of red and white wine on endothelium-dependent vasorelaxation of rat aorta and human coronary arteries. Am J Physiol Heart Circ Physiol 275: H1183–H1190, 1998.

41. Andriambeloson E, Kleschyov AL, Muller B, Beretz A, Stoclet JC, Andriantsitohaina R: Nitric oxide production and endothelium-dependent vasorelaxation induced by wine polyphenols in rat aorta. Br J Pharmacol 120: 1053–1058, 1997.

42. Zenebe W, Pechanova O, Andriantsitohaina R: Red wine polyphenols induce vasorelaxation by increased nitric oxide bioactivity. Physiol Res 52: 425–432, 2003.

43. Wallerath T, Poleo D, Li H, Forstermann U: Red wine increases the expression of human endothelial nitric oxide synthase: a mechanism that may contribute to its beneficial cardiovascular effects. J Am Coll Cardiol 41: 471–478, 2003.

44. Wallerath T, Deckert G, Ternes T, Anderson H, Li H, Witte K, Forstermann U: Resveratrol, a polyphenolic phytoalexin present in red wine, enhances expression and activity of endothelial nitric oxide synthase. Circulation 106: 1652–1658, 2002.

45. Goldberg DM, Yan J, Ng E, Diamandis EP, Karumanchiri A, Soleas G, Waterhouse AL: A global survey of *trans*-resveratrol concentrations in commercial wines. Am J Enol Viticult 46: 159–165, 1995.

46. Andriambeloson E, Stoclet JC, Andriantsitohaina R: Mechanism of endothelial nitric oxide-dependent vasorelaxation induced by wine polyphenols in rat thoracic aorta. J Cardiovasc Pharmacol 33: 248–254, 1999.

47. Diebolt M, Bucher B, Andriantsitohaina R: Wine polyphenols decrease blood pressure, improve NO vasodilatation, and induce gene expression. Hypertension 38: 159–165, 2001.

48. Derek D, Pearson DA, German JB: Endothelial cell basal PGI2 release is stimulated by wine *in vitro*: one mechanism that may mediate the vasoprotective effects of wine. J Nutr Biochem 8: 647–651, 1997.

49. Ndiaye M, Chataigneau T, Andriantsitohaina R, Stoclet JC, Schini-Kerth VB: Red wine polyphenols cause endothelium-dependent EDHF-mediated relaxations in porcine coronary arteries via a redox-sensitive mechanism. Biochem Biophys Res Commun 310: 371–377, 2003.

50. Corder R, Douthwaite JA, Lees DM, Khan NQ, Viseu Dos Santos AC, Wood EG, Carrier MJ: Endothelin-1 synthesis reduced by red wine. Nature 414: 863–864, 2001.

51. Estruch R, Sacanella E, Badia E, Antunez E, Nicolas JM, Fernandez-Sola J, Rotilio D, de Gaetano G, Rubin E, Urbano-Marquez A: Different effects of red wine and gin consumption on inflammatory biomarkers of atherosclerosis: a prospective randomized crossover trial. Effects of wine on inflammatory markers. Atherosclerosis 175: 117–123, 2004

52. Fielding CJ, Fielding PE: Molecular physiology of reverse cholesterol transport. J Lipid Res 36: 211–228, 1995.

53. Parthasarathy S, Barnett J, Fong LG: High-density lipoprotein inhibits the oxidative modification of low-density lipoprotein. Biochim Biophys Acta 1044: 275–283, 1990.

54. Watson AD, Berliner JA, Hama SY, La Du BN, Faull KF, Fogelman AM, Navab M: Protective effect of high-density lipoprotein associated paraoxonase. Inhibition of the biological activity of minimally oxidized low density lipoprotein. J Clin Invest 96: 2882–2891, 1995.

55. Watson AD, Navab M, Hama SY, Sevanian A, Prescott SM, Stafforini DM, McIntyre TM, Du BN, Fogelman AM, Berliner JA: Effect of platelet activating factor-acetylhydrolase on the formation and action of minimally oxidized low density lipoprotein. J Clin Invest 95: 774–782, 1995.

56. Rajaram OV, Barter PJ: Increases in the particle size of high-density lipoproteins induced by purified lecithin: cholesterol acyltransferase: effect of low-density lipoproteins. Biochim Biophys Acta 877: 406–414, 1986.

57. Christison JK, Rye KA, Stocker R: Exchange of oxidized cholesteryl linoleate between LDL and HDL mediated by cholesteryl ester transfer protein. J Lipid Res 36: 2017–2026, 1995.

58. Araya J, Rodrigo R, Orellana M, Rivera G: Red wine raises plasma HDL and preserves long-chain polyunsaturated fatty acids in rat kidney and erythrocytes. Br J Nutr 86: 189–195, 2001.

59. Perret B, Ruidavets JB, Vieu C, Jaspard B, Cambou JP, Terce F, Collet X: Alcohol consumption is associated with enrichment of high-density lipoprotein particles in polyunsaturated lipids and increased cholesterol esterification rate. Alcohol Clin Exp Res 26: 1134–1140, 2002.

60. Yuhanna IS, Zhu Y, Cox BE, Hahner LD, Osborne-Lawrence S, Lu P, Marcel YL, Anderson RG, Mendelsohn ME, Hobbs HH, Shaul PW: High-density lipoprotein binding to scavenger receptor-BI activates endothelial nitric oxide synthase. Nat Med 7: 853–857, 2001.

61. Frankel EN, Kanner J, German JB, Parks E, Kinsella JE: Inhibition of oxidation of human low-density lipoprotein by phenolic substances in red wine. Lancet 341: 454–457, 1993.

62. Aviram M, Fuhrman B: Wine flavonoids protect against LDL oxidation and atherosclerosis. Ann NY Acad Sci 957:146–161, 2002.

63. Hayek T, Oiknine J, Brook JG, Aviram M: Increased plasma and lipoprotein lipid peroxidation in apo E-deficient mice. Biochem Biophys Res Commun 201: 1567–1574, 1994.

64. Hayek T, Fuhrman B, Vaya J, Rosenblat M, Belinky P, Coleman R, Elis A, Aviram M: Reduced progression of atherosclerosis in apolipoprotein E-deficient mice following consumption of red wine, or its polyphenols quercetin or cate-

chin, is associated with reduced susceptibility of LDL to oxidation and aggregation. Arterioscler Thromb Vasc Biol 17: 2744–2752, 1997.

65. Fuhrman B, Lavy A, Aviram M: Consumption of red wine with meals reduces the susceptibility of human plasma and low-density lipoprotein to lipid peroxidation. Am J Clin Nutr 61: 549–554, 1995.

66. Miyagi Y, Miwa K, Inoue H: Inhibition of human low-density lipoprotein oxidation by flavonoids in red wine and grape juice. Am J Cardiol 80: 1627–1631, 1997.

67. Serafini M, Maiani G, Ferro-Luzzi A: Alcohol-free red wine enhances plasma antioxidant capacity in humans. J Nutr 128: 1003–1007, 1998.

68. Carbonneau MA, Leger CL, Monnier L, Bonnet C, Michel F, Fouret G, Dedieu F, Descomps B: Supplementation with wine phenolic compounds increases the antioxidant capacity of plasma and vitamin E of low-density lipoprotein without changing the lipoprotein Cu^{2+}-oxidizability: possible explanation by phenolic location. Eur J Clin Nutr 51: 682–690, 1997.

69. Fuhrman B, Aviram M: Flavonoids protect LDL from oxidation and attenuate atherosclerosis. Curr Opin Lipidol 12: 41–48, 2001.

70. Ragione FD, Cucciolla V, Borriello A, Pietra VD, Racioppi L, Soldati G, Manna C, Galletti P, Zappia V: Resveratrol arrests the cell division cycle at S/G2 phase transition. Biochem Biophys Res Commun 250: 53–58, 1998.

71. Fukuhara K, Miyata N: Resveratrol as a new type of DNA-cleaving agent. Bioorg Med Chem Lett 8: 3187–3192, 1998.

72. Rosenkranz S, Knirel D, Dietrich H, Flesch M, Erdmann E, Bohm M: Inhibition of the PDGF receptor by red wine flavonoids provides a molecular explanation for the "French paradox". FASEB J 16: 1958–1960, 2002.

73. Oak MH, Chataigneau M, Keravis T, Chataigneau T, Beretz A, Andriantsitohaina R, Stoclet JC, Chang SJ, Schini-Kerth VB: Red wine polyphenolic compounds inhibit vascular endothelial growth factor expression in vascular smooth muscle cells by preventing the activation of the p38 mitogen-activated protein kinase pathway. Arterioscler Thromb Vasc Biol 23: 1001–1007, 2003.

74. Feng AN, Chen YL, Chen YT, Ding YZ, Lin SJ: Red wine inhibits monocyte chemotactic protein-1 expression and modestly reduces neointimal hyperplasia after balloon injury in cholesterol-Fed rabbits. Circulation 100: 2254–2259, 1999.

75. Blanco-Colio LM, Valderrama M, Alvarez-Sala LA, Bustos C, Ortego M, Hernandez-Presa MA, Cancelas P, Gomez-Gerique J, Millan J, Egido J: Red wine intake prevents nuclear factor-kappaB activation in peripheral blood mononuclear cells of healthy volunteers during postprandial lipemia. Circulation 102: 1020–1026, 2000.

76. Sen CK, Bagchi D: Regulation of inducible adhesion molecule expression in human endothelial cells by grape seed proanthocyanidin extract. Mol Cell Biochem 216: 1–7, 2001.

77. Bagchi D, Bagchi M, Stohs S, Ray SD, Sen CK, Preuss HG: Cellular protection with proanthocyanidins derived from grape seeds. Ann NY Acad Sci 957: 260–270, 2002.

78. Carluccio MA, Siculella L, Ancora MA, Massaro M, Scoditti E, Storelli C, Visioli F, Distante A, De Caterina R: Olive oil and red wine antioxidant polyphenols inhibit endothelial activation: antiatherogenic properties of Mediterranean diet phytochemicals. Arterioscler Thromb Vasc Biol 23: 622–629, 2003.

79. Neish AS, Williams AJ, Palmer HJ, Whitley MZ, Collins T: Functional analysis of the human vascular cell adhesion molecule 1 promoter. J Exp Med 176: 1583–1593, 1992.

80. Soleas GJ, Diamandis EP, Goldberg DM: Wine as a biological fluid: history, production, and role in disease prevention. J Clin Lab Anal 11: 287–313, 1997.

81. Rotondo S, Rajtar G, Manarini S, Celardo A, Rotillo D, de Gaetano G, Evangelista V, Cerletti C: Effect of trans-resveratrol, a natural polyphenolic compound, on human polymorphonuclear leukocyte function. Br J Pharmacol 123: 1691–1699, 1998.

82. Demrow HS, Slane PR, Folts JD: Administration of wine and grape juice inhibits *in vivo* platelet activity and thrombosis in stenosed canine coronary arteries. Circulation 91: 1182–1188, 1995.

83. Xia J, Allenbrand B, Sun GY: Dietary supplementation of grape polyphenols and chronic ethanol administration on LDL oxidation and platelet function in rats. Life Sci 63: 383–390, 1998.

84. Keevil JG, Osman HE, Reed JD, Folts JD: Grape juice, but not orange juice or grapefruit juice, inhibits human platelet aggregation. J Nutr 130: 53–56, 2000.

85. Wollny T, Aiello L, Di Tommaso D, Bellavia V, Rotilio D, Donati MB, de Gaetano G, Iacoviello L: Modulation of haemostatic function and prevention of experimental thrombosis by red wine in rats: a role for increased nitric oxide production. Br J Pharmacol 127: 747–755, 1999.

86. Pace-Asciak CR, Hahn S, Diamandis EP, Soleas G, Goldberg DM: The red wine phenolics trans-resveratrol and quercetin block human platelet aggregation and eicosanoid synthesis: implications for protection against coronary heart disease. Clin Chim Acta 235: 207–219, 1995.

87. Ruf JC: Wine and polyphenols related to platelet aggregation and atherothrombosis. Drugs Exp Clin Res 25: 125–131, 1999.

88. Pellegrini N, Pareti FI, Stabile F, Brusamolino A, Simonetti P: Effects of moderate consumption of red wine on platelet aggregation and haemostatic variables in healthy volunteers. Eur J Clin Nutr 50: 209–213, 1996.

89. Mennen LI, Balkau B, Vol S, Caces E, Eschwege E: Fibrinogen: a possible link between alcohol consumption and cardiovascular disease? DESIR Study Group. Arterioscler Thromb Vasc Biol 19: 887–892, 1999.

90. Mukamal KJ, Jadhav PP, D'Agostino RB, Massaro JM, Mittleman MA, Lipinska I, Sutherland PA, Matheney T, Levy D, Wilson PW, Ellison RC, Silbershatz H, Muller JE, Tofler GH: Alcohol consumption and hemostatic factors: analysis of the Framingham Offspring cohort. Circulation 104: 1367–1373, 2001.

91. Mezzano D: Distinctive effects of red wine and diet on haemostatic cardiovascular risk factors. Biol Res 37: 217–224, 2004.

92. Szmitko PE, Verma S: Antiatherogenic potential of red wine: clinician update. Am J Physiol Heart Circ Physiol 288: H2023–H2030, 2005.

93. Bagchi D, Sen CK, Ray SD, Das DK, Bagchi M, Preuss HG, Vinson JA: Molecular mechanisms of cardioprotection by a novel grape seed proanthocyanidin extract. Mutat Res 523–524: 87–97, 2003.

94. Wollin SD, Jones PJ: Alcohol, red wine and cardiovascular disease. J Nutr 131: 1401–1404, 2001.

95. Goldberg IJ, Mosca L, Piano MR, Fisher EA: Nutrition Committee, Council on Epidemiology and Prevention, and Council on Cardiovascular Nursing of the American Heart Association. AHA Science Advisory: Wine and your heart: a science advisory for healthcare professionals from the Nutrition Committee, Council on Epidemiology and Prevention, and Council on Cardiovascular Nursing of the American Heart Association. Circulation 103: 472–475, 2001.

23

Endothelial Dysfunction and Atherosclerosis: Role of Dietary Fats

KANTA CHECHI, PRATIBHA DUBEY, AND SUKHINDER KAUR CHEEMA

Abstract

Atherosclerosis or arteriosclerosis is a generic term that defines a number of diseases in which the arterial wall becomes thickened and loses elasticity. Coronary heart disease (CHD), myocardial infarction, angina pectoris, cerebral vascular disease, and thrombotic stroke are few examples of such disorders. The development of atherosclerosis is a complex process that begins with an accumulation of lipid deposits on the arterial wall called "plaques". The plaque formation in the inner lining of vessel wall progresses to arterial thickening and narrowing of its lumen, which can partially block the flow of blood through the artery. Extensive research is aimed to unravel the cellular and molecular events involved in the initiation and propagation of an atherogenic plaque. However, the exact sequence of events that leads to its development is still unknown. There is an increasing evidence that endothelial dysfunction is the primary event in the progression of atherosclerosis. Endothelial dysfunction is known to appear years before the actual symptoms of atherosclerosis develop, which emphasizes its causal role in the progression of the disease. Oxidative stress and inflammatory processes have been recognized as key mediators of the loss of normal endothelial function. In fact, reactive oxygen species (ROS) generation, and triggering of various inflammatory cascades are now known to be intimately linked with all the phases of development and progression of atherosclerosis. Consumption of high dietary fats can alter these parameters and therefore, play an important role in the etiology of CHDs. However, all fats do not exert similar effects in the development of these disorders. While saturated fat intake is associated with an increased incidence of CHDs, fish consumption has been shown to lower the incidence of these disorders. Considering the role played by different dietary fats in the alteration of lipoprotein metabolism and other inflammatory cascades, they are now looked upon as nutritional tool for the prevention of atherosclerosis and maintenance of normal cardiovascular health. This chapter provides a detailed account of endothelial dysfunction and its role in the development of atherosclerosis. The role of dietary fats in the development and progression of endothelial dysfunction and atherosclerosis has also been discussed.

Keywords: dietary fats; endothelial dysfunction; nitric oxide; oxidized LDL; reactive oxygen species

Abbreviations: AngI, angiotensin I; AngII, angiotensin II; ACE, angiotensin-converting enzyme; ADMA, asymmetric dimethylarginine; AA, arachidonic acid;

BMK1, big mitogen-activated protein kinase 1; CHD, coronary heart disease; cAMP, cyclic adenosine monophosphate; cGMP, cyclic guanidine monophosphate; COX, cyclooxygenase; DHA, docosahexaenoic acid; EPA, eicosapentaenoic acid; ET-1, endothelin-1; EDHF, endothelium-derived hyperpolarizing factor; ERK, extracellular signal-regulated kinase; FGF, fibroblast growth factor; ICAM-1, intercellular adhesion molecule-1; IFN-γ, interferon-γ; IL, interleukin; LA, linoleic acid; LOX, lipoxygenase; LDL, low-density lipoproteins; MMP, matrix metalloproteinases; MAPK, mitogen-activated protein kinase; MUFA, monounsaturated fatty acids; NO, nitric oxide; NOS, nitric oxide synthase; Ox LDL, oxidized low-density lipoproteins; PLA2, phospholipase A2; PAF-1, platelet activating factor-1; PDGF, platelet-derived growth factor; PUFA, polyunsaturated fatty acids; PKC, protein kinase C; ROS, reactive oxygen species; SR, scavenger receptors; SFA, saturated fatty acids; TGF-β, transforming growth factor beta; TNF-α, tumor necrosis factor alpha; VCAM-1, vascular cell adhesion molecule-1; VSMC, vascular smooth muscle cells

Introduction

The vascular endothelium is the innermost lining of the blood vessels. For several decades, it was considered a unicellular layer acting as a semipermeable membrane between the blood and interstitium. However, it is now being recognized as a dynamic heterogeneous organ, which possesses secretory, regulatory, and immunological functions [1]. The endothelium regulates the flow of nutrients, various biologically active molecules, and blood cells through the entire human body. It is selectively permeable, possessing various cell membrane receptors for molecules that include proteins (growth factors, coagulation, and anticoagulation proteins), lipid-transporting particles, metabolites (nitric oxide; NO, serotonin), and hormones (endothelin-1; ET-1). The normal endothelium also regulates the vascular tone and controls inflammation and thrombotic events occurring during hemostasis and repair [2].

   Activation of the endothelium due to injury or various pathological factors leads to the alterations in its different regulatory functions. The endothelium becomes incapable of maintaining vascular homeostasis. This characterizes a condition of endothelial dysfunction, which can be defined as an imbalance between relaxing and contracting factors, between procoagulant and anticoagulant mediators, or between stimulants and inhibitors of cell growth and proliferation, respectively [3]. Some of the important parameters of endothelial function, which get disturbed during the diseased or pathological conditions, are described below.

Regulation of Vascular Tone by the Endothelium

Regulation of normal vascular tone is known to be the most critical function of the endothelium. Vascular endothelium maintains the vascular tone by secreting various vasoconstricting and vasodilating factors. The balance between these factors determines the functional or dysfunctional state of the endothelium.

Vasorelaxing Factors

The principle vasodilating factor secreted by the endothelial cells is NO. It is a free radical that is produced by the oxidation of L-arginine to L-citrulline, by the enzyme nitric oxide synthase (NOS) (Fig. 23.1) [4]. This enzyme has three isoforms: the neuronal NOS (nNOS), the inducible NOS (iNOS), and the endothelial NOS (eNOS). The endothelial cells constitutively express the eNOS, which constantly releases small amounts of NO in systemic and pulmonary circulations [5]. Once released from the endothelial cells, NO interacts with the heme prosthetic group of guanylate cyclase present on the surface of underlying smooth muscle cells. This causes the activation of guanylate cyclase and increased production of cyclic guanidine monophosphate (cGMP). Increased cGMP leads to a decrease in the intracellular

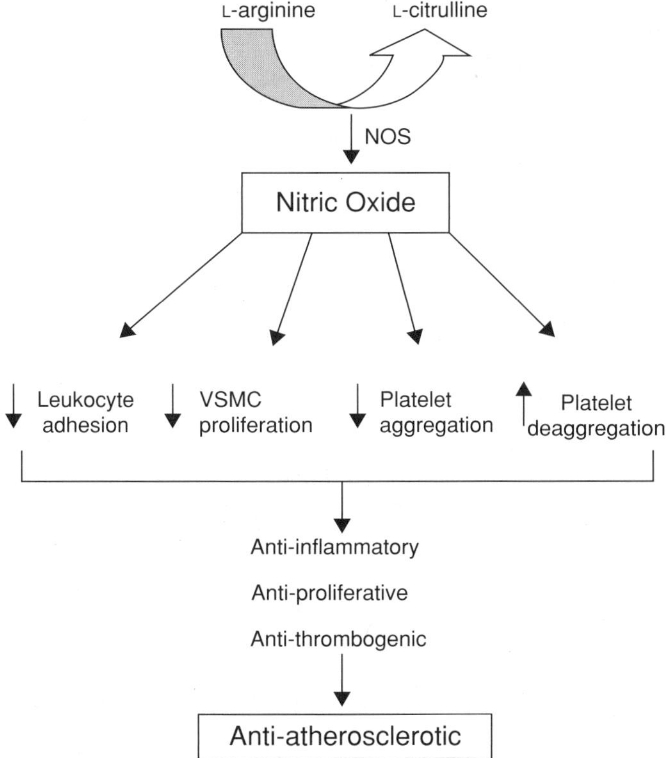

FIGURE 23.1. Various pathways involved in mediating the antiatherosclerotic effects of nitric oxide. NOS, nitric oxide synthase; VSMCs, vascular smooth muscle cells. ↑, increase; ↓, decrease.

calcium concentration, causing relaxation of the underlying vascular smooth muscle cells (VSMC) [6].

Prostacyclins and endothelium-derived hyperpolarizing factor (EDHF) are the other important vasodilators involved in the regulation of vasomotor tone that are secreted by the endothelium. Prostacyclins are produced in response to humoral and hemodynamic factors by the endothelium [7]. These are synthesized by enzyme cyclooxygenase (COX) using arachidonic acid (AA) as a substrate [8]. Prostacyclins stimulate adenylate cyclase leading to increased intracellular concentration of cyclic adenosine monophosphate (cAMP) in VSMC ultimately causing their relaxation [9, 10]. EDHF, on the other hand is known to promote VSMC relaxation by increasing cell membrane conductance of potassium [11].

Vasoconstrictive Factors

For the regulation of normal vascular tone, endothelial cells express variety of constrictive factors, which include endothelins, thromboxane A2, and prostaglandin H. Among these, ET-1 is the most potent factor secreted by the endothelial cells [12]. It is produced in response to stimuli like thrombin, adrenaline, angiotensin II (AngII), hypoxia, and increased shear stress [13]. It antagonizes the action of NO by binding to specific receptors in VSMC causing increased intracellular concentration of calcium leading to vasoconstriction. Interestingly, ET-1 is known to stimulate the production of NO and prostacyclins from the functionally intact endothelium, which modulates vasoconstrictor action and reduces the synthesis of ET-1 itself [14].

Thromboxane A2 and prostaglandin H2, secreted by the endothelial cells activate the thromboxane receptors in VSMC and platelets. These factors also bring about the constriction of VSMC as opposed to the effects of NO and prostacyclins. However, the role of these substances in coronary circulation has not been clearly established. Platelet activating factor-1 (PAF-1) is another vasoconstrictor synthesized and released by endothelial cells in response to humoral and hemodynamic stimuli, which participates in the regulation of vasomotor tone. Finally, the endothelium also expresses the angiotensin-converting enzyme (ACE), which leads to the production of AngII from the angiotensin I (AngI), which directly stimulates the production of ET-1 [1].

Overall, it is the fine interplay of vasoconstrictive and vasorelaxing factors that ultimately determines the vascular tone of healthy endothelium. Any alterations in this balance leads to the development of endothelial dysfunction.

Loss of Vascular Tone Associated with Endothelial Dysfunction

Loss of vascular tone is the primary event occurring during the manifestation of endothelial dysfunction. It is characterized by the reduction in vasorelaxing factors and an increase in vasoconstricting factors. Several studies have reported a strong association between reduction in the bioavailability of the vasodilator NO and progression of endothelial dysfunction [15–20]. This may result either due to reduced activity of eNOS or due to increased degradation of NO. Since endothelial NO is largely responsible for endothelium-dependent dilation of blood vessels, this dilation is severely blunted in coronary and peripheral arteries during atherosclerosis [18, 21, 22].

In addition to the loss of vasodilator NO, endothelial injury is also associated with an increase in the production of vasoconstrictive factors such as ET-1 [23–25]. This leads to further exaggeration of loss of NO and the vasculature becomes prone to unregulated constriction. This can further lead to the development of hypertension and future coronary heart disease (CHD).

Regulation of Inflammation and Thrombosis by the Endothelium

In addition to the maintenance of vascular tone, a normal endothelium possesses anti-proliferative and anti-inflammatory properties. Endothelium-derived vasodilator NO inhibits leukocyte adhesion to the endothelium [26, 27], migration and proliferation of VSMC [28, 29], and stimulates the proliferation of endothelial cells [30]. NO is further known to inhibit platelet aggregation and promote platelet deaggregation [31] (Fig. 23.1). Prostacyclins, another endothelium-derived vasodilator, also interact synergistically with NO, causing inhibition of platelet adhesion and aggregation [32]. Furthermore, the healthy endothelial cells have a negatively charged surface coated with heparans, which exert contact inhibition [33]. Endothelial cells also express anticoagulant factors such as tissue plasminogen activator (tPA), inactivators of thrombin, and thrombomodulin [32]. As a result, leukocytes do not stick to the vascular surface and cell proliferation is tightly controlled [33, 34]. These are among the multiple and redundant healthy defenses that are impaired during the endothelial dysfunction and atherosclerosis.

Loss of Control of Inflammation and Thrombosis

Loss of endothelial function results in an impairment of normal anticoagulant defenses. Reduced bioavailability of NO, abnormal heparans, local thrombin activation, relative lack of tissue PAF-1, and thrombomodulin are the factors that favor local coagulation [35]. Similarly, increased tissue plasminogen activator inhibitor and lack of tissue PAF-1 impair clot lysis [36, 37]. Thrombin activation, membrane-bound platelet adhesion molecules, exposure of collagen, increased production of tissue factor, and loss of NO also favor increased platelet adhesion and aggregation [38, 39].

The various risk factors and pathological stimuli associated with the development of endothelial dysfunction also lead to the abnormal activation of the vessel wall. This leads to the stimulation of certain signaling cascades resulting in the onset of adhesion and inflammation in the vasculature. This results in an abnormal increase in the growth of VSMC, fibroblasts, and matrix within the vessel wall. This process further leads to intimal thickening and plaque formation [35].

Endothelial Dysfunction: Role of Oxidative stress

The development of the endothelial dysfunction is associated with a number of altered physiological markers. Oxidative stress or the presence of reactive oxygen species (ROS) is one such marker that is known to be a key player in its development. The mechanism for the generation of intracellular ROS involves multiple enzyme systems including NAD(P)H oxidase, xanthine oxidase, lipoxygenase (LOX), cytochrome p450, monooxygenase, and COX [40, 41]. Presence of ROS in vasculature leads to quenching of NO, resulting in the formation of peroxynitrite [42]. Peroxynitrite, a cytotoxic oxidant, further enhances the oxidative stress in the system by uncoupling of NOS [43]. It causes a reduction of eNOS cofactor tetrahydrobiopterin BH4, resulting in the formation of BH2. When this occurs, eNOS is uncoupled, thus losing the oxygenase and NO formation activity. This further activates the reductase function of eNOS and more ROS are generated. So, NOS switches from its NO-producing oxygenase activity to ROS producing reductase activity. This leads to a positive feedback loop resulting in enhanced ROS generation, thereby causing endothelial dysfunction [44]. Peroxynitrite is also known to be an important mediator of oxidation of low-density lipoproteins (LDL) [45]. Oxidized LDL (Ox LDL), once formed, can induce endothelial dysfunction and thus significantly contribute to the pathogenesis of atherosclerosis. There is evidence that the degree of endothelial dysfunction relates to the number and density of LDL particles, the level of Ox LDL, the susceptibility of LDL to oxidation in an individual and the titer of autoantibodies to Ox LDL [46]. Ox LDL also affects NO production by several mechanisms like: (a) Ox LDL can inhibit agonist-stimulated arginine transport by endothelial cells, (b) Ox LDL can mediate downregulation of eNOS expression both at mRNA and protein levels, thereby directly inhibiting NO production [47], (c) Ox LDL upregulates asymmetric dimethylarginine (ADMA), which is a competitive antagonist of NOS and may reduce NO production via substrate competition with L-arginine [48, 49], (d) Ox LDL increases synthesis of caveolin-1, which is a membrane protein known to bind and inactivate eNOS, thereby inhibiting production of NO [50]. By reducing NO production, Ox LDL also promotes thrombin generation and platelet aggregation [51]. In addition, Ox LDL alters the balance of other endothelial products, such as prostacyclins and ET-1, further contributing to platelet aggregation [52, 35]. Thus, generation of Ox LDL may mediate endothelial dysfunction by a range of molecular mechanisms (Fig. 23.2).

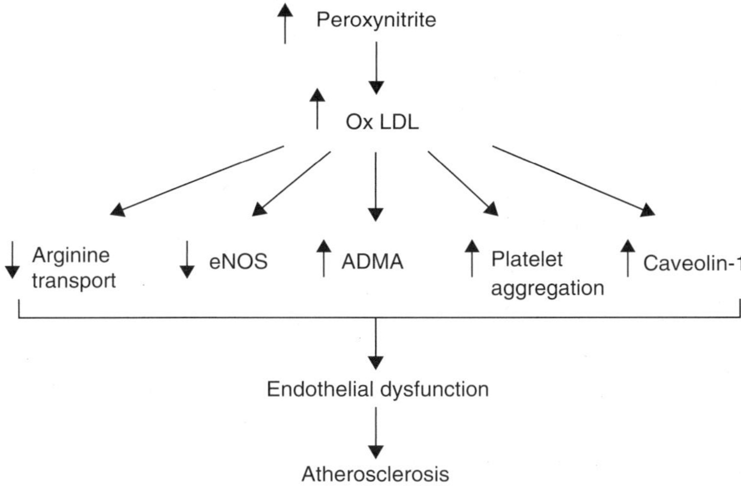

FIGURE 23.2. Different mechanisms by which Ox LDL causes endothelial dysfunction and atherosclerosis. Ox LDL, Oxidized low-density lipoprotein; eNOS, endothelial nitric oxide synthase; ADMA, asymmetric dimethylarginine. ↑, increase; ↓, decrease.

The importance of Ox LDL in atherogenesis has been well appreciated. Ox LDL activates vascular component cells, such as endothelial cells, macrophages, and VSMC. Also, it increases the chemotactic stimulus for monocytes, attracting them to the site of endothelial dysfunction and transforming them into monocyte-derived macrophages. These macrophages take up Ox LDL via the scavenger receptors (CD-36, SR-A, SR-PSOX) and further get transformed to "foam cells", which are a key feature of atherosclerosis [53].

The ROS and their modified target biomolecules (e.g., Ox LDL) are also defined as true second-messenger molecules that regulate various signal transduction cascades upstream of nuclear transcription factors, including modulation of Ca^{2+} signaling, protein kinase, and protein phosphatase pathways [54, 55]. Recently, G-protein Ras has been proposed to act as a mediator of ROS signaling, activating a cascade of kinases, including diverse members of the mitogen-activated protein kinase (MAPK) family [56]. In the case of extracellular signal-regulated kinase 5 (ERK5) or big mitogen-activated protein kinase 1 (BMK1), hydrogen peroxide appears to be an exclusive activator [57]. Different agents that induce oxidative stress have been demonstrated to stimulate tyrosine kinase activity, induce tyrosine phosphorylation events, and activate downstream kinases such as protein kinase C (PKC), c-Src, Raf-1, and MAPK [54, 57–59]. MAPK phosphorylates and enhances the transcriptional activity of redox sensitive nuclear transcription factors such as NFκB and activator protein-1 [54, 60]. Activation of these nuclear factors can further upregulate genes encoding proteins that control adhesion and binding of macrophages and leukocytes to the endothelium.

The increased production of macrophage chemoattractant protein-1 (MCP-1) recruits mononuclear phagocytes and increased expression of intercellular adhesion molecule-1 (ICAM-1), vascular cell adhesion molecule-1 (VCAM-1), and E-selectin leads to an increase in the aggregatory events of these cells to the vessel wall. This upregulation has been proposed to be mediated both directly by Ox LDL and indirectly by changes in the redox state of endothelial cells [61–63]. Decreased NO and increased oxidative stress can also activate matrix metalloproteinases (MMP) namely MMP-2 and MMP-9, which contribute to the progression of atherosclerosis [64, 65]. Thus, endothelial dysfunction with reduced NO bioavailability, increased oxidative stress, and expression of adhesion molecules and cytokines contributes not only to the initiation but also to the progression of the atherosclerotic disease [66].

The vasoconstrictive factor, AngII, also contributes to ROS generation by activating NADPH oxidase leading to the production of superoxide radicals and decreased availability of NO, thus, resulting in impaired vascular function [67]. AngII not only acts as a prooxidant but also stimulates the production of ET-1 [68]. ET-1 and AngII promote proliferation of VSMC and thereby contribute to the development of atherosclerosis [69, 70].

Endothelial Dysfunction: Role of Inflammation

Oxidative stress has been intimately linked to the proinflammatory state of the endothelium. The expression of VCAM-1, ICAM-1, integrins, and E-selectin, in response to oxidative stress, plays an important role in the initiation of the inflammatory process [71]. In fact, the soluble forms of VCAM-1, ICAM-1, and E-selectin in the plasma have been hailed as the markers of endothelial dysfunction [72–75]. Under normal physiological conditions, the endothelium does not allow the circulating immune effector cells to adhere to its surface. However, dysfunction of the endothelium results in the loss of this property of nonadherence. Dysfunctional endothelium expresses a variety of adhesion molecules such as VCAM-1 and ICAM-1 that mediate the recruitment of monocytes, macrophages, T-lymphocytes, and platelets to the endothelium. This results in the production of proinflammatory cytokines such as interferon-γ (IFN-γ), tumor necrosis factor alpha (TNF-α), interleukins (IL-1, 2, 6, 8), and other factors like MCP-1, MCP-4, COX, and MMP by the endothelium, which in turn stimulate the immune cells to secrete these proinflammatory factors [76–77]. A proinflammatory feedback loop is thus established, as these mediators will (a) recruit additional local cytokine releasing immune cells to the endothelium (b) promote LDL receptor expression by the endothelium causing injurious aggregation of Ox LDL particles on its surface, and (c) promote the stimulation of systemic inflammation cascades [78].

The proinflammatory status of endothelium affects its membrane permeability. It allows the passage of adherent Ox LDL into the subendothelial space. Circulating macrophages recognize this immunogenic LDL and migrate into the subendothelial space in order to sequester and phagocytose these lipid antigens [79]. This directed migration is driven by the interaction of MCP-1 with its receptors. Once resident in the intima, the macrophages express scavenger receptors (SR) to capture the internalized Ox LDL particles, which ultimately results in the formation of "foam cells," a hallmark of the arterial lesions [80]. Importantly, these foam cells continue secreting proinflammatory cytokines that results in a disproportionate recruitment of additional immune cells [81]. This process further activates the classical and alternative complement pathways, which in turn stimulates the proliferation of VSMC [82, 83]. Thus, a vicious cycle of inflammatory activity is established in the vessel wall, which exacerbates the endothelial dysfunction and lesion development (Fig. 23.3).

The dysfunctional endothelium is highly synthetic and proliferative in nature and functions like an endocrine tissue in several ways. The production of various growth factors like platelet-derived growth factor (PDGF), fibroblast growth factor (FGF), transforming growth factor beta (TGF-β), IL-1, and

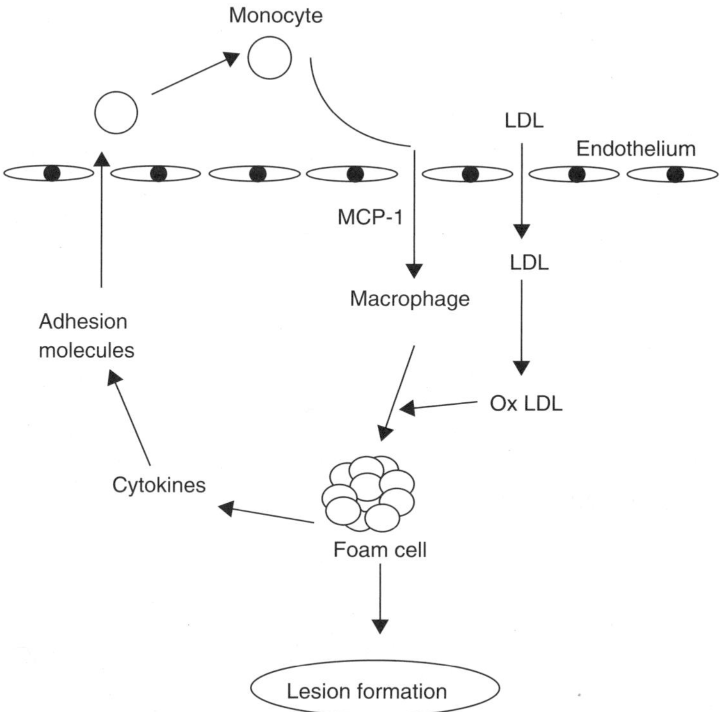

FIGURE 23.3. Involvement of monocyte/macrophages in the formation of foam cells and lesions. Ox LDL, Oxidized low-density lipoprotein; MCP-1, macrophage chemoattractant protein-1 (modified from Ref. [159]).

TNF-α promote VSMC proliferation and migration as well as the proliferation of the endothelium itself [78]. Moreover, other factors like loss of NO, increased platelet adherence, release of vasoactive agents (thromboxanes, ET-1, AngII), and increased expression of MMP leads to enhanced local vasoconstriction, fibroblast-mediated collagen synthesis, and matrix deposition in the endothelium [84–88]. Especially, MMP secreted by the endothelium and other immune cells play an important role in the degradation of extracellular matrix that lends stability to the plaque's fibrous cap. MMP-mediated areas of fissuring or ulceration that subsequently develop in advanced atherosclerotic plaques are particularly vulnerable to platelet-associated vascular hemorrhage, rupture, thrombosis, embolization, and occlusion [78, 89]. Thus, inflammatory molecules play an important role in mediating the phases of the onset and progression of atherosclerosis.

Endothelial Dysfunction: Role of Dietary fats

The development of endothelial dysfunction and its progression to atherosclerosis is a multistep process, where each step is regulated by an array of vasoactive molecules, growth factors, cytokines, and ROS. This multifactorial etiology has known to be modulated through diet. This has sparked an interest in the role of nutritional factors in modulating endothelial function. However, to derive maximum health benefits from dietary fats, it is necessary to possess knowledge of the varied effects of different dietary fats on processes like endothelial dysfunction and atherosclerosis.

Saturated Fatty Acids

Excess consumption of saturated fatty acids (SFA) has always been associated with increased CHD [90–94]. Furthermore, there are numerous epidemiological studies that relate high SFA intake with increased endothelial dysfunction. However, the mechanisms employed by SFA to induce these abnormalities are not completely understood. One of the possible mechanisms is the elevation of plasma LDL levels [95]. SFA like lauric (12:0), myristic (14:0), and palmitic (16:0) acids are associated with increased plasma LDL levels [96, 97], thereby causing endothelial dysfunction. Both *in vitro* and *in vivo* studies in animals have shown that endothelial function may be abnormal within a few hours of exposure to increased levels of LDL cholesterol [98, 99]. Furthermore, increased LDL levels increase their susceptibility to oxidation, resulting in the formation of Ox LDL that affects endothelial function, as discussed in the earlier sections.

In addition to altering the plasma LDL levels, some of the SFA may also affect thrombosis. Stearic acid (18:0) has been shown to induce thrombosis in experimental animals via affecting platelet activity and the activation of coagulation [100]. There is further evidence from human studies that increased dietary intake of stearic acid is strongly associated with enhanced platelet reactivity [101]. Stearic acid has also been reported to result in increased plasma levels of fibrinogen and factor VII (which could increase thrombotic tendency) [102, 103]. However, stearic acid has no effect on the excretion of thromboxane, prostacyclin metabolites, and eicosanoids that modulate platelet aggregation [104].

SFA can further affect relaxation of the vessels by influencing endothelial NO production. Feeding a diet rich in SFA has been shown to impair endothelium-dependent relaxation in pregnant rats and their offspring, suggesting a decrease in NO production by endothelial cells [105]. The possible mechanism behind the decreased endothelial NO synthesis is the enhanced acylation of eNOS by SFA like myristate (C14:0) and palmitate (C16:0). Acylated eNOS interacts strongly with the inhibitor caveolin-1 resulting in decreased NO production [106] thereby altering endothelial function (Fig. 23.4).

As discussed earlier, adhesion molecules like VCAM-1, ICAM, and selectins play an important role in mediating leukocyte–endothelial interactions, which facilitate impairment of the endothelial function. An *in vitro* study reported that SFA does not affect cytokine-induced expression of adhesion molecules like VCAM-1 and thus has no effect on leukocyte–endothelial cell interactions [107]. It has further been shown that SFA has a little effect on lymphocyte proliferation or cytokine production [108]. Though SFA plays an important role in elevating plasma LDL levels, inducing thrombosis, and reducing NO production, however its role in the leukocyte–endothelial cell interactions is yet to be elucidated.

Trans Fatty Acids

Unsaturated fatty acids with *trans* configuration (*trans*-fatty acids) are also known to result in endothelial dysfunction similar to SFA. *Trans*-fatty acids can directly impair endothelial function by altering vasodilation as well as certain inflammatory cascades [109]. However, high consumption of *trans*-fatty acids might also affect endothelial function indirectly by increasing plasma LDL concentrations [110, 111] and affecting LDL oxidation

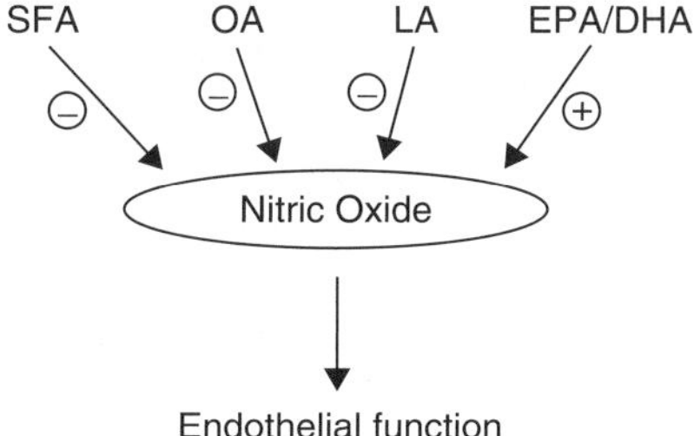

FIGURE 23.4. Effects of different dietary fats on nitric oxide production. The inhibitory effects of different fatty acids are indicated by "−" and the stimulatory effects are indicated by "+". SFA, saturated fatty acids; OA, oleic acid; LA, linoleic acid; EPA, eicosapentaenoic acid; DHA, docosahexaenoic acid.

[112]. Certain studies have examined the relation between the intake of *trans*-fatty acids and plasma concentrations of biomarkers of inflammation and endothelial dysfunction [109, 113]. One of these studies indicated a positive correlation between plasma concentrations of IL-6, soluble ICAM-1, soluble VCAM-1, E-selectin, and dietary *trans*-fat intake [109]. In addition, *trans*-fatty acids can inhibit enzymes of the eicosanoid metabolism, which further results in disturbed prostaglandin balance and increased thrombosis [114].

Monounsaturated Fatty Acids

Consumption of monounsaturated fatty acids (MUFA) has been linked with a reduced incidence of CHD [115, 116]. Among these, oleic acid (18:1) has been shown to reduce plasma LDL levels as well as its oxidative modification [117, 118]. MUFA can decrease dietary thrombogenic potential when substituted for SFA diets. However, it has never been reported to have antithrombotic effects on its own. One study aimed at analyzing the effects of MUFA on the expression of VCAM-1 revealed an increased inhibition of VCAM-1 activity [107]. Thus, the presence of even one double bond appears to be crucial to the effects of modulation of endothelial–leukocyte interactions. It was also predicted that the substitution of SFA in membrane phospholipids by MUFA, renders the endothelial cells less responsive to the stimulation of cytokines [119]. Also, human studies have shown that a high MUFA diet possesses potential beneficial effect on platelet aggregation and postprandial activation of factor VII [120]. However, oleic acid has also been shown to inhibit endothelial NO synthesis by decreasing eNOS activity, resulting in impaired endothelium-dependent vasorelaxation [121] (Fig. 23.4).

Polyunsaturated Fatty Acids

Dietary polyunsaturated fatty acids (PUFA) comprise *n*–6 and *n*–3 PUFA. *N*–6 PUFAs are derived from linoleic acid (LA, 18:2), consumption of which has been promoted in the Western diets based on its plasma LDL lowering properties [122]. However, dietary LA is now being recognized to favor oxidative modification of LDL [123, 124], increase platelet response to aggregation [125], and suppress the immune system [126]. Furthermore, it is known to play an important role in the development of endothelial dysfunction. It can cause an imbalance in cellular oxidative stress of the endothelium thereby resulting in the activation of certain stress responsive transcription factors, inflammatory cytokine production, and the expression of adhesion molecules [127]. An increased consumption of *n*–6 PUFA is also known to result in the accumulation of AA, which results in the enhanced production of eicosanoids. In small quantities, the eicosanoids from AA are biologically active, however, they can contribute to the formation of thrombi and atheromas when formed in large quantities. Thus, high dietary intake of *n*–6 fatty acids shifts the physiological state to prothrombotic and proaggregatory, resulting in progression of atherosclerosis. [122].

In contrast to LA, α linolenic acid (ALA, *n*–3, 18:3) is known to reduce endothelial dysfunction and CHD [127]. ALA is the precursor for other important *n*–3 PUFA, i.e., eicosapentaenoic acid (EPA, 20:5) and docosahexaenoic acid (DHA, 22:6). Fish or fish oil serve as a rich source of these *n*–3 fatty acids. Several studies have reported an association between fish oil consumption and protection against CHD [128–132]. The primary mechanism by which *n*–3 fatty acids are beneficial in cardiovascular disorders is by decreasing plasma LDL levels. However, their role in affecting the oxidative status of LDL is not clear. There are conflicting results among studies on the susceptibility of LDL to oxidation due to *n*–3 PUFA supplementation. There are some human studies which reported enhanced oxidation of LDL [133, 134], whereas some other studies observed no effect of dietary *n*–3 PUFA on LDL oxidation [135–137]. It, therefore, remains to be established whether LDL oxidative status *in vivo* is affected by *n*–3 fatty acids [138].

In addition to the hypolipidemic effects, EPA has been shown to increase endothelium-dependent relaxation *in vitro* through multiple mechanisms, including an increase in intracellular Ca^{2+} concentration and translocation of eNOS [139, 140]. Similarly, treatment of isolated rat aortic rings with DHA was shown to enhance NO synthesis in endothelial cells and relaxation of VSMC [141] (Fig. 23.4).

N–3 fatty acids have also been shown to mediate anti-inflammatory effects by decreasing the expression of adhesion molecules and cytokines, thus affecting leukocyte–endothelium interactions. *N*–3 fatty acids are known to inhibit the synthesis and release of proinflammatory cytokines such as TNF-α, IL-1, and IL-2 that are released during the early course of ischemic heart diseases [142]. Mice fed fish oil were reported to have lower plasma concentrations of TNF-α, IL-1β, and IL-6 following endotoxin injection than mice fed safflower oil [143]. An anti-inflammatory cytokine IL-10 has been shown to be upregulated by *n*–3 fatty acids [144]. There are some studies, which demonstrate that *n*–3 fatty acids can further alter the metabolism of adhesion molecules such as VCAM-1, E-selectin, and ICAM-1 [138, 145]. One study reported that supplementation with a moderate dose of fish oil (1.2 g EPA+DHA per day) for 12 weeks decreased plasma levels of soluble VCAM-1 in older human subjects [146]. Fish oil has further been shown to increase plasma soluble E-selectin [147]. *In vitro* studies have highlighted the potential for *n*–3 PUFA to modulate the expression of adhesion molecules by some cell types. Murine peritoneal macrophages were reported to show less adherent properties when cultured in the presence of EPA or DHA [148]. Similarly, human monocytes revealed a reduction in the expression of ICAM-1 when incubated with EPA, while DHA had no effect [149].

N–3 PUFAs are known to protect the cells from oxidative damage. It is proposed that long chain PUFA can scavenge the ROS undergoing self-peroxidation, thereby preventing the formation of hydrogen peroxide, which mediates cell activation due to oxidative stress [150]. Alternatively, long-chain PUFA can induce the expression of antioxidant enzymes like glutathione peroxidase, superoxide dismutase, and catalase, which maintain the oxidative balance of the cells [151]. Therefore, decreased expression and altered metabolism of adhesion, inflammatory, and antioxidant molecules caused by fish oil results in reduced platelet aggregation and their binding to the endothelial cells, ultimately resulting in an antiatherogenic phenotype.

Another potential antiatherogenic mechanism employed by n–3 fatty acids is their interference with the AA cascade that generates a wide variety of eicosanoids. When ingested, EPA and DHA can partially replace AA in cell membranes of platelets, erythrocytes, neutrophils, monocytes, and liver cells [122]. In addition, EPA and DHA are known to be the competitive inhibitors of COX, which results in reduced production of AA-derived eicosanoids [152, 153]. As a result, ingestion of EPA and DHA from fish or fish oil leads to an increased production of total prostacyclin by increasing PGI3, which is known to promote vasodilation and reduce platelet aggregation. In addition, n–3 fatty acids reduce the production of thromboxane A_2, known to promote platelet aggregation and vasoconstriction. This is associated with an increase in thromboxane A_3, which is a relatively weak vasoconstrictor [154]. N–3 fatty acids also affect the inflammatory processes by decreasing leukotriene B_4, and simultaneously increasing leukotriene B_5 [155, 156].

N–3 fatty acids are further known to possess a novel stabilizing effect on the myocardium itself, resulting in lowering the incidence of arrhythmias [138]. This seems to be due to the ability of n–3 fatty acids to prevent calcium overload by maintaining the activity of L-type calcium channels during periods of stress [157], and to increase the activity of cardiac microsomal Ca^{2+}/Mg^{2+}-ATPase [158]. In addition, n–3 fatty acids (including ALA) are potent inhibitors of voltage-gated sodium channels in cultured neonatal cardiac myocytes, which may contribute to a reduction in arrhythmias [152].

Conclusions

Atherosclerosis is known to be the leading cause of morbidity and mortality worldwide. It is a complex disease that involves exploitation of various cellular pathways for its progression. However, there is increasing evidence that violation of normal endothelial function is the primary event for the initiation of these disorders. Endothelium not only serves as a barrier between blood and the interstitium, it further maintains the cardiovascular health by regulating vascular tone, anti-inflammatory, and anti-aggregatory pathways. Endothelial dysfunction caused due to injury or any pathophysiological stimuli renders it susceptible to the development of atherosclerosis and CHD. Oxidative stress and inflammation are known to play a key role in the development of endothelial dysfunction. Various dietary fats exert different effects on the endothelial function. While, increased consumption of SFA and *trans*-fatty acids has been shown to have deleterious effects, MUFA and *n*–3 PUFA are known to possess cardioprotective effects. Although individual effects of dietary fats are well established, the complete understanding of the employed molecular mechanisms is still lacking. It is crucial to advance our knowledge in this area, as this will not only provide us with primary nutritional management of these disorders, but will further allow the secondary control of these disorders.

Acknowledgments: Research in our laboratory is funded by the Heart and Stroke Foundation of Canada, Natural Sciences and Engineering Research Council (NSERC), Canadian Institutes of Health Research (CIHR), and Canadian Innovation Fund (CFI). SKC is a CIHR New Investigator.

References

1. Caramori PR, Zago AJ: Endothelial dysfunction and coronary artery disease. Arq Bras Cardiol 75: 173–182, 2000.
2. Fishman AP: Endothelium: a distributed organ of diverse capabilities. Ann NY Acad Sci 401: 1–8, 1982.
3. Rubanyi GM: The role of endothelium in cardiovascular homeostasis and diseases. J Cardiovasc Pharmacol 22: S1–S4, 1993.
4. Palmer RM, Ashton DS, Moncada S: Vascular endothelial cells synthesize nitric oxide from L-arginine. Nature 333: 664–666, 1988.
5. Stamler JS, Singel DJ, Loscalzo J: Biochemistry of nitric oxide and its redox-activated forms. Science 258: 1898–1902, 1992.
6. Arnold WP, Mittal CK, Katsuki S, Murad F: Nitric oxide activates guanylate cyclase and increases guanosine 3, 5′-cyclic monophosphate levels in various tissue preparations. Proc Natl Acad Sci USA 74: 3203–3207, 1977.
7. Weksler BB, Marcus AJ, Jaffe EA: Synthesis of prostaglandin I2 (prostacyclin) by cultured human and bovine endothelial cells. Proc Natl Acad Sci USA 74: 3922–3926, 1977.
8. Moncada S, Herman AG, Higgs EA, Vane JR: Differential formation of prostacyclin (PGX or PGI2) by layers of the arterial wall. An explanation for the antithrombotic properties of vascular endothelium. Thromb Res 11: 323–344, 1977.
9. Stamler JS, Vaughan DE, Loscalzo J: Synergistic disaggregation of platelets by tissue-type plasminogen activator, prostaglandin E1, and nitroglycerin. Circ Res 65: 796–804, 1989.
10. FitzGerald GA, Pedersen AK, Patrono C: Analysis of prostacyclin and thromboxane biosynthesis in cardiovascular disease. Circulation 67: 1174–1177, 1983.
11. Garland CJ, Plane F, Kemp BK, Cocks TM: Endothelium-dependent hyperpolarization: a role in the control of vascular tone. Trends Pharmacol Sci 16: 23–30, 1995.
12. Levin ER: Endothelins. N Engl J Med 333: 356–363, 1995.
13. Haynes WG, Webb DJ: Contribution of endogenous generation of endothelin-1 to basal vascular tone. Lancet 344: 852–854, 1994.
14. Pigazzi A, Heydrick S, Folli F, Benoit S, Michelson A, Loscalzo J: Nitric oxide inhibits thrombin receptor-activating peptide-induced phosphoinositide 3-kinase activity in human platelets. J Biol Chem 274: 14368–14375, 1999.

15. Freiman PC, Mitchell GG, Heistad DD, Armstrong ML, Harrison DG: Atherosclerosis impairs endothelium-dependent vascular relaxation to acetylcholine and thrombin in primates. Circ Res 58: 783–789, 1986.

16. Cooke JP, Singer AH, Tsao P, Zera P, Rowan RA, Billingham ME: Antiatherogenic effects of L-arginine in the hypercholesterolemic rabbit. J Clin Invest 90: 1168–1172, 1992.

17. Jayakody L, Senaratne M, Thomson A, Kappagoda T: Endothelium-dependent relaxation in experimental atherosclerosis in the rabbit. Circ Res 60: 251–264, 1987.

18. Ludmer PL, Selwyn AP, Shook TL, Wayne RR, Mudge GH, Alexander RW, Ganz P: Paradoxical vasoconstriction induced by acetylcholine in atherosclerotic coronary arteries. N Engl J Med 315: 1046–1051, 1986.

19. Nabel EG, Selwyn AP, Ganz P: Large coronary arteries in humans are responsive to changing blood flow: an endothelium-dependent mechanism that fails in patients with atherosclerosis. J Am Coll Cardiol 16: 349–356, 1990.

20. Zeiher AM, Drexler H, Wollschlager H, Just H: Modulation of coronary vasomotor tone in humans. Progressive endothelial dysfunction with different early stages of coronary atherosclerosis. Circulation 83: 391–401, 1991.

21. Zeiher AM, Drexler H, Saurbier B, Just H: Endothelium-mediated coronary blood flow modulation in humans: effects of age, atherosclerosis, hypercholesterolemia and hypertension. J Clin Invest 92: 652–662, 1993.

22. Adams MR, Robinson J, McCredie R, Seale JP, Sorensen KE, Deanfield JE, Celermajer DS: Smooth muscle dysfunction occurs independently of impaired endothelium-dependent dilation in adults at risk of atherosclerosis. J Am Coll Cardiol 32: 123–127, 1998.

23. Cernacek P, Stewart DJ: Immunoreactive endothelin in human plasma: marked elevations in patients in cardiogenic shock. Biochem Biophys Res Commun 161: 562–567, 1989.

24. Shichiri M, Hirata Y, Ando K, Emori T, Ohta K, Kimoto S, Ogura M, Inoue A, Marumo F: Plasma endothelin levels in hypertension and chronic renal failure. Hypertension 15: 493–496, 1990.

25. Lopez JA, Armstrong ML, Piegors DJ, Heistad DD: Vascular responses to endothelin-1 in atherosclerotic primates. Arteriosclerosis 10: 1113–1118, 1990.

26. Kubes P, Suzuki M, Granger DN: Nitric oxide: an endogenous modulator of leukocyte adhesion. Proc Natl Acad Sci USA 88: 4651–4655, 1991.

27. De Caterina R, Libby P, Peng HB, Thannickal VJ, Rajavashisth TB, Gimbrone MA Jr, Shin WS, Liao JK: Nitric oxide decreases cytokine-induced endothelial activation. Nitric oxide selectively reduces endothelial expression of adhesion molecules and proinflammatory cytokines. J Clin Invest 96: 60–68, 1995.

28. Marks DS, Vita JA, Folts JD, Keaney JF Jr, Welch GN, Loscalzo J: Inhibition of neointimal proliferation in rabbits after vascular injury by a single treatment with a protein adduct of nitric oxide. J Clin Invest 96: 2630–2638, 1995.

29. Grag UC, Hassid A: Nitric oxide (NO) and 8-bromo-cyclic GMP inhibits mitogenesis and proliferation of cultured rat vascular smooth muscle cells. J Clin Invest 83: 1974–1978, 1989.

30. Fukuo K, Inoue T, Morimoto S, Nakahashi T, Yasuda O, Kitano S, Sasada R, Ogihara T: Nitric oxide mediates cytotoxicity and basic fibroblast growth factor release in cultured vascular smooth muscle cells. J Clin Invest 95: 669–672, 1995.

31. Mendelsohn ME, O'Neill S, George D, Loscalzo J: Inhibition of fibrinogen binding to human platelets by S-nitroso-N-acetylcysteine. J Biol Chem 265: 19028–19034, 1990.

32. Stamler JS, Vaughan DE, Loscalzo J: Synergistic disaggregation of platelets by tissue-type plasminogen activator, prostaglandin E1, and nitroglycerin. Circ Res 65: 796–804, 1989.

33. Simari RD, San H, Rekhter M, Ohno T, Gordon D, Nabel GJ, Nabel EG: Regulation of cellular proliferation and intimal formation following balloon injury in atherosclerotic rabbit arteries. J Clin Invest 98: 225–235, 1996.

34. Rubanyi GM: The role of endothelium in cardiovascular homeostasis and diseases. J Cardiovasc Pharmacol 22: S1–S14, 1993.

35. Adams MR, Kinlay S, Blake GJ, Orford JL, Ganz P, Selwyn AP: Atherogenic lipids and endothelial dysfunction: mechanisms in the genesis of ischemic syndrome. Annu Rev Med 51: 149–167, 2000.

36. Arnman V, Nilsson A, Stemme S, Risberg B, Rymo L: Expression of plasminogen activator inhibitor 1 mRNA in healthy, atherosclerotic and thrombotic human arteries and veins. Thromb Res 76: 487–499, 1994.

37. Hamsten A: Hemostatic function and coronary disease. N Engl J Med 332: 677–678, 1991.

38. Szczeklik A, Musial J, Undas A, Swadzba J, Gora PF, Piwowarska W, Duplaga M: Inhibition of thrombin generation by aspirin is blunted in hypercholesterolemia. Arterioscler Thromb Vasc Biol 16: 948–954, 1996.

39. Davi G, Ganci A, Averna M, Giammarresi C, Barbagallo C, Catalano I, Cala A, Notarbartolo A: Thromboxane biosynthesis, neutrophil and coagulation activation in type IIa hypercholesterolemia. Thromb Haemost 74: 1015–1019, 1995.

40. Abe J, Berk BC: Reactive oxygen species as mediators of signal transduction in cardiovascular disease. Trends Cardiovasc Med 8: 59–64, 1998.

41. Freeman BA, Crapo JD: Biology of disease: free radicals and tissue injury. Lab Invest 47: 412–426, 1982.

42. Koppenol WH, Moreno JJ, Pryor WA, Ischiropoulos H, Beckman JS: Peroxynitrite, a cloaked oxidant formed by nitric oxide and superoxide. Chem Res Toxicol 5: 834–842, 1992.

43. Milstien S, Katusic Z: Oxidation of tetrahydrobiopterin by peroxynitrite: implications for vascular endothelial function. Biochem Biophys Res Commun 263: 681–684, 1999.

44. Landmesser U, Dikalov S, Price SR, McCann L, Fukai T, Holland SM, Mitch WE, Harrison DG: Oxidation of tetrahydrobiopterin leads to uncoupling of endothelial cell nitric oxide synthase in hypertension. J Clin Invest 111: 1201–1209, 2003.

45. Griendling KK, FitzGerald GA: Oxidative stress and cardiovascular injury: Part I: Basic mechanisms and *in vivo* monitoring of ROS. Circulation 108: 1912–1916, 2003.

46. Anderson TJ, Meredith IT, Charbonneau F, Yeung AC, Frei B, Selwyn AP, Ganz P: Endothelium-dependent coronary vasomotion relates to the susceptibility of LDL to oxidation in humans. Circulation 93: 1647–1650, 1996.

47. Hernandez-Perera O, Perez-Sala D, Navarro-Antolin J, Sanchez-Pascuala R, Hernandez G, Diaz C, Lamas S: Effects of the 3-hydroxy-3-methylglutaryl-CoA reductase inhibitors, atorvastatin and simvastatin, on the expression of endothelium-1 and endothelial nitric oxide synthase in vascular endothelial cells. J Clin Invest 101: 2711–2719, 1998.

48. Bode-Böger SM, Böger RH, Kienke S, Junker W, Frolich JC: Elevated L-arginine/dimethylarginine ratio contributes to enhanced systemic NO production by dietary L-arginine in hypercholesterolemic rabbits. Biochem Biophys Res Commun 219: 598–603, 1996.

49. Vallance P, Leone A, Calver A, Collier J, Moncada S: Accumulation of an endogenous inhibitor of nitric oxide synthesis in chronic renal failure. Lancet 339: 572–575, 1992.

50. Feron O, Dessy C, Desager JP, Balligand JL: Hydroxy-methylglutaryl coenzyme A reductase inhibition promotes endothelial nitric oxide synthase activation through a decrease in caveolin abundance. Circulation 103: 113–118, 2001.

51. Radomski MW, Palmer RM, Moncada S: Comparative pharmacology of endothelial-derived relaxing factor and prostacyclin in platelets. J Pharmacol 92: 181–187, 1987.

52. Boulanger CM, Tanner FC, Béa ML, Hahn AW, Werner A, Luscher TF: Oxidied low density lipoproteins induce mRNA expression and release of endothelin from human and porcine endothelium. Circ Res 70: 1191–1197, 1992.

53. Kita T, Kume N, Minami M, Hayashida K, Murayama T, Sano H, Moriwaki H, Kataoka H, Nishi E, Horiuchi H, Arai H, Yokode M: Role of Oxidized LDL in atherosclerosis. Ann NY Acad Sci 947: 199–206, 2001.

54. Palmer HJ, Paulson KE: Reactive oxygen species and antioxidants in signal transduction and gene expression. Nutr Rev 55: 353–361, 1997.

55. Sundaresan M, Yu ZX, Ferrans VJ, Irani K, Finkel T: Requirements for generation of H_2O_2 for PDGF signal transduction. Science 270: 296–299, 1995.

56. Irani K, Xia Y, Zweier JL, Sollott SJ, Der CJ, Fearon ER, Sundaresan M, Finkel T, Goldschmidt-Clermont PJ: Mitogenic signaling mediated by oxidants in Ras-transformed fibroblasts. Science 265: 808–811, 1997.

57. Abe J, Takahaashi M, Ishida M, Lee JD, Berk BC: c-Src is required for oxidative stress-mediated activation of big mitogen-activated protein kinase (BMK1). J Biol Chem 33: 20389–20394, 1997.

58. Staal FJ, Anderson MT, Staal GEJ, Herzenberg LA, Gitler C, Herzenberg LA: Redox regulation of signal transduction: tyrosine phosphorylation and calcium influx. Proc Natl Acad Sci USA 91: 3619–3622, 1994.

59. Suzuki YJ, Forman HJ, Sevanian A: Oxidants as stimulators of signal transduction. Free Radic Biol Med 22: 269–285, 1997.

60. Abe J, Berk BC: Reactive oxygen species as mediators of signal transduction in cardiovascular disease. Trends Cardiovasc Med 8: 59–64, 1998.

61. Cybulsky MI, Gimbrone MA Jr: Endothelial expression of a mononuclear leukocyte adhesion molecule during atherogenesis. Science 251: 788–791, 1991.

62. Khan BV, Parthasarathy SS, Alexander RW, Medford RM: Modified low density lipoprotein and its constituents augment cytokine-activated vascular cell adhesion molecule-1 gene expression in human vascular endothelial cells. J Clin Invest 95: 1262–1270, 1995.

63. Kume N, Cybulsky MI, Gimbrone MA: Lysophosphatidylcholine, a component of atherogenic lipoproteins, induces mononuclear leukocyte adhesion molecules in cultured human and rabbit arterial endothelial cells. J Clin Invest 90: 1138–1144, 1992.

64. Lopez-Garcia E, Schulze MB, Meigs JB, Manson JE, Rifai N, Stampfer MJ, Willett WC, Hu FB: Consumption of *trans* fatty acids is related to plasma biomarkers of inflammation and endothelial dysfunction. Nutrition 135: 562–566, 2005.

65. De Caterina R, Liao JK, Libby P: Fatty acid modulation of endothelial activation. Am J Clin Nutr 71: 213S–223S, 2000.

66. Wallace FA, Miles EA, Evans C, Stock TE, Yaqoob P, Calder PC: Dietary fatty acids influence the production of Th1-but not Th2-type cytokines. J Leuk Biol 69: 449–457, 2001.

67. Griendling KK, Sorescu D, Lassègue B, Ushio-Fukai M: Modulation of protein kinase activity and gene expression by reactive oxygen species and their role in vascular physiology and pathophysiology. Arterioscler Thromb Vasc Biol 20: 2175–2183, 2000.

68. Sowers JR: Hypertension, angiotensin II, and oxidative stress. N Engl J Med 346: 1999–2001, 2002.

69. Drexler H: Factors involved in the maintenance of endothelial function. Am J Cardiol 82: 3S–4S, 1998.

70. Langheinrich AC, Bohle RM: Atherosclerosis: humoral and cellular factors of inflammation. Virchows Arch 446: 101–111, 2005.

71. Griendling KK, FitzGerald GA: Oxidative stress and cardiovascular injury: Part I: Basic mechanisms and *in vivo* monitoring of ROS. Circulation 108: 1912–1916, 2003.

72. Blankenberg S, Rupprecht HJ, Bickel C, Peetz D, Hafner G, Tiret L, Meyer J: Circulating cell adhesion molecules and death in patients with coronary artery disease. Circulation 104: 1336–1342, 2001.

73. Pradhan AD, Rifai N, Ridker PM: Soluble intercellular adhesion molecule-1, soluble vascular adhesion molecule-1, and the development of symptomatic peripheral arterial disease in men. Circulation 106: 820–825, 2002.

74. Ridker PM, Hennekens CH, Roitman-Johnson B, Stampfer MJ, Allen J: Plasma concentration of soluble intercellular adhesion molecule 1 and risks of future myocardial infarction in apparently healthy men. Lancet 351: 88–92, 1998.

75. Endemann DH, Schiffrin EL: Endothelial dysfunction. J Am Soc Nephrol 15: 1983–1992, 2004.

76. Ito T, Ikeda U: Inflammatory cytokines and cardiovascular disease. Curr Drug Targets Inflamm Allergy 2: 257–265, 2003.

77. Strom A, Franzen A, Wangnerud C, Knutsson AK, Heinegard D, Hultgardh-Nilsson A: Altered vascular remodeling in osteopontin-deficient atherosclerotic mice. J Vasc Res 41: 314–322, 2004.

78. Mullenix PS, Andersen CA, Starnes BW: Atherosclerosis as inflammation. Ann Vasc Surg 19: 130–138, 2005.

79. Linton MF, Fazio S: Macrophages, inflammation, and atherosclerosis. Int J Obes Relat Metab Disord 21: S35–S40, 2003.

80. Libby P: Inflammation in atherosclerosis. Nature 420, 868–874, 2002.

81. Jeziorska M, McCollum C, Woolley DE: Mast cell distribution, activation, and phenotype in atherosclerotic lesions of human carotid arteries. J Pathol 183: 248, 1997.

82. Bhakdi S, Torzewski M, Klouche M, Hemmes M: Complement and atherogenesis: binding of CRP to degraded nonoxidized LDL enhances complement activation. Arterioscler Thromb Vasc Biol 19: 2348–2354, 1999.

83. Leskinen MJ, Kovanen PT, Lindstedt KA: Regulation of smooth muscle cell growth, function and death by activated mast cells—a potential mechanism for the weakening and rupture of atherosclerotic plaques. Biochem Pharmacol 66: 1493–1498, 2003.

84. De Caterina R, Zampolli A: From asthma to atherosclerosis—5-lipoxygenase, leukotrienes, and inflammation. N Engl J Med 350: 4–7, 2004.

85. Cracowski JL: Isoprostanes: an emerging role in vascular physiology and disease? Chem Phys Lipids 128: 75–83, 2004.
86. Bombeli T, Schwartz BR, Harlan JM: Adhesion of activated platelets to endothelial cells: evidence for a GPIIbIIIa-dependent bridging mechanism and novel roles for endothelial intercellular adhesion molecule 1 (ICAM-l), ab integrin, and GPIba. J Exp Med 187: 329–339, 1998.
87. Hamsten A, Wiman B, Faire U, Blomback M: Increased plasma levels of a rapid inhibitor of tissue plasminogen activator in young survivors of myocardial infarction. N Engl J Med 313: 1557–1563, 1985.
88. Eberhardt W, Beeg T, Beck KF, Walpen S, Gauer S, Bohles H, Pfeilschifter J: Nitric oxide modulates expression of matrix metalloproteinase-9 in rat mesangial cells. Kidney Int 57: 59–69, 2000.
89. Flak E: Pathogenesis of plaque disruption In: Fuster V, Ross R, Topol EJ (eds), Atherosclerosis and Coronary Artery Disease 2. Lippincott-Raven, Philadelphia, 1996, pp 492–510.
90. Brown AA, Hu FB: Dietary modulation of endothelial function: implications for cardiovascular disease. Am J Clin Nutr 73: 673–686, 2001.
91. McGee DL, Reed DM, Yano K, Kagan A, Tillotson J: Ten-year incidence of coronary heart disease in the Honolulu Heart Program. Relationship to nutrient intake. Am J Epidemiol 119: 667–676, 1984.
92. Posner BM, Cobb JL, Belanger AJ, Cupples LA, D'Agostino RB, Stokes J III: Dietary lipid predictors of coronary heart disease in men. The Framingham Study. Arch Intern Med 151: 1181–1187, 1991.
93. Hegsted DM, Ausman LM: Diet, alcohol and coronary heart disease in men. J Nutr 118: 1184–1189, 1988.
94. Kushi LH, Lew RA, Stare FJ, Ellison CR, el Lozy M, Bourke G, Daly L, Graham I, Hickey N, Mulcahy R, Kevaney J: Diet and 20-year mortality from coronary heart disease. The Ireland-Boston Diet-Heart Study. N Engl J Med 312: 811–818, 1985.
95. Watts GF, Jackson P, Burke V, Lewis B: Dietary fatty acids and progression of coronary artery disease in men. Am J Clin Nutr 64: 202–209, 1996.
96. Mensink RP: Effects of the individual saturated fatty acids on serum lipids and lipoprotein concentrations. Am J Clin Nutr 57: 711S–714S, 1993.
97. Kris-Etherton P, Yu S: Individual fatty acids on plasma lipids and lipoproteins: human studies. Am J Clin Nutr 65(Suppl): 1628S–1644S, 1997.
98. Jayakody L, Senaratne M, Thomson A, Kappagoda T: Endothelium-dependent relaxation in experimental atherosclerosis in the rabbit. Circ Res 60: 251–264, 1987.
99. Cohen RA, Zitnay KM, Haudenschild CC, Cunningham LD: Loss of selective endothelial cell vasoactive functions caused by hypercholesterolemia in pig coronary arteries. Circ Res 63: 903–910, 1988.
100. Connor WE, Hoak JC, Warner ED: Massive thrombosis produced by fatty acid infusion. J Clin Invest 42: 860–866, 1963.
101. Watts GF, Jackson P, Burke V, Lewis B: Dietary fatty acids and progression of coronary artery disease in men. Am J Clin Nutr 64: 202–209, 1996.
102. Mitropoulos KA, Miller GJ, Martin JC, Reeves BEA, Cooper J: Dietary fat induces changes in factor VII coagulant activity through effects on plasma free stearic acid concentration. Arterioscler Thromb 14: 214–222, 1994.
103. Furguson J, Mackay N, McNicol G: Effect of feeding fat on fibrinolysis, Stypven time, and platelet aggregation in Africans, Asians, and Europeans. J Clin Pathol 23: 580–585, 1970.

104. Temme EH, Mensink RP, Hornstra G: Individual saturated fatty acids and effects on whole blood aggregation *in vitro*. Eur J Clin Nutr 52(10): 697–702, 1998.

105. Ghosh P, Bitsanis D, Ghebremeskel K, Crawford MA, Poston L: Abnormal aortic fatty acid composition and small artery function in offspring of rats fed a high fat diet in pregnancy. J Physiol 533: 815–822, 2001.

106. Giovannini L, Migliori M, Longoni BM, Alderton WK, Cooper CE, Knowles RG: Nitric oxide synthases: structure, function and inhibition. Biochem J 357: 593–615, 2001.

107. De Caterina R, Liao JK, Libby P: Fatty acid modulation of endothelial activation. Am J Clin Nutr 71: 213S–223S, 2000.

108. Wallace FA, Miles EA, Evans C, Stock TE, Yaqoob P, Calder PC: Dietary fatty acids influence the production of Th1-but not Th2-type cytokines. J Leuk Biol 69: 449–457, 2001.

109. Lopez-Garcia E, Schulze MB, Meigs JB, Manson JE, Rifai N, Stampfer MJ, Willett WC, Hu FB: Consumption of *trans* fatty acids is related to plasma biomarkers of inflammation and endothelial dysfunction. Nutrition 135: 562–566, 2005.

110. Denke MA: Serum lipid concentrations in humans. In: *Trans* fatty acids and coronary heart disease risk: report of the expert panel on *trans* fatty acids and coronary heart disease. Am J Clin Nutr 62: 693S–700S, 1995.

111. Judd JT, Clevidence BA, Muesing RA, Wittes J, Sunkin ME, Podczasy JJ: Dietary *trans* fatty acids: effects on plasma lipids and lipoproteins of healthy men and women. Am J Clin Nutr 59: 861–868, 1994.

112. Sorensen NS, Marckmann P, Hoy CE, van Duyvenvoorde W, Princen HMG: Effect of fish-oil-enriched margarine on plasma lipids, low density lipoprotein particle composition, size, and susceptibility to oxidation. Am J Clin Nutr 68: 235–241, 1998.

113. De Roos NM, Bots ML, Katan MB: Replacement of dietary saturated fatty acids by *trans* fatty acids lowers serum HDL cholesterol and impairs endothelial function in healthy men and women. Arterioscler Thromb Vasc Biol 21: 1233–1237, 2001.

114. Loi C, Chardigny JM, Cordelet C, Leclere L, Genty M, Ginies C, Noel JP, Sebedio JL: Incorporation and metabolism of *trans* 20:5 in endothelial cells. Effect on prostacyclin synthesis. Lipids 35: 911–918, 2000.

115. Penny M, Kris-Etherton P: Monounsaturated fatty acids and risk of cardiovascular disease. Circulation 100: 1253–1258, 1999.

116. Keys A, Menotti A, Karvonen MJ, Aravanis C, Blackburn H, Buzina R, Djordjevic BS, Dontas AS, Fidanza F, Keys MH, Kromhout D, Nedeljkovic S, Punsar S, Seccareccia F, Toshima H: The diet and 15-year death rate in the Seven Countries Study. Am J Epidemiol 124: 903–915, 1986.

117. Tsimikas S, Tsimikas AP, Alexopoulos S, Sigari F, Lee C, Reaven PD: LDL isolated from Greek subjects on a typical diet or from American subjects on an oleate-supplemented diet induces less monocyte chemotaxis and adhesion when exposed to oxidative stress. Arterioscler Thromb Vasc Biol 19: 122–130, 1999.

118. Mata P, Varela O, Alonso R, Lahoz C, De Oya M, Badimon L: Monounsaturated and polyunsaturated n-6 fatty acid–enriched diets modify LDL oxidation and decrease human coronary smooth muscle cell DNA synthesis. Arterioscler Thromb Vasc Biol 17: 2088–2095, 1997.

119. Carluccio MA, Massaro M, Bonfrate C, Siculella L, Maffia M, Nicolardi G, Distante A, Storelli C, De Caterina R: Oleic acid inhibits endothelial activation:

a direct vascular antiatherogenic mechanism of a nutritional component in the Mediterranean diet. Arterioscler Thromb Vasc Biol 19: 220–228, 1999.

120. Smith RD, Kelly CN, Fielding BA, Hauton D, Silva KD, Nydahl MC, Miller GJ, Williams CM: Long-term monounsaturated fatty acid diets reduce platelet aggregation in healthy young subjects. Br J Nutr 90: 597–606, 2003.

121. Davda RK, Stepniakowski KT, Lu G, Ullian ME, Goodfriend TL, Egan BM: Oleic acid inhibits endothelial nitric oxide synthase by a protein kinase C independent mechanism. Hypertension 26: 764–770, 1995.

122. Simopoulos AP: Essential fatty acids in health and chronic disease. Am J Clin Nutr 70: 560S–569S, 1999.

123. Reaven P, Parthasarathy S, Grasse BJ, Miller E, Almazan F, Mattson FH, Khoo JC, Steinberg D, Witztum JL: Feasibility of using an oleate-rich diet to reduce the susceptibility of low-density lipoprotein to oxidative modification in humans. Am J Clin Nutr 54: 701–706, 1991.

124. Abbey M, Belling GB, Noakes M, Hirata F, Nestel PJ: Oxidation of low-density lipoproteins: intra individual variability and the effect of dietary linoleate supplementation. Am J Clin Nutr 57: 391–398, 1993.

125. Renaud S: Linoleic acid, platelet aggregation and myocardial infarction. Atherosclerosis 80: 255–256, 1990.

126. Endres S, Ghorbani R, Kelley VE, Georgilis K, Lonnemann G, van der Meer JW, Cannon JG, Rogers TS, Klempner MS, Weber PC: The effect of dietary supplementation with n-3 polyunsaturated fatty acids on the synthesis of interleukin-1 and tumor necrosis factor by mononuclear cells. N Engl J Med 320: 265–271, 1989.

127. Ascherio A, Rimm EB, Giovannucci EL, Spiegelman D, Stampfer M, Willett WC: Dietary fat and risk of coronary heart disease in men: cohort follow up study in the United States. Br Med J 313: 84–90, 1996.

128. Hau MF, Smelt AH, Bindels AJ, Sijbrands EJ, Van der Laarse A, Onkenhout W, van Duyvenvoorde W, Princen HM: Effects of fish oil on oxidation resistance of VLDL in hypertriglyceridemic patients. Arterioscler Thromb Vasc Biol 16: 1197–1202, 1996.

129. Stone NJ: Fish consumption, fish oil, lipids, and coronary heart disease. Circulation. 94: 2337–2340, 1996.

130. Kromhout D, Bosschieter EB, de Lezenne Coulander C: The inverse relation between fish consumption and 20-year mortality from coronary heart disease. N Engl J Med 312: 1205–1209, 1985.

131. Kromhout D, Feskens EJ, Bowles CH: The protective effect of a small amount of fish on coronary heart disease mortality in an elderly population. Int J Epidemiol 24: 340–345, 1995.

132. Shekelle RB, Missell L, Paul O, Shryock AM, Stamler J: Fish consumption and mortality from coronary heart disease. N Engl J Med 313: 820, 1985.

133. Dolecek TA, Granditis G: Dietary polyunsaturated fatty acids and mortality in the Multiple Risk Factor Intervention Trial (MRFIT). World Rev Nutr Diet 66: 205–216, 1991.

134. Sorensen NS, Marckmann P, Hoy CE, van Duyvenvoorde W, Princen HM: Effect of fish-oil-enriched margarine on plasma lipids, low-density lipoprotein particle composition, size, and susceptibility to oxidation. Am J Clin Nutr 68: 235–241, 1998.

135. Brude IR, Drevon CA, Hjermann I, Seljeflot I, Lund-Katz S, Saarem K, Sandstad B, Solvoll K, Halvorsen B, Arnesen H, Nenseter MS: Peroxidation of

LDL from combined-hyperlipidemic male smokers supplied with omega-3 fatty acids and antioxidants. Arterioscler Thromb Vasc Biol 17: 2576–2588, 1997.

136. Bonanome A, Biasia F, De Luca M, Munaretto G, Biffanti S, Pradella M, Pagnan A: n-3 fatty acids do not enhance LDL susceptibility to oxidation in hypertriacyl-glycerolemic hemodialyzed subjects. Am J Clin Nutr 63: 261–266, 1996.

137. Higdon JV, Du SH, Lee YS, Wu T, Wander RC: Supplementation of post-menopausal women with fish oil does not increase overall oxidation of LDL ex vivo compared to dietary oils rich in oleate and linoleate. J Lipid Res 42: 407–418, 2001.

138. Kris-Etherton PM, Harris WS, Appel LJ: Fish consumption, fish oil, omega-3 fatty acids, and cardiovascular disease. Circulation 106: 2747–2757, 2002.

139. Okuda Y, Kawashima K, Sawada T, Tsurumaru K, Asano M, Suzuki S, Soma M, Nakajima T, Yamashita K: Eicosapentaenoic acid enhances nitric oxide production by cultured human endothelial cells. Biochem Biophys Res Commun 232: 487–491, 1997.

140. Omura M, Kobayashi S, Mizukami Y, Mogami K, Todoroki-Ikeda N, Miyake T, Matsuzaki M: Eicosapentaenoic acid (EPA) induces Ca^{2+}-independent activation and translocation of endothelial nitric oxide synthase and endothelium-dependent vasorelaxation. FEBS Lett 487: 361–366, 2001.

141. Lawson DL, Mehta JL, Saldeen K, Mehta P, Saldeen TGP: ω-3 polyunsaturated fatty-acids augmented endothelium-dependent vasorelaxation by enhanced release of EDRF and vasodilator prostaglandins. Eicosanoids 4: 217–223, 1991.

142. Das UN: Beneficial effect(s) of n-3 fatty acids in cardiovascular diseases: but, why and how? Prostaglandins Leukot Essent Fatty Acids 63: 351–362, 2000.

143. Sadeghi S, Wallace FA, Calder PC: Dietary lipids modify the cytokine response to bacterial lipopolysaccharide in mice. Immunology 96: 404–410, 1999.

144. Sierra S, Lara-Villoslada F, Olivares M, Jimenez J, Boza J, Xaus J: IL-10 expression is involved in the regulation of the immune response by omega 3 fatty acids. Nutr Hosp 19: 376–382, 2004.

145. De Caterina R, Liao JK, Libby P: Fatty acid modulation of endothelial activation. Am J Clin Nutr 71: 213S–223S, 2000.

146. Miles EA, Thies F, Wallace FA, Powell JR, Hurst TL, Newsholme EA, Calder PC: Influence of age and dietary fish oil on plasma soluble adhesion molecule concentrations. Clin Sci 100: 91–100, 2001.

147. Johansen O, Seljeflot I, Hostmark AT, Arnesen H: The effect of supplementation with omega-3 fatty acids on soluble markers of endothelial function in patients with coronary heart disease. Arterioscler Thromb Vasc Biol 19: 1681–1686, 1999.

148. Calder PC, Bond JA, Harvey DJ, Gordon S, Newsholme EA: Uptake and incorporation of saturated and unsaturated fatty acids into macrophage lipids and their effect upon macrophage adhesion and phagocytosis. Biochem J 269: 807–814, 1990.

149. Hughes DA, Southon S, Pinder AC: n-3 polyunsaturated fatty acids modulate the expression of functionally associated molecules on human monocytes *in vitro*. J Nutr 126: 603–610, 1996.

150. De Caterina R, Spiecker M, Solaini G, Basta G, Bosetti F, Libby P, Liao J: The inhibition of endothelial activation by unsaturated fatty acids. Lipids 34, S191–S194, 1999.

151. Venkatraman JT, Chandrasekar B, Kim JD, Fernandes G: Effects of n-3 and n-6 fatty acids on the activities and expression of hepatic antioxidant enzymes in autoimmune-prone NZBxNZW F1 mice. Lipids 29: 561–568, 1994.

152. Kang JX, Leaf A: Antiarrhythmic effects of polyunsaturated fatty acids: recent studies. Circulation 94: 1774–1780, 1996.
153. Nair SS, Leitch JW, Falconer J, Garg ML: Prevention of cardiac arrhythmia by dietary (n-3) polyunsaturated fatty acids and their mechanism of action. J Nutr 127: 383–393, 1997.
154. Hu FB, Manson JE, Willett WC: Types of dietary fat and risk of coronary heart disease: a critical review. J Am Coll Nutr 20: 5–19, 2001.
155. Weber PC, Fischer S, von Schacky C, Lorenz R, Strasser T: Dietary omega-3 polyunsaturated fatty acids and eicosanoid formation in man. In: Simopoulos AP, Kifer RR, Martin RE (eds). Health Effects of Polyunsaturated Fatty Acids in Seafoods. Academic Press, Orlando, FL, 1986, pp 49–60.
156. Lewis RA, Lee TH, Austen KF: Effects of omega-3 fatty acids on the generation of products of the 5-lipoxygenase pathway. In: Simopoulos AP, Kifer RR, Martin RE (eds). Health effects of polyunsaturated fatty acids in seafoods.Academic Press, Orlando, FL, 1986, pp 227–238.
157. Hallaq H, Smith TW, Leaf A: Modulation of dihydropyridine-sensitive calcium channels in heart cells by fish oil fatty acids. Proc Natl Acad Sci USA 89: 1760–1764, 1992.
158. Grimsgaard S, Bonaa KH, Hansen JB, Myhre ES: Effects of highly purified eicosapentaenoic acid and docosahexaenoic acid on hemodynamics in humans. Am J Clin Nutr 68: 52–59, 1998.
159. Osiecki H: The role of chronic inflammation in cardiovascular disease and its regulation by nutrients. Altern Med Rev 9: 32–53, 2004.

24

Oxidized LDL and Antioxidants in Atherosclerosis

LESLEY MACDONALD-WICKS AND MANOHAR GARG

Abstract

Athersoclerosis is characterized by the development of foam cells from mononuclear phagocytes, which progress to become fatty streaks and further into plaque in the arterial intima. It has long been associated with the development of cardiovascular disease (CVD) and coronary heart disease (CHD). Low-density lipoprotein (LDL), the plasma transport protein for cholesterol, are downregulated by the cellular concentration of cholesterol and therefore are unlikely to be the initiator of the development of foam cells, fatty streaks, or atherosclerosis.

It is clear that the development of foam cells is due to an altered state of metabolism and one of the most popular hypotheses for the development of foam cells from the mononuclear phagocytes is via an oxidative modification of the LDL molecule. This modification allows unregulated uptake of oxidized LDL (Ox-LDL) by the phagocyte scavenger receptor and through this process the development into the foam cell. There is much evidence in the literature to support this hypothesis.

It is unlikely that the oxidative modification of the LDL is an initiator of foam cell development. Therefore it is unlikely that oxidation is the initiator for the development of atherosclerosis. Clearly there is a role for the oxidative process in the ongoing proliferation of atherosclerosis; however, the evidence to date does not support oxidation as essential to the initiation of the process of atherosclerotic development.

Keywords: antioxidants; atherosclerosis; inflammation; oxidation; oxidized LDL

The Oxidized LDL Hypothesis: Atherosclerosis

Atherosclerosis is a disease of the large arteries that is associated with the development of coronary heart disease (CHD) and coronary vascular disease. It is characterized by the accumulation of lipid-laden cells in the intima of the arterial wall. These cells are known as foam cells and are derived from mononuclear phagocytes. The fatty streaks that develop are comprised predominantly of both free and esterified cholesterol.

Low-density lipoproteins (LDL) deliver cholesterol to the peripheral cells where cellular uptake of the cholesterol occurs through the LDL receptor, which recognizes the apo B_{100} moiety on the particle. The LDL receptor is then downregulated in response to cellular cholesterol concentrations [1]. Due to this it is recognized that the native LDL and the LDL receptor cannot be responsible for the creation of foam cells. One avenue of thought is that the LDL is oxidatively modified in order for the aberrant uptake of the cholesterol to occur in sufficient quantities to lead to the development of the foam cell. This hypothesis was first articulated by Steinberg and coworkers [2] when they noted that oxidation of LDL impacts on the pathophysiology of atherosclerosis by facilitating recruitment of monocytes into the arterial intima, preventing resident macrophages exiting the intima, and accelerating uptake of LDL. This resulted in the formation of foam cells and the loss of endothelial integrity due to the cytotoxic nature of the oxidized LDL (Ox-LDL).

Macrophages scavenge oxidatively modified LDL through the scavenger receptors, which is the main route for the uptake and accumulation of cholesterol in the quantities required. When the blood concentration of LDL is high, the intimal LDL concentration also increases. In the intimal location, the LDL is brought into close proximity with endothelial cells, which causes the LDL to be minimally oxidized (MM-LDL). MM-LDL is oxidized but not to the extent that the apo B_{100} molecule is changed [3] therefore the LDL receptor can still recognize MM-LDL. MM-LDL is a chemoattractant for mononuclear phagocytes and encourages their proliferation indirectly through the production of monocyte chemotactic protein-1 (MCP-1) [4]. MM-LDL also stimulates the production of monocyte colony-stimulating factor (M-CSF), which stimulates mononuclear phagocytes to differentiate into tissue macrophages, which are therefore prevented from rejoining the circulation [4]. The presence of the macrophages stimulates further oxidation of the LDL. The macrophages take up the now Ox-LDL in an unregulated manner through the scavenger pathway, causing cholesterol accumulation and eventually the creation of the foam cell.

What is still largely unclear is where the oxidation occurs. It is unlikely to be in the antioxidant-rich environment of the plasma, and more likely to be in the microenvironment of the intima of the arterial wall where the local concentration of antioxidants can be exhausted allowing oxidation to progress [5]. It is generally believed that it is in the intima that the MM-LDL could be further oxidized to Ox-LDL. However, recent studies have shown that the quantities of antioxidants, in particular coantioxidants that could reduce the tocopherol radical, are nearly as high in the intima and in the atherosclerotic lesions as found in plasma [3].

In vitro the LDL can become modified in a number of ways, including enzymatically (principally through myeloperoxidase and 15-lipoxygenase) and through oxidation and due to this modification is retained in the arterial wall. These biologically active compounds stimulate the endothelium and attracted macrophages, T cells, mast cells, and smooth muscle cells to the fatty streak, leading to further oxidation of LDL [6]. This leads to the creation of the fatty streak, which develops into a lesion and eventually atherosclerosis. The Ox-LDL hypothesis requires that oxidation is integral to the process of initiating and developing atherosclerosis and that antioxidants will slow or prevent the development of the atheroma [6].

The Evidence: for and against

There is evidence for the existence of the MM-LDL particle. Berliner et al. [4] have identified a particle in plasma that responds in a way that is consistent with a partially Ox-LDL molecule. Avagaro et al. [7] also identified subfractions of LDL that exhibit characteristics of Ox-LDL but are also sufficiently similar to native LDL.

A number of modifications occur to the LDL when it is oxidized by incubation, for example, with endothelial or smooth muscle cells. There is increased electrophoretic mobility toward the anode and extensive fragmentation occurs leading to the production of short- and long-chain aldehydes such as malondialdehyde (MDA) [8]. Some of these fragments are free, however others become bonded to the apo B_{100} or other lipids causing conjugated dienes. The surface phosphatidylcholine are vulnerable to hydrolysation to lysophosphatidylcholine. The removal of oxidized fatty acids from the surface of the LDL may have a role in the rapid propagation of the oxidation within the LDL itself [8]. The oxidation of cholesterol components can also lead to the fragmentation of apo B_{100}. This fragmentation is not caused by proteolysis, as proteolytic inhibitors do not prevent the fragmentation, though antioxidants have been shown to do so in animals [4, 8, 9].

In Vitro *Studies*

The work of Goldstein and Brown [1] supplied the first evidence that something other than native LDL and the LDL receptor are required for sufficient accumulation of lipid that lead to foam cell formation. Mouse peritoneal macrophages, incubated with LDL *in vitro*, did not accumulate lipids, which is evidence that the macrophages had few LDL receptors. To support this, patients with familial hypercholesterolemia and the Watanabe heritable hyperlipidemic (WHHL) rabbit both develop lipid lesions while having only limited amounts of functional LDL receptors. To further support the lack of LDL receptor involvement in the development of atherosclerosis, in situ hybridization techniques show little or no mRNA for the LDL receptor in macrophage-rich lipid lesion [10].

Goldstein and Brown [11] subsequently showed that the LDL could be modified, and initial modification was through acetylation, which resulted in the uptake by the scavenger receptor of macrophages. In acetylation the ε-amino groups of the lysine residues are affected which causes a neutralization of the positive charge on the protein. This enables recognition by the scavenger receptor on the macrophage. Studies have shown that many such modifications are possible though many are also not likely under physiological conditions [8]. Oxidation of the LDL is a modification that is likely and also prevents recognition by the LDL receptor. It is likely that the lipids of the LDL themselves become oxidized, leading to the generation of peroxides and small- to long-chain aldehydes, such as MDA, which could then modify the apo B_{100}. This modification could cause the Ox-LDL to be recognized by the scavenger receptor.

The scavenger receptor has been shown to be ubiquitously expressed on macrophages of different origin, and on endothelial cells, cultured smooth muscle cells of the rabbit, and on fibroblasts [8]. Via et al. [12] have characterized the scavenger receptor from a murine macrophage cell line, and similar receptor characterizations have followed from a number of studies using a variety of cell types [8]. Although the exact nature of the receptor recognition site is unclear, it is clear that it occurs in the apoprotein moiety of Ox-LDL. Parthasarathy et al. [13] showed that delipidated and resolublized Ox-LDL protein was still recognized by the scavenger receptor and competed with Ox-LDL for uptake. Similarly considerable animal evidence shows that where there is no scavenger receptor, for example mice lacking scavenger receptor-A gene, there is resistance to the development of atherosclerosis [3].

The adhesion of leukocytes to the endothelium, via the vascular cell adhesion molecule-1 (VCAM-1), appears to be influenced by oxidation. Interestingly the genes for the MCP-1 and VCAM-1 appear to be regulated by the redox-sensitive transcription factor NFκβ [14].

It has been shown that LDL incubated with endothelial cells becomes cytotoxic. Vascular cells and macrophages can produce reactive oxygen species (ROS) and reactive nitrogen species (RNS) that Ox-LDL *in vitro* [3]. Henriksen et al. [15] showed that the modifications to the LDL, caused by proximity to the endothelial cells, resulted in uptake by the scavenger receptor, and was competitive with acetyl-LDL for uptake, which indicated a role for the same receptor in both modifications. The oxidative modification of LDL has been shown to act to promote atherosclerosis [6]. However, there is also evidence of another receptor that is not subject to competition by acetyl-LDL [16]. There are two forms of scavenger receptors A1 and A2. In addition to this, other receptors such as CD68, CD36, SR-B1, and LOX-1 also act in ways similar to the classic scavenger receptor [3].

Ox-LDL has been shown to have components that are powerful chemoattractants thought to be monocyte-specific, largely because neutrophils are rarely seen in atherosclerotic aorta [4]. This chemotactic activity is associated with the lysophosphatidylcholine (lyso PtdCho) that is generated during the oxidative modification of the LDL. The β-VLDL, isolated from cholesterol-fed rabbit plasma, has high concentrations of lyso PtdCho and these β-VLDL are also potent chemotactic molecules for human monocytes [4].

Native LDL is not chemotactic for human monocytes. MM-LDL may be active in stimulating the endothelial cells to recruit the monocytes [4]. The monocytes migrate to the area in increasing numbers, and differentiate into tissue macrophages. Ox-LDL prevents both basal and stimulated migration of macrophages from the intima [17, 18].

In Vivo *Studies*

Initially the evidence for the Ox-LDL hypothesis of atherosclerosis only showed that oxidation occurred in atherosclerotic lesions. Studies demonstrated the presence of oxidized lipoproteins in atherosclerotic plaques [4]. These studies showed that the lipid and cholesterol peroxides could be correlated with disease severity.

This initial evidence began in 1971 when human peripheral lymph was found to contain an apoprotein of higher density than that found in plasma [19]. Raymond and Reynolds [20] isolated LDL from rabbit interstitial inflammatory fluid that had properties similar to Ox-LDL. Hoff et al. [21] then isolated apo B lipoproteins from the atherosclerotic aorta that contained peroxides, and Yla-Herttuala et al. [22] showed that isolated lipoprotein had many properties of Ox-LDL and was also chemotactic to monocytes.

Since then autoantibodies to Ox-LDL (aOx-LDL) have been isolated from humans and rabbits. Haberland et al. [23] and Palinski et al. [24] used these antibodies to identify modified LDL in atherosclerotic lesions of WHHL rabbits. Also, immunohistochemical techniques detected material in the macrophage-rich areas of the lesions that were shown to cross-react with these antibodies. Prospective studies have shown aOx-LDL can predict myocardial infarction (MI) [25]. aOx-LDL is elevated in established atherosclerosis and aOx-LDL are present in healthy individuals. However, there was no evidence to show that oxidation was the cause of disease not simply due to the presence of concentrated lipid. What is also unclear is the role of Ox-LDL in different stages and in the development of atherosclerosis. Finally, while it is believed that Ox-LDL is atherogenic, it is unclear whether aOx-LDL is also atherogenic [26].

Some of the most compelling direct evidence for this hypothesis comes from the studies of apo–/– mice, which are a good model of oxidative stress. These mice develop atherosclerosis in a manner similar to human atherosclerosis, and in these mice F_2-isoprostanes (a lipid peroxidation product that is used as a marker of oxidative stress) are elevated in the plasma, tissue, and urine [27, 28].

The evidence to date is quite convincing that oxidation has a role in the development of atherosclerosis, although the evidence is mostly indirect. In summary, it is clear that some form of modification of the LDL is required for foam cell formation and that the most likely modification is through oxidation of the lipids in the LDL molecule. There is significant data on the

Ox-LDL role in the formation of foam cells via the scavenger pathway in macrophages, and that the Ox-LDL exists *in vivo* and that it is present in aortic lesions. It is also clear that there are autoantibodies generated to Ox-LDL and that titers of aOx-LDL are generally correlated with the extent of atherosclerosis.

However, the evidence cannot provide answers at this stage for whether oxidation is an initiating factor for the development of atherosclerosis or whether the oxidation is caused by the inflammatory nature of atherosclerosis, and is therefore subsequent to the development of atherosclerosis. What is also still unclear is the exact location of the oxidation events.

Antioxidants

The definition of antioxidant is given by Halliwell [29] as a substance present in low concentrations when compared to the oxidizable substrate, which significantly delays or prevents oxidation of that substrate.

If oxidation is the cause of atherosclerosis then antioxidants must also be able to delay or prevent the progression of the atheroma development. In support of this idea studies of human populations have shown an association between the intake of antioxidants and the low prevalence of cardiovascular disease (CVD) [6, 30, 31, 32]. Similarly, most but not all animal studies have also been successful in showing antioxidant interventions have slowed the development of atherosclerosis [4, 6] with reductions of up to 60% in atherosclerotic lesions [33].

The most often studied antioxidants include probucol, vitamin E, β-carotene, and vitamin C [34, 35]. Several studies have demonstrated inhibition of atherosclerosis by probucol in animals [36–39]. Successful inhibition of atherosclerotic development has been shown in rabbits [37] and monkeys [38]. The treatment of LDL receptor-deficient rabbits with the hypolipidemic agent and antioxidant, probucol, had an antiatherogenic effect independent of the traditionally recognized cholesterol lowering effect of probucol. Presumably this added effect was due to the protection of the LDL from oxidation [39]. Studies of probucol compared to lovastatin, a drug known to reduce cholesterol to a similar degree to probucol, demonstrated a greater reduction in atherosclerosis from probucol [40]. Similarly, Mao et al. [41] demonstrated that a probucol analog inhibited LDL oxidation and atherosclerosis in the absence of any cholesterol lowering effects, suggesting the effect was mediated by an antioxidant mechanism. In a murine model of atherosclerosis, probucol promoted atherosclerosis in the aortic root, but inhibited disease progression at distal sites [42].

Subsequent tests in humans appear to have equivocal results, and are controversial. One study by Walldius et al. [43] showed no improvements in hypercholesterolemic subjects, while another study by Sawayama et al. [44] showed significant reduction in the progression of atherosclerosis in the carotid artery as assessed by intima-to-media thickness. It is unlikely that antioxidants would have no role in the prevention of atherosclerosis as the basic etiology of atherosclerosis in humans and in animals is similar, therefore it would be difficult to understand why antioxidants work so effectively in animals but not in humans [45].

From the vitamin E studies it is clear that vitamin E alone is insufficient to prevent LDL oxidation, and it requires the synergistic effect of a number of lipid and aqueous soluble antioxidants to protect LDL from oxidation, termed coantioxidants. Studies have clearly shown that the lag time to oxidation ex vivo can be extended by the addition of vitamin E, and that only when the vitamin E is fully expended to form the tocopheryl radical can oxidation continue [35]. At best vitamin E can only delay the onset of peroxidation by 15–20 min, a small amount of time compared to an antioxidant, such as probucol, which can delay oxidation of the LDL fats by up to 20 h [45]. *In vivo* supplementation of 100 IU to 1600 IU of vitamin E lead to reduced ex vivo copper-induced oxidation of LDL in a dose-dependent manner. Although the oxidizability of LDL was reduced there was no reduction in plasma LDL autoantibodies, indicating the aOx-LDL may not be a sensitive *in vivo* marker of the susceptibility of LDL to oxidation [46]. Wen et al. [46] speculated that this may be due to the long half-life of aOx-LDL or, that once initiated, there is a continuous immunological response to Ox-LDL contained in the atherosclerotic lesions. Dietary antioxidants appear to reduce *in vivo* markers of inflammatory chemokines and cytokines, as well as reducing the levels of autoantibodies to oxidatively modified proteins. This may well indicate a long-term attenuation of oxidation. Antioxidants also appear to ameliorate the proliferation of smooth muscle cells in culture whereas oxidants appear to promote proliferation. Also human studies appear to show that intimal–medial thickness is reduced upon antioxidant supplementation [33].

In support of these and many other animal, cell culture and ex vivo studies, epidemiological studies have shown an inverse relationship between vitamin E intake and ischemic heart disease. The positive results from animal studies and the presumed safety of naturally occurring antioxidant lead to large, prospective cohort studies. In particular two of these studies reported an association between antioxidant intake, serum concentration of antioxidants, and improved cardiovascular outcomes [47, 48]. As these were observational studies, no cause and effect relationship could be concluded. These favorable results were from the Health Professionals Follow-up Study [31, 32] and the Nurses Health Study [30]. These two studies could be considered to be primary prevention studies as individuals with existing disease were excluded from the analysis. Contrary to these findings, the Rotterdam Study [47], which focused on the elderly found a reduced risk of MI with β-carotene intake, but none with vitamin E or vitamin C. This may be due to the low rates of supplementation in the Rotterdam Study in comparison to the Health Profession Follow-up Study where the highest quintiles of vitamin E intake were achieved largely through supplementation.

The concern is that, although antioxidant defenses are comprehensive, there is still evidence of lipid peroxidation products in the plasma, although there is no indication where the oxidation occurred.

The Evidence against

The evidence for the oxidation hypothesis is mounting though it is important to note that there is evidence that contradicts these findings and does not support the oxidative hypothesis under discussion. Williams and Fischer [6] argue that the following evidence indicates caution when interpreting the clinical role of Ox-LDL in CVD:

— Some of the animal studies failed to show improvement from antioxidant supplementation, this was shown even though there was reduction of LDL oxidation *in vitro* or in the vessel wall [3, 48].
— It appears that the oxidized material does collect in the vessel wall, but only after nonoxidized material accumulates initially, this questions the requirement of oxidation to initiate the atherosclerotic process [49]. Also, high-density lipoprotein (HDL) retards the progress of atheroma development but does so without altering oxidative processes.
— In some studies it has been shown that the retardation of the scavenger receptor increases the development of the atheroma process [50] and the overexpression of this receptor decreases the development of the atheroma [51]. This is contrary to the expectations of the oxidative hypothesis.
— Studies that have genetically manipulated the two enzymes believed to be involved with the oxidation of LDL, myeloperoxidase, and 15-lipoxygenase, in an effort to ameliorate their oxidative effect on LDL, have instead shown a protective role of these enzymes in some animal models of atherosclerosis. This is contrary to their believed role in the oxidative modification of LDL and the subsequent chain of events that leads to lesion development and atherosclerosis [6].
— Lastly, the correlation between autoantibodies to Ox-LDL and CVD in humans have not been consistent [3] indicating that there may be another explanation for the correlation when it does occur. Williams and Fischer [6] state that from their studies hyperlipidemia and other factors correlate with the antibody concentration, but vascular disease does not.

Perhaps the most damning evidence against the hypothesis is the lack of efficacy of the myriad of antioxidant clinical trials that have sought to show that antioxidant supplementation halted or prevented the development of atherosclerosis. However you term the criticism of the antioxidant trials they did not achieve a clinical outcome and, for antioxidant supplementation to be effective, a clinical outcome that is measurable is required [3,6,52]. As atherosclerosis is essentially an inflammatory illness, any antioxidant used must perhaps also have an anti-inflammatory effect.

Antioxidant Trials

The body of evidence that was mounting for the effectiveness of antioxidants in atherosclerosis in animals, and the support provided by the epidemiological evidence in favor of antioxidant protection of LDL from oxidation has been followed up with large randomized clinical trials of antioxidant supplementation. These trials were largely secondary prevention trials where the patients recruited had known vascular disease, active tobacco use, asbestos exposure, or documented previous malignancies.

The results of the large clinical antioxidant supplementation trials appeared to disprove the role of antioxidants in preventing or halting CHD.

The conflicting evidence from the animal, *in vitro*, and clinical trial appears to have been resolved when a meta-analysis of these key studies was undertaken by Vivekananthan et al. [52], involving 138,113 persons enrolled in the β-carotene trials and 81,788 enrolled in the vitamin E trials using all cause mortality as the clinical end point in common. This analysis showed no clinical benefit from the supplementation of antioxidants in these randomized placebo-controlled clinical trials. In the vitamin E analysis the authors divided the trials into primary prevention and secondary prevention trials and this also showed no clinical benefit from the supplementation of antioxidants. This lack of clinical benefit was seen even when the antioxidant supplement was shown to increase antioxidant plasma concentration, to decrease oxidation lag time ex vivo, and to lower the levels of circulating Ox-LDL [53].

This is indeed damning evidence against the oxidation hypothesis of atherosclerosis. However, some authors argue that it is not as damming as it may seem. According to Parthasarathy et al. [33] the failure is more in the clinical trial study design, specifically in the inappropriateness of the outcome measures chosen, than in the oxidation hypothesis itself.

It is important to note that the decision to focus on vitamin E and β-carotene may not have been the appropriate route to take. There is very little evidence to show that either of these naturally occurring antioxidants is the most efficacious in protecting LDL from oxidation [45]. In particular, the stage of atherosclerotic development where plaque rupture occurs, although oxidation could indeed have a role in the lead up to the rupture, prevention of oxidation at this stage would only have limited effect and to achieve that limited effect large amounts of chain-breaking antioxidants would be required.

As most of the clinical trials have been secondary prevention trials, with established coronary artery disease (CAD) beyond fatty streak lesion formation, relating these patients to the oxidized hypothesis may be beyond the scope of the hypothesis. Indeed in the meta-analysis only one of the trials studied was a primary prevention trial of vitamin E alone, therefore the negative/neutral results were driven by the secondary prevention trials perhaps biasing the results [52].

So, the end points of these trials, such as plaque rupture or MI do not relate to oxidation, and are not an appropriate end point to measure the Ox-LDL hypothesis. Markers of oxidative stress (autoantibodies to oxidatively modified proteins) and surrogate markers of arterial and tissue response (VCAM-1 and MCP-1) can be seen to be reduced by antioxidants—perhaps these are more appropriate markers to test the Ox-LDL hypothesis [33]. Again the authors [4, 52] of the meta-analysis clearly state that none of the large clinical trials use oxidative stress end points in their studies. The authors speculate that the vitamins could have been incorrectly dosed, that is given to patients with low oxidative stress, or given too late in the course of CVD to be efficacious in inhibiting LDL oxidation. The authors [52] state in their article that, although they report no benefit from antioxidant vitamins, this does not disprove the Ox-LDL hypothesis, a position supported by Steinberg and Witztum [45]. It is clear from the evidence to date that, in order to prove or disprove the Ox-LDL hypothesis in relation to atherosclerotic development, it is necessary to have biomarkers that would accurately assess the level of oxidative stress the patient is experiencing and also measure the efficacy of the antioxidant used. Until such biomarkers are tried and tested it is not possible to design an effective clinical trial to test the oxidative hypothesis of atherosclerosis.

In some of the antioxidant clinical trials there were small increases in antioxidant potential in the plasma reported due to the supplementation. The studies often relied on "natural"' antioxidants, such as vitamin E or β-carotene, instead of pharmacologically devised, high-powered antioxidants where a clinically significant increase in antioxidant potential might have been able to be achieved. One study by Meagher et al. [69] indicated that despite very high doses of vitamin E, there was still evidence of lipid peroxidation products in plasma indicating that perhaps vitamin E is not as powerful a lipid protecting antioxidant as often purported.

There is building evidence that the studies performed to date have targeted the wrong micronutrients. Perhaps vitamin E and carotenoids do not have a protective role, but rather are markers for other nutrients also present in the foods. Folate may be of interest due to its role in the reduction of homocysteine concentrations, a risk factor of CHD. Although the mechanism whereby homocysteine is a risk factor of CHD is not clearly elucidated, it is believed to be related to modulation of nitric oxide (NO) production in the endothelium. Recently it has been shown that folate could be operating independently of homocysteine concentrations. It appears that the addition of folate to ex vivo LDL increases lag time of Cu^{2+} Ox-LDL [54]. This effect cannot be attributable to NO production or the indirect effect on homocysteine concentrations, but rather to the scavenging ability of folate itself.

Of particular interest are the phenolic compounds found in foods. One explanation for the French paradox (i.e., a seemingly high intake of dietary fat, in particular, saturated fat, combined with low population rates of heart disease) is the phenolic substances present in red wine [55]. Also of particular interest are the polyphenols present in all plants, but not in vitamin supplements. Some studies have shown an inverse relationship between flavonoid intake and CVD [56, 57]. In particular, flavonoids can prevent the oxidation of LDL [58]. Flavonoids and other polyphenolics do have antioxidant behavior, but interestingly they also have other mechanisms whereby they could influence the development and progression of CVD. Flavonoids can inhibit platelet aggregation and adhesion, inhibit enzymes in lipid metabolism, and inhibit endothelium-dependent vasodilation. These molecules are also investigated for their role in modulating immune cell function and act in a anti-inflammatory manner [59]. Many of these studies are not well controlled, thus interpretation of the data must be done with care.

Oxidative Stress: Friend or Foe?

To say that oxidative stress is ubiquitously problematic is to ignore the many important cellular and protective functions that oxidation performs. This can be highlighted by the role of the oxidative burst in the immune system. Recently Williams and Fisher [6] reported a significantly important role for ROS and RNS in the posttranslational destruction of apolipoprotein-B regulation, and as such would have a role in the normal regulation of this potentially atherogenic protein. It is also clear that ROS are required for the downstream phosphorylation of tyrosine, and insulin signaling in hepatocytes indicating that oxidative stress is a normal and necessary function of insulin signaling in hepatocytes [6]. It should be noted that in the pancreas ROS are responsible for the destruction of pancreatic islet cells and due to this the decreased production of insulin. This is not a desirable result. It is clear that in some circumstances increased antioxidant presence would be desirable and lead to the prevention of chronic illness, and equally in other location or circumstance this would not be the case [6].

Susceptibility of LDL to Oxidation

Esterbauer and colleagues [60] have shown that in lipid oxidation within the LDL particle the rate of diene generation is initially slow and the induction time (lag period) to formation may vary but, once this period has been passed, the rate of diene generation reaches a maximum and continues for some time. This leads to questions about what aspects of the LDL particle could influence the lag period and also the maximum rate of peroxidation (or the slope of the curve).

Type of Fat in the Diet

The Ox-LDL hypothesis holds oxidation to be the key process in the initiation and proliferation of atheromas in the intima of the arterial wall. This is in direct contradiction of the evidence that dietary intake of polyunsaturated fats, which are incorporated into the LDL molecule and are highly oxidizable, are protective of heart disease. Similarly, exercise and moderate alcohol consumption are also considered to be protective of heart disease, and themselves are initiators of oxidative stress.

Many studies from the 1990s indicated that the type of fatty acid in the LDL molecule could radically affect susceptibility to oxidation. From these studies it is clear that the presence of monounsaturated fatty acids (MUFA) in LDL particles decreased the presence of oxidation products, indicating the oleic acid in particular is resistant to oxidation in comparison to linoleic acid. In a study by Bonanome et al. [61] the ratio of oleic acid and linoleic acid in the LDL particle was inversely related to the peroxidation rate. The presence of oleic acid increased the induction time somewhat, but the main effect was to reduce the slope of the curve during the peroxidation period [62]. This indicates that when the antioxidant present in the particle is exhausted, the particle with greater oleic acid concentration resist oxidation in comparison with a high polyunsaturated fatty acid (PUFA) intake. The fatty acid composition of the lipoprotein will affect the stability of these lipoproteins independent of antioxidant concentration [61]. In fact in a study by Dimitriadis et al. [63] showed that the LDL from a diabetic population was more susceptible to oxidation than the control equivalent. This study showed differences in the fatty acid composition of the LDL with increased free/esterified cholesterol ratio in diabetic patients.

Studies by Williams and Fisher [6] indicate that *n*–3 fatty acids, which are highly unsaturated and therefore susceptible to oxidation, inhibit Apo-B secretion in the liver through stimulating its degradation by the postendoplasmic reticulum presecretory proteolysis (PERPP) process. This process is dependent on the oxidation of PUFAs and is potentially protective of atherosclerotic development as the Apo-B moiety is very atherogenic. Similarly, saturated fatty acids, such as myristic acid a 14 carbon saturated fatty acid, are not susceptible to oxidation, have been found to stimulate secretion of Apo-B containing lipoproteins and to protect them from the PREPP process [64].

Presence of Antioxidants

Most of the animal experiments looking at the role of antioxidant in protecting the LDL from oxidation were done using probucol. Probucol is a hypolipidemic agent with impressive reducing potential. Due to the success of probucol, it was decided to proceed with large clinical trials on antioxidants, and in fact the decision was made that all antioxidants, whether soluble in lipid or aqueous solutions, could be grouped together due to their antioxidant actions being similar [45].

What is clear is that the addition of antioxidants lengthens the time to induction of LDL oxidation *in vitro*, that is the lag period, but does not affect the maximum slope of the curve during the peroxidation period [62]. Vitamin E was the most frequently used antioxidant in the clinical trials, however the evidence for a role for vitamin E was at best unclear. From animal trials the evidence is equivocal, there was some protection from atherosclerosis in the mouse model [27], but the rabbit models have not been as effective [65]. In ex vivo measurement of LDL oxidation there is only brief increase in lag time (15–20 min), whereas probucol has much greater peroxidation delaying ability (up to 20 h or more). A study by Jialal and Grundy [66] did show that the propagation of LDL was reduced by vitamin E supplementation, and in populations with increased oxidative stress such as cigarette smokers [67] and hemodialysis [68] vitamin E has been effective in protecting LDL from oxidation. This is not the case in a healthy population indicating that there is only benefit from taking vitamin E in situations where there is increased oxidative stress [69], or perhaps in a population that is deficient in vitamin E initially.

Vitamin E has been shown to be reduced by coantioxidants, such as vitamin C and uric acid [60], but in situations where the coantioxidants are lacking vitamin E has been shown to be a prooxidant [3]. Extensive research into the dual ability of vitamin E to be both pro- and antioxidant has lead to the conclusion that in cases of strong oxidative challenge vitamin E acts as an antioxidant with little concomitant lipid peroxide formation. When in mildly oxidizing conditions, vitamin E is slowly consumed and there is a build-up of lipid peroxides in the LDL molecule but there is no conversion to secondary lipid oxidation products (such as aldehydes) [3]. It is these secondary lipid peroxidation products that are highly atherogenic. So, *in vivo*, the vitamin E radical replaces the lipid peroxide radical, and is the promoter of auto-oxidation, but at a slower rate than incurred if the lipid peroxide radical was the initiator. Vitamin E is very reactive with free radicals, and the presence of vitamin E determines the rate of oxidation, the more vitamin E the more oxidation that occurs, and vitamin E acts as a prooxidant [3]. It is likely that the *in vivo* environment reflects the milder oxidation condition. This leads to the conclusion that vitamin E is not necessarily the most efficacious antioxidant to study the protection of LDL from oxidation or to test the LDL oxidation hypothesis of atherosclerosis [45].

β-Carotene is often used as a biomarker of fruit and vegetable intake, and fruit and vegetable intake is inversely related to CHD in many epidemiological studies [70]. It is not unreasonable to assume that β-carotene is protective of CHD itself. β-Carotene is not an effective chain-breaking antioxidant in comparison to vitamin E, it is an effective singlet oxygen quencher [3], so using β-carotene in trials to prevent LDL oxidation is not useful. It is particularly important that β-carotene should not be used to prove or disprove the Ox-LDL hypothesis of atherosclerosis [45].

The large clinical trials show there is no evidence for using vitamin E and β-carotene supplements in established CHD. This indicates that more potent antioxidants are required to test the role of Ox-LDL in atherosclerosis, Parthasarathy et al. [33] suggests antioxidants that are not naturally occurring, such as probucol, may be better suited to testing this hypothesis.

There is some evidence that folate protects against CHD other than through its potential for lowering homocysteine levels. Ex vivo testing of the antioxidant potential of folate suggest there may be a role for folate as an *in vivo* antioxidant [55]. Although it appears that there is little reducing power of the polyphenolics compared to other more traditional antioxidants *in vivo* [3], there have been some successful studies showing their efficacy in protection from development of atherosclerosis. Some studies suggest that the flavonoids present in onions, apples, and tea provide some protection against CHD [58], although the results are not consistent across all studies.

Perhaps the protective effect of the standard antioxidants, such as vitamin E and flavonoids, does not rely solely on antioxidant ability, this is particularly true as vitamin E has been isolated in atherosclerotic lesions [71]. Vitamin E has many antiatherosclerotic properties; it is an antithrombotic, it inhibits smooth muscle cell proliferation [72], and decreases monocyte adhesion to endothelium, decreases platelet adhesion and proliferation, stabilizes membranes, and affect intracellular calcium mobilization [9, 73]. It has also been shown to modulate superoxide formation and leakage from the mitochondria by preventing electron leakage, directly or indirectly impacting on the superoxide-generating systems, and by scavenging the superoxide that is generated [74]. Flavonoids are anti-inflammatory molecules and this is the role they may have in protecting against atherosclerosis, an established inflammatory condition.

Methods of Detecting Oxidized LDL

Indirect Methods

Initially the oxidation of LDL was measured by the ex vivo method, whereby oxidation of the unsaturated lipids in human LDL molecule was catalyzed by copper and the change in absorption at 234 nm measured. The change in absorption was due to the production of conjugated dienes, which absorb at a different frequency. Similarly the presence of thiobarbituric acid-reacting substance (TBARS) formation can be measured in the presence of copper ions. This is a severe form of oxidation that results in an LDL molecule that is capable of forming foam cells. In this case the LDL is completely devoid of antioxidant potential. This is contrary to the findings that LDL in fatty lesions in humans contain adequate concentrations of antioxidants, including vitamin E [75]. It must be remembered that as the LDL is removed from its native environment and then oxidized, it is difficult to draw comparisons to an *in vivo* environment [76].

Direct Methods

A study of an animal model of atherosclerosis (apo-E knockout mice) demonstrated that autoantibodies against Ox-LDL recognize different epitopes of the complex structures that develop due to oxidative modification of the apolipoprotein moiety [77]. Holvoet et al. [78] developed a competition ELISA method to detect Ox-LDL in plasma using monoclonal antibody 4E6. This monoclonal antibody is directed against the apo B_{100} moiety of LDL that is generated when oxidative modification occurs due to the presence of mid- to long-chain aldehydes. Using this method, elevated levels of Ox-LDL have been shown to correlate well with CAD in heart transplant patients [78]. The limitation of this monoclonal antibody is that it binds to MDA-LDL as well as an array of other modified LDL particles [79].

Toshima et al. [80] also developed a simple sandwich enzyme immunoassay for human Ox-LDL. These authors used a monoclonal antibody FOH1a/DHL3 that reacts specifically against oxidized phophatidylcholine (oxPC) but not against native LDL, MDA-LDL, acetylated LDL, or glycated LDL. This monoclonal antibody was developed by using a homogenate of human atheromatous plaques from the aorta as the antigen. The authors were also able to show a relationship between circulating levels of Ox-LDL using this method and CHD, and showed that the relationship is superior to that of other markers of CHD, such as total cholesterol, triglycerides, or HDL levels [80].

Ehara et al. [79] developed a sandwich ELISA method to measure Ox-LDL levels using a mouse anti-Ox-LDL monoclonal antibody (DLH3) and an anti-apolipoprotein B (apoB) polyclonal antibody. DLH3 is specific of Ox-LDL and does not bind to native or acetylated, glycated, or MDA-treated LDL. This method is highly sensitive to detecting minute amounts of Ox-LDL and avoids the interference of other plasma substances, although strictly speaking does not measure the plasma Ox-LDL levels as the LDL is separated from the blood before the testing. Using this ELISA method Ehara et al. [79] showed that Ox-LDL levels are related to the severity of the coronary syndrome, and thus could serve as a marker of cardiovascular events. It may be that the increased presence of circulating Ox-LDL acts as a destabilizing agent on plaque composition, most likely by enhancing the inflammatory processes and surface thrombosis [79].

The interest in the clinical role of Ox-LDL has been growing since the elucidation of its role in the progression of atherosclerosis. Unfortunately the ongoing study of the clinical role of Ox-LDL has been hampered by a lack in sensitive and specific markers of circulating Ox-LDL. It is currently apparent that this has been rectified with the development of direct ELISA methods for the testing of levels of circulating Ox-LDL and enable a direct comparison between different populations in order to speculate on the likelihood of that population developing atherosclerotic complications.

Use of Ox-LDL as a Marker of Vascular Disease

Studies have indicated an association between increases in the concentration of circulating Ox-LDL in patients and CAD [78]. Holvoet et al. [78] speculated that this increase could be due to a back diffusion of Ox-LDL from the atherosclerotic arterial wall and would be independent of plaque stability. Toshima et al. [80] studied the clinical relevance of circulating antibodies against Ox-LDL and found they were significantly higher in patients with CHD, than the control population. Based on their findings the authors [80] concluded that the Ox-LDL could be a possible risk marker for CHD. In a study by Ehara et al. [79] there was a significant positive correlation between the severity of acute coronary syndromes and the level of Ox-LDL, indicating that the levels of Ox-LDL relate to the instability of plaque in atherosclerotic lesions.

The idea that Ox-LDL and aOx-LDL could be used as a biomarker for CAD has been developed in populations that traditionally are vulnerable to increased CVD, such as type 1 and type 2 diabetes mellitus. Diabetes mellitus (DM) is associated with an increased risk of atherosclerosis; studies in these patients have shown evidence of the increased presence of Ox-LDL and aOx-LDL [81, 82]. There is also evidence that there is increased concentration of Ox-LDL in hypertensive patients [26].

There are many diseases that have been studied and found to have evidence of increased Ox-LDL concentrations in populations at increased risk of developing heart disease. In prospective studies, antibodies to Ox-LDL have been shown to predict MI and progression of atherosclerosis in populations without autoimmune disease. As such these antibodies may be seen as markers of determinants of atherosclerosis such as lipid oxidation, a proinflammatory environment, endothelial dysfunction, and impaired vasodilation [25]. They may in fact be a useful marker for indicating individuals at increased risk of CVD [83, 84].

It appears that antibodies to LDL is a much more sensitive marker to atherosclerosis and CHD than the more traditional biomarkers used currently such as LDL concentration [84], sensitive enough that it may be useful in the identification of silent atherosclerosis in clinically healthy individuals [84].

OxLDL and Inflammation

It is clear that the evidence to support the oxidation hypothesis is compelling, it is also clear that there is also equally compelling evidence that oxidation is not a key process in the development of atherosclerosis. When results are equivocal it implies there is another causative factor not originally considered in the hypothesis. For the Ox-LDL hypothesis to remain viable, oxidation must be the initiating factor in atherosclerotic development, and at this stage, the evidence does not support this. There are avenues for foam cell development that do not rely on oxidation [3].

It is likely that this other factor is inflammation. It is clear that this inflammation is a key to atherosclerosis. Initially in the early 1980s, Ross [59] concluded that diets leading to the development of atherosclerosis produced inflammatory cellular adhesion to arteries. Since then many studies have established markers of inflammation, such as C-reactive protein a systemic marker of inflammation, as sensitive and specific markers of the development and progression of atherosclerosis. C-reactive protein may be a more sensitive marker of atherosclerotic development than traditional biomarkers such as cholesterol [85].

It is clear that there is an intertwined role of Ox-LDL and inflammation. The increased entry and decreased exit of inflammatory cells from the arterial intima would be expected to increase arterial inflammation. Similarly, Ox-LDL is able to upregulate a number of genes associated with inflammation such as MCP-1, and has been shown to alter the scavenger receptor (CD36) expression [3]. Due to the antioxidant clinical trials showing little or no efficacy in the protection or prevention of atherosclerosis, it must now be considered that oxidative events are the consequence of atherosclerosis.

It is likely that oxidation is not the causative agent in the development of atherosclerosis, but rather a by-product of inflammation. LDL oxidation, a key step for the development of atherosclerosis, therefore, could be a result of increased inflammation [3, 6].

Conclusion

Some indirect evidence has accumulated to support a link between Ox-LDL and the pathophysiology of atherosclerosis. To date intervention strategies for reduction of cardiovascular risk included increased intake of antioxidant-rich foods and/or supplements for prevention of oxidation of LDL and the rate of atherogenesis. However, large clinical trials involving isolated antioxidants have failed to produce equivocal evidence to support involvement of Ox-LDL in the development of CAD. It is clear from these large trials that there is no evidence to support the use of supplements to treat or prevent CHD, in fact the only epidemiological evidence that has been sustained is that a high intake of fruits and vegetables is protective in CHD.

It is also important to distinguish whether inflammation increases oxidative stress, which further increases atherosclerosis or it is the increased oxidative stress that increases inflammation to accelerate atherosclerosis. This answer is likely to lie in the former. If that is the case, and taking into consideration the role of redox processes in many cell signaling and other normal cellular activities the ubiquitous prophylactic supplementation with antioxidants may in fact be damaging.

Recent studies have implicated a role for inflammation in the development of atherosclerosis, among other chronic lifestyle diseases. It is well recognized that oxidative stress has a role in, and is a product of, the inflammatory process. The evidence does not appear to support the primary role of oxidative stress in atherogenesis, however the evidence is more favorable for inflammation as the initiating factor in the development of atherosclerosis and through this the development and progression of CHD. Therefore, there is a need to separate oxidative stress and inflammation and their individual and synergistic effect on CHD to improve strategies for the prevention of this number one killer.

References

1. Goldstein JL, Brown MS: The low density lipoprotein pathway and its relation to atherosclerosis. Annu Rev Biochem 46:897–930, 1977.
2. Steinberg D, Parthasarathy S, Carew T, Khoo J, Witztum J: Beyond cholesterol:modifications of low density lipoprotein that increase its atherogenicity. N Engl J Med 320: 915–924, 1989.
3. Stocker R, Keaney JF: Role of oxidative modifications in atherosclerosis. Physiol Rev 84: 1381–1487, 2004.
4. Berliner A, Heinecke JW: The role of oxidised lipoproteins in atherogenesis. Free Radic Biol Med 20: 707–727, 1996.
5. Witztum JL: The oxidation hypothesis of atherosclerosis. Lancet 344:793–795, 1994.
6. Williams KJ, Fisher EA: Oxidation, lipoproteins, and atherosclerosis: which is wrong, the antioxidants or the theory? Curr Opin Clin Nutr Care 8: 139–146, 2005.
7. Avagaro P, Bittolo Bon G, Cazzolato G: Presence of a modified low density lipoprotein in humans. Arteriosclerosis 8:79, 1988.
8. Parthasarathy S, Rankin SM: Role of oxidized low density lipoprotein in atherogenesis. Prog Lipid Res 31: 127–143, 1992.
9. Hamilton CA: Low-density lipoprotein and oxidised low-density lipoprotein: their role in the development of atherosclerosis. Pharmacol Therapy 74: 55–72, 1997.
10. Yla-Herttuala S, Rosenfeld ME, Parthasarathy S, Sigal E, Sarkioja T, Witztum J, Steinberg D: Gene expression in macrophage-rich human atherosclerotic lesions:15-lipoxygenase and acetyl low density lipoprotein receptor messenger RNA colocalize with oxidation specific lipid protein adducts. J Clin Investig 87:1146–1152, 1991.
11. Goldstein JL, Ho YK, Basu SK, Brown MS: Binding site on macrophages that mediates uptake and degradation of acetylated low density lipoprotein, producing massive cholesterol deposition. Proc Natl Acad Sci U S A 76: 333–337, 1979.
12. Via DP, Dresel HA, Cheng SL, Gotto AMJ: Murine macrophage tumours are a source of a 260,000-dalton acetyl-low density lipoprotein receptor. J Biol Chem 260:7379–7386, 1885.
13. Parthasarathy S, Fong L, Otero D, Steinberg D: Recognition of solubilized apoproteins from delipidated, oxidized low density lipoprotein (LDL) by the acetyl-LDL receptor. Proc Natl Acad Sci U S A 84:537–540, 1987.

14. Dwivedi A, Anggard E, Carrier M:Oxidised LDL mediated monocyte adhesion to endothelial cells does not involve NFkappaB. Biochem Biophys Res Commun 284:239–244, 2001.

15. Henriksen T, Mahooney E, Steinberg D:Enhanced macrophage degradation of LDL previously incubated with cultured endothelial cells:recognition by receptor for acetylated LDL. Proc Natl Acad Sci U S A 78:6499–6503, 1981.

16. Sparrow CP, Parthasarathy S, Steinberg D:A macrophage receptor that recognizes oxidized low density lipoprotein but not acetylated low density lipoprotein. J Biol Chem 264:2599–2604, 1989.

17. Quinn MT, Parthasarathy S, Fong D, Steinberg D:Oxidatively modified low density lipoproteins: a potential role in recruitment and retention of monocyte/macrophages during atherogenesis. Proc Natl Acad Sci U S A 84:2995–2998, 1987.

18. Quinn MT, Parthasarathy S, Steinberg D: Lysophosphatidylcholine:a chemotactic factor for human monocytes and its potential role in atherogenesis. Proc Natl Acad Sci U S A 85:2805–2809, 1988.

19. Reichl D, Myant N, Pflug J:Concentration of lipoproteins containing apolipoprotein B in human peripheral lymph. Biochim Biophys Acta 489:98–105, 1977.

20. Raymond TL, Reynolds SA:Lipoproteins of the extravascular space: alterations in low density lipoproteins of interstitial inflammatory fluid. J Lipid Res 24:113–119, 1983.

21. Hoff H, O'Neil J: Lesion-derived low density lipoprotein and oxidized low density lipoprotein share a lability for aggregation, leading to enhanced macrophage degradation. Arterioscler Thromb 11:1209–1222, 1991.

22. Yla-Herttuala S, Palinski W, Rosenfeld ME, Parthasarathy S, Carrier M, Butler S, Witztum J, Steinberg D: Evidence for the presence of oxidatively modified low density lipoprotein in atherosclerotic lesions of rabbit and man. J Clin Invest 84:1086–1095, 1989.

23. Haberland ME, Fong D, Cheng L: Malondialdehyde-altered protein occurs in atheroma of Watanabe heritable hyperlipidemic rabbits. Science 241:215–241, 1988.

24. Palinski W, Rosenfeld ME, Yla-Herttuala S, Gurter GC, Socher SS, Butler S, Parthasarathy S, Carew T, Steinberg D: Low density lipoprotein undergoes oxidative modification *in vivo*. Proc Natl Acad Sci U S A 86: 1372–1376, 1989.

25. Vaarala O: Antibodies to oxidised LDL. Lupus 9:202–205, 2000.

26. Frostegard J, Wu R, Lemne C, Thulin T, Witztum J, de Faires U: Circulating oxidised low-density lipoprotein is increased in hypertension. Clin Sci 105: 615–620, 2003.

27. Pratico D, Tangirala RK, Rader DJ, Rokach J, FitzGerald GA:Vitamin E suppresses isoprostane generation *in vivo* and reduces atherosclerosis in ApoE-deficient mice. Nat Med 4:1189–1192, 1998.

28. Pratico D, Tangirala RK, Horkko S, Witztum JL, Palinski W, FitzGerald GA: Circulating autoantibodies to oxidized cardiolipin correlate with isoprostane $F_{2\alpha}$–VI levels and the extent of atherosclerosis in ApoE-deficient mice: modulation by vitamin E. Blood 97: 459–464, 2001.

29. Halliwell B: How to characterize a biological antioxidant. Free Radic Res Commun 9:1–32, 1990.

30. Stampfer MJ, Hennekens CH, Manson JE, Colditz GA, Rosner B, Willett WC: Vitamin E consumption and the risk of coronary disease in women. N Engl J Med 328:1444–1449, 1993.

31. Rimm EB, Stampfer MJ, Ascherio A, Giovannucci E, Colditz GA, Willett WC: Vitamin E consumption and the risk of coronary disease in men. N Engl J Med 328:1450–1459, 1993.

32. Hennekens CH, Buring JE, Manson JE, Stampfer MJ, Rosner B, Cook NP, Belanger C, La Motte F, Gaziano JM, Ridker PM, et al: Lack of effect of long-term supplementation with beta carotene on the incidence of malignant neoplasms and cardiovascular disease. N Engl J Med 333: 1145–1149, 1996.

33. Parthasarathy S, Khan-Merchant N, Penumetcha M, Khan BV, Santanam N: Did the antioxidant trials fail to validate the oxidation hypothesis? Curr Atheroscler Rep 3: 392–398, 2001.

34. Waickus C: Oxidised LDL, atherogenesis and antioxidants. Am Fam Physician 48:718–720, 1993.

35. Esterbauer H, Gebicki J, Puhl H, Jurgens G: The role of lipid peroxidation and antioxidants in oxidative modification of LDL. Free Radic Biol Med 13:341–390, 1992.

36. Kritchevsky D, Kim HK, Tepper SA: Influence of 4,4′-(isopropylide-nedithio)bis(2,6-di-t-butylphenol) (DH-581) on experimental atherosclerosis in rabbits. Proc Soc Exp Biol Med 136:1216–1221, 1971.

37. Tawara K, Ishihara M, Ogawa H, Tomikawa M: Effect of probucol, pentethine and their combination on serum lipoprotein metabolism and on incidence of atheromatous lesions in the rabbit. Jpn J Pharmacol 41:211–222, 1986.

38. Wissler RW, Vesselinovitch D: Combined effects of cholestyramine and probucol on regression of atherosclerosis in rhesus monkey aortas. Appl Pathol 1:89–96, 1983.

39. Kita T, Nagano Y, Yokode M, Ishii K, Kume N, Ooshima A, Yoshida H, Kawai C:Probucol prevents the progression of atherosclerosis in Watanabe heritable hyperlipidemic rabbits, an animal model for hypercholesterolemia. Proc Natl Acad Sci U S A 84:5928–5931, 1989.

40. Carew T, Schwenke DC, Steinberg D: Antiatherogenic affect of probucol unrelated to its hypocholesterolemic effect: evidence that antioxidants *in vivo* can selectively inhibit LDL degradation in macrophage-rich fatty streaks and slow the progression of atherosclerosis in WHH rabbits. Proc Natl Acad Sci U S A 84: 7725–7729, 1987.

41. Mao SJ, Yates MT, Parker RA, Chi EM, Jackson JL: Attenuation of atherosclerosis in a modified strain of hypercholesterolemic Watanbe rabbits with use of a probucol analogue (MDL 29, 311) that does not lower serum cholesterol. Arterioscler Thromb 11:1266–1275, 1991.

42. Witting PK, Pettersson K, Letters J, Stocker R: Site specific anti-atherogenic effects of probucol in apolipoprotein E deficient mice. Arterioscler Thromb Vasc Biol 20:e26–e33, 2000.

43. Walldius G, Olssom AG, Bergstrand L, Hadell K, Johansson J, Kaijser L, Lassvic C, Molgrard J, Nilsson S: The effect of probucol on femoral atherosclerosis: The Probucol Quantitative Regression Swedish Trial (PQRST). Am J Cardiol 74: 875–883, 1994.

44. Sawayama Y, Shimizu C, Maeda N, Tatsukawa M, Kinnkawa N, Koyanagi S, Kashiwagi S, Hayashi J: Effect of probucol and paavastatin on common carotid atherosclerosis in patients with asymptomatic hypercholesterolemia. Fukuoka Atherosclerosis Trial (FAST). J Am Coll Cardiol 39: 610–616, 2002.

45. Steinberg D, Witztum J:Is the oxidative modification hypothesis relevant to human atherosclerosis? Do the antioxidant trials conducted to date refute the hypothesis? Circulation 105:2107–2111, 2002.
46. Wen Y, Killalea S, Norris LA, Cooke T, Feely J:Vitamin E supplementation in hyperlipidemic patients: effect of increasing doses on *in vitro* and *in vivo* low-density lipoprotein oxidation. Eur J Clin Invest 29:1027–1034, 1999.
47. Klipstein-Grobusch K, Geleijnse JM, den Breeijen JH, Boeing H, Hofman A, Grobbee DE, Witteman JCM: Dietary antioxidants and risk of myocardial infarction in the elderly: the Rotterdam Study. Am J Clin Nutr 69:261–266, 1999.
48. Witting P, Pottersson K, Ostlund-Lindqvist A, Westerlund C, Wagberg M, Stocker R:Dissociation of atherogenesis from aortic accumulation of lipid hydro(pero)xides in Watanabe heritable hyperlipidemic rabbits. J Clin Invest 104: 213–220, 1999.
49. Upston JM, Niu X, Brown AJ, Mashima R, Wang H, Senthilmohan R, Kettle AJ, Dean RT, Stocker R:Disease stage-dependent accumulation of lipid and protein oxidation products in human atherosclerosis. Am J Pathol 160:701–710, 2002.
50. de Winter MPJ, Gijbels MJJ, van Dijk KW, van Gorp PJJ, Suzuki H, Kodama T, Frants RR, Havekes LM, Hofke MH: Scavenger receptor deficiency leads to more complex atherosclerotic lesions in APOE3 Leiden transgenic mice. Atherosclerosis 144:315–321, 1999.
51. Whitman SC, Rateri DL, Szilvassy SJ, Cornicelli JA, Daugherty A: Macrophage-specific expression of class A scavenger receptors in LDL receptor –/– mice decreases atherosclerosis and changes spleen morphology. J Lipid Res 43: 1201–1208, 2002.
52. Vivekananthan DP, Penn MS, Sapp SK, Hsu A, Topol EJ: Use of antioxidant vitamins for the prevention of cardiovascular disease:meta-analysis of randomised trials. Lancet 361:2017–2023, 2003.
53. Hodis HN, Mack WJ, LaBree L, Mahrer PR, Sevanian A, Liu C, Liu C, Hwang J, Slezer RH, Azen SP: Alpha-tocopherol supplementation in healthy indivi-duals reduces low-density lipoprotein oxidation but not atherosclerosis: The Vitamin E Atherosclerosis Prevention Study (VEAPS). Circulation 106: 1453–1459, 2002.
54. Nakano E, Higgins JA, Powers HJ: Folate protects against oxidative modification of human LDL. Br J Nutr 86:637–639, 2001.
55. Frankel EN, Kanner J, German JB, Parks E, Kinsella JE: Inhibition of oxidation of human low-density lipoprotein by phenolic substances in red wine. Lancet 341: 454–457, 1993.
56. Hertog MG, Feskens EJ, Kromhout D: Antioxidant flavanols and coronary heart disease risk. Lancet 349:699, 1997.
57. Yochum L, Kushi LH, Meyer K, Folsom AR: Dietary flavanoid intake and risk of cardiovascular disease in post menopausal women. Am J Epidemiol 149:943–949, 1999.
58. Avariam M, Furhrman B:Polyphenolic flavonoids inhibit macrophage-mediated oxidation of LDL and attenuate atherogenesis. Atherosclerosis 137 (Suppl): S45–S50, 1998.
59. Ross R: Atherosclerosis—an inflammatory disease. N Engl J Med 340:115–126, 1999.
60. Esterbauer H, Rotheneder M, Striegl G, Waeg G, Ashy, A, Sattler W, Jurgens G:Vitamin E and other lipophilic antioxidants protect LDL against oxidation. Fat Sci Technol 8: 316–324, 1989.

61. Bonanome A, Pagnan A, Biffanti S, Opportuno A, Sorgato F, Dorella M, Maiorino M, Ursini F: Effect of dietary monounsaturated and polyunsaturated fatty acids on the susceptibility of plasma low density lipoproteins to oxidative modification. Arterioscler Thromb 12:529–533, 1992.

62. Parthasarathy S, Khoo J, Miller E, Barnett J, Witztum JL, Steinberg D:Low density lipoprotein rich in oleic acid is protected against oxidative modification: Implications for dietary prevention of atherosclerosis. Proc Natl Acad Sci U S A 87:3894–3898, 1990.

63. Dimitriadis E, Griffin M, Owens D, Johnson A, Collins P, Tomkin GH:Oxidation of low-density lipoprotein in NIDDM: its relationship to fatty acid composition. Diabetologica 38:1300–1306, 1995.

64. Krauss RM: Hold the antioxidants and improve plasma lipids? J Clin Invest 113:1253–1255, 2004.

65. Stocker R: Dietary and pharmacological antioxidants in atherosclerosis. Curr Opin Lipidol 10: 589–597, 1999.

66. Jialal I, Grundy SM: Effect of dietary supplementation with alpha-tocopherol on the oxidative modification of low density lipoprotein. J Lipid Res 33: 899–906, 1992.

67. Morrow JD, Frei B, Longmire AW, Gaziano JM, Lynch SM, Shyr Y, Strauss WE, Oates JA, Roberts LJ: Increase in circulating products of lipid peroxidation (F_2-isoprostanes) in smokers. N Engl J Med 94:19–25, 1995.

68. Boaz M, Smetana S, Weinstein T, Matas Z, Gafter U, Iaina A, Knecht A, Weissgarten Y, Brunner D, Fainaru M, et al:Secondary prevention with antioxidants of cardiovascular disease in end-stage renal disease (SPACE): randomised placebo controlled trial. Lancet 356:1213–1218, 2000.

69. Meagher E, Barry O, Lawson J, Rokach J, FitzGerald G:Effects of vitamin E on lipid peroxidation in healthy persons. J Am Med Assoc 285:1178–1182, 2001.

70. Temple N:Antioxidants and disease: more questions than answers. Nutr Res 20:449–459, 2000.

71. Upston JM, Terrentis AC, Stocker R:Tocopherol mediated peroxidation (TMP) of lipoproteins:implications for vitamin E as a potential antiatherogenic supplement. FASEB J 13: 977–994, 1999.

72. Boscoboinik D, Szewczyk A, Hensey C, Azzi A: Inhibition of cell proliferation by α-tocopherol. Role of protein kinase C. J Biol Chem 266:6188–6194, 1991.

73. Ferns GAA, Konneh M, Anggard E:Vitamin E:the evidence for an anti-atherogenic role. Artery 20:61–94, 1993.

74. Chow CK:Vitamin E regulation of mitochondrial superoxide generation. Biol Signals Recept 10:112–124, 2001.

75. Steinberecher U:Receptors for oxidised low-density lipoprotein. Biochim Biophys Acta 1436:279–298, 1999.

76. Kritharides L, Stocker R: The use of antioxidant supplements in coronary heart disease. Atherosclerosis 164:211–219, 2002.

77. Palinski W, Horkko S, Miller E, Steinberecher U, Powell HC, Curtiss LK:Cloning of monoclonal autoantibodies to epitopes of oxidized lipoproteins form apolipoprotein E-deficient mice. Demonstration of epitopes of oxidized low density lipoprotein in human plasma. J Clin Invest 98: 800–814, 1996.

78. Holvoet P, Theilmeier G, Shivalkar B, Fameng W, Collen D:LDL hypercholesterolemia is associated with accumulation of oxidized LDL, atherosclerotic plaque growth and compensatory vessel enlargement in coronary arteries of miniature pigs. Arterioscler Thromb Vasc Biol 18:415–422, 1998.

79. Ehara S, Ueda M, Naruko T, Haze K, Itoh A, Otsuka M, Komatsu R, Matsuo T, Itabe H, Takano T, et al: Elevated levels of oxidized low density lipoprotein show a positive relationship with the severity of acute coronary syndromes. Circulation 103:1955–1960, 2001.
80. Toshima S, Hasegawa A, Kurabayashi M, Itabe H, Takano T, Sugano J, Shimamura K, Kimura J, Michishita I, Suzuki T, et al:Circulating oxidized low density lipoprotein levels. A biochemical risk marker for coronary heart disease. Arterioscler Thromb Vasc Biol 20:2243–2247, 2000.
81. Festa A, Kopp HP, Schernthaner G, Menzel EJ: Autoantibodies to oxidised low density lipoproteins in IDDM are inversely related to metabolic control and microvascular complications. Diabetologica 41:350–356, 1998.
82. Bellomo G, Maggi E, Poli M, Agosta F, Bollarti P, Finardi G:Autoantibodies against oxidatively modified low-density lipoproteins in NIDDM. Diabetes 44:60–67, 1995.
83. Saarelainen S, Lehtimaki T, Jaak-Kola O, Poussa T, Nikkila M, Slolakivi T, Nieminen MM:Autoantibodies against low-density lipoprotein in patients with obstructive sleep apnoea. Clin Chem Lab Med 37:517–520, 1999.
84. Wallenfeldt K, Fagerberg B, Wilstrand J, Hulthie J:Oxidized low density lipoprotein in plasma is a prognostic marker of subclinical atherosclerosis development in clinically healthy men. J Intern Med 256:413–420, 2004.
85. Ricker P, Cook N, Rifai N:C Reactive protein, the metabolic syndrome, and risk of incident cardiovascular events—an 8 year follow up of 14719 initially healthy American women. Circulation 107:391–397, 2003.

25

Dietary Fatty Acids and Stroke

GENE R. HERZBERG

Abstract

The relationship of dietary fatty acid intake and the risk of stroke has been the subject of numerous investigations. The results of studies of the effects of saturated fat have been conflicting making it difficult to reach a conclusion about saturated fat and stroke. Intake of *n*6 fatty acids, particularly linoleic acid, appears to be inversely related to stroke risk. Although concerns have been expressed about a possible increase in risk of hemorrhagic stroke related to long chain *n*3 intake caused by changes in the clotting ability of blood, the results suggest that these fatty acids, when consumed at levels found in typical Western diets, reduce the risk of ischemic stroke with no increase in risk of hemorrhagic stroke.

Keywords: blood clotting; diet; fatty acids; hemorrhagic stroke; ischemic stroke; *n*3 fatty acids, *n*6 fatty acids; saturated fatty acids

Abbreviations: EPA, eicosapentaenoic acid 20:5 *n*3; COX-2, cyclooxygenase; TX, thromboxane; PUFA, polyunsaturated fatty acids; DHA, docosahexaenoic acid 22:6 *n*3; P/S, polyunsaturated/saturated ratio; ALA, alpha-linolenic acid, 18:3 *n*3

Introduction

Stroke is a leading cause of cardiovascular deaths. Annual stroke incidence in Europe, Russia, Australasia, and the United States for those 45–84 years of age is between 0.3% and 0.5% [1]. Worldwide, stroke is the second leading cause of death accounting for 4.4 million deaths in 1990 [2] and results in significant morbidity. The death rate due to stroke varies among countries but is between 17% and 38% [3]. Of those who survive, approximately one half will remain permanently disabled and one half will be capable of independent living [4].

Stroke results from an interruption in the blood supply to a portion of the brain. The extent of the injury depends on the location and severity of the stroke. Strokes can be broadly divided into two categories namely, hemorrhagic and ischemic. Hemorrhagic strokes are a result of the rupture of a blood vessel with the resulting loss of blood supply. Ischemic strokes are the result of a clot forming in an artery. The relative proportion of ischemic compared to hemorrhagic strokes is a function of the age of the population studied since the proportion of ischemic strokes increases with age [5]. Studies in eight countries in those aged 45–84 indicate that hemorrhagic strokes account for approximately 85% of all strokes with ischemic strokes, accounting for approximately 10%, the remainder being of undetermined cause [1].

The type of dietary fat has been suggested to modify the risk of stroke. The effect of dietary fatty acids on stroke is complicated by the fact that different fatty acids have different effects on both blood clotting and blood pressure. For example, linoleic acid (18:2, omega-6; *n*6) may lead to increases in platelet aggregability via its conversion to arachidonic acid (20:4 *n*6), which may increase the likelihood of a clot forming, but may decrease blood pressure. The balance between these effects might depend on dose, and thus make it difficult to determine the mechanism for a relationship between a given fatty acid and stroke.

Fatty Acids and Clotting

The effects of fatty acids on hemostasis have been recently reviewed by Knapp [6], who concluded that evidence of effects of saturated or monounsaturated fatty acids on hemostatis in humans has been difficult to demonstrate conclusively. Polyunsaturated fatty acids (PUFA) alter platelet aggregation and other pathways related to coagulation and fibrinolysis.

Saturated Fatty Acids

Several reports have shown that addition of saturated free fatty acids stimulates aggregation of human platelets [7–10]. Various trials both epidemiologic and intervention in humans and feeding trials in animals have shown effects of saturated fatty acids on platelet function. However, it is difficult to separate the effect of removal of saturates from the diet from the effects of the resulting increase in *n*3 and *n*6 fatty acids in the diet [11–13]. Hoak [14] has concluded that there is no evidence that dietary stearate is thrombogenic.

Monounsaturated Fatty Acids

Oleic acid is generally considered to be neutral with regard to coagulation [6] although there are reports of decreased platelet aggregation when oleic acid was replaced with linoleic acid [15].

Polyunsaturated Fatty Acids

Chronic dietary supplementation with large quantities of fish oils has a modest inhibitory effect on blood clotting compared with aspirin [16]. Ingestion of the pure ethyl ester of eicosapentaenoic acid; 20:5 *n*3 (EPA) led to prolonged bleeding times, so it seems likely that EPA or a product derived from it is responsible for the prolongation of bleeding time after fish oil ingestion [17].

Accumulation of long-chain $n3$ fatty acids in platelets after fish oil supplementation is partially in place of arachidonic acid in membrane phospholipids. Consequently, the increased phospholipase activity after stimulation of platelets results in less arachidonic acid being released. A consequence of this is reduced conversion of arachidonic acid to the proaggregatory and vasoconstrictive thromboxane (TX) A_2 due to the competitive inhibitory effect of the long-chain $n3$ fatty acids on cyclooxygenase (COX-2) activity [18].

An increase in the intake of long-chain $n3$ fatty acids, EPA in particular, is associated with a decrease in blood clotting and an increase in bleeding time in humans [19]. This was first reported by Bang and Dyerberg in 1980, who found that Greenland Eskimos had prolonged bleeding times related to their high intake of fat from marine mammals high in long-chain fatty acids [20]. The mechanism for the antithrombotic effect is not certain, but the most commonly proposed mechanism involves changes in TX levels. EPA competes with arachidonic acid for COX-2, which results in an increase in the TXA_3/TXA_2 ratio. Changes in this ratio are associated with reduced platelet aggregation and vasoconstriction and result in prolonged bleeding time [21]. Products of COX-2 action on arachidonic acid such as prostaglandin I_2 and isoprostanes have been associated with reduced blood flow after traumatic cerebral insult [22]. As with clotting, competition for COX-2 by long-chain $n3$ fatty acids could decrease some of these products and lead to increased blood flow after cerebral trauma leading to increased blood flow. Increased $n3$ fatty acid intake has also been associated with several factors in the intrinsic clotting pathway and with reduced blood viscosity [21, 23].

Perhaps of greatest practical relevance to the relationship between long-chain $n3$ fatty acids and bleeding time is the lack of significant bleeding in patients taking high doses of $n3$ fatty acids. These studies have included examination of bleeding after major surgical procedures or childbirth [24–26]. Thus it appears that while long-chain $n3$ fatty acids may prolong bleeding times, they do not contribute to increased incidence of bleeding except when consumed at very high levels.

Studies of the effects of dietary linoleic acid have generally found that measured platelet aggregation is reduced but with no change in bleeding time (summarized by Knapp [6]). Sanders and Hochland [27] compared a fish oil supplement with a supplement high in oleic and linoleic acids and reported similar reductions in collagen-induced platelet aggregation with both supplements. Overall, studies suggest that $n6$ PUFA may reduce thrombotic processes *in vivo* but have generally not found altered bleeding times.

The available data suggest that PUFA, particularly EPA may increase measured bleeding time or decrease platelet aggregation but there is little evidence, except at high levels of intake, that these fatty acids increase the risk of bleeding.

Fatty Acids and Blood Pressure

Grimsgard et al. studying Norwegian men 40–42 years old, found that blood pressure was linearly related to plasma phospholipid saturated fatty acids and inversely associated with linoleic acid [28]. These relationships were independent of body mass index. Studies in several other countries have also found a positive association between the blood level of palmitic acid and blood pressure [29–31]. Dietary intake of linoleic acid [32] and plasma levels of linoleate have been shown to be inversely associated with blood pressure [30, 31, 33].

Dietary supplementation with high doses of fish oil compared to safflower oil or a mixture of oils that approximated the types of fat present in the American diet can reduce blood pressure in men with essential hypertension [34]. Some [35, 36] but not all [37] clinical trials have supported a blood pressure lowering effect of dietary linoleic acid. A fish oil supplement of 15 g/day reduced blood pressure in a six-month study [38]. Meta-analyses of controlled clinical trials of the effects of fish oil on blood pressure concluded that daily intake of EPA + DHA (docosahexaenoic acid 22:6 $n3$) $\geq$ 3 g/day significantly lowered blood pressure [39, 40].

Studies in both spontaneously hypertensive rats [41] and humans [42–44] have shown that alpha-linolenic acid (ALA, 18:3 $n3$) lowers blood pressure. The effects in humans were seen when blood pressure was related to either dietary intake measured directly or indirectly as the adipose tissue content of alpha linolenate.

While the relationship between blood pressure and diet fatty acid content remains somewhat controversial, the majority of the evidence supports a positive correlation of blood pressure with saturated fat and an inverse relationship with polyunsaturated fat and the polyunsaturated/saturated (P/S) ratio.

Dietary Fatty Acids and Stroke

Saturated Fatty Acids

Results of studies of the relationship between saturated fat and stroke have been inconsistent. In an ecological study, Sasaki et al. [45] reported a positive correlation between saturated fat intake and total stroke mortality. Iso et al. [46] found an inverse relationship between the risk of hemorrhagic stroke in women and intake of saturated or *trans*-unsaturated fat intake but no association with PUFA or monounsaturated fat. Similarly, in men, the Framingham study found the risk of ischemic stroke declined with increasing intake of saturated fat and monounsaturated fat but not polyunsaturated fat. Too few cases of hemorrhagic stroke occurred to reach any conclusions [47]. In a study of Japanese men and women, a high consumption of animal fat was associated with a reduced risk of death from cerebral infarction [48].

Results of the Seven Countries Study, after 20 years of follow-up, found a nonsignificant inverse relationship between consumption of saturated fats and stroke [49]. In a study of men of Japanese descent living in Japan, McGee et al. found the percentage of calories as saturated fat was inversely related to stroke mortality [50].

In both the Lyon Diet Heart Trial as well as in the large Finnish intervention study, decreasing the intake of saturated fat was associated with a large reduction in mortality from stroke [51, 52].

Polyunsaturated Fat

N3 Fatty Acids

The first suggestions that dietary *n*3 fatty acids might modify the risk of hemorrhagic stroke came from ecologic studies of Greenland Eskimos who have a very high intake of *n*3 fatty acids and were shown to have an increased risk of hemorrhagic stroke compared with Danish whites [53, 54]. It should be noted that the intake of *n*3 fatty acids by Greenland Eskimos (10.5 g/day) is much higher than that of Danish whites (0.8 g/day) [55] or of US residents (0.1–0.2 g/day) [56]. Increased mortality due to hemorrhagic stroke has also been reported in some traditional Japanese fishing communities with high intakes of long-chain *n*3 fatty acids [57]. However, Yamori et al. [58] found that *n*3 PUFA levels were significantly higher in the inhabitants of Japanese fishing villages with relatively low stroke morbidity compared to those of farming villages with extremely high stroke morbidity, although the type of stroke was not specified.

Iso et al. [46] examined the association between *n*3 PUFA intake and the risk of stroke in women and found that those in the highest quintile of intake had reduced risk of both total and thrombotic strokes with no increased risk of hemorrhagic stroke. The intake of long-chain *n*3 fatty acids by women in the highest quintile was 0.48 g/day.

In the GISISI-prevenzionne trial, subjects were given a supplement of *n*3 PUFA of 1 g/day, which reduced the risk of an acute heart attack but had no significant effect on the risk of stroke [59].

He et al. [60] examined the relationship between long-chain *n*3 PUFA intake and stroke in men and found an inverse relationship between intake and risk of ischemic stroke while there was no association found with hemorrhagic stroke. However, in a subsequent report, He et al. found no evidence that the type of dietary fat affects the risk of either hemorrhagic or ischemic stroke [61].

Mortality rates from stroke among Alaska Natives, who had previously been shown to have elevated concentrations of plasma *n*3 PUFA [62], are higher than rates for US whites [63]. High levels of long-chain *n*3 fatty acids in perirenal adipose tissue, which reflects high dietary intake, was found to be associated with increased incidence of hemorrhagic stroke [64].

A recent meta-analysis concluded that fish consumption is inversely related to risk of ischemic but not hemorrhagic stroke [65]. Fish is the main dietary source of long-chain $n3$ PUFAs, which have been shown to have many favorable effects on factors related to stroke risk such as hypertension and blood clotting as discussed previously and reviewed by Dyerberg et al. [66] and Nestel [67].

Simon et al. [68] found that higher serum levels of ALA were associated with a lower risk of stroke in middle-aged men at high risk for cardiovascular disease while stearic acid was associated with an increased risk of stroke. Similarly, Leng et al. [69] found that ALA was significantly lower in the red blood cell phospholipids of patients with stroke even after controlling for blood pressure.

Vegetable Fat

A multivariate analysis of the relationship between foodstuffs and stroke in 19 countries found both ischemic and hemorrhagic stroke to be positively correlated with vegetable fat intake [70]. A number of case control studies have found that a lower proportion of linoleic acid in tissues that reflect dietary intake is associated with increased risk of total or ischemic stroke [71–74]. Japanese, who have a higher mortality rate from both ischemic and hemorrhagic stroke [75], have lower serum levels of linoleic acid and higher levels of saturated and $n3$ fatty acids [76]. In a more recent case control study of Japanese men and women, Iso et al. [77] found that linoleic acid was inversely associated with the risk of total stroke and ischemic stroke.

Functional Outcome of Stroke and Effect of Dietary Fatty Acids

Studies of the effects of dietary long-chain $n3$ fatty acids on functional outcome after a stroke have been limited to those with experimental animals. Black et al. [78] using an acute model of cerebral ischemia induced by ligation of the middle cerebral artery in cats found that the neurological deficit and volume of brain infarction was less in a group treated with fish oil than the control group. In contrast, Clarke et al. [79] found, in a model of intracerebral hemorrhage in rats, that a diet high in fish oil compared to safflower oil led to significantly greater impairment of forelimb dexterity and fine motor control. There was no difference in infarct volume but animals maintained on a diet enriched with fish oil exhibited increased cerebral blood flow after the stroke. The differences in these two studies suggest that the effects of fish oil may be related to effects on coagulation or blood flow, and that whether or not fish oil is beneficial depends on the type of stroke.

Conclusion

Overall, given the conflicting results, it is difficult to reach a conclusion about saturated fat and stroke. Intake of *n*6 fatty acids, particularly linoleic acid, appears to be inversely related to stroke risk. Despite expressed concerns about a possible increase in risk of hemorrhagic stroke related to long-chain *n*3 intake, the results suggest that these fatty acids, when consumed at levels found in typical Western diets, reduce the risk of ischemic stroke with no increase in risk of hemorrhagic stroke.

References

1. Sudlow CL, Warlow CP: Comparable studies of the incidence of stroke and its pathological types: results from an international collaboration. International Stroke Incidence Collaboration. Stroke 28: 491–499, 1997.
2. Murray CJ, Lopez AD: Mortality by cause for eight regions of the world: Global Burden of Disease Study. Lancet 349: 1269–1276, 1997.
3. Weir NU, Sandercock PA, Lewis SC, Signorini DF, Warlow CP: Variations between countries in outcome after stroke in the International Stroke Trial (IST). Stroke 32: 1370–1377, 2001.
4. Davenport RJ, Dennis MS, Warlow CP: Effect of correcting outcome data for case mix: an example from stroke medicine. BMJ 312: 1503–1505, 1996.
5. Canada Heart and Stroke Foundation: The Growing Burden of Heart Disease and Stroke in Canada 2003. Ottawa, Canada, 2003.
6. Knapp HR: Dietary fatty acids in human thrombosis and hemostasis. Am J Clin Nutr 65: 1687S–1698S, 1997.
7. Hoak JC, Spector AA, Fry GL, Warner ED: Effect of free fatty acids on ADP-induced platelet aggregation. Nature 228: 1330–1332, 1970.
8. Hoak JC, Warner ED, Connor WE: Platelets, fatty acids and thrombosis. Circ Res 20: 11–17, 1967.
9. Kerr JW, Macaulay I, Pirrie R, Bronte-Stewart B: Platelet-aggregation by phospholipids and free fatty acids. Lancet 43: 1296–1299, 1965.
10. Prost-Dvojakovic RJ, Samama M: Clot-promoting and platelet-aggregating effects of fatty acids. Haemostasis 2: 73–84, 1973.
11. Keys A, Karvonen MJ, Punsar S, Menotti A, Fidanza F, Farchi G: HDL serum cholesterol and 24-year mortality of men in Finland. Int J Epidemiol 13: 428–435, 1984.
12. Nordoy A, Goodnight SH: Dietary lipids and thrombosis. Relationships to atherosclerosis. Arteriosclerosis 10: 149–163, 1990.
13. Renaud S, Kuba K, Goulet C, Lemire Y, Allard C: Relationship between fatty-acid composition of platelets and platelet aggregation in rat and man. Relation to thrombosis. Circ Res 26: 553–564, 1970.
14. Hoak JC: Stearic acid, clotting, and thrombosis. Am J Clin Nutr 60: 1050S–1053S, 1994.
15. Burri BJ, Dougherty RM, Kelley DS, Iacono JM: Platelet aggregation in humans is affected by replacement of dietary linoleic acid with oleic acid. Am J Clin Nutr 54: 359–362, 1991.

16. Knapp HR, Reilly IA, Alessandrini P, FitzGerald GA: *In vivo* indexes of platelet and vascular function during fish-oil administration in patients with atherosclerosis. N Engl J Med 314: 937–942, 1986.

17. Tamura Y, Hirai A, Terano T, Yoshida S, Takenaga M, Kitagawa H: Anti-thrombotic and anti-atherogenic action of eicosapentaenoic acid. Jpn Circ J 51: 471–477, 1987.

18. Holub BJ: Clinical nutrition: 4. Omega-3 fatty acids in cardiovascular care. CMAJ 166: 608–615, 2002.

19. Thorngren M, Shafi S, Born GV: Delay in primary haemostasis produced by a fish diet without change in local thromboxane A2. Br J Haematol 58: 567–578, 1984.

20. Bang HO, Dyerberg J: The bleeding tendency in Greenland Eskimos. Dan Med Bull 27: 202–205, 1980.

21. Lox CD: The effects of dietary marine fish oils (omega-3 fatty acids) on coagulation profiles in men. Gen Pharmacol 21: 241–246, 1990.

22. Hoffman SW, Rzigalinski BA, Willoughby KA, Ellis EF: Astrocytes generate isoprostanes in response to trauma or oxygen radicals. J Neurotrauma 17: 415–420, 2000.

23. Rylance PB, Gordge MP, Saynor R, Parsons V, Weston MJ: Fish oil modifies lipids and reduces platelet aggregability in haemodialysis patients. Nephron 43: 196–202, 1986.

24. Bairati I, Roy L, Meyer F: Double-blind, randomized, controlled trial of fish oil supplements in prevention of recurrence of stenosis after coronary angioplasty. Circulation 85: 950–956, 1992.

25. DeCaterina R, Giannessi D, Mazzone A, Bernini W, Lazzerini G, Maffei S, Cerri M, Salvatore L, Weksler B: Vascular prostacyclin is increased in patients ingesting omega-3 polyunsaturated fatty acids before coronary artery bypass graft surgery. Circulation 82: 428–438, 1990.

26. Olsen SF, Sorensen JD, Secher NJ, Hedegaard M, Henriksen TB, Hansen HS, Grant A: Randomised controlled trial of effect of fish-oil supplementation on pregnancy duration. Lancet 339: 1003–1007, 1992.

27. Sanders TA, Hochland MC: A comparison of the influence on plasma lipids and platelet function of supplements of omega 3 and omega 6 polyunsaturated fatty acids. Br J Nutr 50: 521–529, 1983.

28. Grimsgaard S, Bonaa KH, Jacobsen BK, Bjerve KS: Plasma saturated and linoleic fatty acids are independently associated with blood pressure. Hypertension 34: 478–483, 1999.

29. Cambien F, Warnet JM, Vernier V, Ducimetiere P, Jacqueson A, Flament C, Orssaud G, Richard JL, Claude JR: An epidemiologic appraisal of the associations between the fatty acids esterifying serum cholesterol and some cardiovascular risk factors in middle-aged men. Am J Epidemiol 127: 75–86, 1988.

30. Simon JA, Fong J, Bernert JT Jr: Serum fatty acids and blood pressure. Hypertension 27: 303–307, 1996.

31. Uusitupa MI, Sarkkinen ES, Torpstrom J, Pietinen P, Aro A: Long-term effects of four fat-modified diets on blood pressure. J Hum Hypertens 8: 209–218, 1994.

32. Oster P, Arab L, Schellenberg B, Kohlmeier M, Schlierf G: Linoleic acid and blood pressure. Prog Food Nutr Sci 4: 39–40, 1980.

33. Miettinen TA, Naukkarinen V, Huttunen JK, Mattila S, Kumlin T: Fatty-acid composition of serum lipids predicts myocardial infarction. Br Med J (Clin Res Ed) 285: 993–996, 1982.

34. Knapp HR, FitzGerald GA: The antihypertensive effects of fish oil. A controlled study of polyunsaturated fatty acid supplements in essential hypertension. N Engl J Med 320: 1037–1043, 1989.
35. Folsom AR, Ma J, McGovern PG, Eckfeldt H: Relation between plasma phospholipid saturated fatty acids and hyperinsulinemia. Metabolism 45: 223–228, 1996.
36. Iacono JM, Puska P, Dougherty RM, Pietinen P, Vartiainen E, Leino U, Mutanen M, Moisio S: Effect of dietary fat on blood pressure in a rural Finnish population. Am J Clin Nutr 38: 860–869, 1983.
37. Sacks FM, Rouse IL, Stampfer MJ, Bishop LM, Lenherr CF, Walther RJ: Effect of dietary fats and carbohydrate on blood pressure of mildly hypertensive patients. Hypertension 10: 452–460, 1987.
38. Bairati I, Roy L, Meyer F: Effects of a fish oil supplement on blood pressure and serum lipids in patients treated for coronary artery disease. Can J Cardiol 8: 41–46, 1992.
39. Appel LJ, Miller ER 3rd, Seidler AJ, Whelton PK: Does supplementation of diet with 'fish oil' reduce blood pressure? A meta-analysis of controlled clinical trials. Arch Intern Med 153: 1429–1438, 1993.
40. Morris MC, Sacks F, Rosner B: Does fish oil lower blood pressure? A meta-analysis of controlled trials. Circulation 88: 523–533, 1993.
41. Rupp H, Turcani M, Ohkubo T, Maisch B, Brilla CG: Dietary linolenic acid-mediated increase in vascular prostacyclin formation. Mol Cell Biochem 162: 59–64, 1996.
42. Berry EM, Hirsch J: Does dietary linolenic acid influence blood pressure? Am J Clin Nutr 44: 336–340, 1986.
43. Djousse L, Arnett DK, Pankow JS, Hopkins PN, Province MA, Ellison RC: Dietary linolenic acid is associated with a lower prevalence of hypertension in the NHLBI Family Heart Study. Hypertension 45: 368–373, 2005.
44. Salonen JT, Salonen R, Ihanainen M, Parviainen M, Seppanen R, Kantola M, Seppanen K, Rauramaa R: Blood pressure, dietary fats, and antioxidants. Am J Clin Nutr 48: 1226–1232, 1988.
45. Sasaki S, Zhang XH, Kesteloot H: Dietary sodium, potassium, saturated fat, alcohol, and stroke mortality. Stroke 26: 783–789, 1995.
46. Iso H, Stampfer MJ, Manson JE, Rexrode K, Hu F, Hennekens CH, Colditz GA, Speizer FE, Willett WC: Prospective study of fat and protein intake and risk of intraparenchymal hemorrhage in women. Circulation 103: 856–863, 2001.
47. Gillman MW, Cupples LA, Millen BE, Ellison RC, Wolf PA: Inverse association of dietary fat with development of ischemic stroke in men. JAMA 278: 2145–2150, 1997.
48. Sauvaget C, Nagano J, Hayashi M, Yamada M: Animal protein, animal fat, and cholesterol intakes and risk of cerebral infarction mortality in the adult health study. Stroke 35: 1531–1537, 2004.
49. Menotti A, Keys A, Blackburn H, Aravanis C, Dontas A, Fidanza F, Giampaoli S, Karvonen M, Kromhout D, Nedeljkovic S, et al.: Twenty-year stroke mortality and prediction in twelve cohorts of the Seven Countries Study. Int J Epidemiol 19: 309–315, 1990.
50. McGee D, Reed D, Stemmerman G, Rhoads G, Yano K, Feinleib M: The relationship of dietary fat and cholesterol to mortality in 10 years: the Honolulu Heart Program. Int J Epidemiol 14: 97–105, 1985.

51. Renaud S, de Lorgeril M, Delaye J, Guidollet J, Jacquard F, Mamelle N, Martin JL, Monjaud I, Salen P, Toubol P: Cretan Mediterranean diet for prevention of coronary heart disease. Am J Clin Nutr 61: 1360S–1367S, 1995.

52. Vartiainen E, Sarti C, Tuomilehto J, Kuulasmaa K: Do changes in cardiovascular risk factors explain changes in mortality from stroke in Finland? BMJ 310: 901–904, 1995.

53. Kromann N, Green A: Epidemiological studies in the Upernavik district, Greenland. Incidence of some chronic diseases 1950–1974. Acta Med Scand 208: 401–406, 1980.

54. Ostergaard Kristensen M: Increased incidence of bleeding intracranial aneurysms in Greenlandic Eskimos. Acta Neurochir (Wien) 67: 37–43, 1983.

55. Bang HO, Dyerberg J, Sinclair HM: The composition of the Eskimo food in north western Greenland. Am J Clin Nutr 33: 2657–2661, 1980.

56. Raper NR, Cronin FJ, Exler J: Omega-3 fatty acid content of the US food supply. J Am Coll Nutr 11: 304–308, 1992.

57. Hirai A, Hamazaki T, Terano T, Nishikawa T, Tamura Y, Kamugai A, Jajiki J: Eicosapentaenoic acid and platelet function in Japanese. Lancet 2: 1132–1133, 1980.

58. Yamori Y, Nara Y, Iritani N, Workman RJ, Inagami T: Comparison of serum phospholipid fatty acids among fishing and farming Japanese populations and American inlanders. J Nutr Sci Vitaminol (Tokyo) 31: 417–422, 1985.

59. GISSI-investigators: Dietary supplementation with n–3 polyunsaturated fatty acids and vitamin E after myocardial infarction: results of the GISSI-Prevenzione trial. Gruppo Italiano per lo Studio della Sopravvivenza nell''Infarto miocardico. Lancet 354: 447–455, 1999.

60. He K, Rimm EB, Merchant A, Rosner BA, Stampfer MJ, Willett WC, Ascherio A: Fish consumption and risk of stroke in men. JAMA 288: 3130–3136, 2002.

61. He K, Merchant A, Rimm EB, Rosner BA, Stampfer MJ, Willett WC, Ascherio A: Dietary fat intake and risk of stroke in male US healthcare professionals: 14 year prospective cohort study. BMJ 327: 777–782, 2003.

62. Parkinson AJ, Cruz AL, Heyward WL, Bulkow LR, Hall D, Barstaed L, Connor WE: Elevated concentrations of plasma omega-3 polyunsaturated fatty acids among Alaskan Eskimos. Am J Clin Nutr 59: 384–388, 1994.

63. Schumacher C, Davidson M, Ehrsam G: Cardiovascular disease among Alaska Natives: a review of the literature. Int J Circumpolar Health 62: 343–362, 2003.

64. Pedersen HS, Mulvad G, Seidelin KN, Malcom GT, Boudreau DA: N–3 fatty acids as a risk factor for haemorrhagic stroke. Lancet 353: 812–813, 1999.

65. He K, Song Y, Daviglus ML, Liu K, Van Horn L, Dyer AR, Goldbourt U, Greenland P: Fish consumption and incidence of stroke: a meta-analysis of cohort studies. Stroke 35: 1538–1542, 2004.

66. Dyerberg J, Bang HO, Stoffersen E, Moncada S, Vane JR: Eicosapentaenoic acid and prevention of thrombosis and atherosclerosis? Lancet 2: 117–119, 1978.

67. Nestel PJ: Effects of N–3 fatty acids on lipid metabolism. Annu Rev Nutr 10: 149–167, 1990.

68. Simon JA, Fong J, Bernert JT, Jr., Browner WS: Serum fatty acids and the risk of stroke. Stroke 26: 778–782, 1995.

69. Leng GC, Taylor GS, Lee AJ, Fowkes FG, Horrobin D: Essential fatty acids and cardiovascular disease: the Edinburgh Artery Study. Vasc Med 4: 219–226, 1999.

70. Renaud SC: Diet and stroke. J Nutr Health Aging 5: 167–172, 2001.

71. Bucalossi A, Mori S: Fatty acid composition of adipose tissue in ischemic heart disease and stroke. Gerontol Clin (Basel) 14: 339–345, 1972.

72. Ciavatti M, Michel G, Dechavanne M: Platelet phospholipid in stroke. Clin Chim Acta 84: 347–351, 1978.

73. Ricci S, Patoia L, Berrettini M, Binaglia L, Scarcella MG, Bucaneve G, Vecchini A, Carloni I, Agostini L, Parise P, et al.: Fatty acid pattern of red blood cell membranes and risk of ischemic brain infarction: a case-control study. Stroke 18: 575–578, 1987.

74. Tilvis RS, Erkinjuntti T, Sulkava R, Farkkila M, Miettinen TA: Serum lipids and fatty acids in ischemic strokes. Am Heart J 113: 615–619, 1987.

75. Takeya Y, Popper JS, Shimizu Y, Kato H, Rhoads GG, Kagan A: Epidemiologic studies of coronary heart disease and stroke in Japanese men living in Japan, Hawaii and California: incidence of stroke in Japan and Hawaii. Stroke 15: 15–23, 1984.

76. Iso H, Sato S, Folsom AR, Shimamoto T, Terao A, Munger RG, Kitamura A, Konishi M, Iida M, Komachi Y: Serum fatty acids and fish intake in rural Japanese, urban Japanese, Japanese American and Caucasian American men. Int J Epidemiol 18: 374–381, 1989.

77. Iso H, Sato S, Umemura U, Kudo M, Koike K, Kitamura A, Imano H, Okamura T, Naito Y, Shimamoto T: Linoleic acid, other fatty acids, and the risk of stroke. Stroke 33: 2086–2093, 2002.

78. Black KL, Culp B, Madison D, Randall OS, Lands WE: The protective effects of dietary fish oil on focal cerebral infarction. Prostaglandins Med 3: 257–268, 1979.

79. Clarke J, Herzberg G, Peeling J, Buist R, Corbett D: Dietary supplementation of omega-3 polyunsaturated fatty acids worsens forelimb motor function after intracerebral hemorrhage in rats. Exp Neurol 191: 119–127, 2005.

Index

A

ABCA, *see* ATP-binding cassette (ABC), transporters
ABCA1, 3–6, 9, 12–14, 26, 39
  alleles, 24, 40
  degradation, 47
  expression, 40, 45, 97
  flippase activity, 46
  internalization, 43
  mutants, 44
  pathway, 13
  phosphorylation of, 45–47
ABCG, 95
ABCG1, 57, 92, 95–100, 219
ABCG4, 99
ABCG5, 11–12
  hepatic, 57
ABCG8, 11–12
  hepatic, 57
Abdominal aortic aneurysm, 176
Abdominal obesity, 286
Abetalipoproteinemia, 134
Aboriginals, 248
ACAT, *see* Acyl coenzyme A-cholesterol acyltransferase
Acellular lipid core, 457
Acetylation, 521
Acetylcholine, 437, 445, 481
Acetyl-CoA carboxylase (ACC), 221
N-Acetylcysteine, 262
Acetylhydrolase, 482
AcLDL, 75–76
Actinobacillus actinomycetemcomitans, 418–419
ACTION I, 267
ACTION II, 267
Activated protein C (APC), 153
Activating transcription factor 6 (ATF-6), 364–365

Acute coronary syndrome (ACS), 151, 156, 461
Acute lymphoblastic leukemia, 347
Acute myocardial infarction (AMI), 151, 158, 443, 455, 475; *see also* Myocardial infarction
  and LDL size, 176
Acyl-coenzyme A-cholesterol acyltransferase (ACAT), 5, 9, 11, 92, 131–132
Acylglycerol-3-phosphate acyltransferase (AGPAT), 124, 126
1-Acylglycerol-3-phosphate, 124
Acylglycerols, 137
Acyltransferases, 137
Adaptor proteins, 213
Adenosine, 335
Adenosine deaminase, 335
Adenosine kinase, 335
S-Adenosylhomocysteine (SAH), 130, 334
S-Adenosylmethionine (SAM), 130, 329, 333
Adenylate cyclase, 5, 498
Adhesion
  molecules, 439, 447
  proteins, 288
Adipocytes, 213, 216, 256, 291
Adipokines, 216
Adiponectin, 215–217, 219
ADMA, *see also* Asymmetric dimethylarginine
  and vascular disease, 443–444
  therapeutic reduction in transplant arteriopathy, 446–447
Adrenaline, 498
Adult onset diabetes mellitus, 285
Advanced glycation end products (AGE), 257, 290, 294–296
  atherosclerosis and hypertension, 305–319

Advanced lipoxidation end products (ALE), 310
AGE(s), *see also* Advanced glycation end
         products
    accumulation, 250, 267
    adducts, 255
    and endothelial dysfunction, 314–315
    and oxidative stress, 311–313
    and smooth muscle cells, 263–264
    -carboxymethyl lysine (CML), 250
    -epitopes, 251
    formation, 247, 249, 257
       inhibitors of, 266
       due to insulin resistance, 309–310
    -HDL, 255–256
    -LDL, 256–259
       autoantibodies, 255
    proteins, 265
    receptors, 262
    structures, 251–252
    "second generation", 259
    -RAGE pathway, 295
    -RAGE signaling pathways, 293
AGE-modified lipoproteins, 253
AKT pathways, 214
Alagebrium, 296
Aldehydes, 247, 249
    conjugates, 309–310
Aldose reductase, 289, 309
Allograft
    coronary endothelial cells, 436
    eNOS expression, 448
    heart, 439
    vasculature, 445
Alloxan, 287
ALT-711, 265, 268, 295
Alzheimer's disease, 177, 329, 365, 386
Amadori products, 249, 294
"Amadori inhibitor", 268
American Diabetes Association, 206
Aminoguanidine, 254, 265, 267–268, 295
Ammonia, 336
Amphetamines, 461
Amyloid β-peptide, 77, 259
Angina, 475, 479–480
    pectoris, 495
    recurrent, 424
    unstable, 455–456
    variant, 443
Angiotensin (AT1) receptors, 265
Angiotensin
    I (AngI), 498
    II (AngII), 150, 439, 458, 463, 498
Angiotensin-converting enzyme (ACE), 150,
         158, 265, 498

Annexin V, 42
Antiapolipoprotein B, 535
Antiatherogenic phenotype, 508
Anticonvulsants, 346
Antigen-1, 484, 487
Antigen-4, 484
Antigen-presenting cells, 436
Antioxidant(s), 258, 310, 385, 519–527,
         530–537
    enzymes, 508
Antiphospholipid syndrome (APS), 153
Antithrombin III, 157
Aortic allografts, 444
Aortic lesions, 524
Aortic sinus atherosclerosis, 58
AP1, *see* Transcription factors
ApoA
    -I, *see also* Apolipoprotein, A-1
       –ABCA1 interactions, 41–45
       lipidation of, 43
       lipid-free, 93, 99
       Milano, 14, 94
    -II, 26, 40, 44, 210
    -IV, 40, 44, 98
ApoB, 24–30, 132, 137, 174, 207, 226, 531
    /BATless mouse, 211
    100, 123–124, 151, 170, 253, 478
    B48, 124, 227
ApoB-containing lipoproteins, 33, 222–227
ApoC, 40, 98
    -I, -II and -III, 44
ApoE, 8, 13, 26, 30–32, 40, 44, 57–58, 98,
         264, 360, 397–398, 400, 460, 482
    knockout mouse model, 287
    null mice, 383, 419
ApoE/CBS-deficient double knockout
         mice, 362
Apolipoprotein, 3–7, 10, 44–45, 122, 463
    A-1, 4–7, 24–32, 210, 256, 482
    B, *see* ApoB
    B48, 227
    CII, 193
    E, *see* ApoE
    E knockout (ApoE KO) mice, 58, 61,
         158, 252
    moiety, 535
    mutations in, 24
Apoproteins, 25
Apoptosis, 54, 210, 253, 358, 387, 457–461
    caspase-8-mediated, 81
    ER stress-induced, 370
    of VSMC, 440, 446
AR42J cells, 342
Arabinose, 253

Arachidonic acid (AA), 482, 498, 546–547
L-Arginine, 445, 497, 500
Arginine:glycine amidinotransferase (AGAT), 339
"Arginine paradox", 441
Argpyrimidine, 310, 314
ARIC study, 417
Arrhythmias, 508
Arterial intima, 520
Arteriopathy, transplant, 435–448
Arteriosclerosis, 445, 495
Asbestos, 260
Aspirin, 546
Asymmetric dimethylarginine (ADMA), 369, 440–442, 445, 500
ATF6, 292
Atherogenesis, 70, 501, 537–538
   infectious agents' role in, 413–425
Atherogenic dyslipidemia, 205, 286
Atherogenic lipoprotein phenotype, 169
Atherogenic plaque, 495
Atherogenicity, 29
Atheroma, 259, 419, 443, 506, 524, 532
Atherosclerosis, 519
   and LDL size, 176
   aortic, 28–30
   diet-induced, 59
   LCAT in, 28
   macrophage-specific influences on, 100
   protection against, 54
Atherosclerotic aorta, 522–523
Atherosclerotic lesions, 416, 479–480, 523–524, 534, 536
   formation and AGEs, 315–316
Atherosclerotic plaques, 456–457, 479, 523
Atherosclerotic vascular disease, 329
Atherothrombosis, 152
ATP-binding cassette (ABC)
   transporter A1, see ABCA1
   transporters, 5, 12, 40, 57, 92–95, 219; see also ABCA1
Atrial natriuretic peptide, 439
Australian Aboriginals, 248
Autoantibodies, 29, 255
Autophosphorylation, 212
Azithromycin, 420, 424
Azuridine, 347

B
B cells, 404
B lymphocytes, 398–402
Bacteremia, 419
Bacteroides forsythus, 418

BALBc mice, 287
Balloon angioplasty, 444
Bariatric surgery, 217
Beige mutation mice, 400
β-sheets, 259
Betaine, 335
Betaine homocysteine methyltransferase (BHMT), 335, 338, 342, 359
Big mitogen-activated protein kinase 1 (BMK1), 501
Bile acid(s), 4, 9, 11–12, 96, 131, 218
   biosynthetic pathways, 188, 190
   flux, 197
   general structure of, 189
   resins, 15
   sequestrants, 347
Bile salt export pump (ABCB11), 9, 11
Biliary cholesterol, 12
   pathway, 26
Bioactive lipids, 56, 72
Bio-breeding (BB) rat, 287
Blood
   clotting, 545–547, 550
   flow, 71, 416, 439, 445
   glucose, 286
   pressure, 311, 314, 546, 548
Bone marrow, 62, 101, 404
   transplantation (BMT), 77, 100, 104, 403
Bovine arterial endothelial cells (BAEC), 82
Bowes human melanoma cells, 76
Brachial artery, 27
Brain infarction, 550
Bridging properties of HL, 10
Brown adipose tissue (BAT) knockout (KO) mouse, 211

C
Calcium
   channels, 311
   influx, 314
Calnexin, 364
Calreticulin, 292, 364
cAMP response element-binding protein (CREB), 260, 366
Cancer, 346
Captopril, 158
Carbonyl(s), 266–267
   stress, 247, 251, 258, 266
Carboxyethyl lysine (CEL), 255, 268, 310
Carboxymethyl lysine (CML), 250, 255, 259–260, 268, 310
Carcinoma, 222
Cardiac allografts, 447–448

Cardiac transplant recipients, eNOS
    dysfunction in, 444–445
Cardiac transplantation, 446
Cardiovascular disease (CVD), 122, 130, 151,
    248, 284, 519, 524
Cardiovascular disorders,
    hyperhomocysteinemia-induced, 380
Cardiovascular risk marker, 444
CARE, *see* Cholesterol and Recurrent Events
    study
Carnitine palmitoyl transferase (CPT)
    activity, 217
Carnitine, 334
Carotene, 258
β-Carotene, 317, 525, 527, 530, 533–534
Carotenoids, 530
Carotid
    artery
        atherosclerosis, 158
        disease, 175–176
        stenosis, 250
    intima, 158
    plaques, 461
Caspases, 370, 457
Catalase, 508
Catechin, 482–483
Catecholamines, 460
Catechol-*O*-methyltransferase (COMT),
    339
Cathepsins, 252, 459
Cationic amino acid transporter (CAT),
    440–441
Caucasians, 345
Caveolin-1, 500, 505
Cbl C deficiency, 343; *see also* Cobalamine
CBS, *see also* Cystathionine β-synthase
    deficiency, 387
    mRNA, 341
C-C chemokine receptor (CCR2), 252, 380,
    384, 386
CCAAT/enhancer-binding protein, 7
CD1d, 402
    null mice, 403–404
CD36, 10, 70, 74, 77, 261, 288, 417
    -TLR4 interactions, 78
CE, *see also* Cholesterol esters
    from HDL, 8
    neutral, 25
    selective uptake of, 13
Cell
    death, 253, 292
    debris, 477
    signaling pathways, SR-BI in, 80–83
Cell-mediated cytotoxicity, 399

Cellular adhesion molecule (CAM)
    expression, 256, 260–263
Cellular lipid(s)
    export, 43
    efflux of, 46
Ceramide, 56
Cerebral infarction, 548
Cerebral ischemia, 550
Cerebrotendinous xanthomatosis, 190
Cerebrovascular disease, 476, 495
Ceruloplasmin, 367
CETP *see also* Cholesterol ester transfer
    protein
    inhibitors, 8, 14, 94
    promoter, human, 7
c-*fos*, 263
Chaperones, 363–364
Chemoattractants, 520, 522
Chemokine(s), 362, 414, 436, 439,
    447, 458, 525
    expression, in vascular cells, 385–386
Chemotactic proteins, 72
Chenodeoxycholic acids, 188
Chlamydia pneumoniae (Cpn), 413–414, 417,
    419–424, 458
Chlamydia trachomatis, 423
Chlamydial, 459
CHO cells, 80–81
    -K1 cells, 81
3-[(3-Cholamidopropyl) dimethylammonio]-
    1-propanesulfonate (CHAPS), 136
Cholesterol
    7α-hydroxylase (cyp7a), 11, 187–188, 190;
        see also CYP7A1
    absorption, 14–15, 253
    acceptors, 13, 25
    biosynthesis, 131, 382
    catabolism, 186
    crystals, 457, 477
    dietary, 7
    efflux, 4–7, 13, 40, 62, 80, 101, 256
        ABC transporters and apoE in, 92
        ABCA1-mediated, 45–46, 266
        apolipoprotein-mediated, 43, 97
        enhanced, 47
        SR-BI-mediated, 56–57
        in fibroblasts, 26
    esters (CE), 9–11, 24, 92, 247, 457
        transfer protein (CETP), 7, 14, 26, 32,
            171, 463
    esterificaton rate (CER), 32
    free, 4, 8
    HDL, 7, 10–11, 13–14
    homeostasis, 10–12, 70, 92, 218

Cholesterol (*Continued*)
 metabolism, 11–12
 transport, 8, 11–12, 14
  proteins, 12
Cholesterol and Recurrent Events study
  (CARE), 175, 208
Cholesteryl esters, 28, 71, 121, 257, 258
Cholestyramine, 193, 347
Cholic acids, 188
Choline, 340
 kinase, 130
Chondrocytes, 82
Chromophores, 249
Chylomicron(s), 8, 56, 123, 210
 remnants (CR), 227
Chymase, 458–459
Cirrhosis, 222
Citrulline, 441, 443, 497
c-Jun N-terminal protein kinases (JNK) , 75,
  192, 214, 370
c-*jun*, 263
CLA-1 gene, SNP in, 10
CLAMPS, 60
Claudication, 307
CML, *see also* Carboxymethyl lysine
 epitopes, 251
 -protein adduct, 252
CMV, in transplant endothelial dysfunction,
  445–446
Coagulation, 151, 154, 504, 546, 550
Coantioxidants, 525, 533
Cob(I)alamin, 343, 345, 347
Cob(II)alamin, 345, 347
Cobalamine, 345, 347
Cobalamin metabolism, 343
Cocaine, 461
Coenzyme
 A, 336
 Q10, 310, 316
Colesevelam, 15
Colestipol, 347
Collagen, 251, 294, 311, 361, 459, 479, 486
Collaterals, 456, 461
Complete LCAT deficiency (CLD), 27
Coronary artery(ies)
 endothelial vasodilator function of, 437
 diseases (CAD), 27, 208, 330, 476, 527
 atherosclerosis, 94, 175
 culprit lesions, 455
 endothelial vasodilator dysfunction, 444
 heart disease (CHD), 23, 58, 168, 284, 413,
  495, 519
 heart disease and LDL size, 174–175
 stenoses, 456

syndrome, 480, 535–536
 thrombi, 461
 thrombosis, 455, 462
Corticosteroids, 7
COS-7 cells, 77
C-reactive protein, 418–419
Creatine, 329, 334, 336, 339
CREB, *see* cAMP response element-binding
  protein
C-type lectin-like domain (CLD), 83
CXCR-2, 252
Cyclic adenosine monophosphate (cAMP),
  439, 442, 497–498
Cyclin A, 484
Cyclooxygenase (COX), 498
Cyclooxygenase-2 (COX-2), 481, 547
Cyclosporine, 443
 A, 448
CYP7A1, 12
Cyp7a1 gene expression, 191–192
Cystathionine β-synthase (CBS), 335–338,
  341, 343, 359, 383
Cystathionine γ-lyase (CGL), 335, 343
Cystathionine, 335–336
Cystathioninuria, 344
Cystatin-c, 459
Cysteine, 305, 318, 336, 366, 368
Cysteinylglycine, 366, 368
Cytidine diphosphocholine (CDP-choline),
  130–131
Cytokines, 72, 253, 296, 362, 402, 458, 525
 expression, 262
 production, 502, 505
Cytomegalovirus (CMV), 413, 415–417,
  435, 458
Cytosol, 45
Cytosolic free calcium and AGEs, 314

D
Deadly quartet, 208
Dendritic cells, 398, 402, 458
Deoxycoformycin, 347
3-Deoxyglucosone, 250, 267, 290
Dexamethasone, 341–342
Dgat-1 gene, 128
DGAT-1, overexpression of, 132
Dgat-2 gene, 128
Diabetes, 27, 34, 151, 175, 194, 221,
  414, 458
 type 1 (T1D), 253, 256, 263, 285
 type 2 (T2D), 122, 205–208, 213, 265, 330
 and LDL size, 177–178
Diabetes Atherosclerosis Intervention Study
  (DAIS), 209

Diabetes Control and Complications Trial
    (DCCT), 248
Diabetes mellitus (DM), 32, 307, 341, 346,
    441, 536
  model systems, 286–288
  and development of atherosclerosis,
    284–296
Diabetes-induced atherosclerosis, 247
Diabetic atherosclerosis, animal models of,
    264–266
Diabetic dyslipidemia, 208, 216, 254
Diabetic nephropathy, 158
Diacylglycerol (DAG), 46, 124, 293
Diacylglycerol acyltransferase (DGAT), 124,
    127–128; see also DGAT-1
Diacylglycerol transacylase (DGTA),
    128–129
Dialuric acid, 287
Dietary fats, 495, 504, 509
Dietary stearate, 546
Diethyl-*p*-nitrophenyl phosphate, 136
L-3,4-Dihydroxyphenylalanine (L-DOPA),
    339, 347
Diisopropyl fluorophosphates, 136
Dimethylamine, 441, 443
Dimethylarginine dimethylaminohydrolase
    (DDAH), 369, 441, 443, 446
Dimethylarginine dimethylaminohydrolase I
    (DDAH-I) recipients, 447
N, N-Dimethylglycine, 335
Dimethylglycine, 335
Diphosphoglycerate, 309
DNA, 334
  methylation, 338
  oxidation of, 384
  synthesis, 257, 480
Docosahexaenoic acid (DHA, 22:6), 507,
    548
Dopamine, 339
Doppler flow wire, 437
Double knockout mice, 33
Down syndrome, 345
Doxycycline, 420
Drenomedullin, 439
Dysfunctional endothelium, 503
Dyslipidemia, 14, 32, 157, 177, 221, 268,
    307, 435

E
Ectopia lentis, 359
Eicosanoids, 505–506, 508
Eicosapentaenoic acid (EPA), 482, 507
ELAM-1, 422
Elastase, 485

Elastin, 311, 459
Embolization, 504
Endocannabinoid system, 463
Endocarditis, 420
Endocytosis, 9, 152, 250, 257, 259, 261, 421
Endoplasmic reticulum (ER) stress, 291–293,
    358, 363–366, 370, 387
Endosomal pathway, 8
Endothelial
  cell protein C receptor (EPCR), 153
  growth factor receptor (EGFR), 315
  permeability, 477
  thromboxane A2, 481
  vasodilator dysfunction, 435
Endothelial dysfunction, 305, 358, 436, 487,
    498–508, 536
  and AGEs, 314–315
  and atherosclerosis, 495–509, 495
Endothelial lipase (EL), 13–14, 94
Endothelial nitric oxide synthase (eNOS), 54,
    61, 80, 314, 481
Endothelin, 439–440, 460, 498
Endothelin-1 (ET-1), 481
Endothelium-derived hyperpolarizing factor
    (EDHF), 498
Endotoxemia, 75
Endotoxin, 76
eNOS, see also Endothelial nitric oxide
    synthase
  activation, 81
  dysfunction in cardiac transplant
    recipients, 444–445
  gene, 443
Enterohepatic circulatory pathway, 11
Enterohepatic metabolism of cholesterol,
    11–12
Eotaxin, 458
Epicardial coronary vasoconstriction, 445
Epicardial vasodilatation, 437
EPIC-Norfolk study, 250
Epidermal growth factor (EGF), 459
Epinephrine, 334
24, 25-Epoxycholesterol, 96
ER stress response elements (ERSE), 365
ERK, see also Extracellular signal-regulated
    kinase
  signaling, 225
  1/2, 80
  5, 501
Erythrocytes, 508
Erythromycin, 420
Erythrose, 253
E-selectin, 57, 156, 288, 361, 386, 478, 506
Estradiol, 56

Ethinyl estradiol, 337
Ethionine synthase, 342
Ethyl eicosapentaenoic acid (EPA), 546
Eukaryotic initiation factor-2α (eIF-2α),
    292, 365
Extracellular matrix (ECM), 251, 288,
    361, 459
Extracellular molecular targeting, 367–368
Extracellular signal-regulated kinase (ERK),
    214, 220, 263
Ezetimibe, 15

F
F2-isoprostanes, 523
Factor V, 367
Factor Y, 387
Familial hypercholesterolemia (FH), 73,
    177, 210
Farnesoid X receptor (FXR), 12, 186, 191,
    194–197, 218
Farnesyl pyrophosphate, 157
FAT, 77
Fatty acids
    n3-507–508, 545, 547, 549, 550
    n6-, 551
Fatty acid synthase (FAS), 218, 221
Fatty acyl coenzyme A, 11
"Fatty streak", 398, 414, 421, 479, 519, 521
Fc-γ receptors, 255
FED mutations, 27
Fenofibrate Intervention and Event Lowering
    in Diabetes Study (FIELD), 209
Fenofibrate, 14
    treatment, 209
FFA, *see also* Free fatty acid
    efflux, 217
    flux, 205, 215–216, 223, 225, 227
    metabolism, 211
    oxidation, 217, 219
    synthesis, 220–221
Fibrates, 14, 219
Fibrin, 368
Fibrinogen, 157, 505
Fibrinolysis, 151, 154, 157, 546
Fibrinolytic activity, 461
Fibroblasts, 5, 26, 42, 97, 256, 500, 522
Fibroblast growth factor (FGF), 503
Fibrofatty plaques, 436
Fibronectin, 367–368
Fibrosis, 222, 414
"Fibrotic plaque", 414
Fibrous cap, 456, 459
Fibrous plaques, 421
Fish-eye disease (FED), 27

Flavoenzyme, 335
Flavonoids, 483, 531, 534
Flow-mediated dilatation (FMD), 27
Flow-mediated vasodilation (FMVD), 441
Fluorescence photobleaching technique, 41
Fluorophores, 249
Foam cell, 72, 247, 358, 398, 457, 477, 501, 534
    death, 288
    formation, 254, 256–259
Folate, 337, 347, 381, 530
    metabolism, 346
Folic acid, 329, 359, 386
Foxa1, 194–195
Foxo1, 192
Framingham Heart Study, 24, 331, 344
Frank rupture, 460
Free cholesterol (FC), 11, 24–25, 92, 457
    bidirectional flux of, 95
Free fatty acid (FFA), 207, 210, 212, 218
FXR, *see also* Farnesoid X receptor
    target genes, 197
    farnesoid X receptor, 3
Fyn, 77

G
α-Galactosylceramide, 401
α-GalCer, 402–403
Galectin-3, 262
Gallbladder disease, 122, 286
Gallstone, 190
Gangrene, 307
Gemfibrozil, 14
Genetic polymorphisms, 329
Geranyl pyrophosphate, 157
Geranylgeranyl pyrophosphate (GGPP), 97
Gingivitis, 418
Glibenclamide, 42
Glitazones, 209
Glomerulopathy, 262
Glu298Asp, 443
Glucagons, 341, 343
Glucocorticoid(s), 341, 343
    receptor, 194–195
Gluconeogenesis, 33, 197, 205
Glucose, 152, 247, 249, 253, 309
    biosynthetic pathway, 194
    homeostasis, 210, 219
    intolerance, 207
    metabolism, 305, 308
    tolerance, impaired, 210
    transporter 4 (GLUT4), 214, 285, 291
Glucose-6-phosphatase (G6Pase), 194
β-Glucuronidase, 485
GLUT1 transporter, 259

GLUT2 transporter, 287
Glutamine:fructose-6 phosphate
        amidotransferase (GFAT), 290
Glutathione, 310–311, 336, 366, 368, 483
Glutathione peroxidase (GPx), 310–311
    expression, 363
Glutathione peroxidase (GPx-1) activity, 369
Glutathione reductase, 252, 311
Glycated hemoglobin (HbA1c), 248
Glycation, 152, 155, 249
Glycemia, 217, 264–265
Glyceraldehyde, 253
    -3-phosphate (G3P), 309
Glyceraldehyde-3-phosphate dehydrogenase
        (GAPDH), 252, 309
Glycerol, 134, 222
    -3-phosphate pathway, 125
Glycerol-3-phosphate acyltransferase (GPAT)
        125–126, 128
Glycine N-methyltransferase (GNMT), 334
Glycine, 335
Glycogen synthesis, 213, 222
Glycolaldehyde, 249–250, 253, 257–258, 261
    -pyridine, 310
Glycolytic pathway, 309
β-Glycoprotein, 458
Glycosphingolipid, 402
O-linked Glycosylation, 291
Glycoxidation, 247, 249
Glyoxal, 267, 309
GNMT, 338, 342
Goto Kakizaki rat, 287
Gout, 208
GPIV, 77
G-proteins, 75–76, 158
Graft atherosclerosis, 444
Graft ischemia, 448
Granzyme, 399
    A promoter, 400
Grb-2-SOS complex, 213
Growth arrest- and DNA damage-inducible
        gene 153 (GADD153), 292, 365
Growth hormone, 342
GRP78, 292, 364–365
GRP94, 292, 364–365
GTPases, 260
GTP-cyclohydrolase I, 61
Guanidinoacetate methyltransferase, 339
Guanidinoacetic acid (GAA), 339
Guanylate cyclase, 497

H
H4IIE cells, 341–342
Haemopoetic atherogenesis, 362

HbA1c, 250
HCAECs, see Human, coronary artery
        endothelial cells
HDL, see also High-density lipoproteins
    apolipoproteins, 40
    apoproteins, 40, 46
    biogenesis of, 93, 96
    cholesterol, 226
    deficiency, 24, 26, 31
    nascent pre-β-, 7, 13, 96
    receptor, 54
    spherical, 4, 7, 13
    syndromes, low, 33
HDL-SR-BI ligation, 81
Heart
    attack, 121
    disease, 307
    failure, 441
    transplantation, 435
HEART study, 158
Heat shock proteins (HSP), 156–157,
        458–459
    60 (hsp60), 422–423
HEK 293 cells, 99–100
HeLa cells, 80, 385
Helicobacter pylori (Hpy), 413, 417–418
Helix-loop-helix
    -leucine zipper transcription factors,
        220
    zipper proteins, 387
Hematoma, 399
Hematopoietic stem cells, 100
Heme oxygenase-1 (HO-1), 363
Heme proteins, 439
Hemodialysis, 533
Hemoglobin (HbA1c), 254
    glycated (HbA1c), 295
Hemorrhage, 477
Hemorrhagic stroke, 545–546, 548–551
Hemostasis, 486, 496, 546
Heparans, 499
Heparin, 156, 368
    sulfate proteoglycans, 8
Hepatic lipase, 8, 171
Hepatic lipid metabolism, 387–388
Hepatic steatosis, 221, 286, 359
Hepatic VLDL
    assembly, 121
    overproduction, 223
    production, 222
    TG production, 223
Hepatocellular necrosis, 221
Hepatocyte(s), 124, 223, 261, 292, 339,
        382, 531

Hepatocyte(s) (*Continued*)
   nuclear factor-4 (HNF-4), 3; *see also*
      HNF1; HNF-3α and -4α
   nuclear factor-4α (HNF-4α), 12, 194–195
Hepatoma
   cell lines, 132
   H4IIE cells, 342
HepG2 cells, 10, 132, 135, 223, 341
Herp, 365
Herpes simplex virus type 2, 415
Herpesviruses (HPV), 413, 415–417, 445–446
Hexosamine pathway, 249, 290–292
HHcy
   -induced atherogenesis, animal models of,
      360–362
   -induced CVD, molecular and biochemical
      mechanisms of, 380
High-density lipoproteins (HDL), 5, 11, 14,
      39, 475, 526
   cholesterol (HDL-C), 23–24, 27, 29, 40
High-mobility group protein B1 (HMGB1),
      259
High-plasma homocysteine, 414
High-sensitivity C-reactive protein (hs-CRP),
      482
HIV, 415
HL, 8, 13–14
   -60 cells, 370
HMG-CoA, *see* Hydroxy-3-methylglutaryl
      coenzyme A
HMG-CoA reductase, 10, 157, 209–210,
      366, 387
HNF
   1, 192
   -4, 1, 192
   -3α and -4α, 194–195
Holoprotein, 96
Homeostasis, 95, 314
L-^{35}S-Homocysteine, 368
Homocysteine (Hcy), 293, 362, 530
   expression, 385–386
   hepatic lipid metabolism and, 387–388
   metabolism, 329, 333, 336–338, 381–382
   plasma forms of, 331–333
   kidney injury and, 388
   thiolactonase, 333
   thiolactone, 332, 367, 370
Homocysteinemia, 333
Homocysteinuria, 329, 343–344, 347, 359
Homocysteinylated-factor V (Va), 368
Hordaland homocysteine study, 337, 345
Hordaland study, 387
Hormone(s), 329, 496
   sensitive lipase (HSL), 135, 215

HTG, *see* Hypertriglyceridemia
Human
   apoB transgenic mouse, 211
   CMV, 445
   coronary artery endothelial cells
      (HCAEC), 81–82
   factor V, 368
   hepatoma cell (HepG2), 387
      line, 341
   monocyte-derived macrophages (HMDM),
      257
   umbilical endothelial cells, 484
   umbilical vein endothelial cells (HUVEC),
      481
Hydrocortisone, 342–343
Hydrogen peroxide, 310
Hydroxy-3-methylglutaryl coenzyme A
      (HMG-CoA), 3, 150, 476
α-Hydroxyaldehydes, 266
Hydroxycholesterols, 96
   7α-, 187–188
HyperapoB, 227
Hyperbetalipoproteinemia, 170
Hypercholesterolemia, 71, 131, 150, 210, 441,
      476, 487
   familial, 521
Hypercoagulability, 248
Hyperglycemia, 153, 207, 210, 248–254, 267,
      443
   chronic, 285, 295
   streptozotocin-induced, 289
Hyperhomocysteinemia (HHcy), 329–332,
      337, 441, 443
   atherosclerosis and, 362
   endothelial function and, 383
   genetic and nutritional factors, 359–360
   nonphysiological causes of, 343–346
   role in atherosclerosis, 358
Hyperinsulinemia, 205, 210–211, 218, 221, 287
Hyperlipidemia, 28, 122, 170, 208, 287,
      414, 443
Hyperphagia, 211
Hyperplasia, 210
Hypertension, 71, 175, 207, 222, 296, 414, 443
   atherosclerosis, and AGE products,
      305–319
   risk factors, 306
Hyperthemia, 157
Hypertriglyceridemia (HTG), 8, 27, 32–33
      153, 171, 196, 210, 227
Hypoalphalipoproteinemia, 170
Hypochloride, 152
Hypochlorous acid (HOCl), 310
Hypoxia, 343, 498

I

IκB kinases, 385
ICAM, *see* Intercellular adhesion molecule
IHD, *see* Ischemic heart disease
IKK
  -α, 385
  -β, 385
Immune system in atherosclerosis, 397–405
Immunogenicity, 254
Immunoinflammatory disease, 463
Immunosuppression, 443
Immunosuppressive agents, 435, 448
Inducible nitric oxide synthase (iNOS), 384,
    444, 481
Infertility, 209, 286
Inflammation, 71, 150, 221, 305, 362,
    419, 519
Inflammatory cells, 458–459, 461
Inflammatory cytokines, 447
Inflammatory diseases, 156
Inflammatory markers, 81, 216–217, 422
Inflammatory pathways, 80, 293
Innominate artery, 460
iNOS knockout mice, 444
INSIG-1, 220
Insulin, 10, 341, 343, 531
  -dependent diabetes mellitus (IDDM), 285
    dyslipidemia, 221–222
    pathway, 213
  receptor (IR) signaling pathway, 211
  resistance, 28, 32, 206, 286, 435, 458
  -resistant states, 205
  signaling, 220–221
    substrate (IRS), 213–214, 219
    syndrome, 177, 208, 217, 308
Integrin(s), 502
  β2, 288, 485
Intercellular adhesion molecule (ICAM), 288
  -1 (ICAM-1), 293, 417, 422, 478, 502
Interferon (IFN-γ), 7, 97, 288, 399–405,
    458, 502
Interleukin(s), 502
  -α, 10
  -1 (IL-1), 97, 255, 422, 478
  -2 (IL-2), 403
  -4 (IL-4), 402–403
  -6 (IL-6), 76, 263, 417, 419, 422
  -8, 80, 362, 414, 422, 458
    receptor, 252
  -10 (IL-10), 403, 458
  -12 (IL-12), 402, 405
  -18 (IL-18), 405
Intermediate-density lipoprotein (IDL), 30,
    33, 170

International Diabetes Federation (IDF),
    206
Intima, 71–72, 156, 288
Intimal hyperplasia, 436
Intima-medial thickness (IMT), 27, 178, 443
Intracellular adhesion molecule-1 (ICAM-1),
    156, 263
Intracellular mediators, 157
Intracellular molecular targeting, 368–370
Intracerebral hemorrhage, 550
Intravascular thrombosis, 477
IRE(s), 226
  -1, 292, 364–366
    signaling, 370
Ischemia-reperfusion, 435–436
Ischemic heart disease (IHD), 151, 158,
    507
Ischemic stroke, 413, 415, 418, 545, 548–551,
    546
Ischemic sudden death, 455–456
Isoniazid, 347
Isoprostanes, 547

J

Jα18 null mice, 404
JAK/STAT pathway, 221, 260
Janus kinase 2 (JAK2), 3, 5–6, 46, 214
JCR: LA corpulent rat, 210
JTT-705, 14
Juvenile diabetes, 285

K

Kennedy pathway, 340
α-Ketobutyrate, 336
Kidney
  disease, 307
  failure, 316
  injury, 386, 388
KK mouse, 287
Krebs' cycle, 336
Kupffer cells, 137

L

Lactate dehydrogenase (LDH), 252
Laminin, 294
Lauric acid, 504
LCAT, *see also* Lecithin cholesterol
    acyltransferase
  activity, 26, 29, 57
  in atherosclerosis, 28
  deficiency, 23, 30–34
  gene, 24, 26, 28–31
    knockout mice, 23, 30, 34
  mutations, 27, 31

LDL, 24, 30, 33, 93
    acetylated (acLDL), 73
    ECM affinity, 254
    glycation, 256
    glycation/glycoxidation of, 252
    glycoxidation of, 257
    oxidative modification of, 152
    oxidizability, 222
    oxidized, *see* oxLDL
    phenotype(s), 168–170
        A and B, 172–173
    -R pathway, 72–73
    -receptor null mice, 460
    -receptor related protein (LRP), 8
        receptor, 9–10, 56–58, 254, 397
    knockout mice, 252
    removal of, 54
    uptake of, 423
LDLR knockout mouse model, 287
Lecithin cholesterol acyltransferase (LCAT),
    7, 54, 93, 96, 256, 482
Lectin-like oxLDL receptor-1 (LOX-1), 70,
    74, 81
Lens crystallins, 294
Leptin, 209–210, 216
    deficient-obesity, 28
    receptor gene, 210
Leukocyte(s), 422
    adhesion molecules, 362
    CMV DNA in, 446
    recruitment of, 445
Leukotriene, 485
    B4, 508
    B5, 508
Limb ischemia, 455
Linoleate, 32
Linoleic acid, 506, 532, 545–548,
    550–551
α-Linolenate, 548
α-Linolenic acid, 507, 548
Lipase
    endothelial, 13–14, 94
    hepatic, 8, 17
LIPG gene, 13
Lipid(s)
    antigens, 503
    bidirectional flux of, 55
    in Diabetes Study (LDS), 209
    disorders, 216
    efflux, 40–46, 98
    hydrolysis, 457
    metabolism and atherosclerosis, 93–94
    oxidation, 531, 536
    peroxidation, 222, 253

Research Clinic Prevalence Mortality
        Follow-up Study, 24
    Research Clinic Primary Prevention Trial,
        24
Lipid-binding affinity, 43
Lipogenesis, 33, 197, 205, 215, 218, 223
    in insulin resistance, SREBP-1c role,
        220–221
Lipoic acid, 226, 305, 310, 316, 318
Lipolysis, 8, 14, 121, 134–135, 215
Lipopolysaccharide (LPS), 56, 74, 97, 403,
        422, 460, 484
Lipoprotein lipase (LPL), 8, 33, 171, 193,
        210, 226
Lipoprotein(s), 25, 28, 98, 475, 532
    (a), 155
    apoB-containing, 30–31
    components and classification, 122–123
    in insulin-resistant states 205–228
    metabolism, 495
    native, 77
Lipoteichoic acid, 260
Lipotoxicity, 215
Lipoxygenase (LOX), 500
    15-, 483, 526
    5-, 485
    inhibitors, 225
Liver
    cells, 508
    dysfunction, 386
    receptor homolog-1, 7
    X receptor (LXR), 3, 10, 12, 40, 96, 227
        α (LXRα), 33, 196
Long terminal repeat (LTR), 100
Long-Term Intervention with Pravastatin in
        Ischemic Disease Study Group
        (LIPID), 208
Lovastatin, 524
Low-density lipoproteins (LDL), 121, 151,
        415, 475, 500, 520
LOX-1
    expression, 261–262
    gene, 83, 261
LpX vesicles, 27, 31
LRH-1, 192
Luciferase, 341
Luminal stenosis, 436
LXR
    -α, 196
        activators, 191
    -RXR
        agonists, 100
        pathway, 96
Ly49A transgenic mice, 400

Lymphocyte(s), 399, 477
　proliferation, 505
Lyn, 77
Lynglycolipids, 76
Lysine residues, 478, 521
Lysophosphatidic acid, 479
Lysophosphatidic acid acyltransferase, 124
Lysophosphatidylcholine (LPC), 24, 479,
　　521–522
Lysophospholipids, 72
Lysosomes, 8–9

M
Macrophage(s), 151, 247, 385, 397, 420, 459
　ABCA1 and atherosclerosis, 104–106
　ApoE and atherosclerosis, 100–104
　cholesterol efflux pathways, 94
　colony-stimulating factor (M-CSF),
　　252, 458
　foam cell, 252–253, 266
　scavenger receptor
　　A (MSR-A), 257, 260, 264
　　-1 (SR-A), 70, 74–76
Macrophage lipoxygenase, 478
Madin–Darby canine kidney (MDCK)
　　cells, 61
Major histocompatibility complex (MHC)
　　class I specific receptor, 400
Malondialdehyde (MDA), 253, 521
　4-hydroxynonenal, 478
Malvidin, 480
MAP, 205–206, 213–214
MAP kinase(s), 260
　cascade, 225
　pathway, 212, 214, 220
MAP/ERK kinase (MEK), 214
Marek's disease virus, 415
Marijuana, 461
Mast cells, 398–399, 458–459
MAT 1A, 333
MAT 2A, 333
MAT III, 338
MAT mRNA, 341
Matrix
　components, 398
　depletion, 459
　metalloproteases (MMPs), 73, 253, 261,
　　288, 414, 422, 502
　turnover, 455
McArdle RH7777, 132, 341
MCP, *see* Monocyte, chemoattractant protein
MCP-1
　homocysteine-induced expression of, 362
　expression, 262, 380, 385–386

receptor, 252
Mediterranean diet, 476
Mental retardation, 359
Metabolic dyslipidemia, 207–212, 216–217,
　　222–228
Metabolic syndrome, 32, 168, 175, 207, 222,
　　286, 308
Metabolism, 170
　homocysteine, 333–336
Metalloproteinases, 459
Metformin, 347
　adenosyltransferase (MAT), 333, 343
　cycle, 333, 335, 338
Methionine, 343
　metabolism, 334, 336–338
　synthase (MS), 335, 345, 359
Methotrexate, 346
Methyglyoxal, 249
"Methyl trap," 345
Methylation demand, 338–341
Methylcob(III)alamin, 345
Methylcobalamin, 359
$N^{5,10}$-Methylenetetrahydrofolate, 335
$N^{5,10}$-Methylenetetrahydrofolate reductase
　　(MTHFR), 335, 338, 359
Methylglyoxal, 250, 253, 258, 267, 308–315
N^{5}-Methyltetrahydrofolate, 335, 338
Methyltransferases, 334, 338, 339
Mevalonate, 157
　pathway, 97
MGAT, 135
Micelles, mixed, 11
Microemboli, 477
Microsomal carnitine acyltransferase, 129–130
Microsomal triglyceride transfer protein
　　(MTP), 133–134, 205, 210, 223
Microsomes, 130, 136
Mitogen-activated kinases (MAK), 263
Mitogen-activated protein kinase (MAPK),
　　77, 83, 205, 213, 501; *see also* MAP
Mitogenesis, 480
mLDL, 72, 75
MMP-1 overexpression, 460
Molecular targeting hypothesis, 366
Monoacylglycerol acyltransferase (MGAT),
　　129
Monoacylglycerols, (MG)
　2-, 124
　pathway, 129
Monoclonal antibodies, 535
Monocyte(s), 152, 361, 398, 420–425, 501,
　　508
　chemoattractant protein (MCP-1), 156,
　　288, 417, 442, 480, 520

Monocyte(s) (*Continued*)
  colony-stimulating factor (M-CSF), 520
  infiltration and differentiation, 72
  peripheral blood, 386
Monogenic disorders, 26–27
Monooxygenase, 500
Monounsaturated fatty acids (MUFA), 506,
    532, 546
MTHFR
  677TT polymorphism, 330
  mutations, 345
Mtp gene, 134, 223, 225
Multiple risk factor intervention trial, 24
Mutagenesis, targeted, 57
Mycoplasma pneumoniae, 423
Myeloid differentiation factor (MyD88), 458
Myeloperoxidase, 310, 478, 483, 526
Myocardial infarction (MI), 72, 174, 284,
    316, 358, 424, 480, 523
Myocarditis, 420
Myocardium, 508
α-Myosin, 423
Myointimal hyperplasia, 444
Myotoxicity, 158

N
N-acetylcysteine (NAC), 312, 484
NADPH oxidase, 478, 483, 500
Nascent HDL particles, 40, 47
NASH, *see* Nonalcoholic steatohepatitis
National Cholesterol Education Program
    Adult Treatment Panel III, 168–169,
    176, 178
National Health and Nutrition Examination
    Survey (NHANES III), 418
Natural killer (NK) cells, 397–398, 404–405
  and experimental atherosclerosis, 399–400
  T (NKT) cells, 397–399, 405
    in linking innate and acquired immune
      systems, 401–403
nCEH, *see* Neutral cholesterol ester
    hydrolase
Necrosis, 457
Necrotic cells, 415
Necrotic core, 288–289
Neimann-Pick type C gene, 98
Neisseria meningitidis, 260
Neoangiogenesis, 463
Neomycin gene cassette, 59
Neonatal cardiac myocytes, 508
Neovascularization, 458
Nephropathy, 267–268, 295
Nerve growth factors, 10
Neuropathy, 267

Neutral cholesterol ester hydrolase (nCEH),
    11
Neutrophils, 156, 485, 508
NF-κB, *see* Nuclear factor kappa B
    activation, 77–79, 82, 266, 362, 380–384
Niacin, 347
Niemen-Pick disease type C gene, 457
Nitric oxide (NO), 439–440, 475, 495, 530
Nitric oxide synthase (NOS), 435–436,
    440–444, 446, 497
L-Nitroarginine (L-NA), 442
S-Nitrosylation, 343
3-Nitrotyrosine, 363
NK cells, *see* Natural killer cells
Nonalcoholic fatty liver disease (NAFLD),
    221–222
Nonalcoholic steatohepatitis (NASH), 219,
    221–222
Noninsulin-dependent diabetes mellitus
    (NIDDM), 285
Nonobese diabetic (NOD) mouse, 287
Norepinephrine, 481
Normoglycemia, 249
Normolipidemia, 228
NPC disease, 98
Nuclear factor kappa B (NF-κB), 260, 293,
    388, 422, 484
Nuclear factor Y (NF-Y), 226, 366
Nuclear pore proteins, 291
Nuclear receptors (NRs), 189, 191, 218
NZW rabbit, 442

O
Obesity, 122, 196, 205–211, 217–222, 287,
    293, 414, 458
Occlusive coronary artery atherosclerosis, 58
Oleate, 134
Oleic acid, 506, 532, 546
OLR1, 81
Omega-3 (C20:5), 482
O-*N*-acetylglucosamine transferase (OGT), 291
Orlistat, 15
Orphan nuclear receptors (NR), 191
  ARP1, 7
Osteoporosis, 359, 386
Otsuka Long-Evans Toskushima fatty rat
    model, 224
Oxidative markers, 31
Oxidative stress, 150, 157, 222, 305, 383, 476,
    500, 531
  AGEs and, 311–313
  atherosclerosis and, 442–443
  homocysteine-induced, 363
  on intracellular signaling, 440

Oxidatively modified low-density
    lipoproteins, *see* oxLDL
oxLDL, 10, 29, 72, 259, 478, 495, 500, 502
   thrombosis and, 150–159
    antigens, 152
    autoantibodies, 255
α-Oxoaldehydes, 266
oxPC, 535
Oxysterol(s), 96, 187, 191, 218; *see also*
    specific compounds
   nuclear receptors, 138

P
P selectin, 480
p21ras, 260
p38, 214, 260
   mitogen-activated protein kinase (MAPK),
    80, 484
p38 kinase, 225
PAF-AH activity, 30
PAI-1, 157–158, 291
Palm oil, 28
Palmitate, 505
Palmitic acid, blood level of, 548
Pancreatic β-cells, 211, 287
Paraoxonase (PON), 54, 482
   1 (PON1), 29
Parkinson's disease, 338–339, 347, 365
PCAM-1, 288
PDGF-β receptor, 417
PDI, 292
PDZK1, 61
   KO mice, 60
Pemt gene, 130
PEMT, 131
   null mice, 130
   pathway, 340
Penicillamine, 347
PEPCK1, 33
Perforin, 399
Perilipin, 457
Periodontal disease, 418–419
Peripheral arterial disease, 476
Peripheral nerve dysfunction, 286
Peripheral vascular disease, 156, 358
PERK, 292
Peroxidase, 508
Peroxides, 253
Peroxisome proliferator-activated receptors,
    see PPARs
Peroxisome(s), 131
Peroxynitrite, 152, 310, 315, 363, 384,
    388, 500
Peroxynitrous acid, 315

PEST sequence, 47, 97
Phagocytes, mononuclear, 519–520
3-Phenacyl-4, 5-dimethylthiazolium chloride,
    268, 295
N-Phenacylthiazolium bromide (PTB),
    268
Phenylpropanedione, 268
Phosphatase and tensin homolog (PTEN),
    205, 213
Phosphatidate phosphohydrolase (PAPH),
    124, 127
Phosphatidic acid (PA), 124
Phosphatidylcholine (PC), 24, 130, 329,
    339, 521
Phosphatidylcholine phospholipase
    C (PC-PLC), 46
Phosphatidylethanolamines (PE), 130,
    257, 339
Phosphatidylethanolamine
    N-methyltransferase (PEMT), 130, 339
Phosphatidylinositol, 7, 14
Phosphatidylinositol 3-kinase (PI-3-K), 83,
    205, 212, 220, 260, 484
Phosphatidylserine (PS), 41
Phosphocholine, 130
Phosphocholine cytidylyltransferase (CCT),
    130
Phosphodiesterase, 486
Phosphoenolpyruvate carboxykinase
    (PEPCK), 194
Phospholipase(s), 5
   A2, 486
   C, phosphatidylcholine-specific, 5, 42
   D, phosphatidylcholine-specific, 5
Phospholipid(s), 5, 40, 261, 334, 547, 550
   3′-, 213
   efflux, 45–46
   transfer protein (PLTP), 26
Phospholipid flippase, 43
Phosphotyrosines, 6
Phosphotyrosyl-protein phosphatases
    (PTPases), 213
Physician's Health Study, 177
PKB/Akt pathways, 56
PKC-β, 82, 262
PKR like ER kinase (PERK), 364–366
Plaque, 414
   erosion, 461
   formation, 419, 422–423
   growth, 446
   inflammation, 458
    in thrombosis, 460
   neovascularity, 455, 457
   phenotype, 462–463

Plaque (*Continued*)
    rupture, 72–73, 94, 455–463
    stabilization, 462–463
    vulnerability, 455–458
Plasma homocysteine, 336–337, 341–343
Plasmin, 154, 459
Plasminogen, 154, 459
    activator (uPA), 459
    activator inhibitor-1 (PAI-1), 154–155,
        157–158, 422
Plasmodium falciparum, 261
Platelet-activating factor, 480
    factor-1 (PAF-1), 498
Platelet activating factor acylhydrolase
    (PAF-AH), 28, 54
Platelet activation, 151, 461
Platelet-associated vascular hemorrhage,
    504
Platelet-derived growth factor (PDGF),
    479, 503
PLP, 343
Pneumonia, 420
Polymorphisms, 443
Polyol pathway, 249, 289
Polyphenolics, 534
Polyunsaturated fatty acids (PUFA), 32, 478,
    506–508, 532, 546
Polyunsaturated phospholipids, 482
PON1
    activity, 30
    arylesterase activity, 31
Porphyromonas gingivalis (Pgn), 413,
    418–419
Postendoplasmic reticulum presecretory
    proteolysis (PERPP) process, 532
Postheparin lipase activity assay, 33
Postprandial lipemia, 222, 227
PPAR, 3, 12, 96, 138, 227
    -α, 7–14, 96, 192, 219, 266
        KO mice, 219
        agonists, 7, 14, 138, 209, 463
    γ, 96, 215, 219
        agonists, 97, 138, 209, 265–266
        coactivator-1alpha (PGC-1alpha), 218
    δ agonists, 96–97
Prediabetes, 286
Preeclampsia, 311
Pregnane X receptor (PXR), 189
Pressor stimuli, 461
Proanthocyanidines, 484
Proapoptotic gene *p53*, 461
Probucol, 58
Procoagulant proteins, 288
Proliferating SMCs, 477

Proliferator activated receptor, 3
Prooxidant, 533
Propionyl CoA, 336
Prostacyclin(s), 153, 437–440, 481,
    498–500, 508
    metabolites, 505
Prostaglandin(s)
    H, 498
    I2, 547
Protease(s)
    site-1 (S1P), 220, 365
    site-2 (S2P), 220, 365
Protein(s)
    C, 155, 157, 368, 383
    glycation, 247
    homocysteinylation, 332
    pathway, 154
    receptor, 462
    S, 155
    S-APC complex, 153
Protein disulphide isomerase, 364
Protein kinase, 501
    A (PKA), 5, 45, 61, 97
    C (PKC), 5, 46, 76, 83, 211, 249, 293, 501
    Cα (PKCα) signaling pathway, 97
Protein tyrosine kinases (PTK), 76
Protein tyrosine phosphatase 1B (PTP-1B),
    205, 214, 221
Proteoglycans, 10, 174, 255, 459, 477, 479
Proteolysis, 47, 521
Psammomys obesus, 211
P-selectin, 57, 156, 288, 293
Psoriasis, 346
PTP-1B KO mice, 214
PTPase, 214
PXR, 192, 196
"Pyridorin," 268
Pyridoxal, 329
    5′-phosphate (PLP), 335, 347
    phosphate, 267, 359
Pyridoxamine, 268, 295
Pyridoxine deficiency, 381
Pyruvate, 309
Pyruvate dehydrogenase complex, 336

Q
Quebec Cardiovascular Study, 174, 177
Quercetin, 480, 482–483, 485–486

R
Rac, 158
RAGE-null mice, 260
Ramipril, 158
Rap, 158

Ras, 158
RAW cells, 80
R-BI KO mice, 60
RCT, *see* Reverse cholesterol transport
Reactive nitrogen species (RNS), 522
Reactive oxygen species (ROS), 484, 495,
    500, 522
Receptor for advanced glycation end
        products (RAGE), 259–265, 267,
        295–296, 310
Red wine, 475
Related protein, 3
Remethylation pathway, 359, 381–382
Renal failure, 286, 441
Renal function, impaired, 381
Renal insufficiency, 34, 341
Renin-angiotensin system, 158
Resistin, 216–217
Restrain stress, 157
Resveratrol, 480–481, 484–486
Retention factor, 220
Retinal, 267
Retinoic acid, 7
Retinoid X receptor (RXR), 7, 12,
        96, 191
Retinopathy, 286
Reverse cholesterol transport (RCT), 7–14,
        23, 39, 93, 218
    HDL-dependent, 59
    initiation of, 96
    pathway, 13, 24–25, 27, 40, 47, 54
Rheumatoid arthritis, 346
Rho, 5, 158, 260
Rimonabant, 463
RNA, 334
Rosiglitazone, 226, 265
Rouleaux, 26, 29
Roxithromycin, 424
RTP, 365
Russel's viper venom, 460
Ryanodine receptor, type 2, 314

S

S100 proteins, 259, 265
S100/calgranulin epitopes, 264, 296
S-Adenosylhomocysteine (SAH), 130,
        334, 338
S-Adenosylmethionine (SAM), 130, 329, 333,
        338–339
"Sand rat," 211
SAPK/JNK, 260
Sarcosine, 334
Scarb1, 57, 59
SCARB3, 77

Scavenger receptors (SR), 151, 257, 260–262,
        288, 417, 503
    class B type I, *see* SR-BI
Schiff base, 249
Selectin
    E-, 57, 156, 288, 361, 386, 478, 506
    P-, 57, 156, 288, 293, 480
Selenocysteine, 369
    incorporation sequence (SECIS), 369
Sepsis, 260
Septic shock, 343
Serine-threonine kinases, 214
Serotonin, 480
Serum amyloid A (SAA), 56, 80
Shear stress, 315
Shp gene, 192
Sickle cell anemia, 156
Signaling, 70, 74
    factors, 291
Silent plaque rupture, 461
Simvastatin, 158
Single-nucleotide polymorphisms (SNP),
        10, 13
Sinusitis, 420
Site-1 protease (S1P), 365
Site-2 protease (S2P), 365
Small dense low-density lipoproteins,
        168–179, 207
Small PDZK1-associated protein (SPAP),
        60–62
SMC migration, 72
Smoking, 175, 307, 337, 418, 458
Smooth muscle cells (SMC), 151, 295, 361,
        421, 459, 476
Sorbitol, 266, 289–290
Sp1, transcription factor, 7
Sphingomyelin, 46
Sphingomyelinase, 42
Sphingosine-1-phosphate, 56
SR-A, 288
SR-BI, 8–14, 70, 74, 79–81, 261
    atherosclerosis and, 53–59
    deficiency, 57
    expression, 95
    knockout (KO) mice, 57–62
    overexpression, 59
    protein, hepatic, regulation of expression,
        60–61
SR-BI/apoE double KO mice, 58
SR-BIatt/LDL receptor KO mice, 59
SREBP, 6–7, 33, 193, 293, 365, 387
    -1a, 138, 220
    -1c, 194, 205, 220–221, 223
    -2, 220

SREBP (*Continued*)
   cleavage-activating protein (SCAP), 220
   pathway, 218
Stanford Five Cities Project, 174–177
StAR, *see* Steroidogenic acute regulatory
   protein
Statins, 93, 150, 157–158, 178, 209
Stearic acid, 504–505, 550
Stearoyl CoA-desaturase (SCD), 97, 133, 221
Steatohepatitis, 222
Steatosis, 134, 218–219, 221, 456, 461–462
Steroidogenesis, 11
Steroidogenic acute regulatory protein
   (StAR), 11
Steroids, 152, 336, 443
Sterol(s), 7
   homeostasis, 98, 218
   regulatory element binding proteins, *see*
     SREBP
   response elements (SRE), 220, 225–226
Streptozotocin (STZ), 220, 264, 287, 289,
   342
Stress (es)
   circumferential, 457
   endoplasmic reticulum, 291–293
   hemodynamic, 459
   myocardial perfusion scintigraphy, 456
   oxidative, 31, 150, 152
   restrain, 157
Stroke, 71, 121, 288, 307, 316, 358, 455
Substance P, 437
Succinyl CoA, 336
Superoxide dismutase, 508
Superoxide, 363
Symmetric dimethylarginine (SDMA), 369
"Syndrome X", 208, 308
Syrian golden hamster model, 211
Systemic lupus erythematosus (SLE), 153
Systemic thrombogenicity, 461
Systemic vascular resistance, 446

T
T cell(s), 71–72, 425, 458–459
   activation, 436
   death associated gene 51 (TDAG51), 365
   receptor (TCR), 401
T lymphocytes, 288, 362, 397–405, 414,
   479, 502
Tachycardia, 437
Tacrolimus, 448
Tangier disease, 40, 44, 46, 96, 100
   patients with, 5
Tannic acid, 480
Targeted mutation, 60

Taurine, 336
Tenascin-C, 459
Testosterone esters, 337
Tetrahydrobiopterin (BH4), 61, 136, 314–315,
   440, 500
TF, 288, 362, 419, 460, 480
   expression, 262
   secretion, 422
TG, 121, 124–129, 193–194
   accumulation, 217, 257
   lipolysis, 135, 137–138, 171
   metabolism, 122
TG lipase, 135
TGF, *see* Transforming growth factor
TGH, *see* Triacylglycerol hydrolase
Thapsigargin, 370
Theophylline, 347
Thiazolidinediones, 209–210, 217, 219, 226
Thiobarbituric acid, 483
   -reacting substance (TBARS), 534
Thiol(s), 305, 369
   protease inhibitors, 47
THP-1 cells, 46, 76, 97, 386
Thrombi, 506
Thrombin(s), 156, 486, 498–499
   generation, 460
   α-, 368
Thrombocytes, 486
Thrombogenicity, 461
Thrombomodulin, 153, 156, 462, 499
α-Thrombonin, 484
Thrombosis, 71, 151, 362, 475, 485, 504
   and plaque rupture, molecular mechanisms
     of, 455–463
Thrombospondin, 77, 261
Thromboxane (TX), 505, 547
   A2, 153, 480, 498, 508
   A3, 508
Thrombus, 480
Thyroid hormone (T3), 97
Thyroxin, 342
Tissue factor, *see* TF
Tissue factor pathway inhibitor (TFPI), 153,
   156, 158
Tissue inhibitors of metalloproteinases
   (TIMPs), 459
Tissue plasminogen activator (tPA), 154–158,
   499
TLR, *see* Toll-like receptor
TNF-α, *see* Tumor necrosis factor-α
α-Tocopherol, 54, 258
Tolbutamide, 129
Toll-like receptor 4 (TLR-2 and 4), 78–79,
   422, 458

Torcetrapib, 14
"Total plasma homocysteine" (tHcy), 332
TRAF2, 370
Transcription factor(s), 193, 196, 291, 365,
        380, 387
    AP1, 7
Transforming growth factor-β (TGFα and
        β), 291, 458, 484, 503
Transketolase, 266
Transplant arteriopathy, 435, 444–447
Transplant arteriosclerosis, 445
Transplant coronary artery disease, 441
Transplant endothelial dysfunction, role of
        CMV in, 445–446
Transretinoic acid, 7
Transsexuals, 336–337
Transsulfuration pathway, 335, 341–344, 359,
        381–382
Transthyretin, 367–368
Triacylglycerol hydrolase (TGH), 136–138
Triamcinolone, 341
Triglycerides, 133–134, 169, 174, 207,
        222
Trisomy 21, 345
Tromso study, 330
Trypsin, 459
Tryptase, 458–459
Tumor necrosis factor-α (TNF-α), 76, 156,
        255, 383, 414, 478, 507
Tunica media, 398
Tunicamycin, 370
Tyrosine, 531
    phosphorylation, 501
Tyrosine kinase, 501

U
Ubiquitin, 97
Ulceration, 307
Unfolded protein response (UPR), 291,
        363–366
United Kingdom Prospective Diabetes Study
        (UKPDS), 248
Uremia, 27
Urokinase plasminogen activator (uPA),
        76, 154

V
Vα14
    NKT cells, 402, 404
    T cells, 402
Vα14/Vα receptor, 404
Vα3.2-Jα9, 402
Vascular alterations in transplant
        arteriopathy, 436

Vascular cell adhesion molecule-1 (VCAM-1),
        57, 156, 361, 417–422, 478, 502, 522
Vascular dementia, 177
Vascular disease(s)
    ADMA and, 443–444
    LDL size and, 176–177
Vascular endothelium, 480–481, 496
Vascular homeostasis, 439, 496
Vascular inflammation, 436
Vascular lesion formation, 446
Vascular smooth muscle cells (VSMCs), 288,
        293, 437, 475, 479, 498
Vascular tone, 439
Vascular wall, 61–62
Vasculopathy, 267
Vasoconstriction, 437, 439, 445
Vasodilation, 441
Vasodilator substances, 439
Vasopressor infusion, 461
VCAM, 288
    -1 expression, 262–263
Vegetable fat, 550
Venom, Russel's viper, 460
Venous thrombosis, 359
Very low-density lipoprotein, see VLDL
Vessel injury, 71
Veterans Administration High-Density
        Lipoprotein Intervention Trial (VA-
        HIT), 209
VFVNFA motif, 45
Viral infections, 446
Visna virus, 100
Vitamin(s)
    B_{12}, 329, 337, 359, 381
    B_6, 267–268, 316, 337, 359
        deficiency, 381
        metabolism, 347
    C, 305, 310, 316–318
    D receptor, 189, 196
        deficiency, 343–344
    E, 54, 305, 310, 316–318
VLDL, 7, 29–33, 54, 121, 155, 482
    assembly, 132, 135, 137, 211, 222–223
    lipids, 8
    overproduction, 205, 210
    production, 224
    receptors, 193
    secretion, 123–124, 133, 137, 226
    subclasses of, 170
    triglycerides, 134
von Willebrand factor (vWF), 156–157
VSMC, see also Vascular smooth muscle cells
    apoptosis, 440
    growth, 310

W

Watanabe heritable hyperlipidemic (WHHL)
rabbit model, 210, 521
Wistar-Kyoto (WKY) rats, 311–312, 314–315
World Health Organization (WHO), 206
World Hypertension League, 306

X

Xanthine oxidase, 500

X-box binding protein-1 (XBP-1), 365
Xenical, 15

Y

Yeast, 338
Yes, 77

Z

Zucker rats, 210